WITHDRAWN

CHROMATOGRAPHY OF ENVIRONMENTAL HAZARDS
Volume II

CHROMATOGRAPHY OF ENVIRONMENTAL HAZARDS

VOLUME II

METALS, GASEOUS AND INDUSTRIAL POLLUTANTS

LAWRENCE FISHBEIN

Department of Health, Education and Welfare,
Public Health Service,
Food and Drug Administration,
National Center for Toxicological Research,
Jefferson, Ark., U.S.A.
Adjunct Professor of Entomology-Toxicology,
North Carolina State University,
Raleigh, N.C., U.S.A.

ELSEVIER SCIENTIFIC PUBLISHING COMPANY

AMSTERDAM/LONDON/NEW YORK

1973

ELSEVIER SCIENTIFIC PUBLISHING COMPANY
335 JAN VAN GALENSTRAAT
P.O.BOX 1270, AMSTERDAM, THE NETHERLANDS

AMERICAN ELSEVIER PUBLISHING COMPANY, INC.
52 VANDERBILT AVENUE
NEW YORK, NEW YORK 10017

LIBRARY OF CONGRESS CARD NUMBER: 78-180000
ISBN 0-444-41059-7
WITH 367 ILLUSTRATIONS AND 231 TABLES

PRINTED IN THE GERMAN DEMOCRATIC REPUBLIC

PREFACE

This book, the second of four volumes of the series *Chromatography of Environmental Hazards*, is intended to present a survey of the salient features of selected chromatographic procedures (primarily gas–liquid, thin-layer, paper and ion-exchange) that relate to the detection, separation and determination of the principal metal, gaseous and industrial chemical hazards encountered in the environment. Germane features of the occurrence and/or synthesis, areas of utility, chemical, physical and biological properties, metabolism, stability and toxicity as well as instrumental and chemical analytical methodology are included to present as thorough and cohesive a picture as possible of the specific environmental hazard. Although the focus of treatment concerns chromatographic analytical methodology, clearly many diverse areas of expertise are required for the amelioration or abatement of these hazards. Hence, it is the hope of the author that the contents of this series will be of significance to investigators in disciplines that include biochemistry, biology, genetics, toxicology and public health in addition, of course, to the analyst on whose shoulders the major burden of identification and quantification of the hazard must fall.

In the initial volume of the series, the chromatographic as well as the biological aspects of a number of chemical carcinogens, mutagens and teratogens and related compounds of environmental significance were examined. The contents of that volume are listed on pages xi and xii. As will be recognized, a number of chemical agents examined clearly fall within the realm of gaseous and industrial pollutants and hence are not re-examined in Volume II.

Special acknowledgments of gratitude are extended to Mrs. Peggy Sauls and Mrs. Donna Shields for their patience and skill in the assembly, typing and proofing of the manuscript.

L. FISHBEIN

CONTENTS OF VOLUME II

CONTENTS

CONTENTS OF VOLUME I

Chapter 1

INTRODUCTION

This volume is arranged in three major parts for what is desired to be an enhancement of cohesiveness and subject treatment. These are metals, gaseous and industrial pollutants. The gaseous pollutants include both inorganic and organic pollutants. It is admittedly an arbitrary arrangement and it is well recognized that there are overlaps in both subject area and occurrence classification. For example, the halogen gases, *e.g.* chlorine, hydrogen chloride, fluorine and hydrogen fluoride, are clearly industrial agents but are considered under gaseous pollutants. Polynuclear aromatic hydrocarbons, such as benzpyrene, important ubiquitous organic particulate atmospheric pollutants, would deserve a category in their own right, but are treated under gaseous pollutants.

As in Volume I, no attempt has been made to present an exhaustive account of the chromatographic procedures involved in the analysis of chemical hazards. The literature is voluminous and in the area of hydrocarbons alone, for example, the references number well into the thousands. Rather, chromatographic references are chosen that are believed to be of greatest value and potential significance, particularly in the areas of environmental utility (*e.g.* monitoring and analysis of degradation products and residues) and biological and toxicological applications. Information is provided wherever possible concerning the basic equipment, operating parameters, sensitivity, interferences, approximate time required for analysis of a particular contaminant and/or group of related contaminants. Similarly, germane features of the occurrence and/or synthesis, areas of utility, chemical, physical and biological properties are also provided to further enhance the analyst's ability to see the problem in as broad a context as possible and hence more effectively to select and utilize the best choice of procedures.

The wide areas of chromatographic utility, particularly gas, thin-layer and ion-exchange, the speed of separation and detection, and in many cases the exquisiteness of sensitivity all combine to make chromatography an indispensable part of an analyst's weaponry.

The practical aspects of importance for the chromatographic analysis of metals[1] and gaseous pollutants[2-8] have been described and the chromatographic and biological aspects of inorganic[9] and organic mercury[10], polychlorinated biphenyls[11] and phthalate esters[12] and the analysis of polycyclic aromatic hydrocarbons[13-16] and phenols[17] reviewed. Other important subject areas of relevance include the detection and determination of trace elements[18], detection of radionuclides of biological interest[19], metallic contaminants and human health[20-22], toxicity and metabolism of gases and industrial solvents[23,24].

REFERENCES

1 R. W. MOSHIER AND R. E. SIEVERS, *Gas Chromatography of Metals Chelates*, Pergamon Press, Oxford, 1965.
2 P. G. JEFFERY AND P. J. KIPPING, *Gas Analysis by Gas Chromatography*, Marcel Dekker, New York, 1968.
3 A. P. ALTSCHULER, in A. L. STERN (Ed.), *Air Pollution*, Vol. II, Academic Press, New York, 2nd edn., 1968, p. 116.
4 M. KATZ, in A. L. STERN (Ed.), *Air Pollution*, Vol. II, Academic Press, New York, 2nd edn., 1968, p. 105.
5 E. A. SCHUCK AND E. R. STEVENS, in J. N. PITTS AND R. L. METCALF (Eds.), *Advances in Environmental Sciences*, Vol. I, Wiley-Interscience, New York, 1969, p. 73.
6 W. LEITHE, *The Analysis of Air Pollutants*, Ann Arbor Science, Ann Arbor, Mich., 1971.
7 M. W. ANDERS AND G. J. MANNERING, *Progr. Chem. Toxicol.*, 3 (1967) 121.
8 W. LODDING, *Gas Effluent Analysis*, Marcel Dekker, New York, 1967.
9 L. FISHBEIN, *Chromatog. Rev.*, 15 (1971) 195.
10 L. FISHBEIN, *Chromatog. Rev.*, 13 (1970) 83.
11 L. FISHBEIN, *J. Chromatog.*, 68 (1972) 345.
12 L. FISHBEIN AND P. W. ALBRO, *J. Chromatog.*, 70 (1972) 365.
13 C. R. SAWICKI AND E. SAWICKI, in A. NIEDERWESER AND G. PATAKI (Eds.), *Progress in Thin-Layer Chromatography*, Vol. III, Ann Arbor Science, Ann Arbor, Mich., 1972, p. 233.
14 F. A. GUNTHER AND F. BUZZETTI, *Residue Rev.*, 9 (1965) 90.
15 E. O. HAENNI, *Residue Rev.*, 24 (1968) 42.
16 R. E. SCHAAD, *Chromatog. Rev.*, 13 (1970) 61.
17 S. T. PRESTON, *A Guide to Analysis of Phenols by Gas Chromatography*, Polyscience, Evanston, Ill., 1966.
18 M. PINTA, *Detection and Determination of Trace Elements*, Ann Arbor Science, Ann Arbor, Mich., 1962.
19 U. MUTSCHKE AND O. PRIBILLA, in A. STOLCMAN (Ed.), *Progress in Chemical Toxicology*, Vol. III, Academic Press, New York, 1967.
20 D. H. K. LEE (Ed.), Metallic contaminants and human health, *Fogarty International Center Proceedings No. 2*, Academic Press, New York, 1972.
21 E. BROWNING, *Toxicity of Industrial Metals*, 2nd edn., Butterworths, London, 1969.
22 Z. I. IZRAEL'SON, *Toxicology of Rare Metals*, Israel Program for Scientific Studies, Jerusalem, 1967.
23 F. A. PATTY (Ed.), *Industrial Hygiene and Toxicology*, Vol. II, 2nd edn., Interscience, New York, 1963.
24 E. BROWNING, *Toxicity and Metabolism of Industrial Solvents*, Elsevier, Amsterdam, 1965.

Chapter 2

BERYLLIUM AND CHROMIUM

1. BERYLLIUM

Beryllium, one of the lightest of metals, although widely distributed, exists in relatively small quantities, comprising less than 0.006% of the earth's crust. It occurs chiefly as the mineral ore beryl (3 BeO, Al_2O_3, 6 SiO_2) which contains up to 13.9% of beryllium oxide. Beryllium is also obtained as a by-product from felspar, mica and lithium ores found principally in Brazil, Argentina, the U.S. and South Africa. Other sources of beryllium include most coals which can contain varying concentrations of from approximately 1.5 to 2.5 p.p.m. Despite these low concentrations, the amount of beryllium added to the atmosphere by the burning of coal may be significant in areas of large coal consumption such as the United Kingdom and areas of the United States. Beryllium is known to be present in the atmosphere. The daily average atmospheric concentration of beryllium for 100 stations in the United States[1] is less than 0.0005 $\mu g/m^3$ and the maximum value recorded for the 1964–1965 period was 0.008 $\mu g/m^3$.

Beryllium has found a number of important industrial uses including (*a*) in the production of 2% copper–beryllium alloy, as well as alloys with aluminum, magnesium and nickel, (*b*) in neutron reactors for cans and other reactor access oils, (*c*) in the production of X-ray diffraction tubes and electrodes for fluorescent lamps, neon signs and targets for cyclations, and (*d*) in rocket fuels as finely powdered metallic beryllium.

Industries in which beryllium disease has arisen include beryllium extraction, fluorescent lamp production and disposal, phosphor production, "neon" lamp industry, alloys (production and metallurgy), nuclear energy industry, metal working and machining, ceramics (BeO refractory) and electron tube production. Occupational exposure to beryllium compounds may lead to acute or chronic disorders. Acute berylliosis is characterized by conjunctivitis, tracheitis and bronchitis, dermatitis, subcutaneous granulomas and pulmonary disease. Chronic beryllium disease, also known as berylliosis, is a systemic disease due to the inhalation of toxic beryllium compounds. It is marked by a latency or delay in onset which in some cases has exceeded 20 years and is manifest by severe pulmonary damage in most cases. Metabolic disorders such as excessive concentrations of calcium in the serum and urine are also known to occur in beryllium disease.

Almost all the presently known beryllium compounds are acknowledged to be toxic in both the soluble and insoluble forms, depending on the amount of material inhaled and the length of exposure. Soluble beryllium compounds, *e.g.* beryllium

3

sulfate and chloride, commonly produce acute pneumonitis while insoluble compounds such as metallic beryllium and beryllium oxide can produce berylliosis.

Inhalation of beryllium sulfate produces pulmonary adenocarcinomas in rats[2] and it is believed that the physical and chemical properties of the beryllium compound have a bearing on its toxic or carcinogenic potential. The commonly mined beryl ore, beryllium aluminium sulfate, has been shown to produce pulmonary adenocarcinomas when inhaled by rats at an atmospheric concentration of 15 mg/m³ throughout their lifespan. Beryl, however, was not carcinogenic to the hamster or monkey[3].

Aspects of the excretion and storage of beryllium in animals[4-7] and man[8-10] have been described. Beryllium has been determined in microgram amounts by atomic absorption spectroscopy[11-13], arc-emission spectroscopy[14-16], fluorimetry[17-19], and mass spectrometry[20].

Taylor *et al.*[21] described a rapid microanalysis of beryllium in biological fluids by GLC. The method involved the *direct* treatment of biological fluids containing beryllium with trifluoroacetyl acetone in a sealed glass ampoule and subsequent quantitative analysis of the reaction mixture by GLC using an electron-capture detector. (No preparative steps were needed on the samples, nor were concentration techniques required.) As little as 2.95×10^{-7} g Be^{2+}/50 µl sample was routinely quantitated while a sensitivity on the order of 1×10^{-10} g/50 µl sample was feasible with time requirements for an analysis being less than one hour.

A modified Barber-Colman Model 25-C gas chromatograph with a Model 5120 tritium electron attachment detector (300 mCi), a 4 ft. $\times \frac{1}{8}$ in. Teflon column (Polymer Corp., Reading, Pa.) packed with 5% SE-52 coated on 60–80 mesh Gas Chrom Z was used; the ovens were maintained at isothermal temperature as follows: injector port (Teflon), 130°C; column, 110°C; detector, 125°C. Prepurified nitrogen was used as both carrier and detector scavenger gas with flow rates of 27 and 110 ml/min, respectively. Detector anode voltage was maintained at 40 V d.c. and the cell spacing adjusted to 12 mm. These settings resulted in a standing current on the cell of 0.95×10^{-8} A. Attenuation ranges of 3×10^{-9} A and 1×10^{-8} A were used, depending on the Be^{2+} concentration of the sample. The reaction of trifluoroacetylacetone with Be^{2+} results in the formation of a single beryllium chelate, Be(TFA)$_2$(I), which had a retention time of 1.15 min on a column containing 5% SE-52 operated at the conditions described. Unreacted TFA has a retention time of 0.45 min under the same operating conditions. These analyses were performed by chromatographing aliquots of a solution of beryllium (TFA)$_2$ obtained by heating 0.05 ml samples, and a benzene solution of TFA in sealed glass ampoules with average time required per analysis only 15 min.

I

Taylor and Arnold[22] reported additional refinements to the GLC analysis of beryllium in which aqueous and prepared blood samples contained as little as 0.02 μg of beryllium/ml of sample. A Varian Model 2100 gas chromatograph was used equipped with an electron-capture detector (250 mCi tritium, d.c. mode). Columns were 6 ft. × 2 mm U-tubes packed with 10% SE-52 silicone gum on 60–80 mesh Gas Chrom Z. The operating conditions were injection temperature 140°, detector 180°, column 110°C, carrier gas (prepurified nitrogen) at 100 ml/min. Two chelating solutions were used for the analysis of beryllium. For beryllium concentrations greater than 0.2×10^{-6} g Be/ml, a solution was prepared to contain 4 μl TFA/ml plus 8.52×10^{-7} g tetrabromoethane (TBE)/ml used as an internal standard. For concentrations of beryllium less than 0.2×10^{-6} g Be/ml, a solution containing 1 μl TFA/ml and 8.52×10^{-8} g TBE/ml was employed.

The gas chromatograph was calibrated daily by injecting 0.5–1.0 μl of $Be(TFA)_2$–TBE calibration standards. Peak areas of each of the two peaks were calculated by multiplying peak heights by peak widths at one-half the peak height. The response ratio (area $Be(TFA)_2$ peak)/(area TBE peak) was then plotted vs. picograms of Be injected. Under the specified conditions, the retention indices for $Be(TFA)_2$ and TBE were 12:00 and 12:80 methylene units, respectively. The minimum amount of beryllium which could be detected was 8.0×10^{-14} g.

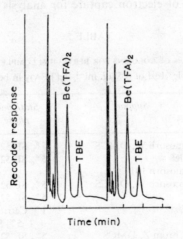

Fig. 1. Chromatograms obtained in analysis of two 50 μl aliquots of whole blood from rat injected with aqueous beryllium. Beryllium found: 1.72 μg/ml blood. Chromatograph attenuation: 16×10^{-9} AFS.

In the preparation of blood samples for analysis, it was found that disodium EDTA was preferred over heparin as an anticoagulant since its use produced samples containing fewer substances interfering with chromatography. Figure 1 illustrates the base line and absence of interfering peaks obtained when Na_2EDTA is used. Disodium EDTA can mask certain metal ions such as iron and prevent their extraction into the organic layer while not interfering with the chelation of beryllium by trifluoroacetylacetone[23].

The determination of traces of beryllium (1 ng/ml) in human urine by electron-capture GLC was reported by Foreman et al.[24]. Beryllium was first isolated by solvent extraction in the form of the volatile, thermally stable trifluoroacetylacetone chelate $Be(TFA)_2$, then determined by GLC. Urine containing beryllium was analyzed by both direct solvent extraction as well as by extraction after wet oxidation. $Be(TFA)_2$ was prepared by dissolving beryllium carbonate in perchloric acid, diluting with water and neutralizing. The solution was buffered with sodium acetate and a solution of trifluoroacetylacetone in benzene added. After shaking the mixture, the bulk of the benzene was then removed by distillation, followed by evaporation with a stream of dry air, the residual dissolved in hexane, recrystallized, sublimed and finally dried under vacuum. GLC analyses were carried out with a Pye 104 chromatograph fitted with an electron-capture or with flame-ionization detectors. Table 1 lists a comparison of columns for $Be(TFA)_2$ analysis. The columns are arranged in order of increasing suitability and illustrate the striking difference between the two di-atomaceous supports Chromosorb W and Gas Chrom Z (containing the same stationary phase 5% SE-52). Figure 2 compares the response of the flame-ionization and electron-capture detectors to solutions of the complex $Be(TFA)_2$ and clearly illustrates the advantages of electron capture for analysis of this complex.

TABLE 1

COMPARISON OF COLUMNS FOR BERYLLIUM COMPLEX ANALYSIS
Sample: 1 μl of 0.5 μg. ml^{-1} $Be(TFA)_2$ in benzene

Column material	Support	Stationary phase	Recovery* (%)
Glass	Chromosorb W, DMCS	5% SE 52	Zero
PTFE	Voltalef	5% SE 52	2.6
PTFE	Chromosorb 101		5.2
PTFE	Chromosorb W, DMCS	5% SE 52	18.2
PTFE	PTFE	5% SE 52	25.0
PTFE	PTFE	1% Cetrimide + 5% SE 52	45.0
PTFE	Gas Chrom Z, DMCS	5% SE 52	100.0 (standard)

* Relative figures assuming complete recovery with the most satisfactory column.

Human urine to which known amounts of beryllium sulfate had been added was analyzed under the following operating conditions: the injection, column, and detector temperatures were 100°, 100°, and 125°C, respectively; column and purge gas flow rates were 67 ml/min (nitrogen gas); the column was of PTFE, 4 ft. × ⅛ in. o.d. containing 5% SE-52 on 72–85 mesh Gas Chrom Z and the electron-capture detector had a pulse period of 150 μsec with 0.75 μsec amplitude, 47–60 V and a source of 10 mCi nickel-63. The use of PTFE columns in conjunction with an electron-capture detector did not cause contamination of the detector provided that both prior to and after packing the column was conditioned for several days at 200°C. Chromatograms of urine extracts are shown in Fig. 3.

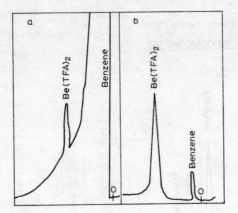

Fig. 2. Chromatograms of Be(TFA)$_2$ solutions in benzene. (a) Response of flame-ionization detector to 0.2 μl of sample containing 50 μg/ml of beryllium (1×10^{-8} g of beryllium); (b) response of electron-capture detector to 0.2 μl of sample containing 1 μg/ml of beryllium (2×10^{-10} g of beryllium).

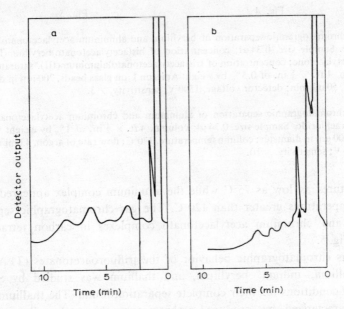

Fig. 3. Chromatograms of urine extracts. (a) 1 μl of human urine extract; (b) 1 μl of human urine extract containing 10 ng/ml of beryllium (10^{-11} g of beryllium).

The GLC analysis of acetylacetonato complexes of beryllium and aluminum has been reported by Biermann and Gesser[25]. A Pye Argon chromatograph equipped with an ionization detector was used with 4 ft. $\times \frac{1}{4}$ in. columns of 0.5% Apiezon L on glass beads 200 μm in diameter. Argon was used as carrier gas at 50 ml/min. Figure 4 shows the gas chromatographic separation of beryllium and aluminum acetylacetonato complexes in acetylacetone. The beryllium complex can be eluted

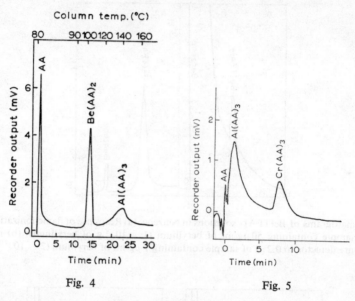

Fig. 4   Fig. 5

Fig. 4. Gas chromatographic separation of beryllium and aluminum acetylacetonato complexes in acetylacetone. Sample size, 0.34 μl; concentration of bis(acetylacetonato)beryllium(II), 0.217 g in 3.00 ml of acetylacetone; concentration of tris(acetylacetonato)aluminum(III), saturated solution at 25°C; column, 4 ft. × ¼ in. of 0.5% by weight Apiezon L on glass beads, 200 μm in diameter; flow rate of argon, 50 ml/min; detector voltage, 1500 V; sensitivity, × 3.

Fig. 5. Gas chromatographic separation of aluminum and chromium acetylacetonato complexes in carbon tetrachloride. Sample size, 0.34 μl; column, 4 ft. × ¼ in. of 1% by weight Apiezon L on glass beads, 200 μm in diameter; column temperature, 170°C; flow rate of argon, 43 ml/min; detector voltage, 2000 V; sensitivity, × 10.

at temperatures as low as 75°C while the aluminum complex appeared only after column temperatures greater than 120°C. The gas chromatographic separation of aluminum and chromium acetylacetonato complexes in carbon tetrachloride is shown in Fig. 5.

The gas chromatographic behavior of the trifluoroacetonates (TFA's) of aluminum, gallium, indium, beryllium, and thallium was studied by Schwarberg et al.[26] and conditions for their complete separation found. The thallium derivative was not characterized, but produced a chromatographic peak well separated from the others.

Using conditions established for complete separations, it was possible to obtain the gas chromatographic analysis of any combination of up to 5 metals by a maximum of two runs which involved only a change in column temperature. The TFA's of scandium(II), chromium(III), copper(II), manganese(III), zirconium(II), hafnium(IV), zinc(II), and thorium(IV) were also studied by GLC, enabling a prediction of interference in this method.

An F & M Model 500 gas chromatograph with a thermal conductivity detector containing either W-1 or W-2 tungsten filaments was used. Borosilicate glass was used for injection port inserts and helical columns. The metal–TFA complexes were prepared[27] by the procedure used for tris(acetylacetonato) Al[II]. Columns used were 4 ft. glass helices with a 4 mm internal diameter. The solid supports and glass columns were silanized according to the procedure of Bohemen et al.[28] with modifications by Ross[29] and involved treatment with 1% chlorotrimethyl silane at room temperature. The columns used were (1) 25 g of glass micro beads (60–80 mesh) coated with 0.5% Silicone 710 oil. It was conditioned at 180° for 24 h while allowing helium to flow at a perceptible rate, (2) same as column (1) except for a silanized column and solid support. It was conditioned in the same way as (1) but at a temperature of 140°, (3) column was packed with 4.2 g of Chromosorb W (30–60 mesh) coated with 5% Silicone 710 oil. It was conditioned at 200° in the same manner as (1). Conditions for gas chromatographic analyses were helium flow rate 79 ml/min for a column temperature of 120°, injection port temperature 135°, block heater temperature setting 775°, actual temperature of detector block 150°, attenuator setting 1 ×, W-2 tungsten detector filaments, and chart speed 0.5 in./min.

Figure 6 shows the partial separation (achieved in 12 min) of the thallium compound and beryllium, aluminum, gallium, and indium TFA's using column (1).

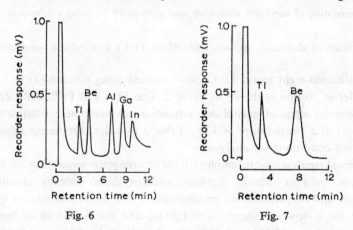

Fig. 6   Fig. 7

Fig. 6. The partial separation of the thallium compound and beryllium, aluminum, gallium, and indium TFA's using column (1). The column temperature was programmed from 85° to 160° at the rate of 7.9°/min. A W–1 tungsten detector filament was used.

Fig. 7. The separation of Be(TFA)$_2$ and the thallium compound at a column temperature of 85°.

The column temperature was programmed from 85° to 160° at the rate of 7.9°/min. A complete separation of the beryllium, aluminum, and gallium chelates and the thallium compound was accomplished on column (*2*), using a column temperature of 85° for the first 3 min, then programming to 116° at the rate of 7.9°/min. (This analysis required 20 min.) The elution time for the complexes of Tl, Be, Al, and Ga were 3.1, 6.6, 11.5, and 16.7 min, respectively. The separation of $Be(TFA)_2$ and the thallium compound at a column temperature of 85° is shown in Fig. 7.

In contrast to the time required for partial separation of the 5 metals by temperature programming, a complete separation of Be, Al, Ga, and In TFA's required 22 min on column (*2*) at 115° (Fig. 7). Conditions for the chromatograms in Figs. 8 and 9 permitted the complete separation and quantitative determination of the maximum number of metals using an isothermal run.

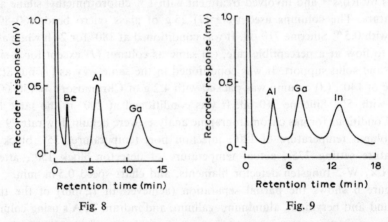

Fig. 8.   Fig. 9.

Fig. 8. The separation of beryllium, aluminum, and gallium TFA's using a column temperature of 115°.

Fig. 9. Separation of aluminum, gallium, and indium TFA's at a column temperature of 120°.

An additional eight metal TFA's were studied using column (*1*) at 125°. Their general order of elution is shown in Table 2. The TFA's of Fe^{III}, Zn^{II}, Zr^{IV}, Mn^{III}, and Hf^{IV} showed signs of partial decomposition or incomplete volatization in the injection port at a temperature of 135°. Table 3 lists the chromatographic columns and operating conditions for analysis.

Isothermal analyses were employed for a complete separation of the TFA's of aluminum, gallium, indium, thallium, and beryllium whereby aluminum, gallium, indium, and beryllium could be determined with an overall relative mean error of 2% for the following ranges: 0.20–1.59 µg and 0.075–0.45 µg of beryllium at column temperatures of 85° and 115°, respectively; 0.37–2.21 µg and 0.67–3.37 µg of aluminum at 115° and 120°, respectively; 2.08–12.48 µg and 3.59–10.77 µg of gallium at 115° and 120°, respectively; and 12.1–36.4 µg of indium at a column temperature of 120°.

TABLE 2

COMPARATIVE ORDER OF ELUTION FOR VARIOUS METAL TFA'S[a]

Metal TFA	Retention time (min)
Al^{III}	2.6
Ga^{III}	4.8
Sc^{III}	6.0
Cr^{III}	7.5
Cu^{II}	7.7
Mn^{III}	8.2
In^{III}	8.5
Zr^{IV}	12.0
Hf^{IV}	12.0
Zn^{II}	12.8
Th^{IV}	[b]

[a] Column 1 at 125°C. Flow rate 83 ml/min at 125°C.
[b] No peak up to 13 min.

TABLE 3

DESCRIPTION OF CHROMATOGRAPHIC COLUMNS AND CONDITIONS FOR GLC ANALYSIS

Conditions for gas chromatographic analyses

Column temperatures: specified for each figure
Helium flow rate: 79 ml/min for a column temperature of 120°C
Injection port temperature: 135°C
Block heater temperature setting: 175°C
Actual temperature of detector block: 150°C
Attenuator setting: IX
W-2 tungsten detector filaments
Chart speed: 0.5 in./min
Column: 1 s

Description of chromatographic columns

Columns were 4 ft. glass helices with a 4 mm i.d.

1   This column was packed with 25 g of glass micro beads (60–80 mesh) coated, 0.5% by weight, with Silicone 710 oil. It was conditioned at 180°C for 24 h while allowing helium to flow at a perceptible rate.

1 s   This was the same as column 1, except for a silanized column and solid support. It was conditioned in the same way as 1, but at a temperature of 140°C.

2   Column 2 was packed with 4.2 g of Chromosorb-W (30–60 mesh) coated, 5.0% by weight, with Silicone 710 oil. It was conditioned at 200°C in the same manner as 1.

The utilization of fluorinated ligands offers significant advantages over the use of the protonated analogues for the formation of volatile metal chelates. For example, the volatility and the excellent response of the electron-capture detector for metal complexes derived from trifluoroacetylacetone, the anion of which may be depicted as $CF_3–\underset{\underset{O}{\|}}{C}–CH=\underset{\underset{O}{|}}{C}–CH_3$, has made this agent one of the most used chelating agents

for GLC analysis of metals.

Ross and Sievers[30,31] described a rapid ultra-trace GLC determination of beryllium as beryllium(II) trifluoroacetylacetone [Be(TFA)$_2$]. The lower limit of detectability was about 4×10^{-13} g of beryllium with calibration plots extending from 8×10^{-13} to 4×10^{-11} g of beryllium. Samples of beryllium in aqueous solution at four concentrations (1.8×10^{-7}, 1.18×10^{-8}, 2.95×10^{-9}, and 8.84×10^{-10} g/ml were analyzed quantitatively by combining solvent extraction and gas chromatography. The distribution of beryllium during the extraction procedures was determined independently by use of radioactive beryllium-7.

An ionization detector (diode type) cell, Model A-4150, was used in a Barber-Colman Model 20 gas chromatograph. This detector was used in a non-pulsed electron-capture mode by reducing the cell potential from a modified power supply. The maximum accessible cell potential was 200 V. The instrument operating conditions were 4 ft. × 0.06 in. bore DuPont Teflon column packed with 5% SE-52 silicone gum rubber on 60–80 mesh Gas Chrom Z, column temperature 80°C, injection port temperature 168°C, detector temperature 200°C, carrier gas prepurified nitrogen at a flow rate of 50 ml/min, column inlet pressure 28 p.s.i.g., scavenger flow rate 214 ml/min prepurified nitrogen, detector voltage 10 V.

Figure 10 illustrates a chromatogram obtained from injection of a 1 μl sample of an extracted organic layer containing Be(TFA)$_2$ in benzene before and after washing with aqueous sodium hydroxide.

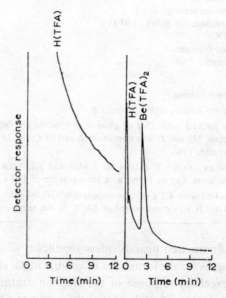

Fig. 10. Chromatogram obtained from injection of a 1 μl sample of the extracted organic layer before (left) and after (right) washing with aqueous sodium hydroxide. The sample contained 2.7×10^{-11} g of beryllium.

Noweir and Cholak[32] described gas chromatographic determination of ultra-trace quantities of beryllium in urine, blood, tissue, and air-borne dust. Except for air-borne dust, the procedure involves a double extraction, first with acetylacetone in benzene and then with a benzene solution of trifluoroacetylacetone. Beryllium in air-borne dust was extracted only by trifluoroacetylacetone in benzene. Levels as low as 1×10^{-10} g (0.001 μg) of beryllium per sample could be determined by this method. Except for Fe^{3+} and Al^{3+}, none of the ions usually present in air-borne dust or biological material interferes with the analysis. Fe^{3+} was removed in a magnetic mercury cathode and Al^{3+} was precipitated with 8-quinolinol in chloroform. Recovery of beryllium ranged from 70 to 90% at levels from 1×10^{-9} g to 1×10^{-7} g per sample.

A Loenco Model 70 Hi-Flex chromatograph (Loe Engineering Co., Altadena, Calif.) was used provided with an electron-capture detector using a tritium source at 22.5 V. The column consisted of 4 ft. $\times$ 0.06 in. i.d. Du Pont Teflon packed with 60–80 mesh Gas Chrom Z impregnated with 5% SE-52 (methylphenyl) silicone gum. The column was operated at a temperature of 80°C with nitrogen provided at an inlet pressure of 24 p.s.i.g. and a rate of 50 ml/min being used as the eluent gas. The temperatures at the injection port and at the detector were maintained at 160° and 200°C, respectively. The injection port was modified by inserting a borosilicate glass liner to reduce likelihood of sample decomposition. A Servoriter II portable 0–1 mV range recorder, Model PS01 WGA (Texas Instruments) was used to record the chromatograms.

Ross and Sievers[33] recently described a GC analysis of ultra-trace concentrations of beryllium in particulate matter collected on air filters and in water. The particulate matter on glass fiber filters was ashed, the residue digested with acid, the resulting solution buffered to the optimum pH for solvent extraction and potential interfering metals eliminated by masking with EDTA. A benzene solution of trifluoroacetylacetone [H(TFA)] was allowed to react with the beryllium in the aqueous solution to form $Be(TFA)_2$ which was analyzed by electron-capture GC. Analysis of air filters showed beryllium concentrations from 0.00016 to 0.00049 μg of Be/m^3 of air in 14 urban samples examined. The results indicated that the GC method was more sensitive and reproducible than any other method now employed for the determination of beryllium in environmental samples.

A Hewlett-Packard Model 402 gas chromatograph was used equipped with a ^3H electron-capture detector and a 2 m $\times$ 3 mm i.d. borosilicate column containing 2.8% W-98 silicone (Union Carbide) on Diatoport (Hewlett-Packard). The carrier gas was 10% methane–90% argon at 54.5 ml/min, column temperature 110°C, detector temperature 200°C, and on-column injection (requiring no additional heat at the site of injection).

Figure 11 shows a chromatogram of an air filter digest analysis for beryllium and Fig. 12 is a chromatogram of a standard solution of $Be(TFA)_2$ (1.0 $\times$ 10^{-11} g Be/μl).

Beryllium can be measured rapidly in air by this GC procedure with sensitivity at least two orders of magnitude below the AEC recommended neighborhood maximum allowable concentration[34] of 0.01 μg Be/m³. The technique required about 30 min for the extraction procedure and less than 10 min for one GC analysis and was suggested to be quite applicable for route analysis of both air and water samples for beryllium, to monitor trends and variations of Be concentrations and, with variations of the method, it could be employed for monitoring biological samples.

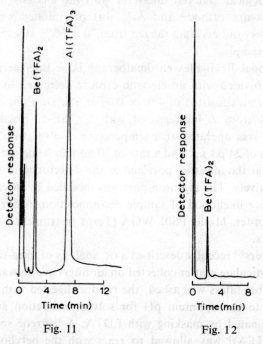

Fig. 11. Chromatogram of beryllium extracted as Be(TFA)₂ from an air filter digest solution, 0.00043 μg Be/m³; 1 μl sample injected.

Fig. 12. Chromatogram of 1 μl of a standard solution of Be(TFA)₂ 1.0×10^{-11} g Be/μl.

The utility of Be(TFA)₂ complexes for electron-capture GLC determination was further demonstrated in the ultra analysis for beryllium in terrestrial, meteoritic and Apollo 11 and 12 lunar samples by Eisentraut et al.[35]. The concentration of beryllium in the crystalline rocks was less than 1 p.p.m. while lunar dust was found to contain as much as five times more beryllium. The detection limit was approximately 4×10^{-14} g beryllium with the sensitivity of the electron-capture detector to Be(TFA)₂ being much greater at higher carrier gas flow rates.

A Hewlett-Packard Model 810 Research Gas Chromatograph equipped with a tritium electron-capture detector was used for the analysis. The chromatograph was modified by incorporation of a Hewlett-Packard Model 5750 injection port assembly. The column used was constructed from thick-walled Teflon (DuPont) tubing (56 cm × 3 mm i.d.) packed with 10% SE-30 on 70–80 mesh Gas Chrom Z. The tubing had a wall thickness of approximately 2 mm. The carrier gas was 90% argon–10% methane at a flow rate of 60 ml/min which was dried by passing through a bed of molecular sieves and Drierite. No purge gas was used and the purge entry opening was sealed off. The column was operated isothermally at 120°C with the electron-capture detector at 176° and the injection port at 161°C. The pulse interval was 150 μsec and the electrometer was operated generally at an attenuation of 1432. (This is equivalent to a current for fullscale display of 1.28×10^{-10} A.) The peak for Be(TFA)$_2$ appeared 1.7 min after injection and the chart speed was 4 min/in. It was found that optimum flow rates for the analysis were between 200 and 300 ml/min which resulted in the enhancement of sensitivity to Be(TFA)$_2$ by nearly 50 times over that at 20 ml/min.

Barrett et al.[36] studied the GLC of beryllium and zinc oxyacetates. Beryllium oxyacetate was prepared by reacting freshly precipitated beryllium hydroxide with anhydrous acetic acid. Beryllium oxypropionates (μ$_4$-oxohexa-μ-propionatotetra-beryllium(II)) were prepared for comparison and their properties investigated by thermogravimetry and GLC. A Pye 104 gas chromatograph with a heated flame-ionization detector was used. Eluates were collected for characterization by means of a 100:1 stream splitter and the "manual preparative unit" available for this instrument. Analytical gas chromatography was performed with Pyrex glass (or Teflon) columns (1 m × 0.4 cm) containing 5% Apiezon L on 85–100 mesh "Universal B" support (a silanized flux-calcined diatonite; Phase Separations Ltd.). The column was conditioned for 12 h at 200° before use. Thermogravimetric analyses were carried out on a Perkin Elmer TGS-1 thermobalance. The analyses were performed in a pure nitrogen atmosphere at a flow rate of 20 ml/min. The temperature scan rate was 20°/min with an upper temperature limit of 650° (sample sizes were 1.5 mg).

Figure 13 illustrates comparative thermograms for beryllium and zinc oxyacetates and beryllium β-diketonates. The three metal oxycarboxylates are considerably less volatile than the beryllium β-diketonates, hence higher column temperatures are essential for gas chromatography. The shapes of the thermograms for all the beryllium samples were very similar and showed negligible decomposition. This study indicated that the thermal stability and volatility of the beryllium oxycarboxylates should favor quantitative GLC, whereas quantitative elution of zinc oxyacetate appeared improbable. The GLC behavior of beryllium oxyacetate as a 10% solution in chloroform was examined on a freshly prepared glass column of Apiezon L on Universal B with a nitrogen flow rate of 60 ml/min. Column, injector port, and detector temperatures were 150, 180, and 180°C, respectively. Figure 14 illustrates a gas chromatogram of beryllium oxyacetate (25 μg) and shows good peak symmetry with little tailing. The initial minor peak (ca. 2%) could not be eliminated by changing operating parameters and was probably due to a little decomposition on the column. A rectilinear relationship with peak area was found over the range of 2–60 μg of beryllium oxyacetate. The elution of beryllium oxy-carboxylates was strongly dependent on the state of the column used. After some weeks of use, the Apiezon L column gave a chromatogram for beryllium oxyacetate with multiple peaks, most of which were of shorter retention time than the authentic beryllium oxyacetate peak (Fig. 14, curve B).

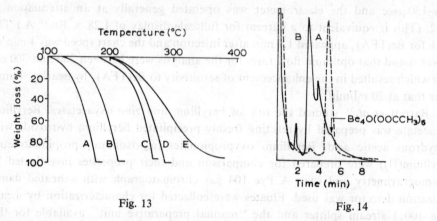

Fig. 13   Fig. 14

Fig. 13. Comparative thermograms of beryllium and zinc oxyacetates and beryllium β-dikentonates. A, Be(TFA)$_2$; B, Be(AA)$_2$; C, Be$_4$O(OOCCH$_2$CH$_3$)$_6$; D, Be$_4$O(OOCCH$_3$)$_6$; E, Zn$_4$O(OOCCH$_3$)$_6$.

Fig. 14. Gas chromatogram of beryllium oxyacetate (25 μg); conditions as in text. – – –, On freshly prepared column (curve A); ———, on aged column (curve B).

Butler[37] described two methods for concentrating beryllium from biological samples. One method was the cation exchange of beryllium from dilute hydro-chloric acid to di-2-ethylhexylphosphoric acid (HDEHP). (This exchange requires 30 min of mixing.) In the other method, beryllium as an oxalate anion complex was anion-exchanged to triisooctylamine (TIOA) in 10 sec. Although neither method was specific for beryllium, microgram quantities of the element could be separated from gram quantities of solids. When 1 liter of urine containing [7]Be tracer was wet-ashed to 12 g of inorganic salts and concentrated by the TIOA method, 95% of the tracer was recovered and only 6 mg of solids accompanied the [7]Be. It was suggested that the TIOA–oxalate exchange in conjunction with other liquid–ion exchange separations may simplify complex methods now in use for ultrasensitive beryllium determinations. Concentration of beryllium to TIOA and subsequent analysis of the beryllium by gas chromatography should thus offer considerably enhanced sensitivity.

Yamane et al.[38] reported the analysis of beryllium, chromium and aluminum in water via oxine, L-methyloxine extraction and by thin-layer chromatography. The extraction procedure involved the initial addition of 2 ml of 2% oxine and 4 ml of 2% 2-methyloxine to a 1 μl sample water containing the above metals; the solution is adjusted to pH 8 with NaOH solution and held at 60° for 5 min. The chelates of Be, Cr, and Al are extracted with 30 ml of $CHCl_3$. The organic phase is then concentrated on a water bath and the metals back-extracted into 0.5 ml 3 N HCl. Oxine and 2-methyloxine are precipitated by bromination with a $KBr–KBrO_3$ mixture since their co-extraction interferes with the subsequent TLC step. The supernatant solution is used for TLC on silica gel with ethanol–$NHNO_3$ (99:1) used as developer. Beryllium and aluminum were identified by a 2% oxine spray and fluorescence in ultraviolet, and chromium was determined by a diphenyl carbazide spray. The detection limit for Be and Cr was 1 p.p.m. and for aluminium, 0.1 p.p.m. Anions such as Cl^-, SO_4^{2-}, BO_3^{3-}, $C_2O_4^{2-}$, and tartrate do not interfere with the extraction, while F^- and PO_4^{3-} do.

Eristavi et al.[39] described the separation and determination of beryllium (in beryl and ML-5 Mg alloys) from iron and aluminum by anion-exchange chromatography. Preliminary separation was accomplished using AV-16 and AV-77 anion exchangers in their F^- form and the subsequent photometric determination with Beryllon II. Beryllium was determined in the alloy by first decomposing 2 g of the sample, precipitating Be, Fe, and Al with NH_4OH as hydroxides to separate from Mg, filtering, dissolving the precipitate in an excess of sulfuric acid, and diluting to 250 ml with water. Passage of a 25–50 ml aliquot at a rate of 1 ml/min through the anion exchange column was followed by washing with 25 ml H_2O thence elution of the Fe with 150 ml 1% NaF, then elution of the Be with 180 ml of 3% NaF at a rate of 5 ml/min. Beryllium in the eluate was then determined photometrically.

Eristavi et al.[40] described the separation of beryllium from nickel, copper, aluminum, iron, titanium, and vanadium on AV-17 anion exchanger (carbonate form). After adsorption of these elements on the anion exchanger, sequential elution was effected with 3 N aq. ammonia (for Ni, Cu, and Al), 1 N sodium hydroxide (for Be), 3 N sodium hydroxide (for Ti and V) and 1.5 N hydrochloric acid (for Fe). Beryllium was then determined spectrophotometrically with arsenazo I. For the determination of 0.01–1 mg of beryllium the error was $\pm 8\%$.

2. CHROMIUM

Chromium occurs in nature chiefly as chrome iron ore (chromite) $(FeOCr_2O_3)$ found mainly in the U.S.S.R., Southern Rhodesia, South Africa, and Turkey. Soils have been found to contain from a trace to 2.4% chromium. Vegetables of 25 botanical families contain 10–1,000 μg of chromium/kg of dry matter with most plants falling within the range of 100–500 $\mu g/kg$. Other sources of chromium include asbestos (content in the most common asbestos mineral, chrysolite, is approx. 1,500 $\mu g/g$), and coal (approx. 7–20 μg and the ash contains 1–270 $\mu g/g$ depending on the origin of the coal).

The industrial uses of chromium include (a) the production of alloys with nickel, molybdenum, vanadium, cobalt, and niobium and of corrosion-resisting steel and iron, (b) in chromium plating, (c) production of catalysts (2.8 million lb. of dichromate equivalent produced[41] in 1958), (d) chrome-tanning industries (15 million lb. of chromic oxide annually), (e) primer paints and dips, pigments (approx. 30,000 tons of sodium dichromate annually), (f) as mordants in the textile industry, and (g) graphic arts (approx. 200,000 lb. of dichromates annually). Other areas of application include fungicides and wood preservatives which consume approx. 2.9 million lb. of chromium chemicals annually. The major fungicide use is for potato and tomato blight control, seed sterilization and lower fungicide control. Chromate wood preservatives are also used in wooden cooling towers. Chromate chemicals, e.g. sodium chromate and complex chromate salts are widely used in cooling-tower recirculating water systems as rust inhibitors as well as wood preservatives. The circulating water usually contains 15–300 p.p.m. chromate ion and such systems continuously discharge about 1% of their flow to waste water and some is lost to the atmosphere. Additional uses of chromium are in paper matches, fireworks, dry cell batteries, and as anti-static additives in aviation and other fuels[42]. Chromates are also used for the oxidation of organic materials in the production of synthetic dyes, saccharin, benzoic acid, anthraquinone, hydraquinone, camphor, and synthetic fibers. The chromates (because of their oxidizing properties) are used extensively as cleaning agents, in the purification of chemicals and in inorganic oxidation. Table 4 lists some chromium compounds and their uses. The uses of chromium that are believed to be the most likely sources of atmospheric pollution are metallurgical and chemical industries

TABLE 4

SOME CHROMIUM COMPOUNDS AND THEIR USES

Compound	Uses
Chromium trioxide CrO_3	Chromium plating, photography, medical and veterinary use (5% solution as topical antiseptic and astringent).
Chromic acetate $Cr(C_2H_3O_2)_3$	Mordant in dyeing; in tanning, photographic emulsions, catalyst for polymerization of olefins.
Chromic bromide $CrBr_3$	In catalysts for polymerization of olefins.
Chromic chloride $CrCl_3$	In chromizing; in manufacture of Cr metal and compounds; weather-proofing agent, corrosion inhibitor; textile mordant.
Chromic hydroxid $Cr(OH)_3$	As pigment; mordant in tanning industry; catalyst.
Chromic oxide Cr_2O_3	In abrasives; as refractory materials; electric semi-conductors; pigment in alloys; catalyst.
Chromium Cr	Manufacture of chrome-steel; chrome-nickel steel, ^{51}Cr isotope as tracer in various blood diseases and determination of blood volume.
Lead chromate $PbCrO_4$	As pigment in oil and water colors; decorating china and porcelain.
Potassium dichromate $K_2Cr_2O_7$	In tanning, leather, dyeing, painting; corrosion inhibitor; oxidizer in manufacture of organic chemicals.
Cupric chromate $CuCr_2O_4$	In fungicides; seed protectants, wood preservatives.

and in products employing chromate compounds as well as its presence in cement and asbestos. Chromium concentrations in urban air average 0.015 µg/m³ and range as high as 0.350 µg/m³.

It is known that chromium (*e.g.* trivalent chromium) is involved in carbohydrate and fat metabolism and hence appears to be an essential trace element. The average total daily intake of 30–140 µg of chromium by man is derived from food (30–100 µg), water (0–40 µg), and air (0–0.3 µg).

A very high proportion of dietary chromium is excreted in the feces while most of what is absorbed is returned to the intestine in the bile and only about 1% enters the general circulation to be excreted in the urine. Traces might be stored, probably in an insoluble form, in the lungs. Regional differences in the chromium content of human tissues have been noted[43], *e.g.* levels of subjects in the U.S. were well below the average in contrast to the exceptionally high level of chromium levels in subjects from Bangkok and Manila.

Chromium(III) is less toxic than chromium(VI)[44] whose toxicity has been noted in humans exposed to industrial dichromates[45–47]. The principal toxic effects of chromium from industrial aspects are exerted on the skin, the nasal membrane and the lungs. Although systemic effects have been rarely described in industrial workers, lesions of the kidneys have been observed in the non-industrial population following ingestion or external application of chromium compounds. Aspects of the potential carcinogenicity of chromium in chrome workers (especially with regard to the incidence of lung carcinoma) have not been unequivocally resolved[48–51]. Intramuscular, subcutaneous, intrapleural, intraperitoneal and intraosseous injections of chromium compounds have been reported to cause the development of local sarcomas in rabbits, rats, and mice[52–55].

A variety of methods for the detection and measurement of chromium in biological materials have been described, and include spectrophotometry[56–58], arc-emission spectrography[59–62], spark source mass spectroscopy[63], atomic absorption spectrometry[64–66], coulometry[67], polarography[68], neutron activation analysis[69,70], X-ray fluorescence[71], and colorimetry[72,73]. The occurrence, functioning, and measurement of chromium in biological systems has been reviewed recently by Mertz[74].

Gas chromatography is a specific and extremely sensitive method for the detection of metals as their β-diketone chelates. β-Diketone chelates of chromium derived from acetylacetone[75,76], trifluoroacetylacetone[75–80], hexafluoroacetylacetone[75,76,81,82], and heptafluorodimethyl octanedione[83] have all been analyzed by GC. Exceptional sensitivity has been obtained from the fluoro-β-diketones using electron-capture detection[75,77,78,81,83].

The GLC of chromium(III) hexafluoroacetonate [$Cr(HFA)_3$] in concentrations in the range $10^{-8}–10^{-3}$ g/ml has been described by Ross and Wheeler[84]. A Barber-Colman Model 20 gas chromatograph equipped with an ionization detector (diode type) cell, Model A-4150 was used with an 11 ft. × $\frac{1}{8}$ in. stainless steel column packed with 20% Dow Corning Silicone Fluid 710R on Gas Chrom Z (60–80 mesh). The operating parameters were: column temperature 90°C, flash heater temperature 170°C, and detector temperature 200°C. Carrier gas was prepurified nitrogen at a flow rate of 67 ml/min.

Savory et al.[85,86] reported a method for measuring toxicological levels of chromium in serum. The procedure involved preliminary decomposition of the organic matter of the sample, thence isolation of chromium from aqueous solutions of the ashed samples *via* chelation–extraction using 0.3 *M* trifluoroacetylacetone in benzene. After removal of excess chelating agent by alkaline washing, the benzene solutions of chromium trifluoroacetylacetone [Cr(TFA)$_3$] were analyzed by gas chromatography using a model 402 Hewlett-Packard chromatograph equipped with a ^{63}Ni electron-capture detector and fitted with a 10 ft. × $\frac{1}{4}$ in. glass column packed with 60–80 mesh WAW Chromosorb pretreated with dimethyldisilazane and impregnated with 5% QF-1. The carrier gas was 95% argon–5% methane, column flow, 80 ml/min, column temperature 150°C, preheater temperature 175°C, and detector temperature 200°C. The ^{63}Ni detector had a pulse interval setting of 150.

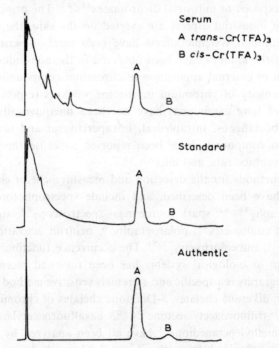

Fig. 15. Chromatograms of *trans*- and *cis*-Cr(TFA)$_3$ from digested serum, digested standard solution of chromium(III), and a solution of authentic Cr(TFA)$_3$ in benzene.

Chromatograms obtained from digested serum, standards, and authentic Cr(TFA)$_3$ are shown in Fig. 15. The identity of the Cr(TFA)$_3$ peaks in chromatograms from serum and urine was ascertained by comparison of retention times of the assigned peaks with authentic samples of *trans* (and *cis*) Cr(TFA)$_3$ under varying conditions of column temperature, carrier gas flow rate and column packing, as well as by the addition of a solution of authentic Cr(TFA)$_3$ isomer mixture to each of the benzene solutions of the serum and urine extracts. (Recoveries of chromium were essentially quantitative.) The retention times of the *trans* and *cis* isomers of Cr(TFA)$_3$ were 10 and 12 min, respectively. Determinations of serum and urine chromium levels by GLC were compared with measurements by the atomic absorption procedure of Feldman *et al.*[64]. The comparison data for 16 serum samples are given in Fig. 16.

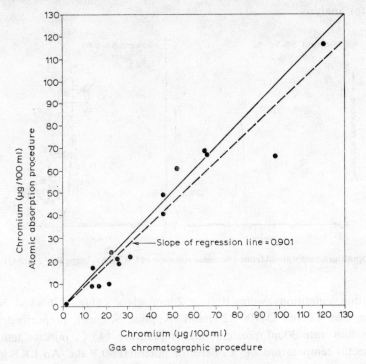

Fig. 16. Comparison of chromium concentration measurements in 16 serum samples by the gas chromatographic and atomic absorption spectrometric procedures. – – – Indicates calculated regression line; —— represents theoretical relationship $x = y$.

The slope of the regression line (m) was 0.901 with a standard error of estimate (s) of 8.2 µg/100 ml and a correlation coefficient (R) of 0.963. A similar comparison of 8 urine samples was made giving $m = 1.043$, $x = 5.8$ µg/100 ml and $R = 0.923$. GLC offers a potential advantage of greater sensitivity over atomic absorption. The limit of sensitivity for the detection of chromium was 0.03 pg injected into the column of the gas chromatograph. This amount, which gave a peak height of 5% above the base but is equivalent to 1 ng chromium/100 ml of final benzene solution. In comparison, the limit of sensitivity of atomic absorption[64] is 0.6 µg of chromium/100 ml in methyl isobutyl ketone.

Savory *et al.*[85] also demonstrated the efficiency of 5% QF-1 over 5% SE-52 for the separation of the *cis* and *trans* isomers of $Cr(TFA)_3$ as shown in Fig. 17.

Booth and Darby[87] described the GLC determination of physiological levels of chromium in biological tissues using a modification of the procedure of Savory *et al.*[86]. The techniques permitted the analysis of diverse materials including liver, fat, plasma, and diet and increased the detection limits of chromium in tissues to values below 20 ng Cr/g tissue (20 p.p.b.). A Varian Aerograph Model 1840 Dual Column Gas–Liquid Chromatograph with dual ^{63}Ni electron-capture detectors was used for analysis.

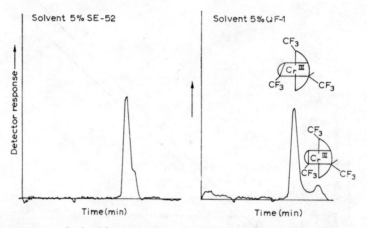

Fig. 17. Chromatograms obtained from a benzene solution of $Cr(TFA)_3$ using Se-52 and QF-1 column packings.

Operating conditions were 10 ft. × 2 mm glass columns packed with 3% OV-225 on Gas Chrom Q 100–120 mesh (Applied Science), prepurified nitrogen carrier gas flow rate 30 ml/min, column temperature 143°C, injector temperature 185°C, detector temperature 245°C, detector potential 90 V d.c. An LKB gas chromatograph–mass spectrometer Type 9000 equipped with a 10 ft. × 4 mm glass column packed with 5% QF-1 on H.P. Chromasorb W (AW-DMCS), 80–100 mesh (Applied Science) operated at 70 meV was used for GLC–MS determinations. Chromium-51 was determined using a Nuclear Chicago Model 4998 gas radiochromatography counting system and Nuclear Chicago scaler Model 8725 with well-type NaI crystal.

$Cr(TFA)_3$ was synthesized and purified by the method of Fay and Piper[88,89]. The acid digestion mixture consisted of a 6:3:1 (v/v/v) mixture of conc. HNO_3, 73% $HClO_4$ and conc. H_2SO_4. The combination of four final 1.0 N NaOH washes together with the use of a 10 ft. 3% OV-225 column produced a relatively clean chromatogram as seen in Fig. 18 (illustrating a gas chromatograph of chromium analysis of liver samples). Under the conditions of analysis, the first $Cr(TFA)_3$ peak (*trans* isomer) emerged with a retention time of 5.6 min. The relative retention time of *trans*-

Cr(TFA)₃ lindane was 0.141 and the time of analysis was approx. 15 min. The peak height measurement of the first Cr(TFA)₃ peak (*trans* isomer) was used for quantitative determinations.

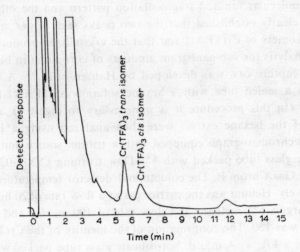

Fig. 18. Typical gas chromatogram of Cr analysis of liver sample.

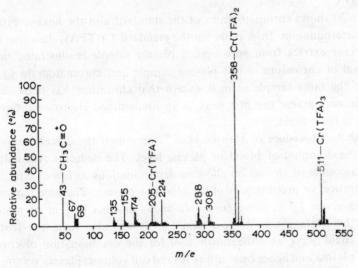

Fig. 19. GLC–mass spectrograph of peak from purified synthetic Cr(TFA)₃.

Analysis by GLC–mass spectrometry produced the mass spectrum shown in Fig. 19 which was identical for both Cr(TFA)₃ peaks. There was a molecular ion at m/e 511, the molecular weight of Cr(TFA)₃. The most abundant ion was at m/e 358, corresponding to Cr(TFA)₂. Prominent peaks were also seen at m/e 205 and 43 corresponding to Cr(TFA) and $CH_3C≡O^+$, respectively. The identity of these fragments was confirmed by the presence of a characteristic isotopic abundance pattern for chromium in the appropriate peaks and by analogy to mass spectra of other chromium complexes which also show sequential loss of chelating units[90-92]. This mass spectrum, its molecular ion and fragmentation pattern and the other physical and chemical data clearly established that the two peaks seen in GLC analysis were the *cis* and *trans* isomers of $Cr(TFA)_3$ and that the crystallized product was pure.

A rapid analysis for sub-nanogram amounts of chromium in blood and plasma using electron-capture GLC was developed by Hansen *et al.*[93]. A 0.050 ml sample was reacted in a sealed tube with a hexane solution of TFA (1,1,1-trifluoro-2,4-pentanedione). (In this procedure it is unnecessary to digest or ash the sample.) Components of the hexane extract were then analyzed using a Hewlett-Packard Model 402 gas chromatograph equipped with a tritium source and a 2 ft. × $\frac{3}{16}$ in. i.d. borosilicate glass tube packed with 5% Dow Corning LSX-3-0295 silicone gum on 60–80 mesh Gas Chrom P. The column and detector temperatures were 132 and 190°C, respectively. Helium was the carrier gas at a flow rate of 20 ml/min, the purge gas was 95% argon–5% methane at a flow rate of 100 ml/min and the pulse mode interval setting was 150. The confirmation of the identity of the $Cr(TFA)_3$ peak was performed using a 4 ft. × $\frac{3}{16}$ in. i.d. borosilicate glass tube packed with 3.8% Union Carbide W-98 silicone gum on 80–100 mesh Anachrom ABS with a column temperature of 133°C.

Figure 20 shows chromatograms of the standard and the hexane extracts from a plasma determination. In A is shown the standard $Cr(TFA)_3$ dissolved in hexane. In B a hexane extract from an unspiked plasma sample is illustrated showing the natural level of chromium in this plasma sample and chromatogram C shows the analysis of the same sample as in B, except that chromium has been added. In all plasma chromatograms the first peak is an unidentified electron-capturing species derived from the plasma.

The above procedure of Hansen *et al.*[93] permitted the determination of chromium at the 50 ng/ml of blood or plasma level. The dedector responded well to picogram amounts of chromium allowing determinations as low as 5 ng/ml without preconcentration or pretreatment of the blood or plasma. The time requirement per sample was about 1.5 h; however, simultaneous anaylsis of four or more samples decreased the average time to less than 1 h. It was also suggested that this GLC technique could serve as a diagnostic tool for the determination of chromium in blood and plasma and hence be of utility for red cell volume, plasma volume measurements, studying red cell survival time and in evaluating blood loss. The use of non-radioactive chromium instead of the usual radioactive materials is of apparent value for certain patients such as pregnant women and children where radioactive methods are contraindicated.

Hansen *et al.*[93] also evaluated the electron–capture GLC determination of chromium in blood and plasma using 1,1,1,5,5,5-hexafluoro-2,4-pentanedione (HFA) and 1,1,1,2,2,2,3,3-heptafluoro-7,7-dimethyl-4,6-octanedione (HFOD) as direct chelating agents. However, the highest extraction efficiency for either agent in hexane never exceeded 50% compared with approximately 90% found for TFA.

Wolf et al.[94] recently reported the application of a modified conventional gas chromatograph–mass spectrometer (GC–MS) for the quantitative determination of picogram amounts of chromium in blood plasma and serum and of beryllium in an environmental residue sample. With the GC–MS technique described, chromatographic conditions were not especially critical for successful analysis since electron-capturing impurities did not interfere with the detection of the metal derivative. Hence, the sensitivity of GC–MS for detection of the metal chelate from the above samples was somewhat greater than that which is attainable with electron-capture GC.

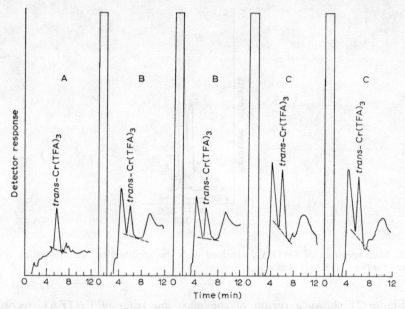

Fig. 20. Chromatograms of A, 1.0 µl standard Cr(TFA)$_3$ solution containing 5.1×10^{-9} g Cr/ml; B, 1.0 µl hexane extract of plasma reaction with no added chromium; and C, 1.0 µl hexane extract of plasma reaction containing added chromium at the 0.052 p.p.m. level. Column, 2 ft. 5% DC LSX-3-0-95; column temperature, 132°C; detector temperatures 190°C; He flow rate, 20 ml/min; CH$_4$–Ar flow rate, 100 ml/min; range, 10; attenuation, 8.

A DuPont 21-491 Double Focusing Mass Spectrometer, coupled through a Biemann-Watson separator to a Loenco Model 160 gas chromatograph, was used. A 2 ft. Teflon column packed with 10% SE-30 on Gas Chrom Z was utilized in the chromatograph with helium as the carrier gas at 40 ml/min and an oven temperature of 130°C. The entire system was silanized periodically by injection of 300 µl of N,O-bis-trimethylsilyl acetamide at a column temperature of 190°C.

A Hewlett-Packard Model 810 gas chromatograph equipped with an electron-capture detector (200 mCi ^{3}H, pulsed mode) and a 6 ft. $\times \frac{1}{4}$ in. o.d. glass coiled column packed with 8% UC-W98 (Applied Science) on Gas Chrom Z was used with a carrier gas of 10% methane–90% argon at 27 ml/min. The oven, injection port, and detector temperatures were 170, 180, and 188°C, respectively. TFA-chelates were prepared by the reaction of 50 μl of the biological material with 500 μl of a benzene solution of trifluoroacetylacetone (9.987 M), for 30 min at 150°C in a flame-sealed tube made from a disposable pipet (Pasteur type, Matheson, Coleman and Bell). The reaction tube was allowed to cool for 5 min, centrifuged, and the benzene layer removed and placed in a vial containing 0.5 ml of 1.0 N sodium hydroxide (to remove excess H(TFA)). After shaking 5 min, the vial was centri-fuged, the benzene layer removed, and aliquots injected into the GC–MS.

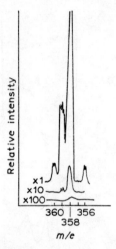

Fig. 21. Mass spectrum of Cr(TFA)$_3$ obtained with the modified GC–MS, illustrating resolution achieved for the Cr(TFA)$_3^+$ peaks.

Figure 21 shows a region of the mass spectrum of Cr(TFA)$_3$ recorded at optimum instrument resolution after the slit dimensions of the instrument were altered. The mass region shown corresponded to that of the Cr(TFA)$_2^+$ species (masses 356–360). The ratio of peak heights from the figure are $I_{356}:I_{358}:I_{359}:I_{360}$ = 0.660:1.000:0.268:0.066. This compared favorably with the calculated inten-sities based upon the naturally occurring isotope abundances (0.052:1000:0.229: 0.055) for the Cr(TFA)$_2^+$ species.

The lower limit of detectability for chromium for this technique is 0.5 pg and thus the chromium in an injection of 40 μl of a solution containing concentrations in the sub-p.p.b. range have been determined in standard solutions with an accuracy of the order of 20%.

Applicability of this technique to the detection and quantitative determination of beryllium present in standard solutions of Be(TFA)$_2$ and in an environmental sample previously analyzed by electron-capture gas chromatography was studied.

The lower limit of detectability of Be from a standard solution was on the order of 1×10^{-12} g when analyzed by GC–MS. GC–MS analysis of beryllium extracted from particulate matter on an air filter sample compared with values obtained by electron-capture gas chromatography, *e.g.* 22–23 pg Be/μl by GC–MS, 20–25 pg Be/μl by electron-capture GC. For GC–MS analysis of beryllium, the peak at m/e 218 was monitored. This corresponded to the fragment ion of largest intensity in the mass spectrum of Be(TFA)$_2$. (No interfering peaks were present at this m/e). It was suggested that the ultimate limit of detectability of metal chelates by mass spectrometry appears to be appreciably lower than the picogram level since mass spectrometers of even higher sensitivity than were employed in the above study are available.

The gas chromatographic behavior of four metal carbonyls, Fe(CO)$_5$, Cr(CO)$_6$, Mo(CO)$_6$, and W(CO)$_6$, on two non-polar phases, squalene and Apiezon L, has been described by Pommier and Guiochon[95]. An IGC-12M gas chromatograph (Intersmat) was equipped with a thermal conductivity detector with two Gow-Mac tungsten rhenium filaments. The columns were 1 m $\times$ 2 mm i.d. stainless steel tubes

TABLE 5

SPECIFIC RETENTION VOLUMES (cm^3/g) OF METAL CARBONYLS ON SQUALANE

Temperature ($^\circ$C)	Fe(CO)$_5$	Cr(CO)$_6$	Mo(CO)$_6$	W(CO)$_6$
63	90.8	177		
70	84.8	142		
79	58.1	101.6	202	
87	47.5	79.3	154	302
97	35.4	59.9	109	202
102				178
112	24.1		68.5	128
120	22.3	32.3	56.9	102

TABLE 6

SPECIFIC RETENTION VOLUMES (cm^3/g) OF METAL CARBONYLS ON APIEZON L

Temperature ($^\circ$C)	Fe(CO)$_5$	Cr(CO)$_6$	Mo(CO)$_6$	W(CO)$_6$
63	40.5	68.1	145	313
70	33.1	55.2	119	159
79	24.5	39.6	76.9	162
87	20.0	31.7	61.7	125
97	14.9	24.8	46.4	87.8
102				78.3
112			28.9	54.6
118	9.3	12.4	23.3	44.0

packed with Chromosorb P (acid-washed, 80–100 mesh) coated with 25% squalene or 20% Apiezon L. Helium was the carrier gas. Tables 5 and 6 give the specific retention volumes of metal carbonyls on squalene and Apiezon L columns at various temperatures, respectively. The compounds were eluted according to the order of decreasing vapor pressure and the relative retentions of the metal carbonyls at 87°C on the two liquid phases were found to be very similar. However, the efficiency of the squalene column was higher and better separations were obtained on this phase as shown in Figs. 22 and 23.

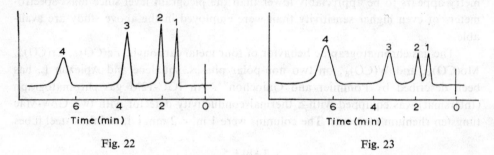

Fig. 22   Fig. 23

Fig. 22. Analysis of a mixture of metal carbonyls on squalene. Temperature 90°C, helium flow rate 33 ml/min. (1) Fe(CO)$_5$; (2) Cr(CO)$_6$; (3) Mo(CO)$_6$; (4) W(CO)$_6$.

Fig. 23. Analysis of a mixture of metal carbonyls on Apiezon L. Temperature: 87°C, helium flow rate 26.9 ml/min. (1) Fe(CO)$_5$; (2) Cr(CO)$_6$; (3) Mo(CO)$_6$; (4) W(CO)$_6$.

Segard *et al.*[96] described GLC and NMR studies of arenetricarbonyl chromium derivatives. Thirteen benzenetricarbonyl chromium derivatives were analyzed by GLC between 80 and 180°C. No decomposition was found to occur in the chromatographic system as shown by the mass spectra of the eluted compounds. A Girdel gas chromatograph (Giravions Dorand) equipped with a flame-ionization detector was used with two columns (1.5 m × 2 mm i.d.) filled with Chromosorb W, 80 to 100 mesh, acid-washed, and coated with 10% SE-30 or Apiezon L grease. The chromium complexes were prepared by the method of Nicholls and Whiting[97]. Table 7 lists the specific retention volumes and relative retention times of arene-

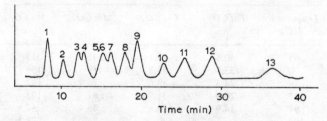

Fig. 24. Analysis of a benzenic solution of the 13 arenetricarbonylchromium complexes on the SE-30 column; conditions: column temperature 140°C, injection and detection block temperature 220°C, helium flow rate 30.6 ml/min (for the peak numbers, see Table 7).

TABLE 7

SPECIFIC RETENTION VOLUMES (V_g) AND RELATIVE RETENTIONS (a) OF ARENETRICARBONYLCHROMIUM
COMPLEXES ON SE-30 AND APIEZON L GREASES

Ligand	Peak number	SE-30		Apiezon L	
		V_g (ml)	a	V_g (ml)	a
Benzene	1	385	1	1150	1
Toluene	2	485	1.25	1450	1.25
p-Xylene	3	510	1.35	1750	1.50
m-Xylene	4	530	1.40	1900	1.65
o-Xylene	5	575	1.50	2160	1.90
Ethylbenzene	6	585	1.52	2270	2.00
Mesitylene (1,3,5-trimethylbenzene)	7	605	1.57	2490	2.15
Pseudocumene (1,2,4-trimethylbenzene)	8	680	1.75	2640	2.30
Cumene (isopropylbenzene)	9	720	1.85	2860	2.50
n-Propylbenzene	10	845	2.20	3380	2.95
tert.-Butylbenzene	11	960	2.50	3880	3.35
Isobutylbenzene	12	1010	2.65	4280	3.70
n-Butylbenzene	13	1290	3.35	5650	4.80

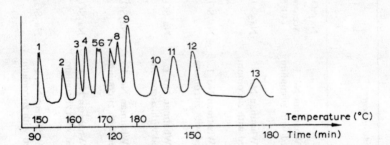

Fig. 25. Analysis of a benzenic solution of the 13 arenetricarbonylchromium complexes on the
Apiezon L column; conditions: column temperature programmed from 80 to 180°C at 0.8°C/min,
program started 3 min after injection; injection and detection block temperature 220°C; helium flow
rate 31 ml/min (for the peak numbers, see Table 7).

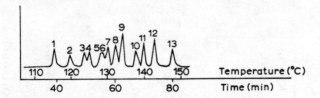

Fig. 26. Analysis of a benzenic solution of the 13 arenetricarbonylchromium complexes on the
SE-30 column. Conditions: column temperature programmed from 80 to 150°C at 0.8°C/min;
injection and detection block temperature 220°C; helium flow rate 30.6 ml/min (for the peak numbers,
see Table 7).

References pp. 33–35

TABLE 8

CHROMATOGRAPHIC SEPARATION OF CR^{III} AND CR^{VI} COMPOUNDS

	Chromium compounds	Adsorbent	Developer	Duration (h)	R_F Cr^{3+}	R_F CrO$_4$$^{2-}$	Observations
1	CrCl$_3$ + K$_2$CrO$_4$	Glass paper	Na$_2$SO$_4$ satd. soln.	2			Not separable
2	CrCl$_3$ + K$_2$CrO$_4$	Filter paper Whatman No. 1 impregnated with zirconium phosphate	Acetone–4 N HCl–H$_2$O (40:5:10)	5	0.00	0.33	CrO$_4$$^{2-}$ is partially reduced to Cr^{3+}
3	^{51}CrCl$_3$ + K$_2$CrO$_4$	Filter paper Whatman No. 1 impregnated with zirconium phosphate	Na$_2$SO$_4$ satd. soln.	2	0.00	0.67	CrO$_4$$^{2-}$ is partially reduced to Cr^{3+}
4	^{51}CrCl$_3$ + CrCl$_3$	Filter paper Whatman No. 1 impregnated with zirconium phosphate	Na$_2$SO$_4$ satd. soln.	2	0.85		
5	Na$_2$^{51}CrO$_4$ + CrCl$_3$	Filter paper Whatman No. 1 impregnated with zirconium phosphate	Na$_2$SO$_4$ satd. soln.	2	0.85	0.36	CrO$_4$$^{2-}$ is partially reduced to Cr^{3+}
6	Na$_2$^{51}CrO$_4$ + K$_2$CrO$_4$	Filter paper Whatman No. 1 impregnated with zirconium phosphate	Na$_2$SO$_4$ satd. soln.	2		0.66	The reduction of ^{51}CrO$_4$$^{2-}$ is decreased
7	^{51}CrCl$_3$ carrier-free	Filter paper Whatman No. 1 impregnated with zirconium phosphate	Na$_2$SO$_4$ satd. soln.	2	0.90		
8	Na$_2$^{51}CrO$_4$ carrier-free	Filter paper Whatman No. 1 impregnated with zirconium phosphate	Na$_2$SO$_4$ satd. soln.	2	0.90	0.38	^{51}CrO$_4$$^{2-}$ is partially reduced to ^{51}Cr^{III}
9	^{51}CrCl$_3$ + CrCl$_3$	Silica Gel thin layer	Acetone–TBP*–H$_2$O– 5% H$_2$SO$_4$ (40:6:10:20)	7			Incomplete separation

No.	Mixture	Medium	Solvent		Rf	Result
10	$^{51}CrCl_3 + K_2CrO_4$	Silica Gel thin layer	Acetone–TBP–H₂O–5% H₂SO₄ (40:6:10:20)	7		Incomplete separation
11	$^{51}CrCl_3 + K_2CrO_4$	Silica Gel thin layer	Acetone–36% HCl–H₂O (40:3.4:10)	5	0.94 0.86	Incomplete separation
12	$^{51}CrCl_3 + K_2CrO_4$	Zirconium phosphate thin layer	Na₂SO₄ satd. soln.	4	0.75	Incomplete separation
13	$^{51}CrCl_3 + K_2CrO_4$	Zirconium phosphate thin layer	Acetone–TBP–H₂O (60:10:30)	4	0.55	Incomplete separation
14	$^{51}CrCl_3 + K_2CrO_4$	Aluminum oxide thin layer	n-BuOH–acetic acid–H₂O (4:1:1)	3		Incomplete separation
15	$^{51}CrCl_3$ carrier-free	Aluminum oxide thin layer	Na₂SO₄ satd. soln.	2	0.00	Complete separation
16	$Na_2\,^{51}CrO_4$ carrier-free	Aluminum oxide thin layer	Na₂SO₄ satd. soln.	2	0.62	Complete separation
17	$^{51}CrCl_3 + Na_2\,^{51}CrO_4$ carrier-free	Aluminum oxide thin layer	Na₂SO₄ satd. soln.	2	0.57 0.00	Complete separation
18	$^{51}CrCl_3 + Na_2\,^{51}CrO_4$ + CrCl₃ + K₂CrO₄	Aluminum oxide thin layer	Na₂SO₄ satd. soln.	2	0.50 0.00	Complete separation
19	$Na_2\,^{51}CrO_4 + CrCl_3$	Aluminum oxide thin layer	Na₂CO₃ satd. soln.	2	0.50 0.00	Complete separation
20	$Na_2\,^{51}CrO_4 + K_2CrO_4$	Aluminum oxide thin layer	Na₂CO₃ satd. soln.	2	0.50 0.00	Complete separation
21	$Na_2\,^{51}CrO_4 + K_2CrO_4$ + CrCl₃	Aluminum oxide thin layer	Na₂CO₃ satd. soln.	2	0.50 0.00	Complete separation
22	$Na_2\,^{51}CrO_4$ carrier-free	Aluminum oxide thin layer	Na₂CO₃ satd. soln.	2	0.62 0.00	Complete separation

* TBP = tributyl phosphate.

tricarbonyl chromium complexes on SE-30 and Apiezon Greases measured at
145°C on both phases. At the same temperature, the retentions on Apiezon L are
about 3–4 times greater than the retentions on SE-30 silicone gum. Figure 24 shows
a chromatogram of a benzene solution of the 13 arenetricarbonyl chromium complexes on the SE-30 column at 140°C. As can be seen in Figs. 25 and 26, improved
separations were obtained by programming the column temperature from 80 to
180°C and 80 to 150°C for both Apiezon L and SE-30 columns, respectively.

The determination of trace quantities of aluminum and chromium in uranium
by electron-capture GLC of their trifluoroacetylacetonates has been accomplished
by Genty et al.[98]. A glass column (3 ft. × 0.08 in. i.d.) was used filled with 60 to
80 mesh silanized glass beads coated with 0.2% DC-710 and operated at $\simeq 115°$
with nitrogen as carrier and scavenger gas. It was possible to determine down to
0.1 p.p.m. of Al or Cr in the samples.

The separation of chromium ions of different valence (Cr^{3+} from CrO_4^{2-}) has
been achieved by a variety of chromatographic techniques. The separation of non-
radioactive chromium ions (Cr^{2+}, Cr^{3+} and CrO_4^{2-}) by paper chromatography
(Whatman No. 1) using n-butanol–acetic acid–ethyl acetate–water (50:10:5:35) as
developer has been described by Bighi[99]. Pollard et al.[100] separated Cr^{3+} from
CrO_4^{2-} on paper using methanol–ethyl ether–conc. HCl–water (30:50:4:15).
Sastri and Rao[101] separated Cr^{3+} from CrO_4^{2-} on paper impregnated with zirconium
phosphate using saturated sodium sulfate as developer. The adsorption of chromate
ions and other ionic species on paper impregnated with ion-exchange resin from
different media has been reported by Lederer et al.[102,103] and Ossicini[104]. The paper
and thin-layer chromatography of radioactive carrier-free chromium(III) from
chromium(VI) has been reported by Galateanu[105] (Table 8). In the course of determining radiochemical purity using the procedures of Bighi[99] and Sastri and Rao[101]
(with carrier and carrier-free chromate and chromium chloride labeled with ^{51}Cr)
a supplementary reduction of chromate ions by the reducing action of the chromatographic paper and/or solvents was noted by Galateanu[105]. When the concentration
of chromate ions in the solution is very low, this reduction was increased and may
be a source of error in the analysis or determination of radiochemical purity of some
labeled compounds of chromium.

The use of TLC using plates of aluminum oxide, Merck Grade, (80 g) with
calcium sulfate (10 g) developed with a saturated solution of sodium sulfate enabled
the separation of $^{51}Cr^{+3}$ ions from $^{51}CrO_4$ ions (R_F 0.00 and 0.57, respectively).

Separation of different valences of chromium by high voltage paper electrophoresis was described by Batasheva et al.[106] using 0.1 N KCl as electrolyte for
5–10 min at 1000–500 V. ^{51}Cr was used to follow the separation by drying the
electrogram, cutting into 5 mm strips and measuring the ^{51}Cr radioactivity by
γ-spectrometry with the peak areas being proportional to the amounts of Cr^{3+} and
Cr^{6+}. Classical methods of precipitation of Cr^{3+} as $CrPO_4$ and Cr^{6+} as $PbCrO_4$
agreed with the electrophoretic method.

The detection of traces of transition metals by formation of peroxo compounds and sorption on a chelating resin was reported by Maura and Rinaldi[107]. The detection of traces of Ti, V, Cr, Mo, Ce, and U by formation of colored peroxo compounds with hydrogen peroxide in the presence of sulfuric acid or alkali was improved by carrying out the test on a spot-plate in the presence of Dowex A-1 resin (30–70 mesh). The concentration ranges for U and Ti were 20 p.p.m. and 0.5 p.p.m., respectively.

REFERENCES

1   *Air Quality Data, National Air Sampling Networks, 1964–1965*, U.S. Dept. HEW, Cincinnati, Ohio, 1966.
2   G. W. H. SCHEPERS, T. M. DURKAN, A. B. DELEHANT AND F. T. CREEDON, *Arch. Ind. Health*, 15 (1957) 32.
3   W. D. WAGNER, D. H. GROTH, J. L. HOLTZ, G. E. MADDEN AND H. E. STOKINGER, *Toxicol. Appl. Pharmacol.*, 15 (1969) 10.
4   F. HYSLOP, E. D. PALMES, W. C. ALFORD, A. R. MONACO AND L. T. FAIRHALL, *U.S. Public Health Serv. Bull. No. 18*, Washington, D.C., 1943.
5   J. T. CROWLEY, J. G. HAMILTON AND K. G. SCOTT, *J. Biol. Chem.*, 177 (1949) 975.
6   C. D. VAN CLEVE AND C. T. KAYLOR, *Arch. Ind. Health*, 11 (1955) 375.
7   F. W. KLEMPERER, A. P. MARTIN AND R. E. LIDDY, *Arch. Biochem. Biophys.*, 41 (1952) 148.
8   F. R. DUTRA, J. CHOLAK AND D. M. HUBBARD, *Am. J. Clin. Pathol.*, 19 (1949) 229.
9   J. N. DENARDI, H. S. VAN ORDSTRAND, G. H. CURTIS AND J. ZIELINSKI, *Arch. Ind. Hyg.*, 8 (1953) 1.
10 W. MACHLE, E. C. BEYER AND H. TEBROCK, *Proc. 9th Intern. Congr. Ind. Med.*, Wright Publishing Co., London, 1949, p. 615.
11 R. H. MULLER, *Anal. Chem.*, 40 (1968) 85A.
12 J. A. ADAM, E. BOOTH AND J. D. H. STRICKLAND, *Anal. Chim. Acta*, 6 (1952) 462.
13 D. L. BOKOWSKI, *J. Am. Ind. Hyg. Assoc.*, 29 (1968) 474.
14 J. CHOLAK AND D. M. HUBBARD, *Anal. Chem.*, 20 (1948) 73.
15 J. CHOLAK AND D. M. HUBBARD, *Am. Ind. Hyg. Assoc. Quart.*, 13 (1952) 125.
16 R. J. POWELL, P. J. PHENNAH AND J. E. STILL, *Analyst*, 85 (1960) 347.
17 J. W. ROBINSON AND C. J. HSU, *Anal. Chim. Acta*, 43 (1968) 109.
18 C. W. SILL, C. P. WILLIS AND J. K. FLYGARE, *Anal. Chem.*, 33 (1961) 1671.
19 C. W. SILL, C. P. WILLIS AND J. K. FLYGARE, *Anal. Chem.*, 31 (1959) 598.
20 J. A. McHUGH AND J. C. SHEFFIELD, *Anal. Chem.*, 39 (1967) 377.
21 M. L. TAYLOR, E. L. ARNOLD AND R. E. SIEVERS, *Anal. Letters*, 1 (1968) 735.
22 M. L. TAYLOR AND E. L. ARNOLD, *Anal. Chem.*, 43 (1971) 1328.
23 W. G. SCRIBNER, M. J. BORCHERS AND W. J. TREAT, *Anal. Chem.*, 38 (1966) 1779.
24 J. K. FOREMAN, T. A. GOUGH AND E. A. WALKER, *Analyst*, 95 (1970) 794.
25 W. J. BIERMANN AND H. GESSER, *Anal. Chem.*, 32 (1960) 1525.
26 J. E. SCHWARBERG, R. W. MOSHIER AND J. H. WALSH, *Talanta*, 11 (1964) 1213.
27 R. C. YOUNG, in W. C. FERNELIUS (Ed.), *Inorganic Synthesis*, Vol. 2, McGraw-Hill, New York, 1946, p. 25.
28 J. BOHEMAN, S. H. LANGER, R. H. PERETT AND J. H. PURNELL, *J. Chem. Soc.*, (1960) 2444.
29 W. D. ROSS, personal communication in ref. 26.
30 W. D. ROSS AND R. E. SIEVERS, in A. B. LITTLEFIELD (Ed.), *Gas Chromatography, Rome, 1966*, Inst. of Petroleum, London, 1967.
31 W. D. ROSS AND R. E. SIEVERS, *Talanta*, 15 (1968) 89.
32 M. H. NOWEIR AND J. CHOLAK, *Environ. Sci. Technol.*, 3 (1969) 927.
33 W. D. ROSS AND R. E. SIEVERS, *Environ. Sci. Technol.*, 6 (1972) 155.

34 L. B. TEPPER, H. L. HARDY AND R. I. CHAMBERLIN, *Toxicity of Beryllium Compounds*, Elsevier,
    Amsterdam, 1961, p. 156.
35 K. J. EISENTRAUT, D. J. GRIEST AND R. E. SIEVERS, *Anal. Chem.*, 43 (1971) 2003.
36 B. S. BARRATT, R. BELCHER, W. I. STEPHEN AND P. C. UDEN, *Anal. Chim. Acta*, 57 (1971)
    447.
37 F. E. BUTLER, *Am. Ind. Hyg. Assoc. J.*, 30 (1969) 559.
38 Y. YAMANE, M. MIYAZAKI AND H. IWASE, *Eisei Kagaku*, 14 (1968) 106; *Chem. Abstr.*, 69 (1968)
    45939d.
39 V. D. ERISTAVI, D. I. ERISTAVI AND F. I. BROUCHEK, *Anal. Khim.*, 23 (1968) 782; *Chem. Abstr.*,
    69 (1968) 40951s.
40 D. I. ERISTAVI, V. D. ERISTAVI AND S. A. KEKELIYA, *Chem. Zvesti*, 25 (1971) 132.
41 R. A. BAKER, SR. AND R. C. DOERR, *J. Air Pollution Control Assoc.*, 14 (1964) 409.
42 J. R. LODWICK, *J. Inst. Petrol.*, 50 (1964) 297.
43 H. A. SCHROEDER, J. J. BALASSA AND I. H. TIPTON, *J. Chronic Diseases*, 15 (1962) 941.
44 A. M. BAETJER, in M. J. UDY (Ed.), *Chromium*, Vol. I, Reinhold, New York, 1950, p. 76.
45 G. E. MORRIS, *A.M.A. Arch. Dermatol.*, 78 (1958) 612.
46 A. M. BAETTER, C. DAMRON AND V. BUDACZ, *Arch. Ind. Health*, 20 (1949) 136.
47 T. F. MANGUSO, *Ind. Med. Surg.*, 20 (1951) 393.
48 H. P. BRINTON, E. S. FRASIER AND A. L. KOVEN, *U.S. Public Health Serv. Rept.*, 67 (1952)
    385.
49 A. M. BAETJER, *Arch. Ind. Hyg.*, 2 (1950) 487.
50 W. MACHLE AND F. GREGORIUS, *U.S. Public Health Serv. Rept.*, 63 (1948) 1114.
51 P. L. BIDSTRUP, *Brit. J. Ind. Med.*, 8 (1951) 302.
52 W. C. HUEPER, *Am. Ind. Hyg. J.*, 20 (1959) 274.
53 W. C. HUEPER, *Arch. Environ. Health*, 5 (1962) 445.
54 F. J. C. ROE AND R. L. CARTER, *Brit. J. Cancer*, 23 (1969) 172.
55 P. P. DVIZHOV AND V. I. FEDDROVA, *Vop. Onkol.*, 13 (1967) 57.
56 H. J. CAHNMANN AND R. BISEN, *Anal. Chem.*, 24 (1952) 1341.
57 D. O. MILLER AND J. H. YOE, *Clin. Chim. Acta*, 4 (1959) 378.
58 P. F. URONE AND H. K. ANDERS, *Anal. Chem.*, 22 (1950) 1317.
59 K. M. HAMBIDGE, *Anal. Chem.*, 43 (1971) 103.
60 R. MONACELLI, H. TANAKA AND J. H. YOE, *Clin. Chim. Acta*, 1 (1966) 577.
61 H. J. KOCH, JR., E. R. SMITH, N. F. SHIMP AND J. CONNER, *Cancer*, 9 (1959) 499.
62 W. B. HERRING, B. S. LEAVELL, Z. M. PAIXAO AND J. H. YOE, *Am. J. Clin. Nutr.*, 8 (1960) 846.
63 W. A. WOLSTENHOLME, *Nature*, 203 (1964) 1284.
64 F. J. FELDMAN, E. C. KNOBLOCK AND W. C. PURDY, *Anal. Chim. Acta*, 38 (1967) 489.
65 J. O. PIERCE AND J. CHOLAK, *Arch. Environ Health*, 13 (1966) 208.
66 E. E. CARY AND W. H. ALLAWAY, *J. Agr. Food Chem.*, 19 (1971) 1159.
67 F. J. FELDMAN, G. C. CHRISTIAN AND W. C. PURDY, *Am. J. Clin. Pathol.*, 49 (1968) 826.
68 L. R. PASCALE, S. S. WALDSTEIN, G. ENGBRING, A. DUBIN AND P. B. SZANTO, *J. Am. Med.
    Assoc.*, 149 (1952) 1385.
69 H. J. M. BOWEN, *Analyst*, 89 (1964) 658.
70 J. PIJEK, J. GILLIS AND J. HOSTE, *Intern. J. Appl. Radiation Isotopes*, 10 (1961) 149.
71 K. BEYERMANN, J. H. ROSE, JR. AND R. P. CHRISTIAN, *Anal. Chim. Acta*, 45 (1969) 51.
72 B. E. SALTZMAN, *Anal. Chem.*, 24 (1952) 1016.
73 J. P. MCKAVENAY AND H. FRIESER, *Anal. Chem.*, 30 (1958) 1965.
74 W. MERTZ, *Phys. Rev.*, 49 (1969) 163.
75 W. D. ROSS, *Anal. Chem.*, 35 (1963) 1596.
76 R. D. HILL AND H. GESSER, *J. Gas Chromatog.*, 1 (1963) 11.
77 D. K. ALBERT, *Anal. Chem.*, 36 (1964) 2034.
78 W. D. ROSS AND R. E. SIEVERS, *Anal. Chem.*, 37 (1965) 598.
79 W. TANIKAWA, K. HIRANO AND K. ARAYAWA, *Chem. Pharm. Bull.*, 15 (1967) 915.
80 H. VEENING AND J. F. K. HUBER, *J. Gas Chromatog.*, 7 (1968) 326.
81 W. D. ROSS AND G. WHEELER, *Anal. Chem.*, 37 (1965) 598.
82 R. S. JUVET AND R. P. DURBIN, *J. Gas Chromatog.*, 1 (1963) 14.

83 R. E. SIEVERS, J. W. CONNOLLY AND W. D. ROSS, *J. Gas Chromatog.*, 5 (1967) 241.
84 W. D. ROSS AND G. WHEELER, *Anal. Chem.*, 36 (1964) 266.
85 J. SAVORY, P. MUSHAK AND F. W. SUNDERMAN, JR., *J. Chromatog. Sci.*, 7 (1969) 674.
86 J. SAVORY, P. MUSHAK, F. W. SUNDERMAN, JR., R. H. ESTES AND N. O. ROSZEL, *Anal. Chem.*, 42 (1970) 294.
87 G. H. BOOTH, JR. AND W. J. DARBY, *Anal. Chem.*, 43 (1971) 831.
88 R. C. FAY AND T. S. PIPER, *J. Am. Chem. Soc.*, 84 (1962) 2302.
89 R. C. FAY AND T. S. PIPER, *J. Am. Chem. Soc.*, 85 (1963) 500.
90 C. G. MACDONALD AND J. S. SHANNON, *Australian J. Chem.*, 19 (1966) 1545.
91 B. R. KOWALSKI, T. L. ISENHOUR AND R. E. SIEVERS, *Anal. Chem.*, 41 (1969) 998.
92 J. L. BOOKER, T. L. ISENHOUR AND R. E. SIEVERS, *Anal. Chem.*, 41 (1969) 1705.
93 A. C. HANSEN, W. G. SCRIBNER, T. W. GILBERT AND R. E. SIEVERS, *Anal. Chem.*, 43 (1971) 349.
94 W. R. WOLF, M. L. TAYLOR, B. M. HUGHES, T. O. TIERNAN AND R. E. SIEVERS, *Anal. Chem.*, 44 (1972) 616.
95 C. POMMIER AND G. GUIOCHON, *J. Chromatog. Sci.*, 8 (1970) 486.
96 C. SEGARD, B. ROQUES, C. POMMIER AND G. GUIOCHON, *Anal. Chem.*, 43 (1971) 1146.
97 B. NICHOLS AND M. C. WHITING, *J. Chem. Soc.*, (1959) 551.
98 C. GENTY, C. HOUIN, P. MACHERBE AND R. SCHOTT, *Anal. Chem.*, 43 (1971) 235.
99 C. BIGHI, *Ann. Chim. (Rome)*, 45 (1955) 1087.
100 F. H. POLLARD, J. F. W. MCOMIE AND A. J. BANISTER, *Chem. Ind. (London)*, 49 (1955) 1598.
101 M. N. SASTRI AND A. P. RAO, *J. Chromatog.*, 9 (1962) 250.
102 M. LEDERER AND L. OSSICINI, *J. Chromatog.*, 13 (1964) 188.
103 G. BAGLIANO, G. GRASSINI, M. LEDERER AND L. OSSICINI, *J. Chromatog.*, 14 (1964) 238.
104 L. OSSICINI, *J. Chromatog.*, 9 (1962) 114.
105 I. GALATEANU, *J. Chromatog.*, 19 (1965) 208.
106 V. K. BATASHEVA, A. M. BABESHKIN AND A. N. NESMEYANOV, *Vestn. Mosk. Univ. Khim.*, 11 (1970) 361; *Chem. Abstr.*, 73 (1970) 105121a.
107 G. MAURA AND G. RINALDI, *Anal. Chim. Acta*, 53 (1971) 466.

Chapter 3

MANGANESE, COBALT, NICKEL AND CADMIUM

1. MANGANESE

Manganese is widely distributed in the combined state ranking 12th in abundance among the elements in the earth's crust. It is found in nature as an oxide (in pyrolusite, braunite, mangonite, and Housmannite), as a sulfide (in manganese blend and hauserite), as a carbonate (in manganesespot), as silicate (in tephroite, knebelite, rhodomite), and in other ores. The worldwide production of manganese ore of 35% manganese or more was about 16 million tons in 1965 and increasing.

The smelting and refining of manganese ores (over 90% of which are used in the iron and steel industry) represents the largest potential pollution source in the United States. The ferromanganese blast furnace production in the United States in 1966 was 651,987 tons[1] which, it is suggested, could contribute from 19,500 to 32,500 tons of manganese to the atmosphere per year. In addition, electric-arc furnaces produced 294,223 tons of ferromanganese in 1966[1] and the manganese emissions from these sources may be even higher than from those producing steel. Additional sources of manganese include coal ash which contains between 0.005 and 1% manganese, depending on the origin of the coal[2], and manganese (as MnO_2) from burning fuel oil (the concentration of MnO_2 in the dust emission particulate is approximately 400 $\mu g/g$). A burner burning 1000 lb. of oil/h would be discharging 0.8 g of MnO_2/h into the air. The oxides of manganese, *e.g.* MnO, MnO_2, Mn_2O_3, or Mn_3O_4, are the most important of the manganese compounds from an air pollution standpoint. However, these oxides may react rapidly with other pollutants such as SO_2 and NO_2 to form water-soluble manganese compounds. The average concentration of manganese in the air in the U.S. in 1967 was 0.10 $\mu g/m^3$ with the maximum value recorded at 10 $\mu g/m^3$. Inhalation of manganese or its compounds may induce manganese poisoning (a disease of the central nervous system) or manganic pneumonia. Industrial poisoning occurs principally where the dust or fumes containing manganese oxides pollute the atmosphere. The higher valence oxides are believed to be more poisonous than manganous oxide.

Approximately 150,000 tons of manganese ore are used by the chemical industry and the production of dry-cell batteries uses approximately 30,000 tons of high-grade manganese ore annually. The disposal of old dry-cell batteries by incineration is a possible source of manganese air pollution. About 10,000 tons of manganese (20,000–25,000 tons of manganese dioxide) are used annually in welding rods and fluxes for iron and steel[3]. Fungicidal agents such as Maneb (manganese ethylene bis-dithiocarbamate) are another potential source of manganese exposure; approximately 10,000 tons of manganese are used in fungicides annually.

Fuel additives containing manganese derivatives represent another potential source of manganese air pollution. These additives fall into two classes, anti-knock compounds and smoke inhibitors. The most important manganese-containing anti-knock compound is methylcyclopentadienyl manganese tricarbonyl which is mixed with tetraethyl lead to increase the octane rating of gasoline. More than 2 ml of this mixture/gallon of gasoline (approx. 36,000 μg/gal. of manganese) is recommended.

Permanganates are widely used oxidants. One of the larger uses of $KMnO_4$ is in water purification and with problems of water shortage and pollution becoming greater, this use is increasing. The largest use for $KMnO_4$ comprises chemical-process oxidations and purifications of liquid organics to produce both intermediates and final products in the pharmaceutical and chemical industries. Other uses include glass and enamel, ceramics, dyes and pigments, catalysts, electronics, in safety matches and flares, in agriculture to enrich manganese deficient soils, and as a trace element in poultry and animal feeds. Figure 1 illustrates the areas of utility for manganese. In addition to industrial exposures to manganese fumes and dusts *per se*, manganese is found in tobacco, *e.g.* an average of 180 p.p.m. of manganese was found in tobacco, but only 0.4 μg in the mainstream of 100 cigarettes[4]. Cereals and their products which are considered rich in manganese contain only slightly more than 100 p.p.m. on a dry matter basis and other foodstuffs range very much lower, mostly below 30 p.p.m.[5]. Manganese is an essential trace element with human requirements estimated between 3 mg and 9 mg/day[6]. Manganese is found throughout the body but is concentrated particularly in liver, bone, and lymph nodes. Chronic reactions may result from prolonged exposure to the dust of the metal and its compounds. The main route of exposure is *via* the lungs, skin absorption being minimal.

Manganese chloride has been reported to be mutagenic in *E. coli*[7,8] and *Serratia marscescans*[9] and manganic sulfate has caused cytogenic damage in red cells, white cells, and megakarocytes following intraperitoneal administration to rats[10].

Manganese has been analyzed in air samples by atomic absorption[11], emission spectroscopy, neutron activation, and colorimetry[12]. A wide variety of chromatographic techniques has been utilized for the separation of the various valence states of manganese from other cations.

The paper separation of the oxidation states of manganese(II, III, and VII), arsenic(III, V), chromium(II, VI), mercury(I, II), antimony(III, V), and thallium(I, III) has been accomplished using tributyl orthophosphate–acetic acid–acetone solvent systems[13].

Manganese(II) was separated from 15 metal ions in micro amounts by circular paper chromatography according to Bariborli and Lipparini[14]. A 45 cm disc of Whatman No. 1 paper was first buffered with a solution containing 2.81 g boric acid and 1.60 g of sodium borate/l at pH 8.0. Ten to 20 μl of an aqueous solution containing 5–80 μg of each metal was chromatographically separated in 14 h with

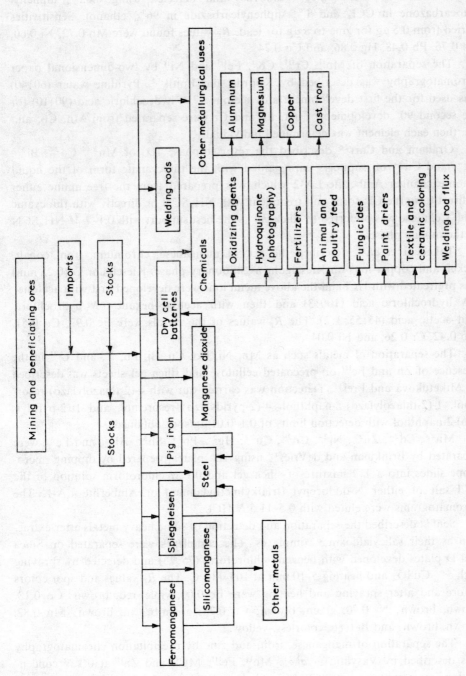

Fig. 1. Manganese areas of utility.

ethylacetate–acetic acid–water (50:30:20) and detected using 0.02% diphenyl thiocarbazone in CCl_4 and 1% diphenylcarbazide in 96% ethanol. Sensitivities varied from 0.5 μg for zinc to 8 μg for lead. R_F values found were Mn 0.72, Ni 0.60, Cd 0.76, Pb 0.48, Hg 0.86, and Co 0.74.

The separation of Mn^{II}, Cr^{III}, Co^{II}, Fe^{III} and Ni^{II} by two-dimensional paper chromatography was described by Koprda and Fojtik[15]. Pyridine–water (60:40) was used for the first development and acetone –7 N hydrochloric acid (90:10) for the second 90° development. First, Fe and Cr were separated from Mn, Co, and Ni, then each element was determined individually.

Graham and Carr[16] described the reversed-phase TLC of Mn^{2+}, Cd^{2+}, Bi^{3+}, Ag^+, and Cu^{2+} on supports impreganted with the thiocyanate form of the liquid anion-exchanger Amberlite LA-2 which was prepared from the free amine either indirectly *via* the Cl^- form and treatment with NH_4SCN or directly with thiocyanic acid. Cellulose was preferred to silica gel as the best support with 0.1–7 M NH_4SCN as the mobile phases.

Takeuchi *et al.*[17] described the TLC of manganese, chromium, iron, cobalt, nickel, and copper ions on Kieselguhr as stationary phase. Kieselguhr G (0.25 mm) was pretreated with HCl and the above metal ions were developed first with acetone–6 N hydrochloric acid (100:3) and then with acetone–hexane–6 N hydrochloric acid–acetic acid (45:55:3:2). The R_F values of the metals were Fe 0.97, Cu 0.51, Mn 0.42, Cr 0.26, and Ni 0.04.

The separation of metals such as Mn, Ni, Co, Cu, Bi, Pb, V, and U^{VI} in the presence of Zn and Fe^{III} on precoated cellulose and silica gel sheets was described by Miketukova and Frei[18]. Detection was carried out with 4-(2-thiazolylazo) resorcinol, 1-(2-thiazolylazo)-2-naphthol, 4-(2-pyridylazo) resorcinol, and 1-(2-pyridylazo)-2-naphthol with detection limits of 0.1–0.01 μg/spot obtained.

Mn^{2+}, Cd^{2+}, Zn^{2+}, Ni^{2+}, Co^{2+}, Cu^{2+}, Hg^{2+}, Pb^{2+}, Sb^{3+}, Al^{3+}, and Fe^{3+} were separated by Brinkman and deVries[19] using TLC plates prepared by dipping microscope slides into a 1:2 mixture of silica gel and 0.1 M chloroform solution of the HCl salt of either N-dodocenyl (trialkylmethylamine) or Amberlite LA-1. The chromatograms were eluted with 0.5–11.5 N HCl.

Senf[20] described the separation and detection of some heavy metals after extraction as their salicylaldoxime complexes. The complexes were separated on Silica Gel D plates developed with benzene–chloroform (75:25) and detected by spraying with 5% $CuSO_4$ and heating 5–10 min at 105–110°C. The R_F values and spot colors before and after spraying and heating were Fe 0.00, violet-red, brown; Co 0.18, brown, brown; Ni 0.70, green, brown; Cu 0.78, green-brown, brown; Mn 0.82, brown, brown; and Bi 0.86, colorless, yellow.

The separation of manganese, iron, and zinc by precipitation chromatography was described by Vasyutin *et al.*[21]. Mn^{II}, Fe^{III}, Mg^{II}, and Zn^{II} at 0.1 N concentrations were separated in 90–95% yields by selective elution following their precipitation on a column of 3 g of AV-17 (hydroxyl form) anion exchange resin. Zinc was eluted first with a 1:4 mixture of ethylenediamine and water, after which the column was rinsed with 1–2 ml of water, and then magnesium was eluted with 10 ml of aq. $NH_4Cl–NH_3$ (3:1), a portion of which was retained on the column for 15 min. To elute manganese, the dioxide was first reduced by treating the column for 15 min with 5 ml of 20% solution of hydroxylamine hydrochloride adjusted to pH 6 with hydroxylamine. After rinsing, iron was extracted with 0.5 N hydrochloric acid.

The separation of manganese from 13 diverse elements by cation-exchange chromatography was studied by Strelow and Baxter[22]. Manganese was separated by elution from an AG50W-X8 (sulfonated polystyrene) cation-exchange resin column by 0.75 M HCl in 90% acetone. The diverse elements remained absorbed on the column and could be eluted with the following eluents: Li and Na, MHCl; Mg, Be, K, and Ni, 2.0 M HCl; Ti^{IV}, 2.0 M HCl containing 0.03% H_2O_2; Al and Ca, 3.0 M HCl; La and Zr, 5.0 M HCl; and Ba, 3.0 M HNO_3. Thorium was removed by ashing the resin. Quantitative separations were attained for several millimoles on 30 ml resin columns. Manganese(II) ($\leq$ 10 mmole) could be separated from 140 γ Li, the least strongly absorbed of the diverse elements, by using a 60 ml resin column.

The separation of a number of metal–EDTA complexes by anion-exchange chromatography was reported by Vanderdeelen[23]. Fe^{3+}, Mn^{2+}, Co^{3+}, Ni^{2+}, Zn^{2+}, Cu^{2+} and Mn^{2+}, Co^{3+}, Zn^{2+}, Ca^{2+}, Cr^{3+}, and Fe^{3+}, each $\leq$0.2 mg/ml, were quantitatively separated in the above given order as their respective EDTA complexes by chromatography on a 17.3–27.2 cm $\times$ 0.95 cm^2 column of Dowex 2-X8 resin (220–400 mesh, Cl^- form) by eluting with 0.1 M KCl and with 0.5 M ammonium acetate, respectively at flow rates of 0.2–0.5 ml/min/cm^2.

The electrophoretic mobilities of azide, hydroxylamine, hydrazine, thiocyanate complexes of 16 metals including manganese were described by Majumdar and Mitra[24]. The complexing ligands served as electrolytes. Ternary and quaternary metal mixtures (10–15 µg) were resolved at 150 V in 3 h. The quaternary mixtures were best resolved using azide ion and hydroxylamine while SCN^-, $S_2O_3^{2-}$, and hydrazine were more suited for ternary mixture separations.

Karayannis and Corwin[25,26] studied the hyperpressure gas chromatography (HPGC) of 15 etioporphyrin II metal chelates of 14 different elements at 145° and 1000–700 p.s.i., with CCl_2F_2 as the solvent gas. A column of 10% Epon 1001 resin on Chromosorb W effected the following separations: Mg^{II}, Cu^{II} or Co^{II}; Ni^{II} or Co^{III}; VO^{2+} or TiO^{2+}; Mn^{II}, Zn or Pt^{II} and Pd^{II}; Ag^{II}, Pt^{II} and Pd^{II}. A 10% silicone gum rubber XE-60 on Chromosorb W column separated Mn^{II} or Fe^{III}, Zn or TiO^{2+}, Pt^{II} and VO^{2+}. All the chelates were eluted intact with the exception of Mg^{II} which was demetallated in the column. A novel method for the insertion of metal ions in the porphyrin ring, involving the reaction of the corresponding metal acetylacetonate with the porphyrin in a suitable solvent was also reported. The apparatus used in this study was a Perkin-Elmer hyperpressure gas chromatograph, with a Hitachi–

TABLE 1

RELATIVE RETENTION VALUES OF ETIOPORPHYRIN II METAL CHELATES

Etioporphyrin II chelate	Monochromator setting during detection (mμ)	Relative retention time (min)	
		Epon 1001[a]	XE-60[b]
Mg^{II}	397	0.90	1.05
TiO	401	2.26	1.16
VO	405	2.39	2.36
Mn^{II}	400	2.13	1.08
Mn^{III}	359	Insoluble	
Fe^{III}	392	Retained	1.07
Co^{II} 1st peak (Co^{II})	391	1.08	0.86
2nd peak (Co^{III})		1.41	1.06
Ni^{II}	390	1.27	1.14
Pd^{II}	395	2.90	2.19
Pt^{II}	380	2.31	1.68
Cu^{II}	398	1.00	1.00
Ag^{II}	405	1.19	1.12
Zn^{II}	400	2.07	1.03
Al^{III}	398	Retained	Retained
Sn	406	Insoluble	
Etioporphyrin II	397	0.95	1.15
Retention time of Cu^{II} etioporphyrin II (min)		29,250	18,5625

[a] Column: 64 in., $\frac{1}{8}$ in. i.d., column of 10% Epon 1001 resin on Chromosorb W, 60–80 mesh. Pressure: 1500 p.s.i. Column temperature: 145°C. Flow rate: 162 ml/min.
[b] Column: 46 in., $\frac{1}{8}$ in. i.d., column of 10% Silicone XE-60 on AW-HMDS-treated Chromosorb W, 80–100 mesh. Pressure: 1350 p.s.i. Column temperature: 145°C. Flow rate 136 ml/min.

Perkin-Elmer Model 139 spectrophotometer as detector as described earlier by Karayannis et al.[27]. The columns were preconditioned at ca. 2000 p.s.i. and temperatures of 150–155°C. Table 1 shows the relative retention values of 15 etioporphyrin II chelates on Epon 1001 and XE-60 columns and Table 2 summarizes the separations that were effected from these columns. HPGC at lower pressures and flow rates or use of longer columns led to better separations, but the peaks obtained under such conditions were usually too broad and could not be measured.

The HPGC technique has the advantage of the possibility of operation at substantially lower temperatures for the analysis of chelates of metals.

2. COBALT

Cobalt is found chiefly in the copper–cobalt area of the Belgian Congo and also as smalite ($FeNiCo)(As_2$), cobalt bloom ($Co(AsO_4)_4$ $2.8H_2O$), and cobaltite or cobalt glance (($CoFe)AsS$). The industrial uses of cobalt include (a) the manu-

TABLE 2

SEPARATIONS OF ETIOPORPHYRIN II METAL CHELATES

Column E = 46 in., ⅛ in. i.d., column of Epon 1001 on Chromosorb W, 60–80 mesh. Column X = 46 in., ⅛ in. i.d. column of XE-60 on AW-HMDS-treated Chromosorb W, 80–100 mesh. c = Complete separation. s = Satisfactory partial separation (small overlapping of the peaks). n = No separation (considerable overlapping of the peaks). p = Poor separation (complete overlapping of the peaks).

Chelate	Mg^{II}		TiO		VO		Mn^{II}		Fe^{III}		Co^{III}		Co^{II}		Ni^{II}		Pd^{II}		Pt^{II}		Cu^{II}		Ag^{II}		Zn^{II}	
	E	X	E	X	E	X	E	X	E	X	E	X	E	X	E	X	E	X	E	X	E	X	E	X	E	X
Mg^{II}			c	p	c	c	c	n	n	n	n	n	s	n	s	p	c	c	c	c	n	n	n	n	c	n
TiO	c	p			n	c	n	n	n	n	c	n	c	p	c	n	s	c	c	n	c	p	c	n	n	n
VO	c	c	n	c			n	c	n	c	c	c	c	c	c	c	c	c	n	s	c	c	c	c	n	c
Mn^{II}	c	n	n	n	n	c			n	c	c	n	c	n	c	c	p	p	n	c	c	n	n	n	c	n
Fe^{III}	n	n	n	n	n	c	n	c			n	p	n	n	s	n	n	c	n	c	n	n	n	n	n	n
Co^{II}	s	n	c	p	c	c	c	n	n	n	s	p			n	n	c	c	c	c	s	n	c	c	c	n
Co^{III}	n	n	c	n	c	c	c	n	n	p			s	p	c	c	c	c	c	c	s	n	n	n	c	n
Ni^{II}	s	p	c	n	c	c	c	c	s	n	c	c	n	n			c	c	c	c	s	p	c	c	o	c
Pd^{II}	c	c	s	c	c	c	p	p	n	c	c	c	c	c	c	c			s	p	c	c	n	n	n	c
Pt^{II}	c	c	c	n	n	s	n	c	n	c	c	c	c	c	c	c	s	p			s	p	c	c	s	c
Cu^{II}	n	n	c	p	c	c	c	n	n	n	s	n	s	c	s	p	c	c	s	p			n	n	n	c
Ag^{II}	n	n	c	n	c	c	n	n	n	n	n	n	c	c	c	c	n	n	c	c	n	n			c	n
Zn^{II}	c		n	n	n	c	c	n	n	n	c	n	c	n	o	c	n	c	s	c	n	c	c	n		

facture of alloys with chrome, nickel, copper, aluminum, beryllium, and molybdenum for use in the automobile, aircraft, and electrical industries; (b) in the tungsten carbide tool industry as a binder; and (c) in the form of a blue pigment in the china and glass industry. Radioactive cobalt-60 has been widely used for the treatment of breast cancer and non-radioactive cobalt salts are used therapeutically for certain types of anemia.

The toxic effects of cobalt in man have included both lung affection and skin lesions following exposures in the tungsten carbide industry[28-30]. The carcinogenic effect of metallic cobalt in rats[31,32] and cobalt oxide and cobalt sulfide in the mouse[33] following intramuscular administration has been reported.

Cobalt is one of the major constituents of fission products and its separation from other elements has been the subject of intensive investigations. The separation of cobalt from nickel has been accomplished on cation-exchange resin KU-1 by selective elution with hydrochloric acid[34]. Strelow[35] achieved similar separation of cobalt from iron, zinc, etc., on Dowex-50 resin. Cobalt has been separated from silver on Amberlite IR-120 resin using a 2% sodium nitrite solution[36]. Katsura[37] separated copper from cobalt on Amberlite IR-120 by first eluting copper with a solution of 0.5 M sodium thiosulfate. A mixture of 0.5 M hydrochloric acid–acetone has been used[38] for the separation of cadmium and zinc from cobalt on Dowex 50W-X8.

A systematic study of the cation-exchange behavior of cobalt on Dowex 50W-X12 (H$^+$ form) with different eluents has been described by Akki and Khopkar[39]. Ions such as zinc, beryllium, cesium, uranium, rubidium, cadmium, lead, and mercury were found to be much more weakly bound to Dowex 50W-X12 than cobalt and it was thus possible to remove first the unwanted cations with 1 M hydrochloric acid and then elute cobalt with 4 M hydrochloric acid. With some cations, other eluents were preferable: for uranium, 0.5 M sulfuric acid; for mercury, 0.5 M nitric acid; and for lead, 0.5 M ammonium acetate solutions. In these cases, cobalt was removed later with the same eluent at a concentration of 4 M.

Table 3 shows the elution behavior of cobalt(II) on Dowex 50W-X12. On the basis of elution constants and volume distribution coefficients, the eluents can be arranged in order of decreasing efficiency as follows:

$$Na_2SO_4 = (NH_4)_2SO_4 = NaCl = NH_4Cl = NaNO_3 < HCl < HNO_3$$

$$< HClO_4 < CH_3COONH_4$$

The quantitative adsorption and elution of microgram amounts of cobalt-60 on columns of cellulose powder, carboxymethyl cellulose, aminoethyl cellulose, diethylaminoethyl cellulose, and cellulose phosphate was described by Muzzarelli[40]. The organic eluting solvents were (a) anhydrous ether, (b) conc. nitric acid–diethyl ether (5:95), (c) NH$_4$NCS (40 mg) dissolved in 20 ml methanol and 80 ml diethyl ether, (d) NH$_4$NCS (5 g) dissolved in 20 ml methanol and 80 ml diethyl ether, and

TABLE 3

ELUTION BEHAVIOR OF COBALT(II) ON DOWEX 50 *W-X* 12

[Co = 25.335 mg; wt. of resin (oven-dried) = 18.5 g]

Eluent	M	V_{max} (peak elution volume) (ml)	V_t (total volume for full recovery) (ml)	Elution (%)	E	D_v
HCl	1.0		200	5.0		
	2.0	125	200	98.5	0.3373	2.965
	3.0	75	150	101.2	0.6322	1.582
	4.0	40	130	100.6	1.629	0.6141
H_2SO_4	1.0		200	43.5		
	2.0	75	175	102.6	0.6322	1.582
	3.0	50	150	101.9	1.123	0.8906
	4.0	40	120	102.3	1.629	0.6141
HNO_3	1.0		200	3.0		
	2.0	125	250	100.4	0.3373	2.965
	3.0	75	200	101.8	0.6322	1.582
	4.0	50	150	100.0	1.123	0.8906
$HClO_4$	3.0	125	250	101.5	0.3373	2.965
	4.0	75	200	100.7	0.6322	1.582
CH_3COONH_4	1.0	150	200	98.9	0.247	3.661
	4.0	50	200	102.1	1.123	0.8906
NaCl	1.0–2.0	75	250	99.6	0.6322	1.582
	3.0–4.0	50	150	100.8	1.123	0.8906
$NaNO_3$	2.0–4.0	75	200–300	100.3	0.6322	1.582
NH_4Cl	1.0	200	200	35.0		
	2.0	75	225	99.6	0.6322	1.582
	3.0–4.0	50	150	100.5	1.123	0.8906
Na_2SO_4	0.5	100	225	100.0	0.4398	2.274
	1.0	75	150	101.4	0.6322	1.582
	2.0	50	125	99.8	1.123	0.8906
$(NH_4)_2SO_4$	0.5	125	275	101.8	0.3373	2.965
	1.0	75	150	101.0	0.6322	1.582
	2.0–3.0	50	125	100.9	1.123	0.8906

(e) conc. hydrochloric acid–methanol (15:85). ^{60}Co was estimated by gamma-ray spectrometry and it was shown that the strength of adsorption of cobalt depended on the special functional groups attached to the modified celluloses.

Mixtures of Co, Ni, Cu, Cd, Fe, U, La, Th, and Zr have been separated by reversed-phase partition chromatography on columns of poly(trifluorochloro-ethylene) (Kel-F) supporting the liquid anion exchange resin Aliquot 336 as the stationary phase[41]. An 11 cm × 1 cm² column was used containing about 5 g ml/ cm²/min. The eluent was collected in 5 ml fractions and the metal ions were determined in the effluent colorimetrically (Cu, Cd, La, Th, and Zr).

The semiquantitative determination of micro amounts of cobalt, nickel, copper, iron, and zinc in blood and tissue by circular thin-layer chromatography was

TABLE 4

SPRAY REAGENTS FOR SENSTIVITY STANDARD OF METAL IONS

Reagent	Solution
Tannic acid (TA)	10% in water
8-Hydroxyquinoline (HQ)	3% in alcohol
Potassium ferrocyanide (PF)	2% in water
2,2'-Diquinolyl (DQ)	Saturated in alcohol followed by spray with 5% aqueous solution of hydroxylamine hydrochloride
Benzoin oxime (BO)	5% in alcohol
Ammonium thiocyanate (AT)	5% in water
Rubeanic acid (RA)	0.05% in alcohol
α-Nitroso-β-naphthol (NN)	0.2% in alcohol
Dimethylglyoxime (DG)	2% in alcohol
Dimercaptothiodiazol (DT)	1% in alcohol
Dithizone (DZ)	0.2% in carbon tetrachloride
Potassium ferricyanide (PFi)	1% in water

TABLE 5

SENSITIVITY STANDARD OF METAL IONS

Cation[a]	Color and spray reagent[b]	Sensitivity (μg)[c]	R_F[d]
Fe^{3+}	Light violet (TA)	0.04	1.00
	Yellow (HQ)	0.06	
	Blue (PF)	0.02	
Cu^{2+}	Pink (DQ)	0.01	0.98
	Light green (BO)	0.40	
Co^{2+}	Blue (AT)	0.06	0.41
	Brown (RA)	0.04	
	Brown (NN)	0.03	
Ni^{2+}	Blue (RA)	0.03	0.38
	Pink (DG)	0.04	
	Light brown (DT)	0.08	
Zn^{2+}	Pink (DZ)	0.02	0.98
	Light violet (PFi)	0.90	

[a] Alumina, DS-5 (Camag) for Fe^{3+}, Cu^{2+}, Co^{2+}, and Ni^{2+}; cellulose powder CC 41 (Whatman) for Zn^{2+}.

[b] See Table 4.

[c] Represents the minimum amount of the metal ion in the volume of the liquid spotted which is visible on the chromatoplate with the corresponding spray reagent provided the diameter of the "circular spot", i.e. R_F value, remains the same. By this method, the amount of metal ions as low as "sensitivity" limit can be determined.

[d] Mixture of acetone–hydrochloric acid (4 M)–acetylacetone (45:3:2) was used as solvent.

described by Hashmi *et al.*[42]. The blood and dried tissue samples were separately ashed for about 12 h in a muffle furnace at 700° using a china crucible. A circular thin-layer chromatographic apparatus as described by Hashmi *et al.*[43] previously was used with special care taken to saturate the developing chamber before running the chromatoplates. Thin layers of uniform thickness (400 µm) were prepared by mixing 5 g of absorbent with 12 ml of water using a Camag applicator (Camag, Muttenz). The plates were dried overnight at room temperature (30–32°). A very fine hole was made with a pin in the prepared surface and a known volume (1–10 ml) of different concentrations of solutions containing iron, copper, cobalt, nickel, and zinc applied with a micrometer syringe or a micro pipette. The chromatogram was developed for 1–2 min using a mixture of acetone–hydrochloric acid (0.4 *M*)–acetylacetone (45:3:2) as solvent. After drying, the chromatoplate was sprayed with different spray reagents (Table 4). Table 5 shows the sensitivity of the Fe, Cu, Co, Ni, and Zn obtained with different spray reagents as well as the R_F values obtained with the developer acetone–hydrochloric acid (4 *M*)–acetylacetone.

Successful separations of cobalt, chromium, rhodium, and platinum complexes have been achieved by chromatography on paper, ion-exchange, and silica gel starch[44]. Microcrystalline cellulose (Avicel) as an adsorbent in inorganic thin-layer chromatography has been successfully applied to the separation of several cobalt complexes in a variety of solvent systems[45]. Mixtures of the geometric isomers of the cobalt(III) complexes were detected as orange spots by spraying with a 1% aqueous

TABLE 6

SOLVENT SYSTEMS FOR THE TLC SEPARATION OF SOME *cis*- AND *trans*-Co[III] COMPLEXES ON MICRO-CRYSTALLINE CELLULOSE (AVICEL)

No.	Solvent system
1	65 ml acetone–30 ml water–5 ml conc. nitric acid
2	70 ml acetone–20 ml water–10 ml conc. hydrochloric acid
3	60 ml methyl alcohol–40 ml dimethylformamide–10 drops perchloric acid
4	60 ml methyl alcohol–40 ml dimethylformamide–10 drops formic acid
5	60 ml methyl alcohol–40 ml dimethylsulfoxide–10 drops perchloric acid
6	60 ml methyl alcohol–40 ml dimethylsulfoxide–10 drops formic acid
7	100 ml methyl alcohol–10 drops perchloric acid
8	100 ml methyl alcohol–10 drops formic acid
9	70 ml methyl alcohol–30 ml propionaldehyde–10 drops perchloric acid
10	70 ml methyl alcohol–30 ml nitromethane–10 drops perchloric acid
11	50 ml alcohol–50 ml nitromethane–10 drops perchloric acid
12	50 ml methyl alcohol–50 ml propionaldehyde–10 drops perchloric acid
13	70 ml methyl alcohol–30 ml propionitrile–10 drops perchloric acid
14	100 ml propionaldehyde–2 g pyridine N-oxide–10 drops perchloric acid
15	50 ml methyl alcohol–50 ml *n*-butyronitrile–10 drops perchloric acid
16	70 ml methyl alcohol–30 ml *n*-butyronitrile–10 drops perchloric acid
17	70 ml methyl alcohol–30 ml tetrahydrofuran–10 drops perchloric acid
18	70 ml methyl alcohol–30 ml acrylonitrile–10 drops perchloric acid
19	50 ml methyl alcohol–50 ml acrylonitrile–10 drops perchloric acid

solution of the disodium salt of 1-nitroso-2-naphthol-3,6-disulfonic acid. Heating the plates at 60–110°C further heightened the spot colors.

Table 6 lists the solvent systems used and Table 7 shows the R_F values of some cis- and trans-Co^{III} complexes obtained by TLC on microcrystalline cellulose. In all separations, the R_F values of the cis complexes were greater than the trans. Stefanovic and Janjic[46] separated the same cobalt(III) complexes using Whatman No. 1 paper (ascending) with acetone–water–acid and found the same relationship in R_F values, e.g. cis > trans.

TABLE 7

R_F VALUES OF SOME cis- AND trans-Co^{III} COMPLEXES BY TLC USING MICROCRYSTALLINE CELLULOSE

Solvent system	$[Co(en)_2Cl_2]Cl$		$[Co(en)_2(NO_2)_2]NO_3$		$[Co(NH_3)_4(NO_2)_2]NO_3$	
	cis	trans	cis	trans	cis	trans
1			0.75	0.45	0.40	0.13
2	0.52	0.21				
3			0.75	0.37		
4						
5	0.73	0.27	0.74	0.59		
6			0.14	0.09		
7			0.45	0.11		
8	0.50	0.34	0.36	0.15		
9					0.16	0.06
10					0.44	0.20
11					0.47	0.15
12			0.31	0.15	0.17	0.03
13					0.35	0.14
14					0.13	0.03
15			0.72	0.07	0.25	0.09
16			0.53	0.08	0.30	0.09
17					0.25	0.13
18			0.59	0.42	0.27	0.11
19					0.34	0.10

Mori et al.[47,48] found trans-[Co(NH$_3$)$_4$(NO$_2$)]$^+$ and trans-[Co(en)$_2$(NO$_2$)$_2$]$^+$ to be more strongly absorbed on an ion-exchange resin than their corresponding cis isomers.

Druding and Hagel[49] studied the TLC silica gel separations of some cobalt(III)–amine complexes and showed that the trans isomers were more mobile than their cis isomers. It was suggested that the cis isomers with their acid groups adjacent would have two linkages to the silica gel resulting in a more tightly held isomer. The trans isomer with the acido group in the opposite side of the ion would be able to form only one linkage to the silica gel, and hence be more mobile. Silica gel with a surface consisting of silanol groups (Si–OH) operates very effectively as a weak acid ion-exchanger in the same manner as a cation-exchange resin. For geometric isomers separated on paper and microcrystalline cellulose TLC, the *cis* isomer, because of its larger dipole moment, is more mobile than the *trans* isomer.

The TLC determination of traces of cobalt, copper, and nickel in cereals has been reported by Frei[50]. The samples were initially ashed by the technique of Conolly and Maguire[51]. Merck Silica Gel G plates were used with 0.5% conc. HCl in acetone as developing solvent. Detection was achieved following drying of the developed plates at 110° for 5 min, then spraying with either a 0.1% w/v rubeanic acid in ethanol solution or with 0.03% w/v ethanolic solution of pyridine-2-aldehyde-2-quinolyl hydrazone (PAQH) reagent. The separation on silica gel occurred in the order of $Ni^{2+} < Co^{2+} < Ca^{2+}$ with complete resolution being achieved within 10–20 min. The separation time was about one-half the time required to separate the ions on cellulose and about one-tenth the time needed on alumina[52]. The visual detection limits for the rubeanic acid complexes were found to be about 0.03 μg/spot (0.5 cm in diam.) for nickel and copper and 0.05 μg/spot for cobalt. The instrumental detection limits for 20 mg samples taken at an accuracy level of 50% were 0.05 μg for nickel and copper and 0.1 μg for cobalt.

Following spraying, the metal complexes were determined by reflectance spectrometry. A Beckman DU spectrophotometer with a standard diffuse reflectance attachment was used for photometric measurements. For rubeanic acid complexes the wavelength settings were 630 nm for nickel, 465 nm for cobalt, and 580 for copper. For the PAQH complexes they were 508 nm for cobalt and 490 nm for nickel. For calibration curves the reflectance values, R, or the values of the Kubelka–Munk function $(1 - R)^2/2R$ were plotted against concentration[52,53].

Linear Kubelka–Munk plots extending up to concentrations of 10 μg/spot for rubeanic complexes and up to 1 μg/spot for PAQH complexes have already been reported by Frei *et al.*[52,53]. It was found that metal concentrations of Co, Cu, and Ni of about 0.02 and 0.03 p.p.m. in oats could be measured with an accuracy of ±20%.

Muzzarelli[54] described the automatic scanning of chromatograms by multi-scaling with a gamma ray spectrometer. This technique was applied to the determination of the chromatographic behavior of nanogram amounts of ^{51}Cr, ^{54}Mn, ^{59}Fe, ^{60}Co, ^{65}Zn, and ^{109}Co thiocyanate complexes on silica gel. Separations were made of Co, Zn, and Cd from Cr, Mn, and Fe. The thin-layers were Eastman Chromagram sheets, type K301R, inert polyester base coated with silica gel and the eluent was 40 mg of ammonium thiocyanate in 20 ml of methanol plus 80 ml of ethyl ether. For automatic scanning, a ND-110 Nuclear Data 128-channel spectrometer connected to a Dumont photomultiplier was used with an NaI (Tl) crystal $1\frac{3}{4}$ in. × 2 in. The X–Y recorder was a Moseley autograph Model 135, set to write the multi-scaling graphs in steps from 1.5 mm. It was driven by the spectrometer, generally with a 1 sec dwell-time, so that gamma rays from a surface of 9 mm × 20 mm of the chromatoplate were counted during 1 sec and the chromatoplate was advanced by 1.5 mm each second. A Teletype model 33C was used to present the results in digital form. The simple and rapid technique permitted many separations and determinations of nanogram amounts of metals by gamma ray spectrometry without destroying the chromatographic layers. Table 8 lists the levels of radioactivity and the amounts of radioisotopes used for each chromatograph. Figure 2 illustrates a scan of the TLC separation of Co from Cr and Fig. 3 shows the geometry of the detector shield employed with the gamma ray spectrometer.

TABLE 8

THE LEVELS OF RADIOACTIVITY AND THE AMOUNTS OF RADIOISOTOPES USED FOR EACH CHROMATOGRAPH

Isotope	^{51}Cr	^{54}Mn	^{59}Fe	^{60}Co	^{65}Zn	^{109}Cd
Quantity of metal (g)	5×10^{-8}	1×10^{-4}	1×10^{-7}	1×10^{-8}	1×10^{-7}	5×10^{-8}
Level of activity (μC)	0.5	0.25	1.0	0.5	0.5	1.0
Mean dispersion of determinations after chromatography	10%	12%	10%	8%	8%	8%

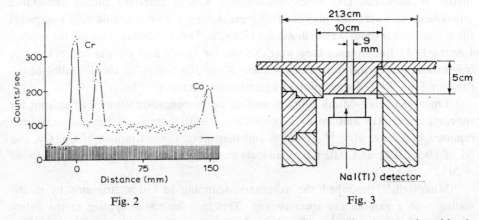

Fig. 2.   Fig. 3

Fig. 2. The separation of Co from Cr. The 128 segments marking the distances are plotted by the recorder when drawing the graph. The counts/sec/step are stored in the memory of the gamma-ray spectrometer, and are plotted by the recorder on the same graph, or read out on the Teletype. The first spot is as long as the window, the second is shorter, and the last is longer.

Fig. 3. The geometry of the detector shield. The window is 9×20 mm and is filled with polyethylene. The chromatographic thin layer is placed on the table and moved from left to right 5.2 cm above the upper surface of the crystal.

Morse and Welford[55] described a radioreagent technique for the determination of cobalt in steels with application to the assay of cobalt added to solutions of NBS steel standards. The technique was based on the equilibration of a constant amount of ^{60}Co with the cobalt in the dissolved steel sample. The cobalt was then isolated by paper chromatography and the density of the tracer in the cobalt area on the paper measured by beta scintillation counting. These results were compared with those obtained by optical density measurements of radioautographs prepared from the chromatogram. Whatman No. 42 paper was used with the developing solvent consisting of hydrochloric acid–distilled water–acetone (10:5:85).

The paper and thin-layer chromatographic behavior of a number of cobalt(III) dioxime chelates has been studied by Soo et al.[56]. Complexes of the type [Co(α-dioxime)$_2$(amine)$_2$]Br (where the α-dioxime is dimethylglyoxime, benzyl α-dioxime or cyclohexane-1,2-dione dioxime, and the amine is aromatic or hetero-cyclic, e.g. a toluidine, a halogenated aniline or picoline) and the more common ethylenediamine or ammine complexes of cobalt were studied on S & S 2043b paper chromatograms developed with methanol–butanol (4:1) and by TLC on kieselgel-gypsum (5:2) with methanol–butanol (10:1) as solvent. Where necessary, spots were rendered visible by spraying with ammonium sulfide solution. This study demonstrated the possibility of differentiating some complexes containing isomeric ligands by TLC.

Cobalt(III), copper(II), and nickel(II) acetylacetonates have been separated by paper chromatography[57] on Whatman No. 1 strips using mixtures of cyclohexane, dioxane, and methanol as developing solvents. Cyclohexane–dioxane–methanol (84:10:6) was the solvent of choice with chromatograms developed in 3 h (25 cm). Table 9 lists the R_F values of metal acetylacetonates of Co^{III}, Cu^{II}, and Ni^{II}. The R_F value of the copper(II) acetylacetonate increased with increasing concentration of dioxane and/or methanol.

The paper chromatography of metal 2-thenoyl-trifluoroacetone (TTA) chelates was reported by Berg and McIntyre[58]. Iron(III), cobalt(III), and nickel(II); iron(III'

TABLE 9

R_F VALUES OF METAL ACETYLACETONATES IN MIXED SOLVENT SYSTEM (CYCLOHEXANE–DIOXANE–METHANOL)

Paper: Whatman No. 1. Development: Ascending

Dioxane (%)	Co^{III} acetylacetonate			Cu^{II} acetylacetonate			Ni^{II} acetylacetonate		
	3%[a]	6%	9%	3%	6%	9%	3%	6%	9%
5	0.58			0.18			0.00		
7		0.57			0.22			0.00	
10	0.69	0.64	0.56	0.25	0.27	0.27	0.00	0.00	0.00
15	0.76	0.72	0.65	0.29	0.34	0.37	0.00	0.00	0.00
20	0.79	0.77	0.73	0.34	0.39	0.44	0.00	0.00	0.02
25	0.83	0.81	0.75	0.43	0.47	0.48	0.00	0.00	0.02
35	0.90	0.88	0.84	0.56	0.57	0.60	0.00	0.01	0.03

[a] Percent methanol in methano–cyclohexane.

TABLE 10

R_F VALUES OF METAL TTA CHELATES IN VARIOUS SOLVENTS

Composition of benzene–methanol–acetic acid solvent by volume	Type paper	R_F values of metal chelates				
		Fe	Co	Ni	Mn	Cu
95:5:0	Plain	0.98	0.04	0.93	0.00	0.98
	NaCl	1.0	0.00	0.77	0.00	1.0
	Al$_2$O$_3$	0.98	0.00	0.00	0.00	0.97
88:10:2	Plain	1.0	0.58	0.95	0.37	1.0
	NaCl	0.98	0.08	0.47	0.07	0.96
	Al$_2$O$_3$	0.99	0.11	0.29	0.00	0.95
68:30:2	Plain	0.95	0.92	0.95	0.53	0.95
	NaCl	0.95	0.53	0.90	0.49	0.93
	Al$_2$O$_3$	0.94	0.10	0.14	0.00	0.96

TABLE 11

SEPARATION OF SOME COBALT COMPLEXES USING ETHER–METHANOL–WATER–CONC. HYDROCHLORIC ACID
(50:30:20:2)

Solute	R_F values	(tailing limits)	Remarks
1 CoCl$_2$	0.43	(0.35–0.49)	Compact spot
2 NH$_4$[Co(NH$_3$)$_2$(NO$_2$)$_2$Ox]	0.74	(0.73–0.75)	Slight tailing to R_F 0.35[a]
3 NH$_4$[Co(NH$_3$)$_2$(NO$_2$)$_4$]	0.31	(0.28–0.33)	Some tailing to R_F 0
4 Co(NH$_3$)$_3$(NO$_2$)$_3$	0.08	(0.05–0.11)	
	0.34	(0.30–0.38)	
5 cis-[Co(NH$_3$)$_4$(NO$_2$)$_2$]Cl	0.23	(0.18–0.27)	Faint tailing to R_F 0.10
6 trans-[Co(NH$_3$)$_4$(NO$_2$)$_2$]Cl	0.74	(0.73–0.75)	Slight tailing to R_F 0[b]
7 [Co(NH$_3$)$_5$Cl]Cl$_2$	0.14	(0.11–0.17)[c]	
8 Co(NH$_3$)$_6$Cl$_3$	0[d]	(0.11–0.20)	Main spot at R_F 0, but a little at
	and 0.15		0.15
9 Na$_3$Co(NO$_2$)$_6$	0.43	(0.35–0.49)	Some tailing to R_F 0.70. Main spot=Co^{II} (due to rapid conversion of the complex during elution)
10 K$_3$CoOx$_3$	0.66	(0.62–0.70)	Some tailing to R_F 0.35 (due to decomposition)
11 cis- or trans-[Co(en)Cl$_2$]Cl	0.27	(0.19–0.33)	Compact spot

The R_F value of the acid front was 0.75. Ox = $(COO)_2{}^{2-}$. en = ethylenediamine.

[a] The tailing back to R_F 0.35 indicates decomposition to Co^{II} during elution.
[b] There was some tailing between spots, and back to R_F 0. The double-spotting suggests either a cis–trans separation, or a separation of the nitro and nitrato isomers.
[c] An intensification of the tailing at R_F 0.34 suggests decomposition to one of the components present in the spot of Co(NH$_3$)$_3$(NO$_2$)$_3$. The original solution contained solid chloro-pentammino-cobaltic chloride (purpureo cobaltic chloride) dissolved in the eluting solvent.
[d] The spot at R_F 0 was very probably due to hydrolysis at the starting line.

nickel(II), and manganese(II); and copper(II), nickel(II), and manganese(II) TTA chelates were separated on paper impregnated with alumina or sodium chloride. Table 10 lists the R_F values of the metal TTA chelates in various solvent systems. The intense color of the iron(III) and copper(II) TTA chelates permitted their detection on the developed chromatogram. Cobalt(II), manganese(II), and nickel(II) chelates were first sprayed with a 1 % solution of sodium sulfide that resulted in the formation of black cobalt sulfide. The chromatograms were next sprayed with a 1 % alcoholic solution of dimethyl glyoxime for the detection of nickel. The Mn^{II} chelate was verified using 0.05 N NaOH then benzidine reagent.

Pollard et al.[59] studied the paper chromatographic separation of cobalt(II) chloride and some ammine and ethylenediamine complexes of cobalt(III). Table 11 illustrates the separation of some cobalt complexes using ether–methanol–water–conc. HCl (50:30:20:2).

3. NICKEL

Nickel is widely distributed, e.g. 0.016% in the earth's crust, with the most common minerals being sulfide ores (pentlandite, $NiFeS_2$; chalcopyrite, $CuFeS_2$; pyrrholite, Fe_7S_8), oxides, silicates, and arsenicals (kupfernickel, Ni_2As_2; nickel glance, $Ni_2(AsS)_2$). Nickel is widely distributed[60] in foods, water, sea water, animal tissues, soils, and miscellaneous materials, the amount ranging from practically none to a few $\mu g/g$, depending on the material and its geographical location. Nickel is concentrated by certain foods such as whole grains and legumes which supply the nickel in the diet of some animals. Other sources of nickel include (1) asbestos (the nickel concentrations in the most common asbestos material, chrysolite, ranges from approx. 1500 to 1800 $\mu g/g$)[61], (2) coal ash, where the concentration of nickel ranges from 3 to 10,000 $\mu g/g$, depending on the origin of the coal[62] (Given[63] reported 100–150 p.p.m. as the average crustal abundance of nickel in Western coal), and (3) fly ash from residual fuel oil contains small amounts of nickel, usually as nickel oxide[64]. Given[63] reported 110 p.p.m. nickel in crude oil and Boldt[65] reported that crude oil contained 55 $\mu g/g$ of nickel while the asphaltene fraction contained 245 $\mu g/g$. Diesel oil has been found to contain 2 p.p.m. nickel while the particulate matter samples collected from diesel engine exhaust emission contained as much as 1 % nickel[66]. Particulate emissions from municipal incinerators[67] have also been shown to contain nickel.

The areas of utility for nickel include (1) the production of alloys with copper, manganese, iron, zinc, chromium, molybdenum, etc. Stainless steels contain 3–35% nickel and the nickel–copper alloys such as Monel contain ca. 70% nickel. Alloys with aluminum and iron, also containing cobalt and other elements, are widely used as permanent magnets, (2) in steel manufacture for the production of special corrosion and heat-resisting steels and cast iron, (3) in nickel plating where large quantities of nickel anodes and nickel salts (chiefly the sulfate) are used in processes such as electroplating, electrodeless plating, electroforming, pressure welding, spraying, vapor-deposition, and chemical deposition, (4) the preparation of catalysts (such as Raney nickel) which are widely used for hydrogenation and dehydrogenation of organic compounds, drying of oils, bleaching, purification of waste water, manufacture of alcohol from gas oil, cracking of ammonia, manufacture of hydrazine from urea, artificial ageing of liquors, and removal of organic sulfur compounds from coal gas, (5) in the production of ground-coat enamels, colored ceramics and glass, (6) in the production of Ni–Cd batteries, and (7) in the preparation of nickel carbonyl which is used as catalyst for carbonylation, polymerization, and miscellaneous reactions as well as to produce powdered nickel for powder metallurgical applications. In the U.S. in 1965 approx. 350 million lb. of nickel were consumed for stainless steels, non-ferrous alloys, miscellaneous steels, electroplating, high-temperature and electrical resistance alloys, cast irons, and miscellaneous salts. The major uses in the U.S. in 1965 of 11,214,000 lb. of nickel in the chemical form included catalysts 4,482,000 lb., electroplating solutions 4,074,000 lb., magnets 1,650,000 lb., and ceramics 1,002,000 lb. More recent suggested areas of nickel utility include nickel complexes (cyclopentadienyl nickel nitrosyl[68]) as fuel additives and nickel chelates of salicylaldehyde nitrites to minimize carbon deposits[69].

Inhalation of nickel or its compounds may cause cancer of the lung or sinus or cause other disorders of the respiratory system or dermatitis. The most likely sources of nickel in urban air are emissions from metallurgical plants, burning coal and oil, nickel plating facilities and engines burning fuels containing nickel additives. Concentrations of nickel in urban air in 1964 averaged 0.032 µg/m^3 and ranged up to a maximum of 0.690 µg/m^3.

Toxicity of nickel and its compounds has been reviewed by Stokinger[70] and Mastromatteo[71]. While nickel metal itself is relatively non-toxic, most nickel salts are highly toxic and nickel carbonyl ($Ni(CO)_4$) is extremely toxic. The effects produced may be cancer of the respiratory system[70-73], respiratory disorders or dermatitis. Sunderman et al.[74-76] reported cancer of the respiratory tract in rats exposed to nickel carbonyl, and it was suggested that nickel carbonyl which has been identified in tobacco smoke (1.5–3.0 µg of nickel per cigarette in 6 common brands of American cigarettes) may be a carcinogen for man[77]. Table 12 summarizes aspects of nickel carcinogenesis in experimental animals.

Nickel carbonyl has been shown to inhibit RNA synthesis in vitro[79] and nickel sulfide inhibited mitotic activity and induced abnormal mitotic figures in cultured rat-embryo muscle cells[80].

Nickel particulates in air have been measured by emission spectroscopy[81], atomic absorption[81], and colorimetric procedures[82] utilizing dimethylglyoxime[83]. Nickel carbonyl in air samples has been determined by spectrographic[84-86] and colorimetric techniques[87], while nickel in biological materials has been analyzed by atomic absorption[88].

TABLE 12

NICKEL CARCINOGENESIS IN EXPERIMENTAL ANIMALS[78]

Species	Compound	Route	Malignant tumors
Rat	Ni dust in lanolin	Intrafemoral intrapleural	Sarcomas
Rat and rabbit	Ni dust in gelatin	Intrafemoral subcutaneous	Sarcomas
Guinea pig	Ni dust	Inhalation	Anaplastic carcinoma
Rat	Ni(CO)$_4$ vapor	Inhalation	Squamous cell and adenocarcinomas
Rat	Ni pellets	Subcutaneous	Sarcomas
Rat	Ni dust in gelatin	Intrapulmonary	Sarcomas
Rat	Ni$_3$S$_2$; NiO dusts in penicillin	Intramuscular	Sarcomas
Rat	Ni$_3$S$_2$ dust, discs, chips	Intramuscular	Sarcomas
Rat	Ni$_3$S$_2$ dust in pencillin	Intramuscular	Sarcomas
Rat	Ni$_3$S$_2$ dust	Intramuscular	Sarcomas
Rat	Ni dust in fowl serum	Intramuscular	Sarcomas
Rat	Ni$_3$S$_2$; NiCO$_3$; NiO dusts in sheep fat	Intramuscular	Sarcomas
Rat	Ni(CO)$_4$ vapor	Inhalation	Squamous cell, anaplastic and adenocarcinomas
Rat	Nickel dicyclopentadienyl (nickelocene)	Intramuscular	Fibrosarcomas

TABLE 13

ABSORPTION MAXIMA (mμ) OF PAQH COMPLEXES OF Co, Ni AND Cu ON SILICA GEL AND CELLULOSE CHROMATOPLATES

	Aqueous solution	Silica gel	Cellulose
Co	500 (pH 7–8)	508 (0.01 N NaOH)	536 (0.01 N NaOH)
Ni	480 (pH 9)	490 (1.0 N NaOH)	520 (1.0 N NaOH)
Cu	475 (pH 9)	485 (1.0 N NaOH)	505 (1.0 N NaOH)

Nickel, cobalt, and copper have been separated from interfering metals by TLC on Silica Gel G and MN-cellulose 300 (Macherey, Nagel) layers by Frei *et al.*[53]. The metals were sprayed with pyridine-2-aldehyde-2-quinolyl hydrazone solution (PAQH) and determined by diffuse reflectance spectroscopy. Under optimum conditions, 0.01 μg/spot could be easily determined and with good reproducibility. The solvent systems for the chromatographic separations consisted of 0.5% conc. hydrochloric acid in acetone for silica gel plates and a freshly prepared mixture of methyl ethyl ketone–conc. HCl–water (15:3:2) for cellulose plates. The absorption maxima (nm) of PAQH complexes of cobalt, nickel, and copper on both silica gel and cellulose plates are shown in Table 13. PAQH reagent has been used previously to develop sensitive and highly selective spectrophotometric methods for palladium, nickel, and cobalt[89,90].

Chelates of Ni, Co, Cu, and Fe (initially extracted with 1-(2-pyridylazo)-2-naphthol) have been separated on Silifol TLC plates and determined by means of scanning reflectance photometry *in situ*[91]. Testa[92] described the separation of a number of cations by means of paper treated with tri-*n*-octylamine (TOA). (Table 14 lists the R_F values found and the solvents and detecting reagents used.)

The utility of hexafluoromonothioacetylacetone (HFAS) as a new ligand for the gas chromatographic separation of d^8 metals was elaborated by Bayer *et al.*[93]. The synthesis of HFAS has recently been described by Bayer and Müller[94] and Brauer[95] and has been shown to proceed according to the sequence

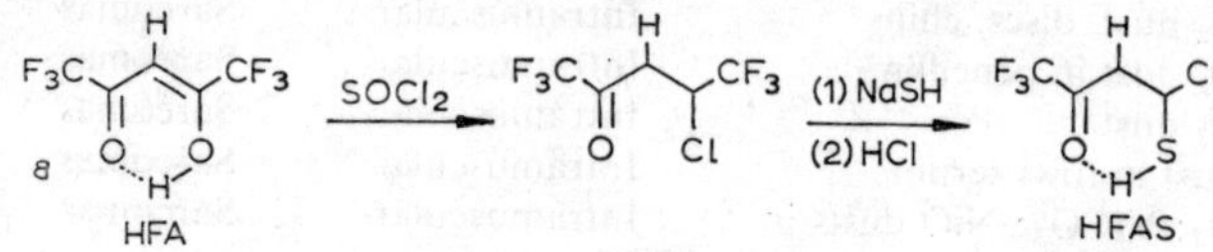

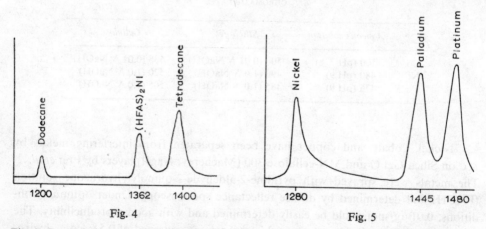

HFAS exhibited completely different analytical selectivity compared with its oxygen analog hexafluoroacetylacetone (HFA). With the divalent metal ions of Cu, Zn, Fe, Ni, Pt, Pd, Cd, and Pb, volatile chelates from HFAS were prepared and characterized by their mass spectra. An LKB 9000 Gas Chromatograph–Mass Spectrometer was used. The total ion current was measured as a detector signal. The accelerating voltage was 3.5 kV and the ion source temperature was 250°C.

Figure 4 shows the gas chromatogram of Ni(HFAS)$_2$ obtained using 1% SE-30 at a column temperature of 100°C. Figure 5 shows the gas chromatographic separation

Fig. 4.

Fig. 5.

Fig. 4. Gas chromatogram of Ni(HFAS)$_2$. 3 m glass column (1% SE-30), column temperature, 100°C; 16 ml He/min carrier gas.

Fig. 5. Gas chromatographic separation of Ni-, Pd-, Pt-HFAS chelates. 1.5 m glass column (3% OV-17), column temperature, 80°C; 18 ml He/min carrier gas.

TABLE 14

R_F VALUES IN THE SEPARATIONS OF SOME CATIONS BY MEANS OF PAPER TREATED WITH TOA

Cation	Solvent	R_F	Developer
Fe^{3+}	4 N HCl	0	KCNS
Co^{2+}	4 N HCl	0.50	8-Hydroxyquinoline
Ni^{2+}	4 N HCl	0.97	Dimethylglyoxime
U^{6+}	10 N HCl	0	$K_4Fe(CN)_6$
Zr	10 N HCl	0.35	Quercetin
Th	10 N HCl	0.95	8-Hydroxyquinoline
Fe^{3+}	1 N HCl	0	$K_4Fe(CN)_6$
Cu	1 N HCl	0.80	$K_4Fe(CN)_6$
Al	1 N HCl	0.98	Alizarin
U^{6+}	3 N HCl	1	$K_4Fe(CN)_6$
V^{5+}	3 N HCl	0.38	Hydrogen peroxide
Ti^{4+}	3 N HCl	0.95	Hydrogen peroxide
Zn	3 N HCl	0.33	8-Hydroxyquinoline
Mn^{2+}	3 N HCl	0.96	8-Hydroxyquinoline
Zn	4 N HCl	0.15	8-Hydroxyquinoline
Mn^{2+}	4 N HCl	0.82	8-Hydroxyquinoline
Ni	4 N HCl	0.97	Dimethylglyoxime
Co^{2+}	4 N HCl	0.45	8-Hydroxyquinoline
Zn	10 N HCl	0.17	8-Hydroxyquinoline
Mn^{2+}	10 N HCl	0.50	8-Hydroxyquinoline
Ni	10 N HCl	0.96	Dimethylglyoxime
Zr	8 N HCl + 5% conc. HNO_3	0.25	Quercetin
Hf	8 N HCl + 5% conc. HNO_3	0.80	Quercetin
U^{6+}	0.2 M H_2SO_4	0.08	$K_4Fe(CN)_6$
Fe^{3+}	0.2 M H_2SO_4	0.66	$K_4Fe(CN)_6$
Cu	0.2 M H_2SO_4	0.68	$K_4Fe(CN)_6$
Ni	0.2 M H_2SO_4	0.97	Dimethylglyoxime
Mo^{6+}	0.5 M H_2SO_4	0	8-Hydroxyquinoline
Fe^{3+}	0.5 M H_2SO_4	0.85	$K_4Fe(CN)_6$
Cu	0.5 M H_2SO_4	0.87	$K_4Fe(CN)_6$
Ni	0.5 M H_2SO_4	0.97	Dimethylglyoxime

Paper: Whatman No. 1. Development: Circular.
Treatment of the paper: After spotting the hydrochloric acid solution (0.02–0.05 ml) of cations to be separated and drying these spots, the paper was immersed in a solution of 0.2 M tri-n-octylamine in kerosene (30–50°).

of Ni-, Pd-, Pt-HFAS chelates on 3% OV-17 at a column temperature of 80°C. The retention indices determined according to Kovats[96] were Ni(HFAS)$_2$ 1280, Pd(HFAS)$_2$ 1445, and Pt(HFAS)$_2$ 1480. Although an exact quantitative assessment was not possible due to the application of a quite selective molecular separation in the gas chromatograph–mass spectrometer, it was estimated that in the gas chromatographic noble metal analysis, microgram amounts could be easily detected and identified.

The ligand HFAS predominantly reacts with metal ions having a d^8 electron configuration (Ni, Pd, Pt). Of the d^{10} elements, zinc and cadmium form HFAS complexes. The Hg complex, however, decomposed shortly after its formation under precipitation of mercury sulfide. All trivalent metals which tend to change their metal states oxidize the ligand to the disulfide, with the exception of gold; no metal chelates were formed by these metal ions.

The gas chromatography of the mixed complexes of nickel(II) and cobalt(II) with trifluoroacetylacetone and dimethylformamide was reported by Jacquelot and Thomas[97]. Ni[II] and Co[II] were separated and determined as $Ni(HFA)_2 \cdot 2DMF$ and $Co(HFA)_2 \cdot 2DMF$ on a 10% XE-60-coated Chromosorb W (KOH and HMDS treated) column at $100°C$ with a flame-ionization detector and helium as carrier gas. The detector response was linear and the lower limit of detection was about $0.4\ \mu g$ complex/$2\ \mu l$ benzene solution of complex injected onto the column.

The most toxic of all nickel compounds, and probably one of the most toxic compounds used in industry, is nickel carbonyl, a low-boiling liquid which vaporizes at ambient temperature and which decomposes at temperatures in excess of $60°C$.

Nickel carbonyl can be prepared by the direct reaction between carbon monoxide and finely divided nickel, or it can also be formed by the reaction between carbon monoxide and either nickel salts in solution or an alkaline suspension of nickel sulfide (in these cases high pressures are required). One of the most important industrial applications of nickel carbonyl is in the Mond process for refining nickel, e.g. nickel carbonyl can readily be decomposed to nickel and carbon monoxide. Nickel carbonyl is also extensively used in the carbonylation of acetylene to acrylic acid[98,99] and as a catalyst in the petroleum, plastic, and rubber industries. The pathological[78,100] and biochemical[101–104] mechanisms of $Ni(CO)_4$ toxicity have been intensively studied and the metabolism of nickel carbonyl-[14]C (ref. 105) and [63]Ni-carbonyl[106] in the rat elaborated.

Nickel carbonyl has been identified in blood following exposure of rats to $Ni(CO)_4$ by inhalation as well as by intravenous injection. It was thus shown that $Ni(CO)_4$ can pass across the pulmonary alveolus in either direction without degradation or metabolic alterations. Sunderman et al.[107] described a gas chromatographic method for the determination of nickel carbonyl in blood and breath and suggested this technique for monitoring the concentrations of $Ni(CO)_4$ in industrial atmospheres and in the blood of workmen who are accidently exposed to inhalation of $Ni(CO)_4$.

A gas–liquid chromatograph equipped for electron-capture detection ($200\ \mu Ci$ tritium) was used with a 6 ft. $\times$ $\frac{1}{4}$ in. i.d. column packed with 5 g Carbowax 20M on 100 g of acid-washed Chromosorb W (60–80 mesh). The injection port, the chromatographic columns, and the electron-capture detector were maintained at ambient temperature ($25°C$). The gas phase was 95% argon–5% methane at a flow rate of 60 ml/min.

Nickel exhaled in the breath of rats following intravenous administration was

trapped in 10 ml of absolute ethanol at $-78°$ and nickel in blood samples was
collected by placing the blood in a side arm flask connected to a high-vacuum pump
via an extraction tube containing 10 ml ethanol cooled to $-78°C$. Table 15 indicates
the retention times of $Ni(CO)_4$ obtained on four liquid phases. Retention times
varied from 19 to 120 sec and the mean ratios of the mobility of $Ni(CO)_4$ to the
mobility of ethanol varied from 0.37 to 0.69. It was demonstrated that the electron-
capture detector was approx. 100,000 times more sensitive to $Ni(CO)_4$ than to ethanol
and the limit of sensitivity for the detection of $Ni(CO)_4$ was approx. 0.001 µl $Ni(CO)_4/$
10 ml of ethanol.

TABLE 15

GAS CHROMATOGRAPHIC DETECTION OF NICKEL CARBONYL IN ETHANOL

Liquid phase	Retention time (sec)		Ratio of retention times of
	$Ni(CO)_4$	C_2H_5OH	$Ni(CO)_4$ and C_2H_5OH
Epon 1001	19	51	0.37
Neopentylglycolsuccinate	24	54	0.43
Carbowax 20 M	37	72	0.51
Silicone DC-560	120	174	0.69

4. CADMIUM

Cadmium occurs in nature as the mineral greenockite or cadmium blende
(CdS) in sparingly mineable amounts of sufficiently high cadmium content. The
main source is zinc and lead ores containing approx. 0.1–0.5% and a few hundredths
percent of cadmium, respectively. World production at the present time is greater
than 31 million lb./year[107] with the U.S. accounting for approx. 40% of the world
output. Next in production are the U.S.S.R., Canada, Japan, Belgium, Mexico,
and Australia.

The commercial production and subsequent manufacturing uses in the U.S.
of metallic cadmium has averaged about 11.5 million lb./year over the last 10 years.
The major uses of cadmium were in the plating processes (55–60%), pigments
(cadmium sulfide and sulfo-selenides) and compounds (30–35%), and in metallurgy
(7–10%). Other principal uses of cadmium include plastic stabilizer (PVC), alloys
and solder (silver solder and bearings), nickel–cadmium batteries, semi-conductors,
photocells, insecticide (as the oxide and hydroxide), turf fungicides, and in nuclear
reactors (as a neutron absorber, either as a coating on graphite or in the form of
rods).

Specific uses of cadmium include cooking utensils, metal pitches, ice-cube
trays, chemicals used in the glass industry, photography, in transistors, as phos-

phors in television tubes, and as dimethyl and diethyl cadmium for polymerization catalysts.

Cadmium toxicity arises from its possible presence in the diet and its occurrence as an industrial and environmental chemical. The dissolution of cadmium from galvanized iron pipes by soft waters and the use of galvanized items in food production is considered a potential source for cadmium contamination. Galvanized iron pipes contain zinc which is contaminated by cadmium in concentrations of 1% or more. Copper used in pipes also contains cadmium as may fittings and solders. Commercial fertilizers, especially the phosphate fractions, contain from 1.4 to 2.3 p.p.m. cadmium[108]. The use of zinc-containing fungicides such as Maneb and Zineb affords an additional source of traces of cadmium. The use of such pesticides may be one explanation for the elevated levels of cadmium in certain vegetables, potatoes, and in tobacco. (A pack of 20 cigarettes contains an average of 30 µg of cadmium[109], and analysis of the cadmium content of whole cigarettes and cigarette ash and filters after smoking indicate that *ca.* 70% of the cadmium content of cigarette tobacco passes into the smoke[110].) Airborne sources of cadmium also include the abrasion of automobile tires compounded with zinc oxide containing traces of cadmium. Cadmium concentrations ranging from 10 to 35 p.p.b. have also been found in gasolines.

The average concentration of cadmium in the atmosphere of the U.S. was 0.002 µg Cd/m^3 while the maximum concentration reported[111] was 0.420 µg Cd/m^3.

Cadmium is found in natural waters as well as in nearly all foods and beverages. Fish and meats (0.79–0.88 p.p.m.) tend to have mean levels considerably higher than those for milk, eggs, cereals, and vegetables (0.07–0.27 p.p.m.). A study[112] of 82 food items (market-basket survey) in the U.S. in 1968 suggested the average daily intake of cadmium to be 0.026 mg (about one-tenth of that estimated by Schroeder[113]). Cadmium was present in all food classes examined and in about 12% of the composite samples.

Aspects of the health effects of man following occupational cadmium poisoning have been intensively studied and reviewed by Stokinger[114], Nilsson[115], Flick et al.[116], Prodan[117], Kazantzis[118], and Friberg[119]. In man, lung and kidney involvement appears to be of primary importance with toxic effects of acute cadmium poisoning including the development of acute bronchiolitis, pneumonitis, pulmonary edema, and, in cases of especially severe exposure, renal cortical necrosis. Chronic intoxication resulting from inhalation of cadmium fumes can cause progressive emphysema.

Chronic low-level effects of cadmium following respiratory or oral intake are usually classified as an industrial hazard. However, one apparently environmentally related disease of cadmium has been reported. Since 1962 Japanese authorities have registered 223 cases of the "Itai-Itai" disease. It is a disease affecting the central nervous system, apparently affecting women more than men. A high degree of contamination by cadmium of the rice and soya-beans grown along the River

Jintsu was found. (The contamination was traced to waste mine tailings from a nearby mine where lead, zinc, and cadmium had been produced.) Rice grown near the entrance showed 0.44–3.36 p.p.m. of cadmium, whereas rice grown in the middle of the field had 0.35–1.32 p.p.m. However, more definitive epidemiological aspects concerning the Itai-Itai disease such as an estimate of the daily intake of the cadmium-enriched rice are lacking.

Cadmium is present in most human organs to the order of 1 p.p.m. (dry weight basis); the concentrations in the lung, kidney, and liver are approx. 0.08, 130 and 6.7 p.p.m., respectively[120,121]. At birth, cadmium is undetectable in human tissue but during maturation, the metal gradually accumulates because intake exceeds excretion by about 3 μg/day[122]. However, cadmium has also been reported in fetal and newborn human liver and kidney and its transplacental passage in man seems certain[123]. It has been indicated that a typical adult American male has accumulated 30 mg of cadmium in his tissues and 4.1 mg of this is stored in the liver[124]. Evidence in experimental animals suggests that cadmium may be an important factor for various pathological processes, including testicular tumors, renal dysfunction, hemorrhagic lesions in sensory ganglia within the central nervous system, arteriosclerosis, hypertension, growth inhibition, chronic diseases of ageing, and cancer. For example, cadmium salts have been shown to be carcinogenic in the rat[125–127] and teratogenic in the hamster[128]. Oral or parenteral administration of cadmium in the rat[129,130] and the rabbit[131] resulted in persistent arterial hypertension.

Cadmium in biological tissues has been analyzed by atomic absorption[132,133] and neutron activation[134–137], in foods by atomic absorption[138], colorimetry[138], and polarography[139], and in alloys by atomic absorption[140], emission spectroscopy[141], polarography[142], and spectrophotometry[143].

The cation-exchange behavior of cadmium on Dowex 50W-X8 and its separation from mixtures was studied by Akki and Khopkar[144]. Cadmium, in milligram amounts, was eluted from Dowex 50W-X8 (H+ form, 20–50 mesh) by the following eluants in various concentrations: HCl, HNO$_3$, HClO$_4$, H$_2$SO$_4$, NH$_4$OAc, KI, NaCl, NH$_4$Cl, NH$_4$Br, NaNO$_3$, and EDTA. Since anions and anionic complexes were not retained by the resin, cadmium could be separated from CrO$_4^{2-}$, TeO$_3^{2-}$, SeO$_3^{2-}$, and AsO$_3^-$ as well as from Zn^{IV} and Ce^{IV} by using 5% citric acid at pH $\simeq$ 2.7, and from nickel by using 5% malonic acid at pH $\simeq$ 4.0 for complexation. Cadmium was also separated from a mixture of Cu, V^V, U^{VI}, Zn, Al, and Mn^{II} by eluting it first with 5% NaCl.

The anion-exchange separation of cadmium, lead, zinc, nickel, cobalt, and manganese in hydrochloric and hydrobromic acid media was demonstrated by Washizuka et al.[145]. Distribution coefficients of Cd, Pb, Zn, Ni, Co, and Mn between the ion exchangers Dowex 50W-X8 and Dowex 1-X8 were first determined at various acid concentrations. Using a 10 × 130 mm column of Dowex 1-X8, 100–200 mesh, the following separations were effected: (1) with hydrochloric acid: Zn–Cd, Zn–Pb,

Ni–Cd, Ni–Pb, Co–Zn, Co–Ni, Mn–Co, and Mn–Ni; (2) with hydrobromic acid: Cd–Zn, Cd–Ni, Ni–Zn, and Ni–Pb. The complete separation of Cd and Pt was effected with a mixture of nitric and hydrobromic acids.

Dithizonates of Cd, Bi, Cu, Pb, Ni, Zn, Co, and Hg have been separated on a column (100 mm × 8 mm i.d.) of potassium citrate or potassium bicarbonate (particle size ranging within 0.02–0.4 mm)[146]. The absorption of dithizonates on potassium citrate (and consequently their separation) depended strongly on the concentration of water present in the sorbent. (The optimum was ca. 0.07 wt. %.) Using carbon tetrachloride as eluent, the elution order was Hg, Co, Zn, Ni, Pb, Cd, Cu, and Bi. Bismuth and copper dithizonates were strongly sorbed on the column and required elution by a mixture of carbon tetrachloride–acetone (10:1). The elution order of dithizonates on a column packed with potassium bicarbonate was Hg, Zn, Ni, Co, Pb, Cd, Cu, and Bi. Dithizonates of Cd, Cu, and Bi were strongly sorbed and were eluted by carbon tetrachloride–acetone (10:1).

Zinc, cadmium, and mercury were resolved as their anionic chloro complexes by an ion-exchange chromatographic procedure[147]. The metal chloro complexes were adsorbed from a 0.01 M hydrochloric acid solution by an anion exchange resin, Dowex-1 (Cl⁻). Zinc and cadmium were eluted separately in that order with 0.01 M hydrochloric acid. Mercury was then removed with a 0.01 M hydrochloric acid–0.1 M thiourea solution. Resolutions were enhanced by the addition of a non-aqueous component (e.g. methanol) to the 0.01 M hydrochloric acid solution. Three separation techniques were proposed: (1) an aqueous system where a mixture of the metals was introduced and eluted in 0.01 M hydrochloric acid on a column 28 cm long, (2) eluant consisted of a 0.01 M hydrochloric acid–10% methanol by volume (18 cm column), and (3) separation was accomplished on a 6 cm column in a 0.01 M hydrochloric acid–25% methanol by volume mixture. All three procedures were conducted with a 60 ml sample volume on −60 +100 mesh Dowex-L at a flow rate of 1.5–2.0 ml/min. Procedures (2) and (3) were the methods of choice for quantitative recoveries. Figure 6 illustrates an elution curve for zinc and cadmium in 0.01 M hydrochloric acid–25% methanol.

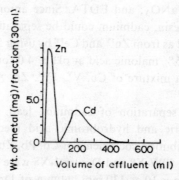

Fig. 6. Elution curve for zinc and cadmium in 0.01 M hydrochloric acid–25% methanol. Column length, 6 cm; resin, Dowex-1 X 10; mesh size, −60 + 100; flow rate, 1.5–2.0 ml/min.

The cation exchange separation of small amounts of metal ions from Cd^{II}, Zn^{II}, and Fe^{III} was described by Fritz and Abbink[148]. Small amounts of Cu^{II}, Mn^{II}, Ni^{II}, and Co^{II} were separated in the form of their chloride complexes from large amounts of Cd^{II}, Zn^{II}, and Fe^{III} on an ion-exchange column containing 100–200 mesh Dowex 50W-X8 cation exchange resin in the H form. Elution from the 1.2×6 cm long column was by $0.5\ M$ HCl containing 50–90% acetone. About 0.2–1.5 mg of Cu, Mn, Ni, or Co were separated from solutions containing 1000 and 10,000 times as much Cd, Zn, or Fe.

The separation of cadmium and zinc as their chloro complexes on the cation exchange resin, Amberlite IR-120, has also been described by Yoshino and Kojima[149].

Zirconium phosphate is a cation exchanger particularly resistant to heat, oxidizing media, and ionizing radiations. It is practically insoluble in nitric and hydrochloric acids up to $8\ M$ and has been used mainly for radiochemical separations[150,151]. The separation of a number of bi- and tervalent cations on zirconium phosphate columns was described by Gal and Peric[152]. On small columns filled with zirconium phosphate in the NH_4^+ form, Zn^{2+}–Cd^{2+} were quantitatively and Ni^{2+}–Co^{2+} partly separated using 0.5–4 N NH_4Cl as eluent. With the exchanger in the H^+ form, UO_2^{2+} was separated from bivalent cations and ferric iron by stepwise elution with 0.1, 0.5, and 4 N HCl solutions.

The determination of cadmium by derivative ion-exchange chromatography was described by Tachikawa[153] which involved the measurement of the conductance difference across a dead space at the column outlet and applied to the determination of 150–500 µg of Cd in the effluent. Amberlite CG-400 (Cl^- form, 100–200 mesh) was used with 0.01 M hydrochloric acid–methanol as eluent.

Cadmium has been separated from dilute solutions with zinc by extraction with 0.1% dithizone in chloroform. The cadmium dithizonate was isolated by TLC on silica gel (using benzene–chloroform–diethylamine (2:2:1) for development), scanned and determined directly by reflectance photodensitometry[154]. The limit of sensitivity was 20 p.p.m. cadmium in solutions with zinc.

A scheme for the TLC of metal dithizonates was developed by Takitani et al.[155]. Pd, Hg, Cu, and Ag were extracted as dithizonates with 0.1% dithizone in chloroform at pH 0–1, then Bi, Cd, Co, Ni, Pb, and Zn at pH 7–8; and Tl, Fe^{2+}, Mn, and Sn^{2+} at pH 9–10. The first and second group metals were separated on MN-Silica S-HR layers with carbon tetrachloride–methylene chloride–benzene (1:7:4) and identified by color. The third group metals were re-extracted with 20% nitric acid, separated with acetone–hydrochloric acid (99:1) and detected by spraying with appropriate reagents. The limits of identification in p.p.m. were 0.2 for Cu, Ni, and Co, 0.3 for Hg, 0.4 for Pd, Ag, and Cd, 0.8 for Zn, 3 for Pb, and 5 for Bi.

The separation and identification of metal ions by reversed-phase TLC was described by Cornett and Gilbert[156]. The adsorbent was prepared by shaking together MN cellulose 300G (15 g), Amberlite LA-2 (1.5 g), and tributylamine (1.5 g) in carbon tetrachloride (70 ml) for 5 min, the mixture then spread on glass plates, dried in air for 15 min and then at 50° (30 min). The chromatograms were developed with 6 M HCl–acetone (3 : 1). Detection was accomplished by successive spraying with ammonium polysulfide solution, 0.1% dithizone solution in chloroform and 0.1% aqueous aluminum. Bi^{IV}, Cd, Co, Cu^{II}, Pb, Mn^{II}, Hg^{II}, Fe^{III}, Al, Ni, Zn, and Ag were subjected to TLC by this technique and only the pairs Ni–Mn and Ag–Bi were not separated.

Fitzgerald[157] described the identification of metals in silver brazing alloys (cadmium, copper, nickel, silver, and zinc) by paper chromatography on Whatman No. 1 paper treated with di-(2-ethylhexyl)orthophosphoric acid. The paper was developed in a chamber previously equilibrated with 0.01 N HCl, dried first in a current of warm air then in an oven for 1 h at 40–50°C. The developed paper was sprayed with dithizone reagent (0.01% in acetone containing 1 ml ammonium hydroxide) to visualize silver, zinc, and cadmium spots and a 0.01% quercetin solution in acetone to detect copper and nickel. Table 16 lists the R_F values, spot colors, and percent composition of Ni, Cu, Cd, Zn, and Hg in brazing alloys.

The separation of 16 cations on styrenated paper using n-butanol saturated with 1 N hydrochloric acid (Table 17) has been described by Anderson et al.[158]

TABLE 16

R_F VALUES[a] AND SPOT COLORS OF METALS IN SILVER BRAZING ALLOYS

Metal	Color	R_F	Composition (%)
Nickel	Orange-pink[b]	0.7–0.9	3–9
Copper	Pink[b]	0.4	15–70
Cadmium	Yellow[c]	0.2	3–25
Zinc	Purple[c]	0.02	15–30
Silver	Orange[b]	0.00	10–90

[a] Values obtained on Whatman No. 1 paper treated with di(2-ethylhexyl) orthophosphoric acid.
[b] Colors obtained with 0.01% quercetin in acetone.
[c] Colors obtained with 0.01% dithizone in acetone containing 1 ml ammonium hydroxide.

TABLE 17

R_F VALUES OF INORGANIC IONS OF STYRENATED PAPER

Solvent: n-Butanol saturated with 1 N HCl. Paper: Whatman No. 41 with 18% (styrene) graft.

Ion	R_F	Ion	R_F
As^{3+}	0.70	Ni^{2+}	0.19
Sb^{3+}	0.72	Cd^{2+}	0.57
Sn^{2+}	0.82	Bi^{3+}	0.72
Fe^{3+}	0.26	Cu^{2+}	0.20
Al^{3+}	0.18	$MoO_4{}^{2-}$	0.42
Zn^{2+}	0.66	Be^{2+}	0.27
Au^{3+}	0.86	Co^{2+}	0.17
Hg^{2+}	0.81	Mn^{2+}	0.20

REFERENCES

1 *Minerals Yearbook*, Bureau of Mines, U.S. Govt. Printing Office, Washington, D.C., 1946–1948.
2 *Air Pollution Survey of Manganese*, Consumer Protection and Environmental Health Service, U.S. Dept. HEW, 1969.
3 A. STANDEN, in KIRK-OTHMER (Ed.), *Encylopedia of Chemical Technology*, 2nd edn., Interscience, New York, 1967.
4 E. G. COGBILL AND M. E. HOBBS, *Tobacco Sci.*, 1 (1957) 68.
5 W. H. PETERSON AND J. T. SKINNER, *J. Nutr.*, 4 (1931) 419.
6 N. L. KENT AND R. A. MCCHANCE, *Biochem. J.*, 35 (1941) 877.
7 M. DEMEREC, G. BERTANI AND J. FLINT, *Am. Naturalist*, 85 (1951) 119.
8 N. N. DURHAM AND O. WYSS, *J. Bacteriol.*, 74 (1957) 548.
9 R. W. KAPLAN, *Naturwissenschaften*, 49 (1962) 457.
10 D. S. MARKARYAN, R. N. MANDZHGALADZE AND V. I. VASHAKIDZE, *Auobshch. Akad. Nauk Gruz. SSR*, 41 (1966) 471.
11 H. H. WILLARD, L. L. MERRITT, JR. AND J. A. DEAN, *Instrumental Methods of Analysis*, 4th edn., Van Nostrand, Princeton, N.J., 1965.
12 *U.S. Public Health Serv. Publ. No. 978*, U.S. Dept. HEW, Washington, D.C., 1962.
13 M. THAKUR, *Separ. Sci.*, 5 (1970) 645.
14 G. BARIBORLI AND L. LIPPARINI, *Rass. Chim.*, 12 (1969) 69.
15 V. KOPRDA AND M. FOJTIK, *Chem. Zvesti*, 20 (1966) 676.
16 R. J. T. GRAHAM AND A. CARR, *J. Chromatog.*, 46 (1970) 293.
17 T. TAKEUCHI, Y. SUZUKI AND Y. YAMAZAKI, *Bunseki Kagaku*, 18 (1969) 459; *Chem. Abstr.*, 71 (1969) 27108.
18 V. MIKETUKOVA AND R. W. FREI, *J. Chromatog.*, 47 (1970) 435.
19 U. A. T. BRINKMAN AND G. DEVRIES, *J. Chromatog.*, 18 (1965) 142.
20 H. J. SENF, *Microchim. Acta*, 3 (1969) 522.
21 V. P. VASYUTIN, N. M. MOROZOVA AND K. M. OL'SHANOVA, *Isv. Vyssh. Ucheb. Zaved. Khim. Khim. Technol.*, 13 (1970) 763.
22 F. W. E. STRELOW AND C. BAXTER, *J. S. African Chem. Inst.*, 22 (1969) 29.
23 J. VANDERDEELEN, *Anal. Chim. Acta*, 49 (1970) 360.
24 A. K. MAJUMDAR AND B. K. MITRA, *Mikrochim. Acta*, 3 (1970) 596.
25 N. M. KARAYANNIS AND A. H. CORWIN, *J. Chromatog.*, 47 (1970) 247.
26 N. M. KARAYANNIS AND A. H. CORWIN, *Anal. Biochem.*, 26 (1968) 34.
27 N. M. KARAYANNIS, A. H. CORWIN, E. W. BAKER, E. KLESPER AND J. A. WALTER, *Anal. Chem.*, 40 (1968) 1736.
28 K. D. LINDGREN AND H. OHMAN, *Arch. Pathol. Anat. Physiol.*, 325 (1954) 259.
29 C. W. MILLER, M. W. DAVIS, A. GOLDMAN AND J. P. WYATT, *Arch. Ind. Hyg.*, 8 (1953) 453.
30 L. SCHWARZ, S. M. PECK, K. E. BLAIR AND K. E. MARKINSON, *J. Allergy*, 16 (1945) 51.
31 J. C. HEATH, *Nature*, 173 (1954) 822.
32 J. C. HEATH, *Brit. J. Cancer*, 10 (1956) 668.
33 J. P. W. GILMAN, *Cancer Res.*, 22 (1962) 158.
34 I. K. SITOVICH, *Zh. Vses. Khim. Obshchestva im. D.I. Mendeleeva*, 6 (1961) 230; *Anal. Abstr.*, 9 (1962) 2304.
35 F. W. E. STRELOW, *Anal. Chim.*, 33 (1961) 994.
36 R. P. BHATNAGER AND R. P. SHUKLA, *Anal. Chem.*, 32 (1960) 777.
37 T. KATSURA, *Japan Analyst*, 10 (1961) 370.
38 J. S. FRITZ AND J. E. ABBINK, *Anal. Chim. Acta*, 22 (1960) 153.
39 S. B. AKKI AND S. M. KHOPKAR, *Anal. Chim. Acta*, 52 (1970) 393.
40 R. A. A. MUZZARELLI, *Talanta*, 13 (1966) 193.
41 M. N. SASTRI, A. P. RAO AND A. R. K. SARMA, *Indian J. Chem.*, 4 (1966) 287.
42 H. H. HASHMI, F. R. CHUGHTAI, N. A. CHUGHTAI, A. RASHID AND M. A. SHAHID, *Mikrochim. Acta*, (1968) 712.
43 M. H. HASHMI, M. A. SHAHID AND A. R. AYAZ, *Talanta*, 12 (1965) 713.
44 V. CARUNCHIO AND G. G. STRAZZA, *Chromatog. Rev.*, 8 (1966) 260.

45 V. T. HAWORTH AND M. J. ZETLMEISL, *Separ. Sci.*, 3 (1968) 145.

46 G. STEFANOVIC AND T. JANJIC, *Anal. Chim. Acta*, 19 (1958) 488.

47 M. MORI, M. SHIBATA AND N. NANASAWA, *Bull. Chem. Soc. Japan*, 29 (1956) 947.

48 M. MORI, M. SHIBATA AND J. AZAM, *Nippon Kagaku Zasshi*, 79 (1955) 1003.

49 L. DRUDING AND R. B. HAGEL, *Anal. Chem.*, 38 (1966) 478.

50 P. W. FREI, *J. Chromatog.*, 34 (1968) 563.

51 J. F. CONOLLY AND M. F. MAGUIRE, *Analyst*, 80 (1955) 172.

52 R. W. FREI AND D. E. RYAN, *Anal. Chim. Acta*, 37 (1967) 187.

53 R. W. FREI, R. LIIVA AND D. E. RYAN, *Can. J. Chem.*, 46 (1968) 167.

54 R. A. A. MUZZARELLI, *Talanta*, 13 (1966) 1689.

55 R. S. MORSE AND G. A. WELFORD, *Anal. Chem.*, 42 (1970) 1100.

56 A. SOÓ, C. VÁRHELYI AND A. SIPOS, *Studia Univ. Babes-Bolyai, Ser. Chem.*, 13 (1968) 57.

57 E. W. BERG AND J. E. STRASSNER, *Anal. Chem.*, 27 (1955) 127.

58 E. W. BERG AND R. T. MCINTYRE, *Anal. Chem.*, 26 (1954) 813.

59 F. H. POLLARD, A. J. BANISTER, W. J. GEARY AND G. NICKLESS, *J. Chromatog.*, 2 (1959) 372.

60 H. A. SCHROEDER, J. J. BACASSA AND I. H. TIPTON, *J. Chronic Diseases*, 15 (1962) 51.

61 J. C. BRIDGE, *Ann. Rept. Chief Inspector of Factories and Workshops, 1932*, Her Majesty's Stationery Office, London, 1939.

62 R. F. ABERNATHY AND F. H. GIBSON, *U.S. Bur. Mines Inform. Circ. 1C-8163*, 1963.

63 P. H. GIVEN, *Advan. Chem. Ser.*, 55 (1966) 1.

64 W. S. SMITH, *U.S. Publ. Health Serv. Publ. 999-AP-2*, 1962.

65 J. R. BOLDT, JR., *The Winning of Nickel*, Longmans, Toronto, 1967.

66 J. W. FREY AND M. CORN, *Am. Ind. Assoc. J.*, 28 (1967) 468.

67 R. L. CHASS AND A. H. ROSE, Paper presented at the 46th Annual Meeting, Air Pollution Control Assoc., Los Angeles, 1953.

68 T. BROWN, *U.S. Pat. 3,249,964*, 1966.

69 E. MIEDZIELSKI, *U.S. Pat. 2,891,853*, 1959.

70 H. E. STOKINGER, in F. A. PATTY (Ed.), *Industrial Hygiene and Toxicology*, Vol. II, 2nd edn., Interscience, New York, 1963, p. 1118.

71 E. MASTROMATTEO, *J. Occupational Med.*, 9 (1967) 127.

72 J. G. MORGAN, *Brit. J. Ind. Med.*, 15 (1958) 224.

73 W. J. WILLIAMS, *Brit. J. Ind. Med.*, 14 (1958) 235.

74 F. W. SUNDERMAN AND A. J. DONNELLY, *Am. J. Pathol.*, 46 (1965) 1027.

75 F. W. SUNDERMAN AND J. F. KINCAID, *J. Am. Med. Assoc.*, 155 (1954) 889.

76 F. W. SUNDERMAN AND F. W. SUNDERMAN, JR., *Arch. Ind. Hyg.*, 20 (1959) 36.

77 F. W. SUNDERMAN AND F. W. SUNDERMAN, JR., *Am. J. Clin. Pathol.*, 35 (1961) 203.

78 H. L HACKETT AND F. W. SUNDERMAN, *Arch. Environ. Health*, 14 (1967) 604.

79 D. J. BEACH AND F. W. SUNDERMAN, JR., *Cancer Res.*, 30 (1970) 48.

80 S. H. H. SWIERENGA AND P. K. BASRUR, *Lab. Invest.*, 19 (1968) 663.

81 R. J. THOMPSON, G. B. MORGAN AND L. J. PURDUE, Paper presented at Instrument Society of America Symp., New Orleans, La., May 5–7, 1969.

82 F. FEIGL AND R. STERN, *Z. Anal. Chem.*, 60 (1921) 1.

83 Y. KOBAYASHI, *J. Chem. Soc. Japan, Ind. Chem. Sect.*, 58 (1955) 728.

84 R. S. BRIEF, F. S. VENABLE AND R. S. AJEMIAN, *Am. Ind. Hyg. Assoc. J.*, 26 (1965) 72.

85 A. A. BELYAKOV, *Zavodsk. Lab.*, 26 (1960) 158.

86 G. A. HUNOLD AND W. PIETRULLA, *Arbeitsschutz*, 8 (1961) 193.

87 G. PITET, *Arch. Mald. Profess.*, 21 (1960) 674.

88 F. W. SUNDERMAN, *Am. J. Clin. Pathol.*, 44 (1965) 182.

89 M. L. HEIT AND D. E. RYAN, *Anal. Chim. Acta*, 34 (1966) 407.

90 S. P. SINGHAL AND D. E. RYAN, *Anal. Chim. Acta*, 37 (1967) 187.

91 A. GALIK AND A. VINCOUROVA, *Anal. Chim. Acta*, 46 (1969) 113.

92 C. TESTA, *J. Chromatog.*, 5 (1961) 236.

93 E. BAYER, H. P. MÜLLER AND R. SIEVERS, *Anal. Chem.*, 43 (1971) 2012.

94 E. BAYER AND H. P. MÜLLER, *Tetrahedron Letters*, (1971) 533.

95 G. BAUER, *Handbuch der präparativen Anorganischen Chemie*, Ferdinand Enke Verlag, Stuttgart, 1960, p. 325.

96   E. Kovats, *Helv. Chim. Acta*, 11 (1958) 1915.
97   P. Jacquelot and G. Thomas, *Bull. Soc. Chim. France*, 2 (1971) 702.
98   J. W. Reppe, *Mod. Plastics*, 23 (1945) 162.
99   J. W. Copenhaver and M. H. Bigelow, *Acetylene and Carbon Monoxide Chemistry*, Reinhold, New York, 1949, p. 246.
100 R. L. Hackett and F. W. Sunderman, Jr., *Arch. Environ. Health*, 16 (1968) 349.
101 F. W. Sunderman, Jr., *Am. J. Clin. Pathol.*, 39 (1963) 549.
102 F. W. Sunderman, Jr., *Am. J. Clin. Pathol.*, 43 (1964) 228.
103 F. W. Sunderman, Jr., *Cancer Res.*, 27 (1967) 950.
104 F. W. Sunderman, Jr., *Cancer Res.*, 27 (1967) 1595.
105 K. S. Kasprzak and F. W. Sunderman, Jr., *Toxicol. Appl. Pharmacol.*, 15 (1969) 295.
106 E. W. Sunderman, Jr. and C. E. Stein, *Toxicol. Appl. Pharmacol.*, 12 (1968) 207.
107 F. W. Sunderman, N. O. Roszel and R. J. Clark, *Arch. Environ. Health*, 16 (1968) 836.
108 H. A. Schroeder and J. J. Balassa, *J. Chronic Diseases*, 4 (1968) 236.
109 K. J. Kingsbury, T. D. Heyes, D. M. Morgan, C. Aylott, P. A. Burton, R. Emmerson and P. J. A. Robinson, *Biochem., J.*, 84 (1962) 124.
110 M. Nandi, H. Jick, D. Slone, S. Shapiro and G. P. Lewis, *Lancet*, 2 (1969) 1329.
111 *Air Quality Data, National Air Sampling Network and Contributing State and Local Networks 1964–1965*, U.S. Dept. HEW, Washington, D.C., 1966.
112 R. E. Duggan and G. Q. Lipscomb, *Pestic. Monit. J.*, 2 (1969) 153.
113 H. A. Schroeder, *J. Chronic. Diseases*, 20 (1967) 179.
114 H. Stokinger, in F. A. Patty, D. W. Fassett and D. D. Irish (Eds.), *Industrial Hygiene and Toxicology*, Vol. 2, Wiley, New York, 1963, p. 1011.
115 R. Nilsson, *Ecol. Res. Comm. Bull. No. 7*, Swedish Natural Science Res. Council, Stockholm, 1970.
116 D. F. Flick, H. F. Kraybill and J. M. Dimitroff, *Environ. Res.*, 4 (1971) 71.
117 L. Prodan, *J. Ind. Hyg.*, 14 (1932) 132.
118 G. Kazantzis, *Brit. J. Ind. Med.*, 13 (1956) 30.
119 L. Friberg, *Acta Med. Scand.*, 138 (Suppl. 240) (1950).
120 I. H. Tipton and B. J. Cook, *Health Phys.*, 9 (1963) 103.
121 S. R. Stitch, *Biochem. J.*, 67 (1957) 97.
122 H. Holden, *Ann. Occupational Hyg.*, 8 (1965) 1.
123 E. J. Underwood, *Trace Elements in Human and Animal Nutrition*, Academic Press, New York, 1962.
124 J. P. Smith, J. C. Smith and A. J. McCall, *J. Pathol. Bacteriol.*, 80 (1960) 287.
125 G. Kazantzis and W. J. Hanbury, *Brit. J. Cancer*, 20 (1966) 190.
126 A. Haddow, F. J. C. Roe, C. E. Dukes and B. C. U. Mitchley, *Brit. J. Cancer*, 18 (1964) 667.
127 S. A. Gunn, T. C. Gould and W. A. D. Anderson, *Proc. Soc. Exptl. Biol. Med.*, 115 (1964) 653.
128 V. H. Ferm and S. J. Carpenter, *Nature*, 216 (1967) 1123.
129 M. Kanisana, *Exptl. Molec. Pathol.*, 10 (1969) 81.
130 H. A. Schroeder, A. P. Nason and R. E. Prior, *Am. J. Physiol.*, 214 (1968) 469.
131 G. S. Thind, G. Karreman and K. F. Stephan, *J. Lab. Clin. Med.*, 76 (1970) 560.
132 G. Lehnert, G. Klavis, K. H. Schaller and T. Haas, *Brit. J. Ind. Med.*, 26 (1969) 156.
133 G. Lehnert, K. H. Schaller and T. Haas, *Z. Klin. Chem.*, 6 (1968) 174.
134 K. W. Lieberman and H. H. Kramer, *Anal. Chem.*, 42 (1970) 266.
135 T. Westermark and B. Sjostrand, *Intern. J. Appl. Radiation Isotopes*, 9 (1960) 78.
136 H. J. Bowen, *Analyst*, 92 (1967) 118.
137 K. Fritze and R. Robertson, *J. Radioanal. Chem.*, 1 (1968) 463.
138 Anal. Methods Committee, *Analyst*, 94 (1969) 1153.
139 J. Cholak and D. M. Hubbard, *Ind. Eng. Chem. Anal. Ed.*, 16 (1944) 333.
140 W. I. Elwell and J. A. F. Gidley, in P. W. West, A. M. G. MacDonald and T. S. West (Eds.), *Analytical Chemistry 1962*, Elsevier, Amsterdam, 1963, p. 291.
141 R. F. Farrell, G. J. Harper and R. M. Jacobs, *Anal. Chem.*, 31 (1959) 1550.

142 W. T. ELWELL AND D. F. WOOD, in P. W. WEST, A. M. G. MACDONALD AND T. S. WEST (Eds.), *Analytical Chemistry 1962*, Elsevier, Amsterdam, 1963, p. 143.
143 G. GHERSINI AND S. MARIOTTINI, *Talanta*, 18 (1971) 442.
144 S. B. AKKI AND S. M. KHOPAR, *Z. Anal. Chem.*, 249 (1970) 228.
145 S. WASHIZUKA, K. ANDO, H. ITO AND T. SAKAMOTO, *Hokkaido Kyoiku Daigaku Kiyo Sect. 2A*, (1967) 39; *Chem. Abstr.*, 70 (1969) 63772u.
146 A. I. BUSEV, I. P. KHARLAMOV AND O. V. SMIRNOVA, *Zavodsk. Lab.*, 35 (1969) 1301.
147 E. W. BERG AND J. T. TRUEMPER, *Anal. Chem.*, 30 (1958) 1827.
148 J. S. FRITZ AND J. E. ABBINK, *Anal. Chem.*, 37 (1965) 1274.
149 Y. YOSHINO AND M. KOJIMA, *Japan Analyst*, 4 (1955) 311.
150 I. GAL AND A. RUVARAC, *J. Chromatog.*, 13 (1964) 549.
151 C. B. AMPHLETT, *Proc. 2nd Intern. Conf. Peaceful Uses At. Energy*, Vol. 28, 1958, p. 17; *Inorganic Ion-Exchangers*, Elsevier, Amsterdam, 1964.
152 I. GAL AND N. PERIC, *Mikrochim. Acta*, (1965) 251
153 T. TACHIKAWA, *Bunseki Kagaku*, 19 (1970) 823.
154 V. MASSA, *Farmaco (Pavia), Ed. Prat.*, 25 (1970) 334.
155 S. TAKITANI, M. SUZUKI, M. YOSHIMURA, S. SATO AND M. SEKIYA, *Eisei Kagaku*, 14 (1968) 324.
156 F. V. CORNETT AND T. W. GILBERT, *Anal. Letters*, 4 (1971) 69.
157 J. W. FITZGERALD, *J. Assoc. Offic. Anal. Chemists*, 52 (1969) 1119.
158 J. R. A. ANDERSON, S. DILLI, J. L. GARNETT AND E. C. MARTIN, *Nature*, 201 (1964) 772.

Chapter 4

TIN AND LEAD

1. TIN

Tin occurs chiefly in the ore cassiterite or tinstone (SnO_2). It is present in traces in most soils and plants. Levels of 1.2 p.p.m. in wheat flour, 1.6 p.p.m. in bran, and 3.9 p.p.m. in the outer pericarp of wheat have been reported[1]. Of the 30 trace metals of possible or probable biological significance, tin ranks 21st in the universe, 17th on the geosphere (3 p.p.m.), 12th in the hydrosphere (3 p.p.b.), and 8th in the body of man (0.4 p.p.m.)[2]. Sizable amounts of tin occur in the air of industrial cities. For example, amounts from 0.003 to 0.3 μg/m^3 were found in 60.6% of samples taken from 22 U.S. cities[3] while 17% of samples from 6 major industrial centers had 0.03–0.3 μg/m^3. The major areas of utility for tin include (*1*) in the manufacture of tin plate, widely used in containers for food and liquids and for electro-deposited coatings employed in electrical, radio, and engineering components, (*2*) in alloys with zinc, nickel, lead, copper, etc., and (*3*) for compounds as opacifiers for vitreous enamels and as reducing agents in chemical processes.

The toxicity of tin is almost exclusively due to its organic compounds[4-6] and most specifically to the alkyl derivatives. The lower alkyl derivatives, especially triethyltin, have a specific effect on the central nervous system, producing cerebral edema.

Organotins are compounds which contain at least one tin–carbon bond and if all radicals attached to tin through carbon are designated R, and all other substituents X, the series R_4Sn, R_3SnX, R_2SnX_2, and $RSnX_3$ is obtained. (R may be a simple aliphatic or aromatic hydrocarbon radical and X halide, hydroxide, OR, SH, SR, or acyl radicals.) Most of the organotins of industrial and pesticidal utilty are the derivatives of quadrivalent tin.

Organotin compounds are used in plastics and polymers as stabilizers of vinyl resins and oxygen-containing polymers and polyamides against degradation by heat and/or UV light; to control pore structure in polyurethane films; preserve the transparency of polyvinyl chloride. Compounds used for this purpose are of the type R_2SnX (*e.g.* dioctylin and dibutyltin dilaurates, maleates, oxides, etc.) and are generally present to the extent of 1–2% of the finished polymer.

The increased use of PVC in the food packaging and disposable medical articles fields has focused the need to elicit both the amount and nature of tin stabilizer residues. Organotin compounds used to stabilize polyvinyl chloride during container-forming operations can migrate into foodstuffs packaged in such containers. The FDA permits the presence in certain foodstuffs, as a result of such migration, of two organotin compounds, namely dioctyltin S,S'-bis-(isooctylmercaptoacetate) and dioctyltin maleate polymer. The concentration of either, or any combination of both, may not exceed 1 p.p.m. which represents 0.158 or 0.259 p.p.m. of tin (as organotin), respectively, in the foodstuffs.

Organotins are employed in a host of other applications that include (a) in rubber products and paints: as antioxidants and anticracking agents; to retard rubber deteriorations and as stabilizers of chlorinated rubbers or chlorinated paints, (b) in transformers, capacitors and cables: to prevent corrosion by serving as scavengers for HCl formed if a short circuit occurs in transformers, etc., using pyranols or chlorinated diphenyls (tetraphenyltin is usefully employed for this purpose), (c) in lubricants and textile oils: as acute oxidants and corrosion-reducing adjuvants for lubricants; as anti-oxidants for textile oils, (d) as activators and catalysts: in oxidation, polymerization (polyesters and silicone elastomers) as Ziegler–Natta type catalysts for polymerization of olefins, (e) tin-containing polymers: numerous tin-containing polymers and macromolecules have been prepared with tin in the main chain or as a substituent as well as tin analogs of silicons (in which carbon and tin alternate), (f) miscellaneous uses of organotins include: treatment of fibreglass with alkyl- and aryltin compounds for adhesion to resins; curing catalysts for application of silicons to textiles, paper.

The biocidal applications of organotin include (a) agricultural fungicides: triphenyltin acetate (I) (Brestan; fentin acetate) and triphenyltin hydroxide (II) (Duter; fentin hydroxide) and bis(tri-n-butyltin) oxide (III) (TBTO), (b) general fungi-

cidal action (e.g. triphenyltin chloride, TBTO): in paints, preservation of manila and sisal ropes, leather, textiles, to confer mildew resistance to fabrics, for protection of jute and jute bags; wood preservative, slimicide; production process paper, (c) bactericides and biostats: disinfectant (triaryltin), bactericides for seeds, (d) anthelmintics: against worms in poultry (dibutyltin laurate, tin oleate, tetra-isobutyltin), (e) nematocide: p-bromophenoxy triethyltin, (f) herbicides: vinyltin compounds (trivinyltin chloride), (g) rodent repellants: protecting food in treated bags (tributyltin chloride, triphenyltin chloride and acetate), (h) molluscides: triphenyltins, (i) ovicides: trialkyl- and triaryltin chlorides (e.g. R_3SnC, R being methyl, ethyl or propyl) as insecticides and ovicides in combination with DDT or pyrethrum.

The mode of action of organotins in mammals can be delineated as the degree of alkylation of the tin compound *per se*. In general, in the whole animal, pharmacological and toxicological effects of trialkyltins are confined to the nervous system[7,8]. For rats, the oral toxicities[5] of the trialkyltins are in the order triethyl > trimethyl > triisopropyl > tri-*n*-butyl. (The decrease of toxicity with increasing length of alkyl chain is analogous to that observed with di- and tetraalkyltins.)

The conversion of tetraalkyltins to trialkyltins[9-11] *in vivo* (as demonstrated for tetraethyltin) accounts for the latent toxicity of the tetraalkyltins, with the site of the conversion being the liver[7,11]. From the toxicological point of view, the tetraalkyltins can thus be considered to behave in a manner analogous to their trialkyltin counterparts[12]. (This conversion of a tetraalkyl to a trialkyl metal has also been shown for tetraethyllead[7,13,14] and may be a general phenomenon.) Once tetraethyltin has been converted to triethyltin, it appears to persist in the body in that form without apparent further reaction to the dialkyl derivative. The trialkyltin ion is a stable entity which is toxic *per se* and persists for some time in the tissues[10]. Long-term feeding experiments with triethyltin have disclosed some testicular atrophy in addition to lesions confined to the central nervous system[8].

Triethyltins and diethyltins ionize in aqueous solution[15,16] with production of the univalent $(C_2H_5)Sn^+$ and divalent $(C_2H_5)_2Sn^{2+}$ cations, respectively. It is reasonable to assume that trialkyl- and triaryltins exert their biological action as R_3Sn^+ ions or as the undissociated hydroxide R_3SnOH formed on dissociation.

The dealkylation of diethyltin by the rat has been reported[17], with diethylation occurring in both the gut and tissues. The induction of biliary and hepatic lesions by dibutyl tin salts in rats has also been described[4].

Triphenyltins, $(C_6H_5)_3SnX$ (where X was halide, hydroxide, alkyl or alkenyl, aryl or alicyclic radicals, and ester groups or organic acids) have been found to be chemosterilants[18,19] when fed to adult houseflies. Triphenyltins are mainly used in agriculture as fungicides (*e.g.* Brestan, TBTO). In the finely-divided state in which triphenyl tins are applied to plants, they are susceptible to light and oxygen[20].

Phenyl groups are gradually split off with step-by-step loss in toxicity. Triphenyltins decompose slowly into diphenyltins and the final stage of non-toxic inorganic quadrivalent tin compounds is probably reached through the unstable intermediate monophenyl tins[21], *viz.*

$$(C_6H_5)_3Sn^- \rightarrow (C_6H_5)_2Sn\big< \ \rightarrow [C_6H_5Sn\big<] \rightarrow \ -\!\!\overset{|}{\underset{|}{Sn}}\!\!-$$

Pate and Hays[22] described several degenerative changes in testicular tissue of male albino rats treated with triphenyltin acetate and chloride. Trialkyltin compounds have also been shown to be inhibitors of oxidative phosphorylation and appear to react with a component of the energy transfer chain leading to the formation of ATP[23].

The paper chromatography of some organotin compounds was described by

Williams and Price[24]. The developing solvents used with Whatman No. 1 paper were n-butanol–pyridine–water (705:3.5 and saturated with water), butanol–ethanol–water (3:1, saturated with water), butanol–ammonia–water and aqueous pyridine (N in ammonia and saturated with water). The organotin compounds were

TABLE 1

R_F VALUES OF ORGANOTIN COMPOUNDS

Compound	Developing solvent			
	Pyridine (60% in water)	Butanol–pyridine–water (7.5:3.5)	Butanol–ammonia–water (N in ammonia)	Butanol–ethanol–water (3:1)
Me$_2$SnCl$_2$	0.36^c	0.55^b	0.03	0.67^b
Me$_3$SnCl	0.35	0.25		0.32
EtSnCl$_3$	Streaks to 0.8	Streaks length of paper	Streaks length of paper	Streaks length of paper
Et$_2$SnCl$_2$	0.36^c	0.80^b	0.16^c	0.98^b
Et$_2$SnOAc	0.40	0.85^b		0.95^b
Et$_3$SnOH	0.88	0.94	0.94	0.95
Pr$_2$SnCl$_2$	0.00^b	0.9^b	Streaks to 0.52	0.98
(Pr$_3$Sn)$_3$O	0.85	0.94		
(iso-Pr$_3$Sn)$_2$O	0.87	0.94		
Pr$_3$SnCH$_2$CH$_2$COONa	0.82	0.90	0.83	
BuSnCl$_3$	Streaks to 0.8	0.99^b	Streaks length of paper	Streaks length of paper
Bu$_2$SnCl$_2$	0.00^a	0.93	0.00^b	0.98^a
Bu$_2$Sn dilaurate	0.00^a	0.95		0.96
Bu$_3$SnCl	0.85	0.92	0.95	0.94
Bu$_3$Sn laurate	0.82	0.92		0.94
(Bu$_3$Sn)$_2$O	0.82	0.97		0.96
Bu$_3$Sn abietate	0.84	0.97	0.94	
Hexabutyl distannane		0.94 (0.8, 0.3^d)	0.95 (0.8^d, 0.3^d)	
Bu$_4$Sn	0.00	0.93	0.94	0.92
n-Hexyl$_2$Sn dilaurate	0.00	0.95		0.95
n-Octyl$_2$Sn dilaurate	0.00			0.94
n-Octyl$_3$SnCl	0.00	0.94	0.93	0.95
n-Octyl$_4$Sn	0.00	0.90	0.94	0.83
PhSnCl$_3$	Streaks to 0.85	0.99^b	Streaks length of paper	Streaks length of paper
Ph$_2$SnCl$_2$	0.00	0.00	0.00	0.95^b
Ph$_3$SnCl	0.85	0.95	0.97	0.97
Ph$_3$SnOAc	0.85	0.95	0.97	0.97
Ph$_4$Sn	0.00	0.00	0.00	0.00
SnCl$_4$	0.00	0.00	0.00	0.00^a

^{a} Slight tail. ^{c} Elongated spot.
^{b} Severe tailing. ^{d} Probably impurity.

Values obtained on Whatman No. 1 paper. Detection reagent: 0.1% catechol violet in 95% ethanol.

detected by spraying with catechol violet (0.1% in 95% ethanol) after initial oxidation of compounds of the types R_4Sn and R_3SnX by ultraviolet irradiation which renders them sensitive to the spray yielding a blue color. Di- or monoorganotin compounds yield a blue color with this reagent. Table 1 lists the R_F values for a number of organotin compounds. Quantitative recoveries of diphenyltin and triphenyltin compounds were achieved following chromatographic separation. After development, irradiation, and spraying as described above, the blue bands obtained were excised with $\frac{1}{2}$ in. margins on both sides and wet ashed with sulfuric and nitric acids. The inorganic tin in the resulting solution was determined turbidimetrically with 4-hydroxy-3-nitrophenyl arsonic acid according to the procedure of Karsten et al.[25].

Adcock and Hope[26] described a method for the determination of tin in the range 0.2–1.6 µg and its application to the determination of organotin stabilizer in certain foodstuffs. Tin was determined spectrophotometrically as its colored complex with catechol violet, the complex being separated from excess of catechol violet by chromatography on asbestos or cellulose. Organotin stabilizer was isolated from the foodstuff (vinegar, orange drink, and cooking oil), separated from inorganic tin by paper chromatography on Whatman No. 1, wet oxidized with sulfuric acid–hydrogen peroxide mixture and the tin determined in the resulting solution.

For the paper chromatographic separation of the mercaptoacetate organotin stabilizer from the foodstuff, chloroform was used as the solvent in a double development and the stabilizer detected by spraying with 0.5% 4-(2-pyridylazoresorcinol disodium salt) (PAR) in ethanol which produced a pink area on a yellow background.

It was also found that the mercaptoacetate in stabilizer (dioctyltin S,S-bis(isooctyl mercaptoacetate)) could be separated from cooking oil (sunflower seed oil) by chromatography on a column of silica gel from which it could be eluted by 2% formic acid in diethyl ether.

Neubert[27] described the utility of TLC with development by isopropyl ether with 1.5% acetic acid and detection with pyrocatechol violet for the separation and identification of organotin stabilizers. It was possible to separate the four stages of alkylation and to distinguish the toxicologically most important dibutyl- and dioctyltin compounds and to also characterize the sulfur-containing stabilizers according to the tin-mercapto compounds. The carboxylates were converted to the diacetates during elution and separation was dependent upon the nature of the alkyl group present. The TLC procedure permitted the detection of 1 µg of organotin derivative. The organotins were detected on the chromatograms by exposure to UV light for 30 min then spraying with freshly prepared 0.5% alcoholic pyrocatechol violet reagent. For the determination of sulfur compounds, a 2.0% alcoholic solution of phosphomolybdic acid was used. Following spraying, the plate was warmed for 10 min at 100–110°C and the spots further detected by the UV fluorescence.

A method was described for the analysis of octyltin compounds and lower alkyltin compounds (mainly butyltin derivatives) occurring therein as impurities[28]. Separation was achieved by TLC on Kieselgel G employing n-butanol–glacial acetic acid–water (300:5:95) as solvent and dithizone for detection. Determination was carried out by spectrophotometry also using 0.012% dithizone in CCl$_4$. A single calibration curve was found sufficient as the color intensity of the different alkyltin compounds (except monoalkyl compounds) was proportional to the tin content. Monoalkyltin compounds were determined volumetrically with EDTA.

The TLC separation of the isomeric di-2-ethylhexyl- and di-(n-octyl)tin derivatives was described by Helberg[29,30]. The developing solvent was isooctane–diisopropyl ether–acetic acid (80:3:8) and detection was accomplished with 0.12% dithizone in carbon tetrachloride. The organotin–dithizone complexes were eluted from the plates with n-propanol–36% hydrochloric acid (10:1) and the eluate taken up in borate buffer (pH 8.4) and determined photometrically at 490 nm.

The TLC determination of a number of organotin compounds by means of the quercetin chelates was described by Wieczorek[31]. The plates were prepared from Silica Gel G or Kieselguhr–Silica Gel G (20:10) and the developing solvents were (1) benzene–ether–hydrochloric acid (150:20:0.3), (2) benzene–ether–acetic acid–water (150:20:5:0.3), (3) hexane–acetic acid (120:35), and (4) benzene–ether–acetic acid–water (70:30:5:0.5). A solution containing R$_n$SnCl$_{4-n}$ (50 mg Sn) in 50 ml of octane–acetic acid (50:1) or isopropyl ether–acetic acid (50:1) was spotted on the plate. After development, the plate was dried at 90° for 20 min and the spots were located with UV light after spraying with 0.1% methanol quercetin solution or with water, 0.1% pyrocatechol violet solution in acetone, and then with 5% malonic acid solution in 50% ethanol. For the determination of RSnCl$_3$ (R = octyl), spots were scraped from the plate, dissolved in 0.1 ml 25% sulfuric acid and 40 ml butanol, treated with 0.1% methanolic quercetin, shaken for 2 min and then filtered. The absorbance of the solution was measured at 443 nm (absorptivity was 25×10^{-4} cm^2/ µg. For 5–15 µg tin, the relative standard deviation was $\simeq 3.2\%$ and the detection limit was 1.3 µg tin.

The TLC of two organotin pesticides and their stability in various soils was studied by Cenci and Cremonini[32]. Triphenyltin acetate (Brestan), triphenyltin hydroxide (Du-Ter) and their degradation products were chromatographed on Silica Gel H with n-butanol–ethanol–water (4:2:1) as the developer of choice. Brestan was found to remain essentially unchanged for 3–10 days following photolysis at different wave lengths while Du-Ter was quickly transformed into several decomposition products.

The TLC of a number of n-alkyltin derivatives on silanized silica gel plates was described by Figge[33]. The silanized material used for the preparation of the plates was Silica Gel PF$_{254}$ (Merck 7751) and Silica Gel HF$_{254}$ (Merck 7750). Following shaking of 40 g of above absorbents for 15 min with 60–80 ml of demineralized water, 250 µm plates were prepared from five 20 × 20 cm glass plates. The plates were allowed to stand for 1 h at room temperature and were then activated by heating for 1 h at 120°C. For development of organotins, the following optimal solvent systems were used: (*1*) methanol–2 *N* hydrochloric acid–tetrahydrofuran (3:1:1), (*2*) ethanol–2 *N* hydrochloric acid–isoamyl alcohol–diisopropyl ether (3:1:1:1), (*3*) 2-chloroethanol–2 *N* hydrochloric acid–tetrahydrofuran–isoamyl alcohol–cyclohexane (5:5:3:2:1). Detection was accomplished using 1.0% ethanolic pyrocatechol violet reagent following a 10 min exposure of the plates to bromine vapor. Figure 1 illustrates the R_F values of octyltin halides obtained on silanized silica gel with 9 solvent systems.

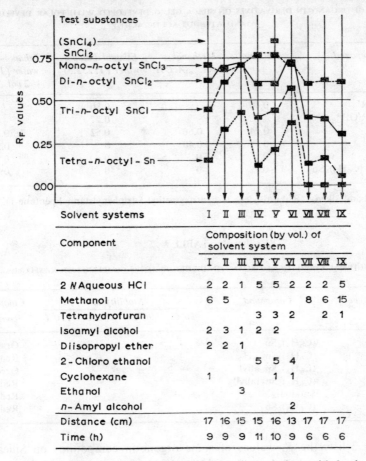

Component	Composition (by vol.) of solvent system								
	I	II	III	IV	V	VI	VII	VIII	IX
2 *N* Aqueous HCl	2	2	1	5	5	2	2	2	5
Methanol	6	5					8	6	15
Tetrahydrofuran				3	3	2		2	1
Isoamyl alcohol	2	3	1	2	2				
Diisopropyl ether	2	2	1						
2-Chloro ethanol				5	5	4			
Cyclohexane	1					1	1		
Ethanol			3						
n-Amyl alcohol						2			
Distance (cm)	17	16	15	15	16	13	17	17	17
Time (h)	9	9	9	11	10	9	6	6	6

Fig. 1. R_F values of octyltin halides obtained on silanized silica gel plates with 9 solvent systems.

Braun and Heimes[34] described the TLC separation of a number of tetraorganotin mixtures. Separations were achieved if the tetraorganotin compounds contained no polar groups and if they differed in the number of phenyl groups per molecule or in the nature of their aliphatic residues (alkyl or allyl). For analytical purposes, mixtures of 80% hexane and 20% benzene by volume were suitable for development.

For preparative purposes (multiple development) pure hexane or hexane containing 10–20% benzene by volume were best suited as mobile phase. Dithizone and silver nitrate were both used for detection. Dithizone permitted the detection of tetra-, tri-, and diorganotin compounds. With silver nitrate, allyltin compounds and polystannones could be distinguished from alkyltin compounds while tetraphenyltin was not colored by this reagent.

TABLE 2

R_F VALUES OF ORGANOTIN DERIVATIVES ON SILICA GEL G DEVELOPED WITH POLAR DEVELOPERS WITH AND WITHOUT ACETIC ACID

Compound	Methanol	Methanol + 5 vol. % AcOH	Ethanol–butanol–water (1:2:2.5)	Ethanol–butanol–water (1:2:2.5) + 2 vol. % AcOH
$(C_6H_5)_3SnR'$	0.70		0.80	
$(C_6H_5)_3SnR'OH^a$			0.77	
$(C_6H_5)_3Sn^+$	0.24	0.66	0.52	0.72
$(C_6H_5)_2^{2+}$	0	0.40	0	0.54
$C_6H_5Sn^{3+}$	0	0	0	0
$(C_6H_5)_4Sn, (C_6H_5)_6Sn_2$	0	0	0	0

^{a} R = alkyl and alkenyl. R' = alkylene (4-hydroxymethyl-5-triphenylstannyl-pentene-1).

TABLE 3

THERMAL STABILITY OF ORGANOTIN DITHIZONE (DTZ) COMPLEXES ON KIESELGEL G

Compound type	Compound	Stability at				Color of DTZ complexes
		50°C	90°C	100°C	110°C	
Sn–C$_6$H$_5$	$(C_6H_5)_4Sn$	+	+	+	+	Orange
Sn–alkyl	$(C_6H_5)_3SnC_3H_7$	+	+	+	−	Red-brown
Sn–allyl	$(C_6H_5)_3Sn$–allyl	+	+	−	−	Green
	$(C_6H_5)_2Sn$–(allyl)$_2$	−	−	−	−	Red
	$Sn(allyl)_4$	−	−	−	−	Red
Sn–Sn	$(C_6H_5)_6Sn_2$	+	+	−	−	Red

Table 2 lists the R_F values of some organotin compounds on Silica Gel G developed with methanol–acetic acid and ethanol–n-butanol–water–acetic acid mixtures and Table 3 depicts the spot colors of the dithizone–tin complexes and their stability from 50 to 110°C.

The organotin compounds have been satisfactorily separated by TLC[35] on silica gel layers as developed by the following solvent systems: (1) n-butanol–acetic acid[36], (2) water–n-butanol–ethanol–acetic acid (48:24.5:24.5:2.5)[36], (3) isopropanol–10% ammonium carbonate–5 N ammonium hydroxide (67:22:11)[35], (4) n-butanol–

2.5% ammonium hydroxide $(80:20)^{35}$, (5) isopropyl ether–acetic acid $(98.5:1.5)^{27}$, and (6) hexane–acetic acid $(92:8)^{37}$. The spots were visualized by spraying the layers with a solution of dithizone in chloroform or by exposure to UV radiation for 30 min and subsequently spraying with pyrocatechol violet solution. Organotin derivatives of the four alkylation stages can be separated and amounts down to 1 μg can be detected. Table 4 lists the R_F values of organotin derivatives obtained with 6 solvent systems on silica gel.

TABLE 4

R_F VALUES OF ORGANOTIN DERIVATIVES OBTAINED WITH 6 SOLVENT SYSTEMS ON SILICA GEL

Compound	Solvents[a]					
	1	2	3	4	5	6
Dimethyltin dichloride	0.0	0.0	0.0	0.0	0.02	0.0
Diethyltin dichloride	0.0	0.0	0.0	0.0	0.09	0.0
Triethyltin chloride	0.02	0.07	0.48	0.17		
Tetraethyltin	1.0	1.0	1.0	1.0		1.0
Tripropyltin acetate	0.09	0.34	0.71	0.29		
Dibutyltin dichloride	0.0	0.0	0.0	0.0	0.38	
Tributyltin chloride	0.21	0.45	0.85	0.41		0.83
Tetrabutyltin	1.0	1.0	1.0	1.0		1.0
Dihexyltin dichloride	0.0	0.0	0.0	0.0	0.57	
Trihexyltin chloride	0.45	0.54	0.92	0.60		
Tetrahexyltin	1.0	1.0	1.0	1.0		1.0
Di-2-ethylhexyltin dichloride	0.0	0.0	0.0	0.0		
Dioctyltin dichloride	0.0	0.0	0.0	0.0	0.68	
Tri-2-ethylhexyltin chloride	0.78	0.69	1.0	0.82		
Tetra-2-ethylhexyltin	1.0	1.0	1.0	1.0		
Diphenyltin dichloride	0.0	0.0	0.0	0.0	0.29	
Triphenyltin acetate	0.55	0.55	0.84	0.60		
Tetraphenyltin	1.0	1.0	1.0	1.0		
Butyltin trichloride						0.0
Butylthiostannic acid	0.0	0.0	0.0	0.0		0.0

[a] Solvents: (1) butanol–pyridine (7.5:3.5), saturated with water; (2) butanol-ethanol (3:1), saturated with water; (3) butanol saturated with 25% ammonium hydroxide; (4) butanol–25% ammonium hydroxide (8:2) (organic phase); (5) isopropanol–sodium acetate (2:1) (1 N acetic acid (1:1)); (6) isopropanol–ammonium carbonate (2:1) (10% ammonium carbonate–5 N ammonium hydroxide (2:1)).

Senf[38] described the separation and detection of pyrrolidinedithiocarbamate complexes of tin, arsenic, and antimony. Table 5 lists the R_F values and limits of detection of the above complexes obtained on silica gel–aluminum oxide plates developed with benzene–chloroform (2:1).

Geisser and Kriegsmann[39] described the GLC analysis of a number of butyltin derivatives using flame-ionization detection and a 2 m column containing 18% phenylmethyl silicone oil OE 4011. The carrier gas was hydrogen at 3.5 l/h and the

TABLE 5

R_F VALUES AND LIMITS OF DETECTION OF PYRROLIDINEDITHIOCARBAMATE COMPLEXES

Complex	System I[a]		System II[b]	
	R_F	Limit of detection (μg)	R_F	Limit of detection (μg)
Sn(PyDTC)$_2$	0.00	0.15	0.00	0.15
As(PyDTC)$_2$	0.12	0.15	0.22	0.3
Sb(PyDTC)$_2$	0.34	0.1	0.66	0.1

[a] System I: Absorbent: Silica Gel D–aluminum oxide (1:1); developer: benzene–chloroform (2:1).
[b] System II: Absorbent: Silica Gel D–aluminum oxide (2:1); developer: benzene–chloroform (2:1).

operating conditions were column temperature 160–180°, detector temperature 210°, and bridge current 300 mA. Figure 2 shows a chromatogram of a mixture of tetra-butyltin, tributyltin chloride, dibutyltin dichloride, dibutyltin dichloride, and butyltin trichloride obtained at a column temperature of 160°C. The retention time for tri-

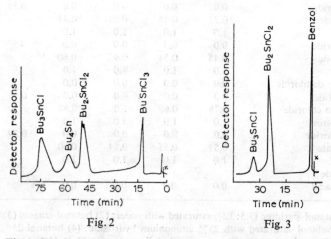

Fig. 2. Gas chromatogram of a mixture of butyltin derivatives at 160°C.
Fig. 3. Gas chromatogram of dibutyltin dichloride (86%) and tributyltin chloride (14%) at 180°.

butyltin chloride was 75 min. When the column temperature was raised to 180°C, the retention time of tributyltin chloride was reduced to 33 min (Fig. 3). Geisser and Kriegsmann[40] extended the above studies to an elaboration of 12 liquid phases for the GLC analysis of butyltin derivatives. A GCl-II gas chromatograph (AWF-DAW, Berlin) was used with a flame-ionization detector, column temperatures of 160–200°C, and hydrogen as carrier gas at 3.5–5 l/h. Table 6 lists the relative retention times of butyltin chlorides on 12 liquid phases.

TABLE 6

RELATIVE RETENTION TIMES OF BUTYLTIN CHLORIDES ON 14 LIQUID PHASES

Liquid phase		Column length (m)	Temp. (°C)	Relative retention time (min)[a]			
				I	II	III	IV
Apiezon L	20%	2	180				
GL 7100 FF	20%	2	175		1	1.88	2.05
	10%	2	175	0.31	1	1.73	1.96
Se-30	10%	1	162	0.34	1	1.74	1.83
	5%	1	163	0.41	1	1.82	1.97
OE-4011[b]	18%	2	178	0.26	1	1.50	1.19
OE-4178[c]	20%	2	176	0.25	1	1.42	1.00
OE-4007[d]	18%	1	194	0.33	1	1.23	0.84
Benzyl diphenyl	20%	2	176		1	1.10	0.56
Reoplex 400	5%	3	163	0.55	1	0.65	0.31
Carbowax 20,000	18%	3	166		1	0.95	0.60
	5%	3	167	0.88	1	0.70	0.30

[a] Compounds: I, n-butyltin trichloride; II, dibutyltin dichloride; III, tributyltin chloride; and IV, tetrabutyltin.
[b] Methylphenyl silicone oil.
[c] Methylcyanopropyl silicone oil.
[d] 1,2,3-Trimethyl-1,1,2,3,3-pentaphenyl trisiloxane.

The gas–liquid chromatographic separation of organotin compounds of type R_6Sn_2 (R = CH_3, C_2H_5, n-C_3H_7, n-C_4H_9), R_3SnH (R = C_2H_5, n-C_3H_7, n-C_4H_9), (n-C_4H_9)$_3$SnR (R = CH_3, C_2H_5, n-C_3H_7), and R_3SnCH_3 (R = C_2H_5, n-C_3H_7 and iso-C_3H_7) has been reported by Faleschini and Doretti[41] using columns of Carbowax and Silicone Oil E-301. A Carlo Erba Fractovap Model B system equipped with a thermal conductivity detector was employed. Columns (2 m × 6 mm) were packed with either 20% Carbowax or 20% Silicone Oil E-301 on Chromosorb W, 60–80 mesh. Helium was the carrier gas at 100 ml/min. The columns and injection port temperatures were 200° and 300°C.

Table 7 lists the relative retention times of a variety of organotin compounds obtained on Carbowax and Silicone Oil E-301. Figure 4 is a chromatogram illustrating the separation of asymmetrical tetraalkyltin derivatives on Carbowax 20M at 200°C.

A gas chromatographic method[42] has been developed for the analysis of butyl-, octyl-, and phenyltin halides, which are intermediates for the manufacture of stabilizers for plastics, fungicides for paints, and certain other biological and agricultural chemicals. A Shandon "Universal" gas chromatograph was used fitted with a hydrogen flame-ionization detector. A 16 cm U-shaped stainless steel section of 4 mm i.d. was packed with 5% Midland Silicone Oil MS 200 supported on Celite 545. The Celite was prepared by the method of James and Martin[43], including a treatment with alkali[44]. Column temperatures of 110° for butyltin halides, 180° for phenyltin halides and 210° for octyltin halides were found to be satisfactory. For the butyltin halides, an injection temperature of 380–400° was sufficient, but the octyltin halides and the phenyltin halides required a temperature of 400–425°. At lower injection temperatures, the organotin halides appeared to be insufficiently volatile for accurate quantitative work.

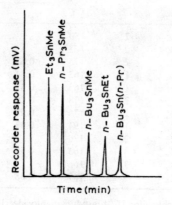

Fig. 4. Chromatograms of asymmetrical tetraalkyltin derivatives on Carbowax 20 M at a column temperature of 200°C; helium carrier gas, 100 ml/min.

TABLE 7

RELATIVE RETENTION TIMES OF ORGANOTIN DERIVATIVES ON CARBOWAX 20 M AND SILICONE OIL E 301
COLUMNS AT 200°C.

Injection port 300°C.

Compound	Carbowax 20 M		Silicone E-301	
	t'_R	$\log t'_R$	t'_R	$\log t'_R$
Me_4Sn	0.67	−0.176	0.89	−0.050
Et_3SnMe	1.22	0.086	2.01	0.303
Et_4Sn	1.50	0.176	2.64	0.422
$n\text{-}Pr_3SnMe$	2.21	0.344	4.72	0.674
$i\text{-}Pr_3SnMe$	2.13	0.328	3.51	0.545
$n\text{-}Pr_4Sn$	3.40	0.531	7.95	0.900
$i\text{-}Pr_4Sn$	3.08	0.488	6.00	0.778
$n\text{-}Bu_3SnMe$	4.01	0.603	12.72	1.046
$n\text{-}Bu_3SnEt$	5.17	0.713	14.42	1.159
$n\text{-}Bu_3Sn(nPr)$	6.17	0.790	18.13	1.258
$n\text{-}Bu_4Sn$	7.50	0.875	24.89	1.396
Et_3SnH	1.87	0.272*		
$n\text{-}Pr_3SnH$	4.36	0.639*		
$n\text{-}Bu_3SnH$	8.85	0.947*		
Me_6Sn_2	1.53	0.185		
Et_6Sn_2	7.73	0.888		
$n\text{-}Pr_6Sn_2$	18.17	1.259		
$n\text{-}Bu_6Sn_2$	44.53	1.649		

* $t = 150°C$.

Figures 5 and 6 illustrate the isolation for typical butyltin and octyltin mixtures. Attention has been drawn to the possible thermal decomposition and disproportionation of organotin halides during exposure to elevated temperatures on the column[45]. No thermal decomposition products were observed in the study of Tonge[42] during the chromatography of a large number of samples of organotin halides, some containing small amounts of organic and inorganic catalysts.

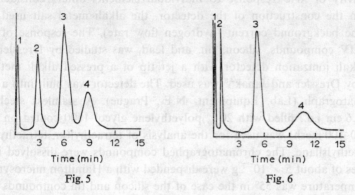

Fig. 5   Fig. 6

Fig. 5. Chromatogram of butyltin bromides. 1, butyl bromide; 2, butyltin tribromide; 3, dibutyltin dibromide; 4, tributyltin bromide.

Fig. 6. Chromatogram of octyltin bromides. 1, octyl bromide; 2, octyltin tribromide; 3, dioctyltin dibromide; 4, trioctyltin bromide.

Devyatykh et al.[46] reported the GLC analysis of a variety of organo tin derivatives. A Tsvet-1 chromatograph was used with a flame-ionization detector. In order to avoid the formation of a film on the detector electrode and to increase the sensitivity, the sample was decomposed thermally in a column 10 cm long × 2.5 mm i.d. at 800° before going to the detector. The thermal decomposition of $(C_4H_9)_4Sn$ formed metallic Sn and C_{1-8} saturated and unsaturated hydrocarbons. Rheoplex 44 and siliconized elastomer E-301 were used as the liquid phase with nitrogen as the carrier gas. Adequate separation was achieved when the length of the chromatographic column was 4 m and 2 m for Rheoplex 400 and E-301, respectively. The analysis of most alkyl derivatives of tin was performed at 180° and that of some of the lower boiling substances at 80°. The products were identified by their retention times. The retention values of octane, dibutyl ether, butyl bromide and butanol were 1, 1, 1.7, and 3.66, respectively, on Rheoplex 400 and 1, 1.68, 0.59, and 0.48 on E-301. The relative retention values for impurities in the heavier fraction for E-301 at 180° were $(C_2H_5)_4Sn$ 1.00, $(C_2H_5)_3SnC_4H_9$ 1.65, $(C_2H_5)_2Sn(C_4H_9)_2$ 3.05, $C_2H_5Sn(C_4H_9)_3$ 5.12, $(C_3H_7)_4Sn$ 2.0, $(C_3H_7)_3SnC_4H_9$ 3.72, $(C_3H_7)_2Sn(C_4H_9)_2$ 4.94, $C_3H_7Sn(C_4H_9)_3$ 6.73, $(C_4H_9)_4Sn$ 8.68, $(C_4H_9)_3SnCl$ 8.68, $(C_4H_9)_2SnCl_2$ 4.43, and $C_4H_9SnCl_3$ 1.91. The sensitivity was about 10^{-15}–10^{-16} M for octane, $(C_4H_9)_4Sn$, $(C_2H_5)_2Sn(C_4H_9)_2$, $C_2H_5Sn(C_4H_9)_3$, $(C_3H_7)_4Sn$, $(C_3H_7)_3SnC_4H_9$,

$(C_3H_7)_2Sn(C_4H_9)_2$, and $C_3H_7Sn(C_4H_9)_3$, about 10^{-9}–10^{-10} M for C_4H_9Br, C_4H_9OH, $(C_4H_9)_2SnCl_2$, $(C_2H_5)_3SnCl$, about 10^{-11} M for $(C_4H_9)_3SnCl$, and 10^{-5} M for $C_4H_9SnCl_3$.

The alkali flame-ionization detector (AFID) in general exhibits a selective response to compounds which contain certain elements. For example, the response of the AFID is known to be selective for nitrogen, phosphorus, arsenic, halogens, and sulfur. The selectivity of AFID response to individual elements differs considerably and depends on the construction of the detector, the alkali metal salt used, and the value of the background current (hydrogen flow rate). The response of the AFID to Group IV compounds, silicon, tin, and lead, was studied by Dressler et al.[47]

An alkali ionization detector with a jet tip of a pressed alkali metal salt as described by Dressler and Janak[48] was used. The detector was built into a Chrom 3 gas chromatograph (Lab. Equipment, N.E., Prague). A stainless steel column, 68 cm × 0.6 cm i.d., filled with 20% polyethylene glycol 1500 coated on Chromosorb W (80–100 mesh) was used for the analysis of tetraethyltin, tetraethyllead and triethoxymethylsilane. The chromatographed compounds were dissolved in hexane and samples of about 5×10^{-7} g were dispended with a Hamilton microsyringe. The column temperature was 55° in the case of the silicon and tin compounds and 85°C for the analysis of the lead compound. The carrier gas flow rate was 60 ml nitrogen/min and the air flow rate 660 ml/min; the hydrogen flow rate was varied in such

TABLE 8

IONIZATION EFFICIENCY OF THE COMPOUNDS CONTAINING DIFFERENT ELEMENTS

Compound	Element	Ionization efficiency (C/mole)	
		Na	K
DIP	P	2.2×10^2	3.5×10^2
Pyridine	N		3.7×10^0
Triethoxymethylsilane	Si	1.9×10^0	6.7×10^0
Bromobenzene	Br	1.9×10^0	
Chlorobenzene	Cl	1.1×10^0	
Iodobenzene	I	6.3×10^{-1}	1.6×10^0
Thiophene	S		-6.0×10^0
Tetraethyllead	Pb	-3.7×10^0	-4.2×10^0
Tetraethyltin	Sn	-8.9×10^0	-4.4×10^1

TABLE 9

RETENTION VALUES OF TIN AND SULPHUR COMPOUNDS AND THEIR STANDARD DEVIATIONS

Compound	Retention value (mm)		σ (mm)	
	FID	AFID	FID	AFID
Tetraethyltin	116.8	127.6	8.4	14.3
Thiophene	126.2	126.2	15.8	15.9

a way that the required value of the background current was attained. Table 8 lists the ionization efficiencies of the compounds containing different elements and Table 9 compares the retention values of tetraethyltin and thiophene as determined by FID and AFID. Figure 7 compares the chromatograms obtained with both FID and AFID systems.

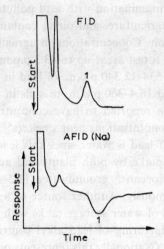

Fig. 7. Chromatograms of tetraethyltin.

2. LEAD

Lead has ubiquitous distribution, being found in the earth's crust (8–20 p.p.m.), food, water, and atmospheric pollution and is the most abundant of the non-essential elements found in the body. Lead ores are widely found in the U.S., England, Mexico, Spain, Germany, South America, and Australia. The most abundant lead-containing ore is galena in which it occurs as the sulfide and from which commercial lead is generally obtained. Other lead-containing ores are cerussite (the carbonate), crocoite (the chromate), wulfenite (the molybdate), matlockite (the fluorochloride), anglesite (the sulfate) and vanadinite (the chlorovanadate).

The industrial uses of lead include (*1*) the production of coverings, pipes, etc., (*2*) in storage batteries (as accumulator plates), (*3*) in various alloys with tin, copper, antimony, etc., (*4*) in paints, pigments, and varnishes (as red lead, litharge, white lead, chromates, sulfate, and titanate), (*5*) in flint glass and vitreous enameling, (*6*) in the manufacture of insecticides (lead arsenate), and (*7*) in the manufacture of tetraethyllead.

Lead is disseminated in the environment *via* mining, smelting, refining, secondary recovery, and use of lead-containing products. In 1969, the mine output in the U.S. was approx. 40% and the gasoline additives approx. 20% of the U.S. total consumption of lead. Red lead and litharge pigments, solder, cable covering, ammunition, and caulking lead together utilized approx. 26%.

Among the factors which may influence the concentration of lead in food are the concentration of lead in the growing environment of plants and food-producing animals, e.g. the uptake from soil and water, excretion, retention and distribution of lead by the organism, contamination with lead pollutants through aerial fallout, use of lead compounds in agriculture, and direct contamination during processing, storage, and final preparation. Concentrations in agricultural soils can range from 13.9 to 95.7 p.p.m.[49] and in forest areas up to 43.3 p.p.m., while the distillation and burning of coal contributes 534–12,340 p.p.m. of lead in the soil[50]. The use of lead-based paints has contributed 16.4–360 p.p.m. to soils in older urban areas[51]. Areas around lead ores have been reported to have concentrations varying from 20 to 10,000 p.p.m. with similar concentrations near smelters[49].

An important source of lead is water since this lead is in a soluble and more easily assimable form for uptake by both plants and animals. The lead levels in sea water (0.08–8 μg/l)[52], oceans, groundwater (1.5–60 μg/l)[53], surface waters (0–55 μg/l)[54] can vary enormously. Another source of lead in drinking water is the result of painting the inside of water storage tanks with lead chromate[55,56].

In the U.S., the manufacturing of lead alkyl contributes approx. 810 tons of lead/year to the atmosphere. (Nationally this represents only 0.43 percent of the total, but the emissions are concentrated and can present localized health problems.) Lead alkyls used as gasoline additives provide an important source of lead in the atmosphere (and ultimately to soil and water), e.g. an average gasoline contains 3 g of lead/gallon and approx. two-thirds of the lead is consumed in exhausts. Combustion of lead alkyl anti-knock compounds, since their introduction about 50 years ago, has been estimated to have contributed 10 mg lead/m^2 to the soil[57]. It has been suggested that airborne lead contributes approx. 0.5–1.5% of the lead content of the U.S. diet[58].

The pesticide lead arsenate (although greatly restricted in use) is also a source of lead residues in food and tobacco. Other sources of lead include storage batteries, lead–tin solder for sealing cans and lead pipe for service use. A minor source of lead is from glazed pottery for acid drinks. High values of extractable lead have been found in stoneware and china (presumably where a lead glaze had been subsequently applied) in addition to the expected source, non-vitreous earthenware.

Available evidence indicates that lead is widely distributed in all natural foods. In the United States, lead intake in food and beverages ranges from 100 to 2000 μg/day for an individual, with long-term averages of 120–350 μg/day[59]. Lewis[60] has indicated that the level of lead in milk is slightly higher than 20–30 years ago and it was suggested that since the level of lead in milk is a function of the body burden of the element rather than that in the early diet, then the lead concentrations in milk might also serve as a useful index to human environmental exposure to lead. (The levels of lead in milk in the U.S. at present are in the order of 0.05 $\pm$ 0.025 p.p.m.[60].)

Severe exposure of the pregnant human female to high levels of lead has resulted in abortion[61,62]. Lead is known to cross the human placenta in cases of high-level exposure[63] and the occurrence of teratogenic effects at a sublethal level was noted following excessive industrial exposure. Offspring with neurological symptoms have been described[64] and growth (both intrauterine and postnatal) may also be retarded[65]. A high incidence of teratogenic malformations in the sacral and tail regions of the offspring have been observed following intravenous administration of lead nitrate, lead chloride, or lead acetate in doses of 50 mg/kg to hamsters on day 8 of pregnancy[66]. However, no evidence of teratogenic activity has been observed in the cow[67] or the sheep[68]. Ferm and Carpenter[66] have noted that administration of lead salts to hamsters may result in congenital malformations, particularly of the central nervous system.

Lead salts such as lead acetate and phosphate have been shown to induce a variety of carcinomas following administration by dietary and parenteral routes[69-71].

Catizone and Gray[72] demonstrated the toxic effect of lead chloride on the morphogenesis of the lead primordium of the chick. The treatment of chick embryo with lead salts has also been shown to cause hydrocephalus and anterior meningoceles[73,74].

Muro and Goyer[75] reported that chromosomes from leukocyte cultures from mice fed a diet containing 1% lead acetate showed an increased number of gap-break type aberrations. These chromosome abnormalities largely involved only single chromatids which suggested that damage occurred after the DNA synthesis phase of the cell cycle. The authors speculated that the chromosomal damage inducible by administration of lead salts may contribute to reduction in fertility and the formation of congenital malformations.

The toxicology[76-80] and biological effects[81,82] of inorganic lead as well as the toxicity[83,84], metabolism[85-88], and chemistry[84,89] of the organic lead compounds have been described.

Analysis of lead in air and biological materials have been achieved by atomic absorption spectroscopy[90-94] and colorimetric (dithizone) techniques[95-98].

Kasteim[99] described the chromatographic determination of lead in air dust. Lead in HCl extracts of air particulates was isolated by TLC on precoated Silica Gel N–HR or cellulose powder MN300 films by developing for 2 h with dioxane–1.5 N HCl–acetylacetone (100:20:0.5) or dioxane–n-butanol–1.5 N HCl–acetylacetone (50:50:20:0.5), respectively, and determined by densitometry after spraying with 0.2% KSCN or 1% Na_2S. The sprayed zones were soaked in Silicone Oil A until transparent and mounted between two glass plates for densitometry. An air particulate sample was analyzed to contain 0.435% Pb with a ±1.15% relative deviation from the mean.

The TLC of cobalt, nickel, copper, bismuth, lead, manganese, cadmium, vanadium(V), and uranium(VI) was investigated with various solvent systems on cellulose and silica gel plates and sheets by ascending technique[100]. The separation of lead, which had a strong tendency to tail, was carried out successfully using methanol–25% nitric acid–water (8:1:1). New spray reagents such as 4-(2-thiazolylazo)-resorcinol, 1-(2-thiazolylazo)-2-naphthol, 4-(2-pyridylazo)-resorcinol and 1-(2-pyridylazo)-2-naphthol were used for the detection of the metals as distinctly colored complexes with detection limits ranging from 0.1 to 0.01 μg/spot.

The circular paper chromatographic separation of lead, mercury, and silver ions has been reported by Murthy et al.[101]. Whatman No. 1 paper was used with the following developers: (1) butanol–pyridine–water (100:20:20), (2) butanol saturated with 4 N acetic acid, and (3) collidine saturated with 0.4 N nitric acid. Detection was accomplished with ammoniacal hydrogen sulfide. Specific reagents for the metallic ions were as follows: lead, mercury, and silver were identified by sodium rhodizonate (blue), diphenyl carbazone (violet) and p-dimethylaminobenzylidine-rhodamine (red-violet), respectively. Table 10 lists the R_F values of Ag^+, Hg^+, Pb^{2+} as determined by paper chromatography. The developing solvent of choice was butanol–pyridine–water (100:20:20).

TABLE 10

CIRCULAR PAPER CHROMATOGRAPHY OF Ag^+, Hg^+ AND Pb^{2+}

Solvent	Time of irrigation for the solvent to travel 9 cm (h)	R_F values of metallic ions			Remarks
		Ag^+	Hg^+	Pb^{2+}	
1. Butanol–pyridine–water (100:20:20)	2	0.86 0.87	0.62 0.65	0.34 0.33	All three ions clearly separated
2. Butanol saturated with 4 N acetic acid	2.25	0.44 0.43	0.75 0.77	0.48 0.45	Silver and lead travel very close to each other. Mercury moves faster with the solvent.
3. Collidine saturated with 0.4 N nitric acid	3.5	0.97	Very little movement		Both Hg and Pb give a diffused patch which is not clearly moved and separated from the center.
		0.95			Ag moves practically along with the solvent and can be separated from the other two.

Paper strips of Whatman No. 1 paper 250 × 50 mm and 250 × 20 mm with 10 mm wide "barriers" obtained by impregnating them with a mixture of equal solutions of 0.5% 8-mercaptoquinoline in 0.1 N HCl and 0.1% H_2SO_4 (a) at the starting line and (b) at 5.2 cm level were used. One drop of 0.5, 0.25, 0.1, and 0.1% aqueous solution of lead acetate, cadmium nitrate, barium chloride, manganese nitrate, and cupric nitrate was applied to the starting line, the strips were dipped in buffer of pH 11.0 or 0.1 N KOH solution and developed at 20° to a distance of 15 cm within 25–30 min. Only lead and cadmium gave yellow spots at the starting line and at 5.2 cm, respectively. The sensitivity of the method was 4γ for Pb and 1γ for Cd, and mineralization of Ba, Mn, Cu, or urine did not interfere with the reaction[102]. The paper and anion-exchange chromatographic separation and determination of some heavy metals in admixture with lead has been studied by Pollard et al.[103]. The solvent system for the separation of lead, zinc, and mercury(II) on acid-washed Whatman No. 1 paper in descending development was diethyl ether–methanol–water–nitric acid (50:30:20:2). Table 11 shows the R_F values of the metal ions chromatographed and the spray reagents employed for their detection.

TABLE 11

R_F VALUES AND SPOT COLORS AND DETECTING REAGENTS FOR THE PAPER CHROMATOGRAPHY OF 14 HEAVY METALS ON WHATMAN NO. 1 PAPER DEVELOPED WITH DIETHYL ETHER–METHANOL–WATER–NITRIC ACID (50:30:20:2)

Metal ion	Spray	Color of spot	R_F value (head and tail of spot)
Tl^I	0.5% w/v 8-hydroxyquinoline in absolute alcohol. Hold over ammonia and under UV light.	Yellow-green fluorescence	0.25–0.17
Ag^I	5% w/v tannic acid in warm 60% v/v aqueous methylated spirits. Warm the damp strip.	Brown stain	0.36–0.27 Dark hydrolysis products also present at starting line
Pb^{II}	Aqueous solution of rhodizonic acid and held over ammonia.	Red	0.41–0.22
Zn^{II}	0.5% w/v 8-hydroxyquinoline in	Yellow fluorescence	0.66–0.55
Cd^{II}	absolute alcohol. Hold over ammonia and under UV light	Yellow fluorescence	0.66–0.55
Cu^{II}	0.1% w/v rubeanic acid in	Green	0.66–0.56
Co^{II}	absolute alcohol. Hold over	Yellow-brown	0.68–0.55
Ni^{II}	ammonia.	Blue	0.68–0.55
Mn^{II}	4% v/v salicylaldehyde in 50% aqueous ethyl alcohol. Hold over ammonia and under UV light.	Dark spot against a yellow-green fluorescent background	0.70–0.56
Fe^{II}	0.5% w/v aqueous potassium	Blue	0.70–0.58
Fe^{III}	ferrocyanide.	Blue	0.70–0.58
Bi^{III}*	10% w/v freshly prepared aqueous sodium dithionite. Warm.	Brown	Hydrolyzed all down the paper to 0.8
U^{VI}	0.5% w/v aqueous potassium ferrocyanide.	Brown	0.78–0.72
Hg^{II}	0.05% w/v dithizone in chloroform.	Pale pink	0.95–0.87

* Bismuth may be chromatographed successfully only if the original solution contained 5% v/v aqueous nitric acid, so that no hydrolysis products are formed, and the R_F value of Bi is 0.78–0.67.

Pollard *et al.*[103] utilized Amberlite IRA-400 resin (Cl^- form), 150 mesh to separate Fe^{III}, Cu^{II}, Zn^{II}, Cd^{II}. The column was first eluted with 3 M hydrochloric acid to elute all the copper(II), and then 0.1 M hydrochloric acid to elute all the ferric iron. By changing to 1 M caustic soda containing 20 g of sodium chloride/l, the zinc was quantitatively eluted and finally cadmium was eluted employing 1 M nitric acid. The interference of both bismuth and lead in this method was not studied since they were removed in the initial paper chromatographic separation. When nickel, cobalt, or manganese were also present, they were extracted with elements iron, copper, and zinc from the paper chromatogram and eluted from the resin in 3 M hydrochloric acid and hence accompany the copper(II). However, in the spectrophotometric method used for the estimation of copper, they caused no interference.

For the estimation of lead in natural waters, Carritt[104] passed the sample down a column composed of a solution of dithizone (diphenyl thiocarbazone) in carbon tetrachloride absorbed on cellulose acetate. Heavy metals were absorbed as their respective dithizonates, the column then stripped of cadmium, zinc, manganese, and lead by treatment with 1 M hydrochloric acid and these metals were simultaneously estimated polarographically.

The measurement of lead in biological samples (blood, urine, and tissue) by combined anion-exchange chromatography and atomic absorption chromatography was described by Lyons and Quinn[105]. It had been shown by Nelson and Kraus[106] that Pb^{II} was poorly adsorbed from HCl onto a quaternary amine anion-exchange resin (Dowex-1) if the HCl is dilute (distribution coefficient $= 1$ in 0.05 N HCl) but that adsorption rapidly increases to a maximum (distribution coefficient about 25) in 1–2 N HCl, and then decreases with increasing HCl concentrations. In the technique of Lyons and Quinn[105] the anion-exchange resin consisted of AG-1-X8, 50–100 mesh Cl form (Bio-Rad) and the adsorption of lead was determined using the 283.3 mm resonance line of a hollow-cathode lamp operated with a Model 303 absorption spectrometer (Perkin-Elmer).

The separation of lead(II) from bismuth(III), thallium(III), cadmium(II), mercury(II), gold(III), platinum(IV), palladium(II), and other elements by anion-exchange chromatography was described by Strelow and Toerien[107]. Lead and the other elements were absorbed from between 0.1 and 4.0 N hydrobromic acid solution on a column of AG 1-X8 anion-exchange resin in the bromide form. The following elements were eluted with 0.1 N HBr: U^{VI}, Th^{IV}, Zr^{IV}, Hf^{IV}, Ti^{IV}, Sc^{III}, Y^{III}, La^{III}, and the rare earths, Al^{III}, Ga^{III}, In^{III}, Fe^{III}, Be^{II}, Mg^{II}, Ca^{II}, Sr^{II}, Ba^{II}, Zn^{II}, Mn^{II}, Co^{II}, Cu^{II}, Ni^{II}, Cr^{III}, Sb^{III}, Ge^{IV}, Li^{I}, Na^{I}, K^{I}, Pb^{I}, and Cs^{I}. Then lead was eluted selectively with 0.3 N HNO_3 plus 0.025 N HBr and after evaporation of the acid, it could be determined by EDTA titration or by mass spectrometric isotope dilution method. Bi^{III}, Tl^{III}, Cd^{II}, Hg^{II}, Au^{III}, Pt^{IV}, and Pd^{II} were retained by the column quantitatively. No element which could interfere with the EDTA titration accompanied lead.

The method described has been successfully employed for the determination of lead in a variety of minerals and materials. The separation could also be applied successfully to the determination of a few micrograms of lead in minerals using the mass spectrometric isotope dilution method for determination.

Bonelli and Hartman[108] described the determination of lead alkyls by electron-capture gas chromatography employing a Bairall–Ballinger type column operated isothermally and with a silver nitrate scrubber column. The five lead alkyls were completely separated in 10 min with good quantitative recovery and an overall accuracy of the method being about 4%. A Wilkens Aerograph Hy-Fi Model A-610-B with electron capture was fitted with a detector heater and Model A-630 voltage control. The output of the chromatograph was recorded on a Leeds & Northrup Speedomax H 1-mV recorder, and a Disc Chart Integrator, Model 207 was used for area determination. The analytical method consisted of separating the lead alkyls on a 10 ft. $\times$ $\frac{1}{8}$ in. stainless steel column of 10% TCEP on 80 to 100 mesh Chromosorb W, HMDS treated. The scrubber section, a 6 in. $\times$ $\frac{1}{8}$ in. stainless steel column, contained 20% Carbowax 300 (saturated with silver nitrate) on 30–60 mesh Chromosorb W precoated with 8% KOH attached between the analytical column and the detector. The operating conditions were column, injector, and detector temperatures, 72, 95, and 150°C, respectively, nitrogen as carrier gas at a flow rate of 27 ml/min, cell potential, 22 V, and attenuation, 10–4 $\times$.

Figure 8 illustrates a chromatogram of the separated lead alkyls in a standard mixture containing 6.25 mole-% of $(CH_3)_4Pb$, 25 mole-% $(CH_3)_3C_2H_5Pb$, 37.5 mole-% $(CH_3)_2(C_2H_5)_2Pb$, 25 mole-% $CH_3(C_2H_5)_3Pb$ and 6.25 mole-% $C_2H_5)_4Pb$. (The mixture was diluted with spectrograde hexane to 4 ml/gal. as

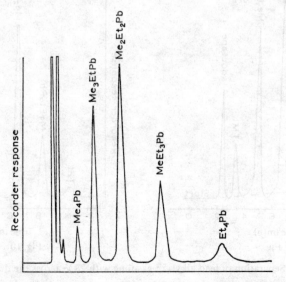

Fig. 8. Sample: 4.0 ml/gal. as Et$_4$Pb diluted 1:200; 0.4 µl injection.

($C_2H_5)_4$Pb, the legal maximum.) As shown in this figure, the alkyl isomers of lead were separated according to increasing molecular weight. The relative electron-capture sensitivity of the lead alkyls is shown in Table 12, which relates the electron affinities in disc units/μg of the five lead additives.

TABLE 12

RELATIVE ELECTRON-CAPTURE SENSITIVITY OF LEAD ALKYLS

Lead alkyl	Disc units/μg IX–X, 22 V
Tetramethyllead (Me_4Pb)	465,000
Trimethylethyllead (Me_3EtPb)	463,000
Dimethyldiethyllead (Me_2Et_2Pb)	505,000
Methyltriethyllead ($MeEt_3Pb$)	470,000
Tetraethyllead (Et_4Pb)	245,000

Tetramethyl, tetraethyl, and three methylethyl lead alkyls were determined in gasoline directly by temperature programmed gas chromatography with an electron-capture detector[109]. The lead scavengers, ethylene dichloride and ethylene dibromide, were removed in the gas chromatograph column to avoid interference. A Jarrell-Ash 76-755 electron-capture detector was thermostated in a 26–750 oven. A Gyra

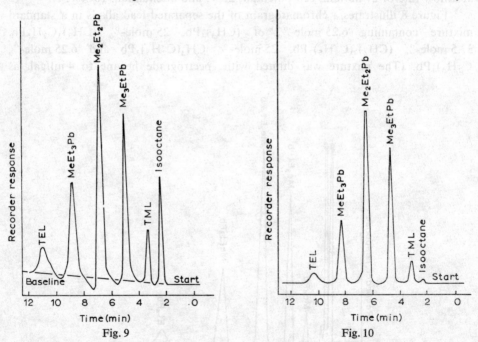

Fig. 9. Separation of lead alkyls in gasoline with 28 V across the detector.

Fig. 10. Separation of lead alkyls in gasoline with 37 V across the detector.

Electronics E-202 electrometer amplifier, a 1 mV recording potentiometer with 1 sec pen response and a Perkin-Elmer 194 printing integrator were also used. The column consisted of 30 in. × $\frac{1}{8}$ in. stainless steel sheathed in an 18 mm i.d. glass jacket. The packing consisted of two sections separated by a glass wool plug: an inlet section, 24 in. long, in which the boiling point separation occurs and an exit section, 6 in. long, in which the scavengers were removed. For both sections, the solid support was 30–60 mesh Chromosorb W coated with 8 % sodium hydroxide. The inlet section packing was 15 % SE-30 silicone rubber and the exit section packing was 20 % of a saturated solution of silver nitrate in Carbowax 400 on the solid support. The column was preconditioned at 200°C with a nitrogen flow of 100 ml/min for 12 h before connecting to the detector. The best sensitivity established for detector operation was in the range 26–30 V. The GLC operating parameters were: detector temp., 180°; column to detector block temp., 160°; ambient sample valve temp.; 37 V to the detector; 100 ml of eluting gas/min and 150 ml of dilution gas/min. Figures 9 and 10 illustrate the separation of lead alkyls in gasoline with 28 and 37 V across the detector, respectively. It was found that standard deviations of the 25 min GLC analysis varied from ±0.01 to ±0.08 g of lead/gal. for the individual lead alkyls.

The simultaneous determination of lead alkyls and halide scavengers in gasoline by gas chromatography with flame-ionization detection was described by Soulages[110]. These compounds were separated in a partition column and hydrogenated using a nickel catalyst and the resulting methane and/or ethane was separated from gasoline hydrocarbons on an adsorption column. Figure 11 illustrates the catalytic

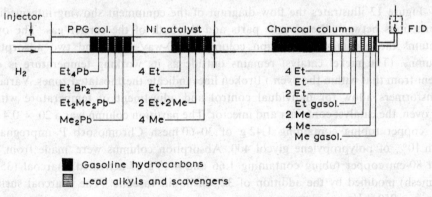

Fig. 11. Catalytic chromatographic process for separation of lead alkyls and scavengers in gasoline.

chromatographed process for the separation of lead alkyls and scavengers in gasoline and shows the three stages of the process. In the first stage the separation of lead alkyl and scavengers is accomplished on a partition GLC column using polypropylene glycol 400 (PPG) as stationary phase. In this step, additives are simultaneously separated from methane and ethane potentially present in the gasoline, thus avoiding

further interference. In the second stage catalytic hydrogenolysis led to the formation
of methane, ethane, or both according to the original compound and simultaneous
hydrogenation of ethylene dichloride and ethylene dibromide to ethane. (Catalysis
was accomplished by passing the carrier gas, hydrogen, over nickel at a suitable
temperature to avoid hydrocarbon cracking.) The third stage consists of the separation of methane and ethane thus formed from normal gasoline hydrocarbons (C_3 and
higher) on a chromatographic column packed with activated charcoal with final
detection by flame detection.

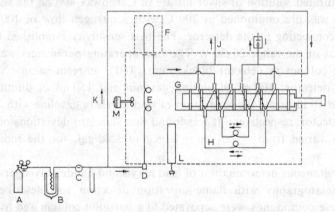

Fig. 12. Flow diagram of apparatus. A, Hydrogen cylinder; B, purification train; C, pressure regulator; D, injector; E, PPG column; F, nickel catalyst; G, ten-way valve; H, charcoal column;
I, flame-ionization detector; J, backflushing outlet; K, backflushing inlet; L, heater; M, fan.

Figure 12 illustrates the flow diagram of the equipment showing intermediate
connections between the various parts and the path of the carrier gas. The oven
contains the preliminary partition column, a 10-way valve and two adsorption
columns. (The nickel catalyst remains outside as its working temperature is different from that within the oven.) Broken lines indicate thermostated zones. Variable
transformers allow for individual control and adjustment of temperature within
the oven, the catalyst container and injector. The partition column was a 20 × 0.4 cm
i.d. copper tubing containing 1.42 g of 30–60 mesh Chromosorb P impregnated
with 10% of polypropylene glycol 400. Absorption columns were made from 30-
and 80-cm copper tubing containing 1.66 and 4.40 g of activated charcoal (35 to
80 mesh) modified by the addition of 3% of liquid vaseline. (The charcoal surface
area was 950 m^2/g.)

Table 13 shows the operational conditions for the determination of tetramethyl
and/or tetraethyl lead and methylethyl lead alkyl in gasolines. Figures 13 and 14
illustrate gas chromatographic separations of $(CH_3)_4Pb$ and $(C_2H_5)_4Pb$ and scavengers in gasoline, respectively. As can be seen from the chromatograms, $(CH_3)_4Pb$
appeared within the first minute, $(C_2H_5)_4Pb$ requires 11 min, and the total mixed
alkyls were resolved in 20 min.

TABLE 13

	Me$_4$Pb and/or Et$_4$Pb	Methylethyl lead alkyls
PPG column (cm)	20	20
Charcoal column (cm)	30	80
Column temperature (°C)	70	60
Catalyst temperature (°C)	140	140
Injector temperature (°C)	90	90
Hydrogen flow (ml/min)	43	40
Inlet pressure (mm Hg)	54	63
Sample volume (μl)	5	5

Lead anti-knock additives have well-defined manufacturer formulations. They are generally composed of (1) one, two, or five lead alkyls responsible for the anti-knock action, (2) scavengers such as ethylene dichloride or ethylene dibromide which function to clean out the cylinder by preventing deposits of lead oxides, and (3) toluene to act as a diluent in certain cases.

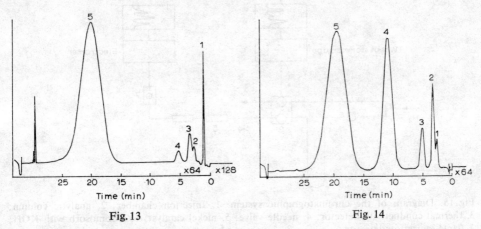

Fig. 13   Fig. 14

Fig. 13. Gas chromatographic separation of Me$_4$Pb and scavengers in gasoline. 1, CH$_4$ from Me$_4$Pb; 2, gasoline C$_2$H$_6$; 3, C$_2$H$_6$ from EtCl$_2$; 4, C$_2$H$_6$ from EtBr$_2$; 5, gasoline C$_3$H$_8$.

Fig. 14. Gas chromatographic separation of Et$_4$Pb and scavengers in gasoline. 1, Gasoline C$_2$H$_6$; 2, C$_2$H$_6$ from EtCl$_2$; 3, C$_2$H$_6$ from EtBr$_2$; 4, C$_2$H$_6$ from Et$_4$Pb; 5, gasoline C$_3$H$_8$.

The analysis of lead anti-knock additives by gas chromatography has been described by Soulages[111]. In earlier work[110], Soulages found that with specially constructed equipment it was possible to analyze for a number of lead alkyls and the halogenated scavengers simultaneously but the sole limitation was the impossibility of determining toluene.

To obtain the true composition of the additive, the chosen technique was to convert the compounds to methane prior to detection using a nickel catalyst in the presence of hydrogen[112,113], which at the time served as carrier gas. The catalyst at the exit of a chromatograph equipped with a thermal conductivity detector was connected to a flame-ionization detector, with provision for having the reaction products first pass through a tube of Chromosorb impregnated with 5% KOH before they reached the detector. The purpose of the KOH is to retain the HCl and HBr produced during decomposition of the halogenated compounds and thus prevent their interference with the response of the flame-ionization detector. With this arrangement it was possible to record simultaneously the signals from the two detectors and to compare the area of a compound after transformation to methane (in the ionization detector) with that of the same compounds detected as such (in the conductivity detector), thus making possible the determination of response factors.

Figure 15 illustrates a diagram of the chromatographic system. Figure 16 illustrates the system for catalytic conversion. A Carlo Erba Fractrap D apparatus with an analytical column connected to a catharometer detector was used in this

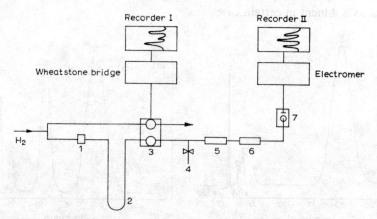

Fig. 15. Diagram of the chromatographic system. 1, Injection chamber; 2, analytic column; 3, thermal conductivity detector; 4, needle valve; 5, nickel catalyst; 6, chromosorb with KOH; 7, flame-ionization detector.

study. Attached to the exit of the chromatograph was the device for catalytic conversion to methane (Fig. 16) consisting of a steel tube, 10×0.4 cm i.d. heated electrically by a 200 W resistance and packed with catalyst[110].

Connected to the catalytic device were, in turn, a 5×0.4 cm i.d. tube filled with alkali-treated Chromosorb and the flame-ionization detector (an Aerograph with a Model 500-D electrometer). The two detectors were connected to a Leeds and Northrup Speedomax G recorder, with 1 mV interval and chart speed 1 in./min, provided with disc integrator. The column used was of copper tubing, 2 m long

× 0.4 cm i.d. packed with 10% PEG (polyethylene glycol 400) and 1% Quadrol (tetrakis(2-hydroxypropyl)ethylenediamine) on non-acid-washed 80–100 mesh Chromosorb W. For additives containing only $(CH_3)_4Pb$ and/or $(C_2H_5)_4Pb$, a column of the same dimensions containing 6.6% PEG on the same support could be used alone.

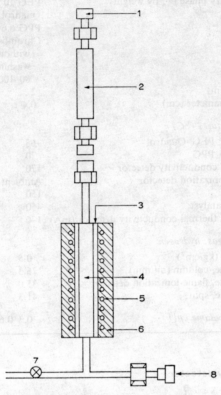

Fig. 16. System for catalytic conversion. 1, Connection to flame-ionization detector; 2, chromosorb with KOH; 3, thermocouple; 4, nickel catalyst; 5, 200 Ω resistance; 6, thermal insulation; 7, needle valve; 8, connection to chromatograph.

Table 14 lists the operating conditions for analysis on the PEG–Quadrol column and on the PPb column. Figure 17 shows the separation on the PEG–Quadrol column of a sample containing five methylethyl alkyl leads and the chromatograms recorded from the two detectors. Figure 18 shows the separation of an additive containing $(CH_3)_4Pb$ on the PPb column. Table 15 lists the relative response factors for the thermal conductivity detector for five lead alkyls, ethylene bromide and ethylene chloride and toluene.

The perfluoroalkanoylpivalylmethanes form complexes with a wide range of metal ions to give volatile, thermally stable chelates that have been employed successfully for gas chromatography[114,115] and mass spectroscopy[116,117] for trace metals.

TABLE 14

OPERATING CONDITIONS FOR ANALYSIS ON THE PEG–QUADROL COLUMN AND ON THE PPG COLUMN

Column

Stationary phase (% by weight)	PEG 10%
	Quadrol 1%
	PPG 6.6%
Support	Chromosorb W without acid washing, 80–100 mesh
Length (m)	2
Inside diameter (cm)	0.4

Temperatures (°C)

Column: PEG–Quadrol	63
PPG	70
Thermal conductivity detector	170
Flame-ionization detector	Ambient
Injector	150
Nickel catalyst	450
Current, thermal conductivity detector (mA)	170

Carrier gas: hydrogen

Pressure (kg/cm^2)	0.5
Flow rate, column (ml/min)	76.3
Flow rate, flame-ionization detector	35.0
Flow rate, split	41.3
Sample volume (μl)	0.3–0.6

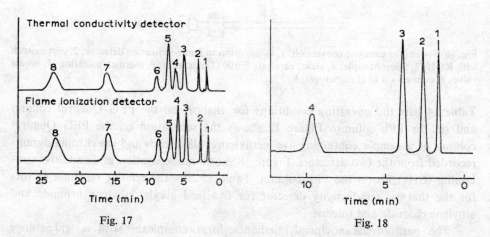

Fig. 17. Separation of methylethyl alkyls of lead on a PEG–Quadrol column. 1, PbMe$_4$; 2, PbMe$_3$Et; 3, PbMe2Et$_2$; 3, PbMe2Et$_2$; 4, toluene; 5, Cl$_2$Et; 6, PbMeEt$_3$; 7, PbEt$_4$; 8, Br$_2$Et.

Fig. 18. Separation of tetramethyl lead on a PPG column. 1, PbMe$_4$; 2, Cl$_2$Et; 3, toluene; 4, Br$_2$Et.

TABLE 15

RELATIVE RESPONSE FACTORS FOR THE THERMAL CONDUCTIVITY DETECTOR: HYDROGEN OR HELIUM
CARRIER GAS

Compound	Relative response ($Cl_2Et = 1$)			
	By weight		Molar	
	Triangulation	Disc integrator	Triangulation	Integrator
PbMe$_4$	1.98	1.95	0.733	0.722
PbMe$_3$Et	1.85	1.82	0.651	0.641
PbMe$_2$Et$_2$	1.79	1.78	0.600	0.597
PbMeEt$_3$	1.83	1.82	0.586	0.582
PbEt$_4$	1.75	1.75	0.536	0.536
Cl$_2$Et	1.00	1.00	1.00	1.00
Br$_2$Et	1.65	1.63	0.869	0.858
Toluene	0.866	0.853	0.932	0.919

The analytical potential in gas chromatography and mass spectrometry of fluorinated lead beta-diketonates was examined by Belcher et al.[118]. The lead chelates were prepared by direct reaction of aqueous solutions of lead(II) with trifluoroacetylpivalylmethane (TPM, pentafluoropropanoylpivalylmethane (PPM), and heptafluorobutanoylpivalylmethane (HPM). A Philips PV 4000 (Pye R) chromatograph equipped with a hydrogen flame-ionization detector was used. Eluates were collected for characterization by introducing a 1:1 post-column splitter and passing the vapor through a heated Teflon tube ($\frac{1}{8}$ in. o.d.) into traps cooled in liquid air. Teflon columns (2 ft. and 4 ft. × $\frac{3}{16}$ in. o.d.) were used with liquid phase loadings. of 5, 10, or 15% Apiezon L on "Universal B" (60–85 mesh) (Phase Separations Ltd.) All columns were pre-conditioned for 24 h at 220° and nitrogen was the carrier gas. The mass spectra were recorded in an AEI MS-9 double-focusing mass spectrometer, operating at an ionizing voltage of 70 eV and an accelerating potential of 8 kV. The source temperature was maintained at 250 ± 10° and samples were evaporated into the source with a direct insertion probe.

TABLE 16

GAS CHROMATOGRAPHIC DATA FOR LEAD PERFLUOROALKANOYLPIVALYLMETHANATES
(Injection temperature 230°; column temperature 210°; gas flow rate 60 ml/min)

Chelate	Retention time (sec)	Solution concentration in acetone (%)	Attenuation setting
Pb(TPM)$_2$	110	10	20×10^2
Pb(PPM)$_2$	100	50	1×10^4
Pb(HPM)$_2$	96	50	1×10^4

Figure 19 illustrates chromatograms of three lead perfluoroalkanoyl pivalyl-methanates obtained at 210°C on a 2 ft. Teflon column of 15% Apiezon L on "Universal B" support. Retention data and chromatographic parameters are given in Table 16. Thermal analysis, GC and mass spectral studies all indicated high thermal stability, but strong column interaction made successful quantitative gas chromatography difficult. The integrated ion-current technique was applied to determine lead heptafluorobutanoylpivalylmethanate in the range $10^{-9}–10^{-7}$ g of lead, but a lower limit of *ca.* 10^{-14} g was indicated.

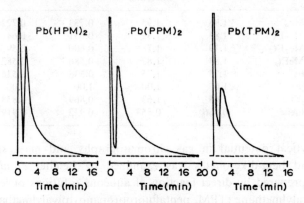

Fig. 19. Chromatograph of lead perfluoroalkanoylpivalylmethanates. 2 ft. Teflon column ($\frac{3}{16}$ in. o.d.) filled with 15% Apiezon L on "Universal B" (60–85 mesh). Column temperature 210°; nitrogen flow rate 60 ml/min.

Quantitative analysis of mixtures of organolead compounds has been reported by Barbieri *et al.*[119]. Compounds of the type R_4Pb, R_3PbCl, and R_2PbCl_2 ($R = CH_3$, C_2H_5 or C_6H_5) were initially separated by paper chromatography, extracted with methanol or chloroform, and then converted to lead iodide with saturated alcoholic iodine. The solvent and excess iodine were removed by heat, the residue dissolved in 4 M KI to give a concentration of 2–16γ Pb/ml as PbI_4^{2-} and the absorbance measured at 357 nm against 4 M KI. When lead chloride was present, it was determined as PbI_4^{2-} in an aqueous solution of the mixture which had been made 4 M in KI. The organolead compounds did not absorb at 357 nm under these conditions and the method was found accurate to within 2%.

REFERENCES

1 N. L. KENT, *J. Soc. Chem. Ind.*, 61 (1942) 183.
2 H. A. SCHROEDER, J. J. BALASSA AND I. H. TIPTON, *J. Chronic Diseases*, 17 (1964) 483.
3 E. C. TABOR AND W. V. WARREN, *Arch. Ind. Health*, 17 (1958) 145.
4 J. M. BARNES AND P. N. MAGEE, *J. Pathol. Bacteriol.*, 75 (1958) 267.
5 J. M. BARNES AND H. B. STONER, *Brit. J. Ind. Med.*, 15 (1958) 15.
6 J. M. BARNES AND J. M. STONER, *Pharmacol. Rev.*, 11 (1959) 211.

7   H. B. STONER, J. M. BARNES AND J. I. DUFF, *Brit. J. Pharmacol.*, 10 (1955) 16.
8   P. N. MAGEE, H. B. STONER AND J. M. BARNES, *J. Pathol. Bacteriol.*, 73 (1957) 107.
9   J. E. CREMER, *Biochem. J.*, 68 (1958) 685.
10 J. E. CREMER, *Biochem. J.*, 67 (1957) 87.
11 J. E. CREMER, *Biochem. J.*, 67 (1967) 28P.
12 R. LECOQ, *Compt. Rend.*, 239 (1954) 678.
13 J. E. CREMER, *Brit. J. Ind. Med.*, 16 (1959) 191.
14 W. ZEMAN, E. GADERMAN AND K. HARDEBECK, *Deut. Arch. Klin. Med.*, 198 (1951) 713.
15 J. G. A. LUIJTEN AND G. J. M. VANDERKIRK, *A Survey of the Chemistry and Applications of Organo Tin Compounds*, Tin Research Institute, 1952.
16 J. G. A. LUIJTEN AND G. J. M. VANDERKIRK, *Investigations in the Field of Organo Tin Chemistry*, Tin Research Institute, 1951.
17 J. W. BRIDGES, D. S. DAVIES AND R. D. WILLIAMS, *Biochem. J.*, 98 (1966) 14P.
18 E. E. KENAGA, *Chem. Week.*, 95 (1964) 64.
19 E. E. KENAGA, *12th Intern. Congr. Entomology, London, July 8–16, 1964*.
20 A. K. SIJPESTEIJN, *Mededel. Landbouwhogeschool Opzoekingsstat. Gent*, 24 (1959) 850.
21 E. KROLLER, *Deut. Lebensm. Rundschau*, 56 (1960) 190.
22 P. D. PATE AND R. L. HAYS, *J. Econ. Entomol.*, 61 (1968) 32.
23 W. N. ALDRIDGE, in J. MANICOFF, J. R. COLEMAN AND M. W. MILLER (Eds.), *Effects of Metals on Cells, Subcellular Elements and Macromolecules*, Thomas, Springfield, Ill., 1970, p. 255.
24 D. J. WILLIAMS AND J. W. PRICE, *Analyst*, 85 (1960) 579.
25 P. KARSTEN, H. L. KIES AND J. J. WALRAVEN, *Anal. Chim. Acta*, 7 (1952) 355.
26 L. H. ADCOCK AND W. G. HOPE, *Analyst*, 95 (1970) 868.
27 G. NEUBERT, *Z. Anal. Chem.*, 203 (1964) 265.
28 B. HEROLD AND K. H. DROEGE, *Z. Anal. Chem.*, 245 (1969) 295.
29 D. HELBERG, *Deut. Lebensm. Rundschau*, 63 (1967) 69.
30 D. HELBERG, *Deut. Lebensm. Rundschau*, 62 (1966) 178.
31 H. WIECZOREK, *Deut. Lebensm. Rundschau*, 65 (1969) 74.
32 P. CENCI AND B. CREMONINI, *Ind. Saccar. Ital.*, 69 (1969) 313.
33 K. FIGGE, *J. Chromatog.*, 39 (1969) 84.
34 D. BRAUM AND H. T. HEIMES, *Z. Anal. Chem.*, 239 (1968) 6.
35 K. BÜRGER, *Z. Anal. Chem.*, 192 (1963) 280.
36 M. TÜRLER AND O. HÖGL, *Mitt. Lebensm. Hyg.*, 52 (1961) 123.
37 R. F. VON DER HEIDE, *Z. Lebensm. Untersuch. Forsch.*, 124 (1964) 348.
38 H. J. SENF, *Mikrochim. Acta*, (1968) 954.
39 H. GEISSER AND H. KRIEGSMANN, *Z. Chem.*, 4 (1964) 354.
40 H. GEISSER AND H. KRIEGSMANN, *Z. Chem.*, 5 (1965) 43.
41 F. FALESCHINI AND L. DORETTI, *Ann. Chim. (Rome)*, 60 (1970) 597.
42 B. L. TONGE, *J. Chromatog.*, 19 (1965) 182.
43 A. T. JAMES AND A. J. P. MARTIN, *Biochem. J.*, 50 (1952) 679.
44 A. T. JAMES, A. J. P. MARTIN AND G. H. SMITH, *Biochem. J.*, 52 (1952) 238.
45 W. P. NEUMANN, *Angew. Chem. Intern. Ed. Engl.*, 2 (1963) 165.
46 G. G. DEVYATYKH, V. A. UMILIN AND Y. N. TSINOVOI, *Tr. Khim. Khim. Teknol.*, 2 (1968) 82.
47 M. DRESSLER, V. MARTINU AND J. JANAK, *J. Chromatog.*, 59 (1971) 429.
48 M. DRESSLER AND J. JANAK, *J. Chromatog. Sci.*, 7 (1969) 451.
49 L. C. HUFF, *Econ. Geol.*, 47 (1952) 517.
50 C. H. MANLEY, *Analyst*, 62 (1937) 544.
51 J. CHOLAK, L. J. SCHAFER AND T. D. SPERLING, *J.A.R. Pollution Control Assoc.*, 11 (1961) 281.
52 R. A. KEHOE, *J. Roy. Inst. Public Health*, 24 (1961) 81.
53 K. BAGCHI, H. D. GANGULY AND J. N. SUDOR, *Indian J. Med. Res.*, 28 (1940) 441.
54 W. H. DURUM AND J. HAFFTY, *U.S. Geol. Surv. Circ. 445*, 1961.
55 J. O. JACKSON, *J. Am. Water Works Assoc.*, 52 (1960) 1370.
56 D. W. CHRISTOFFERSON, *J. Am. Water Works Assoc.*, 53 (1961) 725.
57 T. J. CHOW AND M. S. JOHNSTONE, *Science*, 147 (1965) 502.
58 G. TER HAAR, *Environ. Sci. Technol.*, 4 (1970) 226.

59   R. E. ENGEL, D. I. HAMMER, R. J. M. HORTON, N. M. LANE AND L. A. PLUMLEE, Environmental lead and public health, *Environmental Protection Agency, Publ. 3*, March, 1971.
60   K. H. LEWIS, Symposium on environmental lead contamination, *U.S. Public Health Publ. No. 1440*, Washington, D.C., 1966.
61   A. T. WILSON, *Scot. Med. J.*, 11 (1966) 73.
62   F. J. TAUSSIG, *Abortion, Spontaneous and Induced*, Vol. III, Klimpton, London, 1936, p. 354.
63   D. BARL TROP, *Mineral Metabolism in Pediatrics*, Blackwell, Oxford, 1968.
64   C. R. ANGLE AND M. S. McINTIRE, *Am. J. Diseases Children*, 108 (1964) 436.
65   P. A. PALMISANO, R. C. SNEED AND G. CASSIDY, *J. Pediat.*, 75 (1969) 869.
66   V. H. FERM AND J. J. CARPENTER, *Exptl. Mol. Pathol.*, 7 (1967) 208.
67   V. H. FERM, *Science*, 141 (1963) 42.
68   L. F. JAMES, V. A. LAZAR AND W. BINNS, *Am. J. Vet. Res.*, 27 (1966) 132.
69   B. ZAWIRSKA AND K. MEDRAS, *Zentr. Allgem. Pathol. Pathol. Anat.*, 111 (1968) 1.
70   G. J. VANESCH AND R. KROES, *Brit. J. Cancer*, 23 (1969) 765.
71   F. J. C. ROE, E. BOYLAND, C. E. DUKES AND C. V. MITCHLEY, *Brit. J. Cancer*, 19 (1965) 860.
72   O. CATIZONE AND P. GRAY, *J. Expl. Zool.*, 807 (1941) 71.
73   D. A. KARNOVSKY AND L. P. RIDGEWAY, *J. Pharmacol. Exptl. Therap.*, 104 (1952) 176.
74   E. M. BUTT, H. E. PEARSON AND D. J. SIMONSEN, *Proc. Soc. Exptl. Biol. Med.*, 79 (1952) 247.
75   L. A. MURO AND R. A. GOYER, *Arch. Pathol.*, 87 (1969) 660.
76   C. P. McCORD, *Ind. Med. Surg.*, 23 (1954) 27.
77   R. E. LANE, *Brit. J. Ind. Med.*, 6 (1949) 125.
78   R. A. KEHOE, *J. Air Pollution Control Assoc.*, 19 (1969) 690.
79   Symposium on environmental lead contamination, *U.S. Public Health Publ. No. 1440*, Washington, D.C., 1966.
80   *American Petroleum Inst. Monograph No. 69-7*, New York, 1969.
81   A. DE BRUIN, *Arch. Environ. Health*, 23 (1971) 249.
82   R. L. ZIELHUIS, *Arch. Environ. Health*, 23 (1971) 299.
83   M. W. GOLDBLATT AND J. GOLDBLATT, in E. R. A. MEREWETHER (Ed.), *Industrial Carcinogenesis and Toxicology*, Butterworth, London, 1956.
84   F. W. FREY AND H. SHAPIRO, *Fortschr. Chem. Forsch.*, 16 (1971) 243.
85   R. A. KEHOE AND F. THAMANN, *Am. J. Hyg.*, 13 (1931) 478.
86   W. BOLANOWSKA, J. PIOTROWSKI AND H. GARCZYNSKI, *Arch. Toxikol.*, 22 (1967) 278.
87   W. BOLANOWSKA AND J. M. WISNIEWSKA-KNYPL, *Biochem. Pharmacol.*, 20 (1971) 2108.
88   W. BOLANOWSKA, *Brit. J. Ind. Med.*, 25 (1968) 203.
89   R. W. LEEPER, L. SUMMERS AND H. GILMAN, *Chem. Rev.*, 54 (1954) 101.
90   C. D. BURNHAM, C. E. MOORE, T. KOWALSKI AND J. KRASNIEWSKI, *Appl. Spectry.*, 24 (1970) 411.
91   C. L. CHAKRABARTI, J. W. ROBINSON AND P. W. WEST, *Anal. Chim. Acta*, 34 (1966) 269.
92   M. D. AMOS, P. A. BENNETT, K. G. BRODIE, P. W. Y. LUNG AND J. P. MATOUSEK, *Anal. Chem.*, 43 (1971) 211.
93   J. Y. HWANG, P. A. ULLUCCI, S. B. SMITH, JR. AND A. L. MALENFANT, *Anal. Chem.*, 43 (1971) 1319.
94   S. H. OMANG, *Anal. Chim. Acta*, 55 (1971) 439.
95   J. CHOLAK, *Arch. Environ. Health*, 8 (1964) 222.
96   L. J. SNYDER, *Ind. Eng. Chem. Anal. Ed.*, 19 (1947) 684.
97   E. B. SANDELL, *Colorimetric Determination of Trace Metals*, Interscience, New York, 1969, p. 555.
98   S. L. TOMPSETT, in A. S. CURRY (Ed.), *Methods of Forensic Science*, Vol. 3, Interscience, New York, 1964, p. 8.
99   H. KASTEIM, *Helv. Chim. Acta*, 53 (1970) 2231.
100 V. MIKETUKOVA AND R. W. FREI, *J. Chromatog.*, 47 (1970) 427.
101 A. R. V. MURTHY, V. A. NARAYAN AND M. R. A. RAO, *Current Sci. India*, 24 (1955) 158.
102 H. ROMANOWSKI AND W. HERMANOWSKA, *Farm. Polska*, 21 (1965) 574.
103 F. H. POLLARD, J. F. W. McOMIE AND G. NICKLESS, *J. Chromatog.*, 2 (1959) 284.

104 D. E. CARRITT, *Anal. Chem.*, 25 (1953) 1927.
105 H. LYONS AND F. E. QUINN, *Clin. Chem.*, 17 (1971) 152.
106 F. NELSON AND K. A. KRAUS, *J. Am. Chem. Soc.*, 76 (1954) 5916.
107 F. W. E. STRELOW AND F. V. S. TOERIEN, *Anal. Chem.*, 38 (1966) 545.
108 E. J. BONELLI AND H. HARTMAN, *Anal. Chem.*, 35 (1963) 1980.
109 H. L. DAWSON, JR., *Anal. Chem.*, 35 (1963) 542.
110 N. L. SOULAGES, *Anal. Chem.*, 38 (1966) 28.
111 N. L. SOULAGES, *J. Gas Chromatog.*, 6 (1968) 356.
112 A. ZLATKIS AND J. A. RIDGEWAY, *Nature*, 182 (1958) 130.
113 A. ZLATKIS, J. F. ORO AND A. P. KIMBALL, *Anal. Chem.*, 32 (1960) 162.
114 R. E. SIEVERS, J. W. CONNOLLY AND W. D. ROSS, *J. Gas Chromatog.*, 5 (1967) 241.
115 W. D. ROSS AND R. E. SIEVERS, *Anal. Chem.*, 41 (1969) 1109.
116 B. R. KOWASKI, T. L. ISENHOUR AND R. E. SIEVERS, *Anal. Chem.*, 41 (1969) 998.
117 R. BELCHER, J. R. MAJER, R. PERRY AND W. I. STEPHEN, *Anal. Chim. Acta*, 43 (1968) 451.
118 R. BELCHER, J. R. MAJER, W. I. STEPHEN, I. J. THOMPSON AND P. C. UDEN, *Anal. Chim. Acta*, 50 (1970) 423.
119 R. BARBIERI, N. BELLUCO AND G. TAGLIAVINI, *Ric. Sci.*, 30 (1960) 1671.

Chapter 5

ARSENIC AND PHOSPHORUS

1. ARSENIC

Arsenic is present in small amounts on the earth's crust (2–5 p.p.m.) and is found chiefly combined with minerals such as arsenolite (As_2O_3), realgar (As_2S_2), orpiment (As_2S_3), arsenopyrites ($FeAs_2$), and mispickel (FeS_2 and $FeAs_2$). Virtually all of the arsenic produced, however, is as a byproduct in the smelting of lead, copper, and gold ores (containing up to 3% arsenic). When ores or concentrates of these metals are smelted, arsenic, which sublimates at 218°C, is liberated in the flue dust and is then separated by filters or electrostatic precipitators as an oxide (chiefly as the trioxide). The world production of arsenic trioxide in 1962 was 55,000 tons[1]. Arsenic is also a contaminant in some nickel and cadmium ores and must be removed to improve the quality of the metal. Other sources of arsenic include coal, which contains 0.08–16 μg of arsenic per gram of coal[2]. It is possible for about 327 to 6400 tons of arsenic to be emitted into the atmosphere each year due to the consumption of approx. 409 million tons of coal in the United States[3]. In England, air levels of 0.04–0.14 μg As_2O_3/m^3 have been recorded[4]. Arsenic is also found in sea water (10–100 p.p.b.) with the usual daily intake of arsenic from water estimated[5] to be in the range of 10–20 μg. The average arsenic content of soils of various countries is about 5 p.p.m. (Under conditions of oxidation such as the presence of oxygen and Fe_2O_3, trivalent arsenic in soil is oxidized to the pentavalent form[6].

The industrial uses of arsenic include (*1*) as an addition to alloys in amounts of 0.3–0.5% to increase hardening and heat resistance, (*2*) as As_2O_3 in the manufacture of glass, Paris green, enamels, textile mordants and weed killers, (*3*) as metallic arsenic for preserving hides, and in sheep dips, weed killers, and rodenticides, (*4*) as As_2O_5 in the manufacture of colored glass, adhesives for metals, wood preservatives, and in weed control, and (*5*) as As_2S_3 in the manufacture of oil cloth, linoleum, in electrical semiconductors, photoconductors, as pigment, and for depilitating of hides. Arsenical pesticides formerly constituted a primary use of arsenic but has declined since 1945 with the advent of organic insecticides. The production of calcium arsenate and lead arsenate in 1967 was 2500 and 6000 lb., respectively, compared with 41,349 and 59,569 lb., respectively, in 1939. (Although the use of lead arsenates in the U.S. has been decreasing, it has increased abroad.) In 1964, arsenic acid was the largest volume product in the defoliant–desiccant category (*ca.* 5.0 million lb. were used on *ca.* 1.2 million acres of cotton[7]). Cacodylic acid ($CH_3AsO(OH)_2$) has been used extensively for weed control around industrial sites, fence rows, rights-of-way, etc., to control vegetables. Arsenic-containing compounds, such as arsanilic acid, sodium arsanilate, and 4-hydroxy-3-nitrophenyl arsenic acid, are widely used in rations for poultry, cattle, and swine. (It is usual to incorporate these arsenic derivatives at a level of 90–250 g per ton of feed.) Other sources of arsenic include commercial 20% superphosphate fertilizer which contains *ca.* 5 p.p.m. Arsenic in levels of 10–70 p.p.m. has been found in several common presoaks and household detergents in the U.S.[8] and hence can be expected to increase the arsenic content in water. Of additional environmental significance in this regard is the recent finding of McBride and Wolfe[9] that methano-bacteria act upon a variety of arsenic compounds to produce dimethylarsine. (Methyl cobalamine serves as the methyl donor in the reaction system.) Figure 1 illustrates the proposed schemes for the methylation of arsenic by microorganisms in waterways.

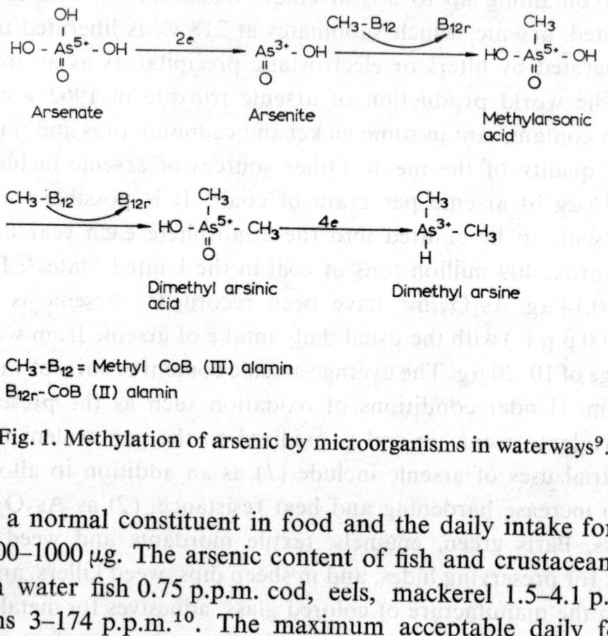

Fig. 1. Methylation of arsenic by microorganisms in waterways[9].

Arsenic is a normal constituent in food and the daily intake for humans is in the range of 400–1000 µg. The arsenic content of fish and crustaceans is frequently high, *e.g.* fresh water fish 0.75 p.p.m. cod, eels, mackerel 1.5–4.1 p.p.m. mollusks and crustaceans 3–174 p.p.m.[10]. The maximum acceptable daily food intake of arsenic suggested by WHO is 0.05 mg/kg[11].

Two forms of arsenic exist, pentavalent and trivalent. Pentavalent arsenic, as arsenate, is non-toxic in normal concentrations and is excreted rapidly largely through the kidneys and probably does not accumulate in human tissues. Trivalent arsenic, the principal form produced commercially, is toxic and accumulates in the mammalian body. It is a contaminant of soils and foods through its use in herbicides and pesticides and performs no known physiological function. Human carcinogenic aspects of arsenic, *e.g.* as a respiratory carcinogen[12,13], as well as recognized sites such as skin, lung, liver and suspected sites such as mouth, esophagus, larynx, and bladder, have been reported by Hueper[14]. (However, the carcinogenicity of arsenic in man has been questioned by Frost[15].)

Marked teratogenic and embryocidal effects of inorganic arsenic (sodium arsenate) in the golden hamster have been reported[16].

Arsenic in foods and biological materials has been determined by the colorimetric methods of Kaye[17] and Sunshine[18], the silver diethyldithiocarbamate coupling of arsenic reduced to arsine[19,20], molybdenum blue procedures[21], activation analysis[22], radioisotope dilution[23,24] and atomic absorption[25].

The separation by paper electrophoresis of arsenic(III), arsenic(V), germanium(IV), and of germanium and tervalent and quinquevalent arsenic-77 was described by Genet and Ferradini[26] with special emphasis on the utility of paper electrophoresis for the kinetic study of the oxidation of radioarsenic by hydrogen peroxide. Conditions of electrophoresis were studied on inactive solutions of Se^{IV} (0.2 M), As^{III} (0.01 M), and As^V (0.01 M) in 1.4 M KOH. The solutions were prepared immediately before use to minimize the formation of CO_3^{2-} which interfered with the separation. Twenty microliter portions of this solution were placed on 39 × 3 cm strips of Whatman 3 MM paper previously impregnated with 0.05 M KOH and a potential of 600 V was applied for 1 h. Germanate was detected with quercetin and arsenite and arsenate with aqueous silver nitrate. The migrations were germanate 11.5 cm/h, arsenite 15 cm/h, and arsenate 20 cm/h. The separation was applied to the study of the oxidation state of the [77]As formed on irradiation of SeO_2 in a neutron flux of 2.2 × 10^{12} neutrons/cm²/sec. The sample (20 mg) was dissolved in 1 ml of 1.4 M KOH containing As^{III} and As^V as carriers (each 0.01 M). The presence of As^{III} was necessary to avoid oxidation of the [77]As^{III}. The mixture was separated as before and the spots were located and recovered by scanning with a Geiger-Müller tube. Eighty-five percent of the [77]As was found to be present as As^{III}.

The paper chromatographic separation and detection of arsenic from a large number of other metallic compounds was described by Elbeih et al.[27]. The best developing solvent was pentanol and water containing 1% ammonium borate, 1% ammonium tartrate, and 0.5% mannitol. The chromatogram was sprayed with a mixture of alcohol containing 1% conc. nitric acid and 5% glycerol to make the paper acid, and then after drying, sprayed with commercial silver nitrate. The arsenic spot was presumably converted to silver arsenate (yellow) which on irradiation with UV light while still wet changes to silver arsenate (brown) and metallic arsenic (black). The spot was then easily visible against the purplish background from the silver nitrate spray. The limit of sensitivity was 0.3 μg of arsenic.

The separation of arsenite and arsenate ions has been achieved on Whatman No. 1 paper with a 140 min ascending development with methanol–1 N ammonium hydroxide (4:1)[28]. (R_F values for AsO_2^- and AsO_4^{3-} were 0.54 and 0.23, respectively.) Detection was by 2% silver nitrate in 10% aq. NH_3. (Yellow and pale brown spots were given by AsO_2^- and AsO_4^{3-}, respectively.) The method was intended for application in toxicology.

The identification and estimation of the arsenic residues in livers of rats ingesting arsenical feed additives was reported by Winkler[29]. It was found in the initial development of analytical methodology that both inorganic and organic As³⁺ (4-aminophenylarsine oxide) could be extracted from an acid solution by sodium ethyl xanthate in carbon tetrachloride. It was also shown that aminophenylarsine oxide, arsanilic acid, and arsenate (inorganic) could be separated from each other on an alumina column. The first was eluted in neutral aqueous solution, the second by 0.1 N NaOH, and the third eluted with 0.5 N NaOH. The salient findings of the arsenical feeding study could be summarized as follows: (*1*) trivalent inorganic arsenic was largely changed to the pentavalent form in the rat (pentavalent arsenic was not reduced), (*2*) when arsanilic acid was fed to rats, no trivalent arsenic was found in the liver, and little or no arsenic was found in the sodium ethylxanthate extracts, (*3*) arsanilic acid was found mainly unchanged in the livers of the rats where the arsenical present was separated as the free compound, (*4*) a protein of the arsenical was still bound to other matter, whose nature was not determined, when arsanilic acid was fed. It is highly probable that this arsenical is arsanilic acid since it forms a strong bond with tissue, (*5*) there was some evidence that small quantities of inorganic As⁵⁺ were present in the tissues of rats that had been fed arsanilic acid.

There are two major questions concerning the metabolism of organic arsenicals that have not been unequivocally answered: (*1*) Are they metabolically reduced to the arsenoxide? and (*2*) Is there metabolic degradation to inorganic arsenate or arsenite? Figure 2 shows the possible pathways of degradation or alteration of arsanilic acid. The possible metabolic products are inorganic arsenate, inorganic arsenite, 4-aminophenyl arsenoxide, and the arsonic acid or arsenoxide with modified ring structures. Previous studies by Overby and Straube[30] with double-labeled

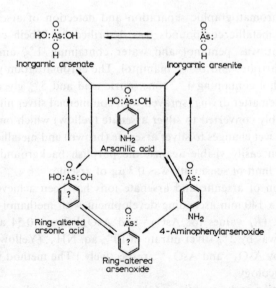

Fig. 2. Pathways of possible metabolic alterations of arsanilic acid.

arsanilic-1-^{14}C-^{74}As acid indicated little or no cleavage of the C–As bond by chickens.

Paper chromatographic, paper electrophoretic and ion-exchange methods were developed by Overby *et al.*[31] for the qualitative and quantitative estimation of arsenate-^{74}As, arsenite-^{74}As, and arsanilic-^{74}As acid. The methods were also found applicable to the analysis of closely related organic arsenicals. The high energy of the isotopic arsenic permitted the detection and identification of trace quantities of the arsenicals present in biological specimens.

The pentavalent compounds were dissolved in aqueous NaOH (pH 8.5) and arsenite and 4-aminophenylarsenoxide in aq. HCl (pH 2–3) and chromatographed with 0.003–0.03 μCi labeled compound on Whatman No. 1 paper using acetonitrile–nitric acid–water (78:2:20) for ascending development. The areas of radioactivity were quantified with a G–M tube paper scanner, coupled to a ratemeter and a strip chart recorder. The labeled and unlabeled organic arsenical spot were visualized under 2537 Å light (dark blue quenched areas). After about 30 min irradiation with this light, or 24–48 h under commercial "Cool-Light" fluorescent bulb (4000 to 6000 Å), the outlined areas showed pronounced spots of light green fluorescence under a 3600 Å light source. Only unsubstituted phenyl arsonic acid failed to show this fluorescence.

For electrophoresis, Whatman No. 3 paper was used in a RSCO #CC-9270 apparatus with water-cooled table and #CC-9268 power supply (Schaar & Co.,

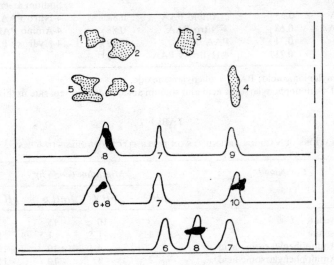

Fig. 3. Tracings of fluorescent spots and scans of radioactivity of paper chromatography strips, showing separations of mixtures of closely related arsenicals. X = point of application; vertical line = solvent front. 1, 2-Aminophenylarsonic acid; 2, 4-aminophenylarsonic acid (arsanilic acid); 3, aminophenylarsonic acid; 4, 4-hydroxyphenylarsonic acid; 5, 4-aminophenylarsenoxide; 6, arsenite-^{74}As; 7, arsenate-^{74}As; 8, arsanilic-^{74}As acid; 9, phenylarsonic-^{74}As acid; 10, 4-nitrophenylarsonic-^{74}As acid. Developing solvent was acetonitrile–HNO$_3$–H$_2$O (87:2:20, by vol.). A, B, C, and D = ascending development. E = descending development.

References pp. 122–124

Chicago) and the following buffers: (*a*) for pH range 1.0–2.0, 0.2 *M* KCl + HCl, (*b*) for pH range 2.0–6.0, 0.1 *M* citrate, (*c*) for pH range 6.0–8.0, 0.1 *M* phosphate, and (*d*) for pH range 8.0–10.0, 0.1 *M* borate.

For ion-exchange chromatography, Dowex 50, X-2, 50–100 mesh resin was used with (*a*) 3% HCl for column regeneration and (*b*) 3% trichloroacetic acid, 10% NaCl, and 5% NaCl in 0.1% NaOH for column development.

TABLE 1

PAPER CHROMATOGRAPHIC R_F VALUES FOR VARIOUS ARSENICALS

The compounds within braces may be separated from compounds in other such groups, but not from each other

Acetonitrile–HNO₃–water (78:2:20 by vol.) ascending development		Acetonitrile–HNO₃–water (78:2:20 by vol.) descending development		Isopropyl alcohol–water (70:30 by vol.) descending development	
Compound[a]	R_F	Compound	R_F	Compound	R_F
3-Amino PAA	0.24⎫	Sodium arsenite	0.52	4-Amino PAA	0.36⎫
Sodium arsenite	0.26⎬			Sodium arsenate	0.40⎬
4-Amino PAO	0.35⎭	3-Amino PAA	0.61⎫	PAA	0.46⎭
		4-Amino PAO	0.62⎬		
		4-Amino PAA	0.65⎭		
4-Amino PAA[b]	0.40			3-Amino PAA	0.50
Sodium arsenate	0.51	Sodium arsenate	0.74	2-Amino PAA	0.60⎫
				Sodium arsenite	0.62⎬
2-Amino PAA	0.55	2-Amino PAA	0.84	4-Nitro PAA	0.64⎭
4-Hydroxy PAA	0.68⎫	4-Nitro PAA	0.88⎫	4-Amino PAO	0.80⎫
PAA	0.71⎬	PAA	0.89⎬	4-Hydroxy PAA	0.85⎭
4-Nitro PAA	0.71⎭	4-Hydroxy PAA	0.90⎭		

[a] PAA = phenylarsonic acid; PAO = phenylarsenoxide.

[b] A mixture of 4-aminophenylarsonic acid and sodium arsenite did not separate in this system.

TABLE 2

AVERAGE MIGRATIONS OF VARIOUS ARSENICALS ON PAPER ELECTROPHORESIS IN 6 h AT 370 V (25 mA)

Compound	Migration (cm/6 h)[a]			
	pH 2	pH 4	pH 6	pH 10
Arsenate	+3	+10	+15	+22
Arsenite	−1.5	−1.5	−1.5	−1.5
4-Aminophenylarsonic acid	−2	+4	+10	+16
3-Aminophenylarsonic acid	−2	+4	+10	+16
2-Aminophenylarsonic acid	−2	+4	+10	+16
4-Nitrophenylarsonic acid	+2	+5	+11	+17
4-Hydroxyphenylarsonic acid	+1	+3	+9	+17
4-Aminophenylarsenoxide	−6	−1.5	−1.5	−1.5
Phenylarsonic acid	+2	+6.0	+11	+16

+ = migration toward anode; − = migration toward cathode.

Figure 3 shows tracings of chromatographic strips and illustrates the separation of closely related arsenicals. The R_F values are recorded in Table 1 and were established by a combination of isotopic scanning and fluorescence visualization. Paper electrophoresis was found most useful for separating and identifying arsenite, arsenate, and arsanilic acid. The migrations at pH 2, 4, 6, and 10 are shown in Table 2. Tracings of typical separations at several pH values are shown in Fig. 4.

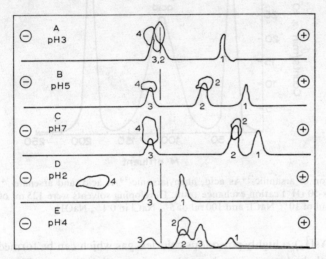

Fig. 4. Tracings of fluorescent spots and scans of radioactivity on paper electrophoresis strips, showing separations of mixtures of arsenicals at different pH. Center line was point of application. 1, Arsenate-^{74}As; 2, arsanilic-^{74}As acid; 3, arsenite-^{74}As; 4, p-aminophenylarsenoxide; 5, phenyl-arsonic-^{74}As acid.

Arsenate migrates rapidly towards the anode at pH 2 and above. Arsenite is a neutral molecule at pH 2–10 and drifts to the true neutral point. The migration of aromatic organic arsenicals is a function of the substituent groups and buffer pH. It was found possible to identify qualitatively the substituted phenyl arsonic acid by careful selection of pH.

It was suggested that the major advantage of the electrophoresis method in arsanilic acid metabolism studies was the clear separation of arsenite, arsenate, and arsanilic acid at pH 4 and above and the separation of 4-aminophenyl arsenoxide and arsenite at pH 2.

Ion-exchange chromatography separated arsanilic acid, non-cationic phenyl arsonic and arsenate or arsenite. Arsanilic acid was quantitatively retained on the cation exchange resin in 3% trichloroacetic acid while 100% of the inorganic arsenicals passed through with the solvent front. Phenylarsonic acid was held up by the resin and was eluted as a broad peak with acid (total recovery was 98–100%). Arsanilic acid was eluted from the resin with 5% NaCl in 0.1% NaOH with recoveries of

92–98%. Figure 5 shows the separation of arsanilic-^{74}As acid–phenylarsonic-^{74}As acid and arsenate-^{74}As or arsenite-^{74}As on Dowex-50 (H$^+$) cation-exchange resin.

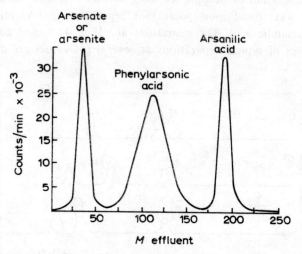

Fig. 5. Separation of arsanilic-^{74}As acid, phenylarsonic-^{74}As acid, and arsenate-^{74}As or arsenite-^{74}As on Dowex-50 (H$^+$) cation exchange resin. Developing solvents were 125 ml of 3% trichloroacetic acid, 25 ml of 10% NaCl, and 100 ml of 5% NaCl in 0.1% NaOH.

Arsine (AsH$_3$), a highly poisonous colorless gas which can be formed on reaction of arsenic with hydrogen, is a byproduct of metallurgical, smelting, and refining industries. It has been shown that arsenic can interact with water and base metals such as aluminum, tin, zinc, or antimony to generate arsine. Arsine poisoning, manifested by acute hemolysis and a variable degree of hepatic and renal damage is an industrial hazard which has long been recognized[32,33].

A variety of industrial processes and circumstances can give rise to the evolution of arsine[33,34]. Arsine is formed by the reaction of nascent hydrogen with some arsenic compounds and can be evolved from reactive arsenides, e.g. aluminum or calcium, by reaction with acids or water. Aluminum, as used by the metal refining industries to remove certain impurities such as arsenic, antimony, and tin, has caused fatalities and hence must be considered as a potential industrial hazard[35,36]. The generation of arsine from sodium arsenite was postulated[37] to proceed according to

$$NaAsO_2 + H_2O \rightarrow NaOH + HAsO_2$$

$$3\ NaOH + Al \rightarrow NaAlO_3 + 3\ H^0$$

$$6\ H^0 + HAsO_2 \rightarrow 2\ H_2O + AsH_3$$

Arsenic-containing herbicides (e.g. sodium acid methane arsenate[38]) have been shown to be a source of arsine when in accidental contact with aluminum. Aspects of arsine hemolysis and human poisoning have been described by Levinsky et al.[37], Uldall et al.[39], Hocken and Bradshaw[40], and DePalma[38].

The GLC of arsine was accomplished by Iguchi et al.[41] using a 3 m column of dioctylphthalate and a 0.75 m column of polyethylene glycol with hydrogen as the carrier gas at 10 and 60 ml/min respectively. The limit of detection was 0.001 mg of AsH_3. Figure 6 illustrates a diagram of the chromatographic apparatus and gas

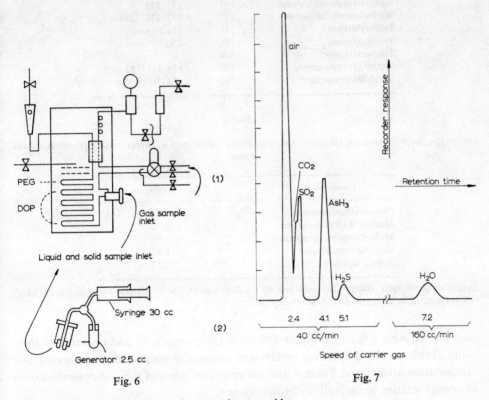

Fig. 6. 1, Diagram of apparatus; 2, generating assembly.
Fig. 7. Gas chromatogram of a mixture of arsine and some related gases. Column, DOP 3 m + PEG 0.75 m; carrier gas, H_2 40 and 160 ml/min; bridge current, 190 mA, span voltage, 2 mV.

generating assembly used and Fig. 7 shows a gas chromatogram of a mixture of arsine, hydrogen sulfide, carbon dioxide, and water.

Gudzinowicz and Martin[42] presented qualitative gas chromatographic data for eight arsines varying in molecular weight from 156 (trivinylarsine) to 306 (triphenylarsine). Nearly linear relationships between the logarithm of the net retention time and either the boiling point or the molecular weight of each component of a homologous series was reported. A modified Barber–Colman Model 10 argon ionization detector chromatograph was employed for the qualitative studies of organo- and organobromoarsenic mixtures. A 6 ft. × ¼ in. o.d. stainless steel column was

References pp. 122–124

TABLE 3

ORGANO- AND ORGANOBROMOARSINES

Compound	Molecular weight	Boiling point (°C)
Dimethylbromoarsine	185	128–130/720 mm
Methylethylbromoarsine	199	152–155
Methylbutylbromoarsine	227	172–178/720 mm
Tributylarsine	246	114/10 mm
Trivinylarsine	156	130
Triphenylarsine	306	
Methyldibromoarsine	250	179–181/720 mm
Vinyldibromoarsine	262	74–76/14 mm

TABLE 4

RETENTION TIMES OF SPECIFIC ORGANO- AND ORGANOBROMOARSINES RELATIVE TO METHYLBUTYLBROMO-
ARSINE

Compound	Relative retention time
Dimethylbromoarsine	0.43
Methylethylbromoarsine	0.70
Methylbutylbromoarsine	1.00
Vinyldibromoarsine	1.53
Tributylarsine	2.13

Operating conditions: column temperature, 95°C; flash heater, 165°C; detector temperature, 195°C; cell voltage, 1250 V; argon flow rate, 40 cc/min.

used packed with 5% SE-30 on 80–100 mesh Chromosorb W and operated isothermally. Table 3 lists the boiling points and molecular weights of the organo- and organobromoarsines and Table 4 lists the retention times of a 5 component mixture of arsines relative to methylbutylbromoarsine.

Figures 8 and 9 are chromatograms of homologous compounds where either the organo and/or bromo content in each series was varied. The peak between the mono- and trivinylarsines in Fig. 8 is believed to be due to divinylbromoarsine.

Quantitative studies involving a binary arsine mixture (a 2:1 ratio of dimethylbromoarsine to trivinylarsine) were determined using a Perkin-Elmer Model 154 D chromatograph and a 6 ft. × $\frac{1}{4}$ in. o.d. copper column operated at 66°C and packed with 10% squalene on 35–80 mesh Chromosorb W. With a helium inlet pressure of 25 p.s.i. and a flow rate of 350–355 ml/min, dimethylbromoarsine and trivinylarsine were eluted in 4.1 and 5.2 min, respectively.

The separation of alkyl/aryl and perfluorinated organoarsenic compounds by GLC was studied by Gudzinowicz and Driscoll[43]. A Perkin-Elmer Model 154C Vapor tractometer was used with helium as carrier gas. The columns used were (a)

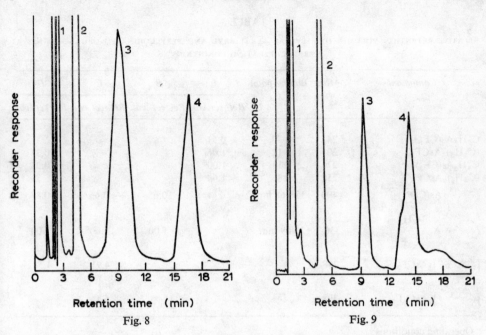

Fig. 8. Gas chromatogram of 1, benzene; 2, trivinylarsine; 3, divinylbromoarsine; and 4, vinyl-dibromoarsine. Operating conditions: column temperature 80°C; flash heater, 167°C; detector temperature, 235°C; cell voltage, 125 V; argon flow rate, 38 ml/min.

Fig. 9. Gas chromatogram of 1, methylene chloride; 2, dimethylbromoarsine; 3, methylethylbromo-arsine; and 4, methylbutylbromoarsine. Operating conditions: column temperature, 68°C; flash heater, 132°C; detector temperature, 200°C; cell voltage, 1250 V; argon flow rate, 44 ml/min.

5.5 ft. × $\frac{1}{4}$ in. o.d. copper tube packed with 15% SE-30 on Fluoropak 80 and (b) 25 ft. × $\frac{1}{4}$ in. o.d. copper tube with 33% Kel-F wax 400 on 30–60 mesh Chromosorb W. Table 5 lists the relative retention volumes for the organic compounds investigated as well as the operating GC parameters. Figure 10 illustrates a gas

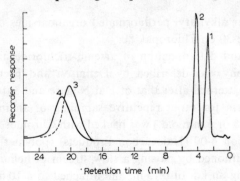

Fig. 10. Gas chromatogram of 1, $C_6H_5As(C_3F_7)_2$; 2, $C_6H_5As(CH_3)C_3F_7$; 3, $(C_6H_5)_2AsCF_3$; and 4, $(C_6H_5)_2AsC_3F_7$. Operating conditions: 5$\frac{1}{2}$ ft. column packed with 15% by weight SE-30 on Fluoropak 80; column temperature, 200°C; helium carrier gas pressure, 20 p.s.i.; helium flow rate, 52 ml/min.

TABLE 5

RELATIVE RETENTION VOLUMES FOR SEVERAL ALKYL/ARYL AND PERFLUORINATED ORGANOARSINES AT VARIOUS OPERATING CONDITIONS[a]

Compound	Mol. wt.	Boiling point (°C)	A	B	C	D
			Rel. ret. vol.	Rel. ret. vol.	Rel. ret. vol.	Rel. ret. vol.
$C_2H_5As(CF_3)_2$	242	77	0.51			
$(C_2H_5)_2AsCF_3$	201	112	1.06			0.33
$C_4H_8As(CF_3)_2$	270	118	1.00[b]			0.23
$(C_4H_8)_2AsCF_3$	258	186	1.64			
⬡—$As(C_3F_7)_2$	490	128/68 mm		0.65	0.66	0.84
⬡—$As\begin{smallmatrix}CH_3\\C_3F_7\end{smallmatrix}$	336	123/69 mm		1.00[c]	1.00[d]	1.00[e]
$(⬡—)_2AsCF_3$	298				4.91	4.58
$(⬡—)_2AsC_3F_7$	398				5.16	

[a] Operating conditions:
A. $5\frac{1}{2}$ ft. column with 15% by weight SE-30 on Fluoropak 80; column temperature 100°C; helium flow rate 35 ml/min at 1 atm and 25°C.
B. $5\frac{1}{2}$ ft. column with 15% by weight SE-30 on Fluoropak 80; column temperature 150°C; helium flow rate 25.4 ml/min at 1 atm and 25°C.
C. $5\frac{1}{2}$ ft. column with 15% by weight SE-30 on Fluoropak 80; column temperature 200°C; helium flow rate 52 ml/min at 1 atm and 25°C.
D. 25 ft. column with 33% by weight Kel-F Wax 400 on Chromosorb W; column temperature 200°C; helium flow rate 88 ml/min at 1 atm and 25°C.
[b] Retention time = 9.75 min.
[c] Retention time = 22.40 min.
[d] Retention time = 4.15 min.
[e] Retention time = 14.42 min.

chromatogram of four alkyl/aryl perfluorinated organoarsines obtained at 200° on a 5.5 ft. column of SE-30 on Fluoropak 80.

The separation and determination of arsenic trichloride and stannic chloride by gas chromatography was described by Dennison and Freund[44]. An all-glass apparatus was used patterned after that of Dal Nogare and Safranski[45] along with a new sampling system for rapid repetitive sampling of an inert atmosphere. A 24.5 ft. glass column (5 or 6 mm o.d.) was packed with Chromosorb W (50–60 mesh) with 24.5% Halocarbon 6-00 (Halocarbon Products) (polymer of chlorotinfluoro-ethylene) and preconditioned by passing a slow stream of helium at 120°C for 6 h. A Teflon needle having an i.d. of 0.13 in. and attached to a 10 ml syringe was used to inject the sample onto the column. The sampling chamber (Fig. 11) was made by attaching a screw-cap culture tube to a three-way Teflon stopcock which was in turn sealed through the other arms to the sample inlet of the column and to the

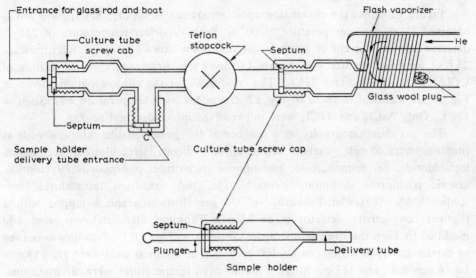

Fig. 11. Sampling chamber and attachment to sample inlet.

sample holder (not shown in Fig. 11). An inter-cooled jacket was incorporated in the sample inlet of the column to provide a sharp temperature change at the point where the Teflon needle enters the flask vaporizer to prevent premature vaporization of the sample and the corresponding broadening of the peaks.

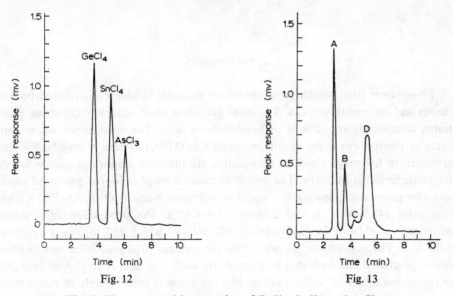

Fig. 12. Chromatographic separation of $GeCl_4$, $SnCl_4$, and $AsCl_3$.

Fig. 13. Chromatographic separation of A, CCl_4; B, $SnCl_4$; C, $AsCl_3$; D, $TiCl_4$.

Figure 12 shows the chromatographic separation of $GeCl_4$, $SnCl_4$, and $AsCl_3$ obtained at an even temperature of 107°C, flash vaporizer temperature of 225°C, detector filament current of 100 mA, corrected flow rate 82.5 ml/min, inlet pressure 25 p.s.i., and sample size 4 μl. Figure 13 shows the chromatographic separation of CCl_4, $SnCl_4$, $AsCl_3$, and $TeCl_4$. The conditions for the chromatogram shown in Fig. 13 are identical to those of Fig. 12, except the oven temperature was raised to 126°C. Only $AsCl_3$ and $TeCl_4$ were not resolved into individual peaks.

The gas chromatography of a number of inorganic volatile chlorides such as titanium tetrachloride, carbon tetrachloride, silicone tetrachloride, germanium tetrachloride, tin tetrachloride, phosphorus trichloride, phosphorus oxychloride, arsenic trichloride, antimony pentachloride, and vanadium tetrachloride was studied[46-48]. A Hewlett-Packard No. 700 gas chromatograph equipped with a thermal conductivity detector (Gow-Mac 4 Tungsten filaments) was used and modified to keep the oven temperature constant within ±0.1°. Nitrogen was used as carrier gas at flow rates of 10–100 ml/min. The columns used were glass (4 mm i.d., 6 mm o.d., and 183 cm long). Column oven temperatures were, at maximum, 30°C below the maximum recommended temperature limit at isothermal conditions. The injection port temperature was maintained at 50°C above oven temperature and the detector temperature was maintained equal to that of the oven. The packing materials were (1) Kel-F Oil No. 10, 10% on Celite 545, (2) Kel-F wax, 10% on Celite 545, (3) Silicone Oil DC 550, 10% on Celite 545, (4) Silicone Rubber UCW 98, 10% on Diatoport S (80–100 mesh), (5) Apiezon L, 10% on Chromosorb R (45–60 mesh), (6) Phasepak P (30–60 mesh).

2. PHOSPHORUS

Phosphorus (the twelfth most abundant element) is widely distributed both in igneous and sedimentary rocks. The most prevalent of all naturally occurring phosphorus compounds are salts of orthophosphoric acid. The most important mineral source of phosphorus is the apatite minerals $Ca_5X(PO_4)_3$ where X may be fluorine, carbonate, or hydroxyl. Calcium phosphates are the most important source for the phosphate fertilizer industry. The major portion of all phosphorus produced comes from phosphate rock, among the major constituents being P_2O_5 (27.5–33.0%), calcium oxide (41.9–46.1%), and fluorine (2.9–4.4%). The production of *elemental* phosphorus worldwide in 1965 was *ca.* 977 tons of which 612 tons was produced in the U.S. The majority of the elemental phosphorus produced is burned to phosphorus pentoxide and hydrated to phosphoric acid. The acid is employed primarily for the manufacture of sodium salts, although some is used directly in metal treatment and in carbonated beverages. Phosphoric acid is also used to make liquid and suspension fertilizers, diammonium phosphate, ammonium phosphate nitrate, and solid ammonium polyphosphate. Other uses of elemental (red) phosphorus include application for wooden and paper safety matches and fireworks, while white phosphorus is used in roach and rodent poisons which usually contain 1–4% phosphorus. Phosphorus derivatives have wide application as animal feed supplements, detergents, surfactants, alloys, gasoline additives (tricresyl phosphate), plasticizers, dyes, lube oil additives, flotation agents, and pesticides (*e.g.* parathion, malathion).

There are three different types of elemental phosphorus (*e.g.* yellow (white), red, and black) each possessing different physicochemical and biological properties. Elemental yellow phosphorus is an extremely toxic substance and protoplasmic poison. Chronic exposure to small quantities of phosphorus can result in epithelial damage of the bone capillaries with thrombosis and bone necrosis[49], while heart damage[50] has resulted from acute phosphorus poisoning. Phosphorus vapor is believed to be the chief industrial cause of poisoning although phosphorus can be absorbed through the skin or by ingestion. Red phosphorus is considered relatively non-toxic unless it contains the white form as an impurity.

Phosphorus is usually determined and estimated as phosphoric acid (expressed as P_2O_5) either gravimetrically as magnesium pyrophosphate, titrimetrically, or colorimetrically[51].

Phosphorus has also been determined by fluorescence quenching[52], neutron activation[53], and spectrophotometric[54–57] techniques.

Addison and Ackman[58] described the direct determination of elemental phosphorus by GLC. Phosphorus levels as low as 10^{-12} g corresponding to about 2 parts in 10^{12} in concentrated extracts could be measured in several minutes with application of this method to the analysis of water, mud, and biological samples. Five samples of these three types were treated as follows: (*a*) *water:* phosphorus was extracted by vigorous shaking with an organic extractant (2:1, v/v) for a few minutes. Samples in which large amounts of phosphorus were expected were re-extracted once, (*b*) *mud:* up to 5 g of mud were extracted with 50 ml extractant by swirling in a stoppered Erlenmeyer-type flask with a few 5 mm glass beads for 10–15 min; samples were then filtered through Whatman No. 1 paper and the residue re-extracted once where necessary. The filtrates were allowed to separate and the organic layers were recovered for analysis, and (*c*) *tissue:* up to 10 g of tissue was homogenized for 2 min with 50 ml extractant in a stainless steel blender, cooled in ice. The blend was filtered and the residue re-extracted if necessary. The filtrate was allowed to separate and the organic layer taken for analysis. Standard solutions of phosphorus were prepared by weighing yellow phosphorus under water, drying it by a 10 sec immersion in acetone, and dissolving it in benzene. The solutions were stored at 0° in the dark under argon (solutions were stable over periods of several weeks).

Analyses were carried out using either a MicroTek 220 or a laboratory-built instrument. Both were fitted with a Melpar flame photometric detector (FPD) (Tracor, Inc., Austin, Texas) employing a 526 mm filter. A Honeywell Elektronik 1 mV recorder was used fitted with disc integrators (Disc Instrument Co.). Operating conditions for the two instruments were

Instrument:	MicroTek	Lab. built
Column:	2 m × 3 mm U-glass	2 m × 3 mm coiled glass
Packing:	3% OV-1 on Chromosorb W	3% Se-30 Chromosorb W
Column temp.:	100°	120°
Injector temp.:	200°	200°
Detector temp.:	200°	200°
Carrier gas flow:	He, 80 ml/min	He, 80 ml/min

The specificity for elemental phosphorus is based on the characteristic retention time of phosphorus and on the relative responses of FPD and FID to phosphorus. The FPD with a 526 mm filter is highly sensitive to P-containing compounds but is considerably less sensitive to purely organic compounds.

By using the FPD and FID simultaneously (drive channel electrometer and two pen recorders) it was possible to distinguish between elemental phosphorus and any organic material emerging with the same retention time. This specificity was demonstrated in the analysis of industrial treatment of waste colloidal solution of phosphorus involving flocculation at a high pH. During this process some phosphine (PH_3) was formed. Since PH_3 is also extractable with organic solvents it can interfere with the determinations of elemental phosphorus by the conventional method. However, GLC with both FPD and FID allowed the isolation and independent determination of elemental phosphorus as well as indicating the presence of phosphine (Fig. 14).

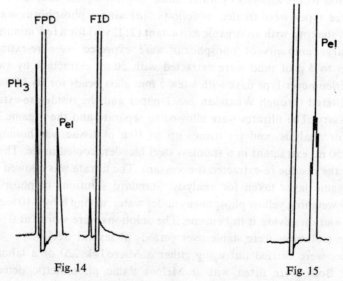

Fig. 14.

Fig. 15.

Fig. 14. FPD and FID responses to a mixture of phosphine and elemental phosphorus.

Fig. 15. FPD response to 10^{-8} g elemental phosphorus.

It was also found that samples of phosphorus exceeding 10^{-8} g were eluted as multiple peaks (Fig. 15). It was suggested by Addison and Ackman[58] that in very low concentrations (e.g. injection of 10^{-12}–10^{-9} g phosphorus) phosphorus vapor may exist monoatomically whereas at higher concentrations there may be recombination. Hence under GLC conditions species of molecular weight 31, 62, and 124 were partially separated.

Drawe and Wendenburg[59] described the radio gas chromatography of white phosphorus and phosphorus derivatives using a thermal conductivity detector (Gow-Mac) and a counting tube for measuring the beta activity of ^{32}P. A 2 m column was used containing 10% Emulphor (Badische Anilin and Sodafabrik AG) or 10% silicone pelastomer (Wacker Chemie) on kieselguhr with hydrogen as carrier gas.

The detection of toxic phosphorus such as in the form of metallic phosphides, e.g. zinc, phosphide, or white phosphorus, in liver or other biological materials is of toxicological importance. The determination of nanogram amounts of phosphides and white phosphorus in biological materials has been achieved by neutron activation[53]. The technique consisted of distillation of phosphorus as phosphine and collection of the phosphine as silver phosphide by reaction with silver nitrate. The silver phosphide was then oxidized to silver phosphate by chlorine and the amount of phosphorus determined by thermal neutron activation analysis. A radiochemical separation procedure involving ion-exchange (Dowex 50-cation exchange) and precipitation methods was used for the separation of ^{32}P activity induced during the neutron activation.

Phosphine (hydrogen phosphide, PH_3) is a reactive highly toxic gas that yields numerous chemical derivatives and has been extensively employed as an effective fumigant against insects that infest cereal grains, tobacco, spices, cocoa, and other stored products[60]. Phosphine has been determined by spectrophotometric[61,62] and colorimetric[63] techniques.

Dumas[64] initially determined phosphine in air by gas chromatography of sample sizes of 0.05–1 ml for concentrations of 0.5–10 mg of PH_3 per liter. A Perkin-Elmer 154-D gas chromatograph with thermistor cell detector was used and a 160 in. × $\frac{1}{4}$ in. stainless steel column packed with 30% Apiezon L on 40–60 mesh firebrick. The column temperature was 35°C and the flow rate of helium carrier gas was 25 ml/min (Under these conditions phosphine had a retention time of ca. 3.5 min.) In a subsequent study, Dumas[65] utilized flame-ionization gas chromatography with phosphorus detector for the determination of phosphine in concentrations of 0.05 to 0.5 mg/l in one limit, sampling in a range of 100 µl. The detection limit was 0.0005 mg/l for 250 µl sample size. An Aerograph 600D was used with a flame-ionization phosphorus detector and a 96 × $\frac{1}{8}$ in. diam. stainless steel column filled with 30% Apiezon L on 30–60 mesh Chromosorb W. The column temperature was 45°C and the flow rate of nitrogen carrier gas was 10 ml/min.

The utility of flame photometric detectors highly selective to phosphorus has been elaborated by Guiffrida[66,67], Karmen[68], and Brody and Chaney[69].

Berck et al.[70] compared microcoulometric, thermionic, and flame photometric (phosphorus mode) detectors (FPD) for minimum detectability, accuracy, reproducibility, and rapidity for the determination of p.p.b. and p.p.t. levels of phosphine in foodstuffs, air, and water. Based on a response at 10% of recorder scale with a reproducibility to within ±10%, the lower limits of detectability were microcoulometric (peak height) 20 pg, flame photometric (peak height) 5 pg. With a 10 g sample of foodstuff or water, these amounts corresponded to a relative minimum detectability of 500, 2, and 0.5 p.p.t. for the microcoulometric, thermionic, and flame-photometric detection methods, respectively. Phosphine could be determined within 30 sec by the appropriate choice of operational parameters. Relatively large concentrations of SO_2, H_2S, CH_3SH, and other sulfur gases or vapors could interfere and steps to curtail this were described.

For the microcoulometric method, a Coulson T-300 titration cell and a C-200 coulometer (Dohrman Instruments) were used. (Since PH_3 could be measured directly in the titration cell, use of GLC and auxiliary apparatus was omitted in this method.)

For the thermionic detection method, an Aerograph Model 1520B gas chromatograph was used equipped with a pelleted cesium bromide thermionic detector attachment (Hartman[71]), and a 5 ft. × $\frac{1}{8}$ in. i.d. stainless steel column packed with 4.5% QF-1 on 70–80 mesh Chromosorb G. The column, detector, and injection port temperatures were 65, 205, and 190°C, respectively. With a nitrogen carrier flow rate of 30 ml/min, the retention time of the PH_3 peak was 16.2 sec.

An Aerograph Model 1520B gas chromatograph equipped with a MicroTek flame phosphorus assembly was used for the FPD method equipped with an 18 in. × $\frac{1}{4}$ in. o.d. stainless steel column packed with 3% Carbowax 20M on 60–80 mesh Gas Chrom Q. Flow rates of hydrogen, oxygen, and nitrogen were 158, 25, and 100 ml/min, respectively. The column, injection port, and detector temperatures were 65, 190, and 200°C, respectively. Under these conditions, the retention of the PH_3 peak was 6.0 sec.

A rapid method of measuring sorption affinity of granular or powdered substrates toward PH_3 in the p.p.b.–p.p.t. range was developed by Berck and Gunther[72]. The method employed rapid fumigation (0.5–6 sec contact time) within a gas chromatographic column. A small amount of substrate (0.15–2 g) was packed in a short Teflon or glass column (1–6 in. × $\frac{1}{4}$ in.) attached to an 18 in. empty Teflon or stainless steel column. PH_3 (500 pg) was injected into the nitrogen carrier gas stream and the response measured by the flame-photometric detection method of Berck et al.[70] was compared with that obtained with an empty column. Uptake by the substrate could be measured to 5 pg (1% level) with a detector response within 15 sec after injection of the 500 pg sample of PH_3. When column temperatures were raised, chemisorption was indicated if the response was significantly lowered and conversely physisorption was indicated if the response was increased. The substrates used included ground wheat, wheat, rye and soy flours, bran, seminolina, wheat germ, starch, gluten powder, aluminum oxide, three types of soil, and glass beads.

It was suggested that the method could be adopted to investigate sorption affinity of other gases applied to virtually any granular or powdered substrate that would be stable under the test conditions.

A number of organoderivatives of phosphorus have been shown to have utility as synthetic intermediates. Triphenylphosphine, for example, is used for the production of phosphorus ylides, important intermediates for the synthesis of complex olefins and aldehydes. Gudzinowicz and Campbell[73] described high-temperature gas chromatographic separations of aryl phosphines and phosphine oxides. Using triphenylmethane as internal standard, triphenylphosphine in mixtures could be quantitatively analyzed. For this study, an F & M Model 124 gas chromatograph was used with a 6 ft. × $\frac{1}{4}$ in. copper column packed with 20% Apiezon L on 40 to 60 mesh acid-washed C-22 firebrick and operated at 275°C with a helium inlet pressure and flow rate of 18 p.s.i. and 107 ml/min. Aryl phosphine mixtures (e.g. triphenylphosphine, p-methoxyphenyldiphenyl phosphine, di(p-methoxyphenyl) phenylphosphine and tri(p-methoxyphenyl)phosphine were separated on a 6 ft. × $\frac{1}{4}$ in. stainless steel column packed with 5% SE-30 on 80–100 mesh Chromosorb W and an argon flow rate of 39 ml/min with temperature programmed from 248 to 331°C. A Barber-Colman Model 10 instrument was used with argon detector maintained at 400°C, cell voltage of 1250 V, and flash heater at 405°C.

The TLC of some aromatic phosphorus and arsenic compounds on Silica Gel G and neutral alumina was described by Berei[74]. The TLC of some aromatic phosphorus compounds (triphenylphosphine, triphenylphosphine oxide, triphenylphosphate, triphenylphosphite, and tricresylphosphate) as well as triphenylarsine on Silica Gel G and neutral alumina was described by Berei[74]. Tables 6 and 7 list the R_F values

TABLE 6

$R_F \times 100$ VALUES ON SILICA GEL G ADSORBENT

Compound	$R_F \times 100$								
	S_1	S_2	S_3	S_4	S_5	S_6	S_7	S_8	S_9
Triphenyl phosphate	93	79	9	24	94	87	79	91	93
Triphenyl phosphite	95	81	18						
Triphenylphosphine oxide	84	54	0	0	55	9	16	70	64
Triphenylphosphine	94	73	21	92	95	95	94	95	95
Tricresyl phosphate	95	90	0	84	90	88	87	89	94
Triphenylarsine	96	75	39	89	95	94	93	91	94

Solvent systems: S_1 = acetone; S_2 = acetone ("oversaturated"); S_3 = petroleum ether; S_4 = benzene; S_5 = benzene–acetone (1:1); S_6 = benzene–acetone (9:1); S_7 = chloroform; S_8 = chloroform–acetone (6:4); S_9 = chloroform-acetone (9:1).

found on Silica Gel G (Merck) and neutral alumina (Woelm) using 9 solvent systems. Table 8 lists the R_F values of triphenyl phosphine oxide on various adsorbents with acetone as solvent.

TABLE 7

$R_F \times 100$ VALUES ON NEUTRAL ALUMINA ADSORBENT

Compound	$R_F \times 100$								
	S_1	S_2	S_3	S_4	S_5	S_6	S_7	S_8	S_9
Triphenyl phosphate	93	73	0	36	94	90	35	90	91
Triphenyl phosphite	92	70	17						
Triphenylphosphine oxide	85	53	0	0	68	13	65	84	75
Triphenylphosphine	95	74	54	93	94	91	91	92	92
Tricresyl phosphate	90	70	0	32	88	88	91	93	91
Triphenylarsine	94	78	75	91	94	94	90	93	94

For the solvent systems see Table 6.

TABLE 8

$R_F \times 100$ VALUES OF TRIPHENYLPHOSPINE OXIDE ON VARIOUS ADSORBENTS WITH ACETONE SOLVENT

Adsorbent	$R_F \times 100$
Silica Gel G	87
Silica Gel G (activated at 160°)	81
Silica Gel + 5% gypsum	82
Silica Gel + 3% starch	76
Alumina + 3% starch	80
Silica Gel–alumina–gypsum (14:14:2)	84

REFERENCES

1 *Minerals Yearbook*, Bureau of Mines, Div. of Minerals, U.S. Govt. Printing Office, Washington, D.C., 1963.
2 R. F. ABERNATHY AND F. H. GIBSON, *U.S. Bur. Mines, Inform. Circ. No. 8163*, 1963.
3 W. S. SMITH AND C. W. GRUBER, *An Inventory Guide*, U.S. Dept, HEW, PHS, Div. of Air Pollution, April, 1966.
4 B. VALLEE, D. D. ULMER AND W. E. C. WALKER, *Arch. Ind. Health*, 21 (1960) 132.
5 H. A. SCHROEDER AND J. J. BACASSA, *J. Chronic Diseases*, 19 (1966) 85.
6 V. M. GOLDSCHMIDT, in A. MUIR (Ed.), *Geochemistry*, Clarendon Press, Oxford, 1958, p. 464.
7 *Agr. Econ. Rept. No. B 1*, Economic Res. Serv., U.S. Dept. of Agriculture, Washington, D.C., 1968.
8 E. E. ANGINO, L. M. MAGNUSON, T. C. WAUGH, O. K. GALLE AND J. BREDFELDT, *Science*, 168 (1970) 389.
9 M. MCBRIDE AND L. S. WOLFE, *Chem. Eng. News*, 49 (1971) 29.
10 D. U. FROST, *Poultry Sci.*, 32 (1953) 217.

11 *Joint FAO/WHO Expert Committee on Food Additives, 10th Rept., FAO Nutrition Meetings Series, No. 43*, 1967, p. 14.
12 H. S. OSBURN, *Central African J. Med.*, 3 (1957) 215.
13 W. BRAUN, *Ger. Med. Monthly*, 3 (1958) 32.
14 W. C. HUEPER, *Ann. N.Y. Acad. Sci.*, 108 (1963) 963.
15 D. V. FROST, *Federation Proc.*, 26 (1967) 194.
16 V. H. FERM AND S. J. CARPENTER, *J. Reprod. Fertility*, 17 (1968) 199.
17 S. KAYE, *Am. J. Clin. Pathol.*, 14 (1944) 83.
18 I. SUNSHINE, *Manual of Analytical Toxicology*, The Chemical Rubber Co., Cleveland, Ohio, 1971, p. 32.
19 *Official Methods of Analysis*, Assoc. of Offic. Agr. Chemists, Washington, D.C., 1965, 10th edn., Sect. 24.001, p. 357.
20 H. K. HUNDLEY AND J. C. UNDERWOOD, *J. Assoc. Offic. Anal. Chemists*, 53 (1970) 1176.
21 L. HOFFMAN AND A. D. GORDON, *J. Assoc. Offic. Agr. Chemists*, 46 (1963) 245.
22 B. SJÖSTRAND, *Anal. Chem.*, 36 (1964) 814.
23 A. ARNOLD, S. DAVIS AND A. L. JORDAN, *Analyst*, 94 (1969) 664.
24 A. ZEMAN, J. RUŽIČKA, J. STARY, E. KLEČKOVA, *Talanta*, 11 (1964) 1143.
25 G. DEVOTO, *Boll. Soc. Ital. Biol. Sper.*, 44 (1968) 425.
26 M. GENET AND C. FERRADINI, *J. Chromatog.*, 37 (1968) 527.
27 I. I. M. ELBEIH, J. F. W. McOMIE AND F. H. POLLARD, *Discussions Faraday Soc.*, 7 (1949) 183.
28 V. MIKETUKOVA, J. KOHLICEK AND K. KACL, *J. Chromatog.*, 34 (1968) 284.
29 W. O. WINKLER, *J. Assoc. Offic. Anal. Chemists*, 45 (1962) 80.
30 L. R. OVERBY AND L. STRAUBE, *Toxicol. Appl. Pharmacol.*, 7 (1965) 850.
31 L. R. OVERBY, S. F. BOCCHIERI AND R. C. FREDRICKSON, *J. Assoc. Offic. Anal. Chemists*, 48 (1965) 17.
32 D. HUNTER, *The Diseases of Occupations*, English Universities Press, London, 2nd edn., 1957, p. 301.
33 W. D. BUCHANAN, *Toxicity of Arsenic Compounds*, Elsevier, Amsterdam, 1962.
34 D. MACAULAY AND D. A. STANLEY, *Brit. J. Ind. Med.*, 13 (1956) 217.
35 S. S. PINTO, *Arch. Hyg. Occupational Med.*, 1 (1950) 437.
36 K. M. MORSE AND A. N. SETTERLIND, *Arch. Ind. Hyg.*, 2 (1950) 148.
37 W. J. LEVINSKY, R. V. SMALLEY, P. N. HILLYER AND R. L. SHINDLER, *Arch. Environ. Health*, 20 (1970) 436.
38 A. E. DePALMA, *J. Occupational Med.*, 11 (1969) 582.
39 P. R. ULDALL, H. A. KAHN, J. E. ENNIS, R. L. McCALLUM AND T. A. GRIMSON, *Brit. J. Ind. Med.*, 27 (1970) 372.
40 A. G. HOCKEN AND G. BRADSHAW, *Brit. J. Ind. Med.*, 27 (1970) 56.
41 M. IGUCHI, A. NISHIYAMA AND Y. NAGASE, *J. Pharm. Soc. Japan*, 80 (1960) 1408.
42 B. J. GUDZINOWICZ AND H. F. MARTIN, *Anal. Chem.*, 34 (1962) 648.
43 B. J. GUDZINOWICZ AND J. L. DRISCOLL, *J. Gas Chromatog.*, 1 (1963) 25.
44 J. E. DENNISON AND H. FREUND, *Anal. Chem.*, 37 (1965) 1776.
45 S. DAL NOGARE AND L. W. SAFRANSKI, *Anal. Chem.*, 30 (1958) 894.
46 D. VRANTI-PISCOU, J. KONTOYANNAKOS AND G. PARISSAKIS, *J. Chromatog. Sci.*, 9 (1971) 499.
47 G. PARISSAKIS AND D. VRANTI-PISCOU, in E. KOVATS (Ed.), *Column Chromatography, 5th Intern. Symp. Separation Methods, 1969*, Sauerlander, 1970, p. 223.
48 G. PARISSAKIS, D. VRANTI-PISCOU AND J. KONTOYANNAKOS, *J. Chromatog.*, 52 (1970) 461.
49 S. MOESCHLIN, *Poisoning, Diagnosis and Treatment*, Grune and Stratton, New York, 1965.
50 R. S. DIAZ-RIVERA, *Medicine*, 29 (1950) 269.
51 M. B. JACOBS, *The Analytical Toxicology of Industrial Inorganic Poisons*, Interscience, New York, 1967.
52 D. B. LAND, *Mikrochim. Acta*, 6 (1966) 1013.
53 S. S. KRISHNAN AND R. C. GUPTA, *Anal. Chem.*, 42 (1970) 557.
54 R. N. ROGERS, *Anal. Chem.*, 32 (1960) 1050.
55 K. P. GUINLAND, *Anal. Chem.*, 27 (1955) 1626.
56 A. S. CURRY, E. R. RUTTER AND C. H. LIM, *J. Pharm. Pharmacol.*, 10 (1958) 635.

57 S. K. NIYOGI, in I. SUNSHINE (Ed.), *Manual of Analytical Toxicology*, Chemical Rubber Co., Cleveland, 1971, p. 285.
58 R. F. ADDISON AND R. G. ACKMAN, *J. Chromatog.*, 47 (1970) 421.
59 H. DRAWE AND J. WENDENBURG, *J. Chromatog.*, 18 (1965) 39.
60 W. H. DIETERICH, G. MAYR, K. HILD, J. B. SULLIVAN AND J. MURPHY, *Residue Rev.*, 19 (1967) 135.
61 J. G. HUGHES AND A. T. JONES, *Am. Ind. Hyg. Assoc. J.*, 24 (1963) 164.
62 R. DECHANT, *Am. Ind. Hyg. Assoc. J.*, 27 (1966) 75.
63 E. J. KING, *Biochem. J.*, 26 (1932) 292.
64 T. DUMAS, *J. Agr. Food Chem.*, 12 (1964) 257.
65 T. DUMAS, *J. Agr. Food Chem.*, 17 (1969) 1164.
66 L. GIUFFRIDA, *J. Assoc. Offic. Anal. Chemists*, 47 (1964) 293.
67 N. F. IVES AND L. GIUFFRIDA, *J. Assoc. Offic. Anal. Chemists*, 50 (1967) 1.
68 A. KARMEN, *Anal. Chem.*, 36 (1964) 1416.
69 S. BRODY AND J. E. CHANEY, *J. Gas Chromatog.*, 4 (1966) 42.
70 B. BERCK, W. E. WESTLAKE AND F. A. GUNTHER, *J. Agr. Food Chem.*, 18 (1970) 143.
71 C. H. HARTMANN, *Bull. Environ. Contamination Toxicol.*, 1 (1966) 159.
72 B. BERCK AND F. A. GUNTHER, *J. Agr. Food Chem.*, 18 (1970) 148.
73 B. J. GUDZINOWICZ AND R. H. CAMPBELL, *Anal. Chem.*, 33 (1961) 1510.
74 K. BEREI, *J. Chromatog.* 20 (1956) 407.

Chapter 6

SELENIUM AND TELLURIUM

1. SELENIUM

Selenium is widely distributed in the earth's crust at a concentration of about 0.09 p.p.m. It is mainly concentrated in sulfide minerals and in the soil of dry plains (*e.g.* Midwestern U.S.), and some plants can absorb and accumulate selenium in large amounts from the soil (*e.g.* varieties of *Astragalus* (milk vetch), woody aster, and golden weed). These plants contain large amounts of selenium, generally 1000 to 10,000 p.p.m. The consumption of these plants by livestock produces the disease syndrome of blind staggers or acute selenium poisoning.

Selenium is found in coal, in ores (as selenides), hydrothermal deposits (where it is associated with silver, gold, antimony, and mercury), and in small concentrations associated with copper deposits. Practically all selenium produced commercially is obtained as a byproduct of precious metals recovered from electrolytic copper refinery slimes. The world production of selenium in 1966 was approx. 3 million lb. of which 620,300 and 575,000 lb., respectively, were produced in the U.S. and Canada. Other sources of selenium are in the making of paper, *e.g.* when pyrites is used as the source of sulfur in the manufacturing process.

The major uses of selenium and its compounds are as follows: as rectifiers (used in electroplating, welding, direct-current motor operation), photoelectric cells (about 0.5–25 g of selenium are used per rectifier and photoelectric cells, depending on cell size), pigments (*e.g.* compounded with cadmium sulfide to form cadmium sulfoselinide pigments used to color plastics, paints, enamels, inks, rubber), glass, lubricants, blasting caps, chromium plating, stainless steel, in xerography photocopiers (amorphous selenium used for paper coating), and medicines (in form of sodium selenate solution to control certain animal diseases[1] in sheep, cattle, pigs, and as a buffered suspension of selenium sulfide for the control of seborrhea dermatitis of the scalp. Table 1 lists some selenium compounds and their uses.

Sources of atmospheric selenium include the combustion of industrial and residential fuels (raw and heavy petroleum contain approx. 0.92 and 1.0 µg/g of selenium)[2]. The amounts of selenium in automobile tires, coal, and soil samples are fairly comparable, *e.g.* 1.33 µg/g. Additional sources of atmospheric selenium include refinery waste gases and fumes and incineration of wastes, including paper products which contain as much as 6 p.p.m. selenium. Selenium content in cigarette papers is less than 0.05 p.p.m., while pipe and cigarette tobacco contain approx. 0.03–0.13 p.p.m. and cigar tobaccos 0.33–1.01 p.p.m. selenium[3].

In animals, selenium exhibits toxic properties when present in food sources at the p.p.m. level. However, selenium-deficient diseases appear if selenium is at

SELENIUM AND TELLURIUM

TABLE 1

SOME SELENIUM COMPOUNDS AND THEIR USES

Compound	Use
Aluminum selenide Al_2Se_3	Preparation of hydrogen selenide for semi-conductors
Ammonium selenite $(NH_4)_2SeO_3$	Manufacture of red glass
Arsenic semiselenide As_2Se	Manufacture of glass
Cadmium selenide $CdSe$	Photoconductors, photoelectric cells, rectifiers
Cupric selenate $CuSeO_4$	In coloring copper and copper alloys
Selenium disulfide SeS_2	In veterinary medicine
Selenium monosulfide SeS	In veterinary medicine
Selenium hexafluoride SeF_6	As gaseous electric insulator
Sodium selenate Na_2SeO_4	As insecticide
Sodium selenite Na_2SeO_3	In glass manufacturing

low levels in the diet of animals. The cycling of selenium has been investigated for soil–plant–animal systems[4]. Significant concentrations of selenium also exist in both fresh and salt water fish[5]; thus selenium apparently cycles through the aquatic ecosystem and accumulates in the higher forms of life, namely fish.

The carcinogenicity of selenium in rats has been reported[6,7]. Selenium has been shown to cross the placenta in rats and cats[8] and the offspring of rats[9], pigs[10], and sheep[11] have shown abnormalities following selenium administration during pregnancy. The passage of selenium can be inferred in other species from the symptoms of selenium poisoning observed in the offspring of animals grazing on seleniferous rangeland. (Foals and calves in affected areas are sometimes born with deformed hoofs.)

It has been suggested[12] that selenium may be a possible teratogen in man. Out of one possible and four certain pregnancies among women exposed to selenite, only one pregnancy went to term and the infant showed bilateral clubfoot. Of the other pregnancies, two could have terminated because of other clinical factors. The potential danger of high concentrations of selenium in food grown on soils with a high selenium content has been cited by Hadjimarkos[12]. Clegg[13] has commented on the special need to observe rural populations in this regard although it would seem unlikely that food contamination with selenium would be a major problem in cities because of the wide variations in the source of foodstuffs. Holmberg and Ferm[14], however, have reported that 2 mg sodium selenite/kg (an almost lethal dose) administered intravenously had no teratogenic or embryotoxic effects in the hamster.

The chief industrial injury caused by inorganic selenium is dermatitis. The greatest hazard of systemic poisoning arises from exposure to hydrogen selenide and certain halogen compounds, such as methyl and ethyl selenide. To some extent, selenium resembles arsenic physiologically. Compounds such as the reactive selenides can be absorbed *via* the lungs or skin. Aspects of industrial selenosis have been described[15-17].

Analyses of selenium (in air samples) have been performed using colorimetry[18], spectrophotometry[19], photometry[20], and neutron activation[21]. Selenium in biological samples has been determined by gravimetry[22], titrimetry[23], colorimetry[24], spectrophotometry[25], atomic absorption[26], neutron activation[27,28], and fluorimetric techniques[29-31].

A number of paper chromatographic and electrophoretic methods have been suggested for effecting a preliminary separation (and identification) of selenium and tellurium in different valency states from one another, or from associated materials[32-37]. Detection is generally carried out by spraying with stannous chloride, resulting in reddish and blackish spots for selenium and tellurium, respectively.

The TLC separation of Se^{IV}, Te^{IV}, V^V, and Mo^{VI} as ternary mixtures on Silica Gel G by a 30–40 min ascending development using $Et_2C_2O_4$–HCl (60:1) and butyl acetate–hydrochloric acid (40:0.6) was described by Johri *et al.*[38]. The ions were detected by spraying with 0.1 M K_2CS_3 and the limits of identification were 1.27–2.04 µg.

The TLC of Se^{IV} and Te^{IV} in alcohol solutions was described by Gaibakyan and Aturyan[39]. Se and Te were separated on Al_2O_3 by using solutions of methanol, ethanol, propanol containing ≥25 ml HCl/100 ml solution. Spraying with $SnCl_2$ solution revealed Se^{IV} and Te^{IV} as light brown and black spots, respectively.

The separation and determination of the valence states of selenium, tellurium, and tin by TLC (using silica gel and cellulose powder supports) combined with Weisz

TABLE 2

SPOT COLORS AND LIMITS OF IDENTIFICATION OF VALENCE STATES OF Se, Te, AND Sn ON TLC PLATES

Ions	Detecting reagent	Spot color	Limits of detection (µg)
Se^{IV}	KI	Red-brown	0.46
Se^{VI}	KI + HCl	Red-brown	0.54
Te^{IV}	10% $FeSO_4$ + H_3PO_4	Black	0.62
Te^{VI}	10% $FeSO_4$ + H_3PO_4	Black	0.62
Sn^{II}	5% Phosphomolybdic acid	Blue	0.02
Sn^{IV}	0.01 M Quercetin	Yellow	0.12

ring oven[40] technique was described by Mehra *et al.*[41]. Se^{IV} and Se^{VI} were separated on Silica Gel G developed with *n*-butanol–3 *N* HCl (5:1). The spots were visualized with sprays of HCl + KI and then heating for 5 min at 80°C. Te^{IV} and Te^{VI} were separated on cellulose powder (Camag plates developed with acetone–3 *N* HCl–2% tartaric acid (4:1:1). Detection was achieved with sprays of 10% $FeSO_4$ + H_3PO_4 and heating for 10 min at 80°C. Sn^{II} and Sn^{IV} were separated on cellulose powder plates developed with *n*-butanol–glacial acetic acid (2:1). Sn^{II} was located using 5% phosphomolybdic acid while Sn^{IV} was detected with an alcoholic solution of 0.01 *M* quercetin. Table 2 lists the results of the TLC separation. Semi-quantitative determinations of the Sn, Se, and Te were carried out using fluorescent TLC supports (*e.g.* Silica Gel G + 2% fluorescent indicator green (Woelm) and cellulose powder DSF (Camag)). Following development and visualization under UV (254 nm), the fluorescent spots were removed and determined by the Weisz ring oven technique The rings were developed with chromogenic agents as shown in Table 3.

TABLE 3

RING OVEN SEPARATION

Ions	Developing agents	Color of ring
Se^{IV}	0.05 *M* Potassium thiocarbonate + 1% $AgNO_3$	Black
Se^{VI}	KI + HCl	Red-brown
Te^{IV}	5% $SnCl_2$ + 25% NaOH	Black
Te^{VI}	5% $SnCl_2$ + 25% NaOH	Black
Sn^{II}	5% Phosphomolybdic acid	Blue
Sn^{IV}	0.01 *M* Quercetin	Green fluorescence under UV light

The application of anion-exchange resins for the separation of the Group VIB elements has been studied by Sasaki[42] who found that selenium was adsorbed on anion-exchange resins from conc. hydrochloric acid solution, while sulfate passed through; selenium was then eluted with 6 *N* HCl. The separation of selenite, sulfate, and iron by cation-exchange resin (Dowex 50W-X8, 100–200 mesh) was reported by Yamamoto and Sakai[43]. Selenite was determined spectrophotometrically with 3,3′-diaminobenzidine as described by Cheng[44]. Figure 1 shows an elution curve of selenite, sulfate, and iron(III). Sulfate in the effluents was determined spectro-photometrically with barium chloranilate[45].

Peterson and Butler[46] described paper chromatographic and electrophoretic systems for the identification of sulfur and selenium amino acids. Whatman 3MM paper was used in descending developments with the following solvent systems: (*1*) *n*-butanol–pyridine–water (1:1:1); (*2*) *n*-butanol–acetic acid–water (25:6:25); (*3*) *n*-butanol–ethanol–water (2:2:1); and (*4*) *tert.*-butanol–formic acid–water (14:3:3). For two-dimensional papers, solvent (*1*) was followed by solvent (*2*).

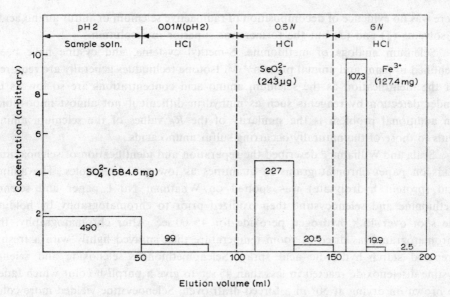

Fig. 1. Elution curve of sulfate, selenite, and iron. The figures in brackets are the amounts added and those assigned to each column represent the amount found.

Anion exchange chromatography (descending) was carried out using diethylamino-ethyl cellulose (DEAE) Whatman DE 20 by modifications of the method of Knight[47]. Paper electrophoresis was carried out on Whatman 3MM paper in the apparatus of Markham and Smith[48] at 20 V/cm for 4 or 8 h. Preliminary separation into neutral, acidic, or basic amino acids was carried out in a volatile pyridine–acetic acid buffer, pH 6.0 for 4 h. After elution and re-application to other electrophoretic strips, further separation was effected in phosphate–citrate buffer at pH 2.7 for 4–8 h. Color development with ninhydrin reagent[49] was carried out at 20°C for 24 h. The colors produced on the DEAE paper for the various compounds were very characteristic and more stable than the corresponding compounds on 3MM paper. On DEAE paper the sulfur amino acids were most clearly separated with 0.02 M acetate buffer at pH 4.7. With decreasing acidity, the amino acids tended to group into 2 bands at approx. R_F 0.25 and 0.50.

The chromatographic and electrophoretic behavior of selenocystine, Se-methyl selenocysteine and selenomethionine in all systems was indistinguishable from the corresponding sulfur compounds. A number of ^{75}Se-labeled compounds extracted from plant tissues also behaved identically with the sulfur amino acids in these systems[50].

The only differences noted in the behavior of the seleno and sulfur analogs were for the substances having a high mobility upon electrophoresis at pH 2.7, e.g. the relative migration of ^{75}Se-selenocysteic acid and cysteic acid over a 4 h period were 14.9 cm and 15.5 cm, respectively. In regard to utility of solvent systems, there was no evidence of decomposition of radioactive selenium or sulfur amino acids in solvents (*1*) and (*3*) but the former was superior in resolution.

Selenium analogs of methionine, S-methyl cysteine, and cystine have been identified in plant and animal protein[51,52]. Isotope techniques generally are required for the identification as the selenium amino acid concentrations are so low as to render detection by reagents such as ninhydrin difficult if not almost impossible. An additional problem is the similarity of the R_F values of the selenium amino acids to those of the naturally occurring sulfur amino acids.

Scala and Williams[53] described the separation and identification of selenoamino acids on paper chromatograms in quantities as low as 0.5 mμmoles. The amino acid (protein hydrolysate) was spotted on Whatman No. 1 paper and seleno-methionine and selenocystine then oxidized prior to chromatography by holding the spot over 15% hydrogen peroxide for 45–60 sec. After chromatography, the chromatogram was dried at room temperature then sprayed lightly with a freshly prepared starch–hydriodic acid spray. Selenomethionine selenoxide and seleno-cystine diselenoxide reacted in less than 45 sec to give a purplish color which faded to brown on drying at 50° in a forced draft oven. Selenocystine yielded more color per mole than selenomethionine. This was suggested to be due to the formation of the diselenoxide, or possibly two selenic acids by peroxide oxidation. Preservation of the chromatograms was accomplished by placing them in a manila envelope in which they are pressed against a paper which had been similarly sprayed and dried. In this manner 0.5 mμmole spots have been clearly visible after 2 weeks. The sulfur amino acids were not oxidized by the described conditions above and hence did not interfere with the identification of the seleno amino acids. This technique allowed the detection of 0.5 mμmoles of selenomethionine in the presence of 200 mμmole; of methionine. The sulfur amino acids could be identified by the ninhydrin reactions

TABLE 4

R_F VALUES ($\times$ 100) OF OXIDIZED SELENOMETHIONINE, SELENOCYSTINE, METHIONINE AND CYSTINE

Solvent	$R_F \times 100$			
	Oxidized Se-methionine	*Oxidized Se-cystine*	*Methionine*	*Cystine*
Butanol–acetic acid–H$_2$O (4:1:1)	22	8	55	4
Ethanol–1 *M* NH$_4$OH–H$_2$O (90:5:5)	20	5	38	6
Phenol–H$_2$O–NH$_4$OH (100:20:0.03)	95	26	76	21

The selenoamino acids were oxidized prior to chromatography and were identified by the starch–hydriodic acid spray described in the text. All chromatography was done on Whatman No. 1 paper.

however, before spraying it was first necessary to immerse the chromatogram in an ammonia atmosphere for 1 h. (This procedure decolorized the paper and clear ninhydrin spots were obtained.)

Table 4 shows some R_F values obtained with selenomethionine and seleno-cystine treated as described above. Methionine and cysteine were identified by ninhydrin.

TABLE 5

R_F VALUES FOR MIXTURES OF METHIONINE AND SELENOMETHIONINE ON SILICA GEL G
Detector: 0.5% ninhydrin in n-butanol

Solvent	R_F values	
	Methionine	Seleno-methionine
Ethanol–isobutanol–water (1:7:2)	0.22	0.30
Ethanol–n-butanol–water (1:7:2)	0.28	0.36
Isopropanol–n-butanol–water (1:3:1)	0.32	0.41
Methanol–n-butanol–water (1:7:2)	0.32	0.37
Ethanol–n-butanol–water (1:3:1)	0.37	0.41

Millar[54] described the separation of selenomethionine and methionine by TLC on Silica Gel G. Table 5 lists the R_F values found for these constituents when developed with 5 solvent systems. The spots were detected by spraying with 0.5% ninhydrin in n-butanol, the limit of detection being 0.02 mg. The most satisfactory solvent was a mixture of isopropanol–butanol–water (1:3:1) which effected complete separation in 10 cm.

The elution rates of selenomethionine and selenocystine from ion-exchang e resins and their position on the chromatograms in relation to other known amino acids was reported by McConnell and Wabnitz[55]. Amino acids and the seleno compounds were determined quantitatively using the methods of Moore et al.[56] and the Beckman/Spinco Model 120 amino acid analyzer with sulfonated styrene–divinyl-benzene (8%) co-polymer resins. Gamma scintillation counting of ^{75}Se was performed using a Picker nuclear spectroscaler III Model 600–333 in conjunction with a Picker scintillation well counter, Type 83F8 (2804-E) and a Picker Model 600-036 printout system. Figure 2 illustrates a flow diagram for simultaneous radioactive amino acid assay and shows how radioactivity of the effluent is measured after it leaves the resin column and before it reaches the ninhydrin mixing manifold. ^{75}Se has two characteristic energy peaks, one at 125 keV and the other at 260 keV. For ^{75}Se assay, the spectrometer was set at 100 keV for the lower level, with a window width at 180 keV. Elution times of selenocystine and selenomethionine from ion-exchange resins and C values (constants per μmole of ninhydrin-positive compound)

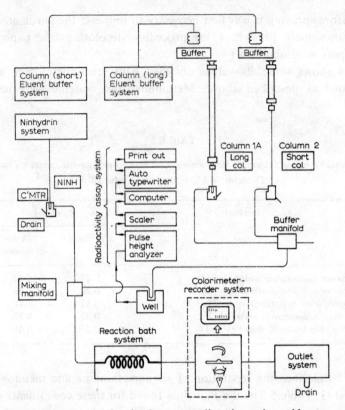

Fig. 2. Flow diagram for simultaneous radioactive amino acid assay.

are shown in Table 6. Selenium analogs of methionine and cystine did not have the same elution rates on the ion-exchange resin as methionine and cystine.

The determination of selenomethionine in biological sources (*e.g.* serum, liver, and muscle) has been achieved by Barak and Swanberg[57] by coupling the technique of paper chromatography and neutron activation analysis. One gram of liver or muscle was homogenized in the presence of 9 volumes of 50% ethanol or, if serum was analyzed, 1 ml was added to 9 volumes of ethanol to obtain a protein-free filtrate which was lyophilized to dryness. The residue from lyophilization was taken into solution in 1 ml of 50% ethanol and 20 µl were chromatographed (descending) on Whatman No. 1 strains developed from 12 h with *n*-butanol–pyridine–water (1:1:1). The strips were sprayed with ninhydrin and developed in an oven at 80° for 10 min. The methionine area (which induced the selenomethionine) was separated from the rest of the ninhydrin-positive areas of tissue (the same separation was effected with liver and serum). Elution and re-chromatography of the solute in this area in other solvent mixtures had shown it to be a single entity, methionine. The methionine area was cut from the paper, placed in a Tygon tube, and irradiated for 20 sec in a Triga Mark T atomic reactor using a rapid transit pneumatic rabbit system and a

TABLE 6

C VALUES AND ELUTION TIMES FOR SELENOCYSTINE AND SELENOMETHIONINE

C values are the constants per μmole of ninhydrin-positive compound. Time from start of all chromatograms to pH 4.25 buffer change is 728 ± 19 min. Flow rate for experiments is 43.6 ± 0.2 ml/h. Elution time in minutes is from pH 4.25 buffer change taken as zero time.

Experiment no. and type	Selenocystine		Methionine		Leucine		Selenomethionine	
	C value	Time (min)	C value	Time (min)	C value	Time (min)	C value	Time (min)
86 Calibration mixture			18.37	22	20.09	80		
98 Calibration mixture + [75Se]selenocystine	8.28[a]	20	61.49[b]	20	20.99	84		
91 [75Se]selenocystine	7.52	18						
97 Stable selenocystine	6.24	20						
102 Calibration mixture + [75Se]selenomethionine			21.77	20	37.62[c]	80	17.08[a]	80
101 Stable selenomethionine							13.50	78
100 [75Se]Selenomethionine							13.30	76

[a] By difference.
[b] 1 μM methionine plus 5 μM [75Se]selenocystine.
[c] 1 μM selenomethionine plus 1 μM leucine.

neutron flux of 1.1×10 in. $n/cm^2/sec$. Following a delay of 20 sec, the sample was analyzed in an RIDL 400 Channel gamma ray spectrometer using a 1 min live time count. This provided a spectrum from 0 to 1 meV, which included the peak due to $17\,m$ Se, a 17.5 sec isotope of selenium peaking at 0.160 meV. The selenium in the sample was estimated by a direct comparison method using selenium standards. The micromoles of selenium within the spot represented the micromoles of seleno-methionine because of the one-to-one stoichiometric relationship between selenium and selenomethionine. Recovery studies indicated that 100% of selenomethionine was recovered from liver and muscle homogenate whereas only 60% of the selenium added as selenomethionine to serum was recovered from the serum methionine area. The remainder of the selenium was associated with the third ninhydrin-positive spot on the chromatogram and suggested a metabolic pathway for selenomethionine at the level of the serum. The lower limit of detectability of this method was 0.1 μmoles of selenomethionine/g of tissue or ml of serum. Levels of natural, free seleno-methionine in rat liver, muscle, and serum and in human serum apparently exist at a lower level than this because more could be detected in these sources by the above technique.

It has been shown by Byard[58] and Palmer et al.[59] that trimethylselenium ion $[(CH_3)_3Se^+]$ is a major selenium product in the urine of rats injected with [75]-Se-selenite. Although McConnell and Portman[60] showed that dimethyl selenide was excreted through the lungs, $(CH_3)_3Se^+$ was the only excretory product of selenium to be isolated and identified. Palmer et al.[61] reported that trimethylselenonium ion is a general excretory product from selenium metabolism in the rat. $(CH_3)_3Se^+$ was shown to be a major excretory product from selenate, selenomethionine, seleno-cystine, methylselenocystine, and seleniferous wheat and its level in the urine was equivalent to 20–50% of the urinary selenium, depending on the selenium source and collection time. A second major unidentified urinary metabolite was excreted from each radioactive selenium source and accounted for 11–28% of the total urinary selenium.

Aliquots of urine (30 ml) were subjected to two-dimensional paper chromato-graphy using n-butanol–acetic acid–water (4:1:1) for the first development and phenol–water (73:27) for the second development. In the experiments where ^{75}Se was used, authentic $(CH_3)_3Se^+$ was co-chromatographed with the urine samples. The trimethyl selenium ion was visualized by spraying the chromatogram with Dragendorff's reagent[62]. Autoradiograms were also developed by exposing a few of the chromatograms to X-ray film for 2 weeks. The finding of the similarities in the excretory products from the various selenium sources suggested that all forms of selenium may be detoxified and excreted by very similar mechanisms. There is apparently much similarity between the toxicities of these forms of selenium[63] which is consistent with the concept of common metabolic pathways.

The GLC characteristics of some alkyl selenium compounds which might prove useful as reference compounds in the identification of selenium metabolites in plant tissues has been reported[64]. The preparative gas chromatograph used in the purification of the organic selenium compounds was a Wilkens Instrument and Research, Inc., Model Aerograph A-90-P, connected to a Leeds and Northrup, Speedomax H, Type S recorder with a disc chart integrator (Disc Instruments, Inc., Model 207). The column used for all purifications was 5 ft. × ¼ in. Silicone Fluid (methyl) SF-96 on 60–80 mesh firebrick. Helium was used as carrier gas with a flow rate of 35–40 ml/min. The analytical gas chromatograph used was a Wilkens Instrument and Research, Inc., Hy-Fi 600-C with either a hydrogen flame-ionization detector or an electron-capture detector using the same recorder as used on the preparative instrument. The three columns used in the Hy-Fi 600-C were (a) 5 ft. × ⅛ in., 20% polymetaphenyl ether (5-ring) on 60–80 mesh Chromosorb W coated with hexamethyl disilazane (HMDS), (b) 10 ft. × ⅛ in., 20% Carbowax 20M

TABLE 7

RETENTION TIMES OF ALKYL SELENIUM COMPOUNDS ON THE POLYMETAPHENYLETHER COLUMN
N_2 carrier gas flow rate: 25 ml/min

	Retention time (min)					
Column temp. (°C)	50	75	100	125	150	175
Injector temp. (°C)	100	180	180	180	225	220
Dimethyl selenide	2.3	1.4	0.8	0.6	0.5	
Diethyl selenide	12.0	5.8	2.4	1.8	1.0	
Dipropyl selenide	35.0	20.0	8.4	4.3	2.2	
Dimethyl diselenide		22.5	10.2	5.2	2.7	2.7
Diethyl diselenide			28.5	12.6	6.0	4.2
Dipropyl diselenide				35.5	12.4	9.0
Ethyl selenocyanate		30.5	11.5	7.0	3.2	2.7

TABLE 8

RETENTION TIMES OF ALKYL SELENIUM COMPOUNDS ON THE CARBOWAX 20M COLUMN
N_2 carrier gas flow rate: 20 ml/min

	Retention time (min)					
Column temp. (°C)	35	45	70	100	120	160
Injector temp. (°C)	100	100	180	180	180	220
Dimethyl selenide	7.0	4.0	2.8	1.7	1.3	0.8
Diethyl selenide		16.0	6.7	3.8	2.5	1.4
Dipropyl selenide			16.2	9.0	5.2	2.3
Dimethyl diselenide			36.5	16.0	8.9	3.8
Diethyl diselenide				32.0	14.7	6.4
Dipropyl diselenide				72.0	30.0	11.0
Ethyl selenocyanate				48.0	19.5	6.5

TABLE 9

RETENTION TIMES OF ALKYL SELENIUM COMPOUNDS ON THE SILICONE OIL DC 550 COLUMN
N$_2$ carrier gas flow rate: 25 ml/min

	Retention time (min)		
Column temp. (°C)	100	125	150
Injector temp. (°C)	180	180	225
Dimethyl selenide	5.2	3.7	2.5
Diethyl selenide	15.5	9.2	5.8
Dipropyl selenide		28.0	14.0
Dimethyl diselenide		28.0	13.0
Diethyl diselenide			29.0
Ethyl selenocyanate		27.5	14.0

on 60–80 mesh HMDS Chromosorb W, and (c) 10 ft. × $\frac{1}{8}$ in., 20% Silicone Oil DC550 on 60–80 mesh Chromosorb W coated with dimethyl dichlorosilane (DMCS). Nitrogen was used as carrier gas at a flow rate between 20 and 30 ml/min.

Tables 7–9 show the retention times of alkyl selenium compounds on the above three columns, respectively, at various column and injector temperatures. The best resolution was achieved on the polymetaphenyl ether column (Table 7), but both Carbowax 20M and Silicone Oil columns gave satisfactory resolution of all the selenium compounds studied (Tables 8 and 9). However, the DC550 column was unsatisfactory for use above 150° as a large amount of continuous column bleeding occurred above this temperature. Figure 3 illustrates the separation of alkyl selenium

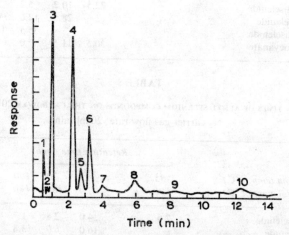

Fig. 3. The separation of alkyl selenium compounds on a polymetaphenyl ether column, with hydrogen flame-ionization detector. Column temperature, 150°; injector temperature, 225°; N$_2$ carrier gas flow rate, 25 ml/min; arrow indicates injection point at time 0 min. 1% solution of each compound in carbon disulfide. 1, Dimethyl selenide; 2, carbon disulfide; 3, diethyl selenide; 4, dipropyl selenide; 5, dimethyl diselenide; 6, ethyl selenocyanate; 7, methylethyl diselenide ?; 8, diethyl diselenide; 9, ethylpropyl diselenide; 10, dipropyl diselenide.

TABLE 10

DETECTOR RESPONSE RATIOS[a]

Compound	Approximate ϕ value
Dimethyl selenide	0.01
Diethyl selenide	0.002
Dipropyl selenide	0.001
Dimethyl diselenide	130.0
Diethyl diselenide	135.0
Dipropyl diselenide	150.0
Ethyl selenocyanate	320.0

[a] Ratio in response given by the electron-capture detector to that of the hydrogen flame detector.

compounds on a polymetaphenyl ether column with hydrogen flame-ionization detector and column and injector temperatures of 150 and 225°, respectively. The detector response ratios for 7 organoselenium compounds are shown in Table 10. The largest response to the monoselenides was obtained with the hydrogen flame-ionization detector whereas the diselenides and ethyl selenocyanate gave the largest responses with the electron-capture detector.

2. TELLURIUM

Tellurium, along with platinum, palladium, and ruthenium, ranks about seventy-first in the order of crustal abundance (average contents in rocks, 0.01 p.p.m.). Tellurium is found chiefly associated with ores such as sylvanite ($(AgAu)Te_2$), black tellurium ($(PbAu)TeS$), hessite (Ag_2Te), and tetraclymite (Bi_2Te_3) as tellurides. Almost all of the commercially produced tellurium is recovered from electrolytic copper refinery slimes. The annual world production of tellurium is about 300,000 lb. of which approx. 85% is produced in the U.S. and Canada (e.g. 199,000 and 72,000 lb., respectively).

The main industrial uses of tellurium include (1) in alloys of lead, copper, and steel for increased resistance to corrosion and stress, machinability, (2) in the rubber industry (as powdered tellurium) as a vulcanizing agent in natural rubber and in styrene–butadiene; and as tellurium diethyldithiocarbamate as an accelerating agent for butyl rubber, (3) in the glass industry as a coloring agent, (4) in cast iron, as a carbide stabilizer for chilling and enhanced surface resistance to wear and corrosion, and (5) as a catalyst in diverse industrial chemical processes.

Industrial poisoning due to tellurium exposure[65-67] and poisoning due to ingestion of sodium tellurite have been reported[68]. Tellurium has been analyzed by gravimetric[69], volumetric[70], polarographic[70], and potentiometric[70] techniques.

The TLC and electrophoretic separation of Te^{IV} and Te^{VI} was described by

Zaitsev *et al.*[71]. Te^{IV} was separated in 5–10 min from Te^{VI} on aluminum oxide plates using (a) 40% aq. sodium hydroxide, (b) 57% nitric acid, or (c) 60% sulfuric acid solutions. R_F values for Te^{IV} and Te^{VI} were, in the respective solvents: (a) 0.65 and 0.00, (b) and (c) 0.0 and 0.95, respectively.

Electrophoretic separation was achieved in a commercial ice-cooled apparatus by using freshly prepared 0.05 N sodium hydroxide as the electrolyte, at a gradient 40 V/cm and with a current of 30 mA; the tellurate(IV) moved 6 cm towards the anode within 10 min, while the tellurate(VI) remained at its initial position. In the elution of Te^{VI} and Te^{IV} from aminoethyl cellulose paper by a 0.05 M Na_2CO_3 solution, the R_F values were 0.15 and 0.65, respectively, and satisfactory separation was achieved within 20 min.

Rai and Kukreja[72] separated tellurium from selenium on a thin layer of silica gel using amyl acetate–conc. hydrochloric acid (100:40) as eluant. Detection was achieved with a stannous chloride spray.

The separation of selenium and tellurium by reversed phase paper chromatography was described by Hu[73]. Tributyl phosphate was used as the stationary phase and HBr or NaBr solutions of various concentrations as the mobile phase. Selenium and tellurium were best separated with 1–2 M HBr or 8 M NaBr and their R_F values decreased with increasing concentration of HBr. While the R_F values of selenium were found to remain practically constant with various concentrations of NaBr, those of tellurium decreased with increasing concentrations of the latter. It was found possible to separate and detect selenium and tellurium at levels of 3–60 μg and 0.5–0.10 μg, respectively, with the estimation being achieved by direct comparison of color intensity with known spots or by conventional colorimetry.

TABLE 11

R_F VALUES OF Se AND Te BY CIRCULAR CHROMATOGRAPHY

Paper: Hsin-Hua No. 1 filter paper. Solvent: Acetone–water–acid (7:2:1). For acid and normality of acid see the table[a].

Acid	Ion	Normality of acid							Temp. (°C)
		1	2	3	4	6	9	12	
H_2SO_4	Se^{IV}	0.75	0.76	0.78	0.78	0.76	0.80	0.78	22
	Te^{IV}	0.08–0.51	0.48	0.50	0.55	0.50	0.60	0.55	
HNO_3	Se^{IV}	0.80	0.79	0.78	0.74	0.73	0.75	0.72	22
	Te^{IV}	0.56	0.58	0.60	0.56	0.55	0.58	0.53	
HCl	Se^{IV}	0.76	0.75	0.80	0.80	0.78	1.00	1.00	21
	Te^{IV}	0.46	0.48	0.62	0.67	0.66	0.86	0.94	
HBr	Se^{IV}	0.74	0.80	0.89	1.00	1.00			25
	Te^{IV}	0.46	0.62	0.73	0.90	0.92			
HI	Se^{IV}	0.09	0.03	0.0	0.0	0.0[a]			25
	Te^{IV}	0.93	1.00	1.00	1.00	1.00[a]			

5 N HI.

The separation of Se^{IV} and Te^{IV} by circular paper chromatography using sulfuric, nitric, hydrochloric, hydrobromic, and hydriodic acid of varying normalities was described by Chih-Hsiang[74] (Table 11).

REFERENCES

1   D. M. HADJIMARKOS, C. W. BONHURST AND J. J. MATTICE, *J. Pediat.*, 54 (1959) 296.
2   Y. HASHIMOTO, J. Y. HWANG AND S. YANAGISAWA, *Environ. Sci. Technol.*, 4 (1970) 157.
3   O. E. OLSON AND D. V. FROST, *Environ. Sci. Technol.*, 4 (1970) 686.
4   O. H. MUTH, *Selenium in Biomedicine, 1st Intern. Symp.*, Avi, Westport, Conn., 1967.
5   V. W. OELSCHLAGER AND K. H. MENKE, *Ernährungswissenschaft*, 9 (1969) 216.
6   A. A. NELSON, O. G. FITZHUGH AND H. O. CALVERY, *Cancer Res.*, 3 (1943) 320.
7   O. G. FITZHUGH, A. A. NELSON AND C. I. BLISS, *J. Pharmacol.*, 80 (1944) 289.
8   B. B. WESTFALL, E. F. STOHLMAN AND M. I. SMITH, *J. Pharmacol. Exp. Therap.*, 64 (1938) 55.
9   I. ROSENFELD AND O. A. BEATH, *Proc. Soc. Exptl. Biol. Med.*, 37 (1954) 295.
10 R. C. WAHLSTROM AND O. E. OLSON, *J. Animal Sci.*, 18 (1954) 141.
11 D. S. F. ROBERTSON, *Lancet*, i (1970) 518.
12 D. M. HADJIMARKOS, *Lancet*, i (1970) 721.
13 D. J. CLEGG, *Food Cosmet. Toxicol.*, 9 (1971) 195.
14 R. E. HOLMBERG AND B. H. FERM, *Arch. Environ. Health*, 8 (1969) 873.
15 R. F. BUCHAN, *Occupational Med.*, 3 (1947) 439.
16 K. HALTER, *Arch. Derm. Syph.*, 178 (1939) 340.
17 A. HAMILTON, *Industrial Poisons in the U.S.*, Macmillan, New York, 1925.
18 P. W. WEST AND C. CIMERMAN, *Anal. Chem.*, 36 (1964) 2013.
19 M. KAWAMURA AND K. MATSUMOTO, *Japan Analyst*, 14 (1965) 789.
20 B. B. MESMAN AND H. A. DOPPELMAYR, *Anal. Chem.*, 43 (1971) 1346.
21 Y. HASHIMOTO AND J. W. WINCHESTER, *Environ. Sci. Technol.*, 1 (1967) 338.
22 A. GUTBIER AND G. METZNER, *Glastech. Ber.*, 12 (1934) 117.
23 C. TOMICEK, *Collection Czech. Chem. Commun.*, 11 (1939) 449.
24 A. RUDRA AND S. RUDRA, *Current Sci. India*, 21 (1952) 229.
25 T. E. GREEN AND M. TURLEY, in I. M. KOLTHOFF AND P. J. ELVING (Eds.), *Treatise on Analytical Chemistry*, Vol. 7, Part 2, Interscience, New York, 1961.
26 H. J. M. BOWEN AND P. A. LAWSE, *Analyst*, 88 (1963) 721.
27 J. P. F. LAMBERT, O. LEVANDER, L. ARGRETT AND R. E. SIMPSON, *J. Assoc. Offic. Anal. Chemists*, 52 (1969) 915.
28 E. STEINNES, *Intern. J. Appl. Radiation Isotopes*, 18 (1967) 731.
29 R. C. EWAN, C. A. BAUMANN AND A. L. POPE, *J. Agr. Food Chem.*, 16 (1968) 212.
30 J. H. WATKINSON, *Anal. Chem.*, 38 (1966) 93.
31 J. B. WILKIE AND M. YOUNG, *J. Agr. Food Chem.*, 18 (1970) 944.
32 F. H. BURSTAL, G. R. DAVIES, R. P. LINSTEAD AND R. A. WELLS, *J. Chem. Soc.*, (1950) 516.
33 C. BIGITI AND I. MANTOVANI, *Boll. Sci. Fac. Chim. Ind. Bologna*, 13 (1955) 102.
34 E. G. WEATHERLEY, *Analyst*, 81 (1956) 404.
35 M. LEDERER, *Chem. Ind. (London)*, (1954) 1481.
36 A. MURATA, *Nippon Kagaku Zasshi*, 78 (1957) 541.
37 F. VESEZY, F. SMIROUS AND J. V. SISKA, *Chem. Listy*, 49 (1955) 1661.
38 K. N. JOHRI, N. K. KAUSHIK AND K. SINGH, *Mikrochim. Acta*, 4 (1969) 737.
39 D. S. GAIBAKYAN AND M. M. ATURYAN, *Arm. Khim. Zh.*, 21 (1968) 1015; *Chem. Abstr.*, 71 (1969) 27109.
40 H. WEISZ, *Micro Analysis by Ring Oven Technique*, Pergamon Press, 2nd edn., London, 1970.
41 H. C. MEHRA, I. P. MITTAL AND K. N. JOHRI, *Chromatographia*, 4 (1971) 532.
42 Y. SASAKI, *Bull. Chem. Soc. Japan*, 28 (1955) 89.
43 M. YAMAMOTO AND H. SAKAI, *Anal. Chim. Acta*, 32 (1965) 370.
44 K. L. CHENG, *Anal. Chem.*, 28 (1956) 1738.

45 H. FLASCHKA AND H. ABDINE, *Chemist-Analyst*, 45 (1956) 58.
46 P. J. PETERSON AND G. W. BUTLER, *J. Chromatog.*, 8 (1962) 70.
47 C. S. KNIGHT, *Nature*, 184 (1960) 1486.
48 R. MARKHAM AND J. D. SMITH, *Biochem. J.*, 52 (1952) 552.
49 E. F. WELLINGTON, *Can. J. Chem.*, 30 (1952) 581.
50 P. J. PETERSON AND G. W. BUTLER, *Australian J. Biol. Sci.*, 15 (1962) 126.
51 K. P. MCCONNELL AND C. H. WABNITZ, *J. Biol. Chem.*, 226 (1957) 765.
52 K. P. MCCONNELL AND A. E. KREAMER, *Proc. Soc. Exptl. Biol. Med.*, 105 (1960) 170.
53 J. SCALA AND H. H. WILLIAMS, *J. Chromatog.*, 15 (1964) 546.
54 K. R. MILLAR, *J. Chromatog.*, 21 (1965) 344.
55 K. P. MCCONNELL AND C. H. WABNITZ, *Biochim. Biophys. Acta*, 86 (1964) 182.
56 S. MOORE, D. H. SPACKMAN AND W. STEIN, *Anal. Chem.*, 30 (1958) 1185.
57 A. J. BARAK AND S. C. SWANBERG, *J. Chromatog.*, 31 (1967) 284.
58 J. L. BYARD, *Arch. Biochem. Biophys.*, 130 (1969) 556.
59 I. S. PALMER, D. D. FISCHER, A. W. HALVERSON AND O. E. OLSON, *Biochim. Biophys. Acta*, 177 (1969) 336.
60 K. P. MCCONNELL AND O. W. PORTMAN, *J. Biol. Chem.*, 195 (1952) 277.
61 I. S. PALMER, R. P. GUNSALUS, A. W. HALVERSON AND O. E. OLSON, *Biochim. Biophys. Acta*, 208 (1970) 260.
62 K. RANDERATH, *Thin-Layer Chromatography*, Academic Press, New York, 1964, p. 129.
63 I. ROSENFELD AND S. A. BEATH, *Selenium*, Academic Press, New York, 1964, p. 171.
64 C. S. EVANS AND C. M. JOHNSON, *J. Chromatog.*, 21 (1966) 202.
65 M. P. SHIE AND F. E. DEEDS, *U.S. Publ. Health Serv. Rept.*, 35 (1920) 939.
66 H. H. STEINBERG, S. C. MASSARI, A. C. MINER AND R. RINC, *J. Ind. Hyg.*, 24 (1942) 183.
67 M. L. AMDUR, *Occupational Med.*, 3 (1947) 386.
68 J. H. H. KEALL, N. H. MARTIN AND R. E. TUNBRIDGE, *Brit. J. Ind. Med.*, 3 (1946) 175.
69 C. DUVAL AND U. M. DOAN, *Anal. Chim. Acta*, 5 (1951) 569.
70 P. W. BENNETT AND S. BARABAS, *Anal. Chem.*, 35 (1963) 139.
71 V. M. ZAITSEV, V. S. GUSEL'NIKOV AND S. M. MAKHOMALKINA, *Radiokhimiya*, 11 (1969) 352.
72 J. RAI AND U. P. KUKREJA, *Chromatographia*, 2 (1969) 18.
73 C. T. HU, *Hua Hsueh Hsueh Pao*, 30 (1964) 426; *Chem. Abstr.*, 61 (1964) 15324h.
74 M. CHIH-HSIANG, *Acta Chim. Sinica*, 31 (1965) 435.

Chapter 7

INORGANIC AND ORGANIC MERCURY

1. INORGANIC MERCURY

Mercury is a comparatively rare element, ranking 16th from the bottom of elements in abundance in the earth. Sufficient concentration of mercury for commercial extraction is found as mercuric sulfide (frequently as the red cinnabar and less often as the black meta cinnabar). Important deposits are located in Spain, Italy, U.S.A., Canada, Mexico, Brazil, Peru, China, Japan, Russia, Hungary, Yugoslavia, and Germany. A less common ore is the mercurous chloride found in Texas. The ore found in Spain has the highest mercury content with an average of 0.5–1.2% mercury with values occasionally as high as 10% mercury.

It has been estimated that the outer 16 km thick layer of the earth's crust contains $2.7 \times 10^{-6}\%$ mercury and the oceans contains $3.0 \times 10^{-9}\%$ or approx. 50 million metric tons of mercury. In soils, it generally ranges from 0.01 to 0.06 p.p.m., in rocks from 0.01 to 0.09 p.p.m., in air from 2 to 5 ng/m³, and in water and beverage from 0.002 to 0.006 p.p.m.

Mercury has been known and utilized since antiquity. Today there are over 3000 recognized applications for mercury and its inorganic and organic derivatives. The world production of mercury amounts to about 10,000 tons per year of which about 3000 tons are used in the U.S. The main areas of utility of mercury include the electrolytic preparation of chlorine and caustic soda, agricultural chemicals, pharmaceuticals (diuretics, cathartics, antibacterial agents), hair dressings and preservatives in cosmetics, pulp and paper making (slimicides and algicides), paint (anti-fouling, pigment), electrical apparatus, catalyst, dental preparations, and amalgamations. Production of chlorine and caustic soda is an electrolytic process where large amounts of mercury are used as a flowing cathode (from 75,000 to 150,000 lb. for a plant with a capacity of 100 tons of chlorine a day)[1]. It is estimated that the chlor-alkali industry loses to the environment approx. 0.45 lb. of mercury per ton of chlorine produced[2]. This loss, based on projected tonnage figures, may be as much as 3300 lb. per day or 1,200,000 lb. per year.

Most of the lost mercury finds its way into streams and lakes; traces of mercury are also carried into the atmosphere with hydrogen gas (20–30 mg/m³) while usually less than 5 p.p.m. of mercury is retained by the caustic soda. Mercury has also been detected in chlorinated hydrocarbons, glycols, acetic acid, carbon dioxide, fertilizers, sulfuric acid, sulfide ores, industrial catalyst wastes (*e.g.* in production of acetaldehyde and vinyl chloride), and in bituminous shales and crude oils. Fossil fuels in the U.S. contain mercury in concentrations ranging from a few parts per billion to several parts per million. The annual consumption of 500 million tons of coal (containing an average concentration of at least 1 p.p.m.) would contribute one million pounds of mercury to the environment, about 450 metric tons[3]. This value does not include the mercury contributed from the refinement and use of products from crude oil which may contain higher concentrations of mercury.

The key inorganic mercury compounds used in agriculture (as insecticides, seed disinfectants, and fungicides) and home, garden, and lawn use are mercuric acetate, chloride, and oxide (red and yellow) and mercurous chloride and nitrate. Elemental mercury is one of the most toxic elements used in the control of crop pests. Quicksilver (metallic mercury) has itself been employed in India to fumigate grain in closed containers.

There is a large variety of miscellaneous sources that present possible mercurial contamination to the environment and including (a) disposal of thermometers, aerometers, barometers, relays, rectifiers, switches, fluorescent tubes, mercury lamps, and batteries, (b) refuse from hospitals, laboratories, and dental clinics, (c) processing or use of raw materials containing mercury such as carbon, coal, chalk, phosphate, and pyrite, (d) use of mercurial compounds to prevent mildewing in commercial laundries, (e) manufacture of and residues from paints and impregnating agents which contain mercury to impart mildew resistance, and (f) refining or redistillation of mercury.

Normal human tissues contain mercury because there is a daily intake in the food. Bread, flour, milk, pork, and beef contain $2-4\,\mu g\%$ mercury and certain

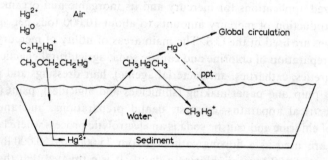

Fig. 1. Conversion of mercury in the environment[7].

vegetables a good deal more, depending on soil conditions and sprays used. Ingestion of fish and fowl with high levels of mercury (as methylmercury) *via* environmental fallout from sources described above introduces yet another dimension to the ever-increasing realization that compounds of mercury present a substantial hazard.

Of major significance has been the recent realization that certain biotransformations[4-7] may take place in the environment (Fig. 1) and in the body, changing

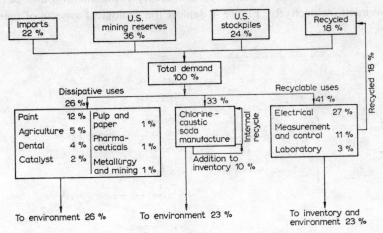

Fig. 2. Dissipative and recyclable uses of mercury in U.S. 1968 (ref. 8).

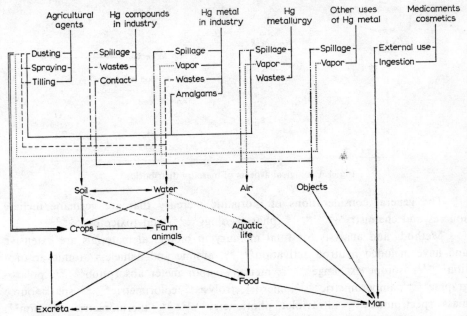

Fig. 3. Principal pathways of mercury contamination and environmental movement.

one form of mercury to another, *e.g.* (*a*) the changing of inorganic mercury to methyl-mercury by microorganisms when oxygen is limited or absent, (*b*) the release of inorganic mercury from phenylmercuric compounds with subsequent conversion to methylmercury, and (*c*) the conversion of phenylmercuric compounds to inorganic mercury in the body.

Figures 2 and 3 illustrate the dissipative and recyclable uses of mercury in the U.S. for 1968 and the principal pathways of mercury contamination and environmental movement, respectively. Figure 4 depicts the ecological aspects of mercury distribution.

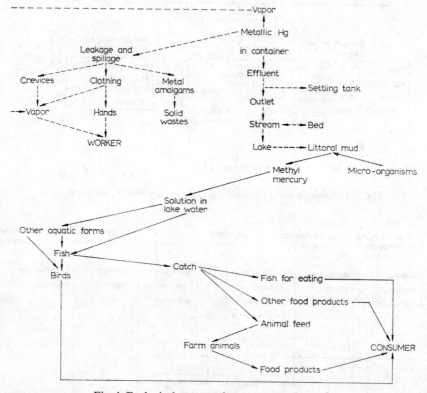

Fig. 4. Ecological aspects of mercury distribution[9].

The general considerations of inorganic mercury that are germane include sources and chemistry[1,3,8,10-12], pharmacology[13,14], and toxicology[15-22].

Methods and analysis for total mercury in biological materials are extensive and have included neutron activation[23-26], atomic and flameless atomic absorption[27-32], isotope exchange[33-35], mercury vapor meter absorption[36-41], polarographic[42,43], amperometric[44,45], microelectrolysis[46], colorimetric[47-50], spark source mass spectrometry[51], ultraviolet photometry[36,52-54], thermometric titration[55], and atomic fluorescence[56].

It is important to distinguish between different forms of mercury (*e.g.* elemental, ionic, alkyl, and aryl) in any evaluation of the potential hazards of mercury residues as well as the development of requisite analytical methodology for their analysis.

Inhaled mercury vapor (in many cases a mixture of vapor and aerosols) is mainly an industrial hazard and there is evidence that the vapor is absorbed through the lungs[57]. The process of absorption involves oxidation of elemental mercury with the subsequent appearance of ionic mercury in the blood, then distribution in the body in analogous manner to injected mercuric chloride, with the kidney as the primary site of deposition. The oxidation of elemental mercury occurs partly in the blood (mainly in the erythrocytes) and partly in the tissues[58]. The higher uptake of mercury in brain following exposure of animals to vapor compared with injections of mercuric nitrate has also been reported[59,60]. Diffusion of elemental mercury into the tissues and across cell membranes is facilitated by its lipid solubility and its lack of charge. Mercuric mercury in the plasma is bound to plasma proteins and in the erythrocytes to hemoglobin.

Although there is some evidence that mercuric mercury can be reduced to elemental mercury in the body[61], the extent of this reduction is not considered of practical importance. Mercurous mercury is probably oxidized to mercuric mercury in the body. Following acute administration of inorganic salts of mercury to man and animals, the highest level of mercury is found in the kidneys and the next highest in the liver. Elimination of mercury from brain, thyroid, and testes is slow[62], permitting accumulation of mercury in these organs. With chronic exposure to mercuric salts, it is uncertain whether mercury levels in brain or testes reach toxic concentrations before the onset of severe renal damage.

Although inorganic mercury salts are more acutely toxic than organic mercury salts, chronic oral studies[63] have shown, for example, that phenylmercury acetate was more toxic to the rat than mercuric acetate. Inorganic mercury salts are poorly absorbed, *e.g.* the absorption has been estimated to be approx. 2% while for methylmercury the available data (man and rat) indicate an intestinal absorption of more than 90%.

Insoluble inorganic mercurous compounds such as calomel (Hg_2Cl_2) can undergo oxidation to more soluble, absorbable compounds. Inorganic mercurials, in suitable vehicles, can be absorbed through the intact skin on passage into the blood; they become firmly bound to plasma proteins and erythrocytes. Mercury readily redistributes to the tissues and within several hours is found in human and animal tissues in the following approximate order of decreasing concentration: kidney, liver, blood, bone marrow, spleen, upper respiratory and buccal mucosa, intestinal wall, skin, salivary glands, heart, skeletal muscle, brain, and lung. Mercury is excreted by the kidney, by the liver *via* bile, by the intestinal mucosa, sweat glands, and salivary glands; urinary and fecal routes are the most important for elimination.

Biotransformation has been shown[64-66] to be important for the excretion of mercury after exposure to methylmercury chloride. The carbon–mercury bond may slowly undergo cleavage *in vivo*, resulting in the release of inorganic mercury and its preferential excretion *via* the gastrointestinal tract (as distinct from the kidney where a relatively higher amount of the intact organomercurial was found in urine). Biotransformation within the intestinal lumen in the cecum was shown to be the major source of inorganic mercury in feces. Clarkson[33] has suggested that the nephrotoxic action of organomercurials may be related to the rate of biotransformation with enzyme systems participating in the breakdown of organomercurials.

A variety of chromatographic techniques has been described for the separation and detection of inorganic mercury.

The utility of paper chromatography and hydrazine sulfate and hydroxylamine hydrochloride as specific reagents for the separation and detection of mercury and arsenic in toxicological analysis was described by Romanowski *et al.*[67]. Mercury and silver were separated using ascending paper chromatography in two solvent systems, (*a*) butyl acetate and (*b*) acetone–conc. hydrochloric acid–isobutanol–glacial acetic acid–water (58:1:20:6.4:14.6); the R_F values were 0.86 and 0.40 for mercury and silver, respectively. The paper was first sprayed with a 5% alcoholic solution of either hydrazine sulfate or hydroxylamine hydrochloride and then heated. Mercury produced a dark brown spot and silver one of yellow-brown. The presence of other metals did not interfere and the elements could be successfully detected in the presence of biological or mineral materials. The limits of detection of both mercury and silver were 0.1γ and 0.3γ with hydroxylamine hydrochloride and hydrazine sulfate, respectively.

The adsorption and development of nineteen mercuric compounds were investigated on Amberlite SB-2 anion exchange paper by using 1.6 N nitric acid as the developing solvent[68]. The R_F values were 0.19–0.33 for alkylmercury compounds; 0.09–0.11 for inorganic salts, and 0.01–0.03 for phenylmercury compounds, hence permitting the differentiation of organic and inorganic mercury compounds.

Anion-exchange paper X-ray emission procedure for the determination of microgram quantities of mercury has been described by Link *et al.*[69]. The procedure permitted the determination of 1γ of Hg in 100 ml of aqueous acid solution with a precision of $\pm 0.25\gamma$. The mercury is first absorbed by anion resin-loaded paper and determined by X-ray emission spectroscopy. The method which was applied to the determination of mercury in inorganic pigments used in foods and drugs gave 75–125% recoveries of 1γ of mercury from acid solutions containing 10 g of Na_2SO_4, NaCl, Fe oxide, $MgCO_3$, and $CaCO_3$ and satisfactory recoveries from the HCl extract of C, $BaSO_4$, chromic oxide, bentanite, kaolin, talc, TiO_2, and Mg stearate.

The paper chromatographic separation of mercury, lead, copper, bismuth, and cadmium cations has been accomplished using Whatman No. 4 paper and *n*-butanol–acetic acid–12 N hydrochloric acid–water and *n*-butanol–hydrochloric acid as developers[70]. Detection was accomplished with dithizone or hydrogen sulfide. Table 1 shows the effect of concentration of hydrochloric acid on the R_F values of the metal cations. (R_F values increase with increasing acid concentrations.) Table 2 depicts the R_F values and spot colors of Hg, Pb, Bi, Cu, and Cd cations developed with n-butanol–acetic acid–12 N hydrochloric acid–water (45:10:1:44).

TABLE 1

R_F VALUES OF Hg, Pb, Bi, Cu, AND Cd CATIONS DEVELOPED WITH n-BUTANOL–HYDROCHLORIC ACID (1:1) (WITH VARYING ACID CONCENTRATIONS)

	R_F Value		
	1 N HCl	2 N HCl	3 N HCl
Mercury	0.75	0.76	0.80
Lead	0.11	0.16	0.28
Bismuth	0.52	0.54	0.58
Copper	0.08	0.11	0.20
Cadmium	0.63	0.67	0.76

TABLE 2

R_F VALUES AND SPOT COLORS OF Hg, Pb, Bi, Cu, AND Cd CATIONS DEVELOPED WITH n-BUTANOL–ACETIC ACID–12 N HYDROCHLORIC ACID–WATER (45:10:1:44)

		Spot colors	
	R_F	H_2S	Dithizone
Mercury	0.78	Black	Rose
Lead	0.052		Rose
Bismuth	0.40	Brown	Violet
Copper	0.13	Brown	Brown
Cadmium	0.20	Lemon-green	Rose-purple

An effective separation of Ag^+, Tl^+, Pb^{2+}, and Hg^{2+} was achieved[71] in approx. 3h at $20 \pm 1°$ by ascending chromatography on Whatman No. 1 paper using a solvent mixture of methanol–11 N nitric acid (60:40). The R_F values were Ag^+ 0.29, Tl^+ 0.39, Pb^{2+} 0.51, and Hg^{2+} 0.75. The spots were detected by dipping the chromatogram in a solution of ammonium sulfide; the lower limit of detection was 1×10^{-5} g. The separation of silver, mercurous, and lead ions by circular paper chromatography was reported by Murthy et al.[72] in a 2 h (9 cm) development with ammoniacal hydrogen sulfide used for detection[73]. The R_F values found were Ag^+ 0.86, Hg^+ 0.62, and Pb^{2+} 0.34.

Blasius and Göttling[74] described the circular paper chromatography of a variety of cations. The following R_F values were found following a 3 h development on Schleicher & Schüll No. 2045 paper with 10% aqueous ammonium acetate:

Pb^{2+} 0.95, Ag^+ 0.80, Tl^+ 0.75, and Hg^+ 0.65. Detection was accomplished with 5% ammonium sulfide (e.g. Tl and Pb, brown and Ag and Hg, green).

The separation and collection of traces of metals on paper impregnated with a few milligrams of a water-insoluble organic reagent was described[75]. The reagent paper was prepared by dipping filter paper, 7 cm in diameter, in 1% dithizone in chloroform or in 0.05% (p-dimethylaminobenzylidene) rhodamine in acetone followed by 0.1 N nitric acid. Approximately 98% of Au, Ag, and Hg in amounts of less than 1–5 μg could be recovered after ashing from 10–250 ml of 0.1–0.5 N mineral acid solution by a single or by repeated filtrations at a rate of 6 ml/min. The method was found applicable for the rapid separation of several p.p.m. of Au, Ag, and Hg in Cu and Pb where the concentration factor of Ag with respect to Cu was approx. 2000. The reagent paper could be stored for more than 80 days without deterioration.

The utility of paper precipitation chromatography for the separation of metal ions was described by Nagai[76]. Hg_2^{2+}, Ag^+, and Pb^+ were separated as their chlorides on filter paper impregnated with 10% sodium chloride with aq. 4 M ammonium acetate and ammonium hydroxide used as developing agents. Pb^{2+} moved with the developing front and was identified with sodium rhodizonate. The Ag^+ band which moved next to Pb^{2+} was identified by a potassium iodide reaction. Hg_2^{2+} was identified by the appearance of a black spot in the center of the chromatogram. Hg^{2+} and Bi^{3+} were separated on potassium iodide impregnated filter paper using 3 M propionic acid as developer.

Sugawara[77] described the separation of a number of heavy metal ions of the Cu group (Hg^{2+}, Bi^{3+}, Cd^{2+}, and Pb^{2+}) and Sn group (As^{3+}, Sb^{3+}) on ion-exchange (Amberlite SA-2 Na^+ type) paper. The groups were developed with 1 M NH_4Cl and 1 N HCl, respectively, and detected with ammonium sulfide plus 4-(2-pyridyl-azo)-resorcinol (for the Cu group) and ammonium sulfide plus palladium chloride (for the Sn group), respectively. The R_F values were Hg^{2+} 0.04, Bi^{3+} 0.11, Cd^{2+} 0.52, Cu^{2+} 0.27, Pb^{2+} 0.47, As^{3+} 0.84, and Sb^{3+} 0.26. The separation of Tl^I, Hg^I, Ag^I, and Cs^I on ammonium molybdophosphate (AMP)-impregnated papers was described by Alberti[78].

Tanaka[79] described a radioisotope dilution method for the determination of inorganic and organic mercury by paper chromatography. Trace amounts of mercuric chloride and phenyl mercuric chloride were first determined chelatometrically with penicillamine. The chelate was satisfactorily separated by paper chromatography from unreacted mercury and the radioactivity of the chelate spot on the paper was measured with a well-type scintillation counter. A known amount of ^{203}Hg-labeled standard solution and sub-stoichiometric amount of penicillamine was then added to an unknown sample solution, the pH adjusted to 6–10 and the mercury chelate separated by paper chromatography on Toyo Filter Paper No. 51A, 2 × 40 cm, with isopropanol–conc. ammonium hydroxide–water (7:1:2) as developer. The amount of mercury in the unknown is treated in an analogous manner and is calculated from the activities measured. Mercury was detected down to 1 μg when $10^{-6} M$ of penicillamine and 10^{-6} M of ^{203}Hg-labeled mercuric chloride or phenylmercuric chloride were used. Ni, Fe^{II}, Cd, Zn, and Co did not interfere with the determination.

Kurayuki and Kusamato[80] described the chromatographic separation of inorganic and organic mercury compounds on anion-exchange paper Amberlite 5B-2 (R-Cl) using 1.6 N nitric acid and 2 M ammonium nitrate as developers and dithizone in chloroform for detection. The R_F values (using 1.6 N HNO$_3$) were inorganic mercury compounds 0.09–0.11, alkyl mercurials 0.19–0.33, and phenyl mercurials 0.01–0.03. The limits of detection were 0.2 μg. In an additional study on the determination of inorganic mercury compounds by measuring the area of the colored zone on strips of Amberlite 5B-2 (R-Cl), the amount of mercury (1–4 mg) was found to be proportional to the colored area at pH 5.0.

Westermark et al.[81] described a concentration electrophoretic method for the separation of charged mercury compounds in aqueous solution.

In electrophoresis, the microcomponents concentrate in very narrow bands on the thin-layer strip; the postulated diffusion forces are counteracted or even balanced by the electrical field strength. This phenomenon was applied to Hg^{2+} and CH$_3$Hg$^+$ in an apparatus consisting of an aqueous mercury containing insert, surrounded by concentration zones and buffer vessels, by using histidine and EDTA ions. The latter move from the cathode and form a negative complex with Hg^{2+}. This complex moves from the inert zone towards the anode, being partly concentrated there. Under suitable conditions CH$_3$Hg$^+$ migrates to the cathode and concentrates behind the insert. At present the method of Westermark and co-workers is limited to 20 μl, but this can apparently be increased. An advantage of this technique compared with gas chromatography is that inorganic mercury and alkylmercury can be measured in the same operation.

The paper electrophoresis of Hg^{II} and 20 other cations was described by Catoni[82]. The electrolytes used were (a) 25 ml 4 N lactic acid and 25 ml 2 N Na$_2$CO$_3$ made up to 1 l, (b) 40.02 ml of 0.2 N HCl and 250 ml of 0.2 M potassium hydrogen phosphate made up to 1 l, (c) 468 ml of 0.1 M sodium citrate and 532 ml of 0.1 N HCl.

Seiler and Seiler[83,84] separated the copper family (Cu^{2+}, Cd^{2+}, Bi^{3+}, Pb^{2+}, and Hg^{2+}) on MN-Silica Gel S-HR using n-butanol–1.5 N HCl–2,5-hexanedione (110:20:0.5) as developer and 2% potassium iodide, ammonia, and hydrogen sulfide reagents for visualization. The addition of the weak complexing agent, hexanedione, reduces tailing. The R_F values increase in the order of Cu^{2+} < Pb^{2+} < Cd^{2+} < Bi^{3+} < Hg^{2+} (Fig. 5).

Cadmium(II), lead(II), and mercury(II) were separated and identified on cellulose MN-300 plates developed with tert.-butanol–acetone–water–6 N HNO$_3$–acetylacetone (4:4:1.1:45:0.45) and detected with 2% alcoholic oxine followed by exposure to ammonia vapors and ultraviolet inspection. Arsenic(III), tin(II),

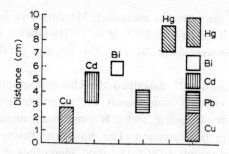

Fig. 5. Separation of the copper group on MN-silica gel developed with n-butanol–1.5 N hydro-chloric acid–2,5-hexanedione (100:20:0.5).

and antimony(III) were separated on silica gel plates with n-butanol–1 M tartaric acid-3 N HCl (10:1:1) as developer using the same detection technique described above. Nickel, cobalt, manganese, and iron ionic mixtures were similarly separated on cellulose MN-300 plates using a hexane–acetone–conc. HCl (8:1:1) solvent system and the identical detection as above.

The TLC of a number of metal dithizonates was described by Takitani et al.[85]. Pd, Hg, Cd, and Ag are extracted as dithizonates with 0.1% dithizone in chloroform at pH 0–1 followed by Bi, Cd, Co, Ni, Pb, and Zn at pH 7–8 and Tl, Fe^{2+}, Mn, and Sn^{2+} at pH 9–10. The 1st and 2nd group metals are separated on a MN-Silica-SHR layer with carbon tetrachloride–methylene chloride–benzene (1:7:4) and identified by color. The 3rd group metals are re-extracted with 20% nitric acid and separated with acetone–hydrochloric acid (99:1) and detected by spraying with appropriate reagents. Identifications limits in p.p.m. are 0.2 for Cu, Ni, Co, 0.3 for Hg, 0.4 for Pd, Ag, Cd, 0.8 for Zn, 3 for Pb, and 5 for Bi.

The utility of TLC in the forensic analysis of metals was described by Kunzi et al.[86]. Silica Gel G (Merck) was used with acetone–benzene (75:25) saturated

TABLE 3

SPOT COLORS OF METALS WITH DITHIZONE AND $(NH_4)_2S$

Metal	Dithizone		$(NH_4)_2S$
	Acid	Ammoniacal	
Copper	Yellow-green	Yellow-brown	Brown
Arsenic	Yellow		Yellow
Cadmium	Lilac	Orange	Yellow
Antimony	Red	Bright brown	Orange
Mercury	Rose	Orange-red	Black
Thallium		Rose	Black
Lead		Rose	Brown
Bismuth	Purple	Orange-red	Brown

with tartaric acid and 6% nitric acid to effect the separation of the metals in the order of R_F: Hg > Bi > Sb > Cd > As > Pb > Cu > Tl.

Both dithizone and ammonium sulfide reagents were used for detection (limits of detection with both reagents was 2 µg). Table 3 illustrates the spot colors of Cu, As, Cd, Sb, Hg, Tl, Pb, and Bi obtained with the above reagents.

Seiler[87] separated Hg^I and Hg^{II} from a cation mixture by TLC on silica gel using n-butanol–4 N nitric acid (80:20) as developing solvent. After separation, mercury was determined by neutron activation analysis by γ-spectrometry of 130 keV ^{197}Hg^m. One µg of mercury was determined in the presence of 150 µg of cation mixture with a precision of ±4%.

Koss and Beisenherz[88] described the determination of heavy metal ions as radioactive sulfides following their initial separation on Silica Gel G-coated slides thence treatment with H$_2{}^{35}$S. Figure 6 illustrates the separation of cadmium, copper, and mercury salts on a Silica Gel G-coated microscope slide developed with n-butanol–1.5 N hydrochloric acid–acetonylacetone (100:20:0.5).

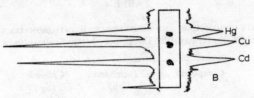

Fig. 6. Separation of cadmium, copper, and mercury salts on a coated microscope slide. Layer, Silica Gel G; solvent, n-butanol–1.5 N HCl–acetonylacetone (100:20:0.5); time of run, 2 h; detection with H$_2{}^{35}$S. The chromatogram was evaluated by means of a Geiger–Müller counting tube (B) and the autoradiograph evaluated photometrically (A).

The separation of the bivalent cations of mercury, copper, cadmium, nickel, and zinc by circular chromatography with dithizone-impregnated silica gel thin-layers was described by Deguchi[89]. The plates were prepared by adding 30 ml of 0.3% dithizone–chloroform solution to 10 g of silica gel. The cations were thus separated as metal chelates yielding colored chromatograms of concentric circles. The best separations of Hg^{II}, Cu^{II}, Cd^{II}, Ni^{II}, and Zn^{II} were obtained by using either 2 N acetic acid–acetone (1:1) or 2–4 N acetic acid–hydrochloric acid (1:0.66) as developers. The cations were developed in the above order, Zn being in the outermost circle.

The separation of a number of metal diethyldithiocarbamates on Silica Gel D plates was described by Senf[90] using a solvent system containing n-hexane–chloroform–diethylamine (20:2:1). After the plate was heated, in order to remove the diethylamine, the spots were developed by spraying with 5% copper sulfate and 0.1γ of each metal was detectable. Interfering metals such as Co, Ni, Fe, and Mn could be removed by controlling the pH range with citrate–phosphate buffer. The R_F values of the metal diethyldithiocarbamates were: Hg, 0.56; Pb, 0.00; Cu, 0.44; Bi, 0.27; and Cd, 0.34.

Takeshita et al.[91] described a reversed-phase TLC method for the separation and detection of the dithizonates of both inorganic mercury and a series of alkylmercury compounds and found the method applicable to the separation and identification of mercury compounds in foods and sewages. The absorbents used were cornstarch containing liquid paraffin and Avicel SF (FMC, American Viscose Div.) containing 20% liquid paraffin. The solvent system consisted of methyl cellosolve–water (75:25) and ethanol and methyl cellosolve treated, respectively, with water in the ratios from 0 to 30%. The study revealed that the best pattern of separation of all the dithizonates was obtained on liquid paraffin–Avicel SF layers with methyl cellosolve–water (7:3), although the development time was longer than with impregnated cornstarch plates (1–1.5 h). No interference of other metals and the residual dithizone was observed. Table 4 lists the detection limits of mercury and alkylmercury compounds (as dithizonates and chlorides) on Avicel SF–liquid paraffin layers with methyl cellosolve–water (75:25) as developer.

TABLE 4

DETECTION LIMITS OF MERCURIC AND ALKYLMERCURIC DITHIZONATES AND CHLORIDES
Layer: Avicel SF–liquid paraffin. Developer: Methyl cellosolve–water (75:25)

Compound	Dithizonate (μg)	Chloride (μg)
Mercury	0.01	0.004
Methylmercury	0.1	0.053
Ethylmercury	0.1	0.055
n-Propylmercury	0.1	0.056
n-Butylmercury	0.1	0.057
n-Amylmercury	0.05	0.029
n-Octylmercury	0.01	0.006
Stearylmercury	0.01	0.007

Tadmor[92] described the gas chromatographic separation of a number of metal halides, e.g. (a) $SnCl_4$, $SnBr_4$, SnI_4, (b) $AsCl_3$ and $GeCl_4$, and (c) $FeCl_3$ and $HgCl_2$ utilizing a laboratory-constructed apparatus. The apparatus (Fig. 7) was comprised of a chromatographic column preceded by a preheater section for the carrier gas. These consist of U-shaped Pyrex glass tubes packed with different stationary phases; the apparatus is placed in a cylindrical electrical furnace (Adamel type T5HT) whose ports are closed with asbestos insulating material in order to thermostat the apparatus. Supplementary heating is provided for the injection port and the end of the column by a heating wire connected to a resistance. The solid stationary phase was Sil-O-Cel brick, −30 to +50 mesh, washed with 2.5 N HNO_3, rinsed with water and dried at 110°. Some of the liquid stationary phases were: n-butanol, n-decane, glycol, Kel-F-oil, nitrobenzene, glycerol, $AlBr_3$, Silicone oil-550, $BiCl_3$, and Woods metal. Samples were labeled with radioactive isotopes, which were

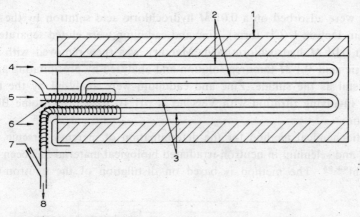

Fig. 7. Home-made gas chromatograph. 1, Electrical furnace; 2, gas preheater; 3, chromatographic olumn; 4, carrier gas; 5, injection port; 6, heating wire; 7, effluent solvent stream; 8, to radioctivity detector and recorder.

distilled in and kept under an atmosphere of dry nitrogen, then introduced into the apparatus as a solution in an organic solvent or as the compound itself. Effluent activity was determined with a liquid beta counter (20th Century Electronic, Model M6) connected to an Atomic Scaler, Model 1091. Figure 8 illustrates the gas–liquid chromatographic separation of $HgCl_2$ and $FeCl_3$ on an 80 cm column coated with 30% $BiCl_3$ on Sil-O-Cel brick (-30 to $+50$ mesh).

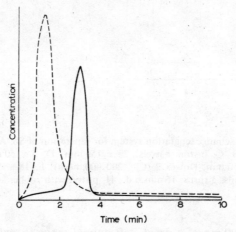

Fig. 8. Gas–liquid chromatograph of $HgCl_2$ and $FeCl_3$; column (80 cm), Sil–O–Cel brick (-30 $+50$ mesh) coated with 30% w/w $BiCl_3$; temperature, 290°; carrier gas (nitrogen) flow, 10 ml/min. $---$ $HgCl_2$; $-$ $FeCl_3$.

Mercury, zinc, and cadmium were quantitatively resolved as their anionic chloro complexes by an ion-exchange chromatographic procedure[93]. The stability of these complexes increases in the order zinc > cadmium > mercury. The metal chloro

complexes were adsorbed on a 0.01 M hydrochloric acid solution by the anion exchange resin Dowex-1 (Cl⁻ form). Zinc and cadmium were eluted separately in that
order with 0.01 M hydrochloric acid. Mercury was then removed with a 0.01 M
hydrochloric acid–0.1 M thiourea solution and analyzed by precipitating and weighing the metal as the sulfide. Zinc and cadmium were analyzed by the titrimetric
procedure involving titration with Versene at pH 10 with Eriochrome Black T as
an indicator.

A radiochemical method for the determination of mercury, arsenic, bromine,
antimony, and selenium in neutron-irradiated biological material has been described
by Samsahl[94,95]. The method is based on distillation of the neutron-irradiated

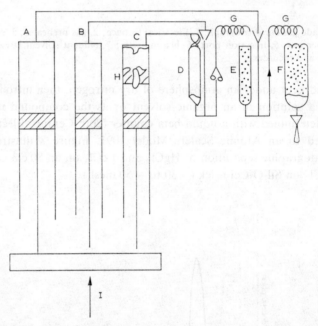

Fig. 9. Scheme of anion-exchange separation system for separation of Sb, As, Hg, and Se. A, piston
barrel, 28×150 mm; B, C, piston barrels, 20×150 mm; D, 5×50 mm, Dowex 2 (HSO₄⁻,
200–400 mesh); E, 7×50 mm, Dowex 2 (Cl⁻, 200–400 mesh); F, 18×50 mm (Br⁻, Cl⁻, 200 to
400 mesh); G, mixing coils, 5 turns, 15 mm o.d.; H, piston with rubber stopper; I, Perspex plate,
15 mm thick.

sample (maximally 200 mg of a dried, soft animal tissue is sealed in a small quartz
tube and irradiated together with standards of As, Br, Hg, Sb, and Se for 1–2 days
with a thermal flux of 2×10^{13} n/cm²/sec) followed after a 2–3 day decaying period
by automated ion exchange and extraction chromatographic methods for separating
the trace elements into 12 groups suitable for gamma-spectrometric measurements.
Figure 9 illustrates the distillation apparatus and scheme of anion-exchange separation system, respectively. The radionuclides ⁷⁶As (0.55 meV), ²⁰³Hg (0.28 meV),

^{122}Sb (0.56 meV), and ^{73}Se (0.14 meV) were separated in the different groups as follows:

Group 1 NaOH absorption: ^{82}Br
Group 2 Dowex 2 (sulfate): ^{197}Hg, ^{203}Hg
Group 3 Dowex 2 (chloride): ^{122}Sb, ^{124}Sb
Group 4 Dowex 2 (bromide, chloride): ^{76}As, ^{75}Se

The simultaneous determination of Cu, Zn, Cd, and Hg with high sensitivity utilizing neutron activation analysis was described[96]. After irradiation, the samples were digested and an initial separation of the four elements made by means of an ion-exchange resin. The elements in the separation fractions were then treated to yield a radiochemical purity, precipitated and their activities measured.

Toribara and Shields[39] described a scheme for the analysis of sub-microgram amounts of mercury in tissues involving (1) the separation of mercury from tissue by a cold digestion in hydrochloric acid–sodium nitrate, (2) collection on an anion exchange column, (3) elution with thiourea solution, (4) collection in cadmium sulfide, volatilization at 550°C, (5) determination photometrically with a General Electric germicidal ultraviolet intensity meter. The ion-exchange separation was accomplished using Amberlite IRA-400 (50–100 mesh) with elution with 0.01 M thiourea.

The separation and detection of picogram quantities of cations in cells and tissues was described by Klimes and Betusova[97]. The sample is first mineralized with HCl–KClO$_4$ and heat, then the cation separated on a strongly basic ion-exchanger (Wofatit 1-150, Veb, Fabrickenwolfen) or by electrolysis and identified by means of classic color reactions in a single grain of silica gel support. The procedure is exemplified by the detection of (a) about 10 pg of mercury in 0.1 mg samples of kidney and in 1 mg samples of adrenal cortex of a rat previously treated with

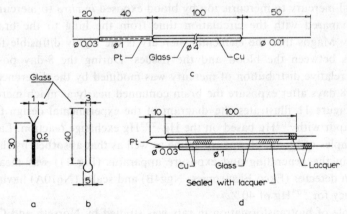

Fig. 10. Apparatus for picogram detection of cations in cells and tissues. (a) Pico test tube; (b) mineralization flask; (c) pico electrode; (d) pico electrode pair.

Agronal and (b) the detection of about 7 pg of iron in a single cell of *Allium cepa*. Figure 10 illustrates the apparatus for the picogram detection of cations in cells and tissues.

Koop and Keenan[98] described an ion-exchange method for the complete separation of sub-microgram and larger quantities of mercury from 0.3 to 0.5 mg of copper, lead, cadmium, thallium, zinc, and nickel.

The mercury in a digest of a biological sample, *e.g.* urine was first absorbed on a 10 mm × 300 mm column containing a strongly basic anionic resin (Dowex 1-X8, Cl^- form, 100–200 mesh) and was then eluted quantitatively with a 0.002 M solution of thiourea in 0.01 M hydrochloric acid. The eluted mercury was then complexed directly with a standard dithizone reagent and read spectrophotometrically at 490 nm, with a sensitivity of 0.3 µg. Experiments with ^{203}Hg and with the stable isotope of mercury in urine yielded recoveries exceeding 92%. The exact chemical composition of the excretory products of mercury in the urine is unknown. There is evidence that organomercurials which have strong C–Hg bonds and do not ionize are excreted in unchanged form[13].

The chromatographic behavior of Zn, Cd, and Hg on columns of natural cellulose and substituted celluloses was studied by using ^{65}Zn, ^{109}Cd, and ^{203}Hg radio tracers[99]. Traces of Zn and Cd were strongly retained by the functional groups attached on the substituted celluloses, while mercury was not retained to any extent. Nanogram amounts of zinc, cadmium, and other metals were separated from 3 g of mercury on cellulose phosphate in ether.

The mercury–blood interaction and mercury uptake by the brain after vapor (^{203}Hg) exposure has been studied by Magos[58]. On the wet basis, the mercury uptake by the brains of mice exposed to 0.003–9.900 g mercury/l of air for 4 h was little more than two-thirds of the total uptake (inhalation plus absorption by fur). This high quotient suggests that the major part of mercury is equilibrated in a highly diffusible form between the blood and the tissues including the brain. The conversion of elemental mercury to mercuric ion by blood exposed *in vitro* to mercury is a slow process compared with the circulation time from the lung to the brain. It was assumed by Magos that the elemental mercury is the highly diffusible form which equilibrates between the blood and the tissues. During the 8-day post-exposure period the relative distribution of mercury was modified by the differences in organ clearance (8 days after exposure the brain contained nearly as much mercury as the kidneys). Figure 11 illustrates the diagram of the experimental design for labeling mercury vapor with ^{203}Hg based on the Hg–^{203}Hg exchange reaction. Labeled mercury in samples of blood, plasma, or saline as well as that absorbed by the hopcalite absorbers in the generating and exposure apparatus (Fig. 11) was measured in a scintillation detector (Ekco Electronics, Ngg4B) and scaler (N610A) having a counting efficiency for ^{203}Hg of 40%.

The role of biotransformation in rats was studied by Norseth and Clarkson[100] by an examination of the intestinal transport of ^{203}Hg-labeled methylmercury chlo-

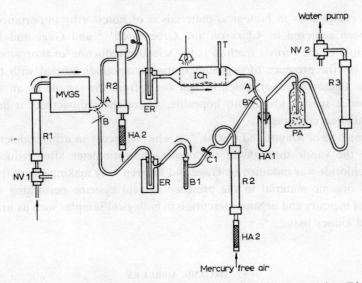

Fig. 11. Diagram of technique for labeling mercury vapour by exchange reaction. Diagram of the experimental design (not to scale). The direction of the flow is from left to right. Connection A is used for inhalation experiments and connection B for exposing blood to mercury vapor. The other symbols are as follows: NV 1 and NV 2, needle valves; R 1, R 2 and R 3, rotameters; MVGS, mercury vapor generating system; HA 1 and HA 2, hopcalite absorbers, type 1 and type 2; ER, exchange reagent; ICh, inhalation chamber; BI, impinger containing the blood; T, thermometer; Cl, clamp; Pa, permanganate absorber.

ride. Inorganic mercury in the presence of methylmercury salts in tissue samples, fluids, and extracts was analyzed by the isotope exchange method[65]. The chemical form of mercury in bile was determined by TLC on Eastman Chromagram 6061 silica gel and electrophoresis on Beckman paper 319328 or 320046. The developing solvents for TLC were n-propanol–water (70:30) and ethanol–34% ammonium hydroxide (70:30). Diethyl barbiturate–sodium barbiturate buffer, pH 8.6, was used for electrophoresis (14 h at 5 mA constant current). Chromatograms were excised for counting or scanned by a gas flow counter and bile was analyzed by column chromatography of Sephadex G-100 to separate diffusible and protein-bound mercury. A Sephadex 15/30 column was used with sodium–potassium phosphate buffer at 0.1 M and pH 8.0 containing 1.0 N sodium chloride and 0.02% sodium azide as eluting and supporting solution.

The detection of mercury in biological material (urine, internal organs) by means of TLC was described by Wysocka[101]. The biological material was first mineralized and the mercury then converted into a dithizonate. Chromatographic analysis was performed on Silica Gel G with n-propanol as the developing agent. After 3 h the separation of the excess dithizone (R_F 0.66) from the Hg dithizonate (R_F 0.82) was completed. Amounts as small as 0.5γ Hg in large amounts of biological material can be detected by this procedure.

The selective determination of inorganic mercury in the presence of organomercurial compounds in biological materials is of noteworthy importance and has recently been achieved by Clarkson and Greenwood[102] and Gage and Warren[41]. The principle of the former method is the selective reduction of inorganic mercury by $SnCl_2$ (in the presence of organomercurial compounds labeled with the ^{203}Hg isotope) to elemental vapor which is then swept from the sample by an air stream collected on a specific absorbent, hopcalite, where the radioactivity is determined by γ-scintillation counting.

The method of Magos and Cernik[54] in which mercury in urine is determined by aspirating the vapor through an ultraviolet absorptiometer after reduction with stannous chloride was modified by Gage and Warren[41] by making use of the varying lability of organic material in the presence of acid cysteine permitting the determination of mercury and organic mercurials in biological samples such as urine, feces, blood, and kidney tissue.

2. ORGANIC MERCURY

The last two decades have dramatized the substantial toxicological significance of organomercury derivatives, particularly methylmercury. As was noted previously (Figs. 1 and 4), the different forms of mercury which get into the environment are usually converted to methylmercury by microorganisms. The reports of environmental contamination in water, foodstuffs, fowl and game[102-105] as well as the toxicological significance of various forms of mercury, primarily methylmercury[106-110], have been well documented.

The adverse effects of mercurial pollution have been extensively reviewed and include (a) Minamata, Japan[106,108,110], where a narcotizing disease of the central nervous system afflicted 111 people of whom 45 died during the period 1953 to 1960, and (b) Nigata, Japan where 26 cases of mercury poisoning and 5 deaths have been documented[108,109,111]. In congenital cases of Minamata disease, cerebral palsy is the major effect. Nineteen of the 111 cases of Minamata disease were encountered in newborn children[112]. Although the mothers of the 19 congenital cases of Minamata disease had all ingested fish or shellfish contaminated with methylmercury, no symptoms had been observed in the majority of cases. (In the few cases where symptoms did occur, they were confined to slight numbness of the fingers.) The lack of symptoms in the mothers was suggested to be due to the relative ease of placental transfer of methylmercury (as also observed in laboratory animals). Tejning[113] reported the passage of methylmercury across the human placenta in normal pregnant women and its possible preferential concentration in the fetus. The human fetal erythrocyte levels showed 28% more mercury than the maternal erythrocytes.

The biological half-life of methylmercury in man is estimated to be about 73 ± 3 days[114,115], but its persistence in nature is believed to be much longer.

The risk of methylmercury accumulation in mammals, the relation between body burdens of methylmercury and toxic effects[116-118], particularly the effects on the central nervous system[109,110,119,120], the embryotoxic and teratogenic effects of methylmercury in mice, rats and cats[109], and in humans, the induction of chromosome breakage with methylmercury[121], as well as the intrauterine effects of methylmercuric dicyandiamide in Sweden[122] and methylmercury in Japan[108], have all been described and attest to the hazardous nature of methylmercury.

Although the current concern for environmental contamination from mercury arose from the Swedish experience showing the agricultural and industrial sources of mercury and the resulting residues in fish and wildlife[10,109], mercury has been reported in foods other than fish[123]. Results from total-diet surveys in the U.S.[124], United Kingdom[125], and Canada[126] have indicated that none of the composites of the twelve food classes contained more than 0.02 p.p.m. The approximate daily intake (ADI) of mercury in the U.S. is 25 μg/day[127,128], although the chemical form of mercury in food is not known.

Aspects of the allowable daily intake for methylmercury compounds in foods have been reviewed by Berglund and Berlin[116] and it was concluded that the no-effect intake should be in the range of 0.6–1.0 mg/day with the corresponding ADIs a factor of 10 lower. The calculations of Berglund and Berlin were based ultimately on neurological effects in adult man recorded during the Japanese epidemics and do not take into account fetal damage by methylmercury compounds. A recent report of an expert committee in Sweden[129] recommended a no-effect intake of 0.3 mg/day and stated that the safety factor of 10 was intended to include effects on the fetus.

Although the importance of methylmercury overshadows the other organo-mercury compounds in terms of toxicity and environmental considerations, it is well to delineate the genetic hazards or aspects of other organomercurials. The genetic effects include mutagenicity of merthiolate in *Drosophila melanogaster*[130], mutagenic activity of Mercuran (fungicide containing 2% ethylmercuric chloride and 12% hexachlorocyclohexane) in germinating apple seeds[131], somatic mutations produced by phenylmercuric hydroxide and phenylmercuric nitrate in flowering plants (seedlings of *Raphanus* and *Zea*) and induction of polyploid nuclei, sticky chromosome and chromosome fragments in root tips of *Allium cepa*[132], cytological effects on root cells of *Allium cepa* of methylmercuric dicyandiamide, methylmercuric hydroxide, phenylmercuric hydroxide and methoxyethylmercuric chloride, and the fungicide Panogen (containing methylmercuric dicyandiamide as the active component)[133,134], cytological effects of inorganic, phenyl- and alkylmercuric compounds (*e.g.* phenylmercuric chloride, butylmercuric chloride, and ethylmercuric chloride) on HeLa cells[135], histological and cytological effects of ethylmercuric phosphate in corn seedlings[136], the C-mitotic action of "Granosan" (fungicide containing ethylmercuric chloride[137,138], and Agrimax M[139] (containing ethylmercuric chloride and phenylmercuric dinaphthyl methanedisulfonate, respectively), and the genetic effects of methylmercuric hydroxide, phenylmercuric acetate, and methoxyethylmercuric chloride in *Drosophila melanogaster*[121,140]. The teratogenicity of phenylmercuric acetate in mice has also been reported[141].

The adverse effects of mercurial pollution in regard to Minamata Bay have been cited above[106,108,110]. The poisoning was due to methylmercuric chloride which was formed in the waste sludges of a plant that used mercuric oxide in sulfuric acid as a catalyst for acetaldehyde production. These solids were discharged into Minamata Bay where the organomercury compound was accumulated by fish and shellfish.

Irukayama *et al.*[142] had earlier reported that an organomercury compound ($CHOCH_2HgCl$) was synthesized from acetaldehyde, mercury sulfate, and sodium chloride. The compound obtained[143] from the Minamata factory sludge CH_3HgCl was considered possibly to have arisen from $CHOCH_2HgCl$ *via* the sequence

$$CHOCH_2HgCl \xrightarrow{\text{O}} COOHCH_2HgCl$$
$$\downarrow -CO_2$$
$$CH_3HgCl$$

A possible source of mercury in Minamata Bay is believed to be the large quantities of mercuric chloride utilized as a catalytic agent in the production of vinyl chloride. (The vinyl chloride output increased from 60 tons in 1949 to 18,000 tons in 1959; this increased production paralleled the increase in cases of Minamata disease until the ban on fishing in 1956.)

$$CH{\equiv}CH + HCl \xrightarrow{HgCl_2} \begin{matrix} H \\ | \\ C=C-Cl \\ | \\ H \end{matrix}$$

At that time it was shown that 60 g of mercury is lost for each ton of vinyl chloride produced. In 1960 it was estimated that over 1000 kg of mercury was lost from the reactors of the Minamata plant in the production of approx. 20,000 tons of vinyl chloride and another 7000 kg of mercury was removed as spent catalyst. In the Minamata plant, the crude vinyl chloride is washed to rid it of impurities and it is not certain whether the mercury removed in the washing process is in the form of inorganic mercury compounds or is converted in the reactor to some stable organic mercury complex. (Irukayama *et al.*[143] believe that the mercury dregs from the vinyl plant consist mainly of activated carbon in which mercuric chloride has been absorbed.)

Sumino[144] described the detailed GLC analysis of organic mercury compounds in various materials related to Minamata disease (both in Kumamoto and Nigata prefectures). The materials subjected to alkylmercury analysis included shellfish, oysters, sludge, organs, and hair of affected patients, experimental animals, by-products from an acetaldehyde manufacturing plant, crops on which alkylmercury pesticides had been sprayed (unhulled rice, strawberries, pimentoes) and for comparison marine products and hair from individuals not exposed to direct contamination. The materials subjected to phenylmercury analysis included crops on which phenylmercury pesticides had been sprayed (rice leaves, raw and polished rice) and experimental animals administered phenylmercury compounds (mice, goldfish). The analytical conditions for GLC were mostly those as described previously[145]. Total mercury was determined by the dithizone method[146].

The salient features of this investigation were (1) the establishment by GLC and TLC of methylmercury from the vicinity of the drainage outlet from an acetaldehyde factory (methylmercury is produced as a byproduct from the initial use of inorganic mercury or a catalyst in the manufacture of acetaldehyde, (2) the results of various animal experiments and examination of organs of affected patients confirmed that the cause of Minamata disease was waste discharge from this factory (methylmercury was found to be distributed almost uniformly throughout each patient's brain), (3) in separate experiments the accumulation of methyl- and ethylmercury in fish kept in alkylmercury solutions was examined and it was found that methylmercury remained within fish for a prolonged period even after transfer to fresh water.

It was suggested that if the concentration of methylmercury remains at a low level, it is still sufficient to allow accumulation in fish and hence could be the causative agent in Minamata disease (in rivers primary accumulations of a high degree occur in the plankton and moss and hence fish eating such microorganisms may accumulate mercury more rapidly). Trace quantities of methylmercury were also found among sea fish from locations where the occurrence of mercury contamination is difficult to envision (e.g. Indian Ocean, Pacific Ocean, Bering Sea, and African Coast), (4) methylmercury was found in the hair of patients affected with Minamata disease as well as in the hair of people not handling mercury, regardless of sex, age, and occupation (the methylmercury content in the hair of control individuals was significantly lower than that of patients with Minamata disease), (5) the residual content of phenylmercury in rice leaves sprayed with phenylmercury pesticides was determined by GLC. Phenylmercury was detected in rice grains and was equivalent to 1/5–1/20 of this total mercury.

Takizawa and Otsuka[147] described the detection of methylmercuric chloride in the moss from water drainage of an acetaldehyde plant. The moss extracted from the drainage of the Showadenko Kanore (synthetic acetaldehyde plant) was examined for methylmercuric chloride by TLC. The chromatogram showed a spot for methylmercuric chloride at R_F 0.47 apart from phenylmercuric acetate (R_F 0.41). The identity of the methylmercuric chloride was further substantiated by GLC techniques.

Takizawa *et al.*[148] investigated the cause of organic mercury poisoning in districts along the Agano River in Japan. Methylmercury was detected by GLC procedures in the moss from the drainage of an acetaldehyde plant and is presumed to be the causative agent responsible for the mercury poisonings reported. A Shimazu gas chromatograph Model GC-1C was used with an ECD-1A (^{3}H) 600 mCi electron-capture detector and a 3 mm × 150 cm stainless steel column containing 15% DEGS on Shimalite (60–80 mesh). The carrier gas was nitrogen at 50 ml/min and the injection, column, and detector temperatures were 180, 190, and 200°, respectively.

In order to illuminate the scope of the mercury problem in terms of the biological properties of the compounds primarily involved, *e.g.* the methylated mercurials, Ostlund described a series of investigations involving the metabolism of ^{203}[Hg]-methyl- and dimethylmercury in the rat[149]. TLC was extensively utilized for the elaboration of radiochemical purity of the methylated mercurials as well as the detection of their metabolic products. TLC offers simple and sensitive methods for the detection of radioactivity. In order to determine the localization of radioactive spots, the quickest method is scanning the chromatographic plate in a radiochromatogram scanner. A more exact method consists of exposing an X-ray film pressed against the plate which yields an autoradiogram of the chromatogram. The adsorbent with the different spots on the plate can be scraped off, transferred to counting vials and the radioactivity of the spots determined by impulse counting.

Three basic TLC procedures were developed by Ostlund in this study.

(1) TLC of organic mercury compounds of low volatility

A combined adsorption–partition chromatography, using water as the stationary phase, was used for the separation of organic compounds which have a moderate volatility, *e.g.* compounds that could be handled at room temperature during the separation process without a noticeable loss of substance. TLC on Silica Gel G (0.4 mm layers on 138 × 138 mm and 46 × 138 mm plates) was used with diethyl ether–toluene (90:10) in 20 min, 120 mm developments for the separation of organic mercury compounds (applied as cyanides by shaking a benzene phase over a water phase containing 1 M KCN). The R_F values obtained were $CH_3OC_2H_5HgCN$ 0.18, CH_3HgCN 0.34, C_2H_5HgCN 0.48, C_6H_5HgCN 0.70, and $C_6H_5HgC_6H_5$ 0.91.

The mode of action of the mercury ion, Hg^{2+}, in this separation was of no interest, since the inorganic mercury(II) cyanide is insoluble in benzene and inorganic mercury was eliminated by the cyanide treatment of the substances before chromatography.

(2) TLC of mercury compounds for the detection of dimethylmercury

For the identification of the most volatile of the mercury compounds of interest in this study, dimethylmercury, the TLC separation was carried out in a freezer in which the temperature varied between −75 and −78°. Silica gel plates (analogous to those in technique (*1*) above were developed with light petroleum–diethyl ether (95:5) for 25 min (120 mm). The mercury compounds were applied as cyanides or chlorides in diethyl ether. After development at $-75°$, the plates were transferred to room temperature and immediately sprayed with a 0.05 M Hg(NO$_3$)$_2$ solution in acetone–ether (80:20) and then with a 0.2% dithizone solution in acetone–ether (80:20). The R_F values obtained were HgX$_2$, MeOEtHgX, MeHgX, EtHgX, C$_6$H$_5$HgX 0, C$_6$H$_5$HgC$_6$H$_5$ 0.38, MeHgMe 0.70 (X = Cn or Cl). Detection was also accomplished with successive sprays of 0.05 M Hg(NO$_3$)$_2$ in acetone–ether (80:20) and with 0.2% dithizone in acetone–ether (80:20).

(3) TLC *of organic and inorganic mercury dithizonates*

This method was used for the determination of the ratio between organic and inorganic mercury in biological material or in water solutions of organic mercury compounds (determination of radiochemical purity). Silica Gel G (0.25 mm layers of 138 × 138 mm and 46 × 138 mm) was used with a solvent system of light petroleum ether–acetone (90:10) in 20 min, 120 mm developments. The mercury compounds were applied as dithizonates obtained by shaking a water solution containing the compound or compounds with a 0.1% solution of dithizone in benzene or in carbon tetrachloride. The R_F values found were dithizonate of Hg^{2+} 0.22, dithizonates of methylmercury and phenylmercury 0.36–0.40.

The metabolism of [^{203}Hg]methylmercuric nitrate in man was studied by Åberg and co-workers[114,150]. The radiochemical purity of CH$_3$^{203}HgNO$_3$ was evaluated by paper chromatography[151] using Munktell S 314 paper with *n*-butanol saturated with 1 N NH$_4$OH (4 h saturation time) in a 15–16 h development. The impurities were less than 10% and consisted of mercuric nitrate. The oral intake of 2.6 µCi by CH$_3$^{203}HgNO$_3$ in three clinically healthy male volunteers aged from 37 to 44 years resulted in an accumulation in the liver and head of the ^{203}Hg. The main activity was localized in the liver (about 50% of the contents of the body) and the head contained about 10% of the total body count. No ^{203}Hg was found in the sperm, and a very rapid uptake was found in the erythrocytes. The main excretory route was the feces but the urinary excretion increased with time up to 30 days after the intake. The biological half-life was found with whole body measurements to be 70–74 days. The decline of ^{203}Hg in the head was less rapid than in the rest of the body. After infinite time, a weekly unit dose of CH$_3$^{203}HgNO$_3$ will result in a whole body burden of 15.2 units of mercury.

In a series of papers, Westöö described the determination of methylmercury in foods[152,153] (*e.g.* fish, egg, meat, and liver) and in various kinds of biological material[154] utilizing both TLC and GLC procedures[153]. A Wilkens Aerograph, Moduline Model 202 with electron-capture detector was used with 5 ft. × $\frac{1}{8}$ in. stainless steel columns containing 10% Carbowax 20M or 10% Carbowax > 1500 (polyethylene glycol, average mol. wt. > 1500) on Teflon-6 (35–60 mesh) or on Chromosorb W, acid-washed DMCS (60–80 mesh). The temperature of the column and injection port were 130–145° add 150–170°, respectively. Nitrogen was the carrier gas at 65 ml/min and the recorder was a Texas Instruments Model PWS-TMVC-05-A25-BT. The peak of each sample solution was compared with that of a standard solution with about the same concentration of methylmercury. Thin-layer chromatography was carried out on aluminum oxide or silica gel plates activated for 1 h at 105° and developed with light petroleum–diethyl ether (70:30). The spraying agent for use with silica gel plates was a saturated solution of Michler's thioketone in ethanol. The transformation of methylmercuric chloride to methylmercuric dithizonate for TLC involved the concentration of 1–10 ml of a benzene solution of methylmercury to 0.1 ml by evaporation of a part of the solvent under reduced pressure. (This results in an obvious loss of mercury compound.) A 0.4% dithizone solution in benzene is added until a green coloration is obtained. The transformation of methylmercuric chloride to bromide, iodide or cyanide for TLC required shaking of a benzene solution of methylmercuric chloride with excess hydrobromic acid, potassium iodide, or potassium cyanide in dilute aqueous solution; the benzene layers are dried over sodium sulfate and concentrated by evaporation under reduced pressure.

The combined GLC and TLC method for the identification of methylmercury in fish[153] was modified by Westöö[152] in order to render it more applicable to a wider range of foods. In animal foodstuffs, methylmercury is probably attached to thiol groups to a great extent. When these foods are extracted according to Gage[155], methylmercuric chloride is formed and dissolves in benzene together with varying amounts of thiol compounds. Purification of the methylmercury in the benzene extracts, e.g. egg yolk or liver, by extraction with aqueous alkali solution did not work, probably because the thio compounds were not volatile and could form alkali-insoluble methylmercury salts. Presumably, methylmercury-S-compounds were reformed at high pH and prevented the formation of water-soluble methylmercuric hydroxide. This clean-up problem was solved by the addition of excess mercuric ions, which expelled the methylmercury from the thio compounds, permitting the analysis of fish, egg-white, egg yolk, meat and liver. The use of an aqueous cysteine solution permitted the extraction of the methylmercury into an aqueous phase, thus resulting in a more facile analytical procedure.

In the original procedure, the methylmercury was extracted with benzene from a homogenate of the fish acidified with hydrochloric acid according to Gage[55], then taken up into ammonium hydroxide solution and finally re-extracted into benzene after acidification with hydrochloric acid. The extraction was incomplete unless the benzene extract was previously concentrated by distillation. (The distillation procedure was assumed to remove volatile thio compounds binding part of the methylmercury and preventing its uptake into ammonia.) Hence, any methylmercury attached to a sulfur atom of non-volatile compounds giving rise to alkali-insoluble methylmercury salts at the purification stage would not be determined.

Figure 12 summarizes the salient steps in the extraction and purification of methylmercury for chromatographic analysis recommended by Westöö[156]. With the cysteine acetate extraction procedure the recovery of methylmercury from fish was $98 \pm 3\%$ (10 samples). (Methylmercury levels of 0.001 mg of Hg/kg of food could be measured by the chromatographic procedures delineated.)

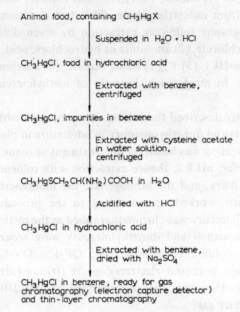

Fig. 12. Extraction and purification of methylmercury in animal foods for chromatographic analysis

Westöö[156] showed that the methylmercury compounds found in eggs, meat, liver (average values < 0.06 mg of Hg/kg), and fish in Sweden in 1966 decreased in 1967 to 1968 to approximately one-third of the average values found earlier following the substitution of methoxyethylmercury for the methylmercury compounds formerly used for seed treatment.

Concerning the mercury content of fish in Sweden, the high mercury content was usually caused by industrial disposal of mercury into the water or the air[157,158]. Downstream from pulp and paper mills which were allowed to use phenylmercury compounds as a fungicide until January 1 1966, and may still use sodium hydroxide and sulfuric acid contaminated with mercury compounds, the mercury content of fish was higher than that of fish caught upstream. Chlorine-alkali factories using mercury electrodes and mercury rectifier factories had a similar effect on fish living downstream[158]. Levels above 1 mg of Hg/kg were usually caused by industrial discharge of mercury compounds. Regardless of the nature of the mercury pollutant, methylmercury has been found in the fish.

It is of interest to note the recent analogous situation in the United States where industrial discharge of mercury from a variety of operations (*e.g.* predominantly chlor-alkali, paper and pulp treatment) have resulted in elevated mercury levels in fish in excess of 0.5 p.p.m. primarily in the Great Lakes region and the banning of sales of fish in at least 33 states[159].

Westöö and Noren[160] reported the analysis of about 550 fish (caught or bought from September 1966 to November 1967) for total mercury and methylmercury to check contamination from industrial waste. The determination of methylmercury was by gas chromatography following extraction by seven different methods, *e.g.* (*1*) benzene, mercuric chloride, (*2*) ammonia or hydrochloric acid, (*3*) 0.20% cysteine-HCl, (*4*) 1.0% cysteine-HCl, (*5*) 1% cysteine acetate, (*6*) combination of the $HgCl_2$ and cysteine methods. In most cases the ratio of methylmercury:mercury in fish was >4:1.

Hartung[161] recently described the solvent separation and subsequent GLC determination of monoethyl- and dimethylmercury in admixture in pheasant, mallard and fish tissues. Initial separation was based on the treatment of tissue with equal volumes of cysteine–borate buffer, pH 8.2, thence extraction with toluene. (The supernatant contained dimethylmercury, and the homogenate the monomethylmercury fraction which was subsequently worked up according to the procedure of Westöö[152].) The extracted dimethylmercury was chromatographed as the methylmercuric bromide derivative. Both monomethyl- and dimethylmercury were separated by GLC using a 5 ft. × $\frac{1}{4}$ in. Pyrex column containing 11% QF-1 + OV-17 on 80–100 Gas-Chrom Q. The injection port and electron-capture (tritium) detector were at 180 and 220°, respectively, the oven was programmed at 100–210°, and nitrogen at 60 ml/min was the carrier gas.

Curley *et al.*[105] reported the first documented episode of indirect organic mercury poisoning in humans in the U.S. caused by the ingestion of contaminated meat from animals that had consumed mercury in their food supply. Atomic absorption spectrophotometry and neutron activation analysis showed the presence of mercury in organic extracts of seed grain and in tissues of hogs fed the contaminated grain. Mercury was also found in the urine, serum, and cerebrospinal fluid of humans who ate contaminated pork. A sample of the contaminated grain was prepared for mass spectral analysis by a modification of the extraction and TLC procedures of Westöö[153] and subsequently examined, using an LKB Model 900 mass spectrometer (ionized at 20 and 70 eV, temperature of direct probe varied from ambient to 120°C; temperature of the ion source, 290°C, accelerating voltage, 3.5 kV; filament current, 4 A, trap current, 60 μA; recording oscillograph 4 cm/sec; scan speed 5).

Bache and Lisk[162] described the gas chromatographic determination of organic mercury compounds by emission spectrometry in a helium plasma with applications of this technique to the analysis of methylmercuric salts in fish. The instrument used for emission spectrometric detection was as described previously[163,164] and the monochromator wavelength was adjusted to isolate the 2537 Å atomic mercury line. A Varian Aerograph Model 705 gas chromatograph was used for confirmatory

TABLE 5

GAS CHROMATOGRAPHIC OPERATING PARAMETERS USING THE EMISSION DETECTOR[a]

Parameter	Dimethylmercury	All other mercury compounds
Column packing	60–80 mesh Chromosorb 101	20% OV-17 and QF-1 (1:1 w/w) on 80–100 mesh Gas Chrom Q (5 μl of Carbowax 200 injected into the column prior to use to coat active sites)
Column length	2 ft., glass	6 ft., glass
Column diameter (i.d.)	5/32 in.	5/32 in.
Carrier gas flow rate	80 ml/min	75 ml/min
Column temperature	100°C	152°C
Injector temperature	140°C	208°C

[a] The electron affinity detector was only used for verifying analysis of methylmercuric chloride in salmon. The operating parameters were identical to those in the right column above except that the temperatures of the column, injector, and detector were, respectively, 150, 200, and 210°C.

TABLE 6

CHROMATOGRAPHIC DATA FOR EMISSION DETECTION OF MERCURY COMPOUNDS

Compound	Retention time (min)	Sensitivity (ng) (50% full-scale deflection)
Dimethylmercury	6.8	0.6
Methylmercuric chloride	2.8	0.6
Methylmercuric dicyandiamide	2.8	0.8
Phenylmercuric acetate	45.0	8.8
Methylmercuric dithizonate	3.0	0.7

analysis of methylmercuric chloride in fish. Table 5 lists the gas chromatographic operating parameters. The Chromosorb 101 column was necessary for slow elution of dimethylmercury which is particularly volatile. Table 6 lists the retention times and estimated sensitivities of several organic mercury compounds analyzed. The sensitivity corresponds to the nanograms of compound required to give a 50% fullscale deflection using a 1 mV recorder. The estimated sensitivity of the method for methylmercuric chloride in fish was about 0.05–0.1.

The selectivity ratio of the 2537 Å atomic emission line for mercury when comparing methylmercuric chloride to eicosane was at least 10,000 to 1, and the detector responded linearly from 0.1 at least 100 ng of injected methylmercuric chloride. The essential lack of hydrocarbon emission in the deep ultraviolet wavelength region (2537 Å) was probably responsible for its selectivity. Although the detector was operated with a low-pressure helium plasma source[163], an argon plasma source could be substituted. It was suggested that the above detection system should be particularly useful for the specific sensitive detection of organic mercury compounds and obviating the difficulty often encountered with the poisoning of tritiated electron-affinity detector foils by eluting mercury compounds[165].

A gas chromatographic separation and detection of dialkylmercury compounds with application to river water analysis was recently described by Dressman[166]. Dimethyl-, diethyl-, dipropyl-, and dibutylmercury were successfully separated using a 6 ft. × 2 mm i.d. glass column packed with 5% DC-200 plus 3% QF-1 (copacked) on Gas Chrom Q (80–100 mesh), with nitrogen carrier gas flow at 50 ml/min. The separation was accomplished by holding the column oven temperature at 70°C for two min (to separate dimethyl- from diethylmercury), then programming the temperature at a rate of 20°C/min to 180°C to elute dipropyl- and dibutyl-

TABLE 7

CHROMATOGRAPHIC RETENTION DATA AND RELATIVE RESPONSE FACTORS FOR SOME DIALKYL MERCURIALS
TO DIPROPYLMERCURY

Dialkyl mercurials	Retention time (min)[a]	Response factor (1/1)	Response factor (1/2)
Dimethylmercury	0.4	2.44	1.20
Diethylmercury	3.0	0.88	0.43
Dipropylmercury	4.9		
Dibutylmercury	6.7		

[a] Conditions as described in text.

mercury. The separated compounds were combusted in a flame-ionization detector and the resultant free mercury was passed into a cold vapor mercury detector (Coleman Mercury Analyzer MAS-50) connected to the exit port of the FID detector of a Perkin-Elmer Model 900 gas chromatograph. The practical absolute sensitivity for mercury was 0.1 ng. Table 7 lists the chromatographic retention data and relative response factors for the dialkyl mercurials to dipropylmercury.

The application to raw river water analysis was performed by dosing a one liter sample with 5 μg each of the dialkyl mercurials. The sample was then extracted twice with 60 ml of a mixture of n-pentane and 20% ethyl ether, the extract then dried by passing it through sodium sulfate, concentrated to 5 ml using a Kudema-Danish apparatus and analyzed as above.

Methylmercury was determined in fish by Newsome[167] using GLC following a modification of existing extraction procedures of Sumino[145] and Westöö[154] The method obviated centrifugation to assist phase separation and yielded overall recoveries of 94% ± 6% for white fish and 98% ± 6% for cod of methylmercury. The modification consisted of incorporating a filtration step and the use of hydrobromic acid rather than hydrochloric acid to enhance the partition ratio and facilitate the extraction of methylmercury from the aqueous phase. Methylmercury was also determined in wheat flour and ground oats by extraction with benzene–formic acid mixture followed by purification and GLC. (Interfering substances were removed from the extracts by column chromatography on silicic acid and partitioning with cysteine acetate solution.) The method was sensitive in the 0.01–0.90 p.p.m. range with a mean recovery greater than 95%.

Methylmercuric chloride (Alfa Inorganics) was ascertained for homogeneity by TLC on silica gel using hexane–diethyl ether (1:1) as developer and a saturated solution of Michler's thioketone in ethanol for visualization. An Aerograph Model 705 fitted with a glass injection insert and tritium foil electron-capture detector was used along with a 40 cm × 4 mm glass column packed with 2% butanediol succinate on 100–120 mesh Chromosorb W, AW, DMCS. The column was conditioned by heating for 24 h at 200°C under a flow of nitrogen carrier gas at 60–80 ml/min. (The conditioning time could be reduced to 2 h using Chromosorb W, HP as a support.) For the determination of methylmercury, typical operating conditions were: injection port, 110°C; oven, 120°C; and detector, 200°C; with carrier gas maintained at 80–100 ml/min (retention time of methylmercuric chloride, approx. 1 min).

The determination of methylmercury in biological materials by GLC and total mercury by cold vapor atomic absorption was described by Langley and Le Blanc[168]. The methylmercury complex in the sample is hydrolyzed by acid and transformed to methylmercuric bromide by sodium bromide solution, viz.

$$R—S—Hg—CH_3 \xrightarrow{H_3O^+, Br^-} CH_3HgBr + H_2O + RSH$$

where R = cysteine group of protein. Any mercaptans or sulfides present are marked by the addition of a copper sulfate solution. The methylmercuric bromide was extracted with benzene and the benzene phase containing the CH_3HgBr then extracted with an aqueous solution of sodium thiosulfate forming the water-soluble thiosulfate complex.

$$CH_3HgBr \xrightarrow{Na_2S_2O_3} CH_3HgS_2O_3^-Na^+ + Na^+Br^-$$

The thiosulfate extract was then reacted with potassium iodide and the resulting methylmercuric iodide extracted with benzene and analyzed by GLC. A Varian Aerograph Model 1800 gas chromatograph with tritium electron capture (90 V) was used with a 1 m × 0.20 cm i.d. glass column packed with 7% Carbowax 20M on 100–120 mesh Varaport-30. The carrier gas was nitrogen (purified by passage through a molecular sieve) at a flow rate of 30–40 ml/min. The operating conditions were: column, injection port, and detector temperatures 155, 195, and 195°C, respectively; standing current: 1.28×10^{-8} to 3×10^{-8} A; attenuation 32/64/128; retention time,

3 min. With optimum chromatographic conditions, a signal-to-noise ratio of ≤ 10 could be obtained with an injected quantity of 10 pg (10 μl of a solution containing 0.0001 ng Hg/μl). The detection limit under the above conditions was 0.0005 mg/kg (0.5 ng/g) when the sample quantity was 1 g.

Kitamura *et al.*[169] reported the GLC behavior of a number of organomercurial chlorides on 3 mm × 187 cm stainless steel columns of 25% DEGS on Shimalite (30–60 mesh) at 170°. Detection was accomplished by flame ionization and electron capture (^{3}H). Figures 13(a) and (b) illustrate the GLC of methyl-, ethyl-, propyl-, and methoxyethylmercuric chlorides using flame-ionization and electron-capture

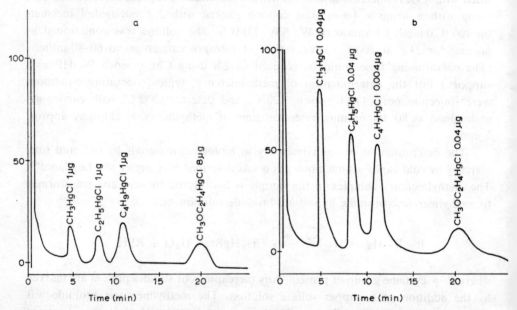

Fig. 13. (a) Gas chromatography of organomercurial chlorides at 170° with FID detection. Column: 3 mm × 1.87 m stainless steel with 25% DEGS coated on Shimalite (30–60 mesh) at 170°; nitrogen carrier gas at 80 ml/min at a pressure of 1.5 kg/cm²; flame-ionization detector at 250°; sensitivity, 1000; range 0.4 V. (b) Gas chromatography of organomercurial chlorides at 170° with electron-capture detection. Column: 3 mm × 1.7 m stainless steel with 25% DEGS coated on Shimalite (30–60) mesh at 170°; nitrogen carrier gas at 80 ml/min at a pressure of 1.5 kg/cm²; electron-capture (^{3}H) detector; sensitivity, 10²; range 0.8 V.

detection respectively. Table 8 lists the retention times of a variety of organomercurials obtained at 170° using a stainless steel column loaded with 25% DEGS on Shimalite (30–60 mesh) with nitrogen as carrier gas at 80 ml/min and detection by electron capture.

The solvent extraction for bio-materials as well as the analysis of organomercury compounds by GLC has been described by Sumino[145]. Shimazu gas chromatographs Mofels GC-1C, 2C, and 3AE were used with both steel and glass columns and

TABLE 8

Column: 2 mm × 1.87 m stainless steel with 25% DEGS coated on Shimalite (30–60 mesh) at $170°$.
Nitrogen carrier gas at 80 ml/min. Electron-capture detector (^{3}H). Sensitivity, 10^2; range, 0.8 V

Compound	Retention time (min)
CH_3HgCl	4.7
CH_3HgI	
$(CH_3Hg)_2SO_4$	
C_2H_5HgCl	8.1
$(C_2H_5Hg)_2HPO_4$	
C_4H_9HgCl	10.8
$CH_3OC_2H_4HgCl$	19.6
⬡—$HgOCOCH_3$	120.0

TABLE 9

COLUMN PACKINGS

The column packings were utilized with U-shaped stainless steel columns, 3 mm × 75 cm, straight stainless steel columns 3 mm × 75 cm, glass columns 4 mm i.d.; 40, 50, 120, 175, and 333 cm lengths were used in the on-column injection techniques

Wt.-%	Liquid phase	Support	Mesh
1.5	SE-30	Chromosorb W	60–80
5	DC-11	Shimalite W	60–80
25	DC-200	Shimalite	60–80
7	Apiezon L	Shimalite	60–80
15	Thermol-3	Shimalite	60–80
1	XE-60	Chromosorb W	60–80
3	SE-52	Chromosorb W	60–80
3	OV-17	Shimalite W	80–100
25	DC-550	Shimalite	30–60
5	QF-1	Chromosorb W	60–80
1	NGS	Chromosorb W	60–80
5	1,4-BDS	Shimalite W	60–80
10	1,4-BDS	Chromosorb W	60–80
10	CHDMS	Chromosorb W	60–80
5	DEGS	Chromosorb W	60–80
15	DEGS	Shimalite	60–80
20	DEGS	Celite 545	60–80
25	DEGS	Shimalite	60–80
25	DEGS	Shimalite	40–50
25	DEGS	Shimalite	30–60
25	PEG 20 M	Chromosorb W	60–80
10	PPE	Celite 545	60–80

sixteen liquid phases as shown in Table 9 for the elaboration of optimal parameters. Both 25% DEGS and 5% 1,4-BDS (butanediol succinate) were the preferred liquid phases on glass columns for the analysis of alkylmercury derivatives. For phenyl-mercury analysis 5% DEGS was used with on-column injection techniques.

Table 10 depicts the operating conditions for the analysis of alkylmercury compounds and Fig. 14(a) and (b), illustrates the chromatograms obtained for a variety of alkylmercury compounds using 25% DEGS and 5% 1,4-BDS columns,

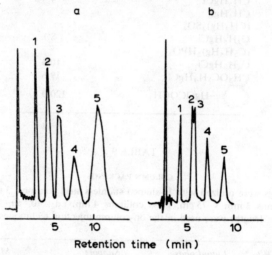

Fig. 14. Chromatograms of alkyl mercury compounds. (a) 25% DEGS 4 mm × 40 cm, 160°, carrier gas (N_2) 80 ml/min. 1, MMC; 2, EMC; 3, BMC; 4, AMC; 5, MtEMC. (b) 5% 1,4-BDS 4 mm × 1.2 m 160°, carrier gas (N_2) 60 ml/min. 1, MMC; 2, EMC; 3, PrMC; 4, BMC; 5, AMC.

TABLE 10

ANALYTICAL CONDITIONS FOR GLC OF ALKYL MERCURY COMPOUNDS

Conditions	Apparatus	
	GC-1 C	GC-3 AE
Column	4 mm × 40 cm glass on-column 160°	4 mm × 1.2 m glass on-column 160°
Packing	DEGS	1,4-BDS
Wt.-%	25	5
Support	Shimalite	Shimalite W
Mesh	60–80	60–80
Carrier gas	N_2	N_2
Flow rate	80 ml/min	60 ml/min
Detector	ECD, 210°	ECD, 160°
Applied volts	Pulse height, 40 V	Pulse height, 40 V
Amplifier range	4×10^{-9} A/f.s.d.	8×10^{-10} A/f.s.d.
Chart speed	10 mm/min	10 mm/min
Detection limit for methylmercuric chloride	2×10^{-12} g	1×10^{-12} g

TABLE 11

RELATIVE RETENTION TIME OF ALKYL MERCURY COMPOUNDS

Column: (A) 25% DEGS, 160°, 40 cm; retention time MMC = 2.35 min, (B) 5% 1,4-BDS, 160°, 1.2 m; retention time MMC = 2.30 min

Compound	Relative retention time	
	A	B
Methylmercuric iodide (MMI)	0.98	0.97
Methylmercuric chloride (MMC)	1.00	1.00
Methylmercuric hydroxide (MM(OH))	1.00	1.03
Methylmercuric acetate (MMA)	1.00	1.06
Methylmercuric sulfate (MMS)	1.00	1.05
Ethylmercuric chloride (EMC)	1.79	1.75
Ethylmercuric phosphate (EMP)	1.80	1.74
n-Propylmercuric chloride (PrMC)	1.80	1.83
n-Butylmercuric chloride (BMC)	2.43	2.54
n-Amylmercuric chloride (AMC)	3.30	3.57
Methoxyethylmercuric chloride (MtEMC)	5.06	4.11

respectively. Table 11 lists the relative retention times compared to methylmercuric chloride of the alkylmercury compounds using the above-mentioned two liquid phases. The compounds eluted from the columns in order of increasing carbon number. MMC, MMI, MM(OH), MMA and MMS, all methylmercury compounds, showed the same retention time. In R–HgX type compounds, the same retention time was again observed, regardless of X, as long as the carbon number was the same. Nishi and Horimoto[170] had previously discussed this phenomenon.

In the comparison of liquid phases for the greatest efficiency of separation, 1,4-BDS had a theoretical plate number of methylmercuric chloride of 900 compared with 500 for DEGS. Under optimum conditions the detection limit for methylmercuric chloride reached 10^{-12} g. This was achieved by controlling the degree of column aging and by using a low column temperature, hence reducing the noise level. The relative peaks for methylmercury did not show any pertinent quantitative variation (MC:MMI:MMA = 1:1.05:0.85).

For the analysis of phenylmercury derivatives, optimum results were achieved using a 40 cm × 4 mm glass column packed with 5% DEGS on Shimalite W (60 to 80 mesh) with nitrogen as carrier gas at 80 ml/min. The applied voltage for the electron-capture detector was 40 V and the column and detector temperatures were 185 and 210°C, respectively. Under these conditions, the detection limit for on-column GLC analysis for phenylmercuric chloride was 1×10^{-11} g.

For the GLC analyses of organomercury compounds, Sumino[145] stressed the following. (1) Electron-capture detection (ECD) should be employed (its sensitivity is several thousand times higher than that of flame ionization), and the apparatus be kept on for stability of ECD conditions (detector voltage of pulse height, 40 V), (2) the on-column injection technique be employed preferentially on glass columns,

(3) the column be aged for over 48 h, (4) nitrogen should be used as carrier gas with a high flow rate and somewhat low column temperatures, (5) ghost peaks occasionally appear after the injection of solvent containing a large quantity of organic and inorganic mercury.

The isolation of organic mercury from different materials was also described by Sumino[145]. The flow charts depicted in Fig. 15 illustrate the extraction procedures for organic mercury compounds from body fluids, water soluble, and other materials. The key features in the extraction procedure involved the separation of organic mercury combined with SH radical in 1 N HCl, the solution then extracted with benzene and this benzene solution re-extracted with glutathione. After acidification of the resulting solution with HCl, final extraction with benzene is performed. In the analysis of alkylmercury, ghost peaks have been found to occasionally appear if extraction solvents such as ether and ethyl acetate are used, hence the utilization of benzene is desirable. However, since this phenomenon was never observed in phenylmercury analysis, mixed solvents of benzene and ether could be employed.

For cases involving the extraction of organomercury compounds from inter-

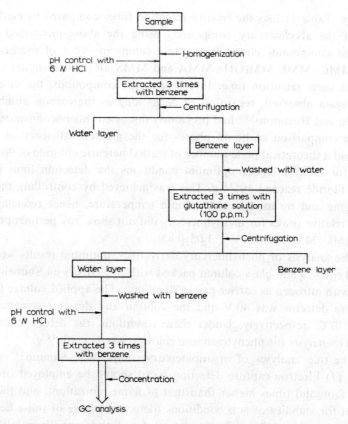

Fig. 15. Method for isolation and estimation of organic mercury compounds.

fering substances which may also dissolve in the extraction solvents and hence interfere with GLC analysis, an ancillary TLC procedure was suggested. After applying the extract to a preparative thin-layer silica gel chromatoplate as a continuous streak, a small amount of the same sample was added to both edges as indicator. The material was developed according to the procedure of Kitamura and Nakayama[171] and the index sample then was visualized with dithizone. Confirmation of an organomercury spot was made and the portions of silica gel containing organic mercury were removed by spatula, extracted with benzene and analyzed by GLC as previously described.

Yamaguchi and Matsumoto[172] recently delineated the GLC of methyl-, ethyl-, propyl-, butyl-, and amylmercuric chlorides using a Shimazu gas chromatograph Model GC-2C with an electron-capture detector ECD-1A (tritium, 300 mCi) in addition to a pulse generator PLS-1A. The following liquid phases and supports were used (1) 3 mm × 75 cm stainless steel column coated with 25% DEGS on Shimalite (60–80 mesh). Injection port and column temperatures were 260 and 165°, respectively. All determinations were performed with nitrogen as carrier gas at 190 ml/min at a pressure of 2.2 kg/cm^2, (2) glass tubing containing 25% DEGS on Shimalite (60–80 mesh). Injection port and column temperatures were 180 and 160°, respectively, with nitrogen as carrier gas at 65.2 ml/min at a pressure of 0.8 kg/cm^2. The concomitant use of a pulse generator with an electron-capture detector remarkably enhanced sensitivity and accuracy. Figure 16 illustrates a gas chromatogram of the chlorides of methyl-, ethyl-, propyl-, butyl-, and amylmercury obtained *via* condition (2) above. The retention times of MMC, EMC, BMC, and

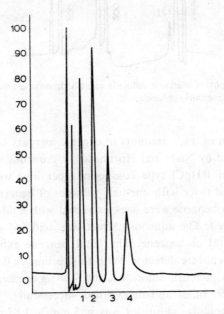

Fig. 16. Gas chromatograph of alkylmercuric chlorides. 1, CH_3HgCl; 2, C_2H_5HgCl; 3, C_4H_9Cl; 4, $C_5H_{11}HgCl$. 1.5 μg of total sample containing 0.6 ng of each alkylmercuric chlorides was injected.

AMC were 1.30, 2.37, 3.58, and 5.15 min, respectively, while PMC had the same retention time as EMC (2.37 min).

In the above study of Yamaguchi and Matsumoto[172], it was found that since the sensitivity of GLC for the detection of MMC was so high, that occasionally unexpected contaminants in reagents (*e.g.* diethyl ether) can affect the determinations and hence its use as a solvent for alkylmercury compounds is to be avoided. It was also disclosed that inorganic mercury (HgCl$_2$) has the same retention time with that of MMC when the concentration of HgCl$_2$ is very high (200 p.p.m.) (Fig. 17).

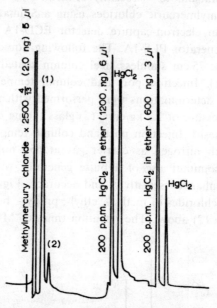

Fig. 17. Gas chromatography of mercuric chloride and alkylmercuric chlorides in ether. 1, Methylmercuric chloride; 2, ethylmercuric chloride.

The determination of trace amounts of organic mercury compounds in aqueous solutions was described by Nishi and Horimoto[173]. Aqueous solutions (100–500 ml) containing >0.4 μg of RHgCl type compounds per liter were acidified to 0.1 N with HCl and extracted twice with one-half volumes of benzene. The mercury compounds extracted into benzene were back-extracted with 8 ml of aqueous 1-cysteine solution containing 1 g/l. The aqueous extract was acidified with 2 ml of 5 N HCl and extracted with 1 ml of benzene. This final benzene extract was analyzed by GLC using an electron-capture detector. With this technique 0.41 μg methylmercuric chloride/l; 0.43 μg ethylmercuric chloride/l, and 0.86 μg of phenylmercuric chloride/l were concentrated ≦50 times and determined successfully. The column used for the analysis of alkylmercuric chlorides was a 3 mm × 1.87 m stainless steel tube packed with 5% diethylene glycol succinate polyester on Chromosorb W at 150°.

For phenylmercuric chloride a 4 mm × 1.05 m glass tube packed with 2% DEGS on Chromosorb W was used at 180°.

Nishi[170,174] described the GLC of a number of organomercury compounds using a Shimazu gas chromatograph Model GC$_6$41E and two columns: (*a*) 2 m × 1 mm stainless steel containing 5% polydiethylene glycol succinate on Chromosorb W (AW) 60–80 mesh, (*b*) glass column containing 1% polyethylene glycol on rock crystal powder at 140–150°. For column (*a*), the columns, injection port and detector temperatures were 140, 140–160, and 150°, respectively. Detection was accomplished with an electron-capture detector with 300 mμCi tritium and the carrier gas was nitrogen of 99.99% purity, flow rate 60 ml/min, and the pressure at the column inlet was 0.8 kg/cm². Nishi[174] recommended the use of Pyrex glass tubings in place of metal columns, and it is advised to wash the inner surfaces with conc. HCl and water before packing the columns. Polyethylene glycol succinate, polyethylene glycol 20M and polybutanediol succinate gave similar results to those described with polydiethylene glycol succinate on Chromosorb W (AW).

Hey[175] described the utility of atomic absorption spectrometry as a mercury-specific detecting system for gas chromatography. The system, especially useful for the detection of mercury compounds in food, consisted of a flame-ionization detector connected with a reaction vessel containing 20% sulfuric acid and 10% stannous chloride, a cold trap, both cooled with ice water, a cuvette in the light path of an atomic spectrometer and a vacuum pump. The detection limit was 50 ng of Hg.

Brodersen and Schlenker[176] described the paper and gas–liquid chromatography of a number of alkyl and aryl mercurial bromides. Table 12 illustrates the R_F values of various organomercurial bromides obtained on Schleicher & Schüll Paper No. 2043 using four solvent systems, *viz.* (*1*) *n*-butanol saturated with 1 *N* ammonia, (*2*) tetra-

TABLE 12

R_F VALUES OF ALKYL AND ARYL MERCURIAL BROMIDES ON SCHLEICHER & SCHÜLL PAPER NO. 2043

Solvent system	Compound	R_F value
n-Butanol with 1 *N* ammonia (satd. solution)	Br–Hg–Br	0.00
	CH$_3$–Hg–Br	0.15
	C$_2$H$_5$–Hg–Br	0.25
	n-C$_3$H$_7$–Hg–Br	0.31
	n-C$_4$H$_9$–Hg–Br	0.86
	p-CH$_3$OC$_6$H$_4$–Hg–Br	0.25
	C$_6$H$_5$–Hg–Br	0.35
	p-Br–C$_6$H$_4$–Hg–Br	0.92
Tetrahydrofuran–*n*-butanol–1 *N* ammonia (15:35:25)	C$_6$H$_5$–Hg–Br	0.01
	p-CH$_3$OC$_6$H$_4$–Hg–Br	0.27
Methyl acetate–*n*-butanol–2 *N* ammonia (47:40:13)	*p*-CH$_3$OC$_6$H$_4$–Hg–Br	0.26
PO[N(CH$_3$)$_2$]$_3$–*n*-butanol–2 *N* ammonia (1:1:1)	C$_6$H$_5$–Hg–Br	0.82
	p-CH$_3$OC$_6$H$_4$–Hg–Br	0.96

hydrofuran–n-butanol–1 N ammonia (15:35:20), (3) methyl acetate–n-butanol–2 N ammonia (47:40:13), and (4) phosphoric acid tri(dimethyl)amide–n-butanol–2 N ammonia (1:1:1). Detection was accomplished with either ultraviolet light or by spraying with an alcoholic dithizone solution. Table 13 records the retention times at 190 and 220° of methyl-, ethyl-, n-propyl-, and n-butylmercuric bromides obtained with a Beckman gas chromatograph Model GC-2 and a 6 ft. column loaded with Silicone Oil DC-550.

TABLE 13

RETENTION TIMES OF ALKYL MERCURIC BROMIDES
Values obtained with a Beckman gas chromatograph GC-2, 6 ft. column containing Silicone DC-550.
Carrier gas: hydrogen at 4.2 atm

Compound	Column temperature (°C)	Retention time (min)
CH$_3$–Hg–Br	190	3.0
	220	1.7
C$_2$H$_5$–Hg–Br	190	5.6
	220	3.0
n-C$_3$H$_7$–Hg–Br	190	6.2
	220	3.5
n-C$_4$H$_9$–Hg–Br	190	9.4
	220	4.7

The chemical methylation of inorganic mercury with methylcobalamin, a vitamin B$_{12}$ analog, has been delineated by Imura et al.[177]. Methylated mercury was detected by GLC and analysis of the products of the reaction by TLC revealed that the methylation proceeded at a very high rate when methylcobalamin and inorganic mercury were mixed. (Dimethylmercury was an initial product of the reaction.) TLC on silica gel was performed using three solvent systems: (a) ether–petroleum ether (30:70); (b) acetone–n-hexane (15:85), and (c) n-hexane–chloroform (10:90). Mercury was detected by spraying with 0.05% dithizone–chloroform solution. With solvent (a) the R_F values of dimethylmercury and methylmercuric chloride were 0.58 and 0.25, respectively.

The gas chromatography apparatus consisted of a Shimadzu gas chromatograph, model GC-4A1E with electron-capture detector and a 100 cm × 0.4 cm column containing 25% polyethylene glycol succinate coated on Chromosorb W (AW-HMDS) (60–80 mesh). The column, injection port, and detector temperatures were 160, 185–190, and 200°C, respectively, with nitrogen carrier gas flow maintained at 80 ml/min. Under these conditions, a linear calibration curve was obtained with 0–2.5 ng of methylmercuric chloride.

Tatton and Wagstaffe[165] described a method for the extraction of a number of organomercurial fungicides from apples, potatoes, and tomatoes and their identification and determination as their dithizonates by TLC and GLC. The following types of organomercury compoinds were examined:

CH$_3$–Hg–X   CH$_3$O–C$_2$H$_5$–Hg–X   Phenyl–Hg–X

C$_2$H$_5$–Hg–X   C$_2$H$_5$O–C$_2$H$_5$–Hg–X   Tolyl–Hg–X

The nature of X affects the properties of the compound and hence dictates the choice of chromatographic solvents to be employed. In general, when X is an anion such as sulfate, nitrate, or acetate, the compound tends to be ionic and water soluble. However, when X is a halogen or dicyandiamide, the compounds tends to be non-

TABLE 14

R_F VALUE $\times$ 100 OBTAINED BY TLC OF DITHIZONATES OF ORGANOMERCURIAL COMPOUNDS
Systems: (*1*) silica gel, hexane–acetone (9:1); (*2*) silica gel, hexane–acetone (19:1); (*3*) silica gel, hexane–acetone (93:7); (*4*) silica gel, light petroleum–acetone (9:1); (*5*) alumina, hexane–acetone (19:1); (*6*) alumina, light petroleum–acetone (19:1). Layer thickness: 250 μm

Dithizonate	System					
	1	*2*	*3*	*4*	*5*	*6*
Methylmercuric	64	48	57	77	89	86
Ethylmercuric	64	51	62	78	91	87
Methoxyethylmercuric	32	16	25	44	58	49
Ethoxyethylmercuric	44	23	34	55	71	67
Phenylmercuric	48	34	46	62	72	69
Tolylmercuric	52	40	53	69	79	76
Mercuric di-dithizonate	19	9	17	28	19	15

polar and hence more soluble in organic solvents. To obviate the difficulties of solvent selections for the variety of organomercurial fungicides screened, the respective dithizonates were prepared. Table 14 depicts the R_F values $\times$ 100 obtained by TLC of dithizonates of organomercury compounds on both silica gel and alumina adsorbents in six solvent systems. The dithizonates are visualized as yellow or red spots with the limits of detection being in the order of 2 μg/spot. The dithizonates of the organomercurials were also found to be successfully separated by GLC using (*1*) 1.5 m $\times$ 3 mm glass columns coated with 2% polyethylene glycol succinate on Chromosorb G (acid-washed, DMCS treated, 60–80 mesh) and (*2*) 1.2 m $\times$ 3 mm column coated with 1% polyethylene glycol succinate on the same support as above.

Retention times for the organomercury dithizonates are depicted in Table 15. The use of columns (*1*) and (*2*), above, permitted the detection of 1 and 0.5 μg respectively, of the aryl mercury compounds. Because mercury compounds are known to "poison" tritiated foil detectors, it was found necessary to maintain an oven temperature of 180° and restrict the mercury content of injections to 100 μg or less.

TABLE 15

GLC RETENTION TIMES FOR ORGANOMERCURIAL DITHIZONATES

(1) 2% polyethyleneglycol succinate on Chromosorb G (acid-washed, DMCS-treated, 60–80 mesh) in glass columns 1.5 m long, 3 mm i.d.; carrier gas, nitrogen

Dithizonate	Column temperature (°C)				
	140	150	160	170	180
Methylmercuric	3.8	2.8	2.2	1.6	1.2
Ethylmercuric	6.6	4.6	3.6	2.7	2.0
Ethoxyethylmercuric	17.0	11.6	8.7	6.2	4.9
Methoxyethylmercuric	17.4	12.0	8.7	6.2	4.9
Tolylmercuric				29.0	19.5
Phenylmercuric				42.0	27.0

(2) 1% polyethyleneglycol succinate on Chromosorb G (acid-washed, DMCS-treated, 60–80 mesh) in glass columns, 1.2 m long, 3 mm i.d.; carrier gas, nitrogen

Dithizonate	Column temperature (°C)	
	170	180
Tolylmercuric	6.4	3.2
Phenylmercuric	10.0	5.0

Takeshita et al.[91] described a reversed-phase thin-layer chromatographic method for the separation and detection of the dithizonates of inorganic mercury compounds and a series of alkylmercury compounds and found the method applicable to the separation and identification of mercury compounds in foods and sewages. The adsorbents used were cornstarch containing liquid paraffin and Avicel SF (FMC, American Viscose Div.) containing 20% liquid paraffin. The solvent systems consisted of methyl cellosolve–water (85:25) and ethanol and methyl cellosolve treated, respectively, with water in the ratios from 0 to 30%. The study revealed that the best pattern of separation of all the dithizonates was obtained on liquid paraffin–Avicel SF layers with methyl cellosolve–water (7:3), although the development time was longer than with impregnated cornstarch plates (1–1.5 h). No interference of other metals and the residual dithizone was observed.

The TLC of alkyl- and alkoxymercury derivatives and the location of mercury in the yolk of hen eggs was described by Rissanen and Miettinen[178]. Methoxymercuric silicate, dimethylmercuric sulfate, and methoxyethylmercuric chloride were separated and detected by TLC on silica gel using n-butanol–acetic acid–water (73:60:13) as solvent and 1% diphenylcarbazone in ethanol as detector. The R_F values found were methoxymercuric silicate 0.46, dimethylmercuric sulfate 0.58, and methoxyethylmercuric chloride 0.69. A ferrocyanate-α,α'-dipyridyl mixture was used as a qualitative spot test for mercury. Alkoxy derivatives were found in the yolk and alkyl derivatives in the white of egg.

Kurayuki and Kusamoto[80] described the chromatographic separation of inorganic and organic mercury compounds on anion-exchange paper, Amberlite SB-2 (R–Cl) using $1.6\,N$ HNO_3 and $2\,M$ ammonium nitrate as developers and dithizone in chloroform for detection. The R_F values were alkylmercury compounds 0.19–0.33, inorganic mercury compounds 0.09–0.11, and phenylmercury compounds 0.01–0.03. The limits of detection of the mercurials by this technique was 0.2 μg. In an additional study on the determination of inorganic mercury compounds by measuring the area of the colored zone on strips of Amberlite SB-2 (R–Cl), the amount of mercury (1–4 mg) was found to be proportional to the colored area at pH 5.0.

The intestinal transport of ^{203}Hg-labeled methylmercuric chloride in the rat was elaborated by Norseth and Clarkson[100]. Mercury was excreted in the bile predominantly as methylmercury cysteine (which was rapidly re-absorbed, not adding significantly to fecal mercury). Protein-complexed methylmercury and protein-bound inorganic mercury were found less extensively. The chemical form of mercury in bile was determined by TLC on silica gel (Eastman Chromagram 6061 Silica Gel) and electrophoresis (Beckman paper No. 319328 or 320046). The developing solvents for TLC were n-propanol–water (70:30) and ethanol–34% ammonium hydroxide (70:30) (for 3–5 h developments). For electrophoresis, diethyl barbiturate–sodium barbiturate buffer pH 8.6 was used for a running time of ca. 14 h with 5 mA constant current. Bile was analyzed by Sephadex G-100 column chromatography to separate diffusible and protein-bound mercury. A Sephadex 15/30 column was used with sodium potassium phosphate buffer at 0.1 M and pH 8.0 containing 1.0 N sodium chloride and 0.02% sodium azide as eluting and supporting solution. The flow rate of the columns varied from 10 to 20 ml/h and the optical transmission of the effluent was monitored at 280 nm.

The utility of a selective chelating resin (Ionac SRXL) for the collection and separation of methylmercury and inorganic mercury was described by Law[179]. Both forms of mercury could be collected from pH 1 to 9 and subsequently eluted with a slightly acid, 5% solution of thiourea. A resin-loaded paper composed of the above resin and cellulose (1:1) exhibited properties similar to those of the loose resin.

Although the predominant environmental concern regarding mercury has centered about methylmercury, it is important to briefly consider the chromatographic aspects of other mercurial sources since they too can serve as methylmercury precursors as discussed earlier.

Phenylmercuric acetate (PMA) is the most important organomercurial of commerce (approx. 500,000 lb. (1964) being produced annually in the U.S.). PMA is made by the reaction of mercuric acetate and benzene, usually in the presence of acetic acid as a co-solvent. The technical grade PMA contains 88% of the pure compound and the remaining 12% is di- and triacetoxy phenylmercury, commonly referred to as "polymercurated benzene". Purified grades of PMA are obtained by recrystallization of the technical product from water. A large proportion of PMA sold in the U.S. is in ammoniacal water solution. In addition to PMA being a potent fungicide and bactericide, it finds application as a herbicide for postemergence control of crabgrass. Phenylmercury compounds are used for the preservation of water-base emulsion paints and for rendering paints resistant to mildew; as "slimicides" in paper mills, as preservatives in adhesives and as bactericides.

Kanazawa et al.[180] described the paper chromatography of a number of organomercurials at 27–28° using n-butanol saturated with 1 N ammonia solution. The R_F values found were phenylmercuric acetate 0.39, phenylmercuric chloride 0.40, ethylmercuric chloride 0.27, diethylmercuric hydrogen phosphate 0.27, ethoxyethylmercuric chloride 0.18, methylmercuric chloride 0.17, and mercuric chloride and mercurous chloride 0.00.

Yamaguchi et al.[181] described the microdetermination of organic mercurials by TLC on silica gel. The separation of phenylmercuric acetate and methylmercuric chloride from mercuric chloride was accomplished using chloroform–ethanol (99:1) while the separation of phenylmercuric acetate (R_F 0.61) from methylmercuric chloride (0.53) was accomplished using benzene–high-boiling petroleum ether (5:1) as developer with 0.05% dithizone in chloroform as the detecting reagent in both cases (the reliability of separation of organic from inorganic mercurials by this method is fairly acceptable in a range 0.2–0.05 μg). Following TLC development, the silica gel zone containing the mercurials is removed and total mercury then measured by ultraviolet absorption which includes fixation of mercury with dithizone decomposition of mercury by heating to produce mercury vapor and subsequent extraction of the mercury by ultraviolet photometry.

Kitamura et al.[182] described the GLC analysis of phenylmercuric acetate and related compounds. Phenylmercuric compounds such as the chloride, iodide, and acetate derivatives in plant specimens were first extracted with hot water, then examined on a 40 cm glass column packed with 5% DEGS on Shimalite W (60 to 80 mesh) at 185° and detected by an electron-capture detector at 210°. Nitrogen was the carrier gas at 80 ml/min and the initial pressure was 0.5 kg/cm². The phenylmercury compounds were detected by an electron-capture type detector at 210°. The retention times of the three compounds studied were essentially the same. The presence of a methylmercury compound (probably methylmercuric acetate) was determined in phenylmercuric acetate by GLC analysis[183].

The peak corresponding to a methylmercury compound was identified in all chromatograms of phenylmercuric acetate such as chemical reagents and agricultural materials. The same peak was also recognized in the chromatograms of benzene or ether-extracted inorganic compounds such as $HgSO_4$, Hg_2SO_4, and $HgCl_2$. This study indicated the possibility of the presence of methylmercuric acetate, a toxic chemical in phenylmercuric actetate, a non-toxic pesticide.

Ishikuro and Yokota[184] described a sensitive and differential determination of phenylmercuric acetate and inorganic mercury. The solution containing both inorganic and organic mercury was shaken with 5 N HCl and 2 × 10^{-4} M dithizone in carbon tetrachloride and the carbon tetrachloride extract chromatographed on an alumina oxide (Merck) column containing 6% water. Successive elution with carbon tetrachloride, carbon tetrachloride–chloroform (19:1), and carbon tetrachloride–chloroform (1:1) yielded first a yellow eluate containing phenylmercuric dithizonate, then an orange eluate of mercuric dithizonate. These eluates were stabilized with anhydrous acetic acid and extinctions measured at 480 and 490 nm for phenylmercuric and mercuric dithizonates, respectively. The method was found useful for analysis of mercurials used extensively as drugs and in agricultural applications. An acidified sample solution was extracted with dithizone–carbon tetrachloride solution and the carbon tetrachloride layer was chromatographed on an alumina column and the separated dithizonate solutions were measured by an Hitachi photo-electric spectrophotometric Model EPU-2A. With this procedure organic and inorganic mercury could be separately determined at levels to 0.02 p.p.m. with an error not exceeding 5% using minute amounts of material and without subjecting it to the tedious procedure of preliminary oxidative degradation.

REFERENCES

1 N. Fimreite, *Environ. Pollution*, 1 (1970) 119.
2 M. Murozumi, *Electrochem. Technol.*, 5 (1967) 236.
3 N. Nelson, T. C. Byerly, A. C. Kolbye, Jr., L. T. Kurland, R. E. Shapiro, W. H. Stickel, J. E. Thomson, L. A. Vandenberg and A. Weissler, *Environ. Res.*, 4 (1971) 1.
4 S. Jensen and A. Jernelöv, *Nordforsk. Biocidinformation*, No. 10 (1967) 4.
5 S. Jensen and A. Jernelöv, *Nordforsk. Biocidinformation*, No. 14 (1968) 5.
6 J. M. Wood, F. S. Kennedy and C. G. Rosen, *Nature*, 220 (1968) 173.
7 A. Jernelöv, in M. W. Miller and G. G. Berg (Eds.), *Chemical Fallout*, Thomas, Springfield, Ill., 1969, p. 68.
8 L. J. Goldwater, *Sci Am.*, 224 (1971) 15.
9 D. K. H. Lee, Personal communication.
10 A. G. Johnels and T. Westermark, in M. W. Miller and G. G. Berg (Eds.), *Chemical Fallout*, Thomas, Springfield, Ill., 1969, p. 221.
11 C. V. King, *Ann. N.Y. Acad. Sci.*, 65 (1957) 360.
12 J. L. Webb, *Enzyme and Metabolic Inhibitors*, Vol. II, Academic Press, New York, 1966, p. 730.
13 H. A. Shoemaker, *Ann. N.Y. Acad. Sci.*, 65 (1957) 504.
14 H. Passow, A. Rothstein and T. W. Clarkson, *Pharmacol. Rev.*, 13 (1961) 185.
15 P. L. Bidstrup, *Toxicity of Mercury and Its Compounds*, Elsevier, Amsterdam, 1964, p. 28.
16 Intern Comm. Rept. on Maximum Allowable Concentrations of Mercury Compounds, *Arch. Environ. Health*, 19 (1969) 891.
17 A. Swensson, *OIKOS*, Suppl. 9 (1967) 27.
18 M. C. Battigelli, *J. Occupational Med.*, 2 (1960) 394.
19 S. C. Harvey, in L. S. Goodman and A. Gilman (Eds.), *Pharmacological Basis of Therapeutics*, Macmillan Co., New York, 3rd edn., 1967, p. 98.
20 T. Sufuki, in M. W. Miller and G. G. Berg (Eds.), *Chemical Fallout*, Thomas, Springfield, Ill, 1969, p. 245.
21 N. Grant, *Environment*, 11 (1969) 18.
22 L. Friberg and J. Vostal, *Mercury in the Environment. A Toxicological and Epidemiological Appraisal*, Karolinska Inst. Stockholm, 1971, Environ. Protection Agency, Research Triangle Park, N.C., 1971.

23   B. UNDERDAL, *Nord. Veterinarmed.*, 20 (1968) 9.
24   B. SJOSTRAND, *Anal. Chem.*, 36 (1964) 814.
25   T. WESTERMARK AND B. SJOSTRAND, *Intern. J. Appl. Radiation Isotopes*, 9 (1960) 1.
26   O. JOHANSEN AND E. STEINNES, *Intern. J. Appl. Radiation Isotopes*, 20 (1969) 751.
27   W. R. HATCH AND W. L. OTT, *Anal. Chem.*, 40 (1968) 2085.
28   E. G. PAPPAS AND L. A. RUSENBERG, *J. Assoc. Offic. Agr. Chemists*, 49 (1966) 792.
29   B. B. MESMAN, B. S. SMITH AND J. O. PIERCE, II, *Am. Ind. Hyg. Assoc. J.*, 31 (1970) 701.
30   A. D. RATHJE, *Am. Ind. Hyg. Assoc. J.*, 30 (1969) 126.
31   M. J. FISHMAN, *Anal. Chem.*, 42 (1970) 1462.
32   J. F. UTHE, F. A. J. ARMSTRONG AND M. P. STAINTON, *J. Fisheries Res. Board Canada*, 27 (1970) 805.
33   T. W. CLARKSON, in M. W. MILLER AND G. G. BERG (Eds.), *Chemical Fallout*, Thomas, Springfield, Ill., 1969, p. 274.
34   T. W. CLARKSON AND M. R. GREENWOOD, *Talanta*, 15 (1968) 547.
35   L. MAGOS, *Brit. Ind. Med.*, 23 (1966) 230.
36   M. B. JACOBS, S. YAMAGUCHI, L. J. GOLDWATER AND H. GILBERT, *Am. Ind. Hyg. Assoc. J.*, 21 (1960) 475.
37   M. M. SCHACHTER, *J. Assoc. Offic. Anal. Chemists*, 49 (1966) 778.
38   W. W. VAUGHN, *U.S. Geol. Surv. Circ. 540*, 1967.
39   T. Y. TORIBARA AND C. P. SHIELDS, *Am. Ind. Hyg. Assoc. J.*, 29 (1968) 87.
40   G. LINDSTEDT, *Analyst*, 95 (1970) 264.
41   J. C. GAGE AND J. M. WARREN, *Am. J. Occupational Hyg.*, 13 (1970) 115.
42   I. M. WEINER AND O. H. MULLER, *J. Pharmacol. Exptl. Therap.*, 113 (1955) 24.
43   I. OKUNO, R. A. WILSON AND R. E. WHITE, *J. Assoc. Offic. Anal. Chemists*, 55 (1972) 96.
44   J. T. STOCK, *Amperometric Titrations*, Interscience, New York, 1965.
45   B. C. SOUTHWORTH, J. H. HODECKER AND K. D. FLEISCHER, *Anal. Chem.*, 30 (1958) 1152.
46   D. PAVLOVIE AND S. ASPERGER, *Anal. Chem.*, 31 (1959) 939.
47   D. M. GOLDBERG AND A. D. CLARKE, *J. Clin. Pathol.*, 23 (1970) 178.
48   V. L. MILLER AND F. SWANBERG, JR., *Anal. Chem.*, 29 (1957) 391.
49   W. H. GUTENMANN AND D. J. LISK, *J. Agr. Food Chem.*, 8 (1960) 306.
50   E. P. LAUG AND K. W. NELSON, *J. Assoc. Offic. Agr. Chemists*, 25 (1942) 309.
51   S. S. C. TONG, W. H. GUTENMANN AND D. J. LISK, *Anal. Chem.*, 41 (1969) 1872.
52   F. N. KUDSK, *Acta Pharmacol. Toxicol.*, 23 (1965) 263.
53   F. N. KUDSK, *Acta Pharmacol. Toxicol.*, 27 (1969) 149.
54   L. MAGOS AND A. A. CERNIK, *Brit. J. Ind. Med.*, 26 (1969) 144.
55   K. C. BURTON AND H. M. N. H. IRVING, *Anal. Chim. Acta*, 52 (1970) 491.
56   V. I. MUSCAT, T. J. VICKERS AND A. ANDREN, *Anal. Chem.*, 44 (1972) 218.
57   J. C. GAGE, *Brit. J. Ind. Med.*, 18 (1961) 287.
58   L. MAGOS, *Environ. Res.*, 1 (1967) 323.
59   M. BERLIN, L. G. JERSELL AND H. VON UBISCH, *Arch. Environ. Health*, 12 (1966) 33.
60   M. H. BERLIN, G. F. NORDBERG AND F. SERENIUS, *Arch. Environ. Health*, 18 (1969) 42.
61   T. W. CLARKSON AND A. ROTHSTEIN, *Health Phys.*, 10 (1964) 1115.
62   M. BERLIN AND S. ULLBERG, *Arch. Environ. Health*, 6 (1963) 589.
63   O. FITZHUGH, A. NELSON, E. P. LAUG AND F. KUNZE, *J. Ind. Hyg. Occupational Med.*, 2 (1950) 433.
64   T. NORSETH, *Ph. D. Thesis, Studies on the biotransformation of methyl mercury salts in the rat*, Univ. of Rochester, Rochester, N.Y., 1969.
65   T. NORSETH AND T. W. CLARKSON, *Biochem. Pharmacol.*, 19 (1970) 2775.
66   T. NORSETH AND T. W. CLARKSON, *Arch. Environ. Health*, 21 (1970) 717.
67   H. ROMANOWSKI, K. IZDEBSKA AND Z. POZECZIEK, *Farm. Polska*, 10 (1961) 452; *Chem. Abstr.*, 57 (1962) 10145f.
68   K. YOSHIO AND K. KUSUMOTO, *Bunseki Kagaku*, 16 (1967) 815.
69   W. B. LINK, K. S. KEINE, JR., J. H. JONES AND P. WATTLINGTON, *J. Assoc. Offic. Agr. Chemists*, 47 (1964) 391.
70   R. INDOVINA, E. DELEO AND B. M. RICOTTA, *Ann. Chim. (Rome)*, 45 (1955) 244.

71   M. B. Celap and Z. Radikovjevic, *Bull. Soc. Chim. Belgrade*, 23–24 (1958–1959) 1; *Anal. Abstr.*, (1960) 3115.
72   A. P. V. Murthy, V. A. Narayan and M. P. A. Rao, *Current Sci. India*, 24 (1955) 158.
73   J. F. W. McOmbie, F. H. Pollard and I. I. M. Elbeih, *Discussions Faraday Soc.*, 7 (1949) 183.
74   E. Blasius and W. Göttling, *Z. Anal. Chem.*, 162 (1958) 423.
75   K. Fukuda and A. Mizuike, *Bunseki Kagaku*, 17 (1968) 65.
76   H. Nagai, *Kumamoto J. Sci. Ser. A.*, 6 (1964) 138.
77   N. Sugawara, *Kagaku Keisatsu Kenkyusho Hokoku*, 20 (1967) 176.
78   G. Alberti, *Chromatog. Rev.*, 8 (1966) 246.
79   H. Tanaka, Y. Sugiura and A. Yokoyama, *Bunseki Kagaku*, 17 (1968) 1424.
80   Y. Kurayuki and K. Kusamoto, *Japan Analyst*, 16 (1967) 815.
81   T. Westermark, D. Hagman, A. Westermark and K. L. Junggren, *Nord. Hyg. Tidskr.*, 50 (1969) 79.
82   G. Catoni, *Atti. Acad. Sci. Torino*, 91 (1956–1957) 23.
83   H. Seiler and M. Seiler, *Helv. Chim. Acta*, 43 (1960) 1939.
84   H. Seiler and M. Seiler, *Helv. Chim. Acta*, 44 (1961) 1753.
85   S. Takitani, M. Suzuki, M. Yoshimura, S. Sato and M. Skeiya, *Eisei Kagaku*, 14 (1968) 324.
86   P. Kunzi, J. Baümler and J. I. Obsersteg, *Z. Gerichtl. Med.*, 52 (1962) 605.
87   H. Seiler, *Helv. Chim. Acta*, 53 (1970) 1893.
88   F. W. Koss and G. Beisenhrz, *Radiochim. Acta*, 3 (1964) 220.
89   T. Deguchi, *Bunseki Kagaku*, 15 (1966) 357.
90   J. H. Senf, *J. Chromatog.*, 21 (1966) 363.
91   R. Takeshita, H. Akagi, M. Fujita and Y. Sakagami, *J. Chromatog.*, 51 (1970) 283.
92   T. Tadmor, *J. Gas Chromatog.*, 2 (1964) 385.
93   E. W. Berg and J. T. Truemper, *Anal. Chem.*, 30 (1958) 1827.
94   K. Samsahl, *Anal. Chem.*, 39 (1967) 1480.
95   K. Samsahl, *Aktiebolaget Atomenergi, Stockholm, AE-82*, 1962.
96   H. D. Livingston, H. Smith and N. Stojanovic, *Talanta*, 14 (1967) 505.
97   I. Klimes and M. Betusova, *Anal. Biochem.*, 23 (1968) 102.
98   J. F. Koop and R. G. Keenan, *Am. Ind. Hyg. Assoc. J.*, 24 (1963) 1.
99   R. A. A. Muzarelli, *Talanta*, 13 (1966) 809.
100 T. Norseth and T. W. Clarkson, *Arch. Environ. Health*, 23 (1971) 568.
101 B. Wysocka, *Dissertationes Pharm.*, 17 (1965) 99.
102 T. W. Clarkson and M. R. Greenwood, *Anal. Biochem.*, 37 (1970) 236.
103 *U.S. Dept. Interior Bibliography on Mercury Contamination on the Natural Environment*, Washington, D.C., 1970.
104 Nat. Inst. of Health, *Nord. Hyg. Tidskr. Suppl.*, 4 (1971) 19.
105 A. Curley, W. A. Sedlak, E. F. Girling, R. E. Hawk and W. F. Barthel, *Science*, 172 (1971) 65.
106 T. Takeuchi, *Proc. 7th Intern. Congr. Neurol., Rome 1961*, p. 1.
107 L. T. Kurland, S. N. Faro and H. Seidlerk, *World Neurol.*, 1 (1960) 1370.
108 H. Matsumoto and T. Takeuchi, *J. Neuropathol. Exptl. Neurol.*, 24 (1965) 563.
109 G. Lofroth, *Ecological Res. Comm. Bull. 4*, Swedish Natural Science Res. Council, Stockholm, 1969.
110 T. Takeuchi, *Intern. Conf. Environ. Mercury Contamination, Ann Arbor, Mich., Sept. 30 to Oct. 2, 1970.*
111 Rept. Intern. Committee, *Arch. Environ. Health*, 19 (1969) 891.
112 K. Irukayama, *Advan. Pollution Res.*, 3 (1969) 153.
113 S. Tejning, *Rept. 68.02.20*, Dept. Occupational Med., Univ. Hospital Lund, Sweden, 1961, p. 396.
114 B. Åberg, L. Ekman, R. Falk, V. Greitz, G. Persson and J. O. Snihs, *Arch. Environ. Health*, 19 (1969) 478.
115 J. K. Miettinen, T. Rahola, T. Hattula, K. Rissane and M. Tillander, *Ann. Clin. Res.*, 3 (1971) 116.

116 F. BERGLUND AND M. BERLIN, in M. W. MILLER AND G. G. BERG (Eds.), *Chemical Fallout*, Thomas, Springfield, Ill., 1969, p. 258.
117 M. BERLIN, *Acta Med. Scand. Suppl*, (1961) 396.
118 M. BERLIN AND S. ULLBERG, *Arch. Environ. Health*, 6 (1963) 610
119 E. SEBE AND Y. ITSUNO, *Nisshin Igaku*, 49 (1962) 407.
120 D. HUNTER, R. R. BOMFORD AND D. S. RUSSELL, *Quart. J. Med.*, 9 (1940) 193.
121 S. SKERVING, K. NANSSON AND J. LINDSTEN, *Arch. Environ. Health*, 21 (1970) 133.
122 G. ENGLESON AND T. HERNER, *Acta Paediat.*, 41 (1952) 289.
123 R. E. JERVIS, D. DEBRUN, W. LE PAGE AND B. TIEFENBACH, *Nat. Health Grant Project, No. 605-7-510*, 1970.
124 P. E. CORNELIUSSEN, *Pestic. Monit. J.*, 2 (1969) 140.
125 D. C. ABBOTT AND J. O'G. TATON, *Pestic. Sci.*, 1 (1970) 99.
126 E. SOMERS AND D. M. SMITH, *Food Cosmet. Toxicol.*, 9 (1971) 185.
127 L. J. GOLDWATER, *Arch. Environ. Health*, 19 (1964) 6.
128 U.S. Dept. of Agriculture, Agr. Res. Service, Household Economics Res. Div., H.H.E. Adm-214 (Dec), Washington, D.C. (1960).
129 L. FRIBERG, *Nord. Hyg. Tidskr. Suppl.*, 3 (1970) 25.
130 I. A. RAPOPORT, *Byul. Mosk. Obshchestva Ispytatelei Prirody, Otd. Biol.*, 69 (1964) 112.
131 V. F. DENISOV, I. M. ZHIRONKIN AND L. N. GORSHKOV, *Tr. Mosk. Obshchestva Isptytatelei Prirady, Otd. Biol.*, 23 (1966) 311; *Chem. Abstr.*, 67 (1967) 81367h.
132 E. W. E. MACFARLANE, *Genetics*, 35 (1950) 122.
133 C. RAMEL, *Hereditas*, 57 (1967) 448.
134 C. RAMEL, *Hereditas*, 61 (1969) 208.
135 M. UMEDA, K. SAITO, K. HIROSE AND M. SAITO, *Japan. J. Exptl. Med.*, 39 (1969) 47.
136 J. SASS, *Phytopathology*, 27 (1937) 95.
137 D. KOSTOFF, *Nature*, 144 (1939) 334.
138 D. KOSTOFF, *Phytopathol. Z.*, 13 (1940) 90.
139 A. BRUHIN, *Phytopathol. Z.*, 23 (1955) 381.
140 C. RAMEL AND J. MAGNUSSON, *Hereditas*, 61 (1969) 231.
141 U. MURAUUMI, Y. KAMEYAMA AND T. KATO, *Ann. Rept. Res. Inst. Environ. Med. Nagoya Univ.*, 195T (1956) 88; *Chem. Abstr.*, 52 (1958) 1449a.
142 K. IRUKAYAMA, M. FUJIKI, F. KAI AND I. KONDO, *Japan. J. Hyg.*, 17 (1962) 33.
143 K. IRUKAYAMA, F. KAI, M. FUJIKI AND T. KONDO, *Kumamoto Med. J.*, 15 (1962) 57.
144 K. SUMINO, *Kobe J. Med. Sci.*, 14 (1968) 131.
145 K. SUMINO, *Kobe J. Med. Sci.*, 14 (1968) 115.
146 K. UZIOKA, *J. Kumamoto Med. Soc.*, 34 (Suppl. 2) (1960) 383.
147 Y. TAKIZAWA AND H. OTSUKA, *Japan. J. Hyg.*, 22 (1967) 475.
148 Y. TAKIZAWA, R. SUGAI, H. KITANO, A. KAWAZI, I. SASAGAWA AND C. SEKIGUCHI, *Japan. J. Hyg.*, 22 (1967) 469.
149 K. OSTLUND, *Acta Pharmacol. Toxicol.*, 27 (Suppl. 1) (1969) 9.
150 R. FALK, J. O. SNIHS, L. EKMAN, U. GREITZ AND B. ÅBERG, *Acta Radiol.*, 9 (1970) 55.
151 B. ÅBERG, personal communication.
152 G. WESTÖÖ, *Acta Chem. Scand.*, 21 (1967) 1790.
153 G. WESTÖÖ, *Acta Chem. Scand.*, 20 (1966) 2131.
154 G. WESTÖÖ, *Acta Chem. Scand.*, 22 (1968) 2277.
155 J. C. GAGE, *Analyst*, 86 (1961) 457.
156 G. WESTÖÖ, in M. W. MILLER AND G. G. BERG (Eds.), *Chemical Fallout*, Thomas, Springfield, Ill., 1969, p. 75.
157 A. G. JOHNELS, T. WESTERMARK, W. BERG, P. I. PERSSON AND B. S. JOSTRAND, *OIKOS*, 18 (1967) 323.
158 K. NOREN AND G. WESTÖÖ, *Var Foda*, 19 (1967) 13.
159 Anon, *Chem. Abstr.*, 48 (1970) 36.
160 G. WESTÖÖ AND K. NOREN, *Var Foda*, 19 (1967) 137.
161 R. HARTUNG, *Intern. Conf. Environ. Mercury Contam.*, Ann Arbor, Mich., Sept. 30–Oct. 2, 1970.
162 C. A. BACHE AND D. J. LISK, *Anal. Chem.*, 43 (1971) 950.

163 C. A. BACHE AND D. J. LISK, *Anal. Chem.*, 39 (1967) 786.

164 C. A. BACHE AND D. J. LISK, *Anal. Chem.*, 40 (1968) 2224.

165 J. O'G. TATTON AND P. J. WAGSTAFFE, *J. Chromatog.*, 44 (1969) 284.

166 R. C. DRESSMAN, Paper presented at *19th Ann. Detroit Anachem. Conf., Detroit, Mich., Oct. 31, 1971.*

167 W. H. NEWSOME, *J. Agr. Food Chem.*, 19 (1971) 569.

168 D. G. LANGLEY AND P. J. LE BLANC, Paper presented at the *Trace Metals Analysis Symp., Univ. of Illinois, Urbana, Ill., 1971.*

169 S. KITAMURA, I. I. SUKAMOTO, K. HAYAKAWA, K. SUMINO AND T. SHIBATA, *Med. Biol. (Tokyo),* 72 (1966) 274.

170 S. NISHI AND Y. HORIMOTO, *Japan Analyst,* 17 (1968) 75.

171 S. KITAMURA AND H. NAKAYAMA, *Japan. J. Ind. Health,* 7 (1965) 276.

172 S. YAMAGUCHI AND H. MATSUMOTO, *Kurume Med. J.,* 16 (1969) 33.

173 S. NISHI AND Y. HORIMOTO, *Bunseki Kagaku,* 17 (1968) 1247.

174 S. NISHI, personal communication.

175 H. HEY, *Anal. Chem.*, 256 (1971) 361.

176 K. BRODERSEN AND V. SCHLENKER, *Z. Anal. Chem.*, 182 (1961) 421.

177 N. IMURA, E. SUKEGAWA, S. K. PAN, K. NAGAO, J. Y. KIM, T. KWAN AND T. UKITA, *Science,* 172 (1971) 1248.

178 K. RISSANEN AND J. K. MIETTINEN, *Ann. Agr. Fenn. Suppl.,* 7 (1968) 22.

179 S. L. LAW, *Science,* 174 (1971) 285.

180 J. KANAZAWA, M. AYA AND R. SATO, *J. Agr. Chem. Soc. Japan,* 31 (1957) 872.

181 S. YAMAGUCHI, H. MATSUMOTO, M. HOSHIDA AND K. AKITAKE, *Kurume Med. J.,* 16 (1969) 53.

182 S. KITAMURA, K. SUMINO AND K. HAYAKAWA, *Igaku To Seibutsugaku,* 73 (1966) 276; *Chem. Abstr.,* 70 (1969) 56503d.

183 K. TERAMOTO, M. KITABATAKE, M. TANABE AND N. YOSHITAKA, *Kogyo Kagaku Zasshi,* 70 (1967) 601; *Chem. Abstr.,* 68 (1968) 94836f.

184 S. I. SHIKURO AND K. YOKOTA, *Chem. Pharm. Bull. (Tokyo),* 11 (1963) 939.

Chapter 8

THALLIUM AND VANADIUM

1. THALLIUM

Thallium is widely distributed with igneous rocks estimated to contain 30 g/ton. Thallium is found mainly in the minerals crooksite [(TlCuAg)$_2$Se], lorandite (TlAsS$_2$) and orboite (TlAs$_2$SbS$_5$). Commercial sources of thallium however are generally flue dusts either from pyrites burners or from lead and zinc smelters and refiners, as a byproduct in the production of cadmium.

Thallium, in the form of its acetate and sulfate salts, is principally used as rodenticides and pesticides. Additional uses of thallium include (*1*) as a stainless alloy with silver and a corrosion-resistant alloy with lead, (*2*) as a fungicide (Tl$_2$CO$_3$), (*3*) in photoelectric cells (thallium oxysulfide), (*4*) in low-range glass thermometers as Tl–Hg alloy (8.5% Tl), and (*5*) in dyes and pigments. Limited uses of thallium include special glasses, selenium rectifiers, insect proofing, and as a phosphor activator.

The toxicology of thallium has been reviewed by Truhaut[1], Heyroth[2] and Downs et al.[3]. Thallium is one of the most toxic elements both acutely and chronically in both man and animals, regardless of the route of entry (toxicity is largely independent of the valence state, Tl$^+$, or Tl^{3+}). The most characteristic symptom of thallium intoxication is the development of alopecia.

The metabolism of thallium has been studied in rats[3], dogs[2], and in rabbit[1]. In man, 15.4% of an oral dose of thallium was excreted in the urine in 5.5 days and the rate of excretion of the remaining body thallium burden was 3.2%/day[4]. The distribution of ^{204}Tl in 57 tissue sites of man following oral administration was also reported[4].

The mechanism of action of thallium salts was described by Truhaut[1]. Reviews of human thallitoxicosis have been published by Cowley[5] and Munch[6] and a number poisonings cited were from thallium rodenticidal and entomological use as well as ingestion of food contaminated with Tl rodenticide. The effect of various substances on the excretion and the toxicity of thallium has been studied[7] and Kamerbeek et al.[8] have cited the dangerous redistribution of thallium in patients with thallium poisoning by treatment with sodium diethyldithiocarbamate. The teratogenicity of thallium sulfate in chick embryo[9] and in rats[10,11] has also been described.

Thallium has been analyzed by atomic absorption spectroscopy[12], coulometric[13], polarographic[14-16], spectrophotometric[1,3], colorimetric[17,18], volumetric[19], gravimetric[20], complexometric[21], and isotope exchange techniques[22].

The paper chromatographic determination of thallium in toxicological analysis has been described by Diller and Rex[23]. Thallium was separated from a mixture of lead, arsenic, mercury(II), copper, and zinc on Schleicher and Schüll No. 2045B paper developed with methanol–25% sulfuric acid–water (70:10:40). The R_F values were lead 0.0, arsenic 0.68, mercury 0.71, copper 0.82, zinc 0.84, and thallium 0.35. Thallium was detected as a yellow spot by sequential spraying with 5% KI then $Na_2S_2O_3$ (limit of sensitivity 0.03 µg). Thallium could also be detected as a yellow complex (to 0.5 µg) using a 5% solution of Thionalid (thioglycolic acid-β-aminonaphthalide) in acetone.

A sensitive test for the detection of thallium was described by Hauck[24]. The organic material (1 ng) containing thallium was first oxidized with sulfuric acid and nitric acid followed by perchloric acid. The mixture was brought to pH 13.5 or higher with sodium hydroxide and shaken with an ethyl acetate solution of dithizone; the ethyl acetate fraction was then concentrated and shaken with 1–2 drops of 2.5% sulfuric acid to dissolve the thallium. The sulfuric acid solution was used for chromatography with 25% methanol–sulfuric acid–water (7:1:9) as developing solvent. After drying, application of 5% KI developed thallium as a yellow spot with an R_F of 0.35 (approx. 1 µg of thallium could be detected). Other metals do not interfere since their complexes with dithizone are insoluble in ethyl acetate.

Adloff[25] separated Tl^I and Tl^{III} on paper impregnated with zirconium phosphate, zirconium oxide, or zirconium tungstate.

The reversed-phase chromatography of Al, Ga, In, Tl, and the transition metals of the iron group on paper treated with di(2-ethylhexyl)orthophosphoric acid (HDEHP) in chloride medium was described by Cerrai and Ghersini[26,27]. Whatman No. 1 chromatographic paper (CRL/1 type) was treated with cyclohexane solutions of HDEHP which had been previously equilibrated with 2.5 M HCl in the usual manner[28] and hydrochloric acid (1.5–8 M) was used as the eluent.

The separation of Tl^I, Hg^{II}, Ag^I and Cu^{II} by ion-exchange chromatography was reported by Bhatnager and Trivedi[29] using Amberlite IR-120 cation exchanger (sodium form). Mercury, silver, or copper were completely eluted with 2% sodium nitrite and thallium was eluted with 10% sodium sulfate.

Bark et al.[30] using pure cellulose TLC plates impregnated with $(C_4H_9)PO_4$ as the stationary phase and various concentrations of hydrochloric acid as mobile phases separated Tl^+, Bi^{3+}, Cd^{2+}, Co^{2+}, Cu^{2+}, Hg^{2+}, Mn^{2+}, Ni^{2+}, Pb^{2+}, Sb^{2+}, SeO_3^{2-}, TeO_3^{2-}, UO_2^{2+}, and Zn^{2+}. Most ions were visualized by spraying with 2-(pyridyl-azo)-2-naphthol (PAN) and exposing to ammonia vapor.

Chitosan has been used as a chelating support for the collection of transition metal traces from salt solution[31,32]. Thallium is often used in alloys which contain very low levels of impurities. Muzzarelli and Tubertini[33] utilized chitosan powder in the column chromatographic separation of cadmium, mercury, iron, and other metals from 1 mM and 38 mM solutions of thallium(I) nitrate. Thallium was recovered with water and other ions were selectively eluted with complexing agents. Precision line (1 cm) Whatman columns, equipped with Teflon diffusors fed by a peristaltic pump, were filled with 100–200 mesh chitosan powder to form a 15 cm bed. A Laben 512 channel gamma-ray spectrometer and a Desage–Barthold radio-chromatogram scanner were used to measure the distribution of metal ions and to follow the chromatographic separation. Table 1 shows the chromatographic separations of trace metals from 1 mM thallium(I) nitrate solutions obtained on a 15 cm column of 100–200 mesh chitosan.

TABLE 1

CHROMATOGRAPHIC SEPARATIONS OF TRACE METALS FROM 1 mM THALLIUM(I) NITRATE SOLUTION
Column length, 15 cm; 100–200 mesh chitosan powder; 78–87% thallium recovered with water

Ion	Eluant	Volume (ml)	Recovery (%)
Cadmium	KCN, 0.1 M	100	100
Mercury	KCN, 0.1 M	100	99
Lead	NH$_4$COOCH$_3$, 2 M	100	100
	KCN, 0.1 M	100	0
Terbium	EDTA, 0.01 M	100	90
Nickel	EDTA, 0.1 M	100	100
Silver	NH$_4$OH + NH$_4$Cl	100	100
Indium	EDTA, 0.01 M	100	95

Sherma and Strain[34] reported the separation of alkali metals from each other, Ag$^+$, Tl$^+$, and Hg$^+$ and from various multivalent cations by electro-migration from water and nitromethane by using a coupled aqueous–nonaqueous system. The aqueous solution was 0.0125 M in nitriloacetic acid, 0.1 M in NH$_4$OH, and 0.05 M in HCN.

The nonaqueous solution was 0.2 M nitromethane in formamide and 0.4 M in trichloroacetic acid. The separated alkali metals were estimated by flame photometry. In addition, Na, K, Rb, and Cs were detected by neutron activation followed by X-ray spectrophotometry.

2. VANADIUM

Vanadium, ranking twenty-second among the elements of the earth's crust, is widely distributed but in rather low abundance (few deposits contain more than 1 or 2% vanadium). It occurs naturally as vanadinite (9PbO$_3$V$_2$O$_5$PbCl$_2$), patronite (V$_2$S$_5$), uranyl potassium vanadate, carnotite, mothamite and other minerals in Colorado, Peru, and Northern Rhodesia and in the mineral oils of Venezuela, Persia, and the Mexican Gulf. For many years, carnotite–roscoelite ores have been the major source of vanadium as well as its co-product uranium. The world production in 1968 of vanadium in ores and concentrates was 12,562 short tons of which the U.S. produced 6483 tons.

The most important use of vanadium is as an alloying element in the steel industry and as an important component in ferrous and non-ferrous alloys, *e.g.*

titanium, aluminum, etc. Vanadium oxide (V_2O_5) is widely used as a catalyst for primarily oxidation reactions such as in the oxidation of SO_2 to SO_3 in the contact process for the manufacture of sulfuric acid; in the oxidation of ammonia in nitric acid manufacture and in the manufacture of phthalic anhydride. Other uses of vanadium include (1) the manufacture of dyes and inks, paints, and varnish dryers, (2) in insecticides, and (3) in photography.

Vanadium (in the form of V_2O_5) has been found in appreciable amounts in the ash of oil-fired boilers (content varies from 2.7% in Texas crude to 63.2% of total ash)[35]. Vanadium concentrations (as V_2O_5) in domestic U.S. coals vary from 16 to 176 p.p.m.[36]. Vanadium is also emitted into the atmosphere from such sources as the industrial production of the metal, its chemical compounds, alloys, and other products. The present concentrations of vanadium in the atmosphere of the U.S.[37] are of the order of a few micrograms/m³. Vanadium is moderately toxic to humans and animals and when inhaled, the chief damage is to the respiratory tract. Other effects may include fatty degeneration of the liver and tubules of the kidneys[38]. The sources of vanadium in foods and its biological effects in man and animals have been reviewed by Schroeder et al.[39]. The oral intake of vanadium by man is in the order of 2 mg daily of which more than half is constantly present in circulating serum (normal levels are 35–48 µg/100 ml).

Vanadium in air has been analyzed by colorimetric[40], atomic absorption[41], polarographic[42] and emission spectrography[37].

The analysis of vanadium in biological materials has been achieved by neutron activation[43,44], autoradiography[45], and low-energy X-ray mass absorption[46].

Seiler[47] separated V^{III}, V^{IV} and V^V (as V^{3+}, VO^{2+}, and VO_2^{2+}) by TLC on Silica Gel MNS-HR plates with n-butanol saturated with 1 N hydrochloric acid as the solvent. The spots were located with 0.2% quercetin in ethanol or ammonia vapor. The separation of a variety of cations on Silica Gel G and visualization with 1 M quercetin was described by Johri and Mehra[48]. The spots were removed for quantitative colorimetric analysis by the Weisz ring-oven technique. The mixtures resolved and developing solvents were U^{VI}–V^V–Mo^{VI}–W^{VI} with tert.-butanol–acetic acid (5:4), Ti^{IV}–Zn^{IV}–Th^{IV}–U^{VI} with n-butanol–hydrochloric acid (4:3), and Re^{VII}–W^{VI}–Mo^{VI} with n-butanol–6 N hydrochloric acid (5:2).

The separation of Se^{IV}, Te^{IV}, V^V, and Mo^{VI} as ternary mixtures on Silica Gel G by ascending TLC for 30–40 min was described[49]. Two solvent systems were used, ethyl oxalate–hydrochloric acid (60:1) and butyl acetate–hydrochloric acid (40:0.6). The ions were detected by spraying with 0.1 M K_2CS_3 with the limits of identification being 1.27–2.04 µg.

Four groups of metals were separated by Rai and Kukreja[50] by TLC on silica gel predried 2 h at 110° and elution with alkyl acetates saturated with hydrochloric acid. Cr^{VI}, V^V, Ti^{IV} were separated with amyl acetate–ethyl acetate–acetic acid (20:20:11) and detected with NaBrO followed by NH_4SCN.

Kiboku[51] separated the diethyldithiocarbamates of VO_3^-, UO_2^{2+}, MoO_4^{2-},

CrO_4^{2-}, Fe^{2+}, Co^{2+}, Ni^{2+}, Cu^{2+}, and Bi^{3+} by TLC using cyclohexane–chloroform (30:70). The R_F values were V 0.10, U 0.0, Fe 0.25, Bi 0.25, Co 0.30, Cr 0.06, Ni 0.37, Mo 0.40, and Cu 0.43. The detection limit for vanadium was 0.05 μg.

The use of organic acids for the electrophoretic separation of vanadium(IV) and (V) compounds was described[52]. Most effective separation of V^{IV} and V^V ions took place at pH 2.0–3.3 in 0.5 M lactic acid and at pH 1.8–3.1 in 0.25 M tartaric acid while 0.5 M lactic acid at pH 2.8 was recommended for best quantitative separations. V^{IV} and V^V spots were made visible by 6% aq. solution of cupferron and 5% hydroxyquinoline solution in acetic acid, respectively.

The gas chromatography of vanadyl trifluoroacetylacetonate was effected on a 5% DC 710/glass bead (hexamethyl disilazane treated) column with helium as carrier gas[53]. The response of the hydrogen flame detector was linear for 4–28 μg $VOL_2/4$ μl C_6H_6 and the use of an electron-capture detector was expected to permit determinations of $\geq 10^{-10}$ gV.

REFERENCES

1 R. TRUHAUT, *Recherchessur la Toxicologie du Thallium*, Institut National Securité Pour la Prevention des Accidents du Travail, Paris, 1959.
2 F. F. HEYROTH, *U.S. Public Health Rept. Suppl. 197*, 1947.
3 W. L. DOWNS, J. K. SCOTT, L. T. STEADMAN AND E. A. MAYNARD, *Am. Ind. Hyg. Assoc. J.*, 21 (1960) 399.
4 R. K. BARCLAY, W. C. PEACOCK AND D. A. KARNOFSKY, *J. Pharmacol. Exptl. Therap.*, 107 (1953) 178.
5 B. E. COWLEY, *J. Am. Med. Assoc.*, 165 (1957) 1566.
6 J. V. MUNCH, *J. Am. Med. Assoc.*, 102 (1934) 1929.
7 A. LUND, *Acta Pharmacol. Toxicol.*, 12 (1956) 260.
8 H. H. KAMERBEEK, A. G. RAUWS, M. TEN HAM AND A. N. P. VAN HEIJST, *Acta Med. Scand.*, 189 (1971) 189.
9 D. A. KARNOFSKY, C. P. RIDGEWAY AND P. A. PATTERSON, *Proc. Soc. Exptl. Biol. Med.*, 73 (1950) 255.
10 J. E. GIBSON, C. P. SIGDESTAD AND B. A. BECKER, *Toxicol. Appl. Pharmacol.*, 10 (1967) 120.
11 J. E. GIBSON AND B. A. BECKER, *Toxicol. Appl. Toxicol.*, 16 (1970) 120.
12 J. SAVORY, N. O. ROSZEL, P. MUSHAK AND F. W. SUNDERMAN, JR., *Am. J. Clin. Pathol.*, 50 (1968) 505.
13 G. D. CHRISTIAN AND W. C. PURDY, *Am. J. Clin. Pathol.*, 46 (1966) 185.
14 G. S. WINN, H. L. GODFREY AND K. W. NELSON, *Arch. Ind. Hyg. Occupational Med.*, 6 (1952) 14.
15 V. P. GLADYSHEV, K. M. NAURYZBAEV AND A. D. AKBASOVA, *Zh. Analit. Khim.*, 25 (1970) 1321.
16 R. A. CULP AND A. F. FINDEIS, *Anal. Chem.*, 42 (1970) 1285.
17 W. B. STAVINOHA, J. B. NASH AND G. A. EMERSON, *Federation Proc. Abstr.*, 391 (1960).
18 E. B. SANDELL, *Colorimetric Determination of Traces of Metals*, Interscience, New York, 1959, p. 834.
19 A. O. GETTLER AND L. WEISS, *Am. J. Clin. Pathol.*, 13 (1943) 374.
20 W. J. WILSON, JR., in I. SUNSHINE (Ed.), *Manual of Analytical Toxicology*, The Chemical Rubber Co., Cleveland, Ohio, 1971, p. 322.
21 A. SZABO AND M. T. BECK, *Acta Chim. Hung.*, 67 (1971) 189.
22 J. STARY, K KRATZER AND A. ZEMAN, *Radiochem. Radioanal. Letters*, 6 (1971) 1.

23 H. Diller and O. Rex, *Z. Anal. Chem.*, 137 (1952) 241.
24 G. Hauk, *Deut Z. Ges. Gerichtl. Med.*, 15 (1961) 570.
25 J. P. Adloff, *J.Chromatog.*, 5 (1961) 366.
26 E. Cerrai and G. Ghersini, *J. Chromatog.*, 16 (1964) 258.
27 E. Cerrai and G. Ghersini, *J. Chromatog.*, 18 (1965) 124.
28 E. Cerrai and C Testa, *J. Chromatog.*, 8 (1962) 232.
29 R. P. Bhatnager and R. G. Trivedi, *J. Indian Chem. Soc.*, 42 (1965) 53.
30 L. S. Bark, G. Duncan and J. R. Graham, *4th Intern. Chromatog. Symp. 1966*, Publ. 1968, p. 207.
31 R. A. A. Muzzarelli and O. Tubertini, *Talanta*, 16 (1969) 1571.
32 R. A. A. Muzzarelli, G. Raith and O. Tubertini, *J. Chromatog.*, 47 (1970) 414.
33 R. A. A. Muzzarelli and O. Tubertini, *Mikrochim. Acta*, (1970) 892.
34 J. A. Sherma and H. H. Strain, *Anal. Chem.*, 34 (1962) 76.
35 J. L. Burpock, Paper presented at *10th Meeting of New England Air Pollution Control Assoc., Hartford, Conn., 1966*.
36 R. F. Abernethy, *U.S. Bur. Mines Circ. No. 8163*, 1963.
37 V. C. Athanassiadis, *Preliminary Air Pollution Summary of Vanadium and Its Compounds*, U.S. Dept. HEW, National Air Pollution Control Admin., Raleigh, N.C., 1969.
38 T. G. F. Hudson, *Vanadium Toxicology and Biological Significance*, Elsevier, Amsterdam, 1964.
39 H. A. Schroeder, J. J. Balassa and I. H. Tipton, *J. Chronic Diseases*, 16 (1963) 1047.
40 M. N. Kuz'micheva, *Gigiena i Sanit.*, 31 (1966) 229.
41 S. L.Sachdev,*Anal. Chim. Acta*, 37 (1967) 12.
42 L. Von Jerman, *Z. Hyg.*, 14 (1968) 12.
43 D. Comar, *Bull. Soc. Chim. Biol.*, 49 (1967) 1357.
44 H. D. Livingston, *Anal. Chem.*, 37 (1965) 1285.
45 R. Soremark, *Acta Odontol. Scand.*, 20 (1962) 225.
46 R. W. Carter, *Health Phys.*, 13 (1968) 593.
47 H. Seiler, *Helv. Chim. Acta*, 53 (1970) 1423.
48 K. N. Johri and H. C. Mehra, *Mikrochim. Acta.* (1971) 317.
49 K. N. Johri, N. K. Kauskik and K. Singh, *Mikrochim. Acta*, (1969) 737.
50 J. Rai and V. P. Kukreja, *Chromatographia*, 9 (1969) 407.
51 M. Kiboku, *Bunseki Kagaku*, 17 (1968) 722.
52 I. A. Tserkounitskaya and V. A. Luginin, *Primenenie Org. Reagentov. Anal. Khim.*, (1969) 165; *Chem. Abstr.*, 73 (1970) 834894.
53 P. Jacquelot and G. Thomas, *Bull. Soc. Chim. France*, 8–9 (1970) 3167.

Chapter 9

RADIONUCLIDES: STRONTIUM-90, YTTRIUM-91
AND CESIUM-137

1. STRONTIUM-90 AND YTTRIUM-91

Strontium-90 is produced in abundance in many nuclear processes. Because it and its daughter product yttrium are deposited in osseous tissue and their excretion is very slow (resulting in chronic irradiation in bone and the contiguous bone-marrow) they are considered to be for man the most biologically hazardous radioactive elements. ^{90}Sr is a β-ray emitter with a physical half-life of 28 years. Approximately 60–70% of the radioactive strontium in fallout is water soluble. Since calcium is known to be the carrier element for ^{90}Sr, the measured values for ^{90}Sr are related to those of calcium. Figure 1 illustrates how radioactive fallouts reach man.

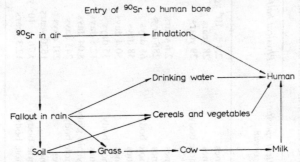

Fig. 1. Pathways through which radioactive fallout enters the body[2].

The absorption of ^{90}Sr is always relative to the quantity of calcium present in foodstuffs. Some ^{90}Sr is lost during the stages of transition from fallout in the atmosphere and on the soil to incorporation into plants and man. The difference between the original concentration and the obtained concentration of ^{90}Sr is known as the "discrimination factor" (DF). The DF from earth to plants and to milk is about 0.09 while that from plants or milk to the human body is 0.25 for ^{90}Sr. Figure 2 illustrates the ecological discrimination against ^{90}Sr with respect to calcium in the U.S. A summary of the discrimination factors of every stage in the ecological series yields a DF of 0.05 for ^{90}Sr from earth to man[1]. (Of the existing ^{90}Sr in the soil, only 5% reaches the human organs through food.) Table 1 is a summary of a number of the radionuclides and their biological importance in terms of the type of radiation, the yield by fission, the critical specific organ, physical half-life, approximate effective half-life, and their maximum permissible concentration (MPC) in air.

TABLE 1

RADIONUCLIDES AND THEIR BIOLOGICAL IMPORTANCE[a]

Radio-nuclide	Type of radiation	Yield by fission (%)	Physical half-life	Critical specific organ	Approximate effective half-life $(T. \text{eff})$[a]	Body MPC[b] in μCi whole population or critical organ
^{239}Pu	α, γ		2.4×10^4 years	Bones	7.2×10^4 years	0.0013
^{90}Sr	β	5.0	28 years	Bones	7.4 years	0.067
^{137}Cs + ^{137}Ba	$\beta+, \gamma$	6.2	29 years	Whole body	70–138 days	0.3
^{147}Pm	α, β	2.6	2.6 years	Bones	140 days	2
^{144}Ce	α, β, γ	5.3	285 days	Bones	180 days	0.167
^{95}Zr	β, γ	6.4	65 days	Whole body	15 days	0.2
^{91}Y	β, γ	5.0	58 days	Bones	51 days	0.167
^{89}Sr	β	4.6	50.4 days	Bones	21 days	0.13
^{95}Nb	β, γ	6.4	35 days	Whole body		0.4
^{140}Ba	β, γ	6.0	12.8 days	Bones	12 days	0.13
^{131}I	β, γ	2.8	8.04 days	Thyroid gland	7.6 days	0.023
^{65}Zn	$\beta+, \gamma$		245 days	Whole body		0.6
^{54}Mn	γ		291 days	Liver		0.67
^{226}Ra	α, β, γ		1620 years	Bones	1.6×10^4 years	0.0033
^{14}C	β		5.76×10^3 years	Whole body		4
^{3}H	β		12.3 years	Whole body	12 days	20

[a] $T. \text{eff} = \dfrac{T_{phys} \times T_{biol}}{T_{phys} + T_{biol}}$

[b] MPC = Maximum permissible concentration in air.

Osteosarcomas have been reported[3] in cats about 1.5 years after a single dose of ^{90}Sr, in dogs 2.5 years after a single dose of ^{90}Sr and continual ingestion in mice and rats in 0.3–2 years. Myelogenous leukemia, lymphosarcoma, and reticulum cell sarcoma have been reported[4,5] in swine and dogs subjected to chronic oral exposure to ^{90}Sr. Chromosomal aberrations and other effects produced by ^{90}Sr–^{90}Y in Chinese hamsters have been described[6].

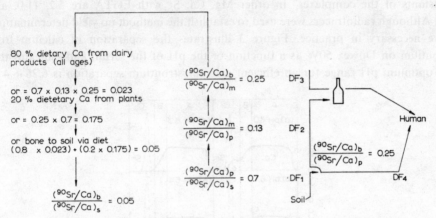

Fig. 2. Ecological discrimination against ^{90}Sr with respect to calcium (U.S.).

The detection and separation of radionuclides in food sources is of paramount importance and as a consequence has been extensively studied. The determination of ^{90}Sr in milk is difficult because of its low concentration (requiring large sample volume) and the limit of detection established by counter background as well as the problems of removal of proteins and the separation of ^{90}Y, daughter of ^{90}Sr.

The use of ion-exchange chromatography has greatly facilitated the determination of ^{90}Sr in milk and diverse biological samples. Porter et al.[7] devised a rapid method for measuring low concentrations of ^{90}Sr in milk with an estimated standard deviation of 1 $\mu\mu$Ci/l in the 1–20 $\mu\mu$Ci/l range. A liter of milk is passed successively through Dowex 50W-X8 (50–100 mesh) cation exchange resin to remove the alkaline and alkaline earth ions, and through a Dowex 1-X8 (50–100 mesh) anion exchange resin (Cl$^-$ form) which retained the ^{90}Y daughter. The yttrium was then eluted with dilute hydrochloric acid, precipitated as the oxalate, and counted in an anti-coincidence beta counter. Calcium, potassium, and total radioactive strontium (strontium-89 and strontium-90) are determined from a 4 M sodium chloride eluant of the cation exchange resin. The columns can be regenerated for repeated use by washing the cation exchange with sodium chloride (4 M) and the anion exchange resin with hydrochloric acid (2 M) and it was suggested that the simplicity of analysis permitted the simultaneous analysis of 8 milk samples of ^{90}Sr in a day by one analyst.

The enrichment (by 10–20 fold) of strontium from milk serum and also separation from interfering ions by application of ethylenebis(oxyethylenenitrilo) tetraacetic acid (EGTA) and Dowex 50W-X8 (50–100 mesh) (NH_4^+ form) has been reported by Brandt and Van't Riet[8]. The described method results in quantitative release of strontium from the resin in a small volume of alkaline EGTA solution and achieves simultaneously separation of strontium from most of the magnesium. The recovery of strontium from all materials investigated was essentially quantitative (98–100%) within the limits of experimental error. The reported log stability constants of the complexes[9] in order Mg, Ca, Sr with EGTA are 5.2, 11.0, and 8.5. Although radiotracers were used to establish the method, no yield determinations were necessary in practice. Figure 3 illustrates the separation of calcium from strontium on Dowex 50W as a function of the pH of the serum and indicates that the optimum pH range for an efficient calcium–strontium separation is 6.26–6.4.

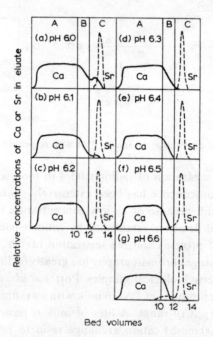

Fig. 3. Separation of calcium from strontium on Dowex 50 W; region A represents serum solutions of various pH values; region B, rinse with 0.05 M ammonium EGTA, 0.05 M ammonium maleate at pH 6.4.; region C, elution with 0.05 M ammonium EGTA, 0.02 M ammonium diethylbarbiturate at pH 8.0.

Milton and Grummitt[10] studied ion-exchange methods for the quantitative separation of the alkaline earths and their application to the determination of ^{90}Sr in milk ash.

The effects of column loading and eluting agent on the Dowex-50 resin (Ion X, 300–400 mesh) separation of beryllium, magnesium, calcium, strontium, barium and radium were studied. The principal eluting agents tested were ammonium lactate, ammonium citrate, and hydrochloric acid. With the column temperature at approx.

80°C, under optimum conditions involving the use of 0.25 g of resin/milliequivalent of sample and with ammonium lactate eluant (0.55 M at pH 5 for Be and 1.5 M at pH 7 for Mg, Ca, Sr, Ba, and Ra) a quantitative separation of all six elements in a single sample was possible in 5 h. This procedure was applied to the routine analysis of ^{90}Sr in bulk milk ash samples containing 20 mg of strontium carrier with yields of 85–95%.

Kornacki et al.[11] studied the characteristics of ion-exchangers and the removal of ^{85}Sr and ^{89}Sr from milk. Polish-made cation-exchanger resins designated MK-2, MK-3, FPC, FHF and SD were compared with Amberlite IR-120 resin in regard to their exchange capacity, bulk density, thermal and chemical resistance, water content, and swelling properties. Milk was in vitro contaminated with 1 μCi of $^{85+89}$Sr-labeled SrCl$_2$/l, kept at 5° for 24 h, and then adjusted to pH 5.4, 6.0, or 6.65 with 0.75 M citric acid. Thirty volumes of milk were passed through the ion-exchange column containing 1 volume of the resin in Ca, Mg, K, and Na form, the radioactivity of the effluent established and the percentage of radioactive Sr retained on the resin calculated. The decrease of pH enhanced the decontamination of milk resulting in 96.24, 91.84, 81.54, 67.30, and 64.86% Sr removed with Amberlite, SD, FPC, MK-2, and MK-3, respectively at pH 5.4.

The ion-exchange chromatography of ^{90}Sr, ^{90}Y, ^{224}Ra, Mg, Ca, and Ba present in milk and bone ash was described by Senegačnik et al.[12] using Dowex 50W-X8 (50–100 mesh) (NH$_4^+$ form) with stepwise elution with buffers, e.g. 1.0 M ammonium lactate (pH 7.5), 0.15 M ammonium citrate (pH 7.5), and 0.3 M ammonium citrate (pH 9–10) as illustrated in Fig. 4.

Rapid radiochemical separation of strontium-90–yttrium-90 and calcium-45–scandium-46 on the cation exchange resin Dowex 50W-X8 resin (200–400 mesh) was

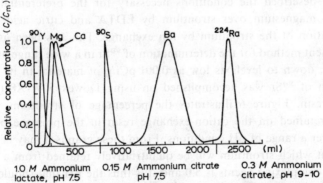

Fig. 4. Ion-exchange chromatography of ^{90}Sr, ^{90}Y, ^{224}Ra and Mg, Ca, and Ba on Dowex 50 W-X8. Column: 1.7 × 55 cm resin bed, 110 ml; ion exchanger: Dowex 50 W-X8 (50–100 mesh), NH$_4^+$ form; buffers: step-wise elution with buffers indicated in the figure; operating conditions: sorption rate 10 ml/min, elution rate 5 ml/min, 20–25°. In the run presented, the column was loaded with a synthetic leaching residue containing 50 mg of Mg, 2000 mg of Ca, ^{90}Sr + ^{90}Y + 30 mg of Sr, 30 mg of Ba, ^{224}Ra in 150 ml of distilled water. Detection: counting of radioactivity. Notes: concentrations of elution peaks: C_m (^{90}Y) = 800 imp./min. ml; C_m(Mg) = 0.32 mg/ml; C_m(Ca) = 9.90 mg/ml; C_m(^{90}Sr) = 0.34 mg/ml; C_m(Ba) = 0.084 mg/ml; C_m(Ra) = 1900 imp./min.ml.

References pp. 209–210

reported by Rane and Bhatki[13]. Strontium-90 and calcium-45 were removed by 1.3–2 M nitric acid solutions and 3 M nitric acid eluted ^{90}Y or ^{46}Sc. Nitric acid (2.1 M) was the optimum concentration of the acid to separate ^{90}Sr and ^{90}Y as well as ^{45}Ca and ^{46}Sc. Figure 5 illustrates the elution of ^{50}Sr–^{90}Y and ^{45}Ca–^{46}Sc from Dowex 50W-X8 resin by dilute nitric acid.

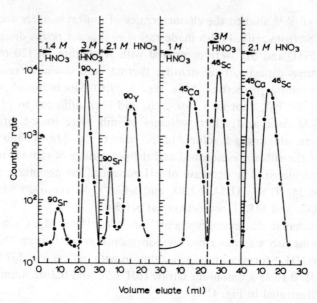

Fig. 5. Elution of Sr^{90}–Y^{90} and Ca^{45}–Sc^{46} from Dowex 50W-X8 resin by dilute nitric acid.

Ibbett[14] described the conditions necessary for the preferential chelation of calcium and magnesium over strontium by EDTA and citric acid and the subsequent isolation of the strontium by ion exchange. This has been used as a basis for a convenient method for the determination of ^{90}Sr in a wide range of environmental materials, down to levels as low as 0.001 pCi/g of material in its natural state. The isolation of ^{90}Sr was accomplished on using Dowex 50W-X8 (50–100 mesh) (H$^+$ form) resin. Figure 6 illustrates the percentage of strontium, calcium, and magnesium retained on the cation-exchange resin in the presence of EDTA and citric acid over a range of pH conditions. From these graphs it may be seen that the highest pH at which strontium will be quantitatively retained from a solution containing both complexing agents is 5.0 and it is thus the pH that should result in the optimum separation.

By the use of 20 g ash samples and overnight counting, total ^{90}Sr contents of 2 pCi could be detected by the above procedure of Ibbett[14] at a standard deviation of ±9%. For some samples this represents levels as low as 0.001 pCi/g of the material in its natural state.

The determination of low levels of ^{90}Sr in a variety of environmental materials, *e.g.* of marine origin, such as sea water, sediments, fish, and sea weeds is often a requirement in regard to the environmental monitoring of nuclear sites and radio-biological and radio-ecological research. It is thus often necessary to separate strontium from naturally occurring and fission-product radioisotopes including radium-226, ruthenium–rhodium-106, zirconium–niobium-95, cerium–praseodymium-144, yttrium-91 and barium–lanthanum-140, in addition to large amounts of calcium, magnesium, and phosphate.

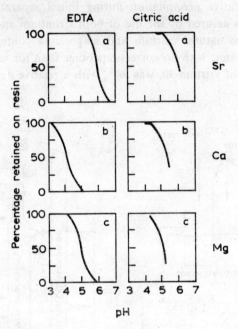

Fig. 6. Relationship between the percentage of (a) strontium, (b) calcium, and (c) magnesium retained on Dowex 50W-X8 resin (H⁺ form), and the pH of the solution.

Most methods for the determination of strontium-90 require a long ingrowth period, usually 10 days to 2 weeks[15-17] after purification of the strontium. Some require[18] that the strontium-90 and yttrium-90 reach equilibrium after the separation of the strontium fraction with the resultant corrections for strontium-89.

The determination of strontium in environmental media has been described by Strong *et al.*[19]. An alkaline fusion of the ashed sample followed by dissolution in dilute acid gives a preliminary separation. Further separation was obtained by sorption on Dowex 50W-X8 (50–100 mesh, Na⁺ form). Cation-exchange resin from an EDTA solution at pH 5.1 followed by elution of the alkali metals with 0.5 M hydrochloric acid and of the strontium with 3 M hydrochloric acid. Strontium was determined by flame photometry (using a Beckman Model DU Spectrophotometer with flame attachment, oxygen–acetylene burner and I, P 2F photo tube). Strontium was determined using the emission line at 460.7 nm and the intensity was enhanced by the addition of *n*-butanol to the solution. The radiochemical determination of strontium-90 in water using ion-exchange has been reported by Talvitie and Demint[20]. The specific determination of strontium-90 was achieved by preferential adsorption of strontium as the cationic resin Dowex 50W-X8 (NH_4^+ form) from an EDTA medium, followed by selective elution of ingrown yttrium-90 with ammonium 2-methyl lactate.

In the analysis of water, yield determinations were considered unnecessary inasmuch as quantitative precipitation during initial separation of strontium-90 as the carbonate was assured by the use of both strontium and calcium carriers in combination with the natural calcium and magnesium content. The recovery of strontium-90 from water, with no corrections other than for reagent blank and for ingrowth and decay of yttrium-90, was 98% with a relative deterioration of 2%.

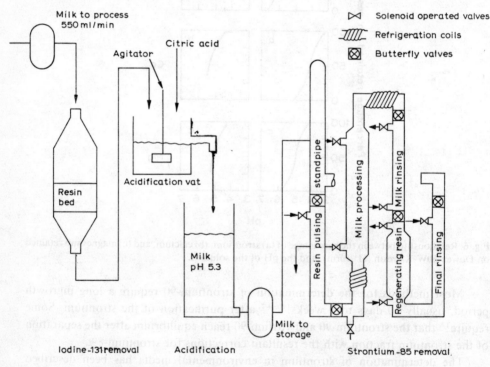

Fig. 7. Flow diagram of combined process for radionuclide removal.

Walter *et al.*[21] described a combined fixed-bed process for the removal of ^{131}I and [85]Sr. Dickerson *et al.*[22] used a modified pulsed-bed contractor for the removal of approx. 97% of the [85]Sr from milk. A combined process to remove ^{131}I and [85]Sr using a fixed-bed anionic contractor and the modified pulsed-bed cationic contractor was described by Stroup *et al.*[23]. Milk labeled *in vitro* containing ^{131}I and [85]Sr was passed through an anionic resin (fixed-bed contractor) in a chloride–

phosphate–citrate cycle, acidified to pH 5.3 with citric acid and passed through a cationic resin (pulsed-bed contractor) in a calcium–magnesium–potassium–sodium cycle. Analyses of the treated milk indicated that approx. 97% of the ^{131}I and 95% of the ^{85}Sr were removed. Figure 7 shows a flow diagram of the combined process for radionuclide removal.

The removal of ^{131}I and ^{90}Sr from milk using ion exchange techniques has been well documented[24-28]. The U.S. Public Health Service and the U.S. Dept. of Agriculture developed a standby ion-exchange system to commercial scale[29,30] for the removal of ^{131}I and ^{90}Sr from raw whole milk. The system has two fixed bed columns, *viz.* the first contained approx. 265 l of Dowex 2-X8 for iodine removal and the second, approx. 1380 l of Amberlite IR-120 for strontium removal. The commercial scale project stripped the ^{131}I from the anion exchange with 2 N HCl at 25°C and ^{90}Sr was stripped from the cation exchanger with a mixture of NaCl, KCl, $MgCl_2$, and $CaCl_2$.

Walker *et al.*[31] described the utility of sodium chloride as an effective stripping agent for the removal of ^{131}I from Dowex 2 X-80, 20–50 mesh resin used to remove radionuclides from milk. This agent did not have the acute corrosive properties of hydrochloric acid and could be used with existing equipment in modern dairies.

A radiochemical determination of ^{90}Sr in fission product samples which have decayed at least 10 days was described by Bryant *et al.*[32]. Such samples contain two radioactive isotopes of strontium, ^{89}Sr with a half-life of 51 days and ^{90}Sr with a half-life of 27.7 years. Strontium-91 with a half-life of 9.7 h, had decayed to an insignificant level in 10 days. Determination of ^{90}Sr in the presence of ^{89}Sr is based on separation and measurement of the radioactive ^{90}Y which is produced by decay of the ^{90}Sr.

Strontium and barium were first adsorbed in a cation-resin column and after a suitable growth period ^{90}Y is selectively eluted. The ^{90}Y was then adsorbed on and eluted from a second cation-resin column and counted. The radiochemical yield was greater than 97% and the gravimetric measurement of the recovery of strontium and yttrium carriers was not required. The method with modifications was also applicable to samples that contain 100 mg of iron or uranium.

The rapid removal of ^{90}Sr in tissue, food, biota, and other environmental media by tributyl phosphate was determined by Baratta and Reavey[33]. For the above types of samples ^{90}Sr and its daughter ^{90}Y are usually known to be in equilibrium. The method was found to be sensitive to less than 1 pCi per sample of tissue or food ash. The samples were prepared by either dry or wet-ashing techniques. then solubilized in nitric acid. ^{90}Y was extracted into equilibrated tributyl phosphate and then further purified by selective stripping and fluoride precipitation to remove the remaining contaminants which may have been carried over. ^{89}Sr could also be determined in the same sample by precipitation of Sr prior to the ^{90}Y extraction.

The rapid separation of ^{89}Sr and ^{90}Sr from ^{60}Co and fission products has been accomplished using a Dowex 2-X8 exchange column[34]. A determination of ^{90}Sr

in urine was described by Testa et al.[35] in which the method depended on the batch-wise separation of ^{90}Y by using the liquid cation exchanger bis(2-ethylhexyl)phosphate supported on microporous polyethylene (Microthene-710).

Carrier-free strontium-90 has been separated from yttrium-90 using paper impregnated with stannic phosphate and ascending development with ammonium nitrate, nitric acid, and hydrochloric acid at various concentrations[36] (Table 2).

TABLE 2

R_F VALUES OF CARRIER-FREE STRONTIUM-90 AND YTTRIUM-90 ON PAPER IMPREGNATED WITH STANNIC PHOSPHATE

Solvents: NH_4NO_3, HNO_3 or HCl at various concentrations
Developments: Ascending
Temperature: 20°

Solvent		Molarity of solvent							
		0.1	0.2	0.4	0.6	0.8	1.0	1.5	2.0
^{90}Sr	NH_4NO_3	0.15	0.39	0.63	0.67	0.73	0.80	0.89	0.96
^{90}Y		0.00	0.00	0.01	0.02	0.02	0.02	0.02	0.03
^{90}Sr	HNO_3	0.16	0.25	0.55	0.62	0.77	0.81	0.85	0.91
^{90}Y		0.00	0.01	0.01	0.01	0.05	0.06	0.02–0.31	0.14–0.49
^{90}Sr	HCl	0.15	0.27	0.53	0.64	0.75	0.78	0.83	0.88
^{90}Y		0.00	0.00	0.02	0.02	0.05	0.07	0.10–0.41	0.13–0.49

The development of polytetrafluoroethylene layers for TLC has been described by Raaen[37,38] who also studied the TLC behavior of the radioisotopes of 31 metal ions on layers of 100% polytetrafluoroethylene layered onto Mylar film. The chromatograms were developed with a methyl ethyl ketone solution of di(2-ethylhexyl)-orthophosphoric acid (HDEHP).

The polytetrafluoroethylene slurry was prepared by dissolving 0.2 g of a fluoro-chemical wetting agent FX-173 (3MCo) in 72 ml of methyl isobutyl ketone, adding 0.8 ml of conc. nitric acid, and then adding to the solution Fluoroglide 200 TWO 218 (Chemplast, Inc.). The mixture was shaken for about 30 sec on an orbital sander. (This amount of slurry was sufficient to layer a 20 × 20 cm area of Mylar polyester film (Kensington Sci. Corp.) to microscope-slide thickness.) Positions of the metal ions on the chromatograms were detected by autoradiography. One of the better separations obtained was that of ^{90}Sr from ^{91}Y. It was also shown that if the HDEHP were located in the layer instead of the developer, the order of migration of the resolved components could be reversed. This reversal was demonstrated for ^{55}Fe and ^{63}Ni. The polyester–fluoroethylene layers were thus suggested to be suitable for either normal or reversed-phase TLC.

2. CESIUM-137

From the standpoint of radiation exposure of population, the most important of the cesium isotopes is ^{137}Cs because of its high yield from nuclear detonations and its introduction into the food chain of animals and man through fallout. In this regard, an important consideration is the consumption of plants by cows which in turn furnish milk and meat for human consumption. Plants and particularly forage crops represent nearly the entire intake of ^{137}Cs by cattle[39,40]. The 30-year physical half-life of ^{137}Cs and its high energy beta disintegration substantiate that the major hazard at present is primarily related to the amount of absorption through the gastro-intestinal tract and subsequent retention in critical organs.

The comparative metabolism of radionuclides (^{137}Cs, ^{90}Sr, ^{226}Ra, ^{65}Zn, ^{239}Pn, ^{131}I, ^{210}Po, ^{210}Pb, and ^{106}Rn) in mammals as well as an appraisal of biological effects and of therapeutic or preventive methods for each radionuclide has been reviewed by Stara et al.[2].

Determination of radio cesium in environmental materials and urine is complicated because large samples must be concentrated to obtain cesium for radiometric assay. High concentrations of alkali and alkaline earth metals in certain biological samples also generally interfere with analytical separations. The use of ammonium molybdophosphate (AMP) as a cation exchanger to concentrate ^{137}Cs in environmental samples has been reported by Collins et al.[41].

Boni[42] has described the rapid ion exchange analysis of radio cesium in milk, urine, sea water, and environmental samples. An inorganic ion exchanger, potassium cobalt ferricyanide (KCFC), was used either alone or with other ion exchange materials.

Figures 8–10 illustrate a disposable polyethylene ion exchange and ion-exchange columns for milk and sea water, respectively. The disposable ion-exchange column consists of a 6 in. long, $\frac{3}{8}$ in. o.d. × $\frac{5}{8}$ in. i.d. polyethylene drying tube that has a serrated tip on one end and a polyethylene funnel with a $4\frac{1}{4}$ in. i.d. top as a reservoir on the other end (Fig. 8). The column is filled to a height of $1\frac{1}{2}$ in. with a water slurry of water-washed 20–30 or 30–60 mesh KCFC supported by a small borosilicate glass wool plug.

The dual ion exchange column used to determine radioiodine and radiocesium in milk consists of two 2 in. o.d. by $1\frac{5}{8}$ in. i.d. polyethylene cylinders, the lower one 2 in. long and the upper one 4 in. long. A neoprene gasket is placed between the cylinders and they are connected by a threaded polyethylene adapter ring. The dual column is closed at both ends with threaded caps with attached stopcocks (Fig. 9). The shorter section of the dual column is filled to a height of $1\frac{3}{4}$ in. with a water slurry of water-washed 20–30 mesh KCFC. The KCFC bed is covered and supported by glass wool pads. The longer section of the dual column is completely filled with a water slurry of 20–50 mesh Dowex 1-X8 anion resin and this bed is also

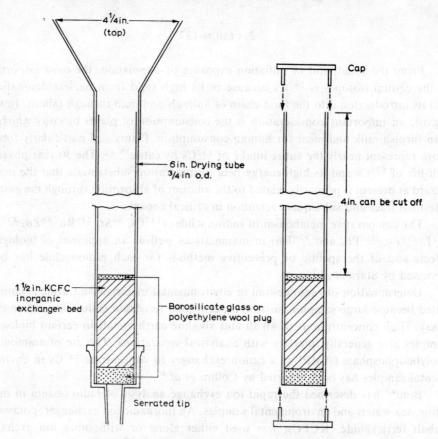

Fig. 8. Disposable polyethylene ion-exchange column.

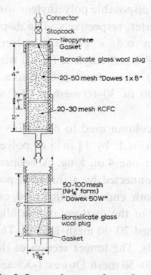

Fig. 9. Ion-exchange column for milk.

covered and supported by glass wool pads. The KCFC concentrates radiocesium and the anion resin concentrates radioiodine from milk[43].

The columns are converted into right circular cylinders for gamma spectrometric analysis by removing the stopcock adapters, separating the two sections and cupping the ends with pressure fitting thin polyethylene protective closures.

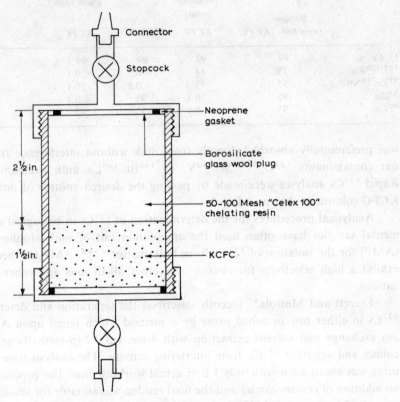

Fig. 10. Ion-exchange column for sea water.

The ion exchange column used to concentrate radiocesium from sea water consists of a 4 in. × 3 in. o.d. by 2½ in. i.d. polyethylene cylinder closed at both ends by threaded caps with attached stopcocks (Fig. 10). The column is filled to a height of 1½ in. with a water slurry of water-washed KCFC supported by a glass wool pad. The remainder of the column is filled with 50–100 mesh B$_{10}$-Rad Celex 100 chelating resin in the calcium form which is used to remove other interfering radionuclides such as ^{59}Co and ^{65}Zn.

Table 3 shows the percent absorption of KCFC of interfering radionuclides. Cesium is exchanged from aqueous samples by replacing the potassium ion in the KCFC crystals. KCFC absorbs cesium from neutral, conc. hydrochloric acid, conc. nitric acid, and highly salted (NaNO$_3$) sodium hydroxide solutions. Cesium-137

TABLE 3

PERCENT ABSORPTION OF KCFC OF INTERFERING RADIONUCLIDES

Sample	Solvent: H_2O Top resin: Bottom resin bed: KCFC	H_2O Celex 100 (Ca^{2+} form, chelating resin) KCFC	1 N HCl–0.5 N HF		10 N HCl–0.5 N HF Dowex 1 X2 (anion resin)
			Dowex 1 X8 (anion resin) KCFC	KCFC	KCFC
^{137}Cs	99	99	99	99	99
^{103,106}Ru	75	44	<0.1	<0.1	<0.1
^{95}Zr–^{95}Nb	95	73	<0.1	<0.1	<0.1
^{65}Zn	95	<0.1	95	<0.1	<0.1
^{60}Co	95	<0.1	95	95	<0.1

was preferentially absorbed directly from milk without interference from the fall-out contaminants ^{131}I, ^{89,90}Sr, ^{90}Y, or ^{140}Br–^{140}La milk sometimes contains. Rapid ^{137}Cs analyses were made by passing the desired volume of milk through a KCFC column as shown in Fig. 9.

Analytical procedures for the determination of ^{137}Cs in biological and environmental samples have often used the ammonium salt of molybdophosphoric acid (AMP) for the isolation of ^{137}Cs from bulk matter[44-46]. AMP has been shown to exhibit a high selectivity for cesium over other alkali ions and other monovalent cations.

Everett and Mottola[47] recently described the separation and determination of ^{137}Cs in either raw or ashed urine by a method which relied upon AMP column ion exchange and solvent extraction with 4-*sec.*-butyl-2-(β-methylbenzyl)phenol to collect and separate ^{137}Cs from interfering activity. The analysis time on 1 liter of urine was about 4.5 h with only 1 h of actual working time. The procedure required no addition of cesium carrier and the final residue was suitable for lew-level β-counting. (A 93% recovery of ^{137}Cs was obtained.)

The distribution of ^{137}Cs in pregnant animals and embryos (rabbits and rats) was studied by Hamada[48]. The distribution of ^{137}Cs in placenta, embryo and amniotic fluid as well as in various organs of the maternal body was examined after administration of ^{137}Cs to rats having 14-day-old embryos in the organ-forming stage and 19-day-old embryos in the growth stage. Comparative examinations were also made on the relation between the amount of K and the distribution of ^{137}Cs in various tissues on the 19th day of gestation and distribution of ^{137}Cs in the organs of pregnant and non-pregnant rats. The results indicated that ^{137}Cs was incorporated into the placenta within a short time after its administration to pregnant animals but its amount decreased gradually thereafter and the decrease became especially marked after 6 h. Levels of ^{137}Cs in the embryo reached a maximum 24 h after its administration to the mother. Transition of ^{137}Cs from the placenta to embryo increased with time during 24 h after its administration and then neared equilibrium thereafter. The concentration of ^{137}Cs in the amniotic fluid was very low being $\frac{1}{7}$ that of the embryo in rats and $\frac{1}{4}$ in rabbits. The concentration of ^{137}Cs was higher in pregnant uteri than in non-pregnant uteri, the difference being statistically significant.

The separation of a number of radioelements (including the pairs $^{137}Cs-^{223}Fr$ and $^{90}Sr-^{90}Y$) on paper impregnated with inorganic ion exchangers has been described by Adloff[49] (Table 4).

TABLE 4

SEPARATION OF SOME RADIOELEMENTS ON IMPREGNATED PAPER[a]

Element mixture	Impregnation medium[b]	Eluant	Development time (min)		R_F
$^{211}Pb-^{207}Tl$	ZP	0.5 M HCl	5	Tl 0.1	Pb 1
	ZO	0.1 M NH$_4$NO$_3$	12	Tl 0.6	Pb 0
$^{204}Tl^+-^{204}Tl^{3+}$	ZP	0.5 M HCl	30	Tl$^+$ 0.2	Tl^{3+} 0.9
	ZO	0.1 M NH$_4$NO$_3$	30	Tl$^+$ 0	Tl^{3+} 0.8
$^{212}Pb-^{212}Bi$	ZO	0.05 M NH$_4$NO$_3$	30	Bi 0	Pb 0.4
$^{140}Ba-^{140}La$	ZP	0.5 M NH$_4$Cl	20	La 0	Ba 0.9
$^{90}Sr-^{90}Y$	ZP	0.5 M HCl	30	Y 0.3	Sr 1.0
	ZO	0.1 M NH$_4$NO$_3$	30	Y 0	Sr 0.95
$^{228}Ra-^{228}Ac$	ZP	0.5 M NH$_4$Cl	30	Ra 0.85	Ac 0.1
$^{223}Ra-^{223}Fr$	ZO	0.1 M NH$_4$NO$_3$	20	Fr 0	Ra 0.85
	ZW	0.1 M NH$_4$Cl	20	Ra 0	Fr 0.5
$^{137}Cs-^{223}Fr$	ZW	1 M NH$_4$Cl	20	Fr 0.5	Cs 0.6

[a] Paper: Whatman No. 1; descending development; temp.: 60° (water vapor atmosphere).
[b] ZP = zirconium phosphate; ZO = zirconium oxide hydrate; ZW = zirconium tungstate.

REFERENCES

1 W. LANGHAM AND E. C. ANDERSON, *Symp. Noxious Effects Low Level Radiation, Lausanne, 1958*, p. 434.
2 F. J. BRYANT, A. C. CHAMBERLAIN, A. MORGAN AND G. S. SPICER, *AERE Rept. HP/R 2353*, 1957.
3 J. F. STARA, N. S. NELSON, R. J. DELLA ROSA AND L. K. BUSTAD, *Health Phys.*, 20 (1971) 113.
4 R. O. McCLELLAN, *Health Phys.*, 12 (1966) 1362.
5 E. B. HOWARD, W. J. CLARKE AND P. L. HACKETT, *3rd Intern. Symp. Comparative Leukemia Res., Paris, 1967.*
6 A. L. BROOKS AND R. O. McCLELLAN, *Intern. J. Radiation Biol.*, 16 (1969) 545.
7 C. PORTER, D. CAHILL, R. SCHNEIDER, P. ROBBINS, W. PERRY AND B. KAHN, *Anal. Chem.*, 33 (1961) 1306.
8 P. J. BRANDT AND B. VAN'T RIET, *Anal. Chem.*, 38 (1966) 1790.
9 G. SCHWARZENBACH AND H. ACKERMAN, *Helv. Chim. Acta*, 40 (1957) 1886.
10 G. M. MILTON AND W. E. GRUMMITT, *Can. J. Chem.*, 35 (1957) 541.
11 K. KORNACKI, S. POZNANSKI AND L. JEDRYCHOWSKI, *Nukleonika*, 14 (1969) 397.
12 M. SENEGAČNIK, S. PALJK AND J. KIRSTIAN, *Z. Anal. Chem.*, 249 (1970) 41.

13 A. T. RANE AND K. S. BHATKI, *Anal. Chem.*, 38 (1966) 1598.
14 R. D. IBBETT, *Analyst*, 92 (1967) 417.
15 F. E. BUTLER, *Anal. Chem.*, 35 (1963) 2069.
16 A. S. GOLDIN, R. J. VELTON AND G. W. FRISHKORN, *Anal. Chem.*, 31 (1959) 1490.
17 D. G. PEPPARD, G. W. MASON AND S. W. MOLINE, *J. Inorg. Nucl. Chem.*, 5 (1957) 141.
18 C. R. PORTER, *Environ Sci. Technol.*, 1 (1967) 745.
19 A. B. STRONG, G. L. REHNBERG AND U. R. MOSS, *Talanta*, 15 (1968) 73.
20 N. A. TALVITIE AND R. J. DE MINT, *Anal. Chem.*, 37 (1965) 1605.
21 H. E. WALTER, A. M. SADLER, D. G. EASTERLY AND L. F. EDMONDSON, *J. Dairy Sci.*, 50 (1967)
    1221.
22 R. W. DICKERSON, JR., A. L. REYES, G. K. MURTHY AND R. B. READ, JR., *J. Dairy Sci.*,
    51 (1968) 1317.
23 W. H. STROUP, A. L. REYES, G. K. MURTHY, R. B. READ, JR. AND R. W. DICKERSON, JR.,
    J. Dairy Sci., 51 (1968) 1500.
24 D. G. EASTERLY, J. Y. HARRIS, L. A. BUNCE AND L. F. EDMONDSON, *J. Dairy Sci.*, 46 (1963)
    1207.
25 L. F. EDMONDSON, *J. Dairy Sci.*, 47 (1964) 1201.
26 G. K. MURTHY, J. E. GILCHRIST AND J. E. CAMPBELL, *J. Dairy Sci.*, 45 (1962) 1066.
27 G. K. MURTHY AND J. E. CAMPBELL, *J. Dairy Sci.*, 47 (1964) 1188.
28 G. K. MURTHY, *J. Dairy Sci.*, 48 (1965) 1429.
29 J. H. FOOKS, J. G. TERRILL, JR., B. H. HEINEMANN, E. J. BALDI AND H. E. WALTER, *Health
    Phys.*, 13 (1967) 279.
30 B. HEINEMANN, E. J. BALDI, R. O. MARSHALL, E. M. SPARLING, H. E. WALTER AND J. H. FOOKS,
    J. Dairy Sci., 50 (1967) 426.
31 J. P. WALKER, B. F. REHNBERG AND I. B. BROOKS, *J. Dairy Sci.*, 51 (1968) 1373.
32 E. A. BRYANT, J. E. SATTIZAHN AND B. WARREN, *Anal. Chem.*, 31 (1959) 334.
33 E. J. BARATTA AND T. C. REAVEY, *J. Agr. Food Chem.*, 17 (1969) 1337.
34 M. STOEPPLER, *Z. Anal. Chem.*, 253 (1971) 35.
35 C. TESTA AND G. SANTORI, *Energia Nucl.* (*Milan*), 17 (1970) 320.
36 Z. SHU-JUN, *Acta Chim. Sinica*, 31 (1965) 549.
37 H. P. RAAEN, *J. Chromatog.*, 44 (1969) 522.
38 H. P. RAAEN, *J. Chromatog.*, 53 (1970) 605.
39 G. M. WARD, H. F. STEWART AND J. E. JOHNSON, *J. Dairy Sci.*, 48 (1965) 38.
40 *Nat. Res. Council, Publ. No. 1092*, National Academy of Sciences, 1963.
41 W. R. COLLINS, JR., D. C. SUTTON AND M. J. SOLAZZI, *U.S. At. Energy Comm. Rept. DP-831*,
    1962.
42 A. L. BONI, *Anal. Chem.*, 38 (1966) 89.
43 A. L. BONI, *Health Phys.*, 9 (1963) 1035.
44 W. W. FLYNN, *Anal. Chim. Acta*, 50 (1970) 365.
45 A. MORGAN AND G. M. ARKELL, *Health Phys.*, 9 (1963) 857.
46 J. V. R. SMIT, *Nature*, 181 (1958) 1530.
47 R. J. EVERETT AND H. A. MOTTOLA, *Anal. Chim. Acta*, 54 (1971) 309.
48 K. HAMADA, *Sanfujinka No Shimpo*, 21 (1969) 355; *Nucl. Sci. Abstr.*, 25 (1971) 16024.
49 J. P. ADLOFF, *J. Chromatog.*, 5 (1961) 366.

Chapter 10

RADIONUCLIDES: ACTINIDES

1. URANIUM

Prior to the discovery of fission in 1939, the uses of uranium or uranium compounds were relatively unimportant. In all of the earlier operations, uranium ores were mined primarily for their radium content. (It is estimated that between 1906 and 1939 the combined production of radium from all of these operations was not more than 1 kg.) The amount of uranium that was concentrated as a byproduct of radium was estimated to be between 4000 and 5000 tons of contained uranium.

There are eleven known isotopes of uranium. Three of these, ^{234}U, ^{235}U, and ^{238}U, exist in nature. All isotopes of uranium are unstable and as they decay, emit alpha or beta particles. ^{238}U is the most stable isotope and it is the most naturally abundant.

The composition of natural uranium is approx. 99.28% ^{238}U, 0.71% ^{235}U, and 0.005% ^{234}U. The half-lives of natural and artificial uranium isotopes are shown in Table 1.

TABLE 1

HALF-LIVES OF NATURAL AND ARTIFICIAL URANIUM ISOTOPES

Isotope	Half-Life	Isotope	Half-Life
Natural		*Artificial*	
^{234}U	235,000 years	^{228}U	9.3 min
^{235}U	700 million years	^{229}U	58 min
^{238}U	4.5 billion years	^{230}U	20.8 days
		^{231}U	4.2 days
		^{232}U	70 years
		^{233}U	160,000 years
		^{237}U	6.8 days
		^{239}U	23.5 min

Uranium compounds are widely distributed in most of the rock types, in ground water, ocean water, and in living matter. The earth's crust contains about 40 times as much uranium as silver and about one-tenth as much uranium as copper. The sedimentary rocks are estimated to contain an average of about 2 p.p.m. of uranium, ocean water about 0.002 p.p.m., and ground water averages about 0.0002 p.p.m. Uranium, as encountered in ore, is nearly always present in the form of the tetravalent ion; when in ground water, it is in the hexavalent state. Pitchblendes ($UO_2 + UO_3$) or uraninite deposits with commercial production records are

located in Australia, Republic of Congo, Czechoslovakia, France, Portugal, and
Canada. Uraninite is the principal uranium mineral in Canada, South Africa, and
in the Front Range of the Rockies in Colorado.

Most of the nuclear reactors built for the generation of electric power in the
U.S. and in many other countries are based on the use of uranium fuel enriched in
isotope ^{235}U. The gaseous diffusion process is the only present method for large-scale
separation of ^{235}U from ^{238}U. The natural uranium power reactors in England
require no isotopic enrichment and the purified natural uranium for direct use in
these reactors may be fabricated in the form of the metal, oxide, or carbide.

Higher uranium concentrations are reported for present-day North American
rivers relative to those found 10–20 years ago for the same rivers or for rivers in
other parts of the world draining less intensively cultivated areas. The increase is
attributed to high concentrations of uranium in phosphate fertilizers[1,2]. Spalding
and Sackett[1] showed a constant phosphorus–uranium ratio for various fertilizers
and for the easily solubilized fraction of 0-46-0 fertilizers. The toxic levels of uranium
in organisms are similar to those of mercury and lead. Uranium concentrations
were determined by a delayed neutron technique.

Radioactive decay of uranium-238 gives, among other products, radium-226,
lead-210, and polonium-210. Polonium-210 and lead-210 also may be absorbed by
plants, hence there is a potential hazard from radionuclide contamination of plants
grown on phosphate-fertilized soils.

The toxic action of uranium and its salts on biological systems, especially on
mammals, has been well documented. Chemical hazards of exposure to uranium
were described in detail by Voegtlin and Hodge[3] and by Tannenbaum[4]. Aspects of
lung cancer and uranium mining have been extensively studied[5–11] and the mechanisms of uranium poisoning have been proposed by Hodge[12].

Many analytical methods have been developed for the determination of uranium
in a wide concentration range and in different materials. These techniques include
spectrophotometric[13,14], fluorimetric[15,16], radiometric[17–19], polarographic[20,21],
electrodeposition[22–24], photochemical[25], mass spectrometric[26,27], and fission track
analysis[26].

TABLE 2

R_F VALUES OF UO_2^{2+} AND THORIUM

Solvent	R_F values		Temp. (°C)
	UO_2^{2+}	Thorium	
Methanol–pyridine–toluene (7:2:1)	0.86	0.16	13
Methanol–pyridine–dioxane (7:2:1)	0.93	0.13	22
Butanol–pyridine–dioxane (7:2:1)	0.69	0.07	15.5
Methanol–chloroform–toluene (7:2:1)	0.72	0.13	20
Pyridine–chloroform–toluene (7:2:1)	0.45	0.13	19

The paper separation of uranium and thorium has been achieved using butanol–$1.0\,N$ nitric acid with 0.5% benzoylacetone[28]; butanol–$3\,N$ HCl or n-butanol–$4\,N$ HCl[29]. (The R_F values for $UO_2{}^{2+}$ are 0.19 and 0.32, respectively, and for the Th 0.05 and 0.13.) The utility of a variety of solvents for the separation of UO_2–Th

TABLE 3

R_F VALUES OF $UO_2{}^{2+}$ AND Pb^{2+} ON PAPERS IMPREGNATED WITH VARIOUS AMOUNTS OF ZIRCONIUM PHOSPHATE AND DEVELOPED AT VARIOUS TEMPERATURES

Paper: Whatman No. 1

Impregnation: The paper strips were drawn uniformly through a solution of zirconium chloride at different concentrations in $4\,N$ HCl. After drying they were dipped into a 60% H_3PO_4 solution in $4\,N$ HCl, dried and washed with distilled water

Development: Ascending

Temp. (°C)	$UO_2{}^{2+}$ ions developed with $3\,N$ $HClO_4$						Pb^{2+} ions developed with $0.4\,N$ $HClO_4$					
	% $ZrOCl_2$ on paper						% $ZrOCl_2$ on paper					
	5	10	15	20	25	30	5	10	15	20	25	30
5	0.70	0.62	0.43	0.38	0.33	0.29	0.67	0.66	0.52	0.51	0.43	0.40
20	0.71	0.63	0.46	0.44	0.38	0.30	0.74	0.68	0.58	0.57	0.52	0.47
35	0.78	0.58	0.46	0.45	0.36	0.33	0.79	0.73	0.62	0.55	0.52	0.45
50	0.80	0.62	0.52	0.40	0.37	0.34	0.86	0.85	0.69	0.66	0.65	0.48

TABLE 4

R_F VALUES OF THE VALENCY STATES OF SOME ELEMENTS ON PAPER IMPREGNATED WITH ZIRCONIUM PHOSPHATE

Paper: Whatman No. 1, impregnated with 10% solution of zirconium oxychloride in $4\,N$ HCl. The dried strips were than passed through a solution of 12% phosphoric acid in $4\,N$ HCl, finally dried and washed with distilled water

Development: Ascending

Valency states of the elements	Composition of the eluant	R_F values
Fe^{II} and Fe^{III}	$0.5\,N$ H_2SO_4	$Fe^{II} = 0.84$; $Fe^{III} = 0.0$
U^{IV} and U^{VI}	$3.0\,N$ HCl	$U^{IV} = 0.0$; $U^{VI} = 0.72$
U^{IV} and U^{VI}	$3.0\,N$ H_2SO_4	$U^{IV} = 0.22$; $U^{VI} = 0.68$
U^{IV} and U^{VI}	$3.0\,N$ HNO_3	$U^{IV} = 0.03$; $U^{VI} = 0.54$
Ce^{III} and Ce^{IV}	$1.0\,N$ H_2SO_4	$Ce^{III} = 0.64$; $Ce^{IV} = 0.06$
Ce^{III} and Ce^{IV}	$1.0\,N$ HCl	$Ce^{III} = 0.68$; $Ce^{IV} = 0.0$
Cr^{III} and Cr^{VI}	Satd. Na_2SO_4 soln.	$Cr^{III} = 0.84$; $Cr^{VI} = 0.3$
As^{III} and As^{V}	$1.0\,N$ HCl	$As^{III} = 0.66$; $As^{V} = 0.14$
As^{III} and As^{V}	$1.0\,N$ HNO_3	$As^{III} = 0.70$; $As^{V} = 0.06$
V^{IV} and V^{V}	Na_2HPO_4–citric acid buffer pH 7	$V^{IV} = 0.06$; $V^{V} = 0.83$
Mo^{V} and Mo^{VI}	$4\,N$ HCl	$Mo^{V} = 0.63$; $Mo^{VI} = 0.35$
Hg^{I} and Hg^{II}	$0.1\,N$ HNO_3	$Hg^{I} = 0.0$; $Hg^{II} = 0.69$

TABLE 5

R_F VALUES OF SOME CATIONS ON PAPER IMPREGNATED WITH ZIRCONIUM PHOSPHATE IN HCl AND $HClO_4$ SOLUTIONS

Paper: Whatman No. 1, impregnated with 3.5 mg/cm^2 of zirconium phosphate
Development: Ascending

Element	Solvent					
	HCl			HClO$_4$		
	1 N	3 N	6 N	1N	3 N	6 N
Fe^{III}	0	0.30	0.51	0	0	0
Bi^{III}	0.83			6	0.16 t^a	0.35 t
Sn^{II}	0	0.42	0.62	0	0	0
UO$_2$^{II}	0.32	0.47	0.52	0.20	0.26	0.27
Th^{IV}	0	0	0	0	0	0
Cs^{I}	0	0	0	0	0	0

^{a} t = tailing.

TABLE 6

R_F VALUES FOR U^{VI} AND Th^{IV}

Paper impregnated with a mixture of HTTA and TOP. The mobile phase is 0.1 M HCl

Mixture of organic extractants		R_F	
		U^{VI}	Th^{IV}
TOP	0.05 M	0.92	0.57
HTTA	0.0025 M	0.85	0.48
TOP	0.0475 M		
HTTA	0.005 M	0.70	0.15
TOP	0.045 M		
HTTA	0.010 M	0.24	0.13
TOP	0.040 M		
HTTA	0.020 M	0.25	0.12
TOP	0.030 M		
HTTA	0.030 M	0.34	0.11
TOP	0.020 M		
HTTA	0.040 M	0.52	0.19
TOP	0.010 M		
HTTA	0.045 M	0.63	0.38
TOP	0.005 M		
HTTA	0.0475 M	0.75	0.40
TOP	0.0025 M		
HTTA	0.05 M	0.97	0.80

The presented curves are rather similar to those obtained by solvent extraction of the above mentioned ions.

has been described by Suchy[30]. Table 2 lists the R_F values of uranyl ion and thorium obtained with five solvent systems.

Grassini and Padiglione[31] described the separation of UO_2^{2+} and Pb^{2+} on papers impregnated with various amounts of zirconium phosphate and developed at various temperatures with 3 N and 0.4 N $HClO_4$ (Table 3). The separation of the valency states of uranium, iron, cesium, chromium, arsenic, vanadium, molybdenum, and mercury on paper impregnated with zirconium phosphate (Table 4) was described by Sastri and Rao[32]. Alberti et al.[33] reported the separation of UO_2^{II}, Th^{IV}, Cs^{I}, Fe^{III}, Bi^{III}, and Sn^{II} on zirconium phosphate-impregnated papers using hydrochloric and perchloric acids of different normality as developers (Table 5).

It had been earlier shown[34] that in the chromatography of $^{241}Am^{III}$, Ce^{III}, and La^{III} on paper treated with a mixture of thenoyl trifluoroacetone (HTTA) and tri-n-octyl phosphate (TOP), the retention of these ions was higher than if each extractant was used separately. Synergism in the reversed-phase partition chromatography of uranium and thorium was studied by Cvjeticanin[35]. In order to obtain the synergistic effect, one of the reagents should be a complexing or chelating agent, capable of neutralizing the charge on the metal ion, while the other is an active donor solvent[36,37]. Whatman No. 1 paper strips impregnated with a mixture of HTTA and TOP[34,38] are developed 4–5 h at 23–25°. UO_2^{2+} was identified with a 3% solution of potassium ferrocyanide and thorium with a 0.05% aqueous solution of thoron. Table 6 shows the R_F values for U^{VI} and Th^{IV} on paper impregnated with a mixture of HTTA and TOP. The mobile phase is 0.1 M HCl. Table 7 depicts the chromatographic separations of U^{VI}–$^{241}Am^{III}$–$^{144}Ce^{III}$; Th^{IV}–Ce^{III}–La^{III}; Th^{IV}–U^{VI}–Ce^{III} and U^{VI}–Ce^{III}–La^{III} on paper treated with a mixture of HTTA + TOP with hydro-

TABLE 7

CHROMATOGRAPHIC SEPARATIONS ON PAPER TREATED WITH A MIXTURE OF HTTA + TOP
Eluant: HCl

Elements	Extractant		HCl (M)	R_F	
	HTTA (M)	TOP (M)			
U^{VI}–$^{241}Am^{III}$–$^{144}Ce^{III}$	0.15	0.05	0.075	U^{VI} $^{241}Am^{III}$ $^{144}Ce^{III}$	= 0.0; = 0.32; = 0.76
Th^{IV}–Ce^{III}–La^{III}	0.15	0.05	0.05	Th^{IV} Ce^{III} La^{III}	= 0.0; = 0.40; = 0.82
Th^{IV}–U^{VI}–Ce^{III}	0.08	0.005	0.05	Th^{IV} U^{VI} Ce^{III}	= 0.0; = 0.21; = 0.98
U^{VI}–Ce^{III}–La^{III}	0.08	0.05	0.025	U^{VI} Ce^{III} La^{III}	= 0.0; = 0.20; = 0.56

chloric acid of varying molarities as developing solvents. Ce, La, and Am formed complexes of the type $M(TTA)_3(TOP)_2$ while UO_2 was found to form the complex $UO_2(TTA)_2TOP$ and thorium the complex $Th(TTA)_4S$.

Paper chromatographic analysis of irradiated uranium in a hydrofluoric acid medium was described by Crouthamel and Fudge[39]. The behavior of the various fission products was studied by irradiating samples of highly purified U_3O_8 in a CP-5 flux of 10^{13} n/cm² sec for 1 h and 1 week. Analysis of the irradiated uranium were made after 24 h cooling and after 1 week's cooling. The movement of individual isotopes was studied employing the chemically purified fission products. Whatman-

TABLE 8

NUCLIDES AND RADIOACTIVE DECAY DATA IDENTIFIED IN THE CHROMATOGRAMS

Nuclide	Gamma spectrum (keV)	Beta (max. keV)	Half-life	Remarks
Cesium-137	661		32.6 years	32 keV K-conversion
and barium-137m			156 sec	from ^{137}Ba
Cerium-141	142		33 days	36 keV K-conversion from ^{141}Pr
Barium-140	537, 165, 30		12.8 days	^{140}La daughter
Niobium-97m	750		60 sec	
and 97	660		74 min	
Zirconium-97	(See ^{97m}Nb)		17 h	
Tellurium-132	230		77 h	27 keV K-conversion from ^{132}I
	265*			
Strontium-90	No γ	610	28 years	
Yttrium-90	No γ	2200	64 h	
Molybdenum-99	0.728, 181		67 h	(See ^{99m}Tc)
Technetium-99m	142		6.0 h	
Neptunium-239	106, 230		2.3 days	106 keV most intense peak
	276, 323			
Plutonium-239	17.8 L-conversion		24,300 years	
	53 and 100 (weak)			
Americium-241	59.8		470 years	32 keV escape peak
Protactinium-233	312		27 days	
	98			
Thorium-234	90		24 days	
	16 L-conversion			
Ruthenium-106	No γ	Very weak	1.0 year	(See ^{106}Rh)
Rhodium-106	513	3500	30 sec	
	624			
	1.04			
Niobium-95	755	160	36 days	
Zirconium-95	730	371	65 days	

* Decay period of 77 h found for both 230 and 265 keV peaks; latter peak has not been associated with ^{132}Te in the literature.

3MM or Whatman No. 1 and No. 2 paper strips were used with aqueous hydrofluoric and methyl ethyl ketone–hydrofluoric acid developing solvents.

Table 8 lists the nuclides and the radioactive decay data identified in the chromatograms. Table 9 summarizes the behavior of radioactive nuclides with HF and HF–methyl ethyl ketone solvents in ascending paper chromatograms. Radioautographs of three irradiated samples are shown in Figs. 1 and 2. Figure 1 shows the chromatograms of relatively green fission products of irradiated uranium. This chromatogram was developed with 60 g of 47% hydrofluoric acid/100 ml of methyl ethyl ketone. Figure 2 shows two chromatograms of a highly irradiated Chalk River uranium sample.

The TLC separation of UO_2^{2+} in a mixture of cations has been described[40,41]. UO_2^{2+} ions migrated towards the upper part of an MN Silica Gel S-HR plate while Fe, Co, Cu, Ni, Al, and Th remained at or near the origin when developed with freshly distilled ethyl acetate–ether saturated with water–tri-n-butyl phosphate

TABLE 9

SUMMARY OF THE BEHAVIOR OF RADIOACTIVE NUCLIDES WITH HF AND HF–MEK* MEDIUM IN ASCENDING PAPER CHROMATOGRAMS

Nuclide	R_F			Remarks
	20 g 49% HF per 100 ml MEK	60 g 49% HF per 100 ml MEK	49% HF	
^{137}Cs, ^{134}Cs	0.3	0.6	1.0	Other alkali and alkali earths behave similarly. Results may be difficult to reproduce as there is a tendency for a fraction to remain fixed in paper.
^{140}Ba	0.0	0.0	0.8	
^{90}Sr	0.0	0.0	0.8	
^{141}Ce, ^{140}La, ^{90}Y, ^{234}Th	0.0	0.0	0.0	
^{93}Nb, ^{97}Nb	1.0	1.0	0.7	
^{181}Ta	1.0	1.0	0.7	
^{95}Zr, ^{97}Zr	0.0	0.05	0.2	
$^{125}Sb^V$	1.0	1.0	0.8	
^{106}Ru		0.2	0.8	Species unpredictable, variable R_F values
$^{132}Te^{VI}$	0.5	0.6	0.8	
^{99}Mo	0.1	0.25	0.7	
^{99m}Tc	0.5	0.6	0.7	
^{233}Pa	0.0	0.05	0.9	
UO_2^{2+}	0.0	0.05	0.8	Considerable tailing
^{239}Np	0.0	0.0	0.2	Considerable tailing
^{239}Pu	0.0	0.0	0.2	Considerable tailing
^{241}Am	0.0	0.0	0.0	

* MEK = methyl ethyl ketone.

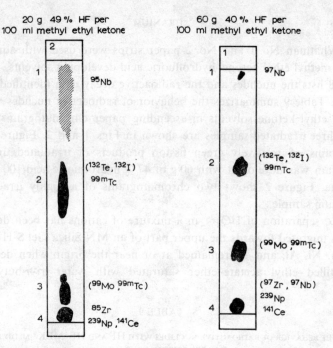

Fig. 1. Uranium irradiated 1 h in CP-5 flux of 10^{13} n/cm² sec and cooled 24 h (strip 1) and 98 h (strip 2) before chromatographing.

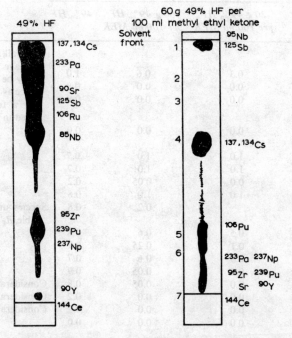

Fig. 2. Chalk River uranium sample irradiated to 10% total uranium atom burn-up and with 2 year cooling period.

(50:50:2) (10–15 min, 15 cm). Detection was achieved with 0.25% solution of pyridylazonaphthol in ethanol. If a $UO_2(NO_3)_2$ solution is applied, the daughter nuclide ^{234}Th can be separated from it and determined at the start by activity measurements. Markl and Hecht[42] separated U^{6+}, Mo^{6+}, and Fe^{3+} from mixtures of other ions using Silica Gel G layers and triisooctylamine as solvent. Layers with exchanger properties can also be obtained by impregnating the adsorbent with triisooctylamine[43] for the separation of mixtures of U, Co, Cu, Zn and Fe, Co, and Ni.

Two anion-exchange separations applicable to the determination of uranium in complex nuclear fuel element solutions was described by Vita *et al.*[44]. In a very selective dual-ion exchange process, uranium was adsorbed on Dowex 1-X8 as the anionic nitrate complex from 1.9 M aluminum nitrate–0.1 M nitric acid and then as an anionic chloride complex from 8 M hydrochloric acid without removal from the column.

In a less selective single ion-exchange process, uranium was adsorbed on Dowex 1-X8 from 8 M hydrochloric acid. Uranium was separated from solutions containing aluminum, beryllium, molybdenum, stainless steel, and zirconium and determined by titration with potassium dichromate[45].

The separation of a variety of cations by anion-exchange chromatography was studied by Strelow and Toerfen[46]. Figure 3 illustrates an elution curve for the separation of U^{VI}, Pb^{II}, and Bi^{III} on AG1-X8 resin. Table 10 lists the analytical methods used for the quantitative determination of various cations following the separation by anion-exchange chromatography.

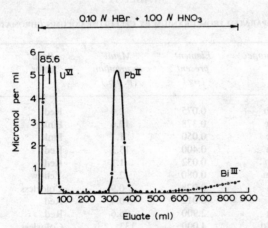

Fig. 3. Elution curve for U^{VI}, Pb^{II}, and Bi^{III}; column of 20 g AG1-X 8 resin in Br^- form; eluant 0.10 N HBr for U^{VI}, 1.00 N HNO$_3$ for Pb^{II} and Bi^{III}; flow rate 3.0 ± 0.2 ml/min.

Submicrogram amounts of thirteen metals have been separated from 50 g of uranyl nitrate by adsorption chromatography on a cellulose column (Whatman cellulose powder, Chromedia Ct II, ashless, standard grade No. 1)[47]. The adsorbed metals were eluted by solutions of potassium thiocyanate and hydrochloric acid in

TABLE 10

ANALYTICAL METHODS

Element	Method
Pb^{II}, Zn^{II}	Titration with EDTA at pH 5.6, xylenol orange indicator
Tl^{III}	Titration with EDTA at pH 8–9 in presence of tartrate, methyl thymol blue as indicator (MTB)
Bi^{III}	Titration with EDTA at pH 2 in presence of malonic acid, MTB indicator
Cd^{II}, Ca^{II}	Titration with EDTA at pH 9, MTB indicator
Pd^{II}	Gravimetrically as dimethylglyoxime complex
Au^{III}, Pt^{IV}	Gravimetrically as the metal
U^{VI}, Fe^{III}, Ti^{IV}, Zr^{IV}	Gravimetrically as the oxides after precipitation with NH_4OH
Th^{IV}, La^{III}, Ce^{III}	Gravimetrically as oxides after precipitation with oxalic acid from 0.05 N HCl
Al^{III}	Gravimetrically as 8-hydroxyquinolate
Be^{II}	Gravimetrically as benzoylacetonate
Sr^{II}, Ba^{II}	Gravimetrically as sulfate
Mg^{II}	Titration with EDTA at pH 9–10, Eriochrome Black T indicator
Mn^{II}	Titration with EDTA at pH 9 in presence of $NH_2OH \cdot HCl$, MTB indicator

TABLE 11

METALS SEPARATED FROM URANIUM BY CELLULOSE COLUMN CHROMATOGRAPHY

Radioisotope	Element present (μg)	Metal in uranium (p.p.b.)	Band color	Eluant[a]
^{114m}In	0.075	2.5	Red	1
^{65}Zn	0.378	12.6	Pink	1
^{60}Co	0.050	1.7	Pink	1
^{64}Cu	0.400	13.3	Red	1
^{51}Cr	0.032	1.1	Red	1
^{54}Mn	0.080	2.7	Yellow	1
^{133}Ba	0.028	0.9	Colorless	1
^{59}Fe	0.012	0.4	Red	1
^{110}Ag	2.900	96.6	Red	1
^{115m}Cd	4.000	133	Colorless	2
^{85}Sr	0.100	3.3	Colorless	2
^{124}Sb	0.034	1.1	Red	3
^{137}Cs	Carrier-free			Colorless	3

[a] Eluting solutions: 1, 40 mg of potassium thiocyanate dissolved in 40 ml of methanol plus 60 ml of ethyl ether; 2, 40 mg of potassium thiocyanate dissolved in 100 ml of acetone; 3, 30 ml of conc. hydrochloric acid dissolved in 70 ml of acetone.

organic solvents. The quantitative recovery of metals was verified by using radio-isotopes and gamma-ray spectrometry. Table 11 illustrates the metals separated from uranium by column chromatography.

In previous studies it has been shown that cadmium, at a concentration of 5 p.p.b., as well as ytterbium and scandium at a concentration of 2 p.p.b., have all been determined in uranium by passing a solution of uranyl nitrate in ethyl ether through cellulose and carrying out emission spectrography on the ashed cellulose column[48].

The separation of uranium and thorium by liquid anion exchangers has been utilized for the analysis of effluents from a laboratory-scale thorium–uranium reprocessing plant and in analysis of low-grade thorium ores[49]. Uranium was separated from sulfate solutions with a tertiary amine (tricuprylamine) (3% in xylene) and thorium from nitrate solutions with a quaternary ammonium salt (aliquot-336), (diluted in xylene, 5%). Uranium was thence determined colorimetrically by means of the sodium carbonate–sodium hydroxide–hydrogen peroxide system[50] when the uranium concentration in the original solution exceeded 10^{-3} M, and by means of the dibenzoylmethane–pyridine system for lower concentrations[51].

Thorium in the aqueous phase was determined colorimetrically with thoran[52] or, for sufficiently high concentrations, by titration with EDTA in the presence of Alizarin S as indicator[53].

Zvarova and Zvara[54] described the separation of rare earth elements by GLC of their chlorides. The analyses were carried out on glass capillary columns at 250° using $AlCl_3$ vapors (40–170 mm Hg) as a component of the carrier gas. $AlCl_3$ forms volatile complexes with rare earth chlorides, and at the same time modifies the surface of the column.

Gas chromatography was performed in the apparatus shown in Fig. 4. Tube furnaces 1, 2, 3, and 4 were maintained at the required stepped temperature distribution along a glass tube 5. This consisted of several welded sections; the spiral one (2.5 m × 1.0 mm i.d.) was placed in a thermostat, 1, which served as the chromato-

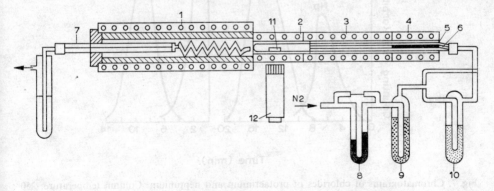

Fig. 4. The chromatographic apparatus [see text for description].

graphic column. Nitrogen (flow rate 12 ml/min) after passing through a flow meter, 8, passed through a molecular sieve desiccant, 9, and was saturated with aluminum chloride vapors in a tube, 6. This was completely filled with the solid chloride and situated in furnace 4, where the temperature was 20–30° higher than in the thermo-

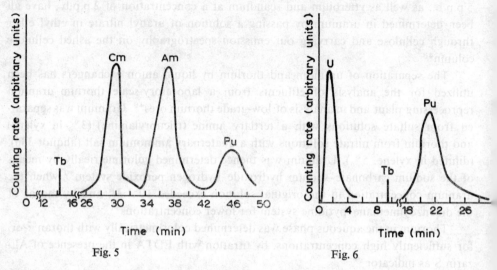

Fig. 5.   Fig. 6.

Fig. 5. Separation of a plutonium–americium–curium mixture. Column temperature, 250°; Al_2Cl_6 partial vapor pressure, 100 mm Hg; helium flow rate, 8 ml/min.

Fig. 6. Separation of a uranium–plutonium mixture. Column temperature, 255°; Al_2Cl_6, partial vapor pressure, 100 mm Hg; helium flow rate, 8 ml/min.

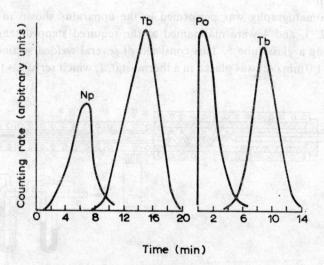

Fig. 7. Chromatograms of chlorides of protactinium and neptunium. Column temperature, 250°; Al_2Cl_6 partial vapour pressure, 100 mm Hg; helium flow rate, 8 ml/min.

stat 3 whose temperature corresponded to the required vapor pressure of Al_2Cl_6. An Al_2Cl_6 partial vapor pressure between 40 and 170 mm Hg was used while maintaining the temperature of furnace 3 at 138–155°.

The separation of transuranium elements by gas chromatography of their chlorides has been described by Zvarova and Zvara[55]. The experiments were carried out in glass capillary columns (10 m × 1 mm i.d.) at 250° and use was made of isotopes ^{231}Pa, ^{233}U, ^{237}Np, ^{239}Pn, ^{241}Am, and ^{244}Cm in amounts not in excess of 1 μg. The chromatograms were measured by collecting fractions of the Al_2Cl_6 condensate at the exit of the column and by measuring their radioactivities. Beta active ^{160}Tb was added to all samples as a monitor. Figures 5–7 illustrate the separation of mixtures of plutonium–americium–curium; uranium–plutonium and the chlorides of protoactinium and neptunium, respectively.

It can be seen from Fig. 5 that curium and americium are eluted from the column in the order of decreasing atomic number and are found on the chromatogram at the position of lighter lanthanide elements. This is analogous to the separation of tervalent ions of Am and Cm by ion-exchange chromatography[56].

The GLC evidence indicates that trichlorides of transuranium elements form with Al_2Cl_6 complexes of the same type as trichlorides of lanthanide elements. The elution of plutonium just after americium and curium (Fig. 5) with similar separation factors of the adjacent elements indicated that under the experimental conditions, plutonium exists in the tervalent state. The elution of plutonium in such a position relative to Am and Cm as shown in Fig. 5 is apparently uncommon in the chromatography of transuranium elements. In ion-exchange and extraction chromatography separations of plutonium are, as a rule found in solution in a higher oxidation state than + 3.

Volatile complexes of some lanthanides and related elements with fluorinated β-ketones and organophosphorus adducts were studied by Mitchell and Banks[57] utilizing both thermogravimetric and gas chromatographic techniques. For the analysis of some chelates of uranium a Beckman Model GC 4 equipped with a flame-ionization detector was used with dual 4 ft. × ¼ in. columns containing (a) 0.1% SE-30 on glass beads and (b) 5% SE-30 on Chromosorb G. Although chelates of uranium(IV) with either hexafluoroacetylacetone (HFA) or trifluoroacetylacetone (HTFA) were not completely volatilized under thermogravimetric conditions, no evidence indicative of thermal decomposition was observed by gas chromatography. Figure 8 illustrates gas chromatograms of some chelates of uranium(IV) with and without the adduct tri-n-butyl phosphate (TBP). Well defined peaks were obtained for the $U(TFA)_4$ and $U(HFA)_4$ chelates while the mixed chelate $U(HFA)_4 \cdot 2$ TBP was eluted with a tailing effect and $U(TFA)_4 \cdot 2$ TBP was not successfully chromatographed. The gas chromatography of chelates of uranium(IV) had not been previously reported.

A gas chromatographic method was reported by Sieck et al.[58] for the separation and determination of $UO_2^{2+} + Th^{IV}$ as mixed-ligand complexes. By using the betadiketone hexafluoroacetylacetone, H(HFA), and the neutral donor, di-n-butyl sulfoxide (Bu₂SO), to form the mixed-ligand complexes, separation and quantitative determination of UO_2^{2+} and Th^{IV} was possible in the range 1–120 mg metal/ml with a relative error of <10%. The detection limits were 0.4 mg/ml for Th and 0.6 mg/ml for uranium

A Hewlett-Packard Model 5756B gas chromatograph was used equipped with a thermal conductivity detector. A 1 mV Bristol recorder equipped with a Disc

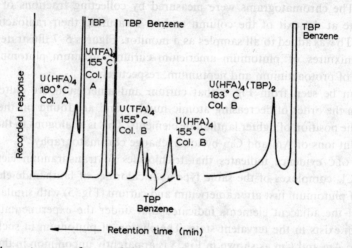

Fig. 8. Gas chromatograms of some chelates of uranium(IV) with and without the adduct TBP. Flow 55 ml/min. Column 4 ft. + ¼ in.; A, 0.1% SE 30 on glass beads; B, 5% SE 30 on Chromosorb G.

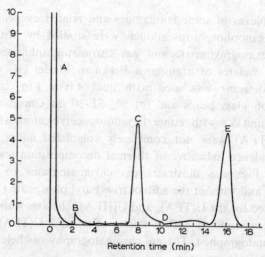

Fig. 9. Gas chromatogram for mixed-ligand complexes of UO_2^{2+} and Th^{IV} with H(HFA) and DBSO. Column B, temperature 192° and 210° C. A, Benzene; B, DBSO; C, Th(HFA)₄ · DBSO; D, temperature to 210° C; E, UO₂ (HFA)₂ · DBSO.

Integrator, Model 202 was used for the quantitative studies. GLC conditions were as follows: inlet temperature 250°, thermal conductivity detector 350°, injection port 260°C, detector bridge current 150 mA. Helium was used as the carrier gas with a flow rate of 48 ml/min for all columns used. Two stainless steel columns (16 in. × ¼ in.) were used, *viz.* column A was packed with Chromosorb W (100–120 mesh) coated with 10 % SE-30. Column B (which gave the best separations) was packed with Chromosorb W (100–120) coated with 17.8 % QF-1. The columns were conditioned at 245°C overnight before use. The chromatogram for a mixture of $UO_2(HFA)_2$ · DBSO and $Th(HFA)_4$ · DBSO is shown in Fig. 9 and illustrates the excellent separation of UO_2^{2+} and Th^{IV} obtained by using this mixed-ligand system. The column temperature for this analysis was held at 192°C until the Th^{IV} complex was eluted, then the temperature was elevated to 210°C for elution of the UO_2^{2+} complex.

The corrosive nature of volatile fluorides such as uranium hexafluoride (UF_6) has required special column materials and modification of gas chromatographic detectors[59–63].

Juvet and Fisher[61] described a direct method for the quantitative analysis of alloys and certain metal oxides, carbides, sulfides, acetates and nitrates by *in situ* reaction with F_2 in a specially designed reactor injection system of a gas chromatograph followed by separation and analysis of the volatile metal fluorides on a chemically conditioned column. The reaction and elution properties of the elements U, S, Se, Te, W, Mo, Re, Si, B, Os, V, Ir, and Pt in various chemical forms were reported. A modified Micro-Tek Model 2000-R research gas chromatograph was modified

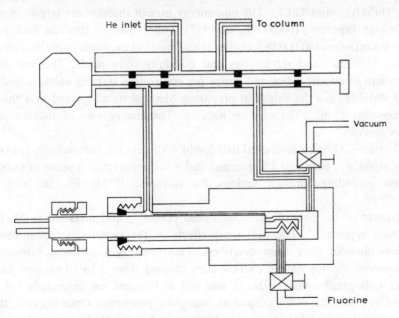

Fig. 10. Reactor injection system for *in situ* fluorinations.

by the use of a Gow-Mac stainless steel thermal conductivity cell equipped with nickel filaments resistant to corrosion by metal fluoride and the addition of the stainless steel reactor injection system as illustrated in Fig. 10. The reaction chamber and gas sampling valve were constructed of Type 316 stainless steel. Tubing connecting the sampling valve to the reactor and leading to the column was heavy-walled Type 316 stainless steel, $\frac{1}{16}$ in. i.d. and vacuum and fluorine inlet lines were $\frac{1}{4}$ in. copper. The total volume of the $\frac{3}{8}$ in. diam. reaction cavity was variable over the range 1–3 ml, depending on the position of the PTFE insert holding the nickel ribbon sample holder–heater assembly.

The column used was a 22 ft. × $\frac{1}{4}$ in. PTFE column packed with 15% Kel F oil No. 10 (Minnesota Mining and Manufacturing Co.) on 40–60 mesh Chromosorb. Packing was performed at liquid nitrogen temperatures[64] and the column was conditioned with ClF_3 or F_2 prior to use to remove moisture and traces of reactive organic matter.

Uranium hexafluoride was found to be readily analyzable by gas chromatography. Owing to its high stability it could be prepared directly from uranium oxides and carbides, e.g. $UO_2(NO_3)_2 \cdot 6H_2O$, $UO_2(C_2H_3O_2)_2 \cdot 2H_2O$, UC_2, and UO_2.

2. THORIUM

The main source of thorium is the phosphate mineral monazite which is reduced by a caustic soda process to yield the principal thorium salts $Th(NO_3)_4$, ThO_2, ThF_4, $Th(SO_4)_2$, and $ThCl_4$. The non-energy uses of thorium are largely in magnesium alloys (approx. 100,000 lb. in 1957). Other uses of thorium include: gas-mantle manufacture (40,000 lb.), chemical and medical products (4000 lb.), electronic products (1000 lb.) and refractories and polishing compounds. Thorium metal is also used in electronic tubes and lamps for controlling starting voltages and maintaining stability, as a deoxidant in preparing Mo and its alloys and in a thoriated-tungsten alloy (3–6% Th) used in welding. The energy uses of thorium are for breeder reactors.

Thorium-232 has a biological half-life of 400 years (its radioactivity is expressed as 90% alpha, 9% beta, and 1% gamma) and it is the parent of a series of radioactive elements extending through various Ra isotopes, ^{208}Th, Po, Bi, and finally stable ^{208}Pb.

Aspects of the toxicology[65], biological effects[66], and metabolism of thorium have been reported[67]. The radiotoxic effects of the radiopaque agent thorotrast (thorium dioxide) have been described. Thorotrast is a colloidal suspension of 25% thorium dioxide having particle sizes ranging from 3 to 10 nm and has been used in radiography since 1928. It was still in frequent use (especially for angiography) well into the 1950's despite its oncogenic properties. Once injected, the thorium material is excreted at a negligible rate and most of the material is retained for life. The radiotoxic effects of this agent have been reported to include severe radiation damage and cancers of the bone, blood vessels, liver, and other organs as a result of its administration 11–20 years previously[68–70].

The analysis of thorium has been generally carried out by colorimetric procedures[71–73].

The separation of thorium from uranium, yttrium, lanthanum, and cerium has been accomplished on Dowex 50W-X8 resin (20–50 mesh, H+ form), in formic acid–dimethyl sulfoxide media[74]. Thorium was separated from U, Yt, La, and Ce by elution first with DMSO and then with 2 M nitric acid. Kuroda et al.[75] have reported Th–Y, La–Th, Th–U separations on the weakly basic anion-exchange resin Amberlyst CG4B in hydrochloric acid media. Korkisch et al.[76,77] have separated thorium from rare earths on a Dowex 1W-X8 column in 75% acetone–25% hydrochloric acid medium, and in a Dowex 50W-X8 column in methanol containing 0.1 M trioctyl phosphine oxide and 5% of 12 M nitric acid.

TABLE 12

R_F VALUES OF FISSION PRODUCTS AND ACTINIDES ON PAPERS TREATED WITH TOA

R_F Values of actinides eluted with
2 M and 8 M HNO$_2$

Actinide	R_F	
	2 M HNO$_3$	8 M HNO$_3$
Pu^{IV}	0.03	0.03
Th	0.18	0.24
U^{VI}	0.68	0.60
Np^V	0.86	(0.20)^a
Pa–Am–Cm	1	1

^{a} Np^{VI}

R_F values of actinides eluted with 0.5 M
LiNO$_3$ + 0.5 M HNO$_3$ (0.1 M amine)

Actinide	R_F
Am–Cm	1
Np^V	0.85
U^{VI}	0.70
Th	0.20
Pu^{IV}–Pa	0

Paper: Schleicher and Schüll No. 2043b MgI. The strips were dipped for 15 min in a solution of TOA (tri-n-octylamine) in xylene.

Knoch et al.[78] described the separation of some fission products and actinides on paper treated with liquid anion exchanger. Schleicher and Schüll paper No. 2034B impregnated with a tri-n-octylamine–xylene solution was used with 2 M and 8 M nitric acid used as solvents in 15 cm developments. The chromatograms were developed either by color reactions and/or counting methods. Table 12 gives the R_F values of actinides (Pu^{IV}, Th, U^{VI}, Np^V, Pa–Am–Cm) obtained on impregnated paper eluted with 2 M and 8 M HNO$_3$.

Figure 11 is a scan plot showing the separation of Ru, Zr, and U obtained on paper with 0.5 M TOA and developed with 0.1 M Al(NO$_3$)$_2$ + 0.5 M HNO$_3$.

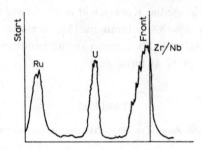

Fig. 11. Scanner plot, separation of Ru, Zr, U. Eluant, 0.1 M Al(NO$_3$)$_3$ + 0.5 M HNO$_3$; paper reated with 0.5 M TOA.

The gas chromatography of metal chelates with carrier gas containing ligand vapor was investigated by Fujinaga et al.[79]. Uden and Jenkins[80] have previously attributed the anomalous peaks in the chromatography of metal chelates to the adsorption of chelates on the solid support and in some cases the presence of isomers of metal chelates must be considered as a cause. However, the main reason for the anomaly could also be the dissociation of metal chelates in the stationary liquid phase. Fujinaga et al.[79] attempted to use a carrier gas containing ligand vapor which would suppress the dissociation. As a typical example, a chromatographic separation of trifluoroacetylacetonates of Th^{IV}, Fe^{III}, and U^{IV} were examined with the vapor of trifluoroacetylacetone as the carrier gas additive.

A Shimadza Model GC-1B gas chromatograph equipped with a thermal conductivity detector was used as the conventional gas chromatograph. Figure 12 shows a schematic diagram of the gas chromatograph equipped with the vapor generator for supplying the carrier gas additive and the resistance column for smoothing the mixing of helium with the additive. The gas chromatographic columns were of 4 mm × 0.75 m stainless steel filled with Gas Chrom-CLH (80–100 mesh) as the stationary solid support and the stationary liquid phases were: Apiezon L, Silicone Se-30, DC-550, XE-60, and Polyethylene Glycol (PEG)-6000 with 1, 0.5, 5, 2.5, and 10% w/w of coating weight, respectively. Helium gas saturated with the vapor of trifluoroacetylacetone (TFA), isobutyl methyl ketone (IBMK), or cyclohexane at 30° was used as carrier gas. The TFA metal chelates of Be, Al, Cr, Fe^{III}, U^{IV}, and Th^{IV} were prepared according to the method of Berg and Truemper[81] and purified by sublimation *in vacuo*. The sample was dissolved in benzene, except for the $(TFA)_4$ which was dissolved in IBMK at a concentration of 10% w/v. Figures 13 and 14 illustrate chromatograms of $Th(TFA)_4$ obtained with the conventional

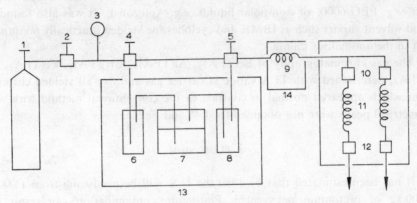

Fig. 12. Schematic diagram of a gas chromatogram using carrier gas containing the vapor of an organic substance. 1, Helium bomb; 2, valve; 3, pressure gauge; 4, 5, valves; 6, 8, buffer tank; 7, generator of organic vapor; 9, resistance column, 4 mm × 0.75 m, filled with Gas Chrom–CLH (80–100 mesh); 10, injection port; 11, volumn; 12, detector; 13, water bath regulated at 30°C; 14, preheater.

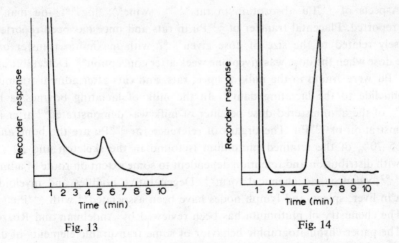

Fig. 13   Fig. 14

Fig. 13. Chromatogram of $Th(TFA)_4$. Carrier gas, helium; sample 10 µl of 10% w/v solution of $Th(TFA)_4$ in IBMK; column, 4 mm × 0.75 m stainless steel U-tube filled with 5% w/w Silicone DC 1550/Gas Chrom-CLH (80–100 mesh); column temperature, 180°C; injection temperature, 250°C; detector temperature, 260°C; flow rate, 65 ml/min at outlet (inlet pressure, 1.07 bar); detector, thermal conductivity; filament current, 150 mA.

Fig. 14. Chromatogram of $Th(TFA)_4$. Carrier gas, helium containing trifluoroacetylacetone vapor; vapor generated at 30°C, other conditions as for Fig. 13.

method using helium alone as carrier gas and helium containing TFA vapor, respectitvely. Figure 14 shows that in the chromatograph of Th(TFA)$_4$ obtained *via* elution with helium gas saturated with TFA at 30°, the anomalous tailing was completely eliminated and a well-defined peak obtained.

As for the choice of stationary phase (in use with ligand vapor), slightly polar liquids, *e.g.* Silicone SE-30, DC-550, and XE-60, were more suitable than strongly polar, *e.g.* PEG-6000, or non-polar liquids, *e.g.* Apiezon L. It was also found that ligand solvent vapors such as IBMK and cyclohexane yielded practically no improvement in the anomalous tailing.

The gas chromatography of Be(TFA)$_2$, Al(TFA)$_3$, Cr(TFA)$_3$, Fe(TFA)$_3$, and U(TFA)$_4$ determined with TFA vapor as carrier gas additive all yielded chromatograms which appeared normal in contrast to the conventional method with which symmetrical peaks were not obtained for U^{IV} and Fe^{III}.

3. PLUTONIUM

It has been estimated that by 1980 the U.S. will be producing from 15,000 to 20,000 kg of plutonium per year[82]. Plutonium contamination can result from weapons testing, fuel reprocessing, nuclear reactor operations, laundry and decontamination wastes, and accidental release during transportation. Most human exposures of the actinides (plutonium, uranium, neptunium, and americium) have been from inhalation of dust or from penetrating wounds contaminated by actinide elements.

Aspects of ^{239}Pu absorption in rats[83,84], swine[85], dogs[86], and man[87] have been reported. Placental transfer of ^{239}Pu in rats and mice has been reported to be inversely related to the size of dose given[88,89] with maximum transfer of 3–8% of the dose when the dose was given one week after conception[88]. Detectable amounts of ^{239}Pu were found in the milk of mice, rats, and cats after administration of the radionuclide to the lactating dam[88]. In the milk of lactating bovines, a level of 10^{-6}% of the administered dose per liter of milk was demonstrated[90] after a single administration of ^{239}Pu. The organs of reference for ^{239}Pu are the bone and liver, *e.g.* 65–70% of the retained plutonium is found in the skeleton and 30% in the liver with distribution and retention dependent to some extent on route of administration[91,92], valence, and chemical form[93]. Degenerative nodules and myeloid metaplasia in liver, spleen and lymph nodes have been associated[94] with ^{239}Pu.

The chemistry of plutonium has been reviewed by Andelman and Rozzell[95].

The paper chromatographic behavior of some transuranic elements of different valence states (*e.g.* Pu^{III}, Pu^{IV}, Pu^{VI}, U^{IV}, U^{VI}, and Am^{III}) was studied by Clanet[96] using *n*-butanol–nitric acid (1:1) of varying molarities (Table 13).

English and Foreman[97] described the paper chromatographic separation of Pu^{IV}–Pu^{VI} by butanol–3 *M* HCl, Pu^{III}–Pu^{IV} by ethyl acetate–conc. HCl, and the separation of Pu^{III} from a number of actinides by ethyl acetate–nitric acid–sodium nitrite. The variation in R_F of plutonium and americium on Whatman No. 1 paper

TABLE 13

R_F VALUES OF SOME TRANSURANIC ELEMENTS

Paper: Whatman No. 1
Solvent: Butanol–aq. HCl (1:1)
Development: Ascending
Temperature: $20 \pm 0.5°C$

Ion	Molarity of HCl							
	1 M	2 M	3 M	4 M	5 M	6 M	8 M	10 M
Pu^{III}	0.019	0.035	0.075	0.179	0.309	0.316	0.231	0.174
Pu^{IV}	0.020	0.024	0.050	0.180	0.260	0.283	0.233	0.176
Pu^{VI}	(0.099	0.189	0.285	0.421	0.541	0.330	0.205	0.181
2 spots	(0.020	0.024	0.051	0.158	0.309			
Am^{III}	0.020	0.024	0.051	0.160	0.293	0.303	0.219	0.117
U^{IV}	$\simeq$0.000	0.014	0.026	0.133	0.281	0.269	0.195	0.110
U^{VI}	0.106	0.169	0.241	0.358	0.469	0.498	0.475	0.555

TABLE 14

R_F VALUES OF SOME TRANSURANIC ELEMENTS

Solvent	Ion			
	Pu^{III}	Pu^{IV}	Pu^{VI}	Am^{III}
n-Butanol equil. with 1 N HCl	0.01	0.01	0.12	0.00
n-Butanol equil. with 2 N HCl	0.02	0.02	0.22	0.02
n-Butanol equil. with 3 N HCl	0.06	0.06	0.30	0.07
n-Butanol equil. with 4 N HCl	0.21	0.22	0.49	0.22
n-Butanol equil. with 5 N HCl	0.41	0.34	0.58	0.36
n-Amyl alcohol equil. with 4 N HCl	0.00	0.01	0.17	

Paper: Whatman No. 1
Development: Descending

Ion	Normality HCl	Ethyl acetate–HCl					
		70:30	60:40	50:50	40:60	20:80	10:90
Pu^{III}	9	0.16	0.17	0.27	0.36	0.52	0.76
Pu^{IV}		0.79	0.77	0.66	0.62	0.57	0.67
Pu^{III}	11	0.10	0.17	0.28	0.39	0.42	
Pu^{IV}		0.80	0.76	0.66	0.56	0.52	

using mixtures of n-butanol–HCl (1:1) has been described by Clanet[96]. The development was carried out at 20 ± 0.5° for 20 h and plutonium was detected with ammoniacal 8-hydroxyquinoline solution. The limits of detection with this reagent were 0.2 μg for Pu^{III} and Pu^{IV} and 0.08 μg for Pu^{VI}. Table 14 lists the R_F values obtained for the paper chromatography of Pu^{III}, Pu^{IV}, Pu^{VI}, and Am^{III} with n-butanol equilibrated with hydrochloric acid of varying normalities, and Pu^{III} and Pu^{IV} developed with ethyl acetate–hydrochloric acid (of varying normalities).

Production of transplutonium elements has increased in the past 5 years posing an increasing potential risk of personal exposure[98]. Programs for the large-scale production of ^{244}Cm (ref. 99) and ^{252}Cf (refs. 100, 101) have also been reported.

Trivalent actinides have been determined by employing the liquid ion-exchanger di-2-ethylhexylphosphoric acid[102].

Butler[103] described rapid, sensitive methods to determine plutonium, neptunium, and enriched uranium in urine. Chloride complexes of the actinides were extracted from oxidized urine salts with the liquid ion-exchanger tri-isooctylamine (TIOA). The actinides were back-extracted and then counted on low-background solid-state counters. All three actinides were exchanged in 10 sec from an 8 N HCl solution of urine salts, with TIOA dissolved in xylene. The actinides could be back-extracted collectively from the organic phase with 0.1 N HCl to determine total alpha emission or they could be stripped individually. When stripped individually, plutonium was adjusted to Pu^{III} and stripped from the TIOA with 8 N HCl containing 0.05 M NH_4I. Np^{IV} was stripped with 0.02 N HF in 4 N HCl and U^{VI} was stripped with 0.1 N HCl. These procedures yielded low-solid mounts (1.5 mg) containing >95% of the actinides with no other radioactive contaminants, thus eliminating lengthy electrodeposition. The mounts were assayed with counters having 30% efficiency and backgrounds less than 5 counts/day. The sensitivity was 0.1 disintegration/min of alpha/1.5 l for a 250 ml urine sample. A procedure for the determination of actinides was developed by Butler and Hall[104] using the bidentate extractant dibutyl-N,N-diethylcarbamyl phosphonate (I). Nine actinides were extracted from 12 N HNO_3, back-extracted to 2 N HNO_3, and counted in a low-background alpha counter. A procedure was also developed for sequential extraction of plutonium, neptunium, and uranium with tri-isooctylamine (TIOA), followed by extraction of horium, americium, cerium, berkelium, californium, and einsteinium with bidentate.

Figure 15 illustrates the TIOA–DBCP procedure for the analysis of actinides. Plutonium, neptunium, and uranium were exchanged from 8 N HCl to TIOA (by the procedure of Butler[103]. Compared with previous methods, this new procedure required less analysis time and gave better recovery. The recovery of Am–Cm–Cf from 250 ml of urine or 20 g of feces was 90% with the sensitivity of analysis being 0.02 ± 0.01 d/min/sample.

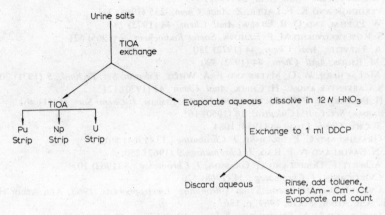

Fig. 15. TIOA–DDCP actinide procedure.

The column chromatographic separation of plutonium of different oxidation states was described by Hoehlein[105]. Pu^{III}, Pu^{IV}, PuO_2^{2+}, and colloidal plutonium were separated and identified by cation-exchange chromatography on Bio-Rad AG50 X8 resin (H^+ form) by elution with hydrochloric acid.

REFERENCES

1 R. F. SPALDING AND W. M. SACKETT, *Science*, 175 (1972) 629.
2 W. S. MOORE, *Earth Planet Sci. Letters*, 2 (1967) 231.
3 C. VOEGTLIN AND H. C. HODGE, *The Pharmacology and Toxicology of Uranium Compounds*, McGraw-Hill, New York, 1951.
4 A. TANNENBAUM, *Toxicology of Uranium*, McGraw-Hill, New York, 1951.
5 R. SELTSER, *Arch. Environ. Health*, 10 (1965) 923.
6 E. LORENZ, *J. Nat. Cancer Inst.*, 5 (1944) 1.
7 F. L. LEITES, *Vop. Onkol.*, 4 (1958) 629.
8 J. K. WAGONER, V. E. ARCHER, B. E. CARROLL, D. A. HOLADAY AND P. A. LAWRENCE, *J. Nat. Cancer Inst.*, 32 (1964) 787.
9 J. K. WAGONER, V. E. ARCHER, F. E. LUNDIN, D. A. HOLADAY AND J. W. LLOYD, *New Engl. J. Med.*, 273 (1965) 181.
10 V. E. ARCHER AND F. E. LUNDIN, *Environ. Res.*, 1 (1967) 370.
11 F. E. LUNDIN, JR., J. W. LLOYD, E. M. SMITH, V. E. ARCHER AND D. A. HOLADAY, *Health Phys.*, 16 (1969) 571.
12 H. C. HODGE, *Arch. Ind. Health*, 14 (1956) 43.
13 J. F. FLAGG, in C. VOEGTLIN AND H. C. HODGE (Eds.), *Pharmacology and Toxicology of Uranium Compounds*, McGraw-Hill, New York, 1949, p. 19.
14 M. JEAN, *Anal. Chim. Acta*, 57 (1971) 440.
15 G. R. PRICE, R. J. FERRETTI AND S. SHWARTZ, *Anal. Chem.*, 25 (1953) 322.

16 J. SOMMER, *Chem. Tech. (Berlin)*, 15 (1963) 38.
17 A. L. BINI, *Health Phys.*, 2 (1960) 288.
18 G. W. ROYSTER, *Health Phys.*, 2 (1960) 291.
19 M. PICER AND P. STROHAL, *Anal. Chim. Acta*, 40 (1968) 131.
20 K. VOLODER, *Clin. Chim. Acta*, 24 (1969) 373.
21 V. VERDINGH AND K. F. LAUER, *Z. Anal. Chem.*, 235 (1968) 311.
22 K. W. PUPHAL AND D. R. OLSEW, *Anal. Chem.*, 44 (1972) 284.
23 V. S. ROSLYAKOV AND M. P. EZHOVA, *Soviet Radiochem.*, 7 (1965) 621.
24 N. A. TALVITIE, *Anal. Chem.*, 44 (1972) 280.
25 W. M. RIGGS, *Anal. Chem.*, 44 (1972) 390.
26 M. MCEACHERN, W. G. MYERS AND F. A. WHITE, *Environ. Sci. Technol.*, 5 (1971) 700.
27 B. S. CARPENTER AND C. H. CHEEK, *Anal. Chem.*, 42 (1970) 121.
28 F. H. POLLARD, J. F. W. MCOMIE AND I. I. M. ELBEIH, *J. Chem. Soc.*, (1951) 461.
29 B. SARMA, *Sci. Cult. (Calcutta)*, 16 (1950) 165.
30 K. SUCHY, *Chem. Listy*, 48 (1954) 1084.
31 G. GRASSINI AND C. PADIGLIONE, *J. Chromatog.*, 13 (1964) 561.
32 M. N. SASTRI AND A. P. RAO, *J. Chromatog.*, 9 (1962) 250.
33 G. ALBERTI, F. DOBICI AND G. GRASSINI, *J. Chromatog.*, 8 (1962) 103.
34 N. CVJETICANIN, *J. Chromatog.*, 34 (1968) 520.
35 N. CVJETICANIN, *5th Intern. Symp. Chromatog. Electrophoresis, 1968*, Ann Arbor-Humphrey Sci., Ann Arbor, Mich., 1969, p. 186.
36 H. IRVING AND D. N. EDGINGTON, *J. Inorg. Nucl. Chem.*, 15 (1960) 158.
37 T. V. HEALY, *Nucl. Sci. Eng.*, 16 (1963) 413.
38 N. CVJETICANIN, *J. Chromatog.*, 32 (1968) 384.
39 C. E. CROUTHAMEL AND A. J. FUDGE, *J. Inorg. Nucl. Chem.*, 5 (1968) 240.
40 H. SEILER AND M. SEILER, *Helv. Chim. Acta.*, 43 (1960) 1939.
41 H. SEILER AND M. SEILER, *Helv. Chim. Acta*, 48 (1965) 117.
42 P. MARKL AND F. HECHT, *Mikrochim. Acta*, (1964) 889.
43 P. MARKL AND F. HECHT, *Mikrochim. Acta*, (1963) 970.
44 O. A. VITA, C. R. WALKER, C. F. TRIVISONNO AND R. W. SPARKS, *Anal. Chem.*, 42 (1970) 465.
45 W. DAVIES AND W. GRAY, *UK At. Energy Authority Reactor Group TRG Rept. T 16(D)*, 1964.
46 F. W. E. STRELOW AND F. V. S. TOERFEN, *Anal. Chem.*, 38 (1966) 545.
47 R. A. A. MUZZARELLI AND L. C. BATE, *Talanta*, 12 (1965) 823.
48 C. A. BERTHELOT, S. HERRMANN, K. F. LAUER AND R. A. MUZZARELLI, *European At. Energy Community, Eur-587-E*, 1964.
49 M. COSPITO AND L. RIGALI, *Anal. Chim. Acta*, 57 (1971) 107.
50 C. J. RODDEN, *Analytical Chemistry of the Manhattan Project*, McGraw-Hill, London, 1950, p. 53.
51 P. BLANQUET, *Anal. Chim. Acta*, 16 (1957) 44.
52 P. F. THOMASON, M. A. PERRY AND W. M. BYERLY, *Anal. Chem.*, 21 (1949) 1239.
53 J. S. FRITZ AND J. J. FORD, *Anal. Chem.*, 25 (1953) 1640.
54 T. S. ZVAROVA AND I. ZVARA, *J. Chromatog.*, 44 (1969) 604.
55 T. S. ZVAROVA AND I. ZVARA, *J. Chromatog.*, 49 (1970) 290.
56 J. BRANDSTETR, T. S. ZVAROVA, M. KRIVANEK AND J. MALY, *Radiokhimiya*, 5 (1963) 694.
57 J. W. MITCHELL AND C. V. BANKS, *Anal. Chim. Acta*, 57 (1971) 415.
58 R. F. SIECK, J. J. RICHARD, K. IVERSEN AND C. V. BANKS, *Anal. Chem.*, 43 (1971) 913.
59 J. F. ELLIS, C. W. FORREST AND P. L. ALLEN, *Anal. Chim. Acta*, 22 (1969) 27.
60 A. G. HAMLIN, G. IVESON AND T. R. PHILLIPS, *Anal. Chem.*, 35 (1963) 2037.
61 R. S. JUVET, JR. AND R. C. FISHER, *Anal. Chem.*, 38 (1966) 1861.
62 H. SHINOHARA, N. ASAKURA AND S. TSUJIMURA, *J. Nucl. Sci. Technol.*, 3 (1966) 373.
63 O. ROCHEFORT, *Anal. Chim. Acta*, 29 (1963) 350.
64 R. J. JUVET AND R. L. FISHER, *Anal. Chem.*, 37 (1965) 1752.
65 R. E. ALLEN, *Thorium—A Bibliography of Published Literature, TID-3044, Suppl. 1*, U.S. At. Energy Comm., Oak Ridge, Tenn., 1959.
66 E. D. HUTCHINSON, *U.S. At. Energy Comm. Res. Div. Rept. UR-561*, 1960.

67 H. C. Hodge, E. A. Maynard and L. J. Leach, *U.S. At. Energy Comm. Res. Div. Rept. UR-562*, 1960.
68 M. L. Janower, V. M. Sidel, W. H. Baker, D. E. P. Fitzpatrick and F. I. Guarino, *New Engl. J. Med.*, 279 (1968) 180.
69 W. B. Looney, *Am. J. Roentgenol. Radium Therapy Nucl. Med.*, 83 (1960) 163.
70 P. Fischer, E. Golub, E. Junze-Mühl and T. Müllner, *Ann. N.Y. Acad. Sci.*, 145 (1967) 759.
71 R. W. Perkins and D. R. Kalkwarf, *Anal. Chem.*, 28 (1956) 1989.
72 G. A. Welford, D. A. Sutton, R. S. Morse and S. Tatrus, *Am. Ind. Hyg. Assoc. J.*, 19 (1958) 464.
73 R. Albert, P. Kevin, J. Fresco, J. Harley, W. Harris and M. Eisenbud, *Arch. Ind. Health*, 11 (1955) 234.
74 M. Qureshi and K. Husain, *Anal. Chim. Acta*, 57 (1971) 387.
75 R. Kuroda, K. Ishida and T. Kiriyama, *Anal. Chem.*, 40 (1968) 1502.
76 J. Korkisch and T. F. Cumming, *Anal. Chim. Acta*, 40 (1968) 520.
77 J. Korkisch and K. A. Orlandini, *Anal. Chem.*, 40 (1968) 1952.
78 W. Knoch, B. Muju and H. Lahr, *J. Chromatog.*, 29 (1965) 122.
79 T. Fujinaga, T. Kuwamoto and S. Murai, *Talanta*, 18 (1971) 429.
80 P. C. Uden and C. R. Jenkins, *Talanta*, 16 (1969) 893.
81 E. W. Berg and J. T. Truemper, *J. Phys. Chem.*, 64 (1960) 487.
82 *IAEA Tech. Rept. 49*, International Atomic Energy Agency, 1965.
83 J. Carritt, R. Fryxell, J. Kleinschmidt and R. Kleinschmidt, *J. Biol. Chem.*, 171 (1947) 273.
84 J. E. Ballou, *Proc. Soc. Exptl. Biol. Med.*, 98 (1958) 726.
85 M. H. Weeks, J. Katz, W. D. Oakley and J. C. Ballou, *Radiation Res.*, 4 (1956) 339.
86 P. E. Morrow, F. R. Bigg, H. Davies, J. Nitola, D. Wood, N. Wright and H. S. Campbell, *Health Phys.*, 13 (1967) 113.
87 W. H. Langham, *Brit. J. Radiol. Suppl.*, 7 (1957) 95.
88 M. P. Finkel, *Physiol. Zool.*, 20 (1947) 405.
89 P. N. Wilkinson and F. E. Hoecker, *Trans. Kansas Acad. Sci.*, 56 (1956) 341.
90 B. F. Sansom, *Brit. Vet. J.*, 120 (1964) 158.
91 J. E. Ballou, *Health Phys.*, 8 (1962) 731.
92 P. W. Durbin, *Health Phys.*, 8 (1962) 665.
93 D. M. Taylor, *Health Phys.*, 8 (1962) 673.
94 R. F. Dougherty, B. J. Stover, J. H. Dougherty, W. S. S. Jee, C. W. Mays, *Radiation Res.*, 17 (1962) 625.
95 J. B. Andelman and T. L. Rozzell, in *Radionuclides in the Environment*, Advances in Chemistry Series, No. 93, American Chemical Society, Washington, D.C., 1970, p. 118.
96 F. Clanet, *J. Chromatog.*, 6 (1961) 85.
97 M. R. English and J. K. Foreman, *Proc. XV Congr. Pure Appl. Chem., Lisbon, 1956*.
98 D. H. Denham, *Health Phys.*, 16 (1969) 475.
99 H. J. Groh, R. T. Huntoon, C. S. Schlea, J. A. Smith and F. H. Springer, *Nucl. Appl.*, 1 (1965) 327.
100 W. C. Reinig, *Nucl. Appl.*, 5 (1968) 24.
101 D. E. Ferguson and J. E. Bigelow, *Actinides Rev.*, 1 (1969) 213.
102 F. E. Butler, *Anal. Chem.*, 37 (1965) 340.
103 F. E. Butler, *Health Phys.*, 15 (1968) 1924.
104 F. E. Butler and R. M. Hall, *Anal. Chem.*, 42 (1970) 1073.
105 G. Hoehlein, *Radiochim. Acta*, 12 (1969) 38.

Chapter 11

NITROGEN GAS (OXIDES OF NITROGEN, AMMONIA)

1. NITRIC OXIDE AND NITROGEN DIOXIDE

Nitric oxide (NO) and nitrogen dioxide (NO_2) are the major oxides of nitrogen produced during combustion. The waste gas of gas turbines contains up to 2000 p.p.m. of nitrogen oxide while that from coal-processing thermal power plants contains between 200 and 1200 p.p.m. of NO depending on the type of heating. The emission of nitrogen oxides has been calculated in tons/day from various cities[1]. For example, 920 tons/day of oxides of nitrogen have been estimated for the Los Angeles County area[2] of which the combustion of fuels and gasoline powered motor vehicles contributed 34.1 and 58.6%, respectively, while petroleum and other sources contributed 4.9 and 2.4%, respectively. Commoner[3] has cited the enormous increase (from the years 1946–1967) in the concentration of nitrogen oxides emitted in engine exhausts as a consequence of the increase in engine compression ratios. It was estimated from the product of passenger vehicle gasoline consumption and p.p.m. of NO_x emitted by engines of average compression ratios 5.9 (1946) and 9.5 (1967) under conditions at 15 in. manifold pressure: 1946, 500 p.p.m. NO_x and 1967, 1200 p.p.m. The ratio of NO to NO_2 emitted from automobile exhausts[4] is generally 99:1. NO is formed by direct combination of oxygen and nitrogen in air at elevated temperatures. NO is slowly oxidized to NO_2 at ordinary temperatures. Nitrogen oxides react in the atmosphere with many organic compounds, particularly hydrocarbons, to yield a spectrum of pollutants including formaldehyde, acrolein, peroxyacetyl nitrate and its analogs (e.g. peroxybutyryl nitrate, peroxyisobutyryl nitrate, and peroxy-benzoyl nitrate), as well as ketones, acids, ozonides and ozononated olefins.

NO_2 and its dimer N_2O_4 are found together at normal environmental temperatures and are encountered as byproducts of many industrial operations. For example, they are found in atmospheres of plants manufacturing nitric and sulfuric acids, in industries in which nitric acid is used for processes such as electroplating, engraving, metal cleaning, and nitration processes as well as occurring in the welding, shipbuilding, and metal repair industries and in the manufacture of dyes, lacquers, and celluloid. NO and NO_2 are also present in concentrations of 14–1000 p.p.m. in cigarette smoke[5-7].

The acute toxicity of nitrogen dioxide has been studied by Gray et al.[8,9] and Carson et al.[10] and the comparative toxicity of NO_2 studied in five species, mice, rats, guinea pigs, rabbits, and dogs by Hine et al.[11]. Table 1 summarizes the toxicologic effects of NO_2 in low concentrations in man, rabbit, rat, and mouse. The odor of NO_2 is characteristic and distinct in concentrations as low as 5 p.p.m. and in concentrations of 10–20 p.p.m. the gas is mildly irritant to the eyes, nose, and upper respiratory mucosa. With man, concentrations above 50 p.p.m. are considered dangerous for short exposure. Deaths resulting from acute exposure to NO_2 have been ascribed to pulmonary edema and not due to the effects of nitrite. For prolonged exposure to concentrations between 25 and 100 p.p.m., the nitrite effect as well as pulmonary irritation are both considerations as a consequence of the absorption of nitrous acid resulting from the hydrolysis of NO_2. The chronic effects of NO_2 in experimental animals have been reported[12-16]. Concentrations

TABLE 1

TOXICOLOGIC EFFECTS OF NITROGEN DIOXIDE AT LOW CONCENTRATIONS[12]

Species	Conc. (p.p.m.)[a]	Duration of exposure	Effect	Comment
Man	5	10 min	Increased airway resistance in 5/5 subjects. Mean increase 92%.	Response was delayed. Maximal increase 30 min after exposure. Time taken to return to normal not investigated.
Rabbit	0.25	6 h/day 6 days	Alteration in structure of lung collagen (by electron microscopy).	Change still apparent in animal sacrificed 7 days after exposure.
Rabbit	1	1 h	Alteration in lung collagen and elastin (by spectroscopy) of animals sacrificed immediately after exposure. Reduced effect observed in animal sacrificed 24 h after exposure.	Chemical changes suggest denaturation of structural protein only partially reversed.
Rats	0.5 or 1	6 h 1 h	Morphologic change in lung mast cells and release of granular substance in animals sacrificed immediately after exposure.	Effect was partially reversed in 24–27 h post-exposure changes are indicative of inflammation.
Rats	1	4 h	Lipoperoxidation of lung lipids.	Effect delayed. Maximum response occurred 24–48 h after exposure.
Rats	2	Continuously for natural lifetime	Pathological changes in bronchiolar epithelium (by electron microscopy), loss of cilia, reduction in normal cellular activity (blebbing)	Changes indicative of a pre-emphysematous condition. Exposure to 0.8 p.p.m. indicates similar process.
Mice	0.5	6–24 h/day for 3 weeks	Distended alveoli, loss of cilia, changes in bronchiolar epithelium.	Indicative of emphasematous condition.
		Continuous for 3 months	Increased susceptibility to bacterial pneumonia.	

[a]By volume.

of NO_2 of 50 p.p.m. or greater can be fatal to most species, depending on the duration of exposure[17]. The phytotoxicity of NO_2 has been cited by Benedict and Breen[18], Taylor and Eaton[19], and Wimmer and Altshuller[20]. Aspects of the oxides of nitrogen and air pollution have been reviewed by Stephens et al.[21-23] and Leighton[24].

Chemical analytical techniques for the determination of nitrogen oxides in gaseous mixtures are based primarily on the Griess–Ilosvay reaction (formation of dye complex formed between sulfanilic acid, nitrite ion, and α-naphthylamine in an acid medium) as modified by Saltzman[25] and Thomas et al.[26]. Other techniques reported include the p-anisidine test paper of Gelman et al.[27], a liquid absorption technique of Christie et al.[28] based on a modified Griess–Ilosvay reaction, and the phenol–disulfonic acid technique[29], ultra-violet spectrophotometry[30], mass spectroscopy[4], and by a nitrate-specific ion electrode[31]. Continuous automatic analyzers for the determination of NO_2 (or of NO after its conversion into NO_2) are mostly based on colorimetric determinations with the Saltzman reagent[26]. The Incometer developed by Fuhrmann[32] in which NO is converted into NO_2 by UV irradiation has also been used for continuous analysis of the oxides of nitrogen.

The determination of NO by gas chromatography has been complicated by the equilibrium existing between NO and NO_2 in the presence of air or oxygen as well as the latter's dimerization to N_2O_4, viz.

$$2\,NO + O_2 \rightleftharpoons 2\,NO_2 \quad \text{and} \quad 2\,NO_2 \rightleftharpoons N_2O_4$$

Another major problem in the GC analysis of nitrogen oxide has been the strong and largely irreversible adsorption of these gases on most materials. The most frequently used stationary phase for the separation of the nitrogen oxides from other inorganic gases has been molecular sieve 5A. Silica gel has also been used for the resolution of these gases.

The quantitative determination of p.p.m. quantities of NO_2 in N_2 and O_2 by electron-capture gas chromatography was described by Morrison et al.[33]. A Loenco 15A gas chromatograph with a Loenco 15B electrometer was used with a Barber–Colman Model A 5042 electron-capture detector (220 mCi tritium source). The column was a 15 ft. $\times \frac{3}{32}$ in. i.d. stainless steel containing 10% SF 96 on Fluoropak 80 operated at 22°C (room temperature) with a flow rate of carrier and scavenger nitrogen at 10 ml/min and 85 ml/min, respectively. The electron-capture cell was operated at 33 V and a background current of 8.6×10^{-9} A. For concentrations of NO_2 from 5 to 150 p.p.m. and for O_2 present to the extent of 9% by volume in N_2, the standard deviation of the best curve showing response vs. concentration was 2 p.p.m. compared with about 3 p.p.m. for chemical techniques[29].

Optimum conditions and variability in use of pulsed voltage in the gas chromatographic determination of p.p.m. quantities of nitrogen dioxide were described by Morrison and Corcoran[34]. Earlier work of Morrison et al.[33] with an opposed flow detector operated in the electron-capture mode showed that the method could be applied to the quantitative analysis of NO_2 between 5 and 150 p.p.m. A Loenco 15A gas chromatograph with a Loenco 15B electrometer was used with a plane-parallel detector containing 180 mCi tritium. The column was a 20 ft. $\times \frac{1}{8}$ in. 304 stainless steel tubing 0.016 in. wall thickness containing 20% SF-96 on 40–80 mesh Fluoro-pak 80 and operated at 22°. Argon was used as both a scavenger and carrier gas. For the d.c. method of operation with the plane-parallel detector, the most sensitive response was with a carrier flow of 10 ml/min of argon, a scavenger flow of 10 ml/min and a detector potential of 4.5 V. With the pulse mode of operation, the response of the plane-parallel detector was relatively independent of voltage between 10 and 50 V. The most sensitive response was with a carrier flow of 10 ml/min and a scavenger flow of 30 ml/min of argon and a fraction on-time of 0.027. Nitrogen dioxide was detected from 3 to 25 p.p.m. with an average deviation of 3.2% and 3.4% from 3 to 75 p.p.m. The standard deviation was 0.59 p.p.m. from 3 to 25 p.p.m. and 1.32 p.p.m. from 3 to 75 p.p.m.

The gas chromatographic determination of nitric oxide on treated molecular sieve was described by Dietz[35]. An F & M Model 810 chromatograph with a 6 ft. $\times \frac{1}{4}$ in. o.d. stainless steel dual column containing molecular sieve 5A (Linde) and a thermal conductivity detector was used with ultra high purity helium (Matheson). The detector was operated at 100°C and a bridge current of 230 mA. The sample column containing molecular sieve 5A was placed in the chromatograph and heated to 300°C under vacuum (1 μm) for 20 h to remove water and fully activate the material. A positive pressure of helium was then introduced at 300°C to minimize the possibility of O_2 adsorption and the flow was then switched to a low flow rate of NO taking care to exclude oxygen. This exclusion of O_2 was necessary to prevent the premature formation of NO_2. After the column was sufficiently saturated with

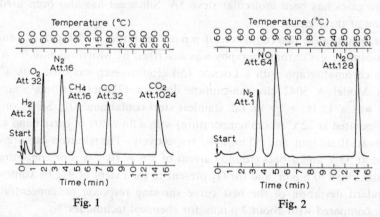

Fig. 1. Separation of NO and N_2O. Sample volume, 0.8 ml: 49.8% NO, 49.8% N_2O, 0.4% N_2.

Fig. 2. Separation of inorganic gases and CH_4. Sample volume, 55 ml; approx. 1% of each component in CO_2.

NO at 300°C (*ca.* 1 h) the temp. was lowered to 20°C, while maintaining the NO flow rate. Following saturation of the sieve for *ca.* 0.5 h at 20°C, the column was flushed with helium (for *ca.* 0.5 h) to remove excess NO then O_2 was introduced to convert the remaining more tightly held NO into permanently bound NO_2. The O_2 flow was maintained for 0.5 h at 25°C and then raised to 100°C for another 0.5 h to insure complete conversion to NO_2. A resulting chromatogram from this treated sieve for 0.4 ml each of NO and N_2O shows no tailing of the NO peak (Fig. 1) and no loss in resolution of the components of a sample containing 1% each of H_2, O_2, N_2, CH_4, and CO in CO_2 (Fig. 2). A full chromatogram was obtained in *ca.* 18 min. The 5 ml sample was injected at ambient conditions using a gas sampling valve and the initial column temperature of 60°C was maintained for 4 min after which the it was increased to 250°C at the constant rate of 30°C/min. The upper limit was held for 8 min to insure the elution of the last component, CO_2, prior to automatic cooling to the initial column temperature. The helium flow rate was 25 ml/min at a delivery pressure of 40 p.s.i.g.

A three-column system for the separation of H_2, O_2, N_2, NO, CO, N_2O, H_2O, and NO_2 was reported by Trowell[36]. Column 1 was 1 ft. × $\frac{1}{8}$ in. o.d. tubing packed with 0.5% Carbowax 1500 on 60–80 mesh silanized glass beads, and operated at 70°C with helium at 30 ml/min for the separation of CO and CO_2. Column 2 was 20 ft. long packed with 40% dimethyl sulfoxide on 60–80 mesh Gas Chrom RZ. This column was operated at 25°C for the separation of N_2O only. Column 3 consisted of 8 ft. of molecular sieve 13X (30–60 mesh), activated at 250°C for 5 min with a helium purge) and was used for the separation of H_2, O_2, N_2, NO, and CO. After all these compounds were eluted from the system, the initial trap was raised from −78°C to +70°C for the release of H_2O and N_2O which were then analyzed on Column 1. The carrier gas flow through this system was 75 ml He/min for column 1 and 30 ml He/min for columns 2 and 3.

Bethea and Adams[37] used a 20 ft. column containing activated charcoal (65 to 80 mesh) impregnated with 2 g of squalene/100 g of the activated charcoal (65 to 80 mesh) at 22°C and helium carrier gas at 66 ml/min for the separation of NO and NO_2 in concentrations of 2–12 mole-%. With thermal conductivity detection and the operating conditions as above, the retention times of nitrogen dioxide and nitric oxide were 7 and 9 min, respectively.

Of the five different columns investigated by Smith and Clark[38] for the separation of N_2O, NO, NO_2, NH_3, O_2, and CO_2, activated coconut charcoal (60 to 120 mesh; washed with 0.1 N sulfuric acid) was the only column which adequately separated CO_2 and N_2O. This was accomplished on a 9 in. column with a helium flow of 104 ml/min. A 4 ft. column of molecular sieve 5 A, 32–80 mesh, with a helium flow of 100 ml/min was used for the separation of NO and NO_2.

Ammonia was analyzed on a 40 in. column containing 33% polyethylene glycol 600 on C 22 firebrick (60–80 mesh, washed with 1 N sodium hydroxide) with a helium flow rate of 112 ml/min. Complete separations of N_2O, NO, NO_2, NH_3,

O_2, and CO_2 were obtained by using these three columns in a multiple column system with a requisite valving arrangement.

Phillips and Coyne[39] separated NO and NO_2 in a variety of nitrogen-containing organic compounds using a 6 ft. column packed with 25% dinonyl phthalate on Chromosorb B at 110°C with a hydrogen flow of 60 ml/min. The nitric oxide was quantitatively scrubbed out of the sample gas by acidified ferrous sulfate and determined by difference from samples taken before and after the scrubber.

Sakaida et al.[40] separated NO and N_2 on a 8 ft. column packed with silica gel (48–60 mesh, Davison) at 28–31°C with helium carrier flows of 40–50 ml/min. Although the column was pre-treated with NO and NO_2, NO_2 was not analyzed.

The analysis of N_2, O_2, NO, N_2O, and NO_2 on a column packed with grade F-20 alumina (activated in an inert atmosphere at 705°F for 4 h) was described by Borland and Schall[41]. The procedure involved the separation of O_2 and N_2 at −78°C, NO and NO_2 at room temperature and finally N_2O at 100°C.

Greene and Pust[42] used a 10 ft. × $\frac{1}{4}$ in. aluminum column containing molecular sieve 5A (20–40 mesh) for the determination of NO and NO_2. The column temperature was 23°C and the carrier gas flow of helium was 60 ml/min. Nitric oxide was directly eluted from this column; however, it was necessary to convert all the nitrogen dioxide to nitric oxide by reaction with water on the column. Detection was achieved with a thermal conductivity cell and the retention volumes were oxygen 180 ml, nitrogen 330 ml, and nitric oxide 480 ml.

Kipping and Jeffrey[43] determined nitric oxide in air using a 5 ft. column of molecular sieve 5A at 100°C with argon ionization detection. Sample sizes ranged from 0.01 to 0.1 μl and the retention time of nitric oxide was dependent on the absolute amount present.

Marvillet and Trenchant[44] separated O_2, N_2, NO, CO, N_2O, and CO_2 on two columns, (a) 5 m × 4 mm i.d. precolumn filled with silica gel (0.125–0.16 mm diam. particles) impregnated with 10% triethanolamine followed by (b) a 4 m × 4 mm i.d. column filled with activated molecular sieve 5A (0.125–0.16 mm diam. particles) at 10°C. Helium was the carrier gas at 30 ml/min, and with thermal conductivity detection, the retention volumes (ml) on columns (a) and (b), respectively, were nitrogen 106 and 139, nitric oxide 137 and 195, carbon monoxide 144 and 202, nitrous oxide 1127 and 4542, and carbon dioxide 1449 and 5030.

Szulczewski and Higuchi[45] separated a mixture of N_2, N_2O, NO_2, CO, and CO_2 using a 6 ft. × 6 mm column of 40–60 mesh silica gel (Davison grade 12). The column was maintained at dry ice–acetone temperature for 18 min and after the emergence of CO, the column was raised to 25°C for the elution of N_2O and CO_2. Helium was the carrier gas at 25 ml/min and with thermal conductivity detection, the retention times (min) obtained were nitrogen 6.0, nitric oxide 14.0, carbon monoxide 16.5, nitrous oxide 46.0, and carbon dioxide 52.0, respectively.

Smith et al.[46] separated the same mixture in 10 min using a 10 ft. × $\frac{1}{4}$ in. copper column containing two layers of 28–200 mesh silica gel (Davison) separated by *ca.* 8 in. of iodine pentoxide powder and *ca.* 0.5 in. of silver metal powder. At elevated temperatures iodine pentoxide will completely oxidize CO to CO_2 (ref. 47) and NO to NO_2 (ref. 48). A Gow-Mac thermal conductivity cell, Model 9193 with TE-11 geometry was used with a Varian Model 6-10 recorder with a full-scale sensitivity of 10 mV. The column was operated at 115°C and helium was the carrier gas at a flow rate of 30 ml/min. The retention times of the gas mixture were nitrogen 2.8 min, carbon monoxide 6.7 min, nitrous oxide 9.3 min, and carbon dioxide 10.3 min. NO_2 did not pass through this column at 115°C.

Meckeev and Smirnova[49] separated a mixture of gases containing H_2, N_2, NO, and CO on NaX molecular sieves (0.4–0.5 mm) at 130°. The sorbent was activated for 3 h at 350° in dry air and then for an additional hour at the same temperature in dry argon, free of oxygen. The mixtures were separated on a 3 m × 6 mm stainless steel column in 5 min with argon as carrier gas at 100 ml/min. Pure nitrogen, previously freed of oxygen by passage over copper and spongy titanium at 400°, as well as nitric oxide plus nitrogen, and carbon monoxide plus nitrogen binary mixtures, were used for calibration. The separated components eluted from the column in the above indicated order. Satisfactory separations were also obtained by using 3 m × 6 mm stainless steel columns filled with KSM-5 silica gel and an additional packing of NaX molecular sieve (to insure the separation of

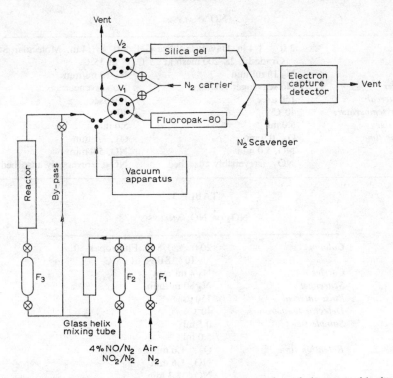

Fig. 3. Schematic diagram of apparatus for the GC determination of nitrogen oxides in air.

N_2 and O_2) as well as with CaA molecular sieves but the analysis achieved with the latter columns was more time consuming.

The gas chromatographic determination of nitrogen oxides in concentrations as low as 10 p.p.m. has been accomplished by Lawson and McAdie[50] using electron-capture detection and employing direct injection of the sample without previous trapping. The apparatus used is shown in Fig. 3. A micrometer valve and flow meter assembly was used to control the concentration of NO in a stream of air. This flowed through a 6 ft. long bypass tube, around a reactor, and then through the series-connected sampling valves of two gas sampling valves. Each valve was connected to an electron-capture detector. Alternatively, the sampling loops of the gas sampling valves could be connected to a vacuum apparatus where calibration samples were made up by partial pressure measurement in a known volume, using N_2 or air as diluent. The sampling loop was evacuated, and filled to one atmosphere with the required sample before injection. A Hewlett-Packard Model 5750 gas chromatograph was used initially with two columns, one for NO and one for NO_2 analysis. For NO analysis, either a 9 ft. $\times$ $\frac{1}{4}$ in. molecular sieve 5A or a 2 ft. $\times$ $\frac{1}{4}$ in. silica gel column was used. Operating conditions for both columns are shown in Table 2. For NO_2 analysis a 20 ft. $\times$ $\frac{1}{8}$ in. column of 10% SF-96 on Fluoropak-80 was used

TABLE 2

NO ANALYSIS

Column	2 ft $\times$ 1/4 in. Davidson Silica Gel, Grade 12, 28–200 mesh at 25°C	9 ft $\times$ 1/4 in. Molecular Sieve 5 A at 35°C
Carrier	N_2 16 ml/min	N_2 50 ml/min
Scavenger	No scavenger gas	No scavenger gas
Pulse interval	150 μsec	150 μsec
Detector temperature	40°C	40°C
Sample size	5.0 ml	5.0 ml
Retention time	O_2, 0.9 min	O_2, 3.5 min
	NO, 5.0 min	NO, 5.9 min
	NO_2, irreversibly adsorbed	NO_2, irreversibly adsorbed

TABLE 3

NO_2 OR NO_x ANALYSIS

Column	20 ft. $\times$ 1/8 in. Fluoropak-80 –10% SF 96 at 25°C
Carrier	N_2 4 ml/min
Scavenger	N_2 80 ml/min
Pulse interval	150 μsec
Detector temperature	40°C
Sample size	0.5 ml
	5.0 ml
Retention time	O_2, 1.6 min
	NO, 1.6 min
	NO_2, 2.3 min

with the operating conditions shown in Table 3. Conditioning of the column for NO analysis was achieved by prolonged injection of NO/air mixtures which subsequently produce NO_2 on the column or preferentially by direct injection of NO_2. For NO_2 analysis on the Fluoropak-80 column, it was necessary to condition the column with several injections of about 5000 p.p.m. NO_2 in air. (If the column was not sufficiently conditioned, the sensitivity of response was low due to excessive adsorption.)

Trowell[51] investigated the reaction of nitrogen dioxide with Porapak Q and Chromosorb 102 and found that NO_2 will react with these materials forming NO, water, and nitration of the aromatic rings of the polymer.

An F & M Model 720 gas chromatograph equipped with a gas sampling valve with 1 ml or less sample loops was used. The column was 8 ft. × 6 mm glass tubing packed with 50–80 mesh Porapak Q (Waters Assoc.). The column was conditioned by heat treating at 230°C for at least 2 h with a helium flow of 60 ml/min. A Perkin-Elmer Model 421 Infrared Spectrometer was used for examining the column packings before and after NO_2 treatment as well as identification of column effluents. It was stressed that if microporous polymer packings are to be used as a column packing material for gases which could contain significant concentrations of NO_2, the limitations due to column interaction with NO_2 be recognized. This would be a primary concern if NO and water were of analytical interest.

2. NITROUS OXIDE

Nitrous oxide, N_2O, occurs naturally in low concentrations. In concentrations of *ca.* 0.3 p.p.m. it is a normal constituent of both unpolluted atmosphere as well as sea water. N_2O is formed upon decomposition of nitrogen-containing inorganic and organic substances and is also found in tobacco smoke (40 µg/g tobacco). It is used as an anesthetic in dental practice and in surgery.

Commercially, nitrous oxide is produced by heating pure ammonium nitrate to a temperature of 245–270°C and allowing it to dissociate exothermically.

$$NH_4NO_3 \rightarrow N_2O + 2\,H_2O + 10.6\,kcal$$

However, higher oxides of nitrogen can result from decomposing ammonium nitrate[52] and the final products can include ammonia, nitric acid, nitrogen, nitric oxide, nitrogen dioxide, and oxygen, and hence serve as a potential source of impurities in nitrous oxide for anesthesia. The reactions can be summarized by the following equations.

$$NH_4NO_3 \rightarrow NH_3 + HNO_3$$
$$2\,NH_4NO_3 \rightarrow N_2 + 2\,NO + 4\,H_2O$$
$$4\,NH_4NO_3 \rightarrow 3\,N_2 + 2\,NO_2 + 8\,H_2O$$
$$2\,NH_4NO_3 \rightarrow 2\,N_2 + O_2 + 4\,H_2O$$

In the normal operating temperature range, these side reactions account for a very small fraction of the decomposition of ammonium nitrate and must be removed by gas washings with fractionation of the liquid gas, or alternatively combining scrubbing of the gas with its passage over finely divided iron to reduce any nitric oxide to nitrous oxide.

The toxicity of nitrous oxide has been cited by Clutton-Brock[53], Von Oettingen[54], and Parbrook[55] and aspects of its leucopenic effects in patients by Lassen *et al.*[56].

The leucopenic effects of nitrous acid have also been correlated with the anti-miotic effects of the gas[57]. The antimiotic and marrow-depressant effects of nitrous oxide have been used therapeutically by Lassen and Kristensen[58] in chronic leukemia and by Eastwood *et al.*[59] in acute leukemia. Nitrous acid has been shown to be lethal to chick embryo[60] and teratogenic in the rat[55,61] and chick embryo[62,63]. The effect of N_2O on RNA and DNA of rat bone marrow and thymus has also been described by Green[64].

A method was described by Buford[65] for quantitatively analyzing gaseous mixtures of N_2, N_2O, CO_2, A, and O_2 by gas chromatography using 3 columns of molecular sieve material at elevated, ambient, and sub-ambient temperatures; with simple modifications, the analysis time of 5 min could be reduced. The columns, all of 3 mm i.d., were packed with Linde molecular sieves. The high-temperature

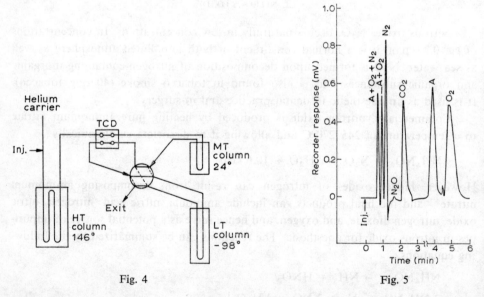

Fig. 4

Fig. 5

Fig. 4. Flow diagram. The full line in the 6-wave stopcock indicates the initial flow path and the broken lines indicates the path after the stopcock has been switched (immediately after the N_2 peak maximum).

Fig. 5. Gas chromatogram of a 5 ml soil atmosphere sample.

(HT) column contained molecular sieve 5A flour (<270 mesh) with non-acid-washed 60–80 mesh Chromosorb[66], the medium-temperature (MT) column molecular sieve 5A (32–60 mesh) and the low-temperature (LT) column, molecular sieve 13X (32–60 mesh). The HT, MT, and LT columns were 225, 38, and 75 cm long, respectively, and all packings were activated by drying in air at 105 °C (16 h) and 350°C (40 h) after the columns had been packed. A Shimadzu GC-1C gas chromatograph was used with a thermal conductivity detector, operated at 220°C, a bridge current of 100 mA, and a recorder of 1 mV range. The carrier gas was helium with an inlet pressure of 3 kg/cm^2 and outlet flow of 75 ml/min. Temperatures were controlled by using the column oven at 146°C (HT column), a water bath at 25°C (MT column), and a freezing methanol bath at −98°C (LT column). Figure 4 illustrates the flow diagram of the apparatus and Fig. 5 shows a chromatogram of a 5 ml soil atmospheric sample.

Table 4 lists the measured values for mass/response of (A + O$_2$), N$_2$, CO$_2$, N$_2$O, A, and O$_2$. An improvement in sensitivity was obtained by increasing the detector current to 170 mA, although a reduction in the sample size to 3 ml was necessary for satisfactory resolution. The minimum detectable amount of N$_2$O was 3–4 p.p.m. in a 3 ml sample. Slight deterioration of the HT and MT columns was found to occur, presumably due to the adsorption of N$_2$O, CO$_2$, and water vapor. All columns were completely reactivated by heating overnight in the carrier gas stream at 350°C.

TABLE 4

MEASURED VALUES FOR MASS/RESPONSE OF (A + O$_2$), N$_2$, CO$_2$, N$_2$O, A, AND O$_2$ (μg/0.01 mV)

Column	Gas	Initial system*	Improved system**
MT	(A + O$_2$)	0.16	0.028
	N$_2$	0.34	0.051
HT	N$_2$O	0.47	0.12
	CO$_2$	1.30	0.35
LT	A	0.77	0.13
	O$_2$	0.93	0.18

* T.C. detector: current 100 mA; temp. 220°C; flow rate 75 ml/min; sample size 5 ml.
** T.C. detector: current 170 mA; temp. 115°C; flow rate 85 ml/min; sample size 3 ml.

Gas adsorption chromatography was used by Rozenberg et al.[67] to determine nitrous oxide in a mixture with nitrogen or nitric oxide. Silica Gel KSK-2.5 (0.25 to 0.5 mm) heated preliminarily for 3 h at 350°C was used as the absorbent. The analysis was performed using a column 163 × 3 mm, a value which gave precise regulation of the gas flow, a manometer, flow meter, sampling apparatus, a katharometer, and an automatic recorder. The rate of hydrogen carrier gas was 30 ml/min. The elution times of N_2 and N_2O were 90 and 216 sec, respectively.

Bock and Schutz[68] analyzed N_2O in air by an initial collection on molecular sieve 5A at room temperature, desorption at reduced pressure at 250–300°C, then determination by gas chromatography. An F & M Model 720 gas chromatograph was used with a thermal conductivity detector and a 1 m × 4 mm column containing molecular sieve 5A (Type OS-0.91, Perkin-Elmer, Bodenseewerk) with helium carrier gas at 50 ml/min.

Bennett[69] described the use of two columns in series to effect complete separation of oxygen, nitrogen, methane, carbon dioxide, and nitrous oxide. The first was packed with porous polymer beads and the second with molecular sieve 5A. A length of copper tubing between the columns enabled the gases which were separated on column I to be eluted before any emerge from column II. This permitted the use of one detector resulting in a continuous chromatogram with no peak overlap. The apparatus is shown diagrammatically in Fig. 6. Column I was a 2 ft. 3 in. length of 0.25 in. o.d. copper tubing filled with 50–80 mesh Porapak Q (Waters Assoc., Stockport). The delay coil was a 7 ft. × ¼ in. o.d. copper tubing housed in the detector oven; all connections were made with compression couplings (Simplifix Ltd., Maidenhead). Column II was a 6 ft. × ¼ in. o.d. copper tubing packed with 30–60 mesh molecular sieve 5A, which was activated prior to packing by heating at 250° for 4 h under vacuum. Both columns I and II were operated at ambient temperature. A Gow-Mac type 9285 thermal conductivity cell fitted with SS-W2 filaments was used in conjunction with a Gas Chromatography Ltd. RY100 bridge unit. The detector was maintained at 113° in a precision air thermostat oven (Griffin and George Ltd.). The bridge output was recorded on a Sunvic 0–5 mV potentiometric recorder and helium was used as the carrier gas at a flow rate of 50 ml/min. The column arrangement used the reference cell of the katharometer to detect gases from column I and the measuring cell to detect gases from column II. Peaks from either cell could be recorded in a continuous chromatogram by reversing the polarity of the detector signal (to achieve this without affecting the baseline, the recorder input signal was first adjusted to zero potential). Samples were introduced into the chromatograph using a Perkin-Elmer gas valve. Column I gave a peak representing a mixture of oxygen and nitrogen followed by individual peaks for methane, carbon dioxide, and nitrous oxide. From the reference cell the gases passed onto column II

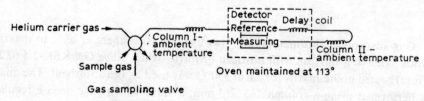

Fig. 6. Diagramatic scheme for the gas chromatographic system employing two columns in series.

which gave individual peaks for oxygen, nitrogen, and methane. Carbon dioxide and nitrous oxide were irreversibly adsorbed on this column. Figure 7 illustrates a chromatogram of the two-column separation of methane, nitrogen, oxygen, nitrous oxide, and carbon dioxide.

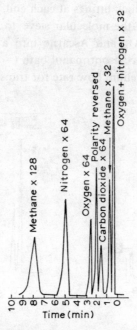

Fig. 7. Gas chromatographic separation of methane, nitrogen, oxygen, nitrous oxide, and carbon dioxide obtained using Porapak Q and molecular sieve 5A columns in series.

The resolution of mixtures of carbon dioxide and nitrous oxide was effected by DeGrazio[70] using a two-column system. An F & M Model 720 gas chromatograph with a thermal conductivity detector was used with a small 4 in. pre-column insert of Linde molecular sieve 13X connected with a $\frac{1}{4}$ in. Swagelock union to a 6 ft. $\times \frac{1}{4}$ in. o.d. stainless steel column packed with 30–60 mesh silica gel. The chromatographic conditions used for the resolution of CO_2 and N_2O were: column, detector, and injection port temperatures, 180, 230, and 210°C, respectively; helium carrier gas flow at 26 ml/min.

The concentration of nitrous oxide in the atmosphere is about 250–300 p.p.b. and the possible sources are believed to be soil bacteria and photochemical reactions. Gas chromatographs employing conductivity detection generally do not have the sensitivity to measure ambient N_2O at levels of 300 p.p.b. with a 5 ml air sample. Attempts to rectify this have mainly centered about the concentration of N_2O on molecular sieves.

LaHue et al.[71] described the GC measurement of atmospheric N_2O using a molecular sieve 5A trap. A sampling train similar to that described earlier by LaHue *et al.*[72] was used with critical orifices controlling the sampling rate. The air was first passed over $CaSO_4$ (8 mesh) and Ascarite (8–20 mesh) to remove most of the atmospheric water and CO_2 prior to reaching the molecular sieve. The molecular sieve tube was 8 in. × $\frac{1}{2}$ in. o.d. (containing 11 g of 5A molecular sieve ($\frac{1}{16}$ in. pellets, Linde)) and fitted with Swagelock fittings at each end. A helium flow saturated with water vapor was passed over the molecular sieve to displace the N_2O and carry the liberated N_2O past $CaSO_4$ and Ascarite into a U-tube filled with activated 8–20 mesh silica gel in a dry ice–isopropanol bath ($-80°C$). Figure 8 illustrates the sample transfer system. The helium flow rate for transfer was 2 l/min and the trans-

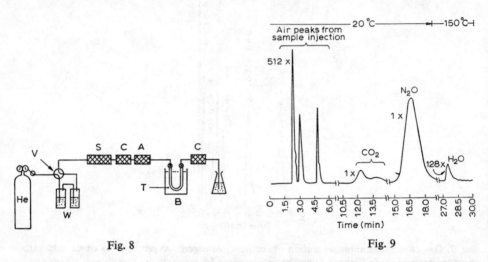

Fig. 8   Fig. 9

Fig. 8. Sample transfer system. N_2O is removed from the molecular sieve and is absorbed by the silica gel. A, $\frac{3}{4}$ in. i.d. × 3 in. long tube filled with 8–20 mesh Ascarite; B, dry ice–isopropyl alcohol bath, $-80°C$; C, $\frac{3}{4}$ in. i.d. × 12 in. long tube filled with 8 mesh $CaSO_4$; S, stainless steel sample tube; T, $\frac{3}{8}$ in. o.d. × 16 in. long thin-walled stainless steel tube packed with activated 8–20 mesh silica gel and fitted with Whitey toggle valves for isolation; V, 4-way stopcock; W, water scrubbing towers for saturation of He.

Fig. 9. Typical gas chromatogram for atmospheric N_2O sample.

fer time was approx. 40 min. The amount of N_2O in the sample was measured by a Perkin-Elmer 900 gas chromatograph with a thermal conductivity detector and a Carle Model 2014 gas sampling valve. The N_2O sample in the U-tube (Fig. 8) was introduced into the gas chromatograph by placing the U-tube in a 140°C oil bath and connecting to the gas sampling valve. After 10 min, the gas sampling valve was switched, the toggle valves were opened and the N_2O transferred onto the gas chromatographic column. The flush time was 2.5 min. Figure 9 shows a gas chromatogram for an atmospheric N_2O sample as well as the instrumental parameters used. The first three peaks in Fig. 9 are a result of the transfer technique. The first two peaks were from dead air above each U-tube valve and the third peak was from switching the gas sampling valve. Measurement of atmospheric N_2O at Boulder, Colorado, over a 3-day period, revealed an average value of 305 p.p.b. with a 1.2% standard deviation. It was suggested that because of the very good precision, accuracy, and simplicity of the above method, the technique could be used to monitor N_2O trends in the atmosphere and to study N_2O sources and sinks.

Leithe and Hofer[73,74] described the gas chromatographic determination of low N_2O concentrations *via* an initial concentration over silica gel at $-70°C$ followed by desorption at room temperature. The separation column was 10.5 m × 6 mm containing 19% propylene carbonate and 16.5% glutaronitrile on 0.2–0.3 mm Sterchamol. The concentration column was a 30 cm × 6 mm U-tube filled with Silica Gel 12 (Davison). (The silica gel was activated before use by heating for 5 h at 160°C). The carrier gas was helium at 60 ml/min, column temperature 20°C, and the detector was a thermistor probe. Under these conditions, the air peak appeared after 6 min, the N_2O peak after 13 min, and the distinct CO_2 peak (levels up to about 400 p.p.m.) after 14 min. When the CO_2 concentration was higher than 400 p.p.m. as in exhaled air, it was expedient to interpose a small soda lime cast ridge (10 cm × 6 mm) before the sample inlet and the beginning of the column. The detection limit for N_2O when concentrated from a 10 l air sample was 0.05 p.p.m. and without the concentration column, 50 p.p.m.

Separations of N_2O and CO_2 have also been accomplished on columns packed with charcoal[38] and dimethyl sulfoxide on Sil–O–Cel[75]. Graven[76] separated nitrous oxide in a gaseous mixture of 6 components using a 10 ft. × ¼ in. column containing 40–80 mesh molecular sieve 5A with the temperature programmed from ambient to 400°C in approx. 25 min and helium as carrier gas. The retention times (min) found were oxygen 2.5, nitrogen 5.0, carbon monoxide 10.0, ethane 15.0, nitrous oxide 19.0, and carbon dioxide 22.0.

3. AMMONIA

Ammonia is principally manufactured by direct synthesis from nitrogen and hydrogen using promoted iron catalysts, although it is also obtained commercially chiefly as a byproduct in the manufacture of coke and gas from coal. Of the total U.S. production of synthetic ammonia, about two-thirds is consumed in agriculture as fertilizer (*e.g.* ammonia and solutions, ammonium nitrate, ammonium sulfate, ammonium phosphates, urea). Ammonia is also used extensively in the production of nitric acid (by oxidation of anhydrous NH_3); in the lead chamber process for manufacturing sulfuric acid, in the production of soda ash from sodium chloride (using the Solvay ammonia–soda process), in the purification and dehydration of sodium hydroxide, in cleaning agents such as household ammonia and in compounding of certain dry cleaning and specialty soaps, in explosives *via* initial conversion to nitric acid thence to basic ingredients of explosives such as nitrocellulose, nitroglycerin, TNT, ammonium nitrate, sodium nitrate, etc.; ammonia is also used in the food and beverage industries, as a fumigant and refrigerant, in metallurgy, petroleum refining, as an intermediate in pharmaceutical manufacture, in pulp and paper industry as a substrate for calcium in the bisulfite pulping of wood, in textiles for the production of synthetic fibers such as cuprammonium rayon, nylon (*via* the production of hexamethylenediamine) and in the manufacture of caprolactam, the monomer for nylon-6, in synthetic resin production to control pH during polymerization of phenol–formaldehyde and urea–formaldehyde synthetic resins and preparation of melamine, and in water purification in combination with chlorine.

Ammonia occurs free or in the form of its salts as traces in air, the level depending on the vicinity of natural or artificial decomposition processes.

The odor threshold for ammonia is 20–40 mg/m^3. Although levels of 100 mg/m^3 are tolerable without adverse effects for a certain time, those of 1500–2500 mg/m^3 are considered highly dangerous after 0.5 h. The mode of action is *via* irritation to mucous membranes of the mouth and nose, as well as the upper respiratory tract. The most frequent cause of death in man from exposure to ammonia is pulmonary edema[77]. The physiological response to ammonia in man at exposures from 100 p.p.m. to 5000–1000 p.p.m. have been described by Henderson and Haggard[78].

The determination of ammonia in air by colorimetric techniques utilizing the Neisler reagent[79–81], the indophenol reaction[82] or by direct UV spectrophotometry at 204.3 mm[83,84] has been described. The enzymatic determination of ammonia in biological medium has also been reported[85,86].

The determination of ammonia content in gas samples by vapor phase chromatographic analysis for nitrogen after catalytic decomposition was described by Diedrich *et al.*[87]. A Dynatronics Chrom-Analyzer Model 100 gas chromatograph equipped with thermal conductivity detector was used with an 8 ft. × ¼ in. diam. stainless steel tube filled with 12.4 g of Linde 5A molecular sieve. The carrier gas was hydrogen at a

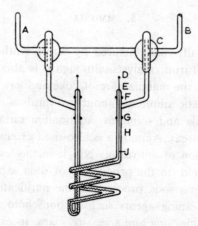

Fig. 10. Ammonia decomposition vessel. A, connection to carrier gas supply; B, connection to chromatograph; C, 3-way stopcock; D, 20 gauge tungsten wire; E, Corning glass (3320); F, Corning glass (5420); G, Pyrex to quartz seal; H, 8 mm o.d. quartz tubing; J, 24 gauge Pt–Rh wire.

flow rate of 40 ml/min, column temperature 30°C, and the thermal conductivity detector current 250 mA. The ammonia decomposition vessel was made of quartz tubing and had a capacity of 25 ml (Fig. 10).

Grune and Chueh[88] separated air and ammonia at 75°C using a 24 ft. × 4 mm column containing silicone grease on C-22 firebrick (24–48 mesh). The detector was a thermal conductivity cell and the carrier gas was helium at 100 ml/min. The separation of ammonia, methylamine, diethylamine, and ethylamine was reported by Amell *et al.*[89] using a 6 ft. column packed with 30% *o*-toluidine on firebrick and a thermal conductivity cell as the chromatographic detector. James et al.[90] used a column containing a mixture of hendecanol and liquid paraffin on Celite for the elution of ammonia and the methylamines in the order of their boiling points. Other stationary phases used for the separation of ammonia from mixtures of amines include polyethylene oxide[91], triethanolamine[92], and a mixture of *n*-hendecanol[93]. For the separation of ammonia and the mono-, di-, and trimethylamines, ethylamines, and *n*-propylamines, Sze *et al.*[94] used a 15 ft. column packed with Chromosorb W containing 5% tetrahydroxyethylethylenediamine with 15% tetraethylenepentamine operated at 58°C.

Smith and Clark[38] compared the sensitivities of a number of gases using katharometer detection (Table 5) and demonstrated the rather poor sensitivity of ammonia for this type of detector.

Liquid anhydrous ammonia is used extensively as a coolant in heat exchange systems because of its chemical stability, low corrosiveness and high latent heat of vaporization. However, anhydrous ammonia is readily contaminated during handling and storage. The gas chromatographic analysis of trace contaminants (O_2, N_2, CO,

TABLE 5

SENSITIVITY OF SOME GASES USING KATHAROMETER DETECTION

Gas	Column	Sensitivity ($\mu g/0.01\ mV\ peak$)	Retention volume (ml)
Oxygen	48 in. molecular sieve 5 A	0.87	96
Nitrogen	48 in. molecular sieve 5 A	1.54	212
Nitric oxide	48 in. molecular sieve 5 A	2.78	324
Nitrogen dioxide	48 in. molecular sieve 5 A (moist)	8.34	324
Carbon dioxide	36 in. silica gel	13.3	340
Nitrous oxide	36 in. silica gel + 36 in. ascarite	6.1	342
Nitrous oxide	36 in. silica gel	6.4	306
Carbon dioxide	9 in. acid-washed charcoal	13.0	456
Nitrous oxide	9 in. acid washed charcoal	9.7	312
Ammonia	40 in. polyethylene glycol 600 on NaOH-washed, but not rinsed firebrick	19.6	298

CH_4, CO_2, and water) in liquid ammonia was described by Mindrup and Taylor[95]. An F & M Model 5750 equipped with a Carle microcavity thermistor detector was used with dual columns for analysis. Table 6 lists the experimental parameters for both columns which were conditioned for a minimum of 12 h at a temperature of 180°C and a 30 ml/min helium flow rate. (The molecular sieve column was conditioned at 250°C.) Detection limits of better than 3 p.p.m. were attained for the above contaminants.

The separation of ammonia from interfering compounds was also based on gaseous diffusion of ammonia from an alkaline medium and absorption by an acidic medium[96-98]. Walker and Shipman[99] described the isolation of ammonia

TABLE 6

EXPERIMENTAL PARAMETERS FOR GC ANALYSIS OF TRACE CONTAMINANTS IN LIQUID AMMONIA

	Analysis of O_2, N_2, CH_4, and CO	Analysis of H_2O	Analysis of CO_2
Column	6 ft. × 1/4 in., o.d., 0.028 in. gage stainless steel: Molecular Sieve 5 A, 80–100 mesh	18 ft. × 1/8 in., o.d. 0.028 in. gage stainless steel: 8 ft. Porapak R, 80–100 mesh; and 10 ft. Porapak R, 80–100 mesh, with 10% polyethylenimine (PEI)	
Detector	Thermal conductivity: Thermistor at 15 mA	Thermal conductivity: Thermistor at 15 mA	Thermal conductivity: Thermistor at 15 mA
Temperature			
column	60°C	80°C	60°C
Detector	Ambient	Ambient	Ambient
Injector	60°C	80°C	60°C
Flow rate	50 ml/min	70 ml/min	50 ml/min
Recorder	0.5 in./min	0.5 in./min	0.5 in./min

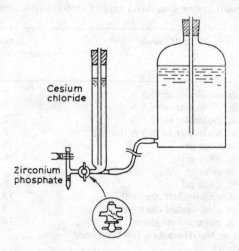

Fig. 11. Zirconium phosphate column for isolation of ammonia.

by use of zirconium phosphate cation exchanger. The adsorbed ammonia was displaced from the column by 1.24 M cesium chloride, then oxidized by hypochlorite, reacted with phenol to form a phenol–indophenol complex which was measured at 395 nm or 625 nm, depending on the concentration range.

The zirconium phosphate (50–100 mesh, BioRad) column (0.8 cm × 0.8 cm) (5 ions connected to two resevoirs by a 3-way stopcock, Fig. 11). Flow rates were controlled by a modification of the Mariotte principle[100] and the column was equipped with a removable cap to facilitate loading. Subsequent to elution of ammonia, the column was regenerated with conc. nitric acid and the cesium (used to displace adsorbed ammonia) recovered for re-use.

REFERENCES

1 W. C. WOHLERS, *Community Air Pollution Sources*, Stanford Res. Inst. Rept., Menlo Park, Calif., 1958.
2 J. N. PITTS, *J. Air Pollution Control. Assoc.*, 19 (1969) 662.
3 B. COMMONER, *Chem. Brit.*, 8 (1972) 52.
4 R. M. CAMPAU AND J. C. NEERMAN, *Automotive Engineering Congress, Detroit, Mich., Jan. 1966*, Society of Automotive Engineers, New York.
5 C. BOKHOVEN AND H. J. NIESSEN, *Nature*, 192 (1961) 458.
6 A. J. HAAGEN-SMIT, M. F. BRUNELLE AND J. HARA, *Arch. Ind. Health*, 20 (1959) 399.
7 V. NORMAN AND C. H. KEITH, *Nature*, 205 (1965) 915.
8 E. L. GRAY, S. B. GOLDBERG AND F. M. PATTON, *Arch. Ind. Hyg. Occupational Med.*, 10 (1954) 409.
9 E. L. GRAY, J. K. MACNAMEE AND S. B. GOLDBERG, J. P. TOLLMAN, L. W. LATOWSKY, E. L. MACQUIDDY AND S. SCHONBERGER, *J. Ind. Hyg. Toxicol.*, 23 (1941) 141.
10 T. R. CARSON, M. S. ROSENHOLTZ, F. T. WILINSKI AND N. H. WEEKS, *Am. Ind. Hyg. Assoc. J.*, 23 (1962) 457.
11 C. H. HINE, F. H. MEYERS AND R. W. WRIGHT, *Toxicol. Appl. Pharmacol.*, 16 (1970) 201.
12 P. K. MUELLER AND M. HITCHCOCK, *J. Air Pollution Control. Assoc.*, 19 (1969) 670.
13 J. P. TOLLMAN, E. L. MACQUIDDY AND S. SCHONBERGER, *J. Ind. Hyg. Toxicol.*, 23 (1941) 269.
14 C. P. MCCORD, G. C. HARROLD AND S. F. MEEK, *J. Ind. Hyg. Toxicol.*, 23 (1941) 200.
15 E. L. GRAY, *Arch. Ind. Health*, 19 (1959) 479.
16 R. EHRLICH AND M. C. HENRY, *Arch. Environ. Health*, 17 (1968) 860.
17 C. W. COOPER AND I. R. TABERSHAW, *Arch. Environ. Health*, 12 (1966) 522.
18 H. M. BENEDICT AND W. H. BREEN, *3rd Natl. Air Pollution Symp., Pasadena, Calif., 1955*.
19 O. C. TAYLOR AND F. M. EATON, *Plant. Physiol.*, 41 (1966) 132.
20 D. B. WIMMER AND A. P. ALTSHULLER, *Environ Sci. Technol.*, 3 (1969) 629.
21 E. R. STEPHENS, *J. Air Pollution Control. Assoc.*, 19 (1969) 181.
22 E. R. STEPHENS, P. L. HANST, R. C. DOERR AND W. E. SCOTT., *Ind. Eng. Chem.*, 48 (1956) 1498.
23 E. A. SCHUCK AND E. R. STEPHENS, in J. N. PITTS AND R. L. METCALF (Eds.), *Advances in Environmental Sciences*, Vol. I, Wiley-Interscience, 1969.
24 P. A. LEIGHTON, *Photochemistry of Air Pollution*, Academic Press, New York, 1961.
25 B. E. SALTZMAN, *Anal. Chem.*, 26 (1954) 1949.
26 M. D. THOMAS, J. A. MACLEOD, R. C. ROBBINS, R. C. GEOTTELMAN, R. W. ELDRIDGE AND L. H. ROGERS, *Anal. Chem.*, 28 (1956) 1810.
27 C. GELMAN, R. GAMSON AND H. KLAPPER, *Proc. 51st Ann. Meeting Air Pollution Control Assoc., 1958*.
28 A. A. CHRISTIE, R. G. LIDZEY AND D. W. F. RADFORD, *Analyst*, 95 (1970) 519.

29 *ASTM-DL-1608-60*, American Society for Testing and Materials, 1960.
30 J. C. MacDonald and L. Haddad, *Environ. Sci. Technol.*, 4 (1970) 676.
31 R. DiMartini, *Anal. Chem.*, 42 (1970) 1102.
32 H. Fuhrmann, *Staub*, 25 (1965) 266.
33 M. E. Morrison, R. G. Rinker and W. H. Corcoran, *Anal. Chem.*, 36 (1964) 2256.
34 M. E. Morrison and W. H. Corcoran, *Anal. Chem.*, 39 (1967) 255.
35 R. N. Dietz, *Anal. Chem.*, 40 (1968) 1576.
36 J. M. Trowell, *16th Ann. Pittsburgh Conf. Anal. Chem. Appl. Spectroscopy, Pittsburgh, Pa.*, 1965.
37 R. M. Bethea and F. S. Adams, *J. Chromatog.*, 10 (1963) 1.
38 D. H. Smith and F. E. Clark, *Soil. Sci. Soc. Am. Proc.*, 24 (1960) 111.
39 L. V. Phillips and D. M. Coyne, *J. Org. Chem.*, 29 (1964) 1937.
40 R. P. Sakaida, R. G. Rinker, R. F. Cuffel and W. H. Corcoran, *Anal. Chem.*, 33 (1961) 32.
41 C. C. Borland and E. D. Schall, *J. Assoc. Offic. Agr. Chemists*, 42 (1959) 579.
42 S. A. Greene and H. Pust, *Anal. Chem.*, 30 (1958) 1039.
43 P. J. Kipping and P. G. Jeffrey, *Nature*, 200 (1963) 1314.
44 L. Marvillet and M. Trenchant, *Proc. 3rd Symp. Gas Chromatog.*, *Edinburgh*, *1960*; R. P. W. Scott (Ed.), Butterworths, London, 1960, p. 321.
45 D. H. Szulczewski and T. Higuchi, *Anal. Chem.*, 29 (1957) 1541.
46 R. N. Smith, J. Swinehart and D. G. Lesnini, *Anal. Chem.*, 30 (1958) 1217.
47 W. H. Welton and N. L. Drake, *Ind. Eng. Chem. Anal. Ed.*, 1 (1929) 20.
48 M. S. Shah and T. M. Oza, *J. Chem. Soc.*, (1931) 32.
49 E. E. Mekeev and G. D. Smirnova, *Izv. Akad. Nauk. Kaz. SSR.*, *Ser. Khim.*, 19 (1969) 87; *Chem. Abstr.*, 72 (1970) 50581x.
50 A. Lawson and H. G. McAdie, *J. Chromatog. Sci.*, 8 (1970) 731.
51 J. M. Trowell, *J. Chromatog. Sci.*, 9 (1971) 253.
52 A. T. Austin, *Brit. J. Anaesthesia*, 39 (1967) 345.
53 J. Clutton-Brock, *Brit. J. Anaesthesia*, 39 (1967) 388.
54 W. F. Von Oettingen, *U.S. Public Health Rept. No. 272*, Federal Security Agency, Washington, 1941.
55 G. D. Parbrook, in *Progress in Anesthesiology, London, Sept. 9–13, 1968*, Excerpta Medica Foundation, Amsterdam, 1970, p. 316.
56 H. C. A. Lassen, E. Hendriksen, F. Neukirch and H. S. Kirstansen, *Lancet*, i (1956) 527.
57 J. Keiler, *Acta Pharmacol.*, 13 (1957) 301.
58 H. C. A. Lassen and H. S. Kristensen, *Danish Med. Bull.*, 8 (1959) 252.
59 D. W. Eastwood, C. D. Green, M. A. Lambdin and R. Gardner, *New Eng. J. Med.*, 268 (1963) 297.
60 G. H. M. Rector and D. W. Eastwood, *Anesthesiology*, 25 (1964) 109.
61 B. R. Fink, T. H. Shephard and R. J. Blandau, *Nature*, 214 (1967) 146.
62 I. G. Mobbs, G. D. Parbrook and J. McKenzie, *Brit. J. Anaesthesia*, 38 (1966) 866.
63 G. D. Parbrook, I. G. Mobbs and J. McKenzie, *Brit. J. Anaesthesia*, 37 (1965) 990.
64 L. D. Green, in B. R. Fink (Ed.), *Toxicity of Anesthetics*, Williams and Wilkins Co., Baltimore, 1968, p. 114.
65 J. R. Buford, *J. Chromatog. Sci.*, 7 (1969) 760.
66 K. J. Bombaugh, *Nature*, 197 (1963) 1102.
67 G. I. Rosenberg and L. T. Kuznetsov-Fetsiov, *Tr. Kazansk. Khim. Teknol. Inst.*, 36 (1967) 1620; *Chem. Abstr.*, 69 (1968) 92723p.
68 R. Bock and K. Schutz, *Z. Anal. Chem.*, 237 (1968) 321.
69 J. Bennett, *J. Chromatog.*, 26 (1967) 482.
70 R. P. DeGrazio, *J. Gas Chromatog.*, 26 (1967) 482.
71 M. D. LaHue, H. D. Axelrod and J. P. Lodge, Jr., *Anal. Chem.*, 43 (1971) 1113.
72 M. D. LaHue, J. B. Pate and J. P. Lodge, Jr., *J. Geophys. Res.*, 75 (1970) 2922.
73 V. W. Leithe and A. Hofer, *Allgem. Prakt. Chem.*, 19 (1968) 78.
74 W. Leithe, *The Analysis of Air Pollutants*, Ann Arbor Science, Ann Arbor, Mich., 1971, p. 176.
75 E. R. Adlard and D. W. Hill, *Nature*, 186 (1960) 1045.

76 W. M. GRAVEN, *Anal. Chem.*, 31 (1959) 1197.
77 M. COPLIN, *Lancet*, 241 (1941) 95.
78 H. HENDERSON AND H. W. HAGGARD, *Noxious Gases*, Reinhold, New York, 1943.
79 M. BUCK AND H. STRATMAN, *Z. Anal. Chem.*, 213 (1965) 241.
80 N. W. HANSON, D. A. REILLY AND H. E. STAGG, *Determination of Toxic Substances in Air*, Heffer, Cambridge, 1965, p. 43.
81 H. B. ELKINS, *The Chemistry of Industrial Toxicology*, Wiley, New York, 2nd edn., 1959, p. 292.
82 W. LEITHE, *The Analysis of Air Pollutants*, Ann Arbor Science, Ann Arbor, Mich., 1971, p. 172.
83 F. A. GUNTHER, J. H. BARKLEY, M. J. KOLBEZEN, R. C. BLINN AND E. A. STAGGS, *Anal. Chem.*, 28 (1956) 1985.
84 M. J. KOLBEZEN, J. W. ECKERT AND C. W. WILSON, *Anal. Chem.*, 36 (1964) 593.
85 H. TACKA AND G. E. SCHUBERT, *Klin. Wochenschr.*, 43 (1965) 174.
86 M. RUBIN AND L. KNOTT, *Clin. Chim. Acta*, 18 (1967) 409.
87 A. T. DIEDRICH, R. P. BULT AND J. M. RAMARADHYA, *J. Gas Chromatog.*, 4 (1966) 241.
88 W. N. GRUNE AND C. F. CHUEH, *Intern. J. Air Water Pollution*, 6 (1962) 283.
89 A. R. AMELL, D. S. LAMPREY AND R. C. SCHIECK, *Anal. Chem.*, 33 (1961) 1805.
90 A. T. JAMES, A. J. P. MARTIN AND G. H. SMITH, *Biochem. J.*, 52 (1952) 238.
91 A. T. JAMES, *Biochem. J.*, 52 (1952) 242.
92 R. E. BUCKS, JR., E. B. BAKER, P. CLARK, J. ERLINGER AND J. C. LACEY, JR., *J. Agr. Food Chem.*, 7 (1959) 778.
93 R. B. HUGHES, *J. Sci. Food Agr.*, 10 (1959) 431.
94 Y. L. SZE, M. L. BORKE AND D. M. OTTERSTEIN, *Anal. Chem.*, 35 (1963) 240.
95 R. F. MINDRUP, JR. AND J. H. TAYLOR, *J. Chromatog. Sci.*, 8 (1970) 723.
96 E. J. CONWAY AND A. BRYNE, *Biochem. J.*, 27 (1933) 419.
97 D. SELIGSON AND H. SELIGSON, *J. Lab. Clin. Med.*, 38 (1951) 324.
98 D. SELIGSON AND K. HIRAHARA, *J. Lab. Clin. Med.*, 49 (1957) 962.
99 R. I. WALKER AND W. H. SHIPMAN, *J. Chromatog.*, 50 (1970) 157.
100 F. A. SCHWERTZ, *Anal. Chem.*, 22 (1950) 1214.

Chapter 12

SULFUR GASES

1. SULFUR OXIDES

Of the several sulfur oxides, sulfur dioxide (SO_2) and sulfur trioxide (SO_3) are of concern in environmental health. The total sulfur entering the atmosphere is estimated to be about one-third man-produced and mostly in the form of sulfur dioxide, about two-thirds is from natural sources such as volcanoes and fumaroles. The estimated U.S. emission of sulfur oxides in 1968 was 33 million tons. Approximately 80% of man-made SO_2 pollution results from the burning of fossil fuels (coal, oil, residual oil) containing sulfur compounds (approx. 46% results from the burning of fossil fuels in electric power plant operations). The largest industrial process contributing to SO_2 production is the smelting of ores. Several important metallic ores such as copper, lead, and zinc are found as sulfides and the processes used in recovery of the primary metals produce SO_2. (Non-ferrous smelters emit approx. 12% of the total emissions of SO_2 in the U.S.) Other sources of SO_2 (and their percentages) include refinery operations (5), coke processing (2), and sulfuric acid manufacture (2). Sulfur dioxide is also used in the manufacture of sodium sulfite and other chemical processes. Large quantities of SO_2 are used in refrigeration, bleaching, fumigating, and preserving. In the U.S. in 1968, sulfur oxides, mostly from stationary fuel consumption and industrial processes, accounted for about 15% by weight of the total of more than 200 million tons per year of pollutants entering the air[1].

The gaseous sulfur compounds cause detrimental effects on the quality and properties of the atmospheric air such as smog formation, production of offensive odors, eye irritation and hydrocarbon photoxidation. Sulfur-containing gas, particularly SO_2, has a very definite detrimental effect, particularly on the respiratory tract of humans. The exposure of man to SO_2 is generally accompanied with digestive ailments, e.g. gastritis, stomach ulcers. In the atmosphere SO_2 is also subject to oxidation of SO_3 and to absorption into water droplets resulting in the formation of sulfuric acid mist which is highly deleterious to textile fibers, building materials, paints, and metals.

The effects of short exposures of SO_2 in animals[2-6] and in man[7-11], as well as chronic inhalation exposure in animals[12-14] and man[15] have been well documented. Laskin[16] has also reported the capacity of inhaled SO_2 to result in benzpyrene tumors which heretofore could not be produced by simple inhalation.

The photochemical[17] and oxidative reactions of SO_2[18] in polluted atmospheres have been reviewed. The analysis of SO_2 in air has been achieved by a variety of spectrophotometric[19-24], titrimetric[25], iodometric–colorimetric[26], polarographic[27], and emission flame photometric techniques[28]. Automatic continuous SO_2 determinations have also been described[29-31].

Adams and Koppe[32] separated SO_2, H_2S, and CH_3SH on a 6 ft. × $\frac{1}{4}$ in. column of 30% Triton X-100 on Chromosorb, 30–60 mesh at 30°C. Helium was the carrier gas at 50 ml/min and detection was with a thermal conductivity detector. Under these conditions, the retention times were H_2S 3.5, CH_3SH 13.0, and SO_2 41.5 min, respectively.

Patrick et al.[33] determined SO_2 and air mixtures on Chromosorb W coated with diisodecyl phthalate. Compositions analyzed (in ca. 3 min) ranged from 50 mole-% to 0.1 mole-% SO_2 and the resulting areas were reproducible to within ±1.0%. A Barber-Colman Series 5000 Selectra chromatograph was used equipped with a thermal conductivity detector and a dual column system and a Barber-Colman Series 8000 recorder. A Perkin-Elmer U-stainless steel column 6 ft. × $\frac{1}{4}$ in. o.d. was packed with 60–80 mesh Chromosorb W (acid-washed) coated with 15% diisodecyl phthalate. The column temperature was 45°C and the carrier gas was helium at 40 ml/min. With the thermal conductivity used, the separation was limited to mixtures with SO_2 concentrations of 0.10% by volume and greater. Use of the detector of Adams and Koppe[32] was suggested to enable the analysis of mixture of much smaller SO_2 concentrations on the same column.

Beuerman and Meloan[34] analyzed SO_2, NO, and NO_2 in 9 min on a 20 ft. × $\frac{1}{4}$ in. column packed with 30% dinonyl phthalate on 40–60 mesh Chromosorb P. The column temperature was 90–95°C and helium carrier was used at 45 ml/min. SO_2, O_2, and CO_2 were also separated on the same column at 90°; thermal conductivity detection was used and the retention times were O_2 3, CO_2 4, and SO_2 8 min, respectively.

Although the GLC of SO_2 has been established, sulfur trioxide is difficult to determine by normal gas chromatographic methods because of its high reactivity and tendency to polymerize. Bond et al.[35] described the GLC of SO_3 by a method which first involved the passage of SO_3 through a tube packed with crystalline oxalic acid. The reaction occurred at room temperature with the formation of CO and CO_2 as follows:

$$COOH + SO_3 \longrightarrow 2\,H_2O + HCOOH + CO_2$$
$$2\,H_2O \cdot COOH$$

$$HCOOH + SO_3 \longrightarrow CO + H_2SO_4$$

The apparatus used for the analysis is shown in Fig. 1. A mixture of sulfur oxides in oxygen was passed continuously over oxalic acid through a calibrated sample loop (10 ml) and then to waste. The carrier gas (oxygen) passed continuously through the rest of the apparatus. The gases passed first through a 6 in. × $\frac{1}{4}$ in. o.d.

column packed with 40% phenyl oxitol (phenyl cellosolve) supported on Sil-O-Cel C22 ($-52 +100$ B.S. mesh sieve) (Column I), and then through one arm of a katharometer; this column resolved SO_2 and yielded a composite peak for the oxides of carbon. The SO_2 was then delayed on a 3 in. column packed with silica gel ("Tell-Tale", $-30 +60$ B.S. mesh sieve) (Column II), while the oxides of carbon were separated on a 20 ft. long column of 40 % phenyl oxitol on Sil-O-Cel. The oxides of carbon were detected as negative peaks by the second filament of the katharometer. The O_2 carrier gas flow rate was 35 ml/min and the oven was thermostated at $46 \pm 1°C$.

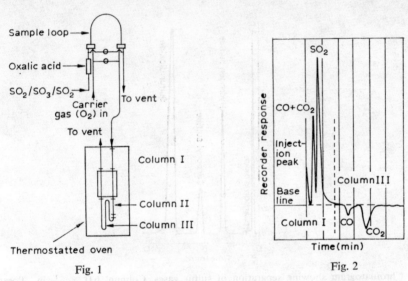

Fig. 1   Fig. 2

Fig. 1. GLC apparatus for the determination of sulfur dioxide and sulfur trioxide.

Fig. 2. Gas chromatographic separation of SO_2, CO, and CO_2.

Figure 2 shows a chromatogram of the separation of SO_2 and SO_3 (as CO_2 and CO following reaction over oxalic acid). The method has been used to determine sulfur dioxides at concentrations of 0.2–0.4 % for the dioxide and about 0.3 % for the trioxide.

The separation of NO, N_2O, CO_2, H_2S, HCN, and SO_2 (in the elution order) was accomplished by Hollis and Hayes[36] using 6 ft. $\times \frac{1}{10}$ in. i.d. stainless steel columns filled with 100–200 μm untreated Porapak at either 26 or 65°C with helium carrier at a flow of 30 ml/min. Porapak Q gave superior results of all the Porapak materials tested. It was necessary to precondition the Porapak columns in an inert atmosphere at 20–30°C higher than the intended operating temperature for 24 h prior to use. The separation of CO_2, H_2S, COS, and SO_2 on the same Porapak column at 68°C is illustrated in Fig. 3.

Wilhite and Hollis[37] investigated a number of Porapak columns for the analysis of NO, CO_2, N_2O, H_2S, CO_2, NH_3, NO_2, SO_2, and HCHO. The best separations were obtained using an 8 ft. Porapak R column (both columns were 0.1 in. i.d. and both Porapaks were 50–80 mesh). The columns were programmed from 25 to 150°C at 12°C/min, helium was the carrier gas at a constant 40 ml/min, and the sample size was 0.1 ml.

A four-column system was used by West[38] for the separation of O_2, N_2, NO, N_2O, SO_2, and H_2S in steam. Two of the columns were 2 ft. long containing 5% Carbowax 1500 on 100–120 mesh Teflon 6, but operated at different temperatures.

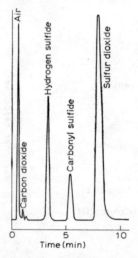

Fig. 3. Chromatogram showing separation of sulfur gases. Column, 6 ft. × $\frac{1}{10}$ in., Porapak Q (117–180 μm); temperature 68°C; flow rate, 30 ml/min; carrier gas, helium.

A 3 in. column containing 4% 2,5-hexanedione on activated charcoal was used to separate H_2S and N_2O. O_2, N_2, and NO were separated on a 4 ft. column packed with 30–60 type 5A molecular sieves. The first column was operated at 160°C and the last three at 25°C. The helium carrier flow was constant throughout at 50 ml/min. The lower detection limits for this system ranged from 0.6% SO_2 to 3% H_2S.

Levchuk et al.[39] analyzed the reaction products of the methane reduction of sulfur dioxide. Two instruments were used with thermal conductivity detectors: (1) SO_2, CS_2, H_2S, COS, and CO_2 were separated on a column of silica gel and (2) CH_4, CO, H_2, O_2, and N_2 were separated on molecular sieve CaA at 105° on a 3 m × 4 mm metal column. The flow rate of argon carrier gas was 32 ml/min; the inlet pressure was 0.2 atm. CO_2 and sulfur compounds were irreversibly adsorbed on the molecular sieve; however, if approx. 10–20 analyses were carried out daily, the adsorptive properties of the packing did not alter after 3 months. Water and sulfur from the CH_4–SO_2 reaction were not estimated by this method.

2. HYDROGEN SULFIDE

Hydrogen sulfide and other sulfur compounds occur to some extent in most petroleum and natural gas deposits. Particularly large and industrially important quantities of H_2S are found in the natural gas and petroleum fields of central and north-central Wyoming and western Texas. (Those in Wyoming alone are estimated to be about 6.5 million tons[40].) Large tonnages of H_2S are also recovered from petroleum refinery operations and very substantial quantities of this gas are also liberated in coking operations or in the production of manufactured gases from coal. Hydrogen sulfide is a byproduct of many industrial operations and in general it is formed whenever sulfur or many sulfur compounds are associated with organic materials at high temperatures. For example, in the process in which sulfur reacts with natural gas at elevated temperatures to produce carbon disulfide, half of the sulfur introduced is consumed in the production of hydrogen sulfide. The process for making thiophene by the reaction of sulfur with butane at elevated temperatures also yields H_2S as a byproduct. In the coagulation of viscose rayon, approx. 6–9 tons of H_2S are formed per 100 tons of rayon produced. Hydrogen sulfide can also be recovered from sulfite waste liquors in the pulp and paper industry as well as from the refining of sulfur by distillation. Hydrogen sulfide can be manufactured by direct synthesis from hydrogen and sulfur vapor at about 500°C in the presence of a catalyst such as bauxite or aluminum hydrosilicate minerals.

The principal use of the hydrogen sulfide recovered as a byproduct is in its conversion into either elementary sulfur or sulfuric acid. Large quantities of hydrogen sulfide are also used in the preparation of sodium sulfide, sodium hydrosulfide, thiophenes, thiols, thioaldehydes, polysulfides, and polymers. Hydrogen sulfide is also used to remove arsenic from the chamber process sulfuric acid produced from pyrites. Other uses of hydrogen sulfide include the manufacture of dyes, pigments, chemicals, in rubber industry products, in tanneries, and the manufacture of glue. Hydrogen sulfide occurs in nature as the putrifaction product of sulfur-containing natural substances. Traces of H_2S are found in the air of slag-slaking plants, foodstuffs plants, and during anaerobic processing of waste waters.

Hydrogen sulfide is one of the more dangerous chemicals in industry because of both its extreme toxicity and its explosive nature when mixed with air or sulfondioxide. The maximum safe concentration of H_2S is about 13 p.p.m. and in its toxic action attacks the nerve centers. The acute[41] and subacute[42,43] effects of hydrogen sulfide have been described. The absorption of hydrogen sulfide is almost exclusively through the respiratory tract.

Hydrogen sulfide in air has been determined by the colorimetric procedures of Jacobs et al.[44], Buck and Stratman[45], potentiometric titration[46], fluorimetric titration[47–49], lead acetate[50], and silver nitrate[51] methods. Hydrogen sulfide in water has been analyzed by a variety of methods, including titrimetric[52], fluorimetric[53], thiomercurimetric[54], and by sulfide ion electrode[55,56] techniques.

The resolution of small amounts of hydrogen sulfide from large quantities of air and water is of importance to industries with wet waste-gas streams containing sulfur gases, *e.g.* pulp and paper and oil-refining industries.

Jones[57] described the separation of H_2, O_2, N_2, CO, CO_2, H_2S, NH_3, H_2O, and C_1–C_4 saturated hydrocarbons in oil-refinery gases using two Porapak Q columns at different temperatures. Cook and Ross[58] recently reported the GC separation of hydrogen sulfide, air, and water in "wet" air and water samples. A Beckman GC-M gas chromatograph was used to analyze gas samples which were either synthetic or extracted directly from the main stack in a Kraft pulp mill chemical recovery plant. A Porapak Q column was used to separate air from H_2S and H_2O, linked to a Carbowax column to increase the resolution. The total column length was 12 ft. (6 ft. of Porapak Q treated with $\frac{1}{4}\%$ Triton 305 joined to 6 ft. of 5% Carbowax 1500 on Teflon 6). The o.d. of the stainless steel columns were $\frac{1}{8}$ in., and the mesh sizes for Porapak Q and Teflon were 80–100 and 40–60, respectively. The carrier gas was helium at 60 ml/min. The operating conditions were: injection inlet temperature 120°C, column compartment 125°C, and detector 175°C. The hot-wire current was kept at 150 mA with a recorder span of 1 mV for both samples.

Figure 4 shows the separation of air, CO_2, H_2S, and H_2O from a vapor phase sample. The chromatogram was obtained from a 1 ml sample injection and the retention time of the H_2S peak was about 30 sec from the air peak. Figure 5 depicts the separation of H_2S from an aqueous sample obtained by bubbling H_2S through a water-filled Dreschel bottle; the sample size was 7.5 μl. The minimum quantity

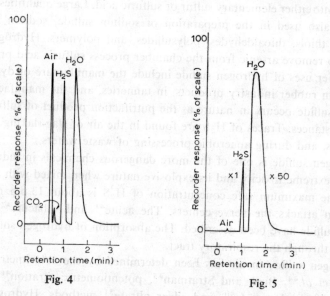

Fig. 4. Separation of air, CO_2, H_2S, and H_2O in "wet" air sample. Hot wire detector; attenuation × 20.

Fig. 5. Separation of air, H_2S, and H_2O in water sample. Hot wire detector.

of H_2S which gave a recorder response was 0.77 µg using an experimental dilution technique of Lovelock[59]. Using a 10 ml sample loop, this corresponded to a minimum detectable concentration of about 50 p.p.m.

Obermiller and Charlier[60] described the gas chromatographic separation of hydrogen chloride, hydrogen sulfide, and water on 5% Carbowax 20M on Fluoropak 80. The apparatus used was built around a Teflon detector block made of conventional design from a single block of Teflon and fitted with two Gow-Mac type W tungsten filaments. The sample gas contacted only Teflon or glass with the exception of the tungsten filament and the liquid substrate. (The sensitivity of the cell was similar to a like cell made from 316 Stainless Steel.) A conventional bridge power supply and output voltage divider were used to operate the cell. A six-way glass stopcock was used as the sampling valve and the cell block and fittings were enclosed in an aluminum heat sink and placed with the column and inlet system within a constant temperature oven. The column used was a 2.5 m × ¼ in. o.d. 0.030 in. wall Teflon tubing packed with 5% Carbowax 20M on Fluoropak operated at 90° with a helium carrier flow of 33 ml/min.

The detector temperature was 100°C and the flash heater and sample valve was at 150°C. Figure 6 shows a chromatogram of the separation of air, H_2S, HCl, and H_2O. The air peak was partially resolved from the H_2S at just under 1 min, followed by HCl and H_2O at 3.2 and 6.2 min, respectively. It was necessary to saturate the system with HCl (use of 5 ml of HCl gas) prior to each determination. The lower limits of detectability of all three components is about 0.1%. The tungsten filaments showed no evidence of corrosion even after a number of determinations.

Ottenstein[61] employed dual columns in series with two separate detectors. To separate CO_2, H_2S, and SO_2 a 9 ft. × ¼ in. column of 10% di-2-ethylhexyl sebacate on 40–60 mesh Teflon-6 was used. It gave a peak for a mixture of O_2, N_2, CH_4, and CO followed by individual peaks of CO_2, H_2S, and SO_2. The other

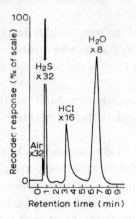

Fig. 6. Separation of air, H_2S, HCl, and H_2O on 5% Carbowax 20 M on Fluoropak 80. Column temperature 90°C; carrier gas, helium at 33 ml/min; sample size, 1.0 ml.

column was arranged in two parts. It was composed of 5 ft. $\times$ $\frac{1}{4}$ in. uncoated 60 to 80 mesh Chromosorb P, followed by 7 ft. $\times$ $\frac{1}{4}$ in. Type 13X molecular sieve (40 to 60 mesh). The uncoated Chromosorb-P "forecolumn" which preceded the molecular sieve made no separation, but delayed the composite mixture from the first column so that the O_2, N_2, CH_4, and CO peaks from the second column would appear at the proper time. A disadvantage of the system was the contamination of the molecular sieve column with CO_2, H_2S, and SO_2 that were separated on the first column.

Gas chromatographic analysis of H_2S was used by Bighi and Saglietto[62] in a study of the decomposition of metal ethylene bis(dithiocarbamates) in acid media between 50 and 100°. A 4 m column of 25% tricresylphosphate on Celite C22 (30–60 mesh) was used at 20° with helium as carrier gas at 5.3 l/hr. A special trap was developed for the condensation and introduction of H_2S samples which permitted its determination at 10^{-7} to 10^{-4} mole/l at 50°, the molar ratio of $H_2S:CS_2$:ethylene thiourea from ethylene bis(dithiocarbamate) was 0.002:1:1; at 80°C it was 0.007:2.5:1; and at 90°C it was 0.03:10.6:1.

3. CARBON OXYSULFIDE

Carbon oxysulfide (carbonyl sulfide), COS, is formed in the destructive distillation of coal, in the purification of petroleum, and is frequently found with hydrogen sulfide and carbon disulfide in coke-oven gas and waste gases of viscose plants.

Carbonyl sulfide (carbonoxysulfide, COS) is always formed when carbon, oxygen, and sulfur or their compounds such as CO, CS_2, and SO_2 are brought together at high temperatures. It is thus found as an impurity in various types of manufactured gases and as a byproduct in the manufacture of carbon disulfide. Carbonyl sulfide reacts with chlorine to form phosgene and sulfur dichloride and with ammonia it forms urea and hydrogen sulfide.

The acute toxic effects of carbonyl sulfide in rabbits[63] and mice[64] have been described. The effects of COS in man may be attributed to the action of H_2S resulting from partial decomposition of COS in the lungs and after absorption into the bloodstream.

Carbonyl sulfide is an impurity in natural gas that reacts irreversibly with the commonly used absorbents and is thus a problem in the natural gas sweetening process.

A quantitative analysis of COS in natural gas by GLC was developed by Schols[65]. The use of a Gow-Mac Model AEL9677 thermistor-type thermal conductivity detector cell permitted the determination of COS concentrations in the range of 25 to 1000 p.p.m. The column used was a 12 ft. $\times$ $\frac{1}{4}$ in. o.d. copper tubing packed with 38 g of Kromat FB support (30–60 mesh) (Burrell Corp.) coated with 30% N,N-di-n-butylacetamide. The column was operated at 28°C with helium as carrier gas at

50 ml/min. The relative retention ratios (compared with n-butane, 9 min) of COS and various hydrocarbons are given in Table 1.

TABLE 1

RELATIVE RETENTION TIMES[a] FOR CARBONYL SULFIDE AND VARIOUS HYDROCARBONS OBTAINED ON A 12 ft. COLUMN OF N,N-DI-n-BUTYLACETAMIDE AT 28°C

Compound	Retention rate
Air	0.115
Methane	0.128
Ethane	0.192
Carbon dioxide	0.209
Propane	0.374
Propylene	0.427
Carbonyl sulfide	0.526
Isobutane	0.705
n-Butane	1.00
Isobutene	1.11
1-Butene	1.11
2,2-Dimethylpropane	1.11
2-Butene	1.35
Isopentane	1.55
n-Pentane	2.97

The GC separation of sulfur-containing mixtures was reported by Berezkina and Mel'nikova[66]. Mixtures of H_2S, SO_2, COS, and Cl_2 were completely separated using a 1 m × 0.4 cm diam. stainless steel column containing silica gel (0.25 to 0.5 mm). The column was used at 100° with helium as carrier gas at 45 ml/min and 0.29 atm pressure at the inlet. The time required for a complete separation was 11 min. H_2S and SO_2 could be separated by GC with a combined glass column 74 cm × 0.5 cm diam. containing a 60 cm layer of Teflon with 10% polyethylene glycol and a 14 cm layer of Teflon with 10% dinonyl phthalate. The column was operated at 34° with helium as the carrier gas at 45 ml/min and an inlet pressure of 0.75 atm with a katharometer detector used in both cases.

The isothermal isolation of COS, H_2S, CS_2, and SO_2 was investigated by Staszewski et al.[67] who used 3 liquid phases on Teflon-6, e.g. squalene, dinonyl phthalate, and polyethylene glycol 400. Squalene did not resolve H_2S and SO_2 over a wide range of temperatures. Partial separation of H_2S and COS was achieved on a 9 ft. column of dinonyl phthalate at 22°C but CS_2 was retained for 78 min. While a 9 ft. column of polyethylene glycol 400 at 46°C provided only partial resolution of COS and H_2S, the SO_2 was retained for 30 min.

The gas chromatographic separation of CO_2, COS, H_2S, CS_2, and SO_2 has been reported by Hodges and Matson[68] who utilized a 1 ft. × $\frac{1}{4}$ in. column of aluminum tubing packed with Davison grade 08, 80–100 mesh silica gel at 100°C with a helium flow of 40 ml/min. An F & M Model 720 dual column programmed-temperature chromatograph equipped with a Gow-Mac W-2 thermal conductivity detector and a 1 mV Honeywell recorder were used. Figure 7 shows the isothermal separation of a 2 ml sample of sulfur gases at 100°C on 1 ft. of 80–100 mesh silica gel in less than 10 min. If the concentration of CO_2, COS, and H_2S were of the same order of magnitude, the separation could be carried out in 5 min by increasing the column temperature to 120°C which caused a decrease in the retention time and improved the peak shape of all components as well as the sensitivity (Fig. 8). However, when the column temperature was maintained at 120°C, the resolution was not as good as that observed at 100° and if the sample contained appreciably more COS than H_2S, the COS interfered with the H_2S peak. A sample containing only air and SO_2 was separated in less than 2 min by using a 1 ft. column at 160° or by using a 6 in. column at 100°C with a limit of detection of approx. 25 p.p.m. of SO_2 for a 5 ml sample.

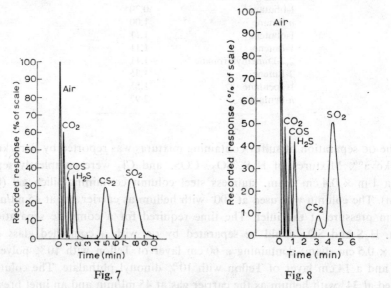

Fig. 7   Fig. 8

Fig. 7. Isothermal separation of a 2 ml sample of sulfur gases on 1 ft. 80–100 mesh silica gel at 100°C, 40 ml of helium/min. Concentrations and attenuations of each component are 5.9% $CO_2 \times 64$, 5.15% COS $\times$ 32, 3.21% $H_2S \times 16$, 1.12% $CS_2 \times 4$, 8.96% $SO_2 \times 8$.

Fig. 8. Isothermal separation of a 2 ml mixture of air, CO_2, COS, H_2S, CS_2, and SO_2 on a 1 ft. 80–100 mesh silica gel column in 6 min at 120°C, 40 ml/min of helium showing partial resolution of COS–H_2S and CS_2–SO_2.

Carbonyl sulfide has been found as a trace impurity in some commercial carbon dioxides in the U.S. and abroad. In contact with water COS slowly hydrolyzes to form H_2S. A 1 p.p.m. COS content in CO_2 produces a discernible off-taste in some carbonated beverages. A quantitative gas–solid chromatographic determination of COS as a trace impurity in CO_2 has been described by Hall[69]. COS at the p.p.m.

level in CO_2 was determined linearly over the range 0.5–2700 p.p.m. with minimum detectable concentrations under these conditions being 0.3 p.p.m. with an accuracy of ± 0.07 p.p.m.

A Pye Argon chromatograph with a Lovelock beta ionization detector and a 0–10 mV range Bristol Dynamaster recorder was used. Detector voltage was 2000 and signal amplification was × 10. A 4 ft. × 4 mm i.d. glass column was filled with 40–60 mesh activated silica gel and was operated at 25°C with argon at 10 p.s.i.

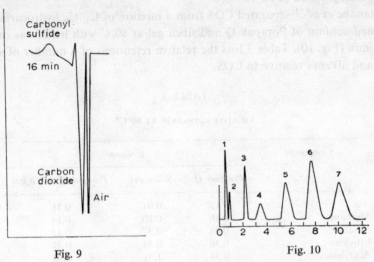

Fig. 9   Fig. 10

Fig. 9. Chromatogram of 3 p.p.m. of carbonyl sulfide in carbon dioxide. Column, 4 ft. silica gel; flow rate, 200 ml/min; temperature, 25°C; sample volume, 114.3 ml; carbonyl sulfide peak height, 9 mm.

Fig. 10. Chromatogram of carbonyl sulfide in C_1–C_3 hydrocarbons. Column, (Porapak Q + silica gel), (75 + 20) cm × 6 mm; temperature, 80°C; carrier gas, He, 60 ml/min. 1, Air; 2, methane; 3, ethane and ethylene; 4, acetylene; 5, carbonyl sulfide; 6, propane; 7, propylene.

TABLE 2

DETERMINATION OF CARBONYL SULFIDE IN CARBON DIOXIDE

Compound	Retention time (min)	Peak and detection characteristics	Minimum detectable concentration (p.p.m.)
(Air) O_2, N_2	1.5 (25°C)	Neg.-sym.-linear	
CO_2	2.7	Neg.-asym.-overload	
SO_2	6.0	Neg.	
COS	16.0	Pos.-sym.-linear	0.3 v/v
C_2H_2	23.6	Pos.-sym.-linear	0.1 v/v
H_2S	32.0	Pos.-linear	<1.0 v/v
Unknown	24.9 (50°C)	Pos.-sym.-linear	
CS_2	46.0 (50°C)	Pos.-sym.-linear	<0.8 w/w

References pp. 287–289

with a corresponding flow rate of 200 ml/min. Figure 9 shows that carbon dioxide and air are detected negatively by the Lovelock detector. Any gas having an ionization potential above the excitation of argon will generally be detected in this manner[70].

Table 2 shows data on the identified contaminant as well as on several others which could possibly be encountered in carbon dioxide manufactured by natural gas combustion processes. The GC separation of air, carbon dioxide, sulfur dioxide, carbonyl sulfide, and acetylene was satisfactory at 25°C while carbon disulfide was best chromatographed at 50°C.

Watanabe et al.[71] separated COS from a mixture of C_1–C_3 hydrocarbons using a combined column of Porapak Q and silica gel at 80°C with helium as carrier gas at 60 ml/min (Fig. 10). Table 3 lists the relative retentions of a number of separated alkanes and alkenes relative to COS.

TABLE 3

RELATIVE RETENTION AT 80°C[a]

Compound	Column		
	Porapak Q	Silica gel	Porapak Q–silica gel
Air	0.12	0.01	0.11
Methane	0.16	0.10	0.14
Ethane	0.45	0.32	0.42
Ethylene	0.36	0.53	0.41
Acetylene	0.34	1.31	0.64
Carbonyl sulfide	1.00	1.00	1.00
Propane	1.54	1.01	1.40
Propylene	1.39	3.03	1.84

[a] Carbonyl sulfide = 1.00.

Carbonyl sulfide has also been determined in various gases by a number of methods including spectrophotometric[72], potentiometric[73], colorimetric[74], gravimetric[75], and selective solvent extraction[76].

4. CARBON DISULFIDE

The traditional method for manufacturing carbon disulfide by the high-temperature reaction of charcoal and sulfur vapor has been largely replaced by the petrochemical process involving the catalytic reaction of natural gas (methane) and sulfur vapor. Another hydrocarbon–sulfur process for the manufacture of CS_2 involves the reaction of petroleum fractions such as gas oils, fuel oils, and heavy residuums with sulfur at elevated temperatures. The U.S. production of carbon disulfide in 1962 was approx. 607 million lb. and that of the major industrial countries approx. 650 million lb.

The major portion of the CS_2 produced today is used as a raw material for the manufacture of regenerated cellulose and the two major products manufactured from regenerated cellulose are viscose rayon and cellophane. (Both materials are formed by steeping cotton linters or wood pulp with caustic soda solution and then xanthating the alkali cellulose with CS_2.) The manufacture of carbon tetrachloride is another principal use for CS_2 with at least 15 % of the CS_2 used in the U.S. employed for this purpose. In 1961 carbon tetrachloride production in the U.S. was 384 million lb. derived from CS_2 and from the chlorination of methane. CS_2 is also extensively used in the cold vulcanization of rubber and in the preparation of rubber accelerators and flotation chemicals. Other areas of utility of CS_2 include fumigants in admixtures with chlorinated hydrocarbons, manufacture of insecticides, the production of resinous materials from condensation reactions with amines, aldehydes, dienes, carboxylic acids, and isocyanates. Formerly, a principal use of CS_2 was as a solvent for the extraction of oils, fats, waxes, and other substances from various materials. Other solvents including chlorinated hydrocarbons such as trichloroethylene have largely replaced CS_2 in this field.

The acute[77,78] and chronic[79,80] effects of exposure of various concentrations of CS_2 on man have been described. The absorption of CS_2 is mainly through the lungs where it enters the bloodstream and is distributed throughout the body. Limited absorption of CS_2 can occur through the skin. Eighty-five to ninety percent of CS_2 is metabolized[81] and eliminated in the urine as inorganic sulfates and other sulfur compounds. The balance of CS_2 is eliminated unchanged (8–13 % in exhaled breath, 0.5 % in urine, and none in the feces).

The present accepted maximum permissible concentration of CS_2 for an 8 h exposure is 20 p.p.m. (0.0624 mg/l). Carbon disulfide has been determined by spectro-photometric[82–85] techniques.

The response of the flame-ionization detector to CS_2 was examined by Douglas and Schaefer[86] using varying conditions of flow rates of fuel and carrier gas and compared with the responses for CH_4, C_2H_6, and CCl_4. The low value of the response for CS_2 (about 1 % of that of CH_4) was attributed to preoxidation of CS_2 in the flame to substances not contributing to the ion forming process (SCO, SO_2, and CO). CS_2 was found to behave like organic compounds with respect to the hydrogen content of the flame and a similar mechanism involving the formations of CH and its oxidation to CHO^+ was proposed for the formation of ions from CS_2. The condition for optimum operation of FID for the analysis of CS_2 was similar to those for hydrocarbons and involved a slight shift to higher hydrogen flow rates. CS_2, CCl_4, and CH_4 possessed a similar response curve as a function of the H_2 content of the unburnt flame gas. The study was carried out using a Beckman GC-2 fitted with FID and a 6 ft. × $\frac{1}{4}$ in. stainless steel column containing 50 % silicone oil on 60–80 mesh firebrick. Column temperatures were selected at 80° or 110° to give suitable elution times, operating at flow rates of air 235, hydrogen 45, and nitrogen 65 ml/min. The background current was within the range of 50–60 × 10^{-12} A

($\mu\mu$A) approx. and the noise variation was $\pm 0.05\,\mu\mu$A. Nitrogen carrier gas was passed through silica gel.

Blades[87] studied the effect of CS_2 concentration on its response to FID. It was shown that ion formation was linearly dependent on concentration below about 0.01 % in the carrier gas, but above this concentration an inhibition of ion formation, second order in CS_2 dominated, this response. Further response was observable above 1 % if the collector was too close to the burner, and the detector was thus only recommended where low concentrations of CS_2 are to be monitored.

5. MERCAPTANS

Mercaptans and organic sulfur compounds such as sulfides play an important role in air pollution problems. Primary concern is generally with the odor associated with some of these compounds. The odor thresholds of a number of these compounds is in the low part per billion range (*e.g.* lower alkyl mercaptans and sulfides).

Methyl mercaptan is formed in the Kraft process as a result of the interaction between hydrosulfide ions in the cooking liquor and the methoxyl groups on the lignin molecule. Methyl mercaptan is extremely malodorous having an odor threshold of about 5 p.p.b. in air and is an important contributor of air pollution from Kraft pulp mills, oil refineries, and sewage plants.

The oxidation reaction to convert mercaptans to disulfides has been reviewed by Tarbell[88] and Oswald and Wallace[89] and recently aspects of the oxidation of methyl mercaptan with molecular oxygen in aqueous solution described by Harkness and Murray[90].

The determination of methyl mercaptan and dimethyl sulfide formation in Kraft pulping with regard to the influence of wood type, temperature of cooking, sulfidity and length of cooking has been studied by Douglas and Price[91] using gas chromatography. Analyses were carried out using a Hy-Fi Model 600-D chromatograph with FID and connected to a Brown Class 15 recorder. The column used was 5 ft. × $\frac{1}{8}$ in. stainless steel packed with 6% tricresyl phosphate on 60–80 mesh Chromosorb G and was operated isothermally at 50°C. Nitrogen and hydrogen flow rates were 25 ml/min, air flow 300 ml/min, and the injection port temperature was 100°C. Under these conditions methyl mercaptan, dimethyl sulfide and dimethyl sulfide had retention times of 30 sec, 54 sec, and 8.5 min, respectively. An increase in temperature or an increase in the length of digestion time was found to increase the total amount of organosulfur compounds produced. Under comparable conditions hardwoods liberated more methyl mercaptan and dimethyl sulfide than soft-wood species. At lower temperatures, the amount of methyl mercaptan formed exceeded the amount of dimethyl sulfide.

The determination of mercaptans in sour natural gases by GLC and microcoulometric titration was reported by Fredericks and Harlow[92]. Sample components were separated on a 25 ft. × $\frac{1}{4}$ in. o.d. aluminum column packed with 30% tricresyl phosphate on acid-washed Chromosorb W, and the individual mercaptans were automatically titrated with coulometrically generated silver ion as they emerged. The large excess of H_2S was bypassed and not determined. Hydrocarbons which entered the titration cell with the mercaptans did not interfere. The method required only 30 ml of sample and was achieved in 1 h with a precision of $\pm 2\%$ at 50 p.p.m. and was sensitive to 1 p.p.m. of individual mercaptan. Table 4 lists the relative retention times of 8 mercaptans and Fig. 11 illustrates a chromatogram of condensate hydrocarbon–hydrogen–mercaptan mixture using both thermal conductivity and microcoulometric detectors. The latter was a Dohrman Model C-100 with a T-200 cell. Helium was the carrier gas at 30 p.s.i.g. (120 ml/min through the column). Under the conditions used in this analysis the microcoulometer was approx. 100 times more sensitive than the thermal conductivity detector.

TABLE 4

RELATIVE EMERGENCE TIMES OF C_1–C_4 MERCAPTANS
n-Pentane $= 1.00$

Compound	B.p. (°C/760 mm Hg)	Relative emergence
(Hydrogen sulfide)	(−59.6)	(0.38)
Methyl mercaptan	5.96	1.70
Ethyl mercaptan	35.00	2.84
Isopropyl mercaptan	52.56	3.16
tert.-Butyl mercaptan	64.22	3.31
n-Propyl mercaptan	67.72	3.64
sec.-Butyl mercaptan	84.98	3.92
Isobutyl mercaptan	88.49	4.00
n-Butyl mercaptan	98.46	4.20

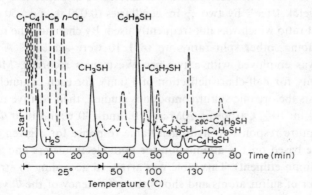

Fig. 11. Chromatogram of condensate hydrocarbon–hydrogen sulfide–mercaptan mixture using two detectors.

Trace components in fuel gases were determined by Reinhardt et al.[93] on a column of 10.1% $NC(CH_2)_2O(CH_2)_2O(CH_2)_2CN$ and 1% lauryl mercaptan or 15.3% dioctylphthalate and 1% lauryl mercaptan on Porolith at column temperature of 50 and 25°C, respectively, with argon carrier gas and a β-argon ionization detector. Quantitative analysis was done by means of a β-argon ionization detector and a vibrating reed electrometer. The linearity range of the detector was 10^{-8}–10^{-7} g. A fuel gas was analyzed and found to contain most of its bound sulfur as mercaptans, thio ethers and disulfides (include CS_2).

A series of partitioning agents were studied by Feldstein et al.[94] to determine their ability to separate organic sulfur compounds by gas chromatography. A number of partitioning agents such as hexatriacontane and benzyldiphenyl tritolyl phosphate, silicone oil, dinonyl phthalate, and β-iminodipropionitrile have been reported to separate mixtures of thiols and sulfides and mixtures of these compounds with lower alcohols. However, a column capable of separating organosulfur compounds from aliphatic hydrocarbons in ambient air is essential since the normal concentration of hydrocarbons is several orders of magnitude greater than trace concentrations of organic sulfur compounds. Table 5 lists the relative retention times of some mercaptans and hydrocarbons obtained on 8 partitioning agents contained in 6 and 25 ft. stainless steel columns and by flame-ionization detectors. Although good separation of the thiols was achieved by a variety of partitioning agents, the presence of atmospheric concentrations of hydrocarbons interfered with the detection of p.p.b. concentrations of the sulfur compounds. Higher concentrations of thiols and sulfides were separated as metallic Cd or Hg salts and identified by GC of the regenerated sulfur compound after addition of acid to the precipitate.

The analysis of sulfur compounds with electron-capture/hydrogen flame dual channel gas chromatography was described by Oaks et al.[95]. The Wilkins Aerograph flame detector was operated with a hydrogen generator at a hydrogen flow of 25 ml/min and 300 ml of air/min. The electron-capture detector was operated at a fixed potential of 90 V. The detectors were connected to a splitter low volume $\frac{1}{16}$ in. modified Swagelok "tee" by two $\frac{1}{16}$ in. capillaries (0.020 in. i.d.) 50 cm long. This gave a 1:1 split ratio which was not frequently used. By changing the relative lengths of capillary tubing, other split ratios up to 1:10 were achieved. A dual channel electrometer was employed with a two-pen Westronics recorder, Model No. 11A with a sensitivity for half-chart deflection of 1.0 mV for each channel used.

Table 6 lists the organic sulfur compounds studied, their relative retention rimes on a 10 ft. × $\frac{1}{8}$ in. 10% Carbowax 20M at 110° and 140°C and their Φ values (ratio of electron capture response to hydrogen flame response for a given peak; it is an empirical value based on 1 × attenuation for both electrometers and assumes a 1:1 split in the column effluent). This table was arranged according to structural group and the number of sulfur atoms and shows the consistency of the Φ values within a particular grouping. The most pronounced increase caused by a single molecular change was observed between the saturated disulfide and the corresponding tri-

TABLE 5

RELATIVE RETENTION TIMES FOR SOME MERCAPTANS AND HYDROCARBONS

	Column							
	Versamid 930	Silicone–triton	Flexol	Tritoyl phosphate	Ucon	Asphalt	TMTS[a]	Apiezon
CH_3SH	1.2	1.25	3.1	1.3	2.4	3.7	5.8	2.3
C_2H_5SH	2.3	2.05	5.0	2.2	4.2	7.9	8.9	4.4
C_3H_7SH		4.2				2.1	19.5	12.0
i-Pentane	2.0	1.6	3.5	1.7	3.1	2.6	5.5	3.3
n-Pentane	2.3	1.6	4.4	2.1	3.9	2.9	6.5	4.8
Isoprene	3.0	2.0	6.4	3.8	6.1	4.6	7.6	8.3
Cyclopentane		3.0	8.5	4.5	7.8	4.1	11.3	
2,2-di-Me-butane	3.5		7.8	3.4	6.2	4.9		
3-Me-pentane	6.0		9.0	7.6	7.4	7.1		
n-Hexane	9.0		13.3	8.6	8.3		17.9	9.6
	6'SS-20%	6'SS-10%	6'SS-20%	6'SS-20%	6'SS-20%	6'SS-10%	25'SS-1%	6'SS-20%

[a] Tetramethylthiuram sulfide.

References pp. 287–289

TABLE 6

RELATIVE RETENTION TIMES AND Φ VALUES OF MERCAPTANS, MONO-, DI-, AND TRISULFIDES AND MISCEL-
LANEOUS SULFUR COMPOUNDS

Compound	Φ	RRT on Carbowax at 110° and 140°C[a]
Mercaptans		
Methyl mercaptan	0.35	0.04
Ethyl mercaptan	0.38	0.05
Allyl mercaptan	0.48	0.07
n-Butyl mercaptan	0.29	0.09
n-Hexyl mercaptan	0.16	0.22
n-Octyl mercaptan	0.13	0.54
Monosulfides		
Dimethyl sulfide	0.01	0.06
Diethyl sulfide	0.02	0.07
Di-*n*-propyl sulfide	0.02	0.15
n-Propyl allyl sulfide	0.02	0.17
Di-allyl sulfide	0.01	0.18
sec.-Butyl allyl sulfide	0.04	0.20
Methyl allyl sulfide	0.04	0.09 (tentative)
Disulfides (saturated)		
Dimethyl disulfide	0.77	0.16
Diethyl disulfide	0.72	0.38
Methyl *n*-propyl disulfide[b]	0.63	0.40
Di-*n*-propyl disulfide	1.1	0.70
Di-*n*-butyl disulfide	1.3	1.52
Di-*n*-amyl disulfide	2.3	3.22
n-Hexyl *n*-amyl disulfide[b]	2.5	4.76
Di-*n*-hexyl disulfide	3.3	7.18
Disulfide (unsaturated)		
Methyl allyl disulfide	8.4	0.46
Ethyl allyl disulfide	10	0.66
n-Propyl allyl disulfide	11.2	0.84
Di-allyl disulfide	20	1.00
n-Hexyl allyl disulfide[b]	8.4	2.67
Trisulfides		
Di-methyl trisulfide	240	0.76
Methyl *n*-propyl trisulfide[b]	180	1.33
Methyl allyl trisulfide[b]	220	1.63
Di-*n*-propyl trisulfide[b]	210	2.28
n-Propyl allyl trisulfide[b]	250	2.83
Di-allyl trisulfide[b]	350	3.42
Others		
Allyl isothiocyanate	22	0.57
Isobutyl sulfoxide	0.01	3.49
Sulfur dioxide		0.04
Hydrogen sulfide		0.04

[a] Diallyl disulfide = 1.00.
[b] Sulfur compounds synthesized.

sulfide (an increase of 200–300 fold). The Φ values were used by Oaks et al.[95] in identifying peaks found in chromatograms of garlic headspace gas and hexane extracts. Of the 18 peaks seen from garlic on the electron-capture channel, nine were represented as the mono-, di-, and trisulfides of methyl–methyl, methyl–allyl, and allyl–allyl. The only other peak identified was methyl propyl trisulfide.

Adams et al.[96] reported the analysis of Kraft-mill sulfur-containing gases employing GLC ionization detectors. The technique included the use of two chromatographic columns in series to separate O_2, N_2, CO, CO_2, H_2O, H_2S, SO_2, and CH_3SH. The chromatograph was a Fisher Partitioner with a four-element thermal detector and a 1 mV Leeds and Northrup Model H recorder. Column 1 was a 6 ft. × $\frac{1}{3}$ in. aluminum tubing packed with 30% Triton X-45 on Chromosorb P 30–60 mesh and column 2 was 6 ft. × $\frac{3}{16}$ in. packed with molecular sieve 13X. Column 1 separated H_2O, H_2S, SO_2, and CH_3SH while column 2 separated O_2, N_2, CO, and CO_2.

The analysis of sulfur-containing gases in the ambient air using selective prefilters and a microcoulometric detector was reported by Adams et al.[97]. A sulfur-sensitive coulometric microtitration cell, originally designed as a gas chromatographic detector, was converted to direct atmospheric analysis by continuously pumping ambient air through the cell. A method was suggested for sequentially separating H_2S, SO_2, CH_3SH, CH_3SCH_3, and CH_3SSCH_3 in ambient air by using a series of chemically impregnated membrane filters. Chemical impregnants were developed which retained one or more of these compounds from mixtures and allowed the others to pass quantitatively. Those sulfur compounds remaining in the air stream emerging from each selective filter were then analyzed with the microtitration cell.

A WSU-1 bromine coulometric microtitration cell was used with a Dohrmann C-200 microcoulometer and a Leeds and Northrup Model H recorder having a 0–1 mV range. The most suitable reagents for the selective filters were (a) for SO_2: $NaHCO_3$ solution (pH 7–11), (b) for H_2S: $ZnCl_2$–H_3BO_3 solution (pH 4.7), (c) for H_2S and CH_3SH: a silver membrane impregnated with $NaHCO_3$ solution, (d) for H_2S, CH_3SH, and CH_3SCH_3: $Hg(NO_3)_2$–tartaric acid solution (pH 1.0), and (e) for H_2S, CH_3SH, CH_3SCH_3, and CH_3SSCH_3: $AgNO_3$–H_3BO_3–tartaric acid solution (pH 2.0).

6. MISCELLANEOUS SULFUR GASEOUS MIXTURES

A gas chromatographic system was described by Robbins et al.[98] for the analysis of CO_2, H_2S, H_2, O_2, N_2, CH_4, CO, and SO_2 from a single gas sample with direct application to the analysis of fixed gases related to the reductive decomposition of gypsum. The system consisted of three columns in series. Column 1 separated the polar gases which were then irreversibly absorbed on column 2. Column 2 delayed but did not absorb the other gases which were subsequently separated on column 3. Eluted peaks from column 1 were detected in the reference side of the thermal conductivity cell and peaks from column 3 in the sample side of the cell. Column 1 was a 20 ft. × ¼ in. o.d., 20 gauge, Type 304 stainless steel tubing filled with 10% dibutyl sebacate on Fluoropak (−20 +80 U.S. standard screen fraction). This column gave a peak of the mixture of H_2, O_2, N_2, CH_4, and CO followed by individual peaks of CO_2, H_2S, and SO_2 in that order. Column 2 was a "fore-column trap",

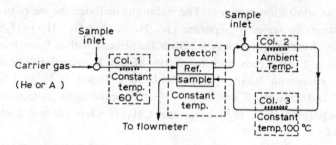

Fig. 12. Gas chromatography system for analysis of CO_2, H_2S, H_2, O_2, H_2, CH_4, CO, and SO_2 from a single gas sample.

11 ft. × ¼ in. o.d., copper tubing filled with 25% potassium hydroxide on Chromosorb W (30–60 mesh). The KOH in the column permanently absorbs CO_2, H_2S, SO_2, and water vapor. Column 3 was a 7 ft. × ¼ in. o.d., 20 gauge, Type 304, stainless steel tubing containing Type 13X molecular sieves (−14 +30, U.S. standard screen, Linde). After the column was activated at 300° for 12 h, it separated H_2, O_2, N_2, CH_4, and CO in that order. The detector was a Model TR-II-B Gow-Mac (Gow-Mac Instrument, Madison, N.J.) thermal conductivity cell, geometry 9193, with a Model 9293-B Gow-Mac power supply control unit. Figure 12 illustrates the GC system for the analysis of the gases. Peaks from either side of the thermal conductivity cell can be recorded in a continuous chromatogram by reversing the polarity of the signal from the detector cell with a double-pole, double-throw switch on the recorder input leads. The recorder input signal must be adjusted to zero potential at the beginning of the chromatogram so reversal of polarity did not affect the baseline. Separation of all components required a helium flow rate of 40 ml/min. Columns 1, 2, and 3 were operated at 60°, ambient and 100°C, respectively. With the polarity of the signal reversed, column 1 gave a peak of a mixture of H_2, O_2, N_2, CH_4, and CO and followed by individual peaks of CO and H_2S. The polarity was then switched to normal to obtain peaks of H_2, O_2, N_2, CH_4, and CO from column 3. Finally, the polarity was reversed again for the SO_2 peak from column 1. Water vapor appeared from column 1 about 5 min after the final peak, but appeared only as a very low hump with a 7 min base width and was trapped on column 2 along with CO_2, H_2S, and SO_2.

Figure 13 shows a chromatogram for the separation of all components in 15 min. The sample contained 50 μl of all components except for 250 μl of hydrogen.

Figure 14 is a chromatogram of a 1 ml sample obtained in 8 min with a helium flow rate of 80 ml/min. Since the sensitivity of hydrogen is poor when helium is used as carrier gas, argon was employed for the determination of low levels of hydrogen. The separation of hydrogen was accomplished on the molecular sieve column. Thus, when hydrogen was analyzed alone, the sample was injected just preceding column 2 (inlet B) (Fig. 12).

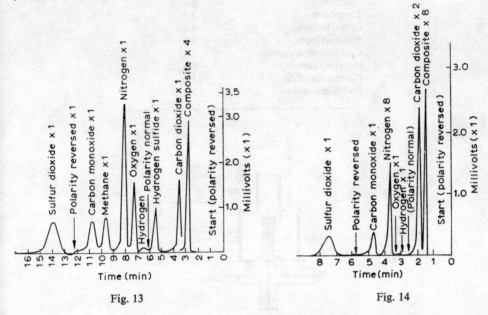

Fig. 13.   Fig. 14.

Fig. 13. Sample chromatogram for all components.

Fig. 14. Sample chromatogram for limited number of components.

Dressler and Janak[99] studied the detection of sulfur compounds utilizing an alkali flame-ionization detector (AFID). The schematic of the detector is shown in Fig. 15. The AFID was incorporated into the Chrom 3 apparatus (Nat. Enterprise Lab. Equip., Prague) and utilized with a 68 cm × 0.6 cm i.d. stainless steel column packed with 25% didecylphthalate on Celite. The carrier gas and air flow rates were kept at 60 and 660 ml/min, respectively, while the hydrogen flow rate was varied to achieve a background ionization current value. H_2S and CS_2 were injected in gaseous state (approx. 1×10^{-8} mole) and other sulfur compounds were first dissolved in appropriate solvents and analyzed in amounts of ca. 1×10^{-8} mole. (The column temperature was 25° in the former and 100°C in the latter case.) Table 7 shows the ionization efficiency found for various sulfur compounds. While all sulfur-containing compounds produced a signal, CS_2 effected a two-fold response. The response of H_2S was, as compared with that of the other compounds, considerably lower.

References pp. 287–289

TABLE 7

IONIZATION EFFICIENCY WITH VARIOUS SULFUR COMPOUNDS

Compound	Ionization efficiency (coulombs/mole)
Thiophene	6.0
Thiophane	5.7
Dibutylsulfide	6.1
Butylmercaptan	5.2
Cyclohexylmercaptan	5.6
Hexylmercaptan	6.1
Carbon disulfide	11.0
Hydrogen sulfide	1.1

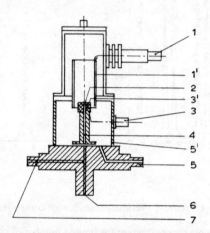

Fig. 15. The alkali flame-ionization detector. 1, Connector of collector electrode; 1', collector electrode; 2, pressed alkali salt tip; 3, connector of the lower electrode; 3', lower electrode; 4, quartz glass jet; 5, air inlet; 5', mesh for air flow control; 6, carrier gas inlet; 7, hydrogen inlet.

Dagnell *et al.*[100] examined the characteristics of a microwave-excited emission detector and its potential use in gas chromatography of various sulfur compounds (carbon disulfide, thiophen, thioglycolic acid, dimethyl sulfoxide, and sulfur dioxide). The column was operated slightly above atmospheric pressure (*ca.* 105 kN/m²) and the microwave detector at a convenient reduced pressure (*e.g.* 13–40 mbar). It was shown that the most sensitive and specific wavelength for analytical purposes was not necessarily the same for all the sulfur compounds examined. The spectra obtained for each compound with argon or helium as carrier gas were characterized and only the atomic lines due to sulfur at 190.0 and 191.5 nm, the C=S system with a band-head around 257.6 nm and the C_2 bandhead at 516 nm were shown to be common to the organic compounds (except CS for thioglycolic acid). CS_2 was the most easily fragmented and gave a limit of detection of 0.2 ng of sulfur at 257.6 nm even when the low luminosity monochromator was used while thioglycolic acid was the least easily fragmented compound.

Figure 16 shows the experimental arrangement used for microwave-excited emissive detector in gas-phase chromatographic analysis. It allowed complete spectra to be recorded for a flowing system and also permitted the evaluation of sensitivity, resolution, etc., following gas chromatographic elution of the compounds. The gas chromatography column could be operated under optimal conditions for resolution (normally slightly above atmospheric pressure) and the microwave discharge run at reduced pressure. This was achieved by using a capillary restrictor with a vent to atmosphere up-stream of this restrictor so that the eluate from the column was stream-split at the vent. The length of the stainless steel capillary restrictor (diameter *ca.* 0.125 mm) was determined by the pressure required in the microwave detector tube which in turn was affected by the capacity of the vacuum pump.

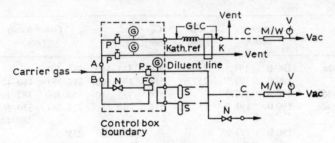

Fig. 16. Experimental arrangement for microwave-excited emissive detector in gas-phase chromatographic analysis. o, On/off taps; C, capillary restriction; G, pressure gauges (0–210 kN/m²); K, GLC katharometer detector; FC, flow controller; M/W, microwave unit, monochromator, and recording system; P, pressure controllers (Watts type 15-2, 0–175 kN/m²); S, samplers; N needle valves; V, vacuum gauges (0–40 mbar). When tap A is open and tap B is closed, the GLC system is operative; when *vice versa*, the sampler system is operative.

A Pye Model Series 104 gas chromatograph was used with a katharometer detector. The microwave excitation apparatus consisted of a "Microtron 200", 2450 MHz generator and a quarter-wave foreshortened radial linecavity (Electro-Medical Supplies, London). A 1 m spectrograph with a diffraction grating blazed at 576 nm and a photoelectric readout system onto a Honeywell recorder was used to examine the spectra emitted from the detector tube.

The column used was 2.7 m × 6.5 mm copper tubing packed with dinonyl phthalate. Argon was used as carrier gas at a column pressure of 105 kN/m². A column temperature of 150°C was found suitable for resolving 1:1 *n*-pentane–CS_2 mixtures using katharometer detection. The retention times were *n*-pentane 190 sec, CS_2 290 sec with well-defined peaks and no tailing. When the gas chromatographic column was connected to the microwave detector under reduced pressure (*via* the katharometer outlet to the capillary restrictor) the retention times of both components were about 30 sec longer, owing to the pressure built up in front of the capillary restrictor. The retention times for thiophen, thioglycolic acid, and dimethyl sulfoxide on the same column were 540 sec, 90 sec, and 110 sec, respectively.

Table 8 lists the prominent spectra observed for carbon disulfide, thiophen, thioglycolic acid, dimethyl sulfoxide, and sulfur dioxide. The sulfur lines at 190 and 191.5 nm were weak in all the spectra and the CS_2 emission at 516 nm was one of the most intense emissions observed. The CS emission at 257.6 nm was prominent for all the carbon–sulfur compounds except thioglycolic acid, for which it was weak as were most of the other bands. The sensitivity of the microwave-excited emission detector, which was estimated to be 100–1000 times that of the katharometer for carbon disulfide, suggests its increasing potential for gas chromatographic analysis.

TABLE 8

SPECTRA OBSERVED PROMINENTLY IN FLOWING SYSTEM

	S (atomic) (nm)		CS (nm)	C_2 (nm)	Other bands (nm)			
Carbon disulfide	190.0	191.5	257.6	516	335–340	382–391		
Thiophen	190.0	191.5	257.6	516	335–340	382–391	410–426[a]	504–522
Thioglycolic acid	190.0	191.5		516	335–340	350–360	382–391	
Dimethylsulfoxide	190.0	191.5	257.6	516	298	326–341	350–360	360–380[b]
							387–408[b]	
Sulfur dioxide	190.0	191.5			250–420	255[c]		

[a] Bandhead at 421, 423, 424, 425, and 426 nm characteristic of thiophen.
[b] Bands characteristic of dimethylsulfoxide.
[c] Bands characteristic of sulfur dioxide.

The use of the flame-photometric detector for the selective detection of phosphorus and sulfur compounds (the Melpar flame-photometric detector) was first reported by Brody and Chaney[101]. The Melpar detector is based on the principle for the detection of phosphorus and sulfur compounds in a hydrogen–air flame as described initially by Draeger and Draeger[102].

The molecular emission due to HPO species from phosphorus compounds at 526 nm and S_2 species from sulfur compounds at 394 nm are measured by a photomultiplier tube through an interference filter that transmits only at one of these wavelengths. Mizany[103] recently studied some characteristics of the Melpar flame-photometric detector (FPD) in the sulfur mode of operation. The effect of the oxygen–hydrogen flow ratio and total gas flow as well as the effect of the oxidation state of the sulfur atom on the emissive response was examined. The flame-photometric responses were normalized by comparison to those obtained with a Dohrmann microcoulometric sulfur detector.

A MicroTek Model MT-220 equipped with a Melpar flame photometric detector and a Dohrmann Microcoulometric detector for sulfur, comprised of a Model S-100 pyrolyzer, T300-R sulfur cell, and C200-A electrometer were used. The needle valve flow controllers for the oxygen, hydrogen, and air on the MicroTek MT-220 were replaced by Millaflow Model 42300080 flow controllers for better reproducibility of the gas flow settings. The GC column used for both detectors was a 120 cm × 4 mm i.d. Pyrex tube packed with a 1:1 mixture of 10% DC-200 and 15% QF-1, both on 80–100 mesh Gas Chrom Q. With the nitrogen carrier gas flow rate at 70 or 80 ml/min the column temperature was varied to obtain an approximately 3 min retention time for each compound. The injector temperature was kept 20–60°C above the column temperature and the detector was kept at 175–180°C. The micro-coulometric detector was operated under the following conditions: furnace temperature 810°C, inlet block temperature 250°C, transfer line temperature 250°C, nitrogen sweep flow 10 ml/min, oxygen flow 20 ml/min, electrometer gain 200, bias 150 mV, range 250 ohms. The peak areas were measured by planimetry and were used in comparing the relative responses of the two detectors. Table 9 shows the sulfur compounds of various oxidative states used for comparison of emissive

TABLE 9

SULFUR COMPOUNDS OF VARIOUS OXIDATION STATES USED FOR COMPARISON OF EMISSIVE RESPONSES

Compound	Structure	S (%)	B.p. (°C)
Diethyl disulfide	Et—S—S—Et	52.5	153
Diethyl sulfide	Et—S—Et	35.6	92
Diethyl sulfoxide	$\overset{\text{O}}{\underset{\parallel}{\text{Et—S—Et}}}$	30.2	90
Diethyl sulfone	$\overset{\text{O}}{\underset{\underset{\text{O}}{\parallel}}{\text{Et—S—Et}}}$	26.2	242
Diethyl sulfite	$\overset{\text{O}}{\underset{\parallel}{\text{Et—O—S—O—Et}}}$	23.2	157
ethyl ethane sulfonate	$\overset{\text{O}}{\underset{\underset{\text{O}}{\parallel}}{\text{Et—O—S—Et}}}$	23.2	208
Diethyl sulfate	$\overset{\text{O}}{\underset{\underset{\text{O}}{\parallel}}{\text{Et—O—S—O—Et}}}$	20.8	208

TABLE 10

COMPARISON OF THE FPD AND MICROCOULOMETRIC RESPONSES FOR THE VARIOUS SULFUR COMPOUNDS

Compound	MCD response peak area for 10 ng S	RFP reponse peak area for 5 ng S	Normalized FPD response
Disulfide	125	280	224*
Sulfide	87	110	181
Sulfone	106	156	139
Sulfite	99	136	139
Sulfonate	107	102	89
Sulfate	47**	100	

* The correction for disulfide is made assuming a linear concentration–response relationship.
** Efficiency of sulfate detection for the MCD is believed to be about 50%.

responses. The responses of both detectors for each compound and the normalized Melpar responses are shown in Table 10. The sulfite was chosen as the arbitrary base for normalization. On the basis of the normalized emissive responses it appeared that the differences in response for the various oxidation states of sulfur were inherent in the Melpar detector.

The normalized FPD responses for different sulfur states were found to be sulfide > sulfone $\simeq$ sulfite > sulfonate. However, the differences attributable to the oxidation state of sulfur were not large and were not believed to detract from the utility of this detector. It was also found that the Melpar flame-photometric detector (in the sulfur mode) produced the highest response with a 0.4–0.5 O_2/H_2 flow ratio.

Reactive sulfur compounds such as SO_2 and H_2S have been particularly difficult to separate chromatographically at levels below 1 p.p.m. due to irreversible reaction with column, support, stationary liquid phase, or carrier gas.

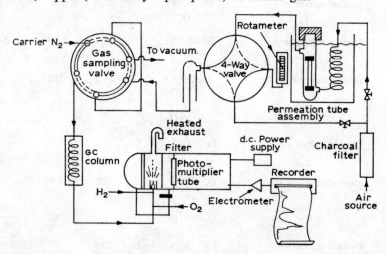

Fig. 17. Automated gas chromatographic–FPD atmospheric sulfur gas analyzer and calibration apparatus.

Stevens *et al.*[104] described the development of the first automated gas chromatographic system that could quantitatively measure the ambient air content of sulfur dioxide, hydrogen sulfide, methyl mercaptan, and dimethyl sulfide down to 0.002 p.p.m. Figure 17 shows the automated gas chromatographic FPD atmospheric sulfur gas analyzer and calibration apparatus. This analyzer consisted of a flame photometric detector, 750 V power supply and electrometer (Micro-Tek Instruments), Varian Aerograph Model 1200 gas chromatograph, a 36 ft. × 0.085 in. i.d. Teflon (DuPont) column packed with 40–60 mesh Teflon coated with a mixture of polyphenyl ether and orthophosphoric acid, and a 6-port rotary Chromatronix automatic gas-sampling valve equipped with a 10 cc Teflon gas loop. The flame-photometric detector was equipped with a narrow-band optical filter that permitted 56% light transmission at 394 nm with a band width of 5 nm. This analytical system was satisfactory for monitoring SO_2 and H_2S in atmospheres that did not contain significant concentrations of sulfur compounds of 3 or more carbon or sulfur atoms (which could eventually elute into the detector causing positive interferences). To prevent these interferences the chromatographic system (Fig. 17) was modified as shown (Fig. 18) by replacing the 6-port valve with a 10-port valve (sliding plate, solenoid-actuated valve, Beckman Instruments, Fillerton, Cal.), equipped with a stripper column. The stripper column consisted of 24 in. × 0.085 in.

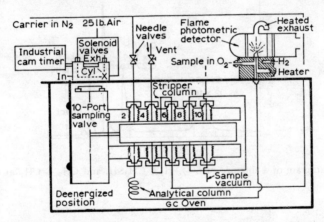

Fig. 18. Automated gas chromatographic–FPD sulfur gas analyzer equipped with precolumn and backflushing modifications.

i.d. Teflon tubing packed with Teflon coated with polyphenyl ether, permitting the passage of SO_2, H_2S, $(CH_3)_2S$ to the analytical column but retaining higher weight compounds. The columns were prepared by lightly packing 36 ft. of $\frac{1}{8}$ in. × 0.085 in. i.d. fluorinated ethylene propylene tubing at 5°C with 40–60 mesh Haloport F (Teflon powder, F & M Scientific, Avondale, Pa.). Fifty milliliters of an acetal solution containing 12 g of polyphenyl ether (5-ringed polymer, F & M) and 500 mg of orthophosphoric acid were forced through the Teflon column at a flow rate of

20 ml/h under nitrogen pressure. Nitrogen was passed through the column until the column packing appeared dry (7–8 h). The column was conditioned at 140°C for 6 h while a nitrogen carrier gas flow of 50 ml/min was maintained through the column. Chromatographic operating parameters detector temperature 105 ± 3°C, hydrogen flow rate of 80 ml/min, and an oxygen flow of 16 ml/min at a carrier gas flow of 100 ml/min gave the most favorable signal-to-noise ratio.

Figure 19 shows a chromatogram of a mixture of SO_2, H_2S, CH_3SH, and $(CH_3)_2S$ at concentrations below 1 p.p.m. The number of theoretical plates for the polyphenyl ether column was calculated to be 1070 which corresponds to an HETP value of 0.97 cm. The analytical system shown in Fig. 18 was calibrated for concentrations between 0.008 and 1.0 p.p.m. (v/v). Sample concentrations in excess of 2 p.p.m. (10 ml sample volume) tended to saturate the photometric detector. Figure 20 illustrates a chromatogram of ambient air taken in Cincinatti, Ohio, showing varying concentrations of SO_2 with time and indicated that gaseous sulfur in that city's atmosphere is largely SO_2. The automated system of Stevens et al.[104] produces a specific quantitative response to SO_2 and H_2S since the FPD has at

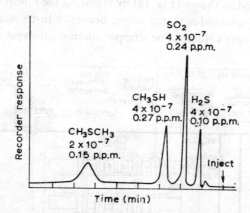

Fig. 19. Chromatogram of a mixture of SO_2, H_2S, CH_3SH, and CH_3–S–CH_3 at concentrations below 1 p.p.m.

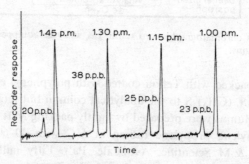

Fig. 20. Chromatogram of ambient air taken in Cincinnati (December 1, 1968, between 1:00 p.m. and 1:45 p.m.) showing varying concentrations of SO_2 with time.

least a 10,000:1 rejection ratio between the sulfur and non-sulfur compounds and the elution order of these compounds was such that it was suggested that no other known sulfur compound in the atmosphere could be eluted before or with these species.

Harrison and Coyne[105] described the flame-photometric determination at the parts per thousand million level of some volatile sulfur compounds in beer headspace samples. A Melpar flame-photometric detector was used at 150°C. A MicroTek dual channel electrometer was used to monitor simultaneously both the flame-ionization and the flame-photometric responses. The oven portion of a Griffin and George flame-ionization gas chromatograph was operated at 50°C. A 300 ft. × 0.02 in. i.d. stainless steel column coated with polyethylene glycol 400 (Merck) was employed with carrier nitrogen, hydrogen, and oxygen used at flow rates of 8, 150, and 20 ml/min. The only volatile found in significant amounts in the headspace above beer was dimethyl sulfide found in amounts varying from 0.002 to 0.006 p.p.m.

REFERENCES

1   Environmental Quality, Council of Environmental Quality Rept., U.S. Govt. Printing Office, Washington, D.C., 1970.
2   O. J. BALCHUM, J. CYBICKI AND G. R. MENEELY, J. Appl. Physiol., 15 (1960) 62.
3   N. R. FRANK AND F. E. SPEIZER, Arch. Environ. Health, 11 (1965) 624.
4   M. O. AMDUR, Intern. J. Air Pollution, 1 (1959) 170.
5   J. A. NADEL, H. SALEM AND B. PAMPLIN, J. Appl. Physiol., 20 (1965) 164.
6   D. M. AVIADO AND H. SALEM, Arch. Environ. Health, 16 (1968) 903.
7   N. R. FRANK, Proc. Roy. Soc. Med. 57 (1964) 1029.
8   N. R. FRANK, M. O. AMDUR AND J. WORCESTER, J. Appl. Physiol., 17 (1962) 252.
9   G. G. BURTON, M. CORN AND J. B. L. GEE, Arch. Environ. Health, 18 (1969) 681.
10 F. E. SPEIZER AND N. R. FRANK, Brit. J. Ind. Med., 23 (1966) 75.
11 M. C. BATTIGELLI, J. Occupational Med., 10 (1968) 500.
12 Y. ALARIE, C. E. ULRICH, W. M. BUSEY, H. E. SWANN AND H. N. MACFARLAND, Arch. Environ. Health, 21 (1970) 769.
13 T. R. VAUGHAN, JR., L. F. JENNELLE AND T. R. LEWIS, Arch. Environ. Health, 19 (1969) 45.
14 T. R. LEWIS, K. I. CAMPBELL AND T. R. VAUGHAN, Arch. Environ. Health, 18 (1969) 596.
15 R. A. KEHOE, W. F. MACHLE, K. KITZMILLER AND T. J. LEBLANC, J. Ind. Hyg., 14 (1932) 159.
16 L. LASKIN, Prelim. Rept. Natl. Cancer Inst. Contract and PH 43-66-962, 1968.
17 P. URONE AND W. H. SCHROEDER, Environ. Sci. Technol., 2 (1968) 611.
18 M. BUFAZINI, Environ. Sci. Technol., 5 (1971) 685.
19 P. W. WEST AND G. C. GAEKE, Anal. Chem., 28 (1956) 1916.
20 B. G. STEPHENS AND F. LINDSTRUM, Anal. Chem., 36 (1964) 1308.
21 F. P. SCARINGELLI, B. E. SALTZMAN AND S. A. FREY, Anal. Chem., 39 (1969) 1709.
22 M. D. THOMAS, J. Air Pollution Control Assoc., 14 (1964) 517.
23 G. F. KIRKBRIGHT, Talanta, 13 (1966) 1.
24 P. URONE, W. E. BOGGS AND C. M. NOYES, Anal. Chem., 23 (1951) 1517.
25 S. KÜNDIG, Chem. Rundschau, 18 (1965) 123.
26 K. ZEPF AND F. VETTER, Mikrochemie, (1950) 280.
27 J. M. KOLTHOFF AND C. S. MILLER, J. Am. Chem. Soc., 63 (1941) 2818.
28 W. L. CRIDER, Anal. Chem., 37 (1965) 1770.
29 H. L. HELWIG AND C. L. GORDER, Anal. Chem., 30 (1968) 1810.
30 W. G. CUMMINGS AND M. W. REDFEARN, J. Ind. Fuel, 30 (1957) 628.

31 W. Leithe, *The Analysis of Air Pollutants*, Ann Arbor Science, Ann Arbor, Mich., 1971, p. 164.
32 D. F. Adams and R. K. Koppe, *Tappi*, 42 (1959) 601.
33 R. Patrick, T. Schrodt and R. Kermode, *J. Chromatog. Sci.*, 9 (1971) 381.
34 D. R. Beuerman and C. E. Meloan, *Anal. Chem.*, 34 (1962) 319.
35 R. L. Bond, W. J. Mullin and F. J. Pinchin, *Chem. Ind.*, (1963) 1902.
36 O. L. Hollis and W. V. Hayes, in A. B. Littlewood (Ed.), *Gas Chromatography, 1961*, The Institute of Petroleum, London, 1967, p. 57.
37 W. F. Wilhite and O. L. Hollis, *J. Gas Chromatog.*, 6 (1968) 84.
38 D. L. West, *Du Pont Rept. DP-380*, Savannah River, 1964, cited in R. M. Bethea and M. C. Meador, *J. Chromatog. Sci.*, 7 (1969) 655.
39 N. N. Levchuk, S. L. Zal'tsman and N. G. Vilesov, *Khim. Prom. Ukr.*, 6 (1969) 40; *Chem. Abstr.*, 72 (1970) 50687m.
40 R. H. Espach, *Ind. Eng. Chem.*, 42 (1950) 2235.
41 Y. Henderson and H. W. Haggard, *Noxious Gases*, Reinhold, New York, 2nd edn., 1943.
42 T. S. Ollman, *A Manual of Pharmacology*, Saunders, Philadelphia, 6th edn., 1944.
43 W. P. Yant, *J. Public Health*, 22 (1940) 598.
44 J. B. Jacobs, M. M. Braverman and S. Hoch-Heiser, *Anal. Chem.*, 29 (1957) 1349.
45 M. Buck and H. Stratmann, *Staub*, 24 (1964) 241.
46 F. Oehme and H. Wyden, *Glas-Instr. Techn.*, 9 (1965) 107; *Staub*, 26 (1966) 252.
47 H. D. Axelrod, J. H. Clary, J. E. Bonelli and J. P. Lodge, Jr., *Anal. Chem.*, 41 (1969) 1856.
48 M. Wronski, *Z. Anal. Chem.*, 305 (1971) 1.
49 B. A. Hardwick, D. K. B. Thistlewayte and R. T. Fowler, *Atmos. Environ.*, 4 (1970) 379.
50 H. Hollings and W. K. Hutchinson, *J. Inst. Fuel*, 8 (1935) 366.
51 A. F. Smith, D. G. Jenkins and D. E. Cunningwort, *J. Appl. Chem.*, 11 (1961) 137.
52 M. Wronski, *Zeszyty Nauk. Uniw. Lodz. Ser. II*, 4 (1958) 181.
53 M. Wronski, *Analyst*, 83 (1958) 314.
54 M. Wronski, *Anal. Chem.*, 43 (1971) 606.
55 G. M. Speyzer abd N. M. Zaydman, *Zavodsk. Lab.*, 31 (1965) 272.
56 T. M. Hseu and G. A. Rechnitz, *Anal. Chem.*, 40 (1968) 1054.
57 C. N. Jones, *Anal. Chem.*, 39 (1967) 1858.
58 W. G. Cook and R. A. Ross, *Anal. Chem.*, 44 (1972) 641.
59 J. E. Lovelock, *Anal. Chem.*, 33 (1961) 162.
60 E. L. Obermiller and G. O. Charlier, *Anal. Chem.*, 39 (1967) 396.
61 D. M. Ottenstein, *13th Conf. Anal. Chem. Appl. Spectry.*, Pittsburgh, Pa., March 1962.
62 C. Bighi and G. Saglietto, *J. Chromatog.*, 17 (1965) 13.
63 F. Flury and F. Zernik, *Schädliche Gase*, Springer, Berlin, 1931.
64 A. Klemenc, *Chem. Ber.*, 76 (1943) 299.
65 J. A. Schols, *Anal. Chem.*, 33 (1961) 359.
66 L. G. Berezkina, S. V. Mel'nikova and N. A. Elefterova, *Khim. Prom.*, 42 (1966) 619; *Chem. Abstr.*, 65 (1966) 16044.
67 R. Staszewski, J. Janak and T. Pompowski, *Chem. Anal. (Warsaw)*, 8 (1963) 897.
68 C. T. Hodges and R. F. Matson, *Anal. Chem.*, 37 (1965) 1065.
69 H. L. Hall, *Anal. Chem.*, 34 (1962) 61.
70 S. J. Clark, *Ionization Detectors*, Jarrell-Ash, New Tonville, Mass., 1968, p. 5.
71 Y. Watanabe, K. Isomura and Y. Tomari, *Japan Analyst*, 42 (1967) 942.
72 F. J. O'Hara, W. M. Keeley and H. W. Fleming, *Anal. Chem.*, 28 (1956) 467.
73 R. M. Creamer and D. H. Chambers, *Anal. Chem.*, 26 (1954) 1098.
74 L. A. Pursglove and H. W. Wainwright, *Anal. Chem.*, 26 (1954) 1835.
75 A. V. Avdeeva, *Zavodsk. Lab.*, 7 (1938) 279.
76 H. Hakewell and E. M. Rueck, *Am. Gas Assoc. Proc.*, 28 (1946) 529.
77 Anon, *U.S. Public Health Rept.*, 56 (1941) 574.
78 F. H. Lewy, *Penna Dept. Labor. Ind. Bull. No. 46*, 1968.
79 H. H. Rubin and A. J. Arieff, *J. Ind. Hyg. Toxicol.*, 27 (1945) 123.
80 H. L. Barthelmy, *J. Ind. Hyg. Toxicol.*, 21 (1939) 141.

81 R. W. McKEE, C. KIPPER, J. H. FOUNTAIN, A. M. RISKIN AND P. DRINKER, *J. Am. Med. Assoc.*, 122 (1943) 217.
82 R. W. McKEE, *J. Ind. Hyg. Toxicol.*, 23 (1941) 151.
83 F. J. VILES, *J. Ind. Hyg. Toxicol.*, 22 (1940) 188.
84 M. P. MATUZAK, *Ind. Eng. Chem. Anal. Ed.*, 4 (1932) 98.
85 N. W. HANSON, D. A. REILLY AND H. E. STAGG (Eds.), *The Determination of Toxic Substances in Air*, Heffer, Cambridge, 1965, p. 86.
86 D. M. DOUGLAS AND B. A. SCHAEFER, *J. Chromatog. Sci.*, 7 (1969) 443.
87 A. T. BLADES, *J. Chromatog. Sci.*, 8 (1970) 414.
88 D. S. TARBELL, in N. KHARASCH (Ed.), *Organic Sulfur Compounds*, Vol. I, Pergamon Press, New York, 1961.
89 A. A. OSWALD AND T. J. WALLACE, in N. KHARASCH AND C. Y. MYERS (Eds.), *The Chemistry of Organic Sulfur Compounds*, Vol. 2, Pergamon Press, New York, 1966.
90 A. C. HARKNESS AND F. E. MURRAY, *Atmos. Environ.*, 4 (1970) 417.
91 I. B. DOUGLAS AND C. PRICE, *Tappi*, 49 (1966) 335.
92 F. W. FREDRICKS AND G. A. HARLOW, *Anal. Chem.*, 36 (1964) 263.
93 M. REINHARDT, R. OTTO AND K. KOCH, *Chem. Tech. (Leipzig)*, 22 (1970) 481.
94 M. FELDSTEIN, S. BALESTRIERI AND D. A. LEVAGGI, *J. Air Pollution Control Assoc.*, 15 (1965) 215.
95 D. M. OAKS, H. HARTMANN AND K. P. DiMICK, *Anal. Chem.*, 36 (1964) 1560.
96 D. F. ADAMS, R. K. KOPPE AND W. N. TRITTLE, *J. Air Pollution Control Assoc.*, 15 (1965) 31.
97 D. F. ADAMS, W. L. BAMESBERGER AND T. J. ROBERTSON, *J. Air. Pollution Control. Assoc.*, 18 (1968) 145.
98 L. A. ROBBINS, R. M. BETHEA AND T. D. WHEELOCK, *J. Chromatog.*, 13 (1964) 361.
99 M. DRESSLER AND J. JANAK, *J. Chromatog. Sci.*, 7 (1969) 451.
100 D. M. DAGNELL, S. J. PRATT, T. S. WEST AND D. R. DEANS, *Talanta*, 16 (1969) 797.
101 S. BRODY AND J. CHANEY, *J. Gas Chromatog.*, 4 (1966) 42.
102 H. DRAEGER AND B. W. DRAEGER, *Ger. Pat. 1,133,918*, 1962.
103 A. I. MIZANY, *J. Chromatog. Sci.*, 8 (1970) 151.
104 R. K. STEVENS, J. D. MULIK, A. E. O'KEEFE AND K. J. KROST, *Anal. Chem.*, 43 (1971) 827.
105 G. A. F. HARRISON AND C. M. COYNE, *J. Chromatog.*, 41 (1969) 453.

Chapter 13

HALOGEN GASES

1. CHLORINE AND HYDROGEN CHLORIDE

Chlorine is produced by the electrolysis of chlorides, either fused or in aqueous solution, and is one of the most important chemicals used in industry. Its spectrum of utility includes (1) the production of plastics such as polyvinyl chloride, (2) insecticides and herbicides (e.g. DDT, lindane, aldrin, phenoxyacetic acids), (3) inorganic chlorine derivatives such as chlorates and perchlorates, (4) bleaching agent in the cellulose and paper industries as well as in laundry plants, (5) disinfectant for water supplies (in amounts of 1.2–3.6 p.p.m.) and swimming pools.

The odor threshold for chlorine is about 0.05–0.1 p.p.m. with irritating symptoms occurring below 1 p.p.m. Concentrations exceeding 3 p.p.m. are considered dangerous. The danger threshold value for damage to vegetation is ca. 0.1–1 p.p.m. The acute effects of chlorine exposure on animals[1-4] and man [1,5] have been described. In man exposure for 0.5–1 h to 14–21 p.p.m. of chlorine is dangerous. Greater concentrations damage the lungs causing pulmonary congestion and edema[6,7]. Chlorine concentrations above 25 p.p.m. have been shown to cause chromosome aberrations in mammalian cells in vitro[8].

Hydrogen chloride (and hydrochloric acid) are prepared commercially by four processes: (1) the salt–sulfuric acid, (2) the Hargreaves (salt, SO_2, air, and water vapor, (3) synthetic or thermal, and (4) byproduct of organic chlorinations. Hydrogen chloride can be synthesized by the combustion of a controlled mixture of hydrogen and chlorine. The chlorine from various types of electrolytic chlorine cells may be wet, or dried cell gas, evaporated chlorine or waste gas from liquid chlorine plants. The hydrogen may come from various sources including chlorine–caustic cells and the hydrocarbon–steam reaction. The total hydrochloric acid production in the U.S. was 2,025,000 short tons in 1971.

Hydrogen chloride and its aqueous solution are widely used in industry (e.g. metal cleaning, chemical manufacture, petroleum-well activation, the production of food and synthetic rubber, the pickling of steel, and oxyhydrochlorination processes. Large quantities of HCl are used in the production of chloroprene and vinyl chloride, alumina for aluminum manufacture and the Dow process for obtaining magnesium from sea water. HCl is used in the manufacture of dextrose and starch, syrups, monosodium glutamate, and in the revivification of bone char and carbon in sugar refining operations. It is also used in many extractive metallurgic processes for treating various high grade ores, among which are those yielding radium, vanadium, tungsten, tantalum, manganese, and germanium. The combustion of plastics (mostly PVC) is a major contributor of hydrogen chloride in local atmospheres.

Gaseous hydrogen chloride of up to 50 p.p.m. can be tolerated for a short period without injury. When inhaled in sufficiently high concentration, hydrogen chloride acts as an irritant to the respiratory tract. The physiological response of hydrogen chloride in animals[9-11] and man[12] has been described.

Chlorine in air has been determined colorimetrically with o-toluidine[13-15] and hydrogen chloride analyzed using a colorimetric procedure with mercuric thio-cyanate[16].

Information regarding the chromatographic analysis of chlorine cell gas is scant. Dudley and Poetker[17] and Murray and Kircher[18] described a chromatographic chlorine cell gas analyzer which analyzed for percent chlorine only (not an analyzer for all gases present). Chlorine cell gas normally contains 90–98% chlorine, 1–3% oxygen, 1–3% carbon dioxide, 0–1% hydrogen, 0–1% carbon monoxide, and 0–3% nitrogen. A gas chromatographic method for the analysis of chlorine cell gas which required a two-column separation was developed by Neely[19]. Two gas chromatographs were used with a manifold in the carrier gas stream so that each chromatograph may be used separately for other types of analyses or can be used with one another for a chlorine cell analysis. The two instruments used were Perkin-Elmer Models 154A and 154C Vapor Fractometers with a Leeds & Northrup 0–5 mV recorder. The hydrogen analyzer consisted of a Gow Mac Model 9677-AEL detector with its necessary circuitry. For sample injection, a Burrell Kromatog glass gas sampling valve was used and all three detectors were wired through a selector switch

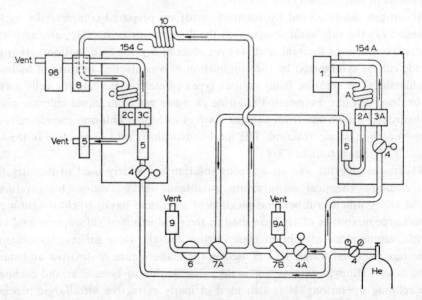

Fig. 1. Flow diagram. 1, Burrell gas sampling manifold (Kromotog); 2, sensing side detector cell; 3, reference side detector cell; 4, pressure regulator and gauge; 5, rotameter; 6, shut-off valve; 7, switch valve; 8, Perkin-Elmer gas sampling valve; 9, Foxboro restricter valve; 10, 1 ft. 5A molecular sieve.

to one recorder. The Model 154A chromatograph was equipped with a 12 ft. column of 30% Fluoroglube grease, Type LG-160 (Hooker Electrochemical Co.) on 30 to 50 mesh Chromosorb. The 154C chromatograph was equipped with an 8 ft. Molecular Sieve column, Type 13X, 30–50 mesh (Linde Air Products Co.).

Figure 1 illustrates the flow diagram of a gas chromatographic system. The two Perkin-Elmer instruments were operated at 40°C primarily for stability of operation and the hydrogen analyzer was operated at room temperature. The Fluorolube column separated the sample into three fractions, *e.g.* the first recorded on the chromatogram contains H_2, O_2, N_2, and CO, the second CO_2, and the third Cl_2. Figure 2 illustrates a chromatogram of chlorine cell gas. Good resolution of all components was obtained in less than 20 min. The same analyses by wet chemical methods[8] required much longer. When compared with values obtained by wet chemical methods, the analyses in volume % obtained chromatographically were accurate within ±0.02%.

A gas chromatographic method for determining up to 1.5% v/v each of hydrogen, nitrogen, oxygen, and carbon dioxide in chlorine gas was described by Lucy and Woolmington[20]. Separation of the components was effected by using two chromatographic columns in series. The first column (silica gel) was used for separating CO_2 and Cl_2 from each other and from the permanent gases. The second column, packed with a molecular sieve, separated H_2, O_2, and N_2.

The apparatus used is shown in Fig. 3. Detection was by means of a Veco thermistor-bead cell at the end of the first column and a Griffin and George thermal conductivity cell at the end of the second column. Each detector was incorporated in a separate Wheatstone-bridge circuit and operated by a 6 V accumulator. Peaks were recorded on a Leeds & Northrup 5 or 10 mV recorder having facilities for

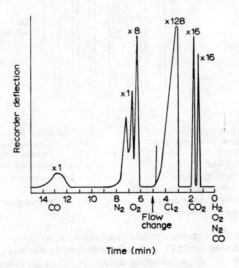

Fig. 2. Chlorine cell gas chromatogram.

switching from one bridge circuit to the other. Chlorine and water vapor were removed from the gas stream by means of a glass U-tube (41 cm × 6 mm i.d.) containing 30–60 mesh Celite impregnated with saturated KI solution containing a little starch. The first column consisted of 114 cm × 6 mm i.d. glass tubing packed with 40 to 60 mesh silica gel (previously dried at 110°C for 24 h); the second column consisted of 198 cm of similar tubing packed with 30–44 mesh Linde 5A molecular sieve that had been activated by drying at 350°C for 5 h. The operating conditions were: temperature of silica gel column 60°C, temperature of molecular sieve column 60°C, and argon carrier flow 43.5 ml/min. The retention times (min) were hydrogen 1.5, oxygen 3, nitrogen 415, carbon dioxide (through column of silica gel only) 8, chlorine (through column of silica gel only) 16.

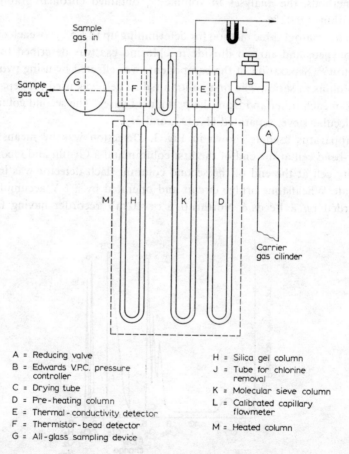

A = Reducing valve
B = Edwards V.P.C. pressure
       controller
C = Drying tube
D = Pre-heating column
E = Thermal-conductivity detector
F = Thermistor-bead detector
G = All-glass sampling device

H = Silica gel column
J = Tube for chlorine
       removal
K = Molecular sieve column
L = Calibrated capillary
       flowmeter
M = Heated column

Fig. 3. Schematic diagram of chromatographic apparatus; A, reducing valve; B, Edwards V.P.C. pressure controller; C, drying tube; D, preheating column; E, thermal conductivity detector; F, thermistor bead detector; G, all-gas sampling device; H, silica gel column; J, tube for chlorine removal; K, molecular sieve column; L, calibrated capillary flowmeter; M, heated column.

Ruthven and Kenney[21] described the gas chromatographic analysis of air, chlorine, and hydrogen chloride with detection achieved using a conventional Gow-Mac katharometer, type 9285 fitted with high-sensitivity W2 (tungsten–rhenium) elements. The column of choice was a 220 cm × 3 mm i.d. PTFE tube containing 4 wt.-% of tritolyl phosphate on 60–80 mesh acid-washed Celite. The carrier gas flow of hydrogen was 75 ml/min and the column was operated at room temperature (with no apparent deterioration with use). Figure 4 illustrates a flow diagram for the chromatograph. The stainless steel Perkin-Elmer gas sampling valve required re-facing every few weeks due to slow corrosion. Accurate pressure control at both column inlet and outlet was essential because of the sensitivity of the detector to small pressure changes. An Edwards diaphragm valve UPCI was used at the inlet and an Edwards Cartesian manostat controlled the outlet to the water pump.

Figure 5 shows a chromatogram of a gas mixture containing hydrogen chloride, chlorine, and oxygen obtained under the conditions specified above. The time

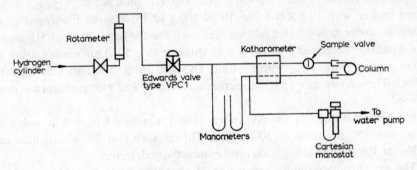

Fig. 4. Flow diagram for the chromatograph.

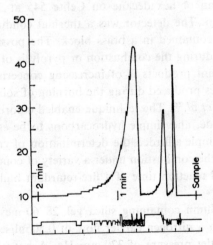

Fig. 5. Chromatogram for typical gas mixture. a, Hydrogen chloride 69.0%; b, chlorine 18.2%; c, oxygen 12.8 %.

References pp. 326–327

required for a complete analysis was about 1.5 min. The retention times (min) of the gases were nitrogen + oxygen 0.262, chlorine 0.354, and hydrogen chloride 0.67. Their relative retention times (to air) were nitrogen + oxygen 1.0, chlorine 1.35, and hydrogen chloride 2.56.

Hydrogen chloride could not be determined to the same degree of accuracy as the other components. For example, the minimum amount that could be detected was about 2% and the accuracy over most of the composition range was about 1–2%. A slight column-conditioning effect was noted with hydrogen chloride and thus several samples were required each day before consistent values were obtained. Although the instrument was developed by Ruthven and Kenney[21] for a kinetic study of the catalytic oxidation of hydrogen chloride to chlorine, it was suggested that with slight modification, it would be applicable to the study of organic chlorination reactions in which the analysis of gases containing wet chlorine and hydrogen can frequently arise.

Ellis and Iveson[22] analyzed Cl_2, HCl, and HF on a 4.5 ft. × ¼ in. i.d. monel column packed with 10 g Kel-F No. 10 oil/100 g of 30–60 mesh Fluoropak at 48°C. An argon carrier flow of 16.5 ml/min was used for the separation of HF and Cl_2.

Ceiplinski[23] detected 3×10^{-12} g of chlorine in a 50 µl air sample using a 1 m column packed with 20% silicone fluid DC-200 on 60–80 mesh Chromosorb W at 50°C. The carrier gas was 1% methane in argon and electron-capture detector was used.

Isbell[24] separated Cl_2 and NO in less than 1 min on a 8 ft. × ¼ in. o.d. column packed with 25% triacetin on 30–60 mesh Chromosorb P at 75° with helium carrier gas flow at 108 ml/min using a thermal conductivity detector.

The gas chromatographic separation of phosgene, hydrogen chloride, chlorine, boron trichloride, and silicon tetrachloride was effected by Turkel'taub et al.[25] using a 3 m × 3.5 mm column of hexadecane on Celite 545 at 28.5° with nitrogen as carrier gas at 40 ml/min. The detector was a thermal conductivity cell with glass-coated platinum wires contained in a brass block. The possibility of formation of toxic gaseous products during the combustion or pyrolysis of new materials such as plastics and heat-resistant products is of increasing concern. A rapid GC method of analysis of toxic gases produced during the burning of solids containing chlorine was developed by Fish et al.[26]. The technique enabled chlorine, hydrogen chloride, phosgene, carbon dioxide, and simple hydrocarbons to be estimated from a single sample (an additional sample afforded the determination of carbon monoxide). The method was applied to the combustion under a variety of conditions of temperature, oxygen availability, and reaction time of a fire-retardant building paper containing about 20% of chlorine.

A 72 in. × ¼ in. column containing silica gel, 28–60 mesh (12 grade, Davison) was used at 56.5°C with a nitrogen carrier flow of 0.50 ml/sec, an inlet pressure of 770 mm Hg and an outlet pressure of 320 mm Hg. A hot-wire type katharometer (Pye) was employed with the block at 62°C and a bridge current and potential of

120 mA and 6 V. Samples having total pressures in the range of 1–20 mm Hg were introduced into the column directly from U-tubes of volume 5–15 ml without interrupting the flow of nitrogen, by using a U-tube and bypass valve. Under these reduced pressure conditions, the retention times (in min) of the various products were hydrogen 1.0, carbon monoxide 2.5, methane 2.9, ethane 7.0, phosgene 9.9, carbon dioxide 11.1, ethylene 12.1, hydrogen chloride 13.3, chlorine 28.5, and acetylene 31.6. The sensitivities of the gases were hydrogen chloride 2.38×10^5, carbon monoxide 3.00×10^5, ethane 3.03×10^5, phosgene 4.80×10^5, ethylene 7.41×10^5, methane 1.13×10^6, acetylene 2.50×10^6, carbon dioxide 2.70×10^6, chlorine 5.20×10^6, hydrogen 1.02×10^8 cm^2/g mole.

Runge[27] used a 9 m × 6 mm i.d. column packed with 1 part Aroclor 1232 to 20 parts by weight of 20–60 mesh Haloport F at 22°C for the separation of CO_2, HCl, Cl_2, and SO_2 (among other acidic gases) in less than 5 min. Helium was the carrier gas at 200 ml/min. The separation of Cl_2, Br_2, NOCl, HCN, SO_2, and H_2S on a 6 ft. column of 50% No. 3 Daifuloil on 30–60 mesh Polifulon powder was described by Araki et al.[28]. Separations of HCl and HBr were also achieved, however, with increase of tailing in the presence of significant amounts of H_2O and NH_3.

Janak et al.[29] and Johns[30] described the use of silica gel and modified silica gel, respectively, for the gas–solid chromatography of chlorine.

Reactive gases such as Cl_2, NOCl, NO_2Cl, and NO_2 were analyzed quantitatively by Huillet and Urone[31] employing diatomaceous earth supports and halogenated compounds as liquid phases contained in nickel, inconel, or stainless steel columns to prevent corrosion. Optimum conditions were established by studying various types of supports and liquid phases at temperatures ranging from -20°C to 50°C. Adsorption of the reactive gases on the support was circumvented by dimethyldichlorosilane treatment and pre-conditioning the column with several injections of the gas mixtures being analyzed. Nitrogen, oxygen, and nitric oxide were analyzed in the same samples using dual-column techniques.

A Consolidated Electrodynamics Model 26-201A gas chromatograph with high nickel alloy filaments and stainless steel internal parts was used for the majority of the study. The column was a 6 in. o.d. coil and was placed in a large Dewar flask filled with a water–ethylene glycol solution. Connections between the carrier exit from the gas sampling valve and the column inlet and between the column exit and the thermal conductivity cell were made of 2 ft. lengths of $\frac{1}{16}$ in. o.d. stainless steel tubing and Swagelock connectors and reducers. A second gas chromatograph used for a portion of the study was constructed from a Perkin-Elmer Model 154C thermistor bead conductivity cell and a Perkin-Elmer gas sampling valve. A Whirlpool thermoelectric cooler was used together with a thermally activated switch to maintain the bath at low temperatures within ± 0.5°C. Columns were 12 ft. × $\frac{3}{16}$ in. or $\frac{1}{4}$ in. o.d. inconel or stainless steel packed straight and then formed into 6 in. diam. coils. Helium was used as the carrier gas. Figure 6 illustrates the calibration

and sample introduction system. A Bourdon vacuum gauge was introduced between
the stainless steel needle valve and the gas sampling valve with a stainless steel "T"
(Fig. 6) for the calibration of the gas chromatograph. Figure 7 illustrates a calibration
curve for NO_2, Cl_2, and NO_2Cl obtained by plotting the curves of peak height
versus millimeters pressure of component.

Tables 1–3 list, in order, effect of solid support on retention volume changes
with NOCl sample size, effect of liquid loading on retention volumes at $-10°C$,
and effect of temperatures on specific retention volumes of Cl_2, NOCl, NO_2Cl, and
NO_2. Decreased surface activity of the solid support and/or increased liquid loading

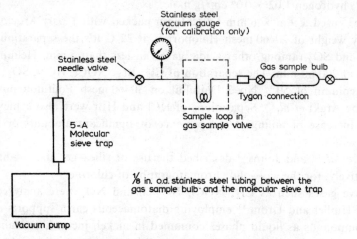

Fig. 6. Calibration and sample introduction systems.

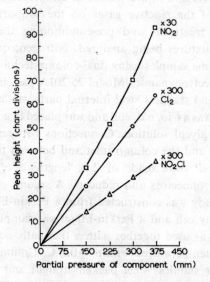

Fig. 7. Typical calibration curves (column 3).

reduced tailing and retention volume changes as the nitrosyl sample size became smaller (Table 2). Of all the columns studied, best results were obtained with columns of 30% Halocarbon 11-14 on Chromosorb P (AW-DHCS), 60–80 mesh or 10% Halocarbon 11-14 on Chromosorb W (AW-DMCS) 60–80 mesh. Aroclor 1232, although briefly investigated, was suggested to be possibly the best stationary liquid for this system while Teflon or Kel-F solid supports were unsatisfactory for this system. From the data on the effect of temperature on retention volumes (Table 3), −10°C was selected as the optimum temperature for gas–liquid partition chromatography of this system.

TABLE 1

EFFECT OF SOLID SUPPORT ON RETENTION VOLUME CHANGES WITH NOCl SAMPLE SIZE
Temperature: 10°C. Liquid Phase: Fluorolube MO-10 on 60–80 mesh supports.

Solid support	Liquid loading (wt.%)	$V'_g NOCl$ (ml/g) Sample size (mg)			Separation factor $V_R (NO_2Cl)/V_R (NOCl)$ Sample size (mg)		
		0.06	0.3	1.5	0.06	0.3	1.5
Firebrick, AW	12		47.7	39.0		0.87	1.01
	21		31.9	26.7		1.16	1.33
Neutraport S	9	57.8	46.7	36.0	0.70	0.83	0.98
Chromosorb P, AW, DMCS	12	30.9	27.2	24.0	1.09	1.25	1.31
	20	29.4	25.6	20.6	1.17	1.30	1.51
	30	28.8	26.8	22.8	1.23	1.32	1.50
Chromosorb W, AW, DMCS	10	22.2	22.2	22.6	1.29	1.29	1.31

TABLE 2

EFFECT OF LIQUID LOADING ON RETENTION VOLUMES
Temperature: −10°C.

Solid support	Liquid phase	Loading (%)	V'_g			
			Cl_2	NOCl	NO_2Cl	NO_2
Firebrick, AW	Fluorolube MO-10	12	14.7	39.0	39.6	167
		21	14.2	26.7	38.4	132
Neutraport S	Fluorolube MO-10	9	13.7	36.0	35.6	140
		16	13.0	26.3		110
Neutraport S	Halocarbon 13-21	13	12.8	32.3	32.3	109
		23	11.9	23.8	31.8	117
Chromosorb P, AW, DMCS	Halocarbon 11-14	20	14.4	25.8	36.5	130
		30	14.1	22.8	36.0	
Chromosorb P, AW, DMCS	Fluorolube MO-10	12	13.6	24.0	34.0	107
		20	12.4	20.6	33.7	107

TABLE 3

EFFECT OF TEMPERATURE ON SPECIFIC RETENTION VOLUMES

Solid support	Liquid phase	Loading (%)	Temp. (°C)	$V'_g (ml/g)$			
				Cl_2	$NOCl$	NO_2Cl	NO_2
Chromosorb P,	Halocarbon	12	52	4.1			6.9
AW, DMCS	11-14	12	22	6.8			28.5
		12	0	12.1			86.3
		12	−12	17.5			164
Chromosorb W,	Halocarbon	10	25	7.3	8.5	12.1	25.2
AW, DMCS	11-14	10	−10	13.7	23.6	36.6	148
Neutraport T	Halocarbon	10	25	6.5	8.9	15.0	20.6
	11-14	10	−10	14.5	22.9	40.8	145
Neutraport S	Fluorolube	9	−10	13.7	47.0	35.6	140
	MO-10	9	−18	16.4	51.1	43.1	182
Chromosorb P,	Fluorolube	12	−10	13.0	24.0	34.0	107
AW, DMCS	MO-10	12	−20	18.1	31.1	50.4	230
Firebrick AW,	Fluorolube	11	−10	12.3	22.2	34.4	109
RIC	MO-10	11	−20	14.7	27.4	43.2	212
Chromosorb P,	Arochlor	12	25	23.4	32.8	27.8	33.4
AW, DMCS	1232	12	−10	61.8	95.5	79.6	260

Figures 8 and 9 illustrate chromatograms of typical process samples and effect of liquid loading on $NOCl$–NO_2Cl separations, respectively. Results were calculated as follows: volume % = mm of component (from calibration curve) × 100/atmospheric pressure (mm).

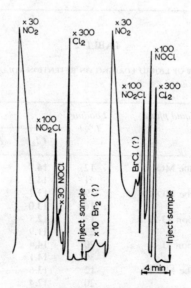

Fig. 8. Typical process samples.

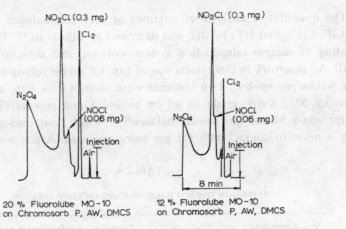

Fig. 9. Effect of liquid loading on NOCl–NO$_2$Cl separation.

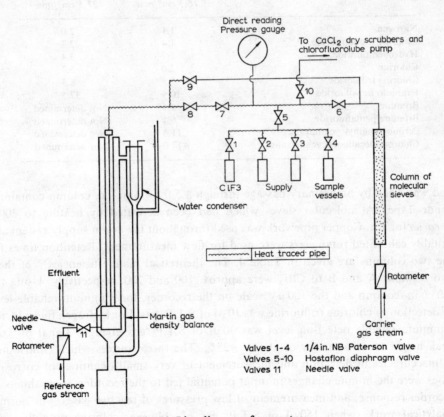

Fig. 10. Line diagram of apparatus.

The quantitative analysis of mixtures of corrosive halogen gases (*e.g.* UF_6, Cl_2, ClF, ClF_3, and HF) by GLC was described by Ellis *et al.*[32]. The line apparatus consisting of sample introduction system, column, and detector is illustrated in Fig. 10. All pipework in this system was of $\frac{1}{4}$ in. i.d. nickel tubing and all joints were either welded or brazed. Two columns were used, 4 ft. × 6 in. and 11 ft. × 6 in. containing 50% Kel-F grade 10 oil on ground Fluon powder (30–60 mesh). The detector was a Martin gas density balance[33,34] which contained a sensing element which is never in contact with the gas being analyzed. Argon was the carrier gas

TABLE 4

RETENTION TIMES OF HALOGEN GASES ON KEL-F COLUMNS

Compound	Retention time for elution peak maximum (min)	
	A 4 ft. 6 in. column with argon rate 16.5 cm³/min	B 11 ft. 6 in. column with argon rate 27.5 cm³/min
Nitrogen	1.1	2.0
Chlorine monofluoride	1.4	2.7
Hydrogen fluoride	1.7	3.4
Chlorine	2.0	5.7
Chlorine trifluoride	2.8	8.3
Uranium hexafluoride	10.5	37.5
Bromine	4.55	Not determined
Bromine pentafluoride	6.1	Not determined
Perfluoromethylcyclohexane	11.5	Not determined
Chloroundecafluorocyclohexane	17.0	Not determined

and was dried by preliminary passage through a 5 ft. × $\frac{1}{2}$ in. i.d. column containing Linde Type 4A molecular sieves which had been activated by heating to 3000° *in vacuo* for 1 h. Copper pipework was used throughout the argon supply system and suitably calibrated rotameters were used for flow measurement. Retention times for the two columns are given in Table 4. The theoretical plate efficiencies[35] of these two columns A and B to ClF_3 were approx. 100 and 200, respectively. Using the 4 ft. 6 in. column and the 100 μV scale on the recorder, the minimum reliable level of detection of chlorine trifluoride was 10 μl of S.T.P. gas. For hydrogen fluoride, the minimum reliable detection level was 30 μl of S.T.P. gas giving a signal of 7 μV peak height to an accuracy of about ±25%. The three sources which contributed to inaccuracies in detection and measurement of very small quantities of corrosive gases were the minute changes in input potential fed to the recorder, sluggishness of recorder response, and measurement of low pressures of reactive gases. For normal analytical work, when 150 μl of S.T.P. gas of chlorine, chlorine monofluoride, chlorine trifluoride, or uranium hexafluoride were present in any gas mixture sample, the amount present can be measured to within 1% of the determined value. For hydrogen fluoride, 150 µl samples could be determined within an accuracy limitation of ±5% of determined value. Accuracies ±1% with hydrogen fluoride could only be achieved with quantities exceeding 500 µl. A chromatogram of a synthetic mixture containing 25% each of ClF, Cl$_2$, ClF$_3$, and UF$_6$ obtained on an 11 ft. 6 in. column of Kel-F oil is illustrated in Fig. 11.

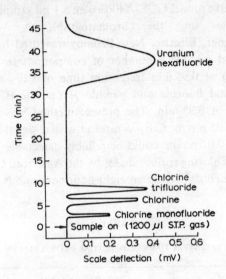

Fig. 11. Chromatogram of halogen gases separated on 11 ft. 6 in. column of Kel-F. Trace obtained from synthetic mixture (25%) of each component.

The gas chromatographic analysis of hydrogen chloride for trace impurities of organic substances was carried out by Dudorov and Angluilov[36]. A column 4 m × 4.5 mm in diameter was packed with 20% dinonyl phthalate on 1NZ-600 brick and used with nitrogen carrier gas at 2.7 l/h. Gaseous hydrochloric acid obtained by the reaction of H$_2$SO$_4$ and HCl (solution) contained 1–2 × 10^{-3} vol.% of methane, ethane, propane, isobutane, and methyl chloride.

Talbot and Thomas[37] separated NOCl and HCl using a 12 ft. column containing 50% dimethyl phthalate on Celite 545 with hydrogen as carrier gas.

The applicability of GC to the detection of the propellant oxidizer chlorine trifluoride in air has been studied by Zolty and Prager[38]. Sub-p.p.m. concentrations in an 1 ml air sample introduced directly into the chromatograph (without concentration procedures) could be detected rapidly by electron-capture detection. A linear response was obtained in the range of 0.05–1.0 p.p.m. volume and detectability of less than 0.01 p.p.m. was indicated. A sample introduced by an automatic gas sampling valve was detected by monitoring a peak with a retention time of 0.95 min.

A MicroTek Model 2000R gas chromatograph was equipped with a Lovelock type parallel plate design electron-capture detector containing a 130 mCi ^{3}H source and operated at 50 V Rf. without purge gas, and a pulse rate of 8 μsec on, 72 μsec off. The column was 10 ft. × $\frac{1}{4}$ in. stainless steel packed with 5% DC-200 (350 cS) on 60–80 mesh Chromosorb G used at room temperature with 5% methane–95% argon carrier gas at 200 ml/min. The carrier gas was dried by passage through a 5 ft. × $\frac{1}{4}$ in. o.d. copper tubing packed with molecular sieve 13X ($\frac{1}{16}$ in. pellets). The detector temperature was 180°C. A MicroTek seven-port air-operated automatic gas sampling valve model GCL-A4007 with a 1 ml sample volume was installed to introduce samples into the chromatograph. A 3-way solenoid valve (U935DEF2100, Skinner, Electric, New Britain) was used to supply the sampling valve with pressurized air. Of a number of compounds investigated for possible interference (Table 5) at 0.95 min (retention time of ClF_3 under the above conditions), only elemental fluorine and perchloryl fluoride (ClO_3F) were found to have retention times of 0.95 min. The present method was shown to be suitable for the detection of 0.05 p.p.m. ClF_3 in air and with a signal-to-noise ratio of 28 at 0.05 p.p.m., less than 0.01 p.p.m. could be reliably detectable. The maximum allowable concentration of chlorine trifluoride set by the American Conference of Governmental Industrial Hygienists[39] from an eight-hour continuous exposure is 0.1 p.p.m.

TABLE 5

RETENTION TIMES OF POTENTIAL INTERFERENCES

Substance	Boiling point (°C)	Retention time of largest peak (min)
Air (oxygen)	(−183)	0.8
Carbon dioxide*	−78	0.8
Tetrafluoromethane (Freon 14)*	−128	0.8
Chlorotrifluoromethane (Freon 13)	−81	0.8
Fluorine	−188	0.95
Perchloryl fluoride	−48	0.95
Chlorine trifluoride	11.75	0.95
Chlorodifluoromethane (Freon 22)	−41	1.12
Dichlorodifluoromethane (Freon 12)	−30	1.15
Chlorine	−34.5	1.4
Dichlorofluoromethane (Freon 21)	9	2.2
Trichlorofluoromethane (Freon 11)	24	4.3
Trichlorotrifluoroethane (Freon 113)	48	4.5
Water	100	8.6
Hydrogen fluoride	19	No response
Hydrogen chloride	−85	No response
Sulfur dioxide	−10	No response to 100 p.p.m.
Gasoline		No response

* Very poor response.

2. FLUORINE AND HYDROGEN FLUORIDE

Fluorine is the most reactive element of the periodic table, combining with all of the elements, even with some of the noble gases. Combined fluorine is widely found in nature and is estimated to be approx. 0.065% by weight of the earth's crust. Among the elements it is about thirteenth in order of abundance and is 5–10 times more abundant than zinc and copper and approx. 30 times more so than lead. The most important industrial source is the mineral fluorspar (fluorite, CaF_2) which contains about 49% fluorine and is widely distributed. Cryolite, Na_3AlF_6, is another fluorine-rich mineral containing about 54% fluorine. Variable amounts of fluorine (3–4%) are also contained in phosphate rock (fluorapatite) $(Ca_{10}F_2(PO_4)_6)$. With a 3% fluorine content, phosphate rock constitutes the world's largest fluorine reserve. Traces of fluorine are found in most natural water and in cases such as sea water the levels are approx. 0.3 mg/l.

Fluorine is commercially produced by the electrolysis of a solution of anhydrous potassium bifluoride electrolyte containing various concentrations of free hydrogen fluoride. The fluoride ion is anodically oxidized to elemental fluorine while hydrogen is liberated at the cathode. The principal raw material for fluorine production is high-purity anhydrous hydrofluoric acid which is obtained by reacting fluorspar with sulfuric acid.

The first major use of elemental fluorine was for the separation of uranium-235 from uranium-238 by gaseous diffusion of uranium hexafluoride, UF_6. The use of fluorine in the form of UF_6 is also the basis for the production of enriched fuel in the nuclear electric-power industry. Elemental fluorine is also used in the manufacture of sulfur hexafluoride, SF_6, which is used as a gaseous insulator for electrical and electronic equipment. Potential uses exist for the application of fluorine derivatives as chlorine trifluoride in the areas of cutting of oil-well casings and along with oxygen difluoride (and fluorine itself) as rocket fuel oxidizers.

Fluorine gas and liquid fluorine are extremely corrosive and irritant to skin tissue. It has been reported[40] that fluorine is more toxic than HF with chronic exposure "tolerance" levels of 1 and 7 p.p.m., respectively. The chronic effects of assimilation of fluorine and fluorides as normally encountered are limited to fluorosis.

Hydrogen fluoride is produced from the reaction of acid-grade fluorspar (calcium fluoride, CaF_2) and sulfuric acid. In volume of production, hydrogen fluoride (hydrofluoric acid, HF) is the most important manufactured compound of fluorine (both the anhydrous and aqueous acids are used directly and as intermediates in forming other fluorine-containing products). The estimated U.S. production of hydrogen fluoride (based on 100% HF content) in 1951 was 62,800 short tons. Hydrogen fluoride has been used in aluminum production for making the aluminum fluoride and synthetic cryolite used in the electrochemical reduction of aluminum oxide. The demand for hydrogen fluoride in the manufacture of fluorinated organics for aerosol propellants, foaming agents, cleaning fluids, and plastics has continued to rise steadily during the last 10 years. Hydrogen fluoride is also used in the petroleum industry as an alkylation catalyst, to produce high octane blending stock for automobile gasolines. Other uses of hydrogen fluoride include: in pickling acids for stainless steel production, etching and polishing of glass, preparation of fluorides, fluoroborates, fluorosilicones, manufacture of fluorine and products derived therefrom, special dyes and pharmaceutical preparations of microelectronic circuits, and manufacture of uranium hexafluoride. The most important sources of fluorine-containing air pollution are the following: (*1*) plants for the production of aluminum by dissolution in fused cryolite and subsequent electrolysis. HF and SiF_4 are generated as waste gas at the anodes *via* reaction with the anodic materials coke and coal-tar pitch. The amount of fluorine produced is *ca.* 6–8 kg/ton aluminum, (*2*) HF can be discharged into the atmosphere at iron works when fluorite (CaF_2) is added to the scrap in order to produce low-boiling slags. In the U.S., iron ores themselves have been found to contain appreciable amounts of fluorides, (*3*) plants for phosphate fertilizers where the crude phosphate consists mainly of fluorapatite $[3Ca_3(PO_4)_2 \cdot CaF_2]$ and contains up to 4% fluorine. When the crude phosphate is decomposed with sulfuric acid, a portion of the fluorine escapes as HF or SiF_4. It has been estimated that over 70,000 tons of fluoride per year, largely as SiF_4, escape into the atmosphere where they react with water to form hydrogen fluoride and hydrofluorosilicic acid[41], (*4*) the use of fluoride-coated welding rods constitutes a definite hazard[12], (*5*) in magnesium founding where fluorides inhibit oxidation when sprayed upon cores or mixed with the core sand to the extent[43] of 4–10%. Fluorides also act as fluxes when added to the melting pots.

The highest concentration of hydrogen fluoride that can be tolerated by man for 1 min is 100 mg/m³ of air (which causes some degree of conjunctivitis and respiratory irritation[44,45]. Chronic injuries can be caused by HF or SiF_4 or by fluorine-containing dusts. The symptoms are weakening of the bone substance (osteomalacia) and teeth caused by disturbances in calcium metabolism. Fluorine in the air (even present in p.p.b. amounts) may also lead to appreciable damage to vegetation.

It has been shown that exposure of *Drosophila melanogaster* to low hydrogen fluoride concentrations can cause genetic damage and that damage is accumulative as the exposure period is increased[46,47]. The cytological effects of hydrogen fluoride on tomato chromosomes have also been reported[48].

Hydrogen fluoride in air has been determined colorimetrically[49,50] and by titrimetric[54] methods. An automatic recorder for very small fluorine concentrations (*e.g.* 1–35 μg/m³) in the atmosphere has been described[52,53] which is based on the spectrophotometric recording of the decolorizing reaction of the color lake of zirconium eriochrome cyanine by fluorine. The determination of fluoride in biological materials has been achieved by spectrophotometric[54,55], fluorimetric[56], polarographic[57], and enzymic techniques[58].

Rochefort[59] studied the gas chromatography of corrosive fluorine compounds such as UF_6, HF, and F_2 using one or several chemical precolumns placed in series. The "precolumn" system retained or destroyed the corrosive compounds and the chromatography column separated the components formed. One precolumn was of 10 cm containing anhydrous sodium chloride dried under vacuum at 350 to 400° for 3 h. The chromatographic column was 2 m of Voltalef No. 3 oil on Voltalef 300 LD resin (40–70 mesh). The columns were maintained at 60–70° with helium as carrier gas and detection was accomplished with a katharometer.

Lysyj and Newton[60] evaluated 5 gas chromatographic columns in respect to the separation of 14 fluorinated materials (*e.g.* SiF_4, CF_4, NF_3, HF, SF_6, $CClF_3$, F_2, CBF_3, ClF_3, Cl_2, CCl_2F_2, C_2F_3Cl, CBr_2F_2, and $CHCl_2F$). A gas chromatograph resistant to corrosive fluorinated materials was designed and custom built using only monel and stainless steel parts. In addition, Teflon-clad hot wires were used in the thermoconductivity detector. (No distortion of the elution peaks was observed, although the sensitivity of the detection system was reduced by about a factor of 4.) Figure 12 illustrates the gas chromatograph used for analysis of fluorinated materials. A Beckman two-way gas sampling valve (with 2 vertical stainless steel U-tubes, 4 in. × $\frac{1}{8}$ in. diam.) was used for sample introduction. The collecting manifold incorporated 3 stainless steel spirals made of $\frac{1}{4}$ in. diam. tube each 2 ft. long. The detector was a Gow-Mac 4 hot wire thermoconductivity detector, Model 9285 (Pretzel) with a conventional Wheatstone bridge electrical circuit arrangement (the hot wires were coated with Teflon). A single-stream flow system (consisting of $\frac{3}{16}$ in. stainless steel tubing) was used. The carrier gas passed through the reference side of the detector, through the Beckman sampling valve, the column, and then the detecting side of the thermal conductivity cell. The columns (except the gel column)

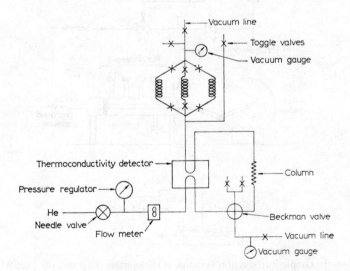

Fig. 12. Gas chromatograph used for analysis of fluorinated materials.

were 10 ft. × ¼ in. diam. stainless steel packed with 30% liquid phase on 70% Chromosorb W, 80–100 mesh.

The best separation of fluorinated materials was achieved using a gel column containing 50% Kel-F polymer plasticized with 50% Halocarbon oil. Using this column, relative retention volumes were obtained which were approximately one order of magnitude higher than the retention volumes for the same compounds obtained on a conventional column under the same operating conditions.

The quantitative separation of F_2, HF, Cl_2, ClF_3, and Freon-114 was accomplished by Million et al.[61] using a 5 ft. × ½ in. o.d. column of 20% Kel-F No. 10 oil on 40–50 mesh Kel-F powder at 60°C. Air was used as the carrier gas at a flow of 24 ml/min and the lower detection limits were 0.2–0.5 micromole of each individual gas when using a sampling chamber with a volume of 31.1 ml; a gas density balance was used as a detector.

Grudzinowicz and Smith[62] used a radioactivity detector (krypton-85 quinol clathrate) to determine 1–20 p.p.m. of fluorine in air. Other gases that possess a similar reactivity to the krypton clathrate include chlorine, bromine, chlorine dioxide, nitrogen dioxide, OF_2, NO_2F, and NO_2Cl.

The analysis of volatile inorganic fluorides by GLC has been studied by Hamlin et al.[63] using specialized equipment made from nickel plate[64]. (The corrosion resistance imparted by this plate was as good as that of pure nickel.) Figures 13 and 14 illustrate the automatic sampling system, the schematic diagram of the com-

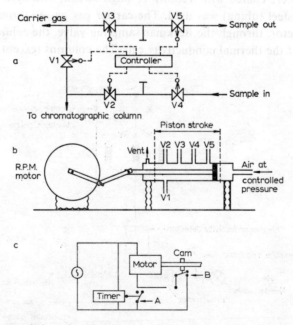

Fig. 13. Automatic sample introduction systems. (a) Schematic diagram; (b) pneumatic operator; (c) electrical control.

plete chromatograph and ancillary valves and the schematic diagram of the split-column chromatograph, respectively. The 14 ft. column operated at 90°C contained (1:5) Kel-F 40 oil on Kel-F 300 low density molding powder (178–211 μm). Nitrogen was the carrier gas at 10 ml/min. Detection was achieved with a specially constructed katharometer containing a nickel wire sensing element. Table 6 lists the retention times of halide gases relative to chlorine. The table shows that, in general, the retention time was related to the volatility of the substance concerned with the exception of hydrogen fluoride. The mixtures usually analyzed with this equipment consisted

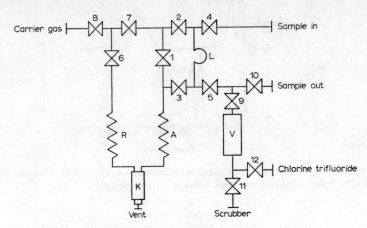

Fig. 14. Schematic diagram of complete chromatograph and ancillary valves.

TABLE 6

RETENTION TIMES OF HALIDE GASES RELATIVE TO CHLORINE

Operating conditions: column temperature, 60°C; carrier gas, nitrogen at 48 ml/min; column length, 14 ft.

Substance	Relative retention time	Boiling point/sublimation temperature (°C)
Uranium hexafluoride	6.0	+56.5
Bromine	2.6	+63
Chlorine monoxide	1.76	+2
Chlorine trifluoride	1.45	+11
Chlorine dioxide	1.45	+11
Chloryl fluoride	1.28	−6
Chlorine	1.0	−33.7
Perchloryl fluoride	0.85	−46
Hydrogen fluoride	0.60	+19
Chlorine monofluoride	0.48	−100
Fluorine monoxide	0.40	−144
Fluorine	Not retained	−188

of uranium hexafluoride, chlorine, chlorine trifluoride, hydrogen fluoride, chlorine monofluoride, and permanent gases. Of these, only hydrogen fluoride and chlorine monochloride were difficult to resolve and if one or the other was present as a trace, quantitative estimation of this was difficult.

Figure 15 illustrates chromatograms of samples containing uranium hexafluoride obtained both with an ordinary and a split-column chromatograph (Fig. 16).

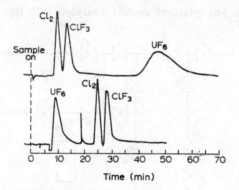

Fig. 15. Chromatograms of samples containing uranium hexafluoride. Above, obtained with ordinary chromatograph; below, obtained with split column chromatograph.

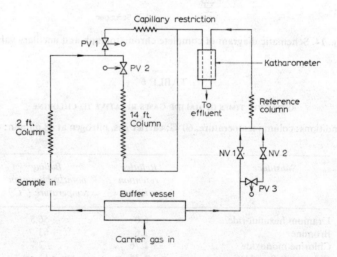

Fig. 16. Schematic diagram of split column chromatograph.

The analysis of F_2, HF, NF_3, trans-N_2F_2, and N_2F_4 mixtures by gas chromatography has been reported by Spears[65]. An F & M Model 720 gas chromatograph with nickel block and Teflon-coated thermistor filaments and a 0–1 mV strip chart recorder with a 1 sec full-scale response was used. Helium, 99.95% pure, was the carrier gas and before entering the GC column it was passed through a 10 ft. × $\frac{1}{4}$ in.

TABLE 7

RETENTION TIMES

Col.	Helium flow rate (ml/min)	Col. temp. or program rate	Retention time				
			Air	F_2	NF_3	N_2F_4	HF
1	50	3.2°C/min	5.5	5.7	2.2	29	70.4
2	50	10.0°C/min	a	0.2	5.9	11.6	31.0
	(both cols.)						
3	65	40°C	0.8	0.8	1.0	a	2.4
4	60	60°C	3.7	b	6.5	b	b
5	45	−72°C	7.0	7.0	7.5	c	c
6	50	Ambient	3.9	4.0	4.3	a	6.3
	(sample col.)						
	77						
	(ref. col.)						

a No data recorded for this gas.
b Reacted with silica gel.
c Boiling points too high.

TABLE 8

PERFORMANCE OF 7 ft. MONEL COLUMN

Same column conditions as No. 2 in Table 7.

Compound	No. of theoretical plates	Column efficiency HETP (cm)	Retention time (min)	Lower limits of detection (p.p.m.)
F_2	0.64	334	0.2	210
NF_3	843	0.253	5.9	180
trans-N_2F_2	675	0.316	7.8	270
N_2F_2	537	0.396	11.6	250
HF	656	0.325	31.0	220

TABLE 9

PERFORMANCE OF 70 ft. COLUMN

Same column and conditions as No. 6 in Table 7.

Compound	No. of theoretical plates	Column efficiency HETP (cm)	Retention time (sec)	Lower limits of detection (p.p.m.)
F_2	995	2.15	240	93.5
NF_3	1250	1.72	260	86.5
trans-N_2F_2	388	5.52	295	88.4
HF	776	2.76	380	105

o.d. copper tubing filled with molecular sieves and then through another 10 ft. $\times \frac{1}{4}$ in. copper tube coil immersed in liquid nitrogen. The six columns used for analyzing mixtures of F_2, HF, NF_3, trans-N_2F_2, and N_2F_4 were: column 1: 16 ft. $\times \frac{1}{4}$ in. o.d., passivated monel tube, used for cryogenic programming; column 2: two 7 ft. $\times \frac{1}{4}$ in. o.d. monel tubes, one for a reference and one for a sample column. The sample column was passivated and both columns were used for cryogenic programming; column 3: 20 ft. $\times \frac{1}{4}$ in. o.d. monel column packed with 50% Halocarbon oil No. 13-21 on 30–50 mesh shredded Teflon; column 4: 10 ft. $\times \frac{1}{4}$ in. o.d. alumina column packed with 30–60 mesh silica gel; column 5: 50 ft. $\times \frac{1}{8}$ in. o.d. Kel-F tubing, packed with 50% Halocarbon oil No. 13-21 on 30–50 mesh Kel-F molding powder; and column 6: a combination of columns 3 and 5. A 7 ft. $\times \frac{1}{4}$ in. o.d. monel tube was used as a reference column. Table 7 gives retention times for mixtures of air, F_2, NF_3, N_2F_4, and HF. Column 2 appeared to be suitable for cryogenic programming and column 6 for conventional analysis. Silica gel (column 4) was useless because F_2, N_2F_4, and HF reacted with it during analysis. Tables 8 and 9 show in more detail the performances of columns 2 and 6, respectively. The numbers of theoretical plates (n) shown in these tables were calculated using the formula

$$n = 16T_R/(T)^2$$

where T_R is the absolute retention time and T the width of peak base in time. The column efficiencies (HETP) (in cm) were calculated using the formula

$$\text{HETP} = L/n$$

where L is the length of the column (cm).

Columns 2 and 6 were recommended for the fast analysis of gases such as F_2 and HF. Column 2 was selected to separate and trap the various electrolysis products of ammonium bifluoride in anhydrous hydrogen fluoride for IR and mass spectral analysis, while column 6 was used for rapid quantitative analysis of the volatile electrolysis products formed as a function of time.

The gas chromatographic analysis of reactive gases such as HF and Cl_2 is generally performed on very inert supports such as the fluorine-containing polymers (Kel-F, Teflon, and Fluoropak). Although these materials have the advantage of high chemical stability, they possess a number of inherent properties which limit their utility. These properties are poor wettability which hinders the formation of a uniform layer of the liquid phase and the presence of electric charges on their surfaces which make regular column packing difficult.

The GC properties of graphitized carbon blocks as solid adsorbents for the separation of some reactive gases were studied by DiCorcia et al.[66]. It was found that by treating graphitized carbon blocks with hydrogen at 6000°C, very polar compounds could be eluted and peak tailing and "ghosting" phenomena could be avoided permitting the GLC of gases such as halogen hydrides and boron trifluoride.

A Carlo Erba gas chromatograph equipped with a thermal conductivity detector with gold filaments was used. Every contact of the reactive gases with metals was avoided by inserting Teflon tubes into the gas lines connecting the samples with the column and the latter with the detector block. Graphon, a graphitized carbon black with a surface area of about 100 m^2/g was treated according to the procedure of DiCorcia and Bruner[67]. This procedure eliminated the active centers constituted by surface oxygen complexes which are always present on an adsorbent surface. HCl and HF were separated from air on a 2 m × 3 mm column of Graphon coated with 40% w/w of a 1:1 mixture of polyfluoroether (Y/25-Montecatini Edison, Milan)–benzophenone. Benzophenone, which is a weak Lewis base, acts as the adsorbate and enhanced the separation between HCl and HF while the fluorinated

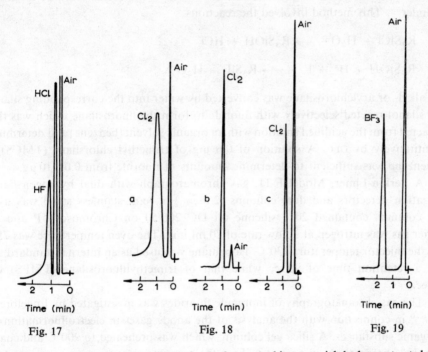

Fig. 17. Elution of HCl and HF. Column, 2 m × 3 mm; packing material, hydrogen-treated Graphon +20% polyfluoroether +20% benzophenone; temperature, 70°; carrier gas, helium; flow rate, 40 ml/min.

Fig. 18. Elution of Cl_2. Column, 2 m × 3 mm; packing material, (a) Graphon + 1% H_3PO_4; (b) Graphon + 5% H_3PO_4; (c) Graphon + 10% H_3PO_4; temperature, 60°; carrier gas, helium; flow rate, 50 ml/min.

Fig. 19. Elution of BF_3. Column 2 m × 3 mm; packing material, Graphon + 15% polyfluoroether + 5% benzophenone, temperature, 96°C; carrier gas, helium.

liquid was used to disperse the benzophenone on the support surface. Chlorine in air was analyzed on a column of Graphon coated with phosphoric acid at various percentages. (No hydrogen treatment of the Graphon was necessary.)

Boron trifluoride has been extremely difficult to analyze by GC because of its high reactivity and corrosiveness. BF_3 is a very strong Lewis acid and forms very stable complexes with molecules containing electron donor atoms such as oxygen, nitrogen, and sulfur. BF_3 was successfully eluted from a Graphon column coated with 15% polyfluoroether and 5% benzophenone. (The Graphon was previously treated with hydrogen at 1000°.) The satisfactory elution of BF_3 suggests its use with suitably efficient columns for the enrichment of boron isotopes.

Figures 17–19 illustrate chromatograms of the elution of HCl and HF, Cl_2, and BF_3, respectively, on 2 m × 3 mm columns of coated Graphon.

A gas chromatographic method was developed by Fresen et al.[68] for the determination of fluoride in biological materials based on the earlier work of Bock and Semmler[69]. This method involved the reactions

$$R_3SiCl + H_2O \longrightarrow R_3SiOH + HCl$$

$$R_3SiOH + H^+ + F^- \longrightarrow R_3SiF + H_2O$$

The alkyl- or arylchlorosilane was converted by water into the corresponding silanol. The silanol reacted selectively with fluoride to form the fluorosilane which was then extracted from the acidified solution with an organic solvent (benzene) and determined quantitatively by GLC. A solution of 0.6 mg of trimethyl chlorsilane (TMCS)/ml of benzene was sufficient to determine amounts of fluoride from 0.01–10 μg.

A Perkin-Elmer, Model F-11, gas chromatograph with dual hydrogen flame-ionization detectors and dual columns (2 m × $\frac{1}{8}$ in. o.d. stainless steel) was used. The columns contained 20% silicone oil DC 200/50 on Chromosorb P and the carrier gas was nitrogen at a flow rate of 20 ml/min. The oven temperature was 75°C and the injector temperature 130°C. Isopentane was used as an internal standard and had a retention time of 90 sec while that of trimethylfluorosilane (TMFS) was 65 sec.

The gas chromatography of inorganic fluorides was investigated by Engelbrecht et al.[70] in connection with the analysis of the anode gases in electrofluorinations of inorganic substances. A silica gel column, which was preheated to 900°C and coated with 30% halocarbon oil proved to be sufficiently resistant to most corrosive gases while still affording good separations. A Perkin-Elmer Type 116E gas chromatograph was used with a flame-ionization detector. Table 10 lists the corrected retention volumes for 15 gases obtained on a 4 m × 6 m column containing 30% Halocarbon oil on silica gel. The column temperature was 30° and the carrier gas was hydrogen at 150 ml/min. Figure 20 illustrates a chromatogram of a mixture of SO_2F_2, SiF_4, SF_4, NF_3, OF_2, N_2, and O_2 obtained with the operating conditions above.

TABLE 10

RETENTION VOLUMES OF GASES OBTAINED ON A 4 m × 6 mm COLUMN OF SILICA GEL COATED WITH 30%
HALOCARBON OIL
Column temperature, 30°C.

Gas	Corrected retention volume (ml)
NO	12.5
OF_3	21.0
NF_3	32.5
SiF_4	57.4
PF_5	96.0
HCl	110
CO_2	121
N_2O	155
SF_4	155
H_2S	220
SO_2F_2	356
ClO_3F	357
SF_4, SOF_2	484
Cl_2	509
SO_2	1050

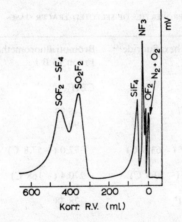

Fig. 20. Chromatogram of a mixture of SO_2F_2, SiF_4, NF_3, OF_2, N_2, and O_2 obtained with a silica gel–30% Halocarbon oil column at 30°C.

Saltzman et al.[71] described the stability and ultrasensitive analysis of a number of halogenated compounds used as gaseous meteorological tracers, e.g. sulfur hexafluoride, bromotrifluoromethane and octafluorocyclobutane. Two systems were used: (A) to provide as much sensitivity as possible, and (B) to study the stability of the materials at higher concentrations under simulated atmospheric conditions. Both systems utilized MicroTek 7-port sampling values with Teflon inserts. The columns used consisted of: (A) 14 ft. × ⅛ in. o.d. stainless steel tubing packed with 40–60 mesh Baymal (colloidal alumina)[72], maintained at 50°C in a MicroTek Model 2500 gas chromatograph. The electron-capture detector was a Barber–Colman Model 5120 with adjustable anode spacing. This device was modified by capping the scavenger gas connection to the cathode and bypassing the linearizing resistor. System B utilized a column consisting of two sections of ⅛ in. o.d. stainless steel tubing. The first section 3.9 ft. long was packed with 40–60 mesh 3A molecular sieve (to remove the relatively large quantities of interfering water vapor from some samples). The second section, 10 ft. long, was packed with 50–60 mesh Baymal and maintained at 58°C in a Barber–Colman bath, Model 5060. The electron-capture detector for system B was also a Barber–Colman Model 5120 with the electrode spacing set at 1 cm. Scavenger gas (5% hydrogen in argon) was used at 50 ml/min. The carrier gas was prepurified nitrogen at 75 ml/min. The gas was purified by passage through a 150 ml cylinder containing 5A molecular sieve. The electron-capture detector was operated with 14 V applied across the detector circuit (which included the linearizing resistor); the operating standing current was 5.3 nA. Sensitivity of 10^{-5} p.p.m. was achieved for sulfur hexafluoride without concentration of the sample and the components in air were determined in a single 10 min run. Table 11 lists the properties of the

TABLE 11

PROPERTIES OF SELECTED TRACER GASES

Name	Sulfur hexafluoride	Bromotrifluoromethane	Octafluorocyclobutane
Synonym		Freon 13 B 1	Freon-C 318, Perfluorocyclobutane
Formula	SF_6	$CBrF_3$	$CF_2CF_2CF_2CF_2$
Molecular weight	146.07	148.93	200.04
Vapor pressure at 70°F (p.s.i.g.)	310	190	25
Boiling point at 1 atm (°F)	−82.8[a] (−63.8°C)	−72.0 (−57.8°C)	21.1 (−6.04°C)
Freezing point at 1 atm (°F)	−59.4 (−50.8°C)	−270.4 (−168°C)	−42.5 (−41.4°C)
Heat of vaporization at normal b.p. (cal/g)	38.61[b]	28.38	27.67
Critical temperature (°F)	114 (45.5°C)	152.6 (67.0°C)	239.5 (115.3°C)
Critical pressure (p.s.i.a.)	545.5 (37.1 atm)	574.8 (39.1 atm)	393 (26.7 atm)
Uses	Gaseous insulating medium	Chemical intermediate, fire-extinguishing agent	Gaseous dielectric, foam-producing agent, propellant

[a] Sublimes.
[b] Heat of sublimation.

selected tracer gases (sulfur hexafluoride, bromotrifluoromethane, and octafluoro-cyclobutane). Figure 21 depicts a chromatogram of a ternary tracer mixture containing SF_6, $CBrF_3$, and C_4F_8 and Fig. 22 illustrates chromatograms of responses of systems A and B to a 0.25 ml sample of ternary tracer mixture in air.

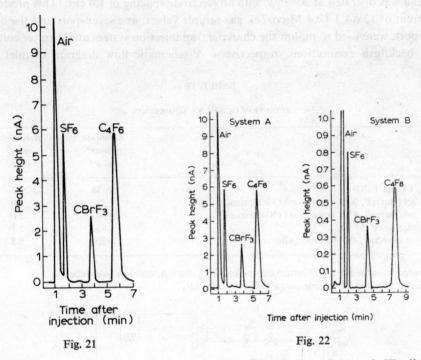

Fig. 21
Fig. 22

Fig. 21. Chromatogram of a ternary tracer mixture containing approximately 6.9 p.p.b. SF_6, 62 p.p.b. $CBrF_3$, and 183 p.p.b. C_4F_8.

Fig. 22. Responses of system A and system B to a 0.25 ml sample of ternary tracer mixture in air. Approximate mixture concentrations (p.p.b.) were: SF_6, 6.9; $CBrF_3$, 62; C_4F_8, 183. Note the differing electrometer scale settings (vertical scales).

Dilutions of SF_6 in air were recently determined by Clemons et al.[73] down to a sensitivity of 1 part of SF_6 in 10^{14} parts of air. The samples were concentrated by an adsorption–desorption process using an activated charcoal trap. The desorbed SF_6 was determined by means of GC using an electron-capture detector. The ultra-sensitivity of the method required that the carrier and sample diluent gases be free of SF_6 and other electron-absorbing impurities. A MicroTek 2500 series gas chromatograph was equipped with a Barber–Colman Model 5120 adjustable anode electron-capture detector, or Keithley Model 417 Picoammeter, a MicroTek polarizing voltage power supply providing 0–50 V d.c. and a Minneapolis-Honeywell Series 1531 mV recorder. The column in two sections consisted of 15 ft. $\times \frac{1}{8}$ in. o.d. stainless steel tubing packed with 40–60 mesh (Baymal). The first 3 ft. of the column was used as a cutter column and the remaining 12 ft. as an analytical column. The system was optimized as previously described by Clemons et al.[73] The column and detector were operated at ambient temperature. The carrier gas was 5 % hydrogen in argon at 75 ml/min; the electron-capture detector contained a 500 mCi tritium foil and was operated at 20–25 V with an electrode spacing of 1.4 cm. (This produced a current of 12 nA.) Two MicroTek gas sample valves, one seven-port and the other nine-port, were used to mount the charcoal trap injection system and a cutter column with backflush connections, respectively. A schematic flow diagram of inlet and

TABLE 12

RETENTION OF SF_6 BY ADSORBENTS

Adsorbent	Retention volume[a] (ml/g)	Retention mass[b] (ng/g)
Al_2O_3 (Alcoa F-10)	88	0.2
Silica gel (Burrell, 300° activation) (50–60 mesh)	333	1.0
Silica gel (Burrell, 450° activation) (50–60 mesh)	667	2.0
Molecular sieve (13×)	833	2.5
Coconut charcoal, (50–60 mesh) (Burrell)	3200	9.5

[a] Adjusted to value for 1 g of adsorbent packed in $\frac{1}{8}$ in. o.d. stainless steel tubes.
[b] Adjusted to value for sample containing 0.5 p.p.b. SF_6.

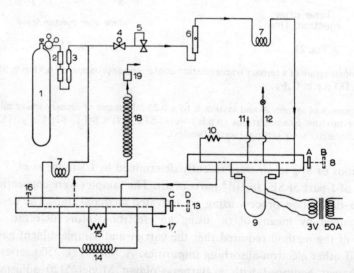

Fig. 23. Flow schematic of inlet and trapping system. 1, Carrier gas; 2, canisters of BTS catalyst; 3, canisters of molecular sieve 5A; 4, 5, flow controls; 6, flowmeter; 7, charcoal traps for purification of carrier; 8, seven-port valve; 9, sample trap with heater; 10, matched flow resistance; 11, sample entry; 12, exit to vacuum source through needle valve control; 13, nine-port valve; 14, cutter column; 15, flow resistance matched to cutter column; 16, capped port; 17, exit for backflush; 18, analytical column; 19, detector.

trapping system is shown in Fig. 23. Table 12 lists the retention values for a 0.5 p.p.b. SF_6 mixture by 5 adsorbents.

The separation of some fluorocarbon and sulfur fluoride compounds by GLC was studied by Campbell and Gudzinowicz[74]. Perkin-Elmer Model 154B and 154C Vapor Fractometers were employed with the following columns: column A, 6 ft. × ¼ in. containing diisodecyl phthalate as stationary phase and 6-, 9-, and 20 ft. × ¼ in. copper columns packed with 33% Kel-F No. 3 loaded on 35–80 mesh Chromosorb W. Helium was the carrier gas at 37 ml/min.

Table 13 lists the relative retention volumes corrected for void volume and pressure drop for several fluorocarbons and S_2F_{10} compared at the various operating conditions shown. The relative retention volumes of a number of S–F and C–F columns using a 20 ft. column of 33% Kel-F on 35–80 mesh Chromosorb W at 25° with a helium flow of 37 ml/min are shown in Table 14. The longest retention time of the sulfur–fluoride compounds studied was 17.0 min for S_2F_{10}. Figure 24 shows a gas chromatogram of typical components in SF_5Cl production obtained on a 20 ft. column of 33% Kel-F at 25° and Fig. 25 illustrates a chromatogram of sulfur halide compounds (obtained as in Fig. 24) identified by trapping and infrared techniques. The chromatographic operating conditions suitable for S–F mixtures were also applied to the pyrolytic degradation products of Teflon. The separation of a representative crude sample of such C_1–C_4 fluorocarbons is shown in Fig. 26. The separation of several of the lower-boiling components of a mixture (SF_6, SF_4, C_3F_6, cyclic-C_4F_8, and S_2F_{10}) are shown in Fig. 27.

Juvet and Fisher[75] described the gas chromatography of fluorinated derivatives of S, Mo, Se, W, Te, and U which were prepared *in situ* by reaction of the elements.

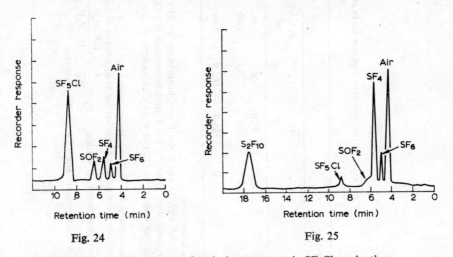

Fig. 24

Fig. 25

Fig. 24. Gas chromatogram of typical components in SF_5Cl production.

Fig. 25. Gas chromatogram of sulfur–halide compounds identified by trapping and infrared techniques.

TABLE 13

RELATIVE RETENTION VOLUMES FOR SEVERAL FLUOROCARBONS AND S_2F_{10} AT VARIOUS OPERATING CONDITIONS

$CF_3CF{=}CF_2 = 1.00$

Compound	A^a				B^a				C^a			
	t_m^b	V_R^c	$V_R^{0\,d}$	Rel. ret. vol.	t_m	V_R	V_R^0	Rel. ret. vol.	t_m	V_R	V_R^0	Rel. ret. vol.
$CF_2{=}CF_2$	2.5	81.0	64.6	0.86	3.8	112.5	77.6	0.79	3.9	76.5	55.8	0.86
$CF_3CF{=}CF_2$	2.9	94.2	75.1	1.00	4.8	142.5	98.3	1.00	4.5	88.5	64.6	1.00
Cyclic-C_4F_3					5.4	160.5	110.7	1.13	5.0	98.5	71.9	1.11
S_2F_{10}	7.0	229.5	182.9	2.44	11.4	340.5	234.9	2.39	9.0	178.5	130.3	2.02

[a] Operating conditions:

A. 9 ft. column with 33 wt.-% Kel-F on Chromosorb W; 33°C column temperature; 33 ml/min helium flow rate at 1 atm and 25°C.

B. 15 ft. column with 33 wt.-% Kel-F on Chromosorb W; 33°C column temperature; 30 ml/min helium flow rate at 1 atm and 25°C.

C. 6 ft. column with 33 wt.-% Kel-F on Chromosorb W plus 6 ft. column A; 29°C column temperature; 20 ml/min helium flow rate at 1 atm and 25°C.

[b] t_m = Observed time for peak maximum (min).

[c] V_R = Total retention volume at column temperature.

[d] V_R^0 = Corrected retention volume corrected for void volume at column temperature.

$$V_R^0 = \text{Corrected retention volume} = V_R \times \frac{3}{2} \times \frac{[(P_1/P_0)^2 - 1]}{[(P_1/P_0)^3 - 1]}$$

where P_1 = inlet pressure and P_0 = outlet pressure.

The fluorinated compounds were separated on a 22 ft. × ¼ in. o.d. PTFE column containing 15 % Kel F-10 on 40–60 mesh Chromosorb T at dry ice–acetone column temperatures with helium as carrier gas at 28 ml/min. Since Kel-F oil No. 10 begins to solidify at −6°C, separation of SF_6. SeF_6, and TeF_6 at dry ice–acetone temperatures was suggested to be due to gas–solid adsorption rather than to partitioning.

The responses of a number of halogenated compounds to electron-capture detection were reported by Clemons and Altschuller[76]. A Model 2500 Micro-Tek dual-column gas chromatograph was equipped with an electron-capture detector mounted in parallel with a flame detector by use of a column effluent splitter. The

TABLE 14

RELATIVE RETENTION VOLUMES FOR SOME S–F AND C–F COMPOUNDS
$SOF_2 = 1.00$

Compound	t_m	V_R	V_R^o	Rel. retention volume
SF_4	5.40	198.3	126.5	0.80
SF_6	4.95	181.7	115.9	0.74
SOF_2	6.70	246.4	157.2	1.00
SF_5Cl	8.50	313.0	199.7	1.27
S_2F_{10}	17.00	627.5	400.3	2.55
CF_4	4.50	165.0	105.3	0.67
CHF_3	4.60	168.7	107.6	0.68
C_2F_4	5.00	183.5	117.1	0.74
$C_2H_4F_2$	6.20	227.9	145.4	0.92
C_3F_6	6.20	227.9	145.4	0.92
cyclic-C_4F_5	7.10	261.2	166.6	1.06
iso-C_4F_2	9.60	353.7	225.7	1.44

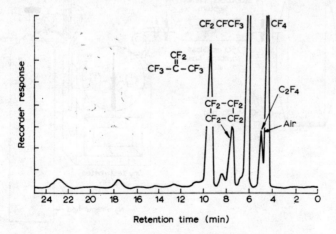

Fig. 26. Gas chromatogram of Teflon thermal degration products.

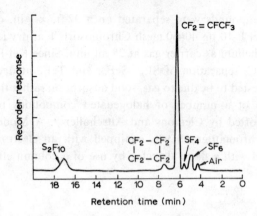

Fig. 27. Gas chromatogram of mixture of fluorocarbons and sulfur–halide compounds.

electron-capture detector was a Micro-Tek detector with a 300 mCi tritium source substituted for a conventional 130 mCi source and was equipped with a heater and with various controls for maintaining sample and scavenger flows. A Micro-Tek polarizing voltage supply modified to provide 0–50 V d.c. calibrated in 0.05 V increments was used to apply a potential across the detector and a Gyra Model E-302 electrometer completed the assembly. The prepurified nitrogen carrier gas, as well as the prepurified hydrogen and the air supply for the flame-ionization detector, were passed through a molecular sieve to remove water and other impurities. The flow rate to the electron-capture detector was 90 ml/min and the detector was maintained at 100°C. Polarization voltage of 11–14 V was based on the optimization of response from the measurements on sulfur hexafluoride and bromotrifluoromethane,

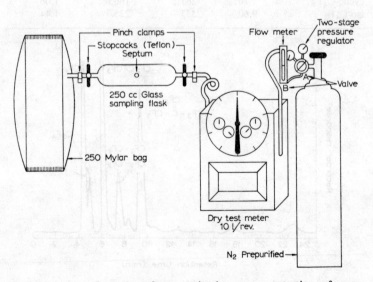

Fig. 28. Diagram of apparatus for preparing known concentrations of vapors.

and the sampling volume was 0.3 μl. An 8 ft. × ⅛ in. o.d. column of 80–100 mesh Baymal (a colloidal alumina)[72] was used for low molecular weight Freons and a 12 ft. × ⅛ in. o.d. column of 10% SF-96 on 50–60 mesh siliconized Sil-O-Cel C-22 firebrick was used for the more polar Freons and other highly polar materials.

TABLE 15

RESPONSES OF THE ELECTRON-CAPTURE DETECTOR TO VARIOUS HALOGENATED COMPOUNDS

Compound	Test concn. (p.p.m.)	Response (in.^{2}/p.p.m.)
$CHF_3{}^a$	4000	$<10^{-4}$
$CF_4{}^a$	4000	3×10^{-4}
$CH_3CHF_2{}^b$	20	9×10^{-3}
$CF_3CF_2CF_3{}^a$	4000	3×10^{-4}
$CF_3CF=CFCF_3{}^b$	0.4	40
$(CF_3)_2C=CF_2{}^b$	0.02	90
$CF_2–CF_2–CF_2–CF_2{}^{b,c}$	0.02	30, 40
$(CF_3)_2–CF_2–CF_2–CF_2CF_2{}^c$	0.13	90
CHF_2Cl^b	20	3×10^{-3}
CF_3Cl^c	4000	1×10^{-3}
$CF_3CF_2Cl^b$	20	5×10^{-2}
$CF_2=CFCl^b$	15	3×10^{-2}
$CHFCl_2{}^b$	20	5×10^{-2}
$CF_2Cl_2{}^c$	0.5	9
$ClF_2C–CF_2Cl^b$	0.5	2
$CF_2=CCl_2{}^b$	0.04	0.2
$CHCl_3{}^c$	0.06	10
$CFCl_3{}^c$	0.004	370
$ClF_2C–CFCl_2{}^c$	0.05	50
$ClHC=CCl_2{}^c$	0.04	20
$CCl_4{}^c$	0.005	650
$CF_3Br^{a,b}$	0.03	40, 12
$F_3C–CHClBr^c$	0.017	120
$CH_2Br_2{}^c$	0.02	300
$CF_2Br_2{}^b$	0.01	500
$BrF_2C–CF_2Br^c$	0.009	230
$CF_3CF_2CF_2I^c$	0.007	180
$IF_2C–CF_2I^c$	0.004	1200
$C_6F_6{}^c$	0.42	1000
$SF_6{}^a$	0.003	580
$SF_5(CF_2)_2Cl^c$	0.002	520
$SF_5(CF_2)_4Cl^c$	0.005	430
$SF_5(CF_2)_6Cl^c$	0.004	460
$SOF_2{}^b$	0.25	5
$SOFCl^b$	0.5	0.2

[a] Baymal column at 65°C.
[b] SF-96 column at 30°C.
[c] SF-96 column at 65°C.

References pp. 326–327

Figure 28 illustrates the method for preparing dilute vapor concentrations of the various halogens in nitrogen. The system allowed a sample dilution of 250×10^6 to 1 when 1 μl of gas was diluted with 250 l of nitrogen. Responses of the electron-capture detector are summarized in Table 15. The responses were found to vary about seven orders of magnitude with low responses exhibited by saturated and vinyl-type fluorinated hydrocarbons including those containing one chlorine atom. Compounds with a chlorine atom attached to a vinyl carbon gave lower responses than the corresponding saturated compounds. The response of the flame-ionization detector was in the range of about 0.1–1 in.^{2}/p.p.m. Thus the flame-ionization response of compounds such as CHF_2Cl, $CF_3ClCF_3CF_2Cl$, $CH_2=CF_2$, CH_3CHF_2, and even some of the dichloro derivatives such as $CHFCl_2$ or $CF_2=CCl_2$ were 10–1000 times greater than their electron-capture responses. It was suggested that if *sensitivity* is the major consideration, flame-ionization detection is preferred for many halogenated substances. However, if *specificity* is critical, the electron-capture detector is preferred even for halogenated substances showing low responses to this detector.

Petrova et al.[77] determined admixtures of air, CO_2, F_3CCl, $FCCl_3$, and CCl_4 in F_2CCl_2 by using gas–liquid and gas adsorption chromatography. Best results

TABLE 16

RETENTION DATA ON PORAPAK Q

Compound	^{130}I	dI/dT (per $10°C$)	Temp. range ($°C$)
CH_3F	174.9	−0.10	43–99
CH_3Cl	307.2	+1.2	117–157
CH_3Br	372.8	+2.0	117–157
CH_2Cl_2	446.4	+1.1	131–183
CH_2ClF	321.5	+0.10	129–183
CH_2F_2	173.1	−0.66	51–100
$CHCl_3$	550.1	−2.2	169–218
$CHCl_2F$	411.1	+0.31	162–210
$CHClF_2$	277.6	−1.1	83–127
CHF_3	160.2	−1.4	51–100
CCl_4	610.1	−3.4	169–218
CCl_3F	458.0	+1.0	162–210
CCl_2F_2	318.2	−0.34	123–183
$CClF_3$	188.9	−0.65	51–100
CCl_2FCClF_2	477.2	−0.21	131–183
$CClF_2CClF_2$	393.9	−1.1	131–179
$CClF_2CF_3$	268.0	−1.3	83–127
CF_3CF_3	161.4	−1.0	51–100
C_2H_4	181.2	+0.14	43–99
$CH_2=CHF$	212.6	−0.43	83–127
$CH_2=CHCl$	352.2	+0.79	131–179
$CH_2=CHBr$	417.1	+1.5	131–183
$CH_2=CF_2$	191.9	−0.55	43–99

were obtained by using two connecting columns. The first column was prepared with dibutyl phthalate and separated fluorotrichloromethane and carbon tetrachloride. The second column (the gas absorber) contained aluminum silicate catalyst and separated the other gases. Optimum conditions were obtained with a 2.5 m × 4 mm i.d. column at 40–42°C with helium as carrier gas at 28–30 ml/min.

Ratcliffe and Targett[78] determined organohalogen impurities (including $CHCl_3$ and CCl_4, 1–10%) in dichlorofluoromethane ($CHCl_2F$) using a modified flame-ionization gas chromatograph on columns of Chromosorb W coated with 6% silicone oil (MS550) and 6% Bentone 34 with argon as carrier gas. The following organohalogens and retention times were found: CF_4 0.51, CHF_3 0.52, $CClF_3$ 0.52, $CH_2{=}CF_2$ 0.54, CCl_2F_2 0.58, $CClF{=}CF_2$ 0.58, $CClF_2CClF_2$ 0.63, $CClHF_2$ 0.64, CH_2F_2 0.69, $CFCl{=}CFCl$ 0.83, CCl_3F 0.94, $CHCl_2F$ 1.00, CH_2ClF 1.24, CH_2Cl_2 3.10, $CHCl_3$ 3.30, and CCl_4 3.62.

The GC of a number of halogenated hydrocarbons (Freons) was reported by Foris and Lehman[79] utilizing Porapak Q (80–100 mesh) at 130°C with helium as carrier gas. An aerograph Model 1521-1B gas chromatograph equipped with nickel filaments (Gow-Mac, Mount 9225) was used in conjunction with 2 m, 3 m, and 4 m × $\frac{1}{8}$ in. stainless steel columns. The Porapak Q columns were conditioned at 240° for 14 h under helium flow. All measurements were taken with a detector temperature of 225°C, a detector current of 200 mA, and an injection port temperature of 200 to 225°C. Table 16 lists the retention data of some Freons obtained using the conditions above. Within each series of halocarbons, the retention index increased with chlorine content, decreased with fluorine and hydrogen content and was directly proportional to molecular weight and boiling point.

Bromotrifluoromethane ($CBrF_3$) has been suggested as a replacement for bromochloromethane (CH_2BrCl) as a fire-extinguishing compound because $CBrF_3$ is essentially non-toxic and exhibits greater thermal stability than most other halogenated fire-extinguishing materials. The pyrolysis products of CH_2BrCl and $CBrF_3$ as well as their toxicity in rats were studied by Haun et al.[80]. Chlorine and phosgene were determined by the GC technique of Priestly[81] while individual halogenated hydrocarbons were identified and their concentrations determined by using a combination of GC and mass spectrometry described by Biemann[82]. Bromotrifluoromethane was determined using thermal conductivity and an 8 ft. column containing 5A molecular sieve (60–80 mesh). The column was operated at 120° with helium as carrier gas.

Nitrogen trifluoride (NF_3) is a stable gas with strong oxidizing properties. It has been shown by Torkelson et al.[83] to have moderate or high toxicity by inhalation in single doses and that the compound caused methemoglobinemia.

The separation of nitrogen trifluoride from carbon tetrafluoride by gas chromatography was described by Richmond[84]. The separation was accomplished in less than 10 min using a column of 10 ft. × $\frac{1}{4}$ in. o.d. copper tubing containing 10% Halocarbon oil (Series 13-21, Halocarbon Corp., Hackensach, N.J.) on 60–80 mesh silica gel (Davison Grade 12) operated at 25°C. The detector consisted of two matched $2000\,\Omega$ thermistors (Veco A-111). The detector block was held at 40°C and helium was the carrier gas at 30 ml/min. As little as 2% NF_3 in mixture with CF_4 has been detected and trace amounts of air were well separated from the CF_4 which eluted before NF_3 under the conditions used.

Nachbaur and Engelbrecht[85] had previously reported the separation of CF_4 and NF_3 in 70 min on long columns held at low temperatures.

REFERENCES

1 F. FLURY AND F. ZERNIK, *Schädliche Gase*, Springer, Berlin, 1931, p. 118.
2 J. A. GUNN, *Quart. J. Med.*, 13 (1920) 121.
3 E. SCHAFER, *Brit. Med. J.*, II (1915) 245.
4 H. G. BARBOUR, *J. Pharmacol. Exptl. Therap.*, 14 (1919) 47.
5 E. B. VEDDER, *The Medical Aspects of Chemical Warfare*, Williams and Wilkins, Baltimore, 1925, p. 70.
6 H. G. BARBOUR, *J. Pharmacol. Exptl. Therap.*, 14 (1919) 65.
7 R. A. KEHOE AND K. V. KITZMILLER, *Cincinatti J. Med.*, 23 (1942) 423.
8 G. H. MICKEY AND H. HOLDEN, JR., *Environ. Mutagen Soc. Newsletter*, 4 (1971) 39.
9 W. MACHLE, K. V. KITZMILLER, E. W. SCOTT AND J. F. TREON, *J. Ind. Hyg. Toxicol.*, 24 (1942) 222.
10 K. B. LEHMAN AND A. BURCH, *Arch. Hyg.*, 72 (1910) 343.
11 E. LEITZ, *Arch. Hyg.*, 102 (1929) 91.
12 W. LUDEWIG, *Arch. Gewerbepathol. Gewerbehyg.*, 11 (1942) 296.
13 L. E. PORTER, *Ind. Eng. Chem.*, 18 (1926) 731.
14 M. B. JACOBS, *The Analytical Toxicology of Industrial Inorganic Poisons*, Interscience, New York, 1967, p. 636.
15 A. WALLACH AND W. A. McQUARY, *Am. Ind. Hyg. Assoc. Quart.*, 9 (1948) 63.
16 J. IWASAKI, S. UTSUMI, K. HAGINO AND J. OZAWN, *Bull. Chem. Soc. Japan*, 29 (1956) 860.
17 W. G. DUDLEY AND B. F. POETKER, *U.S. Pat. 2,850,640*, 1958.
18 R. L. MURRAY AND W. S. KIRCHER, *Trans. Electrochem. Soc.*, 86 (1944) 7.
19 E. E. NEELY, *Anal. Chem.*, 32 (1960) 1382.
20 J. LUCY AND K. G. WOOLMINGTON, *Analyst*, 86 (1961) 350.
21 D. M. RUTHVEN AND C. N. KENNEY, *Analyst*, 91 (1966) 603.
22 J. F. ELLIS AND C. IVESON, in D. H. DESTY (Ed.), *Gas Chromatography, 1958*, Butterworths, London, 1958, p. 300.
23 E. W. CEIPLINSKI, *Perkin-Elmer Corp. Rept., No. GC-DS-003*, 1964.
24 R. S. ISBELL, *Anal. Chem.*, 35 (1963) 255.
25 N. M. TURKEL'TAUB, S. A. AINSHTEIN AND S. V. SYANTSILLO, *Zavodsk. Lab.*, 28 (1962) 141.
26 A. FISH, N. H. FRANKLIN AND R. T. POLLARD, *J. Appl. Chem.*, 13 (1963) 506.
27 H. RUNGE, *Z. Anal. Chem.*, 189 (1962) 111.
28 S. ARAKI, T. KATO AND T. ATOBE, *Bunseki Kagaku*, 12 (1963) 450.
29 J. JANAK, M. NEDOROST AND V. BUBENIKOVA, *Chem. Listy*, 51 (1957) 890.
30 T. JOHNS, *Beckman Instruments Data Sheet No. GC-8064*.
31 F. D. HUILLET AND P. URONE, *J. Gas Chromatog.*, 4 (1966) 249.
32 J. F. ELLIS, C. W. FORREST AND Q. L. ALLEN, *Anal. Chim. Acta*, 22 (1960) 27.
33 A. J. P. MARTIN AND A. T. JAMES, *Brit. Med. Bull.*, 10 (1954) 170.
34 A. J. P. MARTIN AND A. T. JAMES, *Biochem. J.*, (1956) 138.
35 A. I. M. KEULEMANS, *Gas Chromatography*, Reinhold, New York, 1957.
36 V. Y. DUDOROV AND N. K. AGLIULOV, *Zh. Anal. Khim.*, 25 (1970) 162.

37 P. J. TALBOT AND J. H. THOMAS, *Trans. Faraday Soc.*, 55 (1967) 533.
38 S. ZOLTY AND M. J. PRAGER, *J. Gas Chromatog.*, 5 (1967) 533.
39 Air sampling instruments, *Am. Conf. Govt. Ind. Hygienists*, 2nd edn., Cincinatti, Ohio, 1962, p. C-1-4.
40 H. E. STOKINGER, O. VOEGTHIN AND H. C. HODGE, *Pharmacology and Toxicology of Uranium Compounds*, McGraw-Hill, New York, 1949, p. 1021.
41 W. H. MACINTYRE, *Ind. Eng. Chem.*, 41 (1949) 2466.
42 P. DRINKER AND K. W. NELSON, *Ind. Med.*, 13 (1944) 673.
43 C. R. WILLIAMS, *J. Ind. Hyg. Toxicol.*, 24 (1942) 277.
44 W. F. MACHLE, F. THAMANN, K. KITZMILLER AND J. CHOLAK, *J. Ind. Hyg. Toxicol.*, 16 (1934) 129.
45 W. F. MACHLE AND K. KITZMILLER, *J. Ind. Hyg. Toxicol.*, 17 (1935) 223.
46 R. A. GERDES, J. D. SMITH AND H. G. APPLEGATE, *Atmos. Environ.*, 5 (1971) 113.
47 R. A. GERDES, J. D. SMITH AND H. G. APPLEGATE, *Atmos. Environ.*, 5 (1971) 117.
48 A. H. MOHAMED, J. D. SMITH AND H. G. APPLEGATE, *Can. J. Genet. Cytol.*, 8 (1966) 575.
49 B. S. MARSHALL AND R. WOOD, *Analyst*, 93 (1968) 821.
50 R. BELCHER, M. A. LEONARD AND T. S. WEST, *J. Chem. Soc.*, (1959) 3577.
51 S. GERICKE AND B. KURMIES, *Z. Anal. Chem.*, 132 (1951) 335.
52 D. F. ADAMS AND R. K. KOPPE, *Anal. Chem.*, 35 (1953) 794.
53 D. F. ADAMS AND R. K. KOPPE, *Anal. Chem.*, 33 (1961) 117.
54 D. REVINSON AND J. H. HARLEY, *Anal. Chem.*, 35 (1953) 794.
55 H. E. BUMSTED AND J. C. WELLS, *Anal. Chem.*, 24 (1952) 1595.
56 W. A. POWELL AND J. H. SAYLOR, *Anal. Chem.*, 25 (1953) 960.
57 B. J. MacNULTY, G. F. REYNOLDS AND E. A. TERRY, *Nature*, 169 (1952) 688.
58 H. STETTER, *Chem. Ber.*, 81 (1948) 532.
59 O. ROCHEFORT, *Anal. Chim. Acta*, 29 (1963) 350.
60 I. LYSYJ AND P. A. NEWTON, *Anal. Chem.*, 35 (1963) 90.
61 J. G. MILLION, C. W. WEBER AND P. A. KUEHN, *Union Carbide Rept. No. K-1639*, 1966.
62 B. J. GUDZINOWICZ AND W. R. SMITH, *Anal. Chem.*, 35 (1963) 465.
63 A. G. HAMLIN, G. IVESON AND T. R. PHILLIPS, *Anal. Chem.*, 35 (1963) 2037.
64 G. GUTZEIT AND E. J. RAMIREZ, *U.S. Pat. 2,658,842*, 1953.
65 L. G. SPEARS, *J. Gas Chromatog.*, 6 (1968) 392.
66 A. DICORCIA, P. CICCIOLI AND F. BRUNER, *J. Chromatog.*, 62 (1971) 128.
67 A. DICORCIA AND F. BRUNER, *Anal. Chem.*, in press.
68 J. A. FRESEN, F. H. COX AND M. J. WITTER, *Pharm. Weekblad*, 103 (1968) 909.
69 R. BOCK AND H. J. SEMMLER, *Z. Anal. Chem.*, 230 (1967) 161.
70 A. ENGELBRECHT, E. NACHBAUR AND E. MAYER, *J. Chromatog.*, 15 (1964) 228.
71 B. E. SALTZMAN, A. I. COLEMAN AND C. A. CLEMONS, *Anal. Chem.*, 38 (1966) 753.
72 J. J. KIRKLAND, *Anal. Chem.*, 35 (1963) 1295.
73 C. A. CLEMONS, A. I. COLEMAN AND B. E. SALTZMAN, *Environ. Sci. Technol.*, 2 (1968) 551.
74 R. H. CAMPBELL AND B. J. GUDZINOWICZ, *Anal. Chem.*, 33 (1961) 842.
75 R. S. JUVET, JR. AND R. L. FISHER, *Anal. Chem.*, 38 (1966) 1860.
76 C. A. CLEMONS AND A. P. ALTSCHULLER, *Anal. Chem.*, 38 (1966) 133.
77 M. P. PETROVA, D. D. ALEKSEEVA AND A. N. MESCHERYAKOVA, *Zh. Anal. Khim.*, 23 (1968) 1101.
78 D. B. RATCLIFFE AND B. H. TARGETT, *Analyst*, 94 (1969) 1028.
79 A. FORIS AND J. G. LEHMAN, *Separ. Sci.*, 4 (1969) 230.
80 C. C. HAUN, E. H. VERNOT, D. L. GEIGER AND J. M. McNERNEY, *Am. Ind. Hyg. Assoc. J.*, 30 (1969) 551.
81 I. J. PRIESTLY, *Anal. Chem.*, 37 (1965) 70.
82 K. BIEMANN, *Mass Spectrometry, Organic Chemical Applications*, McGraw-Hill, New York, 1962, p. 20.
83 T. R. TORKELSON, F. OYEN, S. E. SADEK AND V. K. ROWE, *Toxicol. Appl. Pharmacol.*, 4 (1962) 770.
84 A. B. RICHMOND, *Anal. Chem.*, 33 (1961) 1806.
85 E. NACHBAUR AND A. ENGELBRECHT, *J. Chromatog.*, 2 (1959) 562.

Chapter 14

CARBON MONOXIDE AND CARBON DIOXIDE

1. CARBON MONOXIDE

Carbon monoxide is one of the most important of urban atmospheric pollutants. Its presence is ubiquitous since it is generated by incomplete combustion and occurs in industry, in household heating, in motor vehicle exhaust, and in tobacco smoke. (The amount of CO present in fuel gases is approx. 40% in water gas, 5% in coke oven gas and household gas.) Carbon monoxide occurs in high concentration in cigarette smoke (>2%), but an estimate of the average concentration in smoke inhaled into the lung is about 400 p.p.m. (0.04%). The scope of the motor vehicle exhaust problem alone can be gleaned from the fact that in New York City alone, *each day*, automobile traffic alone produces 8.3 million pounds (3.8×10^6 kg) of carbon monoxide[1]. Each car emits approx. $\frac{1}{8}$ lb. of CO/mile of travel at 25 m.p.h. (40 k.p.h.) and about $\frac{1}{3}$ lb./mile of travel at 10 m.p.h.[1]. Table 1 lists the emission of gases and vapors from urban activities and for natural processes and illustrates the prominence of CO among the other gases (*e.g.* SO_2, NO, CH_4).

Additional potential exposures of carbon monoxide occur in the manufacture of carbide, synthetic methanol, or other chemicals from CO, and in the production and use of illuminating gas, in the distillation of coal or wood, and in operations near furnaces, ovens, stoves, forges, and kilns. Air levels of CO as high as 50 p.p.m.

TABLE 1

EMISSIONS OF GASES AND VAPORS FROM URBAN ACTIVITIES AND FOR NATURAL PROCESSES[2]

Substance	Emissions from urban activities[a] (g/year)	Emissions from natural processes[b] (g/year)
CO_2	2×10^{15}	
CO	7×10^{13}	
SO_2	2×10^{13}	
Terpenes		10^{12}
NO	7×10^{12}	
H_2	$\sim 10^{13}$	$\sim 10^{13}$
HC	1×10^{13}	10^{13}–10^{14c}
CH_4	$<1 \times 10^{12}$	10^{13}–10^{14}
H_2S		2×10^{14}

[a] United States figures.
[b] From total surface of earth.
[c] Biosphere, methane and terpene production.

and blood carboxyhemoglobin (HbCO) levels as high as 4–5% have been record-ed, although these represent maximal levels. Hence. as an environmental pol-lutant CO is unique in that no other pollutant exists at such high concentrations in urban atmospheres with such high toxic potential.

The toxic effect of carbon monoxide is associated with its ability to form carbon monoxide hemoglobin (a comparatively stable addition compound) with hemo-globin. The affinity of CO for hemoglobin is about 200 times as strong as that of oxygen; however, the formation of carbon monoxide hemoglobin is reversible when sufficient air supply is available; it is slowly reconverted to oxygen hemoglobin. Although the resultant oxygen deficiency is a reversible chemical asphyxia, damage done by severe asphyxia from any cause may not be reversible. Aspects of carbon monoxide and human health have been reviewed by Goldsmith and Landaw[3].

The determination of carbon monoxide in the atmosphere has been carried out by colorimetric[4-6] techniques. Various gas detection devices have been described for semi-quantitative determination of carbon monoxide concentrations (e.g. tubes designed for specific concentration ranges utilizing the reduction of ammonium molybdate to molybdenum blue by carbon monoxide under the catalytic effect of palladium chloride). Automatic analyzers utilizing Hopcalite for the oxidation of carbon monoxide have also been described[7-9].

The determination of carbon monoxide in blood has been achieved by the pyro-tannic acid method[10,11], spectrophotometric[12,13], and Van Slyke gasometric[14] methods.

Carbon monoxide in air samples has been analyzed by gas chromatography using two columns in series[15]. The first column was a 6 ft. $\times$ $\frac{1}{4}$ in. o.d. metal helix packed with di-2-ethylhexyl sebacate on 60–80 mesh "Columpak" and the second column a 6.5 ft. $\times$ $\frac{3}{16}$ in. o.d. metal helix packed with 42–60 mesh molecular sieves. After introduction of the CO, it was reduced with hydrogen catalytically to methane using high-purity nickel powder[16,17]. The methanation was carried out at approx. 260° and the amount of carbon monoxide in the sample was proportional to the height of the methane peak determined by means of FID. The analysis time was about 7 min and the lower sensitivity limit was about 2 p.p.m. with high accuracy compared with the coulometric method which was above 3 p.p.m. CO.

Analysis of carbon monoxide in ambient air and respiratory gases has also been accomplished by Lynn and Hackney[18] using the sequence of steps (a) low-temperature trapping of the CO on a molecular sieve at −78°C, (b) further separation of the CO on a second molecular sieve, and (c) analysis with a thermal conductivity detector. Although samples as large as ca. 1 liter could be trapped efficiently by this technique, the sensitivity was suggested to be marginal for atmospheric analysis[2].

Boreham and Marhoff[19] separated a mixture of carbon monoxide and 8 gases using a 4 m $\times$ 4 mm column of 40–60 mesh silica gel at 30°C. A Perkin-Elmer Vapor Fractometer was used with a thermal conductivity cell, thermistor model, with argon as carrier gas at 3.6 l/h. The retention times found were hydrogen 1.5, oxygen and nitrogen 1.8, carbon monoxide 2.1, methane 2.8, ethane 10.5, carbon dioxide 15.5, and ethylene 17.5 min, respectively.

Murakami[20] described a rapid (4 min) GLC analysis of a mixture of oxygen, nitrogen, methane, carbon monoxide, and carbon dioxide utilizing two columns. Column 1 consisted of a 40 cm × 4 mm i.d. spiral copper column containing 50–80 mesh silica gel. Column 2 was a spiral copper column (200 cm × 4 mm i.d.) containing molecular sieve 5A, 50–80 mesh. The columns were operated at 80°C, the inlet pressure was 0.63 kg/cm² gauge, and the outlet pressure atmospheric. The flow rate of hydrogen as carrier gas was 70 ml/min. The test gas was first separated on column 1 into CO_2 and a mixture of the remaining gases which passed through a thermal conductivity cell. The mixture of remaining gases was separated by passage through column 2 and subsequent passage through a second thermal conductivity cell. Figure 1 illustrates a gas chromatogram of a mixture of O_2, N_2, CH_4, CO, and CO_2.

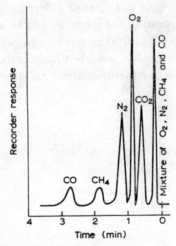

Fig. 1. GLC separation of a mixture of oxygen, nitrogen, methane, carbon monoxide, and carbon dioxide; × indicates switching the poles of the recorder.

The hydrogen flame-ionization detector (FID) is generally regarded as being insensitive to inorganic substances such as carbon monoxide and carbon dioxide[21,22]. The detector has been used[16] for the analysis of CO by converting the CO in the sample to CH_4 by passage with H_2 over a catalyst before entering the FID.

Another method of ionization analysis of inorganic substance has been to raise the level of the background current artificially by addition of a small continuous flow of hydrocarbons to the detector or to the carrier gas then to detect the inorganic gas by the *decrease* in background signal when the sample of inorganic gas passes through the detector[23]. The response of the FID to CO and CO_2 has been recently described by Schaefer and Douglas[24]. Under suitable operating parameters, the FID gave a response to CO and CO_2 which was superior to that of the katharometer. Generally, it was necessary to increase the background current by continuous addition of a trace amount of hydrocarbon to the flame or by operating with a hot burner. Depending on the conditions used, CO and CO_2 may give either positive or negative peaks. An explanation of some of the observed responses was given in terms of expected temperature changes in the diffusion flame of the FID and their influence on the hydrocarbon response.

A Beckman Model GC2 instrument was used with TCD and FID accessory. The FID consisted essentially of a steel tube of 5.5 mm × 0.9 mm o.d., 0.5 mm i.d., as the burner jet. A polarizing electrode set horizontally made contact with the jet. The collector electrode was a "hot-shaped" open cylinder with a flared lower edge and the cylinder was of stainless steel 6 × 6 mm. For most experiments the column used was 6 ft. × ¼ in. o.d. stainless steel filled with 60–80 mesh firebrick coated with 50% silicone oil and operated at about 80°. Porapak Q was also used for some experiments and little quantitative difference was obtained although the Porapak gave a better separation for mixtures of CO and CO_2. Before use both columns were purged at 200°C with 50 ml/min of nitrogen then flushed with frequent samples of SO_2 or H_2S to remove adsorbed hydrocarbons[25]. Controlled amounts of hydrocarbon were added to either the carrier gas stream just prior to the column

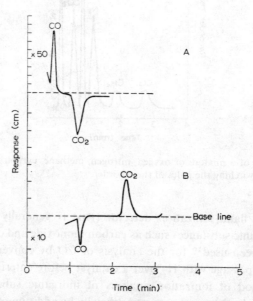

Fig. 2. Sample chromatograms of response to 12.6 μmole of 50% mixture CO/CO_2. Effect of change of H_2% by altering carrier flow rate. Flame contains 16 p.p.m. CH_4 added in hydrogen. Column, Porapak Q 6 ft. 83°C; elution time shifted in B by slower flow rate of carrier. A, at 18.0% H_2 (45 H_2, 93 N_2) CO positive, CO_2 negative, attenuation × 50 = 1.66 pA/cm recorder displacement; B, at 24.2% H_2 (45 H_2, 28 N_2) CO negative, CO_2 positive, attenuation × 10 = 0.34 pA/cm recorder displacement.

or to the fuel supply just before entry to the burner. (The hydrocarbon was supplied from a cylinder containing 1% CH_4 in N_2.)

Figure 2 shows a chromatogram of a mixture of CO and CO_2 (at different equivalent H_2 percentages) when methane was added to the hydrogen fuel line at a constant rate (approx. 12 or 16 p.p.m. of the hydrogen) given a background current of 150 or 200 pA. The flow rate of hydrogen was held constant and that of the carrier was varied. Figure 2(A) at 18% H_2 shows the CO positive and CO_2 as a negative peak. In Fig. 2(B) the composition of the flame was 24.2% hydrogen equivalent and the polarities of the peaks were reversed. The polarity reversal under the conditions described with changes in the flame hydrogen percentage might be used to identify these gases if their relative retention times were not known.

Poli[26] described a method for the rapid monitoring of air samples to give a continuous and unattended record of carbon monoxide and methane content on one analyzer, in the low parts per million range. An event programmer was employed to (1) shut off the sample flow at a predetermined time before the sample was injected into the system, allowing each sample to be equilibrated at the same pressure, (2) inject the sample reproducibly into the pressurizer system, gait component peaks with accurate timing to allow peak height measurement, automatically zero the analog baseline signal before a component measurement is taken and backflush unwanted material in the sample stream, and (3) shift sample streams if a pre-analyzed calibration mixture is to be run. Figure 3 depicts a single channel illustration of such a programmer. Figure 4 illustrates the analyzer block system employed and Fig. 5 shows the complete chromatographic system. The sample valve containing a 7 ml loop introduces ambient air at atmospheric pressure on command from the

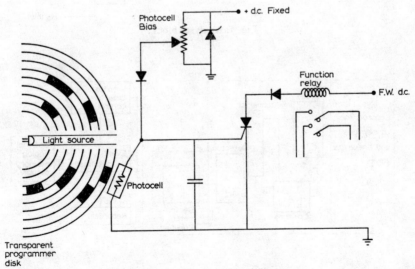

Fig. 3. Programmer functional diagram.

programmer. Chromatographic column no. 1 is used with the backflush valve to strip out unwanted materials (*e.g.* CO_2, hydrocarbons other than methane and any components present) from the sample. Column 2 is the analytical column that separates the air, methane, and carbon monoxide which are then passed through the methanator or converter where only the carbon monoxide is affected (*i.e.* by conversion to methane with hydrogen enrichment over a nickel catalyst) and then transported by the helium carrier to the flame-ionization detector as separate detectable peaks. Using the above system with tight tolerances on flow and temperature reproducibility for a given sample to 2 % of fullscale was obtained. Figure 6 shows a running account of carbon monoxide and methane for a 24 h period during a 59 day overall run.

    The estimation of blood carbon monoxide is required for the accurate determination of the pulmonary diffusing capacity for carbon monoxide. Several methods

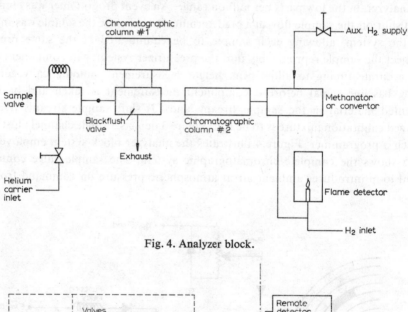

Fig. 4. Analyzer block.

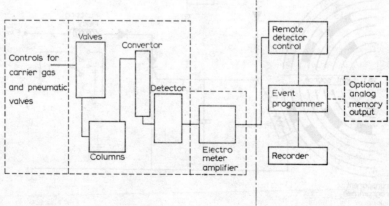

Fig. 5. Complete chromatographic system for air-monitoring of carbon monoxide and methane.

have been described in which CO is liberated by ferricyanide and then measured in either a Hopcalite meter[27] or an infrared analyzer[28,29]. A method was described by McCredie and Jose[30] for the simultaneous measurement of CO and CO_2 in blood. CO and O_2 were dissociated by oxidation with potassium ferricyanide extracted under vacuum in a modified Van Slyke volumetric apparatus and measured by gas chromatography. In routine use the method has a standard deviation for CO content of ± 0.148 ml/100 ml blood and for O_2 content of ± 0.175 ml/100 ml. Extracted gases were injected into the carrier line of a Beckman GC 2 gas chromatograph, passed through CO_2 and water absorbers before entering the chromatograph for separation and measurement. A Van Slyke volumetric apparatus was modified by fitting a glass Y-piece to the waste arm as described by Ramsay[31]. The Y-piece was connected in the carrier gas line of the gas chromatograph with $\frac{1}{4}$ in. o.d. polyethylene tubing. Downstream to it was fitted a 12 in. column of $\frac{1}{2}$ in. o.d. polyethylene tubing containing water and CO_2 absorbers. Upstream to it was inserted a four-way stopcock to provide a loop which bypassed the extraction chamber and the water-absorbing column (Fig. 7). Helium was the carrier gas at a flow of 80 ml/min. Oxygen, nitrogen, and carbon monoxide were separated by a 9 ft. column of 40–60 mesh Linde 5A molecular sieves maintained at 100°C and detected by thermal conductivity cells using filament currents of 350 mA. CO_2 and water were firmly absorbed on molecular sieves and led to zero drift of the instrument and ultimately to loss of resolving power of the column. Daily renewal of water and CO_2 absorbers using

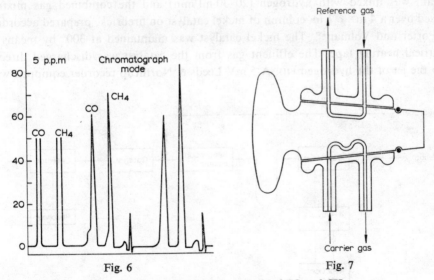

Fig. 6

Fig. 7

Fig. 6. Chromatogram of 24 h analysis of CO and CH_4.

Fig. 7. The stopcock used for calibration; one loop is connected into the carrier gas line while the other is flushed with reference gas, which can be introduced by turning the key through 180°; the volumes of these two loops are 0.1374 ml (SD $\pm$ 0.0004 ml) and 0.0555 ml (SD $\pm$ 0.0004 ml), respectively.

dehydrite and ascarite (A. H. Thomas) prevented the former. Satisfactory column resolution could also be restored by flushing with helium at a temperature of 220°C for 24–48 h. Recordings were made on 1 mV fullscale potentiometric recorder (Minneapolis Honeywell). Oxygen was eluted first (nitrogen elution began before oxygen elution was complete but did not overlap the recorded peak of oxygen). Carbon monoxide was eluted 3 min after nitrogen and was completely separated from it.

The determination of CO content in blood by GLC was described by Collison et al.[32]. CO bound to hemoglobin was released by hemolysis and reaction with $K_3Fe(CN)_6$ in a closed system. The liberated gas was then swept onto a 5A molecular sieve column where CO was separated from other blood gases. The CO after catalytic reduction to methane was detected by flame ionization. The method was specific and sufficiently sensitive to permit analysis of 0.1 ml samples of normal blood. The accuracy of the method expressed as the coefficient of variation (SD × 100/mean) was 1.8% for normal blood.

A Model 154-D vapor fractometer (Perkin-Elmer) was used with a hydrogen FID. Helium was used both as carrier gas at 50 ml/min and to purge the reaction vessel (50 ml/min). A glass U-tube (7 mm i.d. × 20 cm) contained a layer of Ascarite between two layers of Drierite and was placed between the gas valve and the separating column to remove CO_2 and water. The U-shaped separating column (6 ft. × $\frac{1}{4}$ in. stainless steel) was packed with 30–80 mesh 5A molecular sieve and maintained at 100°C. Immediately downstream from the separating column, the carrier gas stream was mixed with hydrogen (20–30 ml/min) and the combined gas mixture passed over a 4 in. × $\frac{1}{8}$ in. column of nickel catalyst on firebrick, prepared according to Porter and Volman[16]. The nickel catalyst was maintained at 300° by means of electrical heating tape. The effluent gas from the catalyst was discharged directly into the jet of the hydrogen FID. A 5 mV Leeds & Northrup recorder equipped with

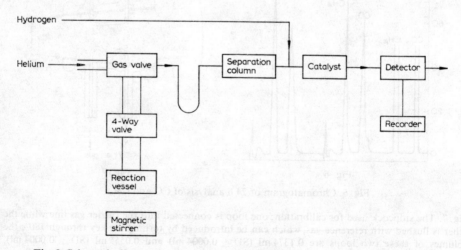

Fig. 8. Schematic diagram of chromatographic system for blood CO determination.

a disc integrator was used to monitor and evaluate the response of the FID. Figure 8 is a block diagram of the system employed. The liberation of CO from blood and its transfer to the chromatograph were accomplished by use of the system illustrated in Fig. 9 which was designed to replace the regular gas sample loop of the chromatograph[33]. In use, the magnet and necessary reagents were placed in the reaction tube, C, which was then seated on the brass body, B, with silicone grease and held in place by a copper wire basket and springs. The vessel was purged with helium through the 4-way valve, A, positioned as shown. The reaction tube, filled with helium, was then isolated by turning the key of the 4-way valve. Blood or standard HbCO (carboxyhemoglobin) (0.1 ml) was injected through the centrally located

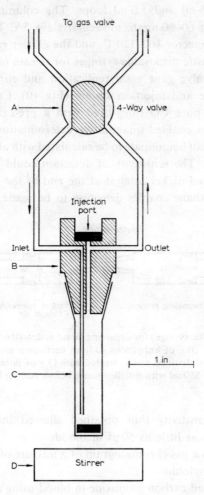

Fig. 9. Reaction system for chromatographic analysis of gases liberated from aqueous solutions includes 4-way valve (A), machined brass body (B) with centrally located silicone rubber injection port, interchangeable reaction tube (C), and external magnetic stirrer (D).

silicone rubber injection port directly into the reagents by use of a gas-tight micro-syringe. At the end of a pre-determined reaction period the gas valve of the chromatograph and the 4-way valve, A, were positioned to pass the carrier gas stream through the reaction tube into the separating column. A sweep-out period of 3 min was used, after which the gas valve on the chromatograph was switched to bypass the reaction system.

Blackmore[34] described the determinations of CO in blood and tissue taken at post-mortem.

GLC was compared with spectrophotometric and differential protein precipitation techniques and found to be the method of choice. A Pye Model 4 gas chromatograph was used fitted with a katharometer detector, linear amplifier, and gas-sampling valve with 0.5 ml and 5.0 ml loops. The column was 5 ft. × ¼ in. o.d. stainless steel containing 60–80 mesh, molecular sieve 5A. Temperatures employed were oven 100–120°C, detector 100–120°C, and the carrier gas was helium at 45 ml/min. The use of 5 ml plastic disposable syringes for release of CO followed by transfer to a gas-sampling valve gave good replication and eliminated the need for a sophisticated gas release and injection system (Fig. 10). Comparison of the peak heights from 0.5 ml of pure CO (measured with a pre-calibrated gas loop) with those of the test samples enabled quantitative determination of the volume of CO liberated from an aliquot of hemolipate to be calculated with allowances for differences in amplifier attenuation. The sensitivity of detection could be greatly increased by the introduction of heated nickel catalyst at the end of the column (which by converting the CO into methane enables detection to be made by flame ionization)[16].

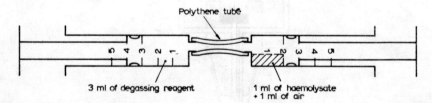

Polythene tube

3 ml of degassing reagent   1 ml of haemolysate + 1 ml of air

Fig. 10. Arrangement of plastic syringes for degassing blood as described in the gas chromatographic method. Degassing reagent: 20 g of anhydrous sodium carbonate and 20 g of sodium hydrogen carbonate dissolved in 400 ml distilled water, treated with 15 g of potassium ferricyanide with constant mixing and made up to 550 ml with distilled water before adding 10 ml of Teepol L or 5 ml of Triton X-100.

The increase in sensitivity thus obtained allowed the accurate quantitative determination of CO in as little as 50 μl of blood.

Figure 11 illustrates a gas chromatogram of a mixture of air and CO from blood containing carboxyhemoglobin.

Carlson[35] determined carbon monoxide in blood using a 2.5 m × 5 mm column containing 30–50 mesh Linde 5A molecular sieve. A Burrell Kromo-Tog II chromatograph equipped with a thermal conductivity detector was employed with the following operating parameters: helium carrier gas at 55 ml/min, column and detector temperatures 95 and 150°C, respectively, and detector current 245 mA. Figure 12 illustrates a chromatogram for the separation of carbon monoxide from contaminating nitrogen and oxygen.

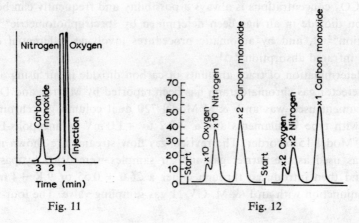

Fig. 11.   Fig. 12.

Fig. 11. Gas chromatogram of a mixture of air and carbon monoxide from blood containing carboxyhemoglobin.

Fig. 12. Optimum conditions for separation of carbon monoxide from contaminating nitrogen and oxygen. Sample size, 1 ml blood; column length, 2.5 m; column packing, Linde 5A molecular sieve 30–50 mesh; column voltage, 45 V; column diameter, 5 mm; carrier gas, helium; column temperature, 95°C; gas flow reference, 10 bubbles/min.; gas flow measuring, 55 ml/min; detector current, 245 mA; detector bath temperature, 150°C.

Dominguez et al.[36] analyzed carbon monoxide in blood using a Perkin-Elmer Model 154C chromatograph equipped with a thermal conductivity detector and a $2 \text{ m} \times \frac{1}{4} \text{ in.}$ column containing Fisher molecular sieve type 5A ($\frac{1}{16}$ in. pellets). Helium was the carrier gas at 135 ml/min and the column temperature was 75°C. Carbon monoxide was eluted in ca. 5 min. Liberation of carbon monoxide from blood has been achieved using a Van Slyke apparatus[37] or a Natelson microgasometer[38].

2. CARBON DIOXIDE

Carbon dioxide is a normal constituent of air, e.g. about 0.03% by volume above the ocean, up to ca. 0.06% in urban areas. However, its level in unpolluted air is not completely uniform. CO_2 levels have increased considerably since the turn of the last century due to the consumption of fossil fuel such as coal, petroleum, and natural gas from about 290 p.p.m. to about 320 p.p.m. on the average. The concentration of CO_2 is higher at the equator than in the arctic (e.g. ca. 350 and 260 p.p.m., respectively). Heating, traffic and industry all can cause enhanced local CO_2 levels which can exceed the average level in open air. Combustion gas can contain up to 20% CO_2, depending on the oxygen content of the air. Aspects of carbon dioxide and the biosphere have been reviewed by Attiwill[39]. Carbon dioxide, except as contributor to oxygen deficiency, does not offer serious industrial exposures except in enclosed spaces where CO_2 poisoning due to the accumulation of higher CO_2 concentrations is always a possibility and frequently can be fatal.

Carbon dioxide in air has been determined by spectrophotometric[40,41], acid–base titration[42,43], and by automatic procedures involving differential conducto-metry and infrared absorption[44,45].

The determination of trace amounts of carbon dioxide in air using a modified hot-wire detector gas chromatograph has been reported by Murray and Doe[46]. The basic instrument used was an F & M Model 720 dual column gas chromatograph equipped with type W filaments and a -0.2 to $+1.0$ mV Minneapolis-Honeywell Electronic Model 15 recorder. The revised gas flow streams are shown in Fig. 13. Helium was used as the carrier gas and air samples were dried by passage over Drierite and then introduced through either a 26.0 ± 0.55 or $9 K \pm 1$ ml sample loop in conjunction with an F & M, GV-11 gas sampling valve. The four-way valve

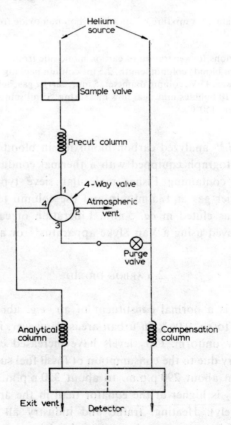

Fig. 13. Revised gas flow streams.

(Fig. 13) is a gas chromatograph selector valve (NO A3-30-2, Republic Valves Co.). The column was 3.5 ft. × ¼ in. o.d. copper tubing containing 30–60 mesh silica gel. The tubing was wrapped with asbestos insulation and an electrothermal heating tape. The temperature of the column was maintained at 60°C. The analytical and compensation columns were 2.5 ft. × ¼ in. o.d. copper tubing containing 30–60 mesh silica gel. These columns were maintained at 40°C. The helium flow rate for the analytical and compensation columns were 50 and 70 ml/min, respectively, and the tungsten (hot-wire) bridge setting was 200 mA.

Lysyi and Newton[47] described the gas chromatography of O_2, N_2, and CO_2 using two columns connected in parallel. One of the columns (1.5 m × 6 mm) was packed with molecular sieve 5A, 60–80 mesh and was used for the separation of oxygen from nitrogen (CO_2 was irreversibly bound on this column). A second column (60 cm × 6 mm) filled with 20–200 mesh silica gel was used for the determination of CO_2. The carrier gas was helium, the separation temperature was 26°C, and a Gow-Mac conductivity cell was used for detection. When air was simultaneously introduced *via* a two-way gas sampling valve into both columns, the O_2–N_2 peak appeared on the silica gel column after 2 min and after 6 min the O_2 and N_2 peaks were separately eluted from the molecular sieve, while the CO_2 peak was eluted after 26 min from the silica gel column.

Bober[48] described the determination of CO_2, O_2, N_2, CO, and CH_4 using two columns in series. The adsorption of CO_2 was effected using a column containing 1.2 m × 5 mm activated carbon while O_2, N_2, CO, and CH_4 were separated on a 1.2 m × 5 mm column containing molecular sieve 5A. The first and second columns were operated at 90 and 70°C with hydrogen as carrier gas and the sample volume was 0.2 ml. O_2, N_2, CH_4, and CO were eluted after 40, 50, 100, and 120 sec, respectively. When separation was completed CO_2 was eluted from the first column in order to prevent its penetration and contamination of the molecular sieve column.

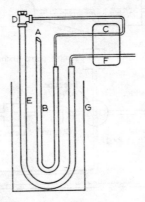

Fig. 14. Schematic diagram of assembled chromatograph. A, Hydrogen carrier gas inlet; B, coiled equilibrium tube; C, reference side of TC detector; D, Swagelok T injection port; E, coiled chromatographic column; F, sample side of TC detector; G, Dewar flask.

The determination of oxygen and carbon dioxide in medicinal gas mixtures by gas–solid chromatography was described by Langham[49]. The method was applicable to the analysis of nitrogen, helium, argon, carbon monoxide, and nitrous oxide. The gas chromatographic assembly utilized is shown in Fig. 14. One end of a gas equilibrating column (empty 17 ft. × $\frac{1}{8}$ in. o.d. aluminum tubing coiled to $1\frac{1}{4}$ in. i.d.) was connected to a hydrogen cylinder and the other end to the reference side of the thermal conductivity detector. (Connections can be made with $\frac{1}{16}$ in. o.d. stainless steel or Teflon tubing.) The outlet of the reference side of the TC detector was connected to a $\frac{1}{4}$ in. Swagelock T which was used as an on-column injection port. The injection port was placed on top of the chromatographic column 16 ft. × $\frac{1}{4}$ in. o.d. aluminum tubing coiled to $2\frac{3}{4}$ in. i.d. packed with 50–80 mesh Porapak Q. The column outlet was connected to the sample side of the TC detector and the gas equilibrating column was positioned inside the coil of the packed column, and both columns were placed in a 2 l wide-mouth Dewar flask. Dry ice conditions were required for the oxygen analysis while carbon dioxide was determined at ambient temperatures. Carbon dioxide had a retention time of ca. 8 min.

The gas chromatographic analysis of mixtures containing oxygen, nitrogen, and carbon dioxide was described by Brenner and Cieplinski[50] using a single sample in a single analysis. A "parallel column" system was used consisting of a column of molecular sieve 5A and a short column of silica gel (Davison type 15, 35–60 mesh). O_2, N_2, and CO_2 were separated in 2 min at 22°C.

The determination of impurities by gas chromatography, with special reference to coolant gas for nuclear reactors, was described by Timms et al.[51]. It was shown that with a 25 ml sample, H_2, A_2, O_2, N_2, CH_4, and CO present at levels as low as 5–20 p.p.m. in CO_2 could be determined using two interchangeable 6 ft. × $\frac{1}{4}$ in. columns containing Linde 5A molecular sieve (36–52 mesh) (activated by heating

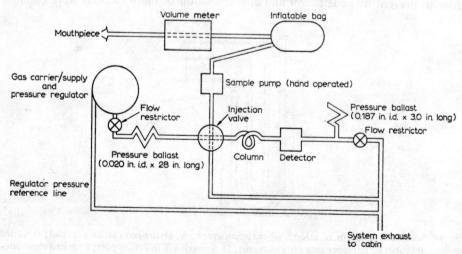

Fig. 15. Schematic flow diagram of the gas meter.

in air to 350°C for 3–4 h). A sensitive katharometer containing tungsten filaments having a resistance of about 30 ohms at 20°C was used for detection.

A small, flight-type gas chromatograph unit incorporating a gas volume meter for monitoring in-flight respiration of astronauts within the Apollo spacecraft was developed by Eaton et al.[52]. The gas analysis system consisted of microcomponents including a packed column capable of separating O_2 and CO_2 with a flow rate of approx. 2 ml/min, a small, light injection valve with zero dead volume, a Gow-Mac thermistor thermal conductivity detector and an amplifier system that integrated the areas of components of interest. The gas meter, 3 in. high × 10 in. wide × 11 in. deep and weighing less than 7 lb., utilized a total power of approx. 3 watts. The schematic flow diagram of the gas meter is shown in Fig. 15. The column was a 6 in. tube packed with Porapak T (120–150 mesh) (conditioned with oxygen, 4 ml/min for approx. 3 h at 100°C). The carrier gas was 10% He–90% N_2. The oxygen-conditioned column when tested at room temperature eluted O_2 and CO_2 within 30 sec at a flow rate of 2 ml/min. There was no change in retention times between 1% and 20% concentrations of CO_2.

Hirt and Palmer[53] described the quantitative determination of carbon suboxide (C_3O_2) by gas chromatography. A Beckman GC-1 gas chromatograph was used with a filament-type sensing cell and 7.5 ft. × $\frac{1}{4}$ in. copper tubing packed with 23% Apiezon grease N on firebrick. Helium was the carrier gas at 15 p.s.i.g., and with the column at room temperature, the retention time of C_3O_2 was 6 min and that of air 1.2 min, followed closely by CO, with CO_2 appearing at approx. 3 min.

REFERENCES

1 A. HELLER, U.S. Public Health Serv. Rept. No. 999, 1967.
2 A. P. ALTSCHULLER, in J. C. GIDDINGS AND R. A. KELLER (Eds.), Gas Analysis by Gas Chromatography, Vol. 5, Marcel Dekker, New York, 1968.
3 J. R. GOLDSMITH AND S. A. LANDAW, Science, 162 (1968) 1352.
4 D. A. LEVAGGI AND M. FELDSTEIN, Am. Ind. Hyg. Assoc. J., 25 (1964) 64.
5 M. B. JACOBS, The Analytical Toxicology of Industrial Inorganic Poisons, Interscience, New York, 1967, p. 703.
6 R. L. BEATTY, U.S. Bur. Mines Bull. No. 557, 1955.
7 C. H. LINDSLEY AND J. H. YOE, Anal. Chem., 21 (1949) 513.
8 C. H. LINDSLEY AND J. H. YOE, Z. Anal. Chem., 131 (1950) 308.
9 J. M. SALSBURY, J. W. COLE AND J. H. YOE, Anal. Chem., 19 (1947) 66.
10 R. R. SAYERS, W. P. YANT AND G. W. JONES, U.S. Public Health Repts., 38 (1923) 2311.
11 R. R. SAYERS AND W. P. YANT, U.S. Bur. Mines. Tech. Paper No. 373, 1927.
12 B. L. HORECKER AND F. S. BRACKETT, J. Biol. Chem., 152 (1944) 669.
13 H. HARTMAN, Physiol. Biol. Chem. Exptl. Pharmacol., 39 (1937) 413.
14 P. B. HAWK AND O. BERGEIM, Pratical Physiological Chemistry, Blakiston, Philadelphia, 11th edn., 1937.
15 M. KATZ, in A. STERN (Ed.), Air Pollution, Academic Press, New York, 2nd edn., 1968, p. 54.
16 K. PORTER AND D. H. VOLMAN, Anal. Chem., 34 (1962) 748.
17 L. DuBois, A. Z. DROJEWSKI AND J. L. MONKMAN, J. Air Pollution Control Assoc., 16 (1966) 135.

18   K. C. Lynn and J. D. Hackney, *5th Conf. Methods in Air Pollution Studies, Los Angeles, Calif.,* Feb., 1963.
19   G. R. Boreham and F. A. Marhoff, *Gas Council Res. Comm. GC 54,* 1958.
20   Y. Murakami, *Bull. Chem. Soc. Japan.,* 32 (1959) 316.
21   D. Jentzsch and K. Friedrich, *Z. Anal. Chem.,* 180 (1961) 96.
22   A. J. Andreatch and R. Feinland, *Anal. Chem.,* 32 (1960) 1021.
23   I. G. McWilliam, *J. Chromatog.,* 6 (1961) 410.
24   B. A. Schaefer and D. M. Douglas, *J. Chromatog., Sci.,* 9 (1971) 612.
25   B. A. Schaefer, *Anal. Chem.,* 42 (1970) 448.
26   A. A. Poli, *Intern. Conf. Inst. Soc. Am., Chicago, October 4–7, 1971,* p. 1.
27   S. M. Siösteen and J. Sjöstrand, *Acta Physiol. Scand.,* 22 (1951) 129.
28   P. J. Lawther and G. H. Apthorp, *Brit. J. Ind. Med.,* 12 (1955) 326.
29   R. F. Coburn, G. R. Danielson, W. S. Blakemore and R. E. Forster, *J. Appl. Physiol.,* 19 (1964) 510.
30   R. M. McCredie and A. D. Jose, *J. Appl. Physiol.,* 22 (1967) 893.
31   J. H. Ramsay, *Science,* 129 (1959) 900.
32   H. A. Collison, F. L. Rodkey and J. D. O'Neal, *Clin. Chem.,* 14 (1968) 162.
33   F. L. Rodkey and H. A. Collison, *U.S. Naval Med. Res. Rept., MR 005.04.0002., Rept. No. 16,* 1967.
34   D. J. Blackmore, *Analyst,* 95 (1970) 439.
35   H. C. Carlson, *Res. Vet. Sci.,* 5 (1964) 186.
36   A. M. Dominguez, H. E. Christensen, L. R. Goldbaum and V. A. Stembridge, *Toxicol. Appl. Pharmacol.,* 1 (1959) 135.
37   S. M. Ayres, A. Criscitiello and S. G. Anneli, Jr., *J. Appl. Physiol.,* 21 (1966) 1368.
38   L. R. Goldbaum, E. L. Schloegel and A. M. Dominguez, *Progr. Chem. Toxicol.,* 1 (1963) 11.
39   P. M. Attiwill, *Environ. Pollution,* 1 (1971) 24.
40   L. W. Loveland, R. W. Adams, H. H. King, Jr., F. A. Nowak and L. J. Cali, *Anal. Chem.,* 31 (1959) 1008.
41   A. N. Spector and B. F. Dodge, *Anal. Chem.,* 19 (1947) 55.
42   H. L. Higgins and W. M. Marriott, *J. Am. Chem. Soc.,* 39 (1971) 58.
43   G. Wagner, *Oesterr. Chem. Ztg.,* 54 (1953) 74.
44   H. Malissa and G. Wagner, *Mikrochim. Acta,* (1962) 332.
45   W. Schmidts and W. Bartscher, *Z. Anal. Chem.,* 181 (1962) 54.
46   J. N. Murray and J. B. Doe, *Anal. Chem.,* 37 (1965) 94.
47   J. Lysyi and P. R. Newton, *J. Chromatog.,* 11 (1963) 173.
48   H. Bober, *Beckman Instrument Rept.,* Feb. 1966, p. 10.
49   W. L. Langham, *J. Assoc. Offic. Anal. Chemists,* 53 (1970) 1268.
50   N. Brenner and E. Cieplinski, *Math. Naturwiss. Unterricht.,* 20 (1967) 705.
51   D. G. Timms, H. J. Kowrath and R. C. Chirnside, *Analyst,* 83 (1958) 600.
52   H. G. Eaton, Jr., V. R. Huebner and C. Mittelman, Jr., *J. Gas Chromatog.,* 6 (1968) 441.
53   T. J. Hirt and H. B. Palmer, *Anal. Chem.,* 34 (1962) 164.

Chapter 15

OXIDANTS (OZONE AND PEROXYACYL NITRATES)

1. OZONE

Ozone (O_3) is a highly reactive, unstable gas formed both in nature at high altitudes as well as by technological processes and equipment. Ozone is formed by natural processes in the upper atmosphere by the photo-dissociative action of solar ultraviolet radiation below 2450 Å on the oxygen molecules present in the stratosphere. Peak atmospheric concentrations of ozone as high as 11 p.p.m. or more have been measured in the stratosphere. The amount of naturally formed ozone at ground level is small, *e.g.* maximum of 1–3 p.p.h.m. (parts per hundred million)[1], depending on the weather conditions and height above sea level.

Ozone has been determined to be one of the principal oxidants found in community photochemical air pollution[2-6]. In the photochemical smog-forming processes, UV radiation from sunlight initiates a series of atmospheric reactions between emissions from sources producing oxides of nitrogen and photochemically reactive organic substances such as the olefinic hydrocarbons and aldehydes which results in the formation of ozone and other compounds. In polluted atmospheres, the peroxide and ozone levels can be as high as 0.6 p.p.m. Ozone is the dominant constituent (up to 90%) of photochemical smog.

Although emissions from motor vehicles constitute a major source of these effluents, other important sources include the combustion of organic fuels for power and heat, evaporation losses from petroleum products, refuse burning, industrial and commercial uses, and losses of organic solvents.

Technological sources of ozone include the use of high voltage ozonizers in the treatment of sewage and in water purification[7] and for control of molds and bacteria in cold storage plants[8]. Ozone is found in appreciable levels in inert gas-shielded arc-welding devices[9,10] and is also formed by high voltage electrical equipment, *e.g.* X-ray apparatus and spectrographs, neon signs, electrical insulators, brushes of motors, ultraviolet-ray quartz lamps[11], and electrically operated office copy equipment.

Ozone plus radiation environment may be associated with nearly any high-intensity radiation field including research accelerators and isotope sources, electron beam irradiators (which are used in industrial polymerization as for paint curing and plywood glue hardening), large-scale isotopic food irradiators, industrial radiography devices and medical therapeutic radiation units. Ozone has increasing applications in a variety of industries, *e.g.* as a bleaching agent for textiles and in various chemical syntheses and analyses involving ozonolysis.

Ozone is a corrosive gas that is one of the most powerful substances known[12]. It combines with many materials including rubber[4,5] and textiles causing the cracking of rubber and deterioration of textiles. Ozone is also highly injurious to vegetation[5,13].

The effects of ozone exposures on man and animals were reviewed by Jaffee[14] with emphasis on the effects of low concentrations of ozone (0.05–0.20 p.p.m.). At high ozone levels such as are found in certain occupational exposures, ozone causes acute lung injury in laboratory animals characterized by pulmonary congestion and edema, while at still higher concentrations hemorrhage and death occur. Ozone may also produce secondary systemic effects on body metabolism and function[15].

The effects of short-term[16-19] and prolonged exposures[20-26] of ozone in man as well as in animals[14,20,27-29] have been reported. Table 1 summarizes the toxicological effects of ozone at low concentrations in man, rabbit, hamsters, and mice. In addition, radiomimetic effects have been reported where ozone simulates the effects of X-irradiation causing such effects as structural damage to myocardial tissue[31] and increased rate in development of lung adenomas and accelerated aging effects[32-34]. The radiomimetic effect of ozone on *Vicia faba* resulting in chromosome breakage was noted by Fetner[35] and the mutagenic activity of inhaled ozone in the Chinese hamster[36] as well as the ozone-induced chromosome breakage in human cell culture also was described[37].

The analysis of ozone has been mainly accomplished by spectrophotometric[38-42], iodometric[43-48], and chemiluminescence[49-54] techniques.

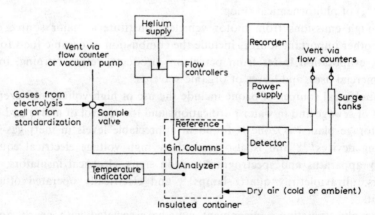

Fig. 1. Low temperature programmed gas chromatograph.

A gas–solid chromatographic technique was developed by Dohohue and Jones[55] for the determination of hydrogen, oxygen, oxygen difluoride, and ozone in gaseous products from the electrolysis of wet hydrogen fluoride. The gases were separated on a silica gel column, temperature-programmed from −75 to −10°C to minimize decomposition. Figure 1 is a schematic diagram of the low-temperature-programmed

TABLE 1

TOXICOLOGICAL EFFECTS OF OZONE AT LOW CONCENTRATIONS[30]

Species	Concentration (p.p.m.)[a]	Exposure period	Effect	Comment
Man	0.02	<5 min	Odor threshold for 9/10 subjects	
Man	0.2–0.5	3 h	Decrease in visual acuity in the dark adaptation and middle vision ranges	Subjective complaints from occupational and experimental exposures
Man	0.3	Unspecified	Nasal and throat irritation, cough	
Man	0.6–0.8	2 h	Lowered DL_{CO} in 11/11 subjects mean decrease = 25%	Transient effect. Almost back to normal 1 h after exposure. Great variability in sensitivity between subjects
Man	0.1–1.0	1 h	Increased airway resistance 1/4 increased at 0.1 p.p.m. 2/4 at 0.4 p.p.m. 2/4 at 0.6 p.p.m. 4/4 at 1.0 p.p.m.	Individual data not given. No effect at 0.2 p.p.m.
Man	0.5	3 h/day; 6 days/week/12 weeks	Significant decrease in average $FEV_{1.0}$ after 10 weeks Returned to normal 6 weeks after final exposure	
Rabbit	1	1 h	Formation of carbonyl groups in lung collagen	Chemical changes are suggestive of crosslinking—changes still apparent 24 h after exposure
Rats, rabbits, hamsters	1	Continuously for at least 1 year	Bronchitis, bronchiolitis, emphysematous and fibrotic changes	
Mice, hamsters	0.08–1.3 or 0.84	3 h	Increased susceptibility to Klebsiella pneumoniae	Applies to ozone exposure before or after K. pneumoniae

gas chromatograph. All parts that came in contact with the sample were made of fluorocarbon plastic or of dry, degreased metals (nickel, copper, or stainless steel) passivated with oxygen difluoride and ozone. Gaseous products from the HF electrolysis cell were swept through the sampling loop and a flow counter (soap bubble meter) by a stream of ordinary helium. The helium carrier gas was purified by passage through molecular sieves (Linde 5A) cooled in liquid and then divided into two streams, one for reference and the other for the sample, each kept at 40 ml/min by its own controller, Model 63BU-L (Moore Products Co.). The sample stream passed through a Perkin-Elmer 1.06 ml sampling valve where it was mixed with the gases to be analyzed. The two streams then passed through twin columns of silica gel (Davison, Grade 923, 100–200 mesh), 6 in. × ⅛ in. diam. contained in an insulated can whose temperature was controlled by a stream of dry air pre-cooled by liquid nitrogen. Effluent streams from the columns were passed onto a thermal conductivity cell (Gow-Mac No. 9285) with matched filaments.

Figure 2 illustrates a chromatographic analysis of gases from wet HF electrolysis together with the range of retention times for the maximum of known peaks. Although ozone was detectable only above 200–300 p.p.m. with the above described equipment configuration, it was suggested by Donohue and Jones[55] that the use of more sensitive detectors and/or larger samples would permit detection down to 1 p.p.m. or less and hence be of possible utility for air pollution studies.

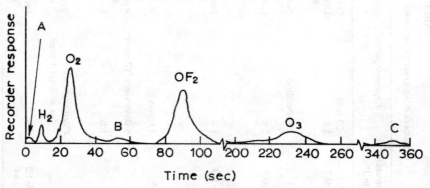

Fig. 2. Typical analysis of gases from wet HF electrolysis. Range of retention time for peak maximum (sec): H_2, 8–9; O_2, 20–35; OF_2, 85–120; O_3, 210–250.

2. PEROXYACYL NITRATES

Peroxyacyl nitrates (PANs), I, are a series of organic nitrogen compounds which are formed when sunlight acts on polluted air containing trace concentrations of nitrogen oxides and organic compounds such as partially oxidized hydrocarbons.

$$R-\underset{\underset{O}{\|}}{C}-OONO_2$$

(I)

Automobile exhaust is considered the most important environmental source of components necessary for peroxyacyl nitrate formation. The five analogs that constitute the family of PANs are where the acyl radical in I is either acetyl, propionyl, butyryl, isobutyryl, or benzoyl. Up to 210 p.p.b. of peroxyacetyl nitrate has been reported in Los Angeles smog and as much as 50–60 p.p.b. in other California cities. (The other analogs are present in lower concentrations but are 4–100 times more reactive.) The benzoyl analog is said to be 200 times more eye-irritating than formaldehyde. Exposures to concentrations of PANs in the order of parts per hundred million are reported to cause visible damage to agricultural crops[56,57], while concentrations in the p.p.m. to p.p.b. range can cause eye irritation[58,59]. The lethality of a 2 h exposure of 100–200 p.p.m. of PANs in mice has been reported[60]. The acute toxicity of PAN in mice[61], the effect on the depression of voluntary activity in mice[62], clinical effects on animals and man[25] as well as aspects of their biochemistry[63], and reactions and physical properties[64] have all been reported.

The analysis of PANs has been accomplished primarily by infrared spectroscopy[65-68]. (These compounds show characteristic bands at 5.4, 7.7, and 12.6 and peroxyacetyl nitrate also shows a band at 8.6 nm; the infrared bands at 8.6 and 12.6 nm ordinarily are used analytically.)

Peroxyacyl nitrates have been analyzed primarily by electron-capture GC which has permitted their detection from the more abundant hydrocarbons.

Darley et al.[69] separated peroxyacyl nitrates on a 3 ft. glass column packed with 5% Carbowax 400 on 100–120 mesh Chromosorb W at 35°C. An electron-capture detector (Wilkins) was used in the d.c. mode; helium was the carrier gas at 25 ml/min. Peroxyacetyl nitrate and peroxypropionyl nitrates (PPN) were eluted after 2 and 3 min, respectively. PAN and PPN were found in samples of smog-laden atmospheres in amounts of 50 and 6 p.p.b., respectively, by this procedure. Subsequently, use of a glass sampling valve with a 9 in. Teflon column packed with the same material and maintained at 22°C was found to give satisfactory results for the analysis of PANs[70].

The quantitative analysis of peroxyacetyl nitrate and methyl nitrate was reported by Bellar and Slater[71]. Adequate separations were obtained with columns packed with polyethylene glycol 400 or 1540 on Gas Chrom Z (60–80 mesh). The electron-capture detector was used in the pulsed mode: frequency 5 kc/sec, peak width 2 μsec and pulse amplitude 60 V. Figure 3 illustrates a chromatogram of electron-capturing substances in urban air and shows the separation of carbon tetrachloride, methyl nitrate, peroxyacetyl nitrate, and 2 unknowns.

The propylene–nitrogen oxide system when irradiated reacts readily to produce oxidant, formaldehyde, acetaldehyde, carbon monoxide, peroxyacetyl nitrate (PAN), and methyl nitrate and causes ozone and PAN-type plant damage and eye irritation. Hence, it is possible to reproduce major "smog" manifestations although not necessarily at the intensities prevailing in the ambient atmosphere. Chemical aspects of the photooxidation of the propylene–nitrogen oxide system have been described

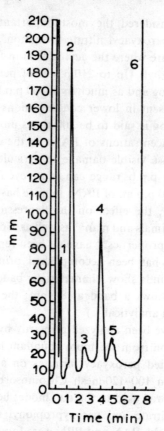

Fig. 3. Chromatogram of electron-capturing substances in Cincinnati air.

by Altshuller *et al.*[72]. The products were measured by colorimetric, gas chromatographic, and various monitoring instruments.

Peroxyacetyl nitrate and methyl nitrate were separated on 8 ft. × ⅛ in. borosilicate glass packed with 10% polyethylene glycol on 60–80 mesh Gas Chrom Z at 28°C and analyzed by means of an electron-capture detector. Carbon monoxide was separated on a molecular sieve column, hydrogenated to methane with 5–10% Raney nickel on a 40–60 mesh C-22 Celite catalyst and analyzed as methane by FID. Acetaldehyde was separated using a 12 ft. × ⅛ in. stainless steel column packed with 10% 1,2,3-tris-(2-cyanoethoxy propane) on 60–80 mesh Gas Chrom Z at 42°C and analyzed by FID.

Stephens *et al.*[73] described the production of pure peroxyacyl nitrates by gas phase reactions at low concentrations, which were purified by preparative-scale GC. PAN was prepared by the photolysis of ethyl nitrite vapor (*ca.* 400 p.p.m.) in oxygen and stored as a vapor diluted with dry nitrogen at about 100 p.s.i.g. and 40°F. The photoreactor used for synthesis of peroxyacyl nitrates is shown in Fig. 4 and the preparative-scale gas chromatograph for PAN purification is illustrated in Fig. 5. Like

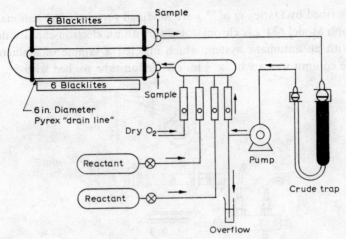

Fig. 4. Photoreactor used for synthesis of peroxyacyl nitrates.

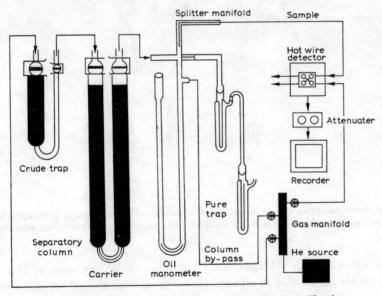

Fig. 5. Preparative-scale gas chromatograph for PAN purification.

the reactor system, the gas chromatographic purifier system was constructed of Pyrex glass and consisted of a crude products freeze-out trap, a 15×900 mm separatory column of 20% Carbowax E-60 on 30–42 mesh C-22 firebrick, a Dow-Corning silicone oil manometer, a $1:10$ ratio 0.05 mm i.d. splitter manifold, two 30×120 mm unpacked pure products freeze-out traps, carrier gas manifold and a hot wire detector, signal attenuator, and Bristol 3 mV recorder.

    An automated gas chromatographic system for the analysis of PAN in the atmosphere has been described by Smith et al.[74]. The procedures for measuring

PAN as described by Darley *et al.*[69] were modified for use in the automated system. An Aerograph Model 681 gas chromatograph with an electron-capture detector was equipped with an automatic system which injected a sample of ambient air every 15 min. The column was a 9 in. × $\frac{1}{8}$ in. i.d. Teflon tube packed with 5% Carbowax

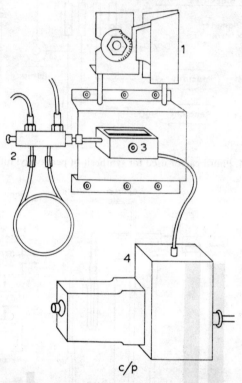

c/p

Fig. 6. Automatic sampling system consisting of: 1, standard timing system; 2, hexaport valve and 2 ml sample loop; 3, solenoid to activate valve; 4, time delay relay to control recorder chart drive motor.

E-400 on 100–120 mesh HMDS-treated Chromosorb W, operated at 25°C with a nitrogen carrier gas flow of 40 ml/min. Under these conditions PAN had a retention time of 60 sec and the minimum detectable quantity was about 1 p.p.b. A stainless steel hexaport valve with an external 2 ml stainless steel sample loop was used instead of the glass sample valve described by Darley *et al.*. With this system there was no detectable decomposition of PAN. The automatic sampling mechanism used is shown in Fig. 6. The chromatograph was calibrated with a flow dilution panel (Fig. 7) from a supply containing PAN at a concentration high enough to be measured in a 10 cm infrared cell (1000 p.p.m.). The automatic system above has been reported in continuous operation 24 h/day for 11 months with only brief interruptions to clean the electron-capture detector. Figure 8 shows a chromatogram produced by the automated PAN analyzer from ambient air samples.

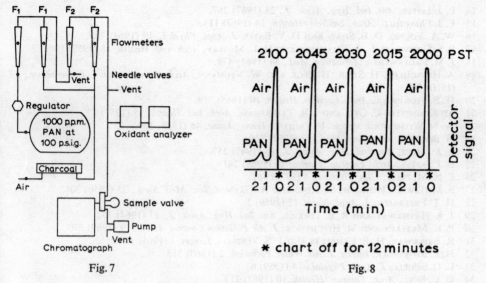

Fig. 7

Fig. 8

Fig. 7. Dynamic calibration system.

Fig. 8. Detection of 19 p.p.b. of PAN riverside air. A chromatogram produced by the automated PAN analyzer from ambient air samples during the period 8.00 to 9.00 p.m. on March 27, 1968; the large air peak is followed by a PAN peak in each of the 15 min periods.

The nitroolefins have been suggested as atmospheric pollutants of biological interest. Stephens and Price[75] described an electron-capture GLC method for the separation of nitroolefins. Separation was achieved on an 18 in. column containing 9 in. of 5% glycerin on 100–120 mesh HMDS-treated Chromosorb W and 9 in. of 5% Carbowax on the same support. Nitroolefins were detected to levels below 1 p.p.b. However, no detectable amounts of nitroolefins were found in atmospheric samples containing from 5–40 p.p.b. of peroxyacetyl nitrate.

REFERENCES

1 A. EHMERT, *J. Atmospheric Terrest. Phys.*, 2 (1952) 189.
2 A. P. ALTSHULLER AND J. J. BUFALINI, *Photochem. Photobiol.*, 4 (1965) 97.
3 L. M. RICHARDS, *J. Air Pollution Control Assoc.*, 5 (1956) 215.
4 A. J. HAAGEN-SMIT, C. E. BRADLEY AND M. M. FOX, *Ind. Eng. Chem.*, 48 (1956) 1484.
5 A. J. HAAGEN-SMIT, *Ind. Eng. Chem.*, 44 (1952) 1342.
6 P. A. LEIGHTON, *Physical Chemistry*, Vol. IX, Academic Press, New York, 1961.
7 R. EHRLICH, *Frontier*, 22 (1960) 16.
8 R. NAGY, *Advan. Chem. Ser.*, 21 (1959) 57.
9 R. FRANT, *Ann. Occupational Hyg.*, 6 (1963) 113.
10 P. J. R. CHALLEN, D. E. HICKISH AND J. BEDORRD, *Brit. J. Ind. Med.*, 15 (1958) 276.
11 M. J. DADLEZ, *Quart. J. Pharm.*, 1 (1928) 99.
12 F. A. PATTY, *Industrial Hygiene and Toxicology*, Vol. II, Wiley-Interscience, 1962, p. 916.
13 L. G. WAYNE, The chemistry of urban atmosphere, *Los Angeles Air Pollution Control Assoc.*, *Tech. Progr. Rept. III*, Dec. 1962.

14 L. J. JAFFEE, *Am. Ind. Hyg. Assoc. J.*, 28 (1967) 267.

15 E. J. FAIRCHILD, *Arch. Environ. Health*, 14 (1967) 111.

16 W. A. YOUNG, D. B. SHAW AND D. V. BATES, *J. Appl. Physiol.*, 19 (1964) 765.

17 S. S. GRISWOLD, L. A. CHAMBERS AND H. L. MOTLEY, *Arch. Ind. Health*, 15 (1957) 108.

18 J. M. LAGERWERFF, *Aerospace Med.*, 34 (1963) 479.

19 A. HENSCHLER, H. STIER, H. BECK AND W. NEUMANN, *Arch. Gewerbepathol. Gewerbehyg.*, 17 (1960) 547.

20 H. E. STOKINGER, *Arch. Environ. Health*, 10 (1965) 719.

21 M. KLEINFELD, C. GEIL AND I. R. TABERSHAW, *Arch. Ind. Health*, 15 (1957) 17.

22 W. N. WITHERIDGE AND C. P. YAGLOU, *Trans. Ashne*, 45 (1939) 509.

23 G. BENNETT, *Aerospace Med.*, 33 (1962) 969.

24 L. J. JAFFEE, *Am. Ind. Hyg. Assoc. J.*, 28 (1967) 257.

25 L. J. JAFFEE, *Arch. Environ. Health*, 16 (1968) 241.

26 P. N. M. NASR, *Clin. Toxicol.*, 4 (1971) 461.

27 S. MITLER, D. HEDRICH, M. KING AND A. GAYNOR, *Ind. Med. Surg.*, 25 (1956) 301.

28 H. T. FREEBAIRN, *J. Appl. Nutr.*, 12 (1959) 2.

29 J. A. HATHAWAY AND R. E. TERRILL, *Am. Ind. Hyg. Assoc. J.*, 23 (1964) 392.

30 P. K. MUELLER AND M. HITCHCOCK, *J. Air Pollution Control Assoc.*, 19 (1969) 670.

31 R. BRINKMAN, H. B. LAMBERTS AND T. S. VENINGA, *Lancet*, i (1964) 183.

32 H. E. STOKINGER, *Intern. J. Air Water Pollution*, 2 (1960) 313.

33 L. D. SCHEEL, *J. Appl. Physiol.*, 14 (1959) 67.

34 G. C. BUEL, *Arch. Environ. Health*, 10 (1965) 213.

35 R. W. FETNER, *Nature*, 181 (1958) 504.

36 R. E. ZELAC, H. L. CROMROY, W. E. BOLCH, JR., B. G. DUNAVANT AND H. A. BEVIS, *Environ. Res.*, 4 (1971) 325.

37 R. H. FETNER, *Nature*, 194 (1962) 793.

38 T. R. HAUSER AND D. W. BRADLEY, *Anal. Chem.*, 39 (1967) 1184.

39 T. R. HAUSER AND D. W. BRADLEY, *Anal. Chem.*, 38 (1966) 1529.

40 D. M. WAGNEROVA, K. ECKSCHLAGER AND J. VEPREKSISKA, *Collection Czech. Chem. Commun.*, 32 (1967) 4032.

41 H. H. BOVEE AND R. J. ROBINSON, *Anal. Chem.*, 33 (1961) 1115.

42 B. E. SALTZMAN AND W. GILBERT, *Am. Ind. Hyg. Assoc. J.*, 20 (1959) 379.

43 A. P. ALTSCHULLER, C. M. SCHWAB, *Anal. Chem.*, 31 (1959) 1987.

44 D. H. BYERS AND B. E. SALTZMAN, *Am. Ind. Hyg. Assoc. J.*, 19 (1958) 251.

45 B. E. SALTZMAN AND A. F. WARTBURG, *Anal. Chem.*, 37 (1965) 779.

46 G. M. MAST AND H. E. SAUNDERS, *Inst. Soc. Am. Trans.*, 1 (1962) 325.

47 B. E. SALTZMAN AND N. GILBERT, *Anal. Chem.*, 31 (1959) 1914.

48 B. E. SALTZMAN, W. A. COOK, B. DIMITRIADES, E. F. FERRAND, *Health Lab. Sci.*, 7 (1967) 152.

49 J. A. HODGESON, K. J. KROST, A. E. O'KEEFFE AND R. K. STEVENS, *Anal. Chem.*, 42 (1970) 1795.

50 R. L. BOWMAN AND N. ALEXANDER, *Science*, 154 (1966) 1454.

51 G. W. NEDERBRAGT, A. VANDER HORST AND J. VAN DUIJN, *Nature*, 206 (1965) 87.

52 V. H. REGENER, *J. Geophys. Res.*, 69 (1964) 3795.

53 D. BERSIS AND E. VASSILIOU, *Analyst*, 91 (1966) 499.

54 I. CHERNIACK AND R. J. BRYAN, Paper presented at *57th Ann. Meeting Air Pollution Control Assoc.*, Houston, Texas, June 1964.

55 J. A. DONOHUE AND F. S. JONES, *Anal. Chem.*, 38 (1966) 1858.

56 E. F. DARLEY, W. M. DUGGER, J. B. MUDD, L. ORDIN, D. C. TAYLOR AND E. R. STEPHENS, *Arch. Environ. Health*, 6 (1963) 761.

57 E. F. DARLEY, C. W. NICHOLS AND J. J. MIDDLETON, *Calif. Dept. Agr. Bull.*, 55 (1966) 11.

58 E. A. SCHUCK, E. R. STEPHENS AND J. J. MIDDLETON, *Arch. Environ. Health*, 13 (1966) 570.

59 E. R. STEPHENS, E. F. DARLEY, O. C. TAYLOR AND W. E. SCOTT, *Intern. J. Air Water Pollution*, 4 (1961) 79.

60 J. T. MIDDLETON, L. O. EMIK AND O. C. TAYLOR, *J. Air Pollution Control Assoc.*, 15 (1965) 476.

61 K. I. CAMPBELL, *Arch. Environ. Health*, 15 (1967) 739.

62 K. I. CAMPBELL, L. O. EMIK, G. L. CLARKE AND R. L. PLATA, *Arch. Environ. Health*, 20 (1970) 22.

63   A. N. M. NASR, *J. Occupational Med.*, 9 (1967) 589.
64   E. R. STEPHENS, in J. N. PITTS AND R. L. METCALFE (Eds.), *Advances in Environmental Sciences*, Vol. I, Wiley, New York, 1969, p. 119.
65   C. S. TUESDAY, *Arch. Environ. Health*, 7 (1963) 188.
66   E. R. STEPHENS, *Anal. Chem.*, 36 (1964) 928.
67   S. L. KOPCZYNSKI, *Intern. J. Air Water Pollution*, 8 (1964) 107.
68   E. R. STEPHENS, P. L. HANST, R. C. DOERR AND W. E. SCOTT, *Ind. Eng. Chem.*, 48 (1956) 1498.
69   E. F. DARLEY, K. A. KETTNER AND E. R. STEPHENS, *Anal. Chem.*, 35 (1963) 589.
70   E. R. STEPHENS AND E. F. DARLEY, *6th Conf. Methods in Air Pollution Studies, Berkeley, Calif., Jan. 1964.*
71   T. A. BELLAR AND R. W. SLATER, *150th Am. Chem. Soc. Mtg., Atlantic City, N. J., Sept. 1965.*
72   A. P. ALTSCHULLER, S. L. KOPCZYNSKI, W. A. LONNELMAN, T. L. BECKER AND R. SLATER, *Environ. Sci. Technol.*, 1 (1967) 899.
73   E. R. STEPHENS, F. R. BURLESON AND E. A. CARDIFF, *J. Air Pollution Control Assoc.*, 15 (1965) 87.
74   R. G. SMITH, R. J. BRYAN, M. FELDSTEIN, B. LEVADIE, F. A. MILLER AND E. R. STEPHENS, *Health Lab. Sci.*, 8 (1971) 48.
75   E. R. STEPHENS AND M. A. PRICE, *J. Air Pollution Control Assoc.*, 65 (1965) 320.

Chapter 16

PHOSGENE, CYANOGEN AND HYDROGEN CYANIDE

1. PHOSGENE

Phosgene is used widely as an intermediate in many chemical syntheses. It is also used in metallurgy to separate ores by chlorination of the oxides and volatilization. Phosgene is prepared commercially by the catalytic chlorination of carbon monoxide. It occurs frequently as one of the products of combustion when a volatile chlorine compound, *e.g.* a chlorinated solvent or its vapor, comes into contact with a very hot metal or flame.

The principal action of phosgene is that of a lung irritant with high concentrations being immediately corrosive to lung tissue resulting in death by suffocation. The action of phosgene and resultant toxic effects in a number of respects resemble those of nitrogen dioxide. The responses to various concentrations of phosgene in air have been reported by Flury and Zernik[1]. Phosgene in air has been determined by colorimetric[2] and gravimetric procedures (*via* reaction with aniline to form diphenylurea)[2,3].

The determination of subtoxic concentrations of phosgene in air by electron-capture GC was described by Priestley[4]. The method can detect and measure concentrations of phosgene of the order of 1000 times less than those of physiological

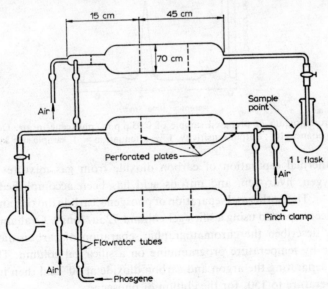

Fig. 1. Diagram of triple dilution system for preparation of phosgene samples.

interest. (The threshold limit value for daily 8 h exposure is 1.0 p.p.m. by volume in air.)

An Aerograph Model A-350-B gas chromatograph was equipped with an electron-capture detector and a Brown Recorder −0.05 to +1.05 mV and a Disc integrator. The column consisted of 2 m × 4.7 mm i.d. aluminum tubing packed with 30% Flexol plasticizer 10–10 (didecyl phthalate, Union Carbide) coated on 100–120 mesh GC-22 Super Support (Coast Engineering Lab.). The column was operated isothermally at 50°C with a flow rate of nitrogen carrier gas at 50 ml/min. The detector potential was 90 V. Samples of known concentrations of phosgene in air were prepared in a dynamic triple dilution system as shown in Fig. 1. The system consisted basically of 3 glass mixing chambers in which the phosgene was progressively diluted with air.

Figure 2 shows a chromatogram of a prepared sample of 0.05 p.p.m. phosgene in air. Concentrations of phosgene as low as 4 p.p.b. were also measured and on the basis of the signal obtained, less than 1 p.p.b. could be measurable.

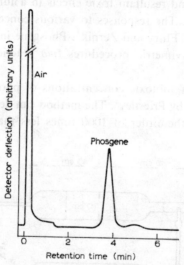

Fig. 2. Gas chromatogram of prepared sample of 0.05 p.p.m. phosgene in air. Detector voltage, 90 V; input impedance, 10^7; output sensitivity, 1 ×, attenuation × 1, sample size 2 ml.

The isothermal separation of carbon dioxide from gas mixtures of chlorine, nitrogen, oxygen, hydrogen, and nitrous acid has been accomplished using silica gel columns[5-7]. The isothermal separation of phosgene, carbon dioxide and other gases at 56.5° has been reported using a silica gel column 72 in. long[8]. Graham and Stevenson[9] recently described the chromatographic separation of carbon dioxide, argon, and phosgene by temperature programming on a silica gel column. The technique consisted of separating the argon and carbon dioxide at 30° and then increasing the column temperature to 150° for the elution of phosgene.

An F & M Model 720 dual-column programmed gas chromatograph with nickel detector was used. The column consisted of 5 in. of A-40 silica gel (F & M) (30 to 60 mesh) in a 6 in. × ⅛ in. o.d. column. The silica gel was activated at 150° for 3 h in a stream of helium. Helium flow rate was 20.5 ml/min; the detector temperature was 45° with a bridge current of 190 mA.

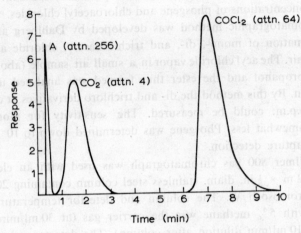

Fig. 3. Chromatogram of argon, carbon dioxide, and phosgene (sample volume 2.5 ml).

Figure 3 is a chromatogram showing the analysis of a 2.5 ml sample containing 33% argon, 2% carbon dioxide, and 65% phosgene. The retention times for argon, carbon dioxide, and phosgene were approx. 20, 100, and 400 sec, respectively, for the conditions as described above. (Chlorine in similar mixtures was eluted at about 270 sec with some interference with the phosgene peak. A 1 in. longer silica gel column resulted in a satisfactory resolution of chlorine.) Figure 4 shows the effect of the column temperature on the phosgene peaks. Increasing the column temperature improved both the sharpness and shape of the phosgene peak.

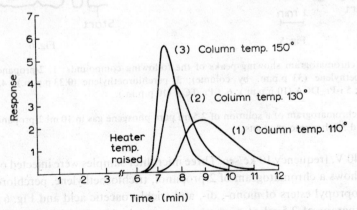

Fig. 4. Effect of column temperature on phosgene retention time (sample volume 1 ml; attenuation 64).

References pp. 363–364

Low-boiling chlorinated hydrocarbons such as tri- and perchloroethylene yield mainly hydrogen chloride on thermal oxidation[10] and chloroacetyl chlorides on photochemical oxidation[11]. Since the chlorinated solvents are used extensively for dry cleaning and degreasing purposes and can, on occasion, come into contact with hot surfaces or be exposed to ultraviolet light, the possibility exists for the formation of hazardous concentrations of phosgene and chloroacetyl chlorides.

A gas chromatographic method was developed by Dahlberg and Kihlman[12] for the determination of mono-, di-, and trichloroacetyl chloride at low concentration levels in air. The acyl chloride vapor in a small air sample (about 1 liter) was absorbed in 2-propanol and the ester thus formed was analyzed using electron-capture detection. By this method the di- and trichloro derivatives at concentrations lower than 1 p.p.m. could be measured. The sensitivity for monochloroacetyl chloride was somewhat less. Phosgene was determined down to 10^{-3} p.p.m. in air using electron-capture detection.

A Perkin-Elmer 800 gas chromatograph was used with an electron-capture detector and a 2 m × ⅛ in. diam. stainless steel column containing 20% silicon oil DC-200 on Chromosorb W. The column and detector temperatures were both 120°C; argon with 5% methane was the carrier gas (at 30 ml/min through the columns, and 210 ml/min dilution after column). The detector voltage was 0.8 µs

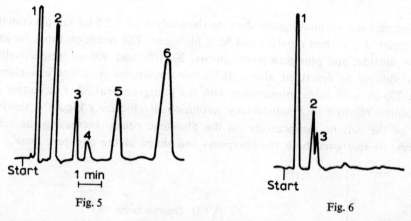

Fig. 5   Fig. 6

Fig. 5. Gas chromatogram showing peaks of the following compounds: 1, 2-propanol (solvent); 2, trichloroethylene (33 p.p.m. by colume); 3, perchloroethylene (0.33 p.p.m.); 4, i-Pr–MCA (33 p.p.m.); 5 i-Pr–DCA (10 p.p.m.); 6, i-Pr–TCA (10 p.p.m.).

Fig. 6. Gas chromatogram of a solution of 2.5 ml pure phosgene gas in 10 ml 2-propanol. 1, 2-Propanol; 2 and 3, phosgene products.

pulses of 40 V, frequency 10 kc/sec. Three microliter samples were injected on column. Figure 5 shows a chromatogram of 2-propanol, trichloroethylene, perchloroethylene, and the isopropyl esters of mono-, di-, and trichloroacetic acid and Fig. 6 illustrates a chromatogram of 2.5 ml of pure phosgene gas in 10 ml of 2-propanol. The suggested parameters for the GC of phosgene were column and detector temperatures of 25 and 120°C, respectively, 1 ml injection with gas-tight syringe on column, carrier gas and detector voltages similar to those described for acetyl chloride above. With these conditions, the retention times of oxygen and phosgene were about 28 and 130 sec, respectively, and trichloroethylene was eluted after about 30 min.

2. CYANOGEN AND HYDROGEN CYANIDE

Cyanogen, $(CN)_2$, is generally prepared by the slow addition of potassium cyanide solution to copper sulfate solution, or by heating mercury cyanide. It is extremely reactive and is used extensively in heterocyclic syntheses. It has also been used as a fumigant and may be formed in processes where there is heating of nitrogen-containing carbon bonds and in blast furnaces, gases, etc.

The physiological response of cyanogen is analogous to that of other cyanides and it is believed to be converted in the body partly to hydrogen cyanide and partly to cyanic acid (HOCN). The toxicity of cyanogen in air for various animal species has been described by Flury and Zernik[1]. Cyanogen is often associated with hydrogen cyanide and their degree of hazard is relatively the same. Derivatives of cyanogen such as cyanogen bromide, CNBr, are used in organic synthesis and as fumigants and pesticides and in gold-extraction processes.

Hydrogen cyanide (HCN, hydrocyanic acid, prussic acid) is manufactured chiefly by the reaction of ammonia, air, and methane in the presence of a platinum catalyst. It is estimated that the production in the U.S. in 1957 was more than 150 million pounds with over half of this used for the preparation of acrylonitrile[13]. Hydrogen cyanide has a broad spectrum of usage involving fumigation, in electroplating and in mining, in the production of various resin monomers such as acrylates, methacrylates, and hexamethylenediamine, in the production of other nitriles, and as a chemical intermediate. Hydrogen cyanide may be generated in such operations as blast furnaces, gas works, and coke ovens. It has been reported to be present (150–300 μg) in the mainstream of cigarette smoke[14]. The chemistry and reactions of hydrogen cyanide have been well documented.[15]

Hydrogen cyanide itself, as well as its simple soluble alkali metal salts, is among the most rapidly acting of all known poisons. The physiological response of animals and man to various concentrations of hydrogen cyanide in air has been described[1,16]. Hydrogen cyanide is toxic by virtue of the cyanide ion and its distribution and metabolism is that of the soluble cyanides (which are rapidly absorbed). Because they are allied both in their physiological and chemical action, cyanogen and hydrogen cyanide are generally estimated together colorimetrically as hydrocyanic acid[17-21].

Hydrogen cyanide, cyanogen, methane, and C_{1-8} amines have been separated[22] by gas chromatography at 125° on a 4.8 m column of Chromosorb G (60–80 mesh) coated with 4.5% Apiezon L. The support was carefully washed before use with HCl, 10% KOH, and a solution of KOH in methanol to remove the active centers.

The symmetry of the peaks was improved by the inclusion of 0.2% Span 80 and 0.8% $C_{18}H_{37}NH_2$ in the stationary phase. Ammonia did not interfere with the determination when the detection was by hydrogen flame ionization. Atmospheric samples were collected on a 15 cm column of 8% Apiezon L on Chromosorb G at liquid nitrogen temperature and then desorbed onto the chromatographic column at 100°. Each gram of cyanogen produced 5.5 millicoulombs of current. For amines, the number of millicoulombs produced/g equaled 44 (no. of C atoms − 0.677 no. of N atoms)/M.W. The log of the retention time relative to methane (log T) was related to the b.p. of the amine (in degrees), thus

$$\log T = 3.65 \times 10^{-3} \text{ (b.p.} +61.8)$$

The separation of hydrogen cyanide, water, and permanent gas has been accomplished by Woolmington[23] using a 7 ft. × 0.3125 in. stainless steel column containing 20% polyethylene glycol 1500 on 30–60 mesh Chromosorb at 90°C with helium as carrier gas at 160 ml/min. A thermal conductivity cell was used for detection and the retention times were permanent gases 1.5, hydrogen cyanide 3.5, and water 9.0 min, respectively.

Isbell[24] separated hydrogen cyanide and cyanogen on 25% triacetin columns. Type 5A or 13X molecular sieve (Wilkins Instrument & Research) was used to separate oxygen, nitrogen, methane, and carbon monoxide (nitric oxide was also

TABLE 1

RETENTION TIME AND VOLUMETRIC FACTORS

| Component | Retention time (min) | | | Volumetric factor (K')[a] with molecular sieve | |
| | Triacetin | Molecular sieve | | | |
		5A	13X	5A	13X
CO_2	0.54	∞^b	∞	1.38	1.25
Cl_2	0.80	∞	∞		
$(CN)_2$	1.02	∞	∞	1.66	1.60
O_2	0.48^c	1.70	1.46	1.00	1.00
CNCl	2.05	∞	∞		
N_2	0.48	3.35	2.06	1.10	1.09
HCN	4.35	∞	∞	1.00^d	1.00^d
NO	0.48	5.33	2.87	1.03	1.06
CH_4	0.48	5.52	3.25	0.99	1.00
CO	0.48	15.3	4.00	1.00	1.18

[a] Determined by method of Johns[25], based on peak area.
[b] Irreversibly adsorbed.
[c] Composite of permanent gases was eluted in 0.48 min.
[d] Determined with mixtures containing 35% HCN or less, $K' = 1.00$.

resolved in the absence of oxygen and methane). The gas chromatograph used was equipped with a stainless steel dual-column detector block with two matched-pair tungsten-filament elements that were operated at 75 mA. The carrier gas, helium, was dried with silica gel and supplied at an inlet gauge pressure of 18 p.s.i. and a rate of 108 ml/min at atmospheric pressure. Two columns, each prepared from $\frac{1}{4}$ in. (5 mm i.d.) copper tubing were used in series. The first column was 8 ft. long and contained non-acid-washed 30–60 mesh Chromosorb P impregnated with 25% triacetin; the second was 9 ft. long and contained a molecular sieve. The triacetin column was operated at 75° and the molecular sieve column at room temperature. These two columns separated mixtures of cyanogen, hydrogen cyanide, oxygen, nitrogen, methane, carbon monoxide, and carbon dioxide. Carbon dioxide was eluted from the first column immediately after the composite of permanent gases so that in the presence of more than about 25% of the permanent gases, CO_2 was not separated completely. Water vapor was absorbed by the triacetin column so that samples saturated with moisture at room temperature were analyzed with no tailing. HCN, $(CN)_2$, and CO_2 were irreversibly adsorbed in the molecular sieve column, but a single set of two columns was used for about 1000 analyses over a period of 6 months with no apparent loss of effectiveness. Table 1 lists the retention times and volumetric factors for the gases analyzed (CO_2, Cl_2, $(CN)_2$, O_2, CNCl, N_2, HCN, NO, CH_4, and CO).

Schneider and Freund[26] determined small quantities of HCN (10^{-6} to 5×10^{-4} M) recovered from aqueous solution using a column of dinonyl phthalate. A measured flow of air (50 ml/min) was first passed through the liquid and the HCN in the air concentrated on a pre-column of dinonyl phthalate on 40–60 mesh firebrick and maintained at dry ice–acetone temperature. The HCN was released by warming the pre-column to *ca.* 57° and the evolved gases passed onto an 18 ft. chromatographic column packed with 20% dinonyl phthalate on Chromosorb W. A thermal conductivity cell was used as the detector.

Cropper and Kaminsky[27] determined hydrogen cyanide in air by using a concentration step in which the HCN from a 1 liter air sample was adsorbed on a silica gel pre-column and subsequently released on heating. Adiponitrile was used as the stationary phase and HCN was detected by FID.

Wolfram and Arsenault[28] separated a mixture of hydrogen cyanide, acetaldehyde, acetone, acrolein, and ethyl acetate on a 1.2 m × 5 mm column containing 40% polyethylene glycol 400 on 30–60 mesh C-22 firebrick at 30°C with helium carrier flow rate of 50 ml/min and detection with a thermal conductivity cell.

REFERENCES

1 F. FLURY AND F. ZERNIK, *Schädliche Gase*, Springer, Berlin, 1931.
2 A. L. CINCH, S. S. LORD, JR., K. A. KUBITZ AND M. D. DeBRUNNER, *Am. Ind. Hyg. Assoc. J.*, 26 (1965) 465.

3   W. P. YANT, J. C. OLSEN, H. H. STORCH, J. B. LITTLEFIELD AND L. SCHEFLAN, *Ind. Eng. Chem. Anal. Ed.*, 8 (1936) 20.
4   L. J. PRIESTLEY, F. E. CRITCHFIELD, N. H. KETCHAM AND J. D. CAVENDER, *Anal. Chem.*, 37 (1965) 70.
5   N. BRENNER AND E. CIEPLINSKI, *Ann. N.Y. Acad. Sci.*, 72 (1959) 705.
6   R. P. DEGRAZIO, *J. Gas Chromatog.*, 3 (1965) 204.
7   J. LACY AND K. G. WOOLMINGTON, *Analyst*, 86 (1961) 350.
8   A. FISH, N. H. FRANKLIN AND R. T. POLLARD, *J. Appl. Chem.*, 13 (1963) 506.
9   R. J. GRAHAM AND F. D. STEVENSON, *J. Chromatog.*, 47 (1970) 555.
10 B. SJÖBERG, *Suensk Kem. Tidskr.*, 64 (1952) 63.
11 J. A. DAHLBERG, *Acta Chem. Scand.*, 23 (1969) 3081.
12 J. A. DAHLBERG AND I. B. KIHLMAN, *Acta Chem. Scand.*, 24 (1970) 644.
13 J. H. WOLFSIE AND B. C. SHAFFER, *J. Occupational Med.*, 1 (1959) 28.
14 A. ARTHO AND R. KOCH, *Beitr. Tabalforsch.*, 5 (1969) 58.
15 V. MIGRDICHIAN, *The Chemistry of Organic Cyanogen Compounds*, Reinhold, New York, 1947.
16 H. C. DUDLEY, T. R. SWEENEY AND J. W. MILLER, *J. Ind. Hyg. Toxicol.*, 24 (1942) 255.
17 C. K. FRANCIS AND W. B. CONNELL, *J. Am. Chem. Soc.*, 35 (1913) 1624.
18 M. B. JACOBS, *The Analytical Toxicology of Industrial Inorganic Poisons*, Interscience, New York, 1967, p. 730.
19 W. LEITHE, *The Analysis of Air Pollutants*, Ann Arbor Science, Ann Arbor, Mich., 1971, p. 227.
20 J. EPSTEIN, *Anal. Chem.*, 19 (1947) 273.
21 W. A. ROBBIE AND P. J. LEINFELDER, *J. Ind. Hyg. Toxicol.*, 20 (1945) 136.
22 M. T. DMITRIEV AND N. A. KITROSSKII, *Gigiena i Sanit.*, 34 (1969) 63; *Chem. Abstr.*, 71 (1969) 244.
23 J. WOOLMINGTON, *J. Appl. Chem.*, 11 (1961) 114.
24 R. E. ISBELL, *Anal. Chem.*, 35 (1963) 255.
25 T. JONES, in Gas chromatographic applications manual, *Bull. 756-A*, Beckman Instruments Co., Fullerton, Calif., 1960, p. 43.
26 C. R. SCHNEIDER AND H. FREUND, *Anal. Chem.*, 34 (1962) 69.
27 F. R. CROPPER AND S. KAMINSKY, *Anal. Chem.*, 35 (1963) 735.
28 M. L. WOLFRAM AND G. P. ARSENAULT, *J. Am. Chem. Soc.*, 82 (1960) 2819.

Chapter 17

HYDROCARBONS (PARAFFINIC AND OLEFINIC)

1. METHANE

The principal source of methane is natural gas which contains from 60 to 98% of methane and provides approx. 35% of the thermal energy produced in the United States. In addition to its major use as a fuel, methane is an important raw material for the manufacture of carbon black, acetylene, hydrogen, halogenated methanes, ethylene, and many other products.

Methane also occurs naturally in coal and is recovered for use as a fuel by destructive distillation (coal gas), as a byproduct from coke-ovens (coke-oven gas), and by the treatment of ventilation exhaust from coal mines (coal gas and coke-oven gas contain approx. 30% methane, the remainder being mainly hydrogen and carbon dioxide). Methane is the anaerobic decomposition product of organic compounds and is contained in large amounts in the waste gases of anaerobic putrefaction processes of wastes such as occur in waste water processing. Because of the possibility of explosions, air–methane mixtures are significant. The explosion limits are between 5 and 14% methane with such mixtures constituting a permanent hazard in coal mines. Methane has no appreciable physiological action except when it lowers the partial pressure of oxygen in the air enough to cause systemic effects due to oxygen deprivation. Hence, methane levels below 1% are believed to have no harmful effect on humans or vegetation. Methane has been determined in air by portable interferometers, Haldane or Orsatt gas apparatus, methane analyzer[1,2], as well as by infrared devices[3]. Absorption in the infrared region, 2–15 μm (corresponding to 5000–660 cm^{-1} wave numbers) is expedient for carbon-containing components such as CH_4, CO, and CO_2.

The separation of methane in a variety of gaseous mixtures has been successfully and rapidly accomplished by gas chromatography. Bombaugh[4] used molecular sieve flour for the separation of methane, hydrogen, oxygen, nitrogen, and carbon monoxide. The packing was prepared by mixing one part by weight of molecular sieve 5A (passing a 200 mesh sieve) with 2 parts by weight of 60–80 mesh red Chromosorb and, after rolling, using the 60–80 mesh fraction. A column prepared with this mixture permitted a six-fold increase in efficiency and an almost two-fold increase in peak height compared with a column packed only with 60–80 mesh molecular sieve.

Ray[5] separated a mixture of methane, carbon monoxide, and hydrogen on a column of activated charcoal. Boreham and Marhoff[6] separated a mixture of methane, ethane, carbon monoxide, carbon dioxide, ethylene, hydrogen, oxygen, and nitrogen on a 4 m × 4 mm column of silica gel at 30° with argon as carrier gas at 3.6 l/h. A Perkin-Elmer Vapor Fractometer gas chromatograph was used with a thermal conductivity cell, thermistor model detector. The retention times were hydrogen 1.5, oxygen and nitrogen 1.8, carbon monoxide 2.1, methane 2.8, ethane 10.5, carbon dioxide 15.5, and ethylene 17.5 min, respectively.

The separation of air, methane, carbon dioxide, and hydrogen sulfide was reported by Grune and Chueh[7] using a column of 35% silicone grease coated on 28–48 mesh firebrick. A short column packed with activated charcoal was placed in the flow system at the exit side of the partition column to improve the separation of methane from the air present. Very low levels of methane in air samples of 0.3 ml have been determined using a 2.2 m × 3 mm column of dioctyl sebacate with helium as the carrier gas at 70 ml/min and detection using FID. The methane peak was eluted in ca. 1.6 min[3].

The use of molecular sieves 5A and 13X for the separation of hydrogen, oxygen, nitrogen, methane, and carbon monoxide have been reported by Kyryacos and Boord[8] and Ellis and Forrest[9]. The separation of mixtures containing oxygen, nitrogen, nitrous oxide, carbon dioxide, carbon monoxide, and methane by temperature programming was reported by Graven[10]. The analysis of a mixture containing oxygen, nitrogen, methane, carbon monoxide, and carbon dioxide using two columns in series was described by Murakami[11]. The gases passed first through a silica gel column to the reference cell of a thermal conductivity detector then through a molecular sieve column and finally through the measuring cell of the detector. By suitable switching of the recorder polarity, a continuous chromatogram was obtained showing a peak for the mixture of oxygen, nitrogen, methane, and carbon monoxide followed by a carbon dioxide peak from the silica gel column, then individual peaks of oxygen, nitrogen, methane, and carbon monoxide from the molecular sieve column. Manka[12] used a similar technique for the separation of hydrogen in addition to the other gases.

The separation of C_1–C_4 hydrocarbons has been achieved using several columns in tandem[13–15]. Liquid-phase chromatography has been employed but the use of this was limited by the volatility and the freezing point of the stationary liquid phase[16–18]. Capillary gas chromatography[19] as well as temperature-programmed GC have also been applied[20] for the analysis of hydrocarbons. The use of porous polymer beads for the separation of C_1–C_2 (ref. 21) and C_1–C_3 hydrocarbons (ref. 22) has been previously reported.

The simultaneous separation of C_1–C_4 hydrocarbons on a single Porapak column was described by Papic[23]. The separation of methane, ethene, ethane, propene, propane, cyclopropane, iso-butane, n-butane, butene-1, methyl propene, and cis- and trans-butene-2 was achieved in 60 min on a 3.15 m Porapak Q (80 to 100 mesh) column at 76°C with helium carrier gas flow rate 100 ml/min. Hydrogen, air, and methane were also separated on a similar column at 0°C. Copper tubing of 4 mm o.d. was used for the columns which were purged before use at 230°C for 1 h. Columns were mounted in a thermostat and 10 μmole samples were introduced by means of a flow of carrier gas through the sample tube and detection of the gas was achieved using a thermal conductivity detector. Figure 1 illustrates the separation of a C_1–C_4 hydrocarbon mixture on Porapak Q (80–100 mesh) at 76°C.

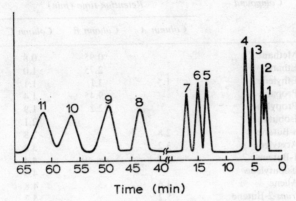

Fig. 1. Chromatogram of the C_1–C_4 hydrocarbon mixture. Column, 315 × 0.4 cm Cu tubing; adsorbent, Porapak Q 80–100 mesh; temperature, 76°C; carrier gas, He flow 100 ml/min; detector, TCD. Assignment of peaks: 1, hydrogen + air; 2, methane; 3, ethene; 4, ethane; 5, propene; 6, propane; 7, cyclopropane; 8, isobutane; 9, methylpropene + butene-1; 10, n-butane + trans-butene-2; 11, cis-butene-2.

The use of ethyl-N,N-dimethyl oxamate (the half dimethylamide of ethyl oxalate) as the liquid phase for the separation of C_1–C_4 hydrocarbons on 3 columns was described by Richmond[24]. Commercially available solid supports were used without special treatment and the columns had a long life if the temperature was maintained at 0°C. Column A consisted of 6 ft. × ¼ in. o.d. copper tubing with 0.030 in. wall packed with 20 wt. % of ethyl-N,N-dimethyl oxamate on 60–80 mesh acid-washed firebrick and operated at 0°C with 40 ml/min helium flow measured at the exit. Column B consisted of 6 ft. × ¼ in. o.d. copper tubing with 0.030 in. wall packed with 20 % ethyl-N,N-dimethyl oxamate on 60–80 mesh silica gel ½ (Davison) operated at 25°C and 60 ml/min helium flow at the exit. Both columns were used with a thermal conductivity detector held at 50°C. Column C was a 10 ft. × ⅛ in. o.d. stainless steel tubing with 0.010 in. wall packed with 10 % ethyl-N,N-dimethyl oxamate on 100–120 mesh acid-washed firebrick. The column temperature was 0°C, helium flow 30 ml/min, and a flame-ionization detector was used. These conditions were used in the nitrogen fixation studies of Hardy et al.[25].

Table 1 summarizes retention times of the hydrocarbons analyzed with the 3 columns. With column A, samples as large as 1 ml of gas containing 90–95 % of one component could be analyzed. Figure 2 shows a typical separation of butene isomers and other low-boiling hydrocarbons. Figure 3 illustrates a separation of ethylene–acetylene as obtained in nitrogen fixation studies.

The liquid phase described in the above study (ethyl-N,N-dimethyl oxamate) has been previously employed (at 20 % level) with silica gel as the solid support

TABLE 1

HYDROCARBON RETENTION TIMES ON 3 COLUMNS

Compound	Retention time (min)		
	Column A	Column B	Column C
Methane		0.95	0.8
Ethane		2.75	1.0
Ethylene	1.5	3.1	1.1
Propane		9.45	1.4
Propylene		12.2	1.9
Isobutane			2.1
n-Butane	2.8		2.8
Acetylene	3.2		3.8
1-Butene	4.2	ca. 40	4.4
Isobutylene			4.5
Allene	4.6		4.8
trans-2-Butene	5.2		5.2
cis-2-Butene	6.0		6.2
Methyl acetylene	8.3		10.4
Butadiene	9.0		

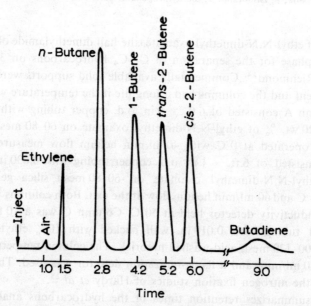

Fig. 2. Typical hydrocarbon separation. Column A, 1 ml gas sample.

for the separation of methane, ethane, ethylene, propane, and propylene at a column temperature of 25°C and helium flow at 60 ml/min[26].

With acid-washed firebrick as the solid support, 0°C column temperature and 40 ml/min helium flow, this material has been used especially for the C_4 hydro-

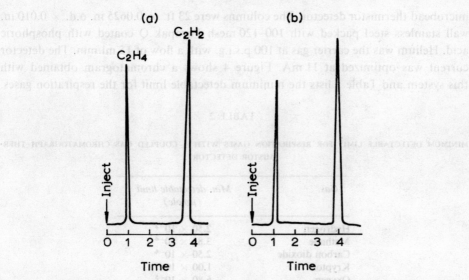

Fig. 3. Ethylene–acetylene separation. Column C 200 μl gas samples, including helium diluent. Nitrogen fixation assay. (a) Standard mix; (b) incubation gas phase N_2-fixation gas assay.

carbons in mixture and for ethylene, acetylene, allene, and methylacetylene in mixture[24].

Carle[27] developed a GC system (with minimal power requirements) for determination of the microbial respiratory gases (H_2, N_2, O_2, CH_4, CO_2) using Kr as an internal standard. Such a system could be used in a life-detection system to be considered for landing on Mars. The system consisted of a column packed with phosphoric acid-treated Porapak Q and maintained at 20°C in conjunction with a

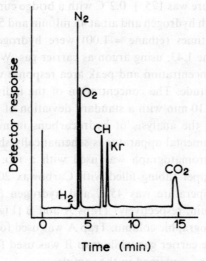

Fig. 4. Typical chromatogram of microbial respiratory gases.

microbead thermistor detector. The columns were 23 ft. × 0.0625 in. o.d. × 0.010 in. wall stainless steel packed with 100–120 mesh Porapak Q coated with phosphoric acid. Helium was the carrier gas at 100 p.s.i.g. with a flow of 15 ml/min. The detector current was optimized at 11 mA. Figure 4 shows a chromatogram obtained with this system and Table 2 lists the minimum detectable limit for the respiration gases.

TABLE 2

MINIMUM DETECTABLE LIMIT FOR RESPIRATION GASES WITH A COUPLED GAS CHROMATOGRAPH–THERMISTOR DETECTOR

Gas	Min. detectable limit (μmole)
Hydrogen	2.50×10^{-2}
Methane	3.85×10^{-4}
Carbon dioxide	2.50×10^{-4}
Krypton	1.00×10^{-4}
Oxygen	6.80×10^{-5}
Nitrogen	5.24×10^{-5}

Durbeck[28] described the gas chromatographic determination of hydrogen, methane, ethane, and ethylene using a Hewlett-Packard Model 5754 gas chromatograph equipped with dual columns and thermal conductivity and FID detectors. The columns were 280 cm × 4 mm i.d. stainless steel containing neutral Al_2O_3 (Woelm) 60–80 mesh. Both nitrogen and argon were investigated as carrier gas (argon at 50 ml/min and nitrogen at 60 ml/min). The oven temperature was 90°C ± 0.5°C and the injection port temperature was 100°C. The thermal conductivity detector temperature was 175 ± 0.2°C with a bridge current of 100 ± 0.05 mA. The FID was at 175°C with hydrogen and air at 45 ml/min and 510 ml/min, respectively. The relative retention times (ethane = 1.00) were hydrogen 0.45, methane 0.58, ethane 1.00, and ethylene 1.43, using argon as carrier gas. With argon a nearly ideal relationship between concentration and peak area response was obtained over more than 3 orders of magnitude. The concentration of the individual components was determined in less than 10 min with a standard deviation of ±1.5%.

Zocchi[29] described the analysis of hydrocarbons in methane in the part per billion range. The experimental apparatus is schematically shown in Fig. 5. A Carlo Erba Fractovap gas chromatograph was used with a FID. The column consisted of 6 m × 4 mm i.d. copper tubing filled with Carbowax 20M on 40–60 mesh alumina. The column temperature was 45°C and hydrogen flow and air flow were 33 ml/min and 300 ml/min, respectively. Traps A and B (Fig. 5) contained the same material as the chromatographic column. Trap A was used for condensing the hydrocarbon impurities in the carrier gas while trap B was used for condensing the trace amounts of hydrocarbons contained in the sample.

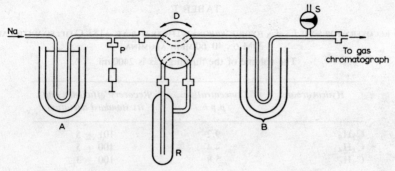

Fig. 5. Schematic drawing of the apparatus for the preconcentration of hydrocarbons.

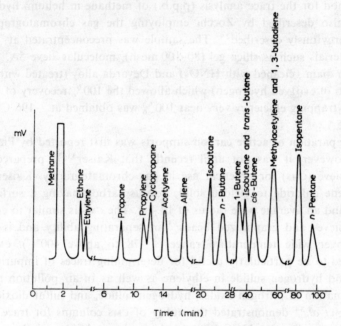

Fig. 6. Typical chromatogram obtained after heating of trap B.

Figure 6 illustrates the elution order and retention time for various hydrocarbons after the heating of trap B. It can be seen that the peak of the methane retained by trap B at the boiling oxygen temperature did not interfere with the analysis of the ethane–ethylene system. The amount of methane retained by trap B after flushing was 10^3–10^4 times less than that introduced with the sample in the apparatus. With the particular column used, the separation of the various C_2–C_5 hydrocarbons was satisfactory except for the *trans*-butene-isobutene system and methylacetylene-1,3-butadiene system whose peaks were completely superimposed. The recovery efficiency of the apparatus was verified for different hydrocarbons and concentrations in the p.p.b. range (Table 3).

References pp. 388–389

TABLE 3

PERCENT RECOVERY FOR SOME C_2–C_3 HYDROCARBONS WITH TRAP B AT $-183°C$ FILLED WITH CARBOWAX 20M ON 40–60 MESH ALUMINA
The volume of the flushing gas is 2400 ml.

Hydrocarbons	Concentrations (p.p.b.)	Recovery efficiency and its standard deviation
C_2H_6	9.2	101 ± 5
C_2H_4	4.4	100 ± 5
C_3H_8	5.8	100 ± 3

A method for the trace analysis (p.p.b.) of methane in helium, hydrogen, and neon was also described by Zocchi employing the gas chromatograph and the apparatus previously described[29]. The sample was preconcentrated at $-215°C$ on various materials such as silica gel (80–100 mesh), molecular sieve 5A, lead pellets (0.6–1.1 mm diam., cleaned with HNO_3) and Devarda alloy (treated with HCl until the cessation of evolved hydrogen) which allowed the 100% recovery of the 8 p.p.b. methane. A trapping efficiency very near 100% was obtained at $-196°C$ with molecular sieve 5A.

The preparation of active carbon supports was first reported by Pierce et al.[30] in 1949. However, it was not until recently that Kaiser[31,32] prepared a carbon molecular sieve (CMS) for practical use in gas chromatography. Kaiser pyrolyzed polyvinylidene chloride to yield a solid, porous carbon having a surface area of 1000 m^2/g and an average pore radius of 12.4 Å. The CMS is similar to conventional molecular sieves and graphitized beads[33] in separating ability and is capable of operating over wide temperature ranges (-78 to above $400°C$). CMS columns are suggested to be particularly useful for determining traces of impurities such as acetylene and hydrogen sulfide in ethylene as well as in air pollution analyses for the determination of nitrogen oxides, hydrogen sulfide, and sulfur dioxide.

Zlatkis et al.[34] demonstrated the utility of CMS columns for trace analysis in gas chromatography. An MCC-100 gas chromatograph (Aptech Instrument Co., Houston) was used with a 5 ft. × 0.023 in. i.d. stainless steel column containing 44–53 μm CMS (Dr. R. Kaiser, BASF, Ludwigshaven). After packing the column, the sieve was activated by heating at $200°C$ for 16 h with helium flowing through at a rate of 40 ml/min. Analyses were performed at three temperatures, -78, 25, and $150°C$. The flow rate of helium carrier gas was 143 ml/min for $-78°C$ operation, whereas the other temperatures required flow rates of 40 ml/min. Detection was accomplished with hot wire thermal conductivity detectors except for the trace analyses of acetylene in ethylene where a FID chromatograph was used.

Table 4 lists the retention data for 12 gases on a carbon molecular sieve column obtained at -78, 25, and $150°C$. All of the major impurities, air, methane, carbon dioxide, and ethane, were resolved in less than 6 min using a 6 ft. column at $150°C$

TABLE 4

RETENTION DATA FOR CARBON MOLECULAR SIEVE COLUMN

Column, 6 ft. × $\frac{1}{8}$ in.; 70–100 mesh CMS; flow rate at −78°C, 143 ml/min; flow rates at 25° and 150°C, 40 ml/min; carrier gas, helium.

Compound	*Retention time (min) at*		
	−78°C	*25°C*	*150°C*
Oxygen	3.60	0.70	0.37
Nitrogen	4.40	0.70	0.37
Nitric oxide	5.45	0.84	0.37
Carbon monoxide	7.60	0.94	0.37
Water			0.50
Methane		3.08	0.50
Carbon dioxide		8.07	0.68
Nitrous oxide		12.88	0.78
Acetylene			1.25
Hydrogen sulfide			1.87
Ethylene			2.06
Ethane			3.11
Sulfur dioxide			4.50

Fig. 7

Fig. 8

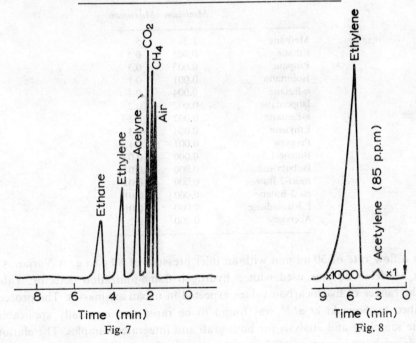

Fig. 7. Carbon molecular sieve separation of light gases. Column: 6 ft. × $\frac{1}{8}$ in., 70–100 mesh; temperature, 150°C; flow rate, 40 ml/min helium.

Fig. 8. Trace analysis of acetylene in ethylene. Column: 2 ft × $\frac{1}{8}$ in. i.d., 70–100 mesh CMS; temperature, 100°C.

References pp. 388–389

as shown in Fig. 7. Rapid determinations were also effected for the analysis of trace acetylene in ethylene using a 2 ft. × $\frac{1}{8}$ in. i.d. column of CMS (Fig. 8).

Air pollutants such as NO, N_2O, CO, H_2S, and SO_2 were resolved with CMS columns using temperature programming. The oxides of nitrogen were eluted at ambient temperatures while the sulfur compounds required elevated temperatures. All the components were eluted within 10 min by programming from 25–150°C. It was also suggested that since CMS columns do not bleed, they would be excellent for use in systems involving the helium ionization detector or a gas chromatograph–mass spectrometer combination.

Sawicki et al.[35] described a procedure for the identification and quantitative analysis of C_1–C_5 atmospheric hydrocarbons; 2.4 m × 3 mm o.d., 1.6 mm copper or stainless steel columns were packed with activated alumina coated with 17% β,β-oxydipropionitrile. The column was maintained at 0°C and the helium carrier

TABLE 5

RANGES OF HYDROCARBON VALUES EXPECTED IN URBAN AIR MASSES

Component	Range (p.p.m.) in air	
	Minimum	Maximum
Methane	1.2	15
Ethane	0.005	0.5
Propane	0.003	0.3
Isobutane	0.001	0.1
n-Butane	0.004	0.4
Isopentane	0.002	0.2
n-Pentane	0.002	0.2
Ethylene	0.004	0.3
Propene	0.001	0.1
Butene-1	0.000	0.02
Isobutylene	0.000	0.02
trans-2-Butene	0.000	0.01
cis-2-Butene	0.000	0.01
1,3-Butadiene	0.000	0.01
Acetylene	0.000	0.2

gas at a flow rate of 30 ml/min with an inlet pressure of 80 p.s.i.g. A Varian Aerograph Model 500B was used with a hydrogen flame-ionization detector. Table 5 lists the ranges of hydrocarbon values expected in urban air masses. The procedure described by Sawicki et al.[35] was found to be rapid and especially applicable to routine sampling and analysis for both grab and integrated samples. The elution of 17 hydrocarbons was accomplished within 16 min (Fig. 9).

Methane, present in both coal and non-coal mines, presents a well-known explosion hazard and lowering of the oxygen content, either by oxidation or dilution with other gases, presents an asphyxiation hazard.

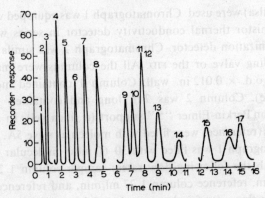

Fig. 9. Sample chromatogram showing position of hydrocarbon elution peaks. 1, Methane; 2, ethane; 3, ethylene; 4, propane; 5, propylene; 6, acetylene; 7, isobutane; 8, *n*-butane; 9, butene-1; 10, isobutylene; 11, 2,2-dimethylpropane; 12, *cis*-2-butene; 13, *trans*-2-butene; 14, isopentane; 15, 3-methylbutene-1; 16, *n*-pentane; 17, 1,3-butadiene.

Freedman *et al.*[36] have developed a gas chromatographic method for the rapid and accurate determination of oxygen, nitrogen, carbon dioxide, methane (in concentrations ranging from 0.01 to 100%) and, if present, carbon monoxide in mine air. This method can replace such time-honored gas reaction techniques as the Haldane[37] or Orsat analyses.

Figure 10 illustrates the gas chromatographic system used consisting of dual detectors and dual columns. Two types of Model C-40 gas chromatographs (In-

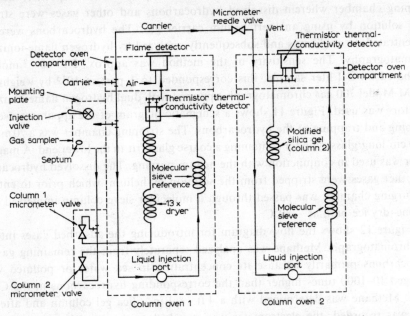

Fig. 10. Gas chromatographic system for analyses of the principal constituents of mine atmospheres.

struments, Inc., Tulsa) were used. Chromatograph 1 was equipped with a gas sampling valve and a thermistor thernal conductivity detector in series with a downstream hydrogen flame-ionization detector. Chromatograph 2 was similar to number 1 but without the sampling valve or the FID. All the columns were made from stainless steel tubing, $\frac{1}{8}$ in. o.d. × 0.012 in. wall. Column 1 contained molecular sieve 5A, 35–48 mesh (Linde). Column 2 was 7 ft. long and contained 1.5% diethylhexyl sebacate (DEHS) on Perkin-Elmer "S" support (a silica gel). In addition, identical 5 ft. coiled columns (reference) were filled with molecular sieve 5A. A 7 in. precolumn dryer for chromatograph 1 was filled with 30–60 mesh molecular sieve. Helium was the carrier gas with the following flow rates: sample column 1 22 ml/min, sample column 2 46 ml/min, reference column 1 27 ml/min, and reference column 2 35 ml/min. The hydrogen flow rate for detectors 1 and 2 was 43 ml/min and the air flow rate for detectors 1 and 2 was 150 ml/min. Currents for detectors 1 and 2 were 16 and 20 mA, respectively. The chromatographic analysis was complete within 7 min. If carbon monoxide was present in excess of 0.01%, it could be determined as a peak separating about 3 min later. Column 1 separated oxygen, nitrogen, and methane and, if indicated, carbon monoxide. (The CH_4 and CO peaks had retention times of 6 and 10 min, respectively.) Column 2 effected the separation of CO_2 from composite gases in about 2 min. Concentrations of the gases were quantitatively determined by comparing the integrated peak areas with the areas of standard gas peaks under similar conditions.

Swinnerton and Linnenbom[38] described a new method for the determination of C_1–C_4 hydrocarbons in sea water. The technique utilized a specially designed stripping chamber wherein dissolved hydrocarbons and other gases were stripped from solution by using an inert helium carrier gas. The hydrocarbons were then concentrated on cold traps and subsequently analyzed by hydrogen flame-ionization chromatography. The sensitivity of the method was approx. 2×10^{-12} moles of gas (based on a 1 liter sample, this corresponded to 1 part in 10^{13} by weight). An F & M Model 700 gas chromatograph equipped with dual hydrogen flame-ionization detectors was used. Figure 11 shows a schematic diagram of the apparatus used for stripping and trapping of the hydrocarbons. The stripping chamber was a 70 mm i.d. × 50 cm long glass cylinder containing a coarse glass frit in the lower end. A magnetic stirrer was used in conjunction with the helium purging. The dissolved hydrocarbons and other gases were stripped from the solution by helium, which prior to entering the purging chamber was passed through a molecular sieve column immersed in an acetone–dry ice bath at −80°C.

Figure 12 shows the flow diagram for introducing the adsorbed gases into the gas chromatograph. Methane was analyzed separately from the remaining gaseous hydrocarbons primarily because its concentration in sea water or polluted water averaged 10–1000 times higher than the corresponding hydrocarbons in the C_2–C_4 range. Methane was determined with a 4 ft. × $\frac{1}{4}$ in. silica gel column and after the peak was recorded, the electrometer was switched to the second column detector

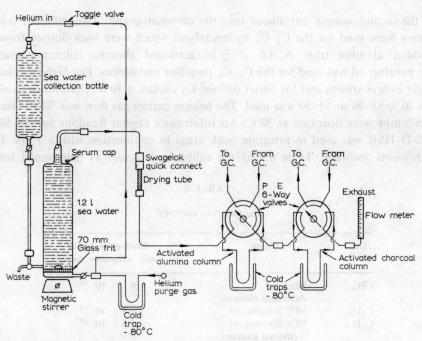

Fig. 11. Sampling and stripping. Flow diagram showing the stripping of sea water by the helium purge gas and the subsequent trapping of the hydrocarbons.

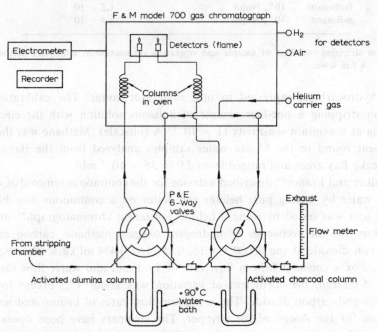

Fig. 12. Injection and analysis. Flow diagram showing the injection of trapped hydrocarbon samples into the chromatographic columns for analysis.

and the second sample introduced into the chromatograph. Two different column systems were used for the C_2–C_4 hydrocarbons which were back-flushed from the activated alumina trap. A 4 ft. × $\frac{3}{16}$ in. activated alumina column containing 10% paraffin oil was used for the C_2–C_4 paraffins and olefins. For higher molecular weight hydrocarbons and for better defined C_4 olefins, a 10 ft. Chromosorb column with 20% SF-96 or SE-30 was used. The helium carrier gas flow was 70 ml/min and all columns were operated at 30°C. An Infotronics Digital Readout System Model CRS-11-HSB was used to integrate peak areas in conjunction with a 1 mV Texas Instruments recorder. Table 6 lists the calibration factors as determined for the

TABLE 6

CALIBRATION FACTORS[a]

Gas	Column	Calibration factor (ml/l integrator count)
CH_4	Silica gel	6.1×10^{-10}
	Activated alumina	
C_2H_6	10% Silicone oil	1.4×10^{-10}
C_2H_4	10% Silicone oil	1.6×10^{-10}
	Activated alumina	
C_3H_8	10% Nujol	1.7×10^{-10}
C_3H_6	10% Nujol	1.6×10^{-10}
Isobutane	10% Nujol	1.2×10^{-10}
n-Butane	10% Nujol	1.4×10^{-10}

[a] Based on stripping one liter of sample and operating on maximum sensitivity of the detector, 1×10^{-11} A full scale.

C_1–C_4 hydrocarbons expressed in ml/l integrator count. The calibrations were based on stripping a one-liter sample of aqueous solution with the electrometer operating at maximum sensitivity (1×10^{-11} A fullscale). Methane was the highest component found in the 65 sea water samples analyzed from the Bahamas and Chesapeake Bay areas and ranged from 2.0 to 36×10^{-5} ml/l.

Walker and France[39] described a device for the continuous removal of dissolved gases in water by mixing pure helium and water on a continuous flow basis. The stripper unit was tested by using a helium ionization chromatograph[40] and found satisfactory for the recoveries of hydrogen, oxygen, methane, carbon monoxide, and carbon dioxide at the 0.1, 0,3, 0.15, 0.18, and 0.454 ml/kg levels, respectively, at 18°C. For a constant helium flow rate of 1 ml/min and water flow rates in the range 1–5 ml/min, the coefficient of variation was about 4%, except for carbon monoxide and carbon dioxide. The maximum flow rates of helium and water were 10 ml/min for the design of the stripper. The strippers have been operated on a continuous basis at concentrations from 0.001 to 1 ml of gas/kg of water. Figure 13 illustrates the schematic diagram of the apparatus.

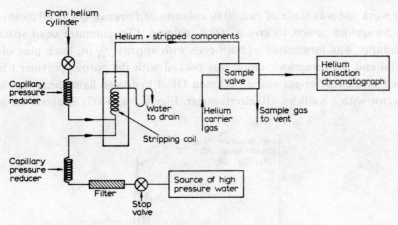

Fig. 13. Schematic diagram of the apparatus.

The GC analysis of methane in expired air was described by Levey and Balchum[41]. Samples of expired air were obtained by having subjects rebreathe 6 l of oxygen contained in a 9.5 × 18.5 in. laminated polyethylene–mylar bag. The determination of CH_4 was carried out with a Barber-Colman Gl-C gas chromatograph equipped with a hydrogen flame-ionization detector. The column was a copper coil, 6 ft. long × ¼ in. o.d. containing molecular sieve 5A. The column was maintained at 90°C and helium was the carrier gas at 15 ml/min through the column. The hydrogen flame was maintained at 7 p.s.i. of hydrogen and the compressed air was at 50 p.s.i. Under these conditions methane eluted in 9 min. The response to methane concentration was linear between 8 and 166 p.p.m. when the instrument was set at a sensitivity of 1000 and an attenuation of 2. (The attenuation was reduced to 1 for the lower concentrations.)

The analysis of organic compounds in human breath by gas chromatography–mass spectrometry was described by Jansson and Larsson[42]. The samples were first collected using a respiratory mask then stored in plastic bags. The whole procedure, sampling and analysis, required 15 min and the following compounds and concentrations in p.p.m. were measured: methane (0–30), isoprene (0.09–0.45), acetone (0.2–0.8), methanol (0.3–3.4), and ethanol (0.05–0.36). The gas chromatographs used (employing flame-ionization detection) were constructed as described by Larsson[43] and Jansson[44] and an LKB 9000 gas chromatograph–mass spectrometer was used for the identifications. A molecular sieve column 5A (Linde) at 70°C was used for the methane analysis, while the other compounds were separated either on a Porapak Q column at 110°C or 20% PEG 1500 at 65°C.

Mixtures of CH_4 and CD_4 have been separated by gas chromatography[45–47] and more recently by use of a porous polymer bead column[46,47]. It was shown in earlier work by Czubryt and Gesser[48] that a 20 ft. × ⅛ in. o.d. (50–80 mesh) Porapak Q column operated at −85°C was effective in CH_4–CD_4 mixture separation.

In later work use was made of two 50 ft. columns of Porapak Q (50–80 mesh) joined with a Swagelock union to give a 100 ft. column. (To minimize dead space each 50 ft. column was terminated at both ends with approx. $\frac{1}{16}$ in. thick plug of G. E. Fonmetal and the Swagelock union was packed with the porous polymer.) Helium was used as the carrier gas and a Beckman GC-4 hydrogen flame head was used as the detector with a Keithley 410 electrometer. Figure 14(A)–(C) illustrate the separa-

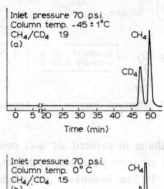

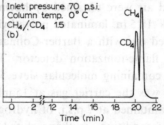

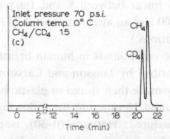

Fig. 14. Gas chromatographic separation of CH_4 and CD_4 on Porapak Q. (a) Inlet pressure, 70 p.s.i.; column temperature, $-45 \pm 1°C$; CH_4/CD_4, 1.9; (b) inlet pressure, 70 p.s.i.; column temperature, $0°C$; CH_4/CD_4, 1.5; (c) inlet pressure, 70 p.s.i.; column temperature, $0°C$; CH_4/CD_4, 1.5.

tion of CH_4 and CD_4 at $-45 \pm 1°C$ and $0°C$, respectively. Although a better separation is obtained at lower temperatures and/or slower flow rates, retention times became much longer such that a slight gain in separation nearly doubles the retention times. Figure 14(C) illustrates an increase in separation at $0°C$ by a sample freezing technique achieved by fuming a small narrow U just past the injection port out of approx. 6 in. of the column. The U was then cooled to liquid nitrogen temperature,

the sample then injected and allowed to freeze out in the U after which the liquid
nitrogen was quickly removed.

Czubryt and Gesser[48] cited a number of advantages of a porous polymer
bead column including relatively short retention times, loss of packing and mecha-
nical stability, no requirement for special treatment (aside from the initial degassing)
nor regeneration, and the column is unaffected by water or by impurities in the
carrier gas.

The complete analysis of gas-phase photolysis and radiolysis product mixtures
by GC was described by Heckel and Hanrahan[49]. Products which were analyzed by
a Duplex Gas Chromatograph included permanent gases such as H_2, CH_4, N_2,
less volatile components with boiling points up to 250°C as well as inorganic components or other components amenable to external analysis by spectrophotometry,
titration, etc. The unit consisted of two separate gas chromatographs and a common
vacuum inlet system which included a removable aliquot tube for external sample
analysis and an automatic Toepler pump. The Toepler pump collected permanent
gases in one sample loop for analysis in an isothermal chromatograph with a thermal
conductivity detector while less volatile components were trapped in a second
sample loop for analysis in a temperature-programmed chromatograph with a
hydrogen flame-ionization detector. The apparatus was equipped with a stream
splitting valve to divert a portion of the sample into a mass spectrometer to permit

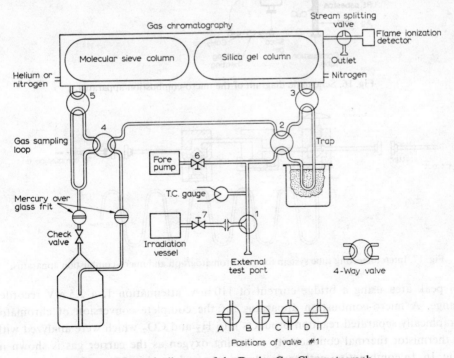

Fig. 15. Schematic diagram of the Duplex Gas Chromatograph.

peak identification. To obtain sensitivity coefficients the sample was similarly diverted into a Miller–Winefordner chromatographic-type micro-combustion apparatus. Figure 15 illustrates the schematic diagram of the Duplex Gas Chromatograph which was designed for analysis of product mixtures from gas phase reactions. The inlet manifold includes a thermocouple vacuum gauge for monitoring the pressure. The gas chromatograph on the left side of the schematic is an isothermal unit with a 3.5 m column of molecular sieve 5A and a Gow-Mac thermal conductivity detector. Using helium carrier gas at a flow rate of 15 to 20 ml/sec and a temperature of 100°, the separation of hydrogen, air, and methane can be accomplished readily. The sensitivity coefficient for H_2 using N_2 flow gas was found to be 3.46×10^{-8} moles/cm^2

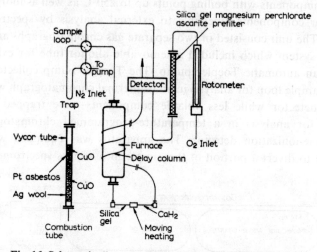

Fig. 16. Schematic diagram of the micro-combustion apparatus.

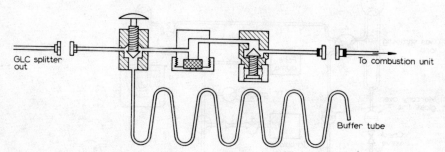

Fig. 17. Interconnecting tube system for gas chromatograph and micro-combustion apparatus.

of peak area using a bridge current of 110 mA, attenuation 1, a 2.5 mV recorder range. A micro-combustion apparatus for the complete conversion of chromatographically separated reaction products into H_2 and CO_2 which were analyzed with a thermistor thermal conductivity cell using oxygen as the carrier gas is shown in Fig. 16. In combination with molecular weight and empirical formula data obtained with a mass spectrometer, quantitative determination of carbon and hydrogen from a particular component allowed calculation of the number of moles, so that the flame-ionization chromatograph could be calibrated and radiolysis yields calculated. Figure 17 illustrates the interconnecting tube system for the gas chromatograph and micro-combustion apparatus.

A gas chromatogram of an irradiated mixture of F-cyclobutane and methane is shown in Fig. 18. The column oven was heated 10 min after injection at 2.1°C/min. After passing the c-C_4F_8 peak, the temperature was brought up to 180° (no com-

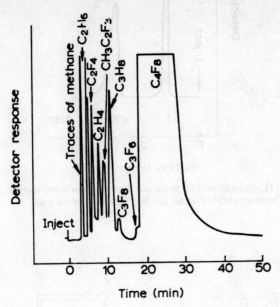

Fig. 18. Analysis of irradiated mixtures of CH_4 and c-C_4F_8 in the flame ionization unit.

pounds higher than F-cyclobutane were observed in this system). Hydrogen and methane were separated in the thermal conductivity unit of the Duplex Gas Chromatograph as shown in Fig. 19.

Seeler and Cahill[50] described a direct gas sampling technique for analyzing trace hydrocarbon radiolysis products in the presence of methane. A Hewlett-Packard Model 5750 gas chromatograph with dual flame detectors was used for the analysis. The instrument was fitted with a Beckman 8-port sampling valve, Model No. 102396. The volumes of the internal and external sample loops were 15 μl and 2 ml, respectively. The pressure of the external sample loop was monitored by either of two pressure transducers ranging from 0–15 p.s.i.a. and 0–2 p.s.i.a. which were installed at point E as shown in Fig. 20, which depicts the gas inlet sampling system for the GC unit. The input–output circuitry was modified as shown in Fig. 21. This modification provided the convenience of adjusting and balancing the circuit between applied input voltage and output voltage. The columns were two matched 12 ft. × ⅛ in. stain-

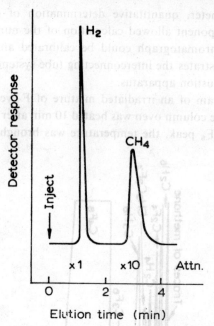

Fig. 19. Separation of H_2 and methane on molecular sieve column in the thermal conductivity section of the Duplex Gas Chromatograph. Carrier gas N_2, flow rate 18 ml/min.

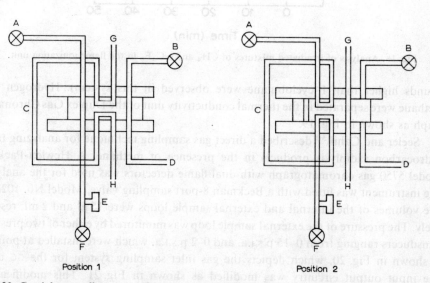

Fig. 20. Gas inlet sampling system for gas chromatography unit. A, Sampling inlet metering valve; B, micro loop inlet metering valve; C, macro sample loop variable; D, micro loop outlet; E, pressure transducer; F, vacuum valve; G, helium from gas chromatograph; H, helium to gas chromatograph.

less steel, Porapak R 80–120 mesh (obtained prepacked, Hewlett-Packard). The columns were preconditioned at 190–200°C with helium flow for 48 h. The program parameters were oven −15°C isothermal for 16 min, program 1°C/min until 35°C was reached and held isothermally until 28 min, reprogrammed at 4°C/min until

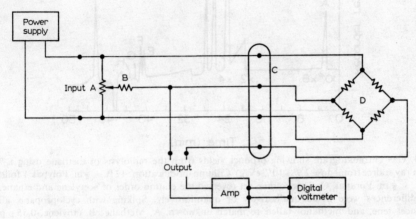

Fig. 21. Input–output circuitry for operating pressure transducers. A, 50 K turn precision potentiometer; B, 10 K fixed resistor; C, cable connector; D, pressure transducer.

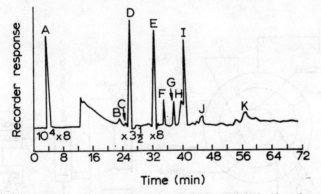

Fig. 22. Gas chromatogram showing product yields from the radiolysis of methane using a pulsed linear accelerator (total dose 8.5×10^{19} eV/g). A, methane; B, ethylene, 12.5 p.p.m.; C, ethane, 54.2 p.p.m.; D, acetylene, 8.8 p.p.m.; E, propylene, 1.1 p.p.m.; F, propane, 1.1 p.p.m.; G, cyclopropane, 0.2 p.p.m.; H, methyl acetylene, 0.2 p.p.m.; I, ghost + n-butane; T, neopentan; K, isopentane.

150°C and held until the last peak was eluted. Helium flow was 35 ml/min at 80 p.s.i.g. An Infotronics integrator Model No. CRS-11 HSB was used to provide digital integration of peaks eluted from the GC. Figure 22 illustrates a chromatogram showing product yields from the radiolysis of methane using a pulsed accelerator. Figure 23 is a chromatogram showing product yields from the radiolysis of methane using a ^{60}Co gamma ray source.

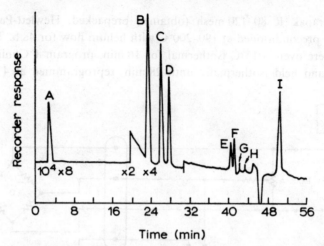

Fig. 23. Gas chromatogram showing product yields from the radiolysis of methane using a ^{60}Co gamma ray source (total dose 2.9×10^{20} eV/g). Column modification: 12 ft. $\times \frac{1}{8}$ in. Polypak 1 followed by 3 ft. $\times \frac{1}{8}$ in. Porapak Q. This column set reversed the elution order of acetylene and ethane. No other differences were noted qualitatively or quantitatively. Spiking with cyclopropane, allene, methylacetylene, and methanol failed to match unknown. A, Methane; B, ethylene, 0.35 p.p.m.; C, acetylene, 0.07 p.p.m.; D, ethane, 180 p.p.m.; E, propane, 22 p.p.m.; F, unknown; G, isobutane, 0.9 p.p.m.; H, ghost; I, n-butane, 3 p.p.m.; J, neopentane, 0.3 p.p.m.; K, isopentane, 0.6 p.p.m.

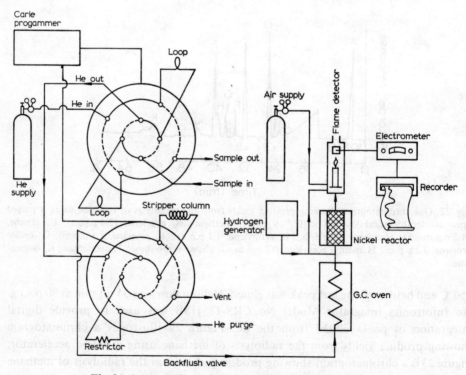

Fig. 24. Automated gas chromatographic CO–CH$_4$ analyzer.

Stevens and O'Keefe[51] have described an automated gas chromatographic carbon analyzer for the analysis of ambient concentrations of CO and CH_4 between 0.010 and 200 p.p.m. A unique pre-column system and a catalytic surface which quantitatively converted CO to CH_4 was incorporated. Figure 24 illustrates the analytical gas chromatographic $CO-CH_4$ analyzer which has been used almost continuously for a year and which required little attention except for changing the pre-column once every 3 months and replacing air and carrier gas cylinders monthly.

2. ACETYLENE

Acetylene is principally produced by the reaction of water with calcium carbide. However, acetylene is also produced on a large scale from hydrocarbons. The U.S. hydrocarbon acetylene capacity in 1963 was more than 700 million lb. compared with about 1125 million lb. for carbide acetylene[52].

Processes for the manufacture of acetylene from hydrocarbons include: (a) electric discharge, (b) regenerative furnace, (c) combustion (one- and two-stage) processes. The annual production of acetylene in the U.S. in 1960 was about 910 million lb. Of this quantity, approx. 270 million lb. (85%) was consumed by the chemical and allied industries in the manufacture of vinyl chloride, neoprene, acrylonitrile, trichloroethylene, perchloroethylene, acetaldehyde, and other miscellaneous acetylene derivatives. The remaining 140 million lb. (15%) was consumed in either oxygen or air in such applications as welding, cutting, and scarfing.

Acetylene is present in polluted air both from vehicle exhausts and as a result of its widespread use as a basic synthetic intermediate. Mixtures of acetylene with air containing more than 3% or less than 65% gas are explosive, the maximum hazard being one volume of acetylene and 12.5 volumes of air. Acetylene acts as a simple asphyxiant with high concentrations causing narcosis and concentrations in excess of 350,000 p.p.m. (35%) causing unconsciousness in 5 min.

The chemistry of acetylene has been reviewed by Nieuwland and Vogt[53] and Miller[54].

The colorimetric determination of acetylene in air at concentrations from 10 p.p.b. to 10 p.p.m. was described by Hughes and Gordon[55]. Acetylene was adsorbed from a gas stream on a column of silica gel, 10 mm × 2 mm diam. cooled to −78°C. The gel was then warmed to room temperature and treated with a solution of ammoniacal cuprous chloride. The presence and quantity of acetylene was indicated by the depth of color produced by cuprous acetylide on the gel. Acetylene has also been determined[56] by infrared spectrophotometry at 13.7 μm.

Klein[57] described the GC analysis of acetylene in gaseous mixtures using a 3.4 m × 4 mm column of silica gel (0.3–0.5 mm) at 50°C with nitrogen as carrier gas at 90 l/min. Acetylene was eluted from a 10 ml air sample 9 sec after methane, ethane, ethylene, and propane, and before propylene and the C_4 hydrocarbons. Acetylene was detected down to 5 p.p.b.

Kuley[58] analyzed a mixture of acetylene and C_3 and C_4 hydrocarbons on a 1.35 m × 5 mm column of activated alumina (60–100 mesh, Burrel Co.) coated with 7% Carbowax 20M. The separation temperature was 40°C with nitrogen as carrier gas and FID; acetylene was eluted 6 min after the C_3 and C_4 hydrocarbons. The detection limit of acetylene was 12 p.p.b.

Gaseous mixtures of acetylene and C_3–C_4 hydrocarbons have been analyzed by Jones and Green[59] utilizing FID and 1.8 m × 5 mm columns containing 3% di(ethylhexyl) sebacate on silica gel (30–60 mesh, Davison). The column was operated at 50°C with helium carrier gas at 40 ml/min. A 5 ml air sample yielded the following gas chromatographic sequence: methane, ethane, ethylene, propane, acetylene, propylene, and C_4 hydrocarbons.

Theer[60] analyzed acetylene, C_3 and C_4 hydrocarbons on a 2 m column of Alusil (Silica Gel A) following enrichment of acetylene on 10 mm × 1 mm absorbers of silica gel (0.125–0.2 mm) maintained at −78°C by CO_2–methanol coolant. Acetylene was desorbed in a hydrogen stream at +40°C and hydrogen at room temperature was the carrier gas. Acetylene was eluted after 2 min between the C_3 and C_4 hydrocarbons. After enrichment, the detection limit was 0.1 p.p.b. hydrocarbons.

The separation of methylacetylene, vinylacetylene, and diacetylene was described by Kontorovitch et al.[61] using columns packed with dibutyl phthalate on firebrick. Vigdergauz et al.[62] used a system of two columns for the separation of alkanes, olefins, and acetylenes. The first column was packed with diisobutyl phthalate on firebrick and the second with 30% sulpholane on firebrick.

REFERENCES

1 G. C. ORTMAN, Anal. Chem., 38 (1966) 644.
2 A. P. ALTSHULLER, G. C. ORTMAN, B. E. SALTZMAN AND R. E. NELIGAN, J. Air Pollution Control Assoc., 16 (1966) 87.
3 W. LEITHE, The Analysis of Air Pollutants, Ann Arbor Science, Ann Arbor, Mich., 1971, p. 200.
4 K. J. BOMBAUGH, Nature, 197 (1963) 1102.
5 N. N. RAY, J. Appl. Chem., 4 (1954) 82.
6 G. R. BOREHAM AND F. A. MARHOFF, Gas Council Res. Comm. GC 54, 1958.
7 W. N. GRUNE AND C. F. CHUEH, Intern. J. Air Water Pollution, 6 (1962) 283.
8 G. KYRYACOS AND C. E. BOORD, Anal. Chem., 29 (1957) 787.
9 J. F. ELLIS AND C. W. FORREST, Anal. Chim. Acta, 24 (1961) 329.
10 W. M. GRAVEN, Anal. Chem., 31 (1959) 1197.
11 Y. MURAKAMI, Bull. Soc. Japan, 32 (1959) 316.
12 D. P. MANKA, Anal. Chem., 36 (1964) 480.
13 H. RODENBUCH, Erdoel Kohle, 19 (1966) 583.
14 T. A. MCKENNA AND J. A. IDLEMAN, Ann. Chem., 32 (1960) 1299.
15 A. RIO, D. RIPA AND S. TRIBASTONE, Chem. Ind. (Milan), 41 (1959) 1185.
16 J. A. KNIGHT AND C. T. LEWIS, Radiation Res., 23 (1964) 319.
17 C. G. S. SCOTT, J. Inst. Petrol., 45 (1959) 115.
18 D. SANDULESCU, O. PETE, L. STANESCU, A. HANES AND M. ENACHE, Rev. Chim., 12 (1961) 297.
19 K. FUJIMOTO, J. Japan Petrol. Inst., 3 (1964) 168.

20 K. ARITA, Y. KUGE AND Y. YOSHIKAWA, *Bull. Chem. Soc. Japan*, 38 (1965) 632.
21 R. A. CROSS, *Nature*, 211 (1966) 409.
22 O. L. HOLLIS AND W. V. HAYES, in A. B. LITTLEWOOD (Ed.), *Gas Chromatography*, Elsevier, Amsterdam, 1967, p. 57.
23 M. PAPIC, *J. Gas Chromatog.*, 6 (1968) 493.
24 A. B. RICHMOND, *J. Chromatog. Sci.*, 7 (1969) 321.
25 R. W. HARDY, R. D. HOLSTEN, E. K. JACKSON AND R. C. BURNS, *Plant Physiol.*, 43 (1968) 1185.
26 A. C. COPE, N. A. LeBEL, H. H. LEE AND W. R. MOORE, *J. Am. Chem. Soc.*, 79 (1957) 4720.
27 G. C. CARLE, *J. Chromatog. Sci.*, 8 (1970) 550.
28 H. W. DURBECK, *Z. Anal. Chem.*, 250 (1970) 377.
29 F. ZOCCHI, *J. Gas Chromatog.*, 6 (1968) 100.
30 C. PIERCE, J. W. WILEY AND R. N. SMITH, *J. Phys. Chem.*, 53 (1949) 669.
31 R. KAISER, *Chromatographia*, 2 (1969) 453.
32 R. KAISER, *Chromatographia*, 3 (1970) 38.
33 W. SCHNEIDER, H. BRUDERRECK AND I. HALASZ, *Anal. Chem.*, 36 (1964) 1533.
34 A. ZLATKIS, H. R. KAUFMAN AND D. E. CURBIN, *J. Chromatog. Sci.*, 8 (1970) 416.
35 E. SAWICKI, R. C. COREY, A. E. DOOLEY, J. B. GISCLARD, J. L. MONKMAN, R. E. NELIGAN AND L. A. RIPPERTON, *Health Lab. Sci.*, 7 (1970) 23.
36 R. W. FREEDMAN, H. W. LAND AND M. JACOBSON, *Bur. Mines Rept. 7180*, U.S. Dept. Interior, Washington, 1968, p. 1.
37 L. B. BERGER AND H. H. SCHRENK, *Bur. Mines Inform. Circ. 7071*, 1938, p. 24.
38 J. W. SWINNERTON AND V. J. LINNENBOM, *J. Gas Chromatog.*, 5 (1967) 570.
39 J. A. J. WALKER AND E. D. FRANCE, *Analyst*, 94 (1969) 364.
40 R. BERRY, in M. VAN SWAAY (Ed.), *Gas Chromatography, 1962*, Butterworths, London, 1962, p. 321.
41 S. LEVEY AND O. J. BALCHUM, *J. Lab. Clin. Med.*, 62 (1963) 247.
42 B. O. JANSSON AND B. T. LARSSON, *J. Lab. Clin. Med.*, 74 (1969) 961.
43 B. T. LARSSON, *Acta Chem. Scand.*, 19 (1965) 159.
44 B. O. JANSSON, *FAO Rept. A-1424-76*, Feb. 1968.
45 J. W. ROOT, E. K. C. LEE AND F. S. ROWLAND, *Science*, 143 (1964) 676.
46 P. L. GANT AND K. YANG, *J. Amer. Chem. Soc.*, 86 (1964) 5063.
47 F. BRUNER, G. P. CARTONI AND A. LIBERTI, *Anal. Chem.*, 38 (1966) 298.
48 J. J. CZUBRUT AND H. GESSER, *J. Gas Chromatog.*, 6 (1968) 41.
49 H. HECKEL AND R. J. HANRAHAN, *J. Gas Chromatog.*, 7 (1969) 418.
50 A. K. SEELER AND R. W. CAHILL, *J. Chromatog. Sci.*, 7 (1969) 158.
51 R. K. STEVENS AND A. E. O'KEEFE, *Anal. Chem.*, 42 (1970) 143A.
52 Anon., *Chem. Week*, 88 (1961) 21.
53 J. NIEUWLAND AND R. VOGT, *The Chemistry of Acetylene*, Reinhold, New York. 1945.
54 S. A. MILLER, *Acetylene*, Academic Press, New York, 1965.
55 E. E. HUGHES AND R. GORDON, JR., *Anal. Chem.*, 31 (1959) 94.
56 W. E. SCOTT, E. R. STEPHENS, P. L. HANST AND R. C. DOERR, *Proc. Am. Petrol. Inst., Sect. III*, 37 (1957) 171.
57 G. KLEIN, *Linde Rept. No. 17*, 1964, p. 24.
58 C. J. KULEY, *Anal. Chem.*, 35 (1963) 1472.
59 K. JONES AND R. GREEN, *Nature*, 205 (1965) 67.
60 J. THEER, *Chem. Technik.*, 14 (1962) 164.
61 L. M. KONTOROVITCH, A. V. IOGANSEN, G. T. LEVCHENKO, G. N. SEMINA, V. P. BOBROVA AND V. A. STEPANOVA, *Zavodsk. Lab.*, 28 (1962) 146.
62 M. S. VIGDERGAUZ, K. A. GOL'BERT, I. M. SAVINA, M. I. AFANAS'EV, R. A. ZIMIN AND N. T. BAKHAREVA, *Zavodsk. Lab.*, 28 (1962) 149.

Chapter 18

MISCELLANEOUS GASEOUS MIXTURES

In the Chapters 11–17, the focus was given as far as possible to the major category pollutant as well as its admixture with a small number of components. It is well recognized that under environmental conditions there is apt to occur complex multi-component mixtures of pollutants. In this chapter consideration is given to a number of such mixtures without attempting further appraisal of the thousands of analytical procedures reported.

Bethea and Meador[1] evaluated 21 columns for the qualitative separation of mixtures of reactive (NO, NO_2, SO_2, HCl, H_2S, Cl_2, NH_3) and other gases (CO_2, N_2O) in air. No single column was found which was effective for the separation of all of these gases. A two-section column of Porapak Q and Porapak R provided adequate separation of NO and air while the separation of air, NO_2, and CO_2 was

TABLE 1

SPECIFICATIONS OF GAS CHROMATOGRAPHIC COLUMNS

Column no.	Partitioning agent	Solid support	Mesh (U.S. std.)	Liquid loading (%)	Length (ft.)
1	Silicone Oil SF-96	Chromosorb T	40–60	10	10
2	Triacetin	Chromosorb T	40–60	10	10
3	Di-*n*-decyl phthalate	Chromosorb T	40–60	10	10
4	Arochlor 1232	Chromosorb T	40–60	10	10
5	QF-1 (FS-1265) fluoro	Chromosorb T	40–60	10	10
6	Silicone XE-60 (nitrile)	Chromosorb T	40–60	10	10
7	Fluorolube HG 1200 grease	Chromosorb W, AW, DMCS	60–80	10	10
8	Kel F 90 grease	Chromosorb W, AW, DMCS	60–80	10	10
9	Halocarbon 11-14	Chromosorb W, AW, DMCS	60–80	10	10
10	Silicone DC 200 oil	Chromosorb W, AW, DMCS	60–80	10	10
11	H_3PO_4	Porapak Q	80–100	3	6
12	None	Polypak 1	80–120	N/A	6
13	None	Polypak 1	80–120	N/A	3
14	Di-*n*-ethylhexyl adipate	Chromosorb W, AW	30–100	10	10
15	None	Porapak Q	30–100	N/A	6
16	None	Porapak QS	80–100	N/A	6
17	None	Porasil B	80–100	N/A	6
18	Arochlor 1232	Porapak Q	80–100	5	8
19	Arochlor 1232	Porapak R	50–80	5	8
20	No. 18 in series with no. 19	Porapak Q + R	50–80	5	16
21	None	Porapak Q	50–80	N/A	8

best carried out on a QF-1 Fluoro column. H_2S, Cl_2, and SO_2 were best analyzed on either silicone oil SF-96 or triacetin columns. These columns were also effective in effecting the air–CO_2–NH_3 separation. Symmetrical, reproducible sharp peaks were obtained for NO_2 on Kel-F 90 grease and Florolube HG 1200 columns and both of these columns completely separated NO_2 from all other compounds tested.

Table 1 lists the specifications of the gas chromatographic columns. All columns were $\frac{1}{4}$ in. o.d., 22 gauge type 304 or 316 stainless steel tubing except columns 12, 13, and 18–20, which were $\frac{1}{8}$ in. o.d., 22 gauge wall thickness, type 304 or 316 stainless steel tubing. A Varian Aerograph Model 1520-1B dual column, dual detector gas chromatograph was used. The detector used in all the evaluations was of the hot wire thermal conductivity type at 150 mA bridge current. The results of the screening studies for the chromatographic columns are shown in Table 2. An entry of nr means that the retention time for that gas on that particular column was not reproducible. No entry in the table means that those gases were either not eluted from the column or were not tested in that column. Table 3 shows the separations actually obtained with these chromatographic systems. The columns generally yielded narrow (less than 0.25 min wide at the baseline), well-resolved peaks for 500 μl samples at × 64 or × 128. Very narrow air peaks were obtained, which would be advantageous for the quantitative analysis of low concentrations of the reactive gases in air. For the optimum analysis of oxides of nitrogen (NO, N_2O, and NO_2) a two-column system was required. For example, a sample was first passed through column 8 (Kel F-90 grease on Chromosorb W, AW, DMCS) which yielded air, NO, and N_2O as a composite peak separated from NO_2. This composite peak was then separated on column 20 (Aroclor 1232 on Porapak Q in series with Aroclor 1232 on Porapak R) for the quantitative analysis of N_2O and NO.

Table 4 shows the lower detection limit for usable separations of mixtures of gases when using thermal conductivity detection.

An electron-capture detector was evaluated in conjunction with column 18 (5% Aroclor 1232 on Porapak Q, 80–100 mesh) for the low-level analysis of SO_2 and chlorine in air. The lower detection limit for SO_2 in air when using 1 ml gas sample was approx. 2.2 p.p.m. and that for chlorine was 0.3 p.p.m.

A condensing system for the determination of trace impurities in gases by GC was described by Brenner and Ettre[2]. The accessory suitable for analysis of trace impurities at the p.p.m. level consists of a condenser column which could be substituted for the sampling volume tube in the Perkin-Elmer precision gas sampling sytem in place of the interchangeable sample volume tubing. The diagram of the system is shown in Fig. 1. The main part of the system is a condenser column, 50 cm × $\frac{1}{4}$ in. o.d. A pressure and flow regulator at the inlet and a flow meter at the terminal end of the system provide for the reproducibility of the conditions of an individual analysis. The condensing system was connected to a Perkin-Elmer Vapor Fractometer Model 154-C which was coupled with a 1 mV Speedomax recorder (Leeds and Northrup). Figures 2–4 illustrate the analysis of 4 p.p.m. acetylene in air

TABLE 2

RETENTION TIMES

Column no.	Temp. (°C)	Flow (ml He/min)	Inlet pressure (p.s.i.g.)	Retention time (min)									
				Air	CO_2	N_2O	NO	NO_2	H_2S	SO_2	Cl_2	HCl	NH_3
1	32	10	20	2.22	2.60	2.73	2.28	nr	3.25	4.13	6.01	nr	2.83
2	32	10	20	0.73	0.94	0.92	0.73	nr	1.47	9.30	2.32	nr	1.99
3	32	10	20	2.25	2.63	2.80	2.29	nr	4.09	11.02	8.80	10.8[a,b]	2.33
4	32	10	20	0.65	0.73	0.80	0.67	0.97	1.07	1.90	2.10	0.88	1.14[b]
5	32	10	20	2.14	2.56	2.74	2.16	4.5–5.1[a,b]	2.86	5.37	3.83	2.72	4.8[a,b]
6	32	10	20	0.65	0.62	0.74	0.65	17.6	0.88	1.97	1.21	1.89[b]	0.87
7	32	10	20	3.04	3.04	3.04	3.04	3.42	3.04	3.18[c]	3.31[c]	3.04	3.67[b]
8	32	10	20	1.10	1.12	1.12	1.12	1.43	1.12	1.25	1.25	1.16	nr
9	32	10	20	3.04	3.32	3.46	3.04	nr	3.50	4.34	5.62	3.32	nr
10	32	10	20	1.08	1.08	1.15	1.08	18.3[a,b,c]	1.23	1.36	1.65	1.37	1.40[b]
11	100	50	50	0.96	1.58	1.81	1.00	1.10	3.65	7.38	3.71	3.65[b]	1.09
12	30	20	20	1.11			1.98	2.42[b]		2.02			
13	32	10	20	0.69	0.93	1.06	0.69	2–2.6[a,b]	2.46[b]	6.9[b]	3.05[a,b]	2.04[b]	1.60[b]
14	50	10	50	0.67	0.78	0.78	0.70	4.8 ± 0.1	1.05	1.92	1.74[b]	2.26[b]	1.02[b]
15	30	40	50	1.20	3.66	4.80[b]		3.75[b]		3.60[b]			
16	78	50	50	0.40	0.60	0.84	0.47	1.70	1.75	3.88	1.49	1.49	0.89
17	100	50	50	0.62	0.62[b]	0.62	0.62	nr	0.62	>20[b]	>10[b]	nr	nr
18	100	10	30	0.81	1.07	1.07	0.81	7.2[b]	2.39	4.95[b]	5.13	2.55[b]	
19	78	20	30	0.64	0.76	0.79	0.60	1.25[b]	1.35	4.26[b]	5.37[b]	nr	
20	35	20	80	1.60	1.65	4.57	1.95	8.05[b]		9.1[b]			1.35
21	100	50	30	0.49	0.73	0.76	0.50	2.9–3.1[b]	1.44	4.32[b]	2.57[b]	2.55[b]	1.28[b]

[a] Peak exhibited considerable leading.
[b] Peak exhibited considerable tailing.
[c] Peak was considerably broader than the rest of the peaks of that column.

TABLE 3

PEAKS IN ORDER OF ELUTION

Column no.	1st Peak	2nd Peak	3rd Peak	4th Peak	5th Peak	6th Peak	7th Peak	8th Peak	9th Peak
1	Air + NO	$CO_2 + N_2O$	NH_3[b]	H_2S	SO_2	Cl_2			
2	Air + NO	$CO_2 + N_2O$	H_2S	NH_3	Cl_2	SO_2	SO_2		
3	Air + NO	NH_3	$CO_2 + N_2O$	H_2S	Cl_2	HCl[a,b]	SO_2		
4	Air + NO	$CO_2 + N_2O$	NH_3[b]	$H_2S + NO_2 + HCl$	$SO_2 + Cl_2$				
5	Air + NO	CO_2	$N_2O + HCl + H_2S$	Cl_2	NO_3[a,b]	NH_3[a,b]	SO_2		
6	Air + NO + N_2O + CO_2	$H_2S + NH_3$	Cl_2	HCl[b] $+ SO_2$	NO_2[c]	SO_2			
7	Air + NO + CO_2 + N_2O + H_2S + HCl	SO_2[c] $+ Cl_2$	NO_2[c]	NH_3[b]					
8	Air + NO + CO_2 + N_2O + H_2S + HCl	$SO_2 + Cl_2$	NO_2						
9	Air + NO + CO_2 + N_2O + H_2S + HCl	CO_2	SO_2	Cl_2					
10	Air + NO + CO_2 + N_2O	H_2S	$SO_2 + HCl + NH_3$[b]	Cl_2	NO_2[a,b,c]	SO_2[c]			
11	Air + NO + NO_2 + NH_3	CO_2	N_2O	$H_2S + HCl$[b]	Cl_2[b]				
12	Air	$NO + SO_2$	N_2O[b]						
13	Air + NO	CO_2	N_2O	NH_3[b]	HCl[b]	NO_2[a,b]	H_2S[b]	Cl_2[a,b]	SO_2[b]
14	Air + NO + CO_2 + N_2O	NH_3[b] $+ H_2S$	Cl_2[b]	SO_2	HCl[b]	NO_2[b]			
15	Air	CO_2	$SO_2 + NO_2$[b]	N_2O[b]					
16	Air + CO_2	CO_2	$NH_3 + N_2O$						
17	Air + CO_2 + N_2O + NO + H_2S	$CO_2 + N_2O$	$NH_3 + N_2O$	$Cl_2 + HCl$	$H_2S + NO_2$	SO_2			
18	Air + NO	$CO_2 + N_2O$	$H_2S + HCl$[b]	SO_2[b]	Cl_2	NO_2[b]			
19	Air + NO	$CO_2 + N_2O$	NO_2[b] $+ H_2S + NH_3$[b]	SO_2[b]	Cl_2[b]				
20	Air + CO_2	NO	N_2O	NO_2[b]	SO_2[b]				
21	Air + NO	$CO_2 + N_2O$	NH_3[b] $+ H_2S$	HCl[b] $+ Cl_2$[b]	NO_2[b]	SO_2[b]			

[a] Peak exhibited considerable leading.
[b] Peak exhibited considerable tailing.
[c] Peak was considerably broader than other peaks on that column.

TABLE 4

LOWER DETECTION LIMITS FOR USABLE SEPARATIONS

Column no.	Lower detection limit (p.p.m.)[a]								
	CO_2	N_2O	NO	NO_2	H_2S	SO_2	Cl_2	HCl	NH_3
1					300	200	500		200
2					1000	2000	300		2000
3					300				400
4				500					
5	400					400			
6							200		
8				400					
9	200					300	300		
10					200		300		
11	200	200							
14						500			
16	400					1000			
18				400	200	300	200	500	
20		500	300						
21						200			

[a] Measured at ambient conditions using 1 ml samples.

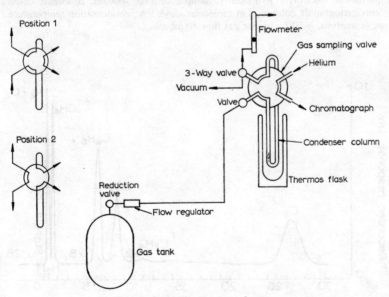

Fig. 1. Schematic diagram of system.

or oxygen, impurities in electrolytic hydrogen I and impurities in electrolytic hydrogen II, respectively.

The analysis of the impurities of electrolytic hydrogen indicated the following levels: oxygen 155 p.p.m., nitrogen 4900 p.p.m., carbon monoxide 10 p.p.m., carbon

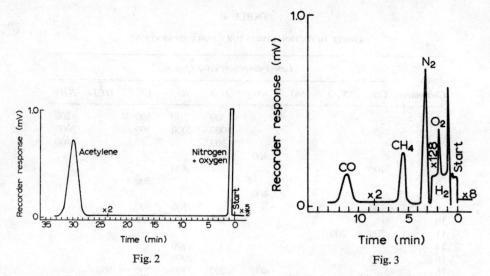

Fig. 2   Fig. 3

Fig. 2. Analysis of 4 p.p.m. acetylene in air or oxygen. Sample volume, 8.90 1; condenser column, 50 cm silica gel; chromatographic column, 2 m silica gel; condensation temperature, $-80°C$; temperature of analysis, $+50°C$; carrier gas flow, 65 ml/min.

Fig. 3. Impurities in electrolytic hydrogen I. Sample volume, 1000 ml; condenser column, 50 cm silica gel; chromatographic column, 2 m molecular sieve 5 A; condensation temperature, $-80°C$; temperature of analysis, $+50°C$; carrier gas flow 80 ml/min.

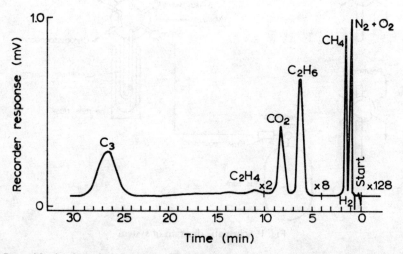

Fig. 4. Impurities in electrolytic hydrogen II. Sample volume, 16 1; condenser column, 50 cm silica gel; chromatographic column, 2 m silica gel; condensation temperature, $-80°C$; temperature of analysis, $+50°C$; carrier gas flow 80 ml/min.

dioxide 10 p.p.m., methane 610 p.p.m., and higher hydrocarbons 20 p.p.m. as well as 100 p.p.m. of water. For the determination of all components three different analyses were necessary. Using large enough sample volumes, the minimum detectability of the system was at or below 0.1 p.p.m. Table 5 summarizes the possibility of determining the impurities in electrolytic hydrogen utilizing several chromatographic and condenser columns.

TABLE 5

POSSIBILITY OF DETERMINATION OF IMPURITIES IN ELECTROLYTIC HYDROGEN

Chromato-graphic column	Condenser column	O_2	N_2	CO	CO_2	CH_4	C_2H_6	C_2H_4	C_3	C_4 and higher
Linde molecular sieve, 5A	Linde molecular sieve 5A or silica gel	+	+	+	−	+	−	−	−	−
Silica gel	Silica gel	One peak		a	+	+	+	+	+	−
Di-2-ethylhexyl sebacate or di-methyl-sulfolane on Chromosorb	Di-2-ethylhexyl sebacate or di-methyl-sulfolane, or poly(ethylene glycol) on Chromosorb	These components superimposed in one peak							+	+

a CO is determinable only if the concentration is similar to N_2 and CH_4. In other cases, it is obscured by these components.

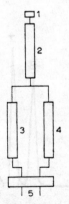

Fig. 5. Three-column system for analysis of light gases. 1, Gas sample valve; 2, primary column: 1.5 m × 4.8 mm i.d. stainless steel tubing packed with 50–80 ASTM mesh Porapak T; 3, secondary column: 3.05 m × 4.8 mm i.d. stainless steel tubing packed with 50–80 ASTM mesh Porapak S; 4, secondary column 3.66 m × 4.8 mm i.d. stainless tubing packed with 44–60 BS mesh 13X molecular sieve; 5, katharometer detector.

Deans *et al.*[3] described a new column system for the isothermal GC analysis of light gases (H_2, O_2, N_2, CO, CH_4, CO, C_2H_4, C_2H_6, and C_2H_2) employing a column switch technique. Analysis of all the columns listed was carried out in 11 min. The system adopted uses three columns connected together as shown in Fig. 5. The primary column of Porapak T served to separate the light components (H_2, O_2, N_2, CO, and CH_4) as a group from the heavier fraction (CO_2, C_2H_4, C_2H_6, and C_2H_2) and these in turn from water and higher hydrocarbons which were back-flushed from the system before emerging from this column. The light components were routed from the primary column to a 13X molecular sieve secondary column for separation and subsequent detection by one arm of the katharometer bridge. Immediately after the light components had emerged from the primary column, the effluent flow was diverted to the other secondary column (Porapak S) where CO_2, C_2H_4, C_2H_6, and C_2H_2 were separated and detected by the other arm of the katharometer bridge. The columns were sized (Fig. 5) so that the detection of the components eluting from the molecular sieve column was complete before the first component emerged from the Porapak S column connected to the other arm of the bridge. An important aspect to achieve this feature was the use of *sparingly* activated 13X molecular sieve. The molecular sieve was first washed in distilled water and the fines removed by decantation; the sieve was then dried at 100–120°C overnight. When packed in the column it was activated for 4 h at 175°C. In this state rapid elution of the permanent gases was achieved in the order H_2, O_2, N_2, CO, and CH_4. The rapidity of elution enabled the detection of methane to be completed before

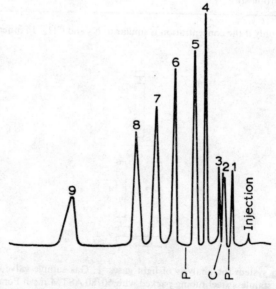

Fig. 6. Isothermal gas chromatographic separation of light gases. Operating conditions: primary Porapak T (see Fig. 5); temperature, 50°C; carrier gas, helium; P, polarity reversal; C, column switch. 1, H_2; 2, O_2; 3, N_2; 4, CO; 5, CH_4; 6, CO_2; 7, C_2H_4; 8, C_2H_6; 9, C_2H_2.

the emergence of CO_2 from the Porapak column. Figure 6 illustrates the isothermal gas chromatographic separation of a 9 component mixture of light gases.

Gas chromatographic analysis for organic compounds in odorous pollution studies requires a sensitivity to 10^{-10} g/l and generally needs sample pre-concentration as well. Dravnieks et al.[4] recently developed a process for the extraction of organic compounds from air at a rate of 4 l/min by adsorption on a high-surface-area styrene–divinyl copolymer (Chromosorb 102) followed by an essentially contamination-free transfer of the sample for gas chromatographic analysis. Air moisture was not collected in amounts sufficient to interfere with the process. The polymeric adsorbent, Chromosorb 102 (60–80 mesh) (Johns-Manville) is highly porous (the pores actually are channels which range up to 30 Å in diam.). Hence, all organic substances analyzable by GC are not subject to size exclusion effects. Figure 7 illustrates

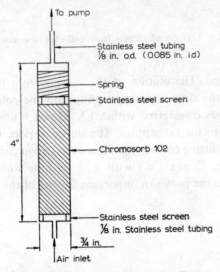

Fig. 7. High-speed organic vapor collector.

the high-speed organic vapor collector which was made of stainless steel and contains 5 g of 60–80 mesh Chromosorb 102 with a total surface area of 1500 m². The collector was preconditioned at 220°C with nitrogen gas for 24 h, then conditioned at 120°C with a pure (zero) helium flow of 60 ml/min overnight.

In sampling, air was drawn through the collector and at 2 p.s.i. (0.14 atm) pressure drop the sampling rate reached 4 l/min. Most organic compounds were completely collected from a 10 l air sample since the retention volumes (in liters) at 25°C were methanol 2.7, ethanol 14, pentane 12, and hexane 30. Since gas chromatographs with FID detectors detect 10^{-9} g of an organic species, an analytical sensitivity of 10^{-10} g/l air was possible.

Figure 8 illustrates sample transfer from the collector to a special injector needle. A copper side tube connects the boiling liquid nitrogen to the covered groove

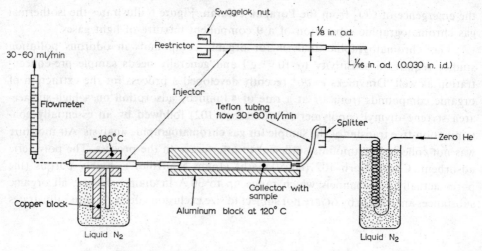

Fig. 8. Transfer of sample from collector to injector.

which holds the injector. The elution of the sample from the collector was in a direction opposite to the direction of sampling into the collector. Approximately 90–95% of a sample was transferred within 1 h. Figure 9 illustrates the system for the sample injection into the GC septum. The injection port was modified by silver brazing a stainless steel fitting containing a reusable Teflon compression plug sealing element (Conax fitting, Conax Co.) with a $\frac{1}{16}$ in. hole which accepted the small end of the injector onto the port. An important feature of the injector was a restric-

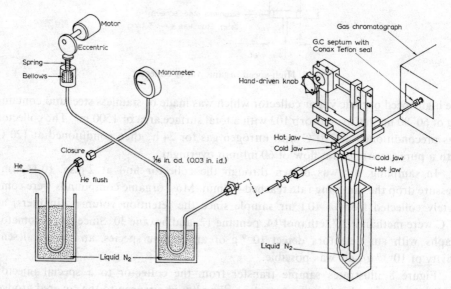

Fig. 9. Sample injection into GC septum.

tion at its downstream end. When the injector was flash-heated to inject the sample into the chromatograph, only a negligible fraction of the sample could travel through the restricted end, thus allowing a 95–98 % complete transfer of the sample into the GC port. In sampling from synthetic mixtures of 9 components in air, the reproducibility of the GC peak areas was within ±3 %.

The chromatographic separation properties of the macroporous sorbent Synachrom (based on a styrene–divinylbenzene bead copolymer) was studied by Dulka et al[5]. for mixtures of gases and a number of other materials. The elution order of several near-boiling substances from different homologous series corresponded to the increasing value of the polarizability of the substances. Thus, Synachrom behaved not only as an adsorbent but also as a liquid phase of non-polar character. Examples of the separation effect of Synachrom for C_2-hydrocarbons together with other gases is shown in Fig. 10. Other examples of separations of mixtures of gases, e.g. H_2, Ne, N_2, O_2, Ar, and CO and N_2, CO, CH_4, CO_2, and N_2O are shown in Figs. 11 and 12. Table 6 lists the Kovats retention indices of 42 substances obtained using a 1.5 m × 5 mm glass column of Synachrom 60–80 mesh at 150°C.

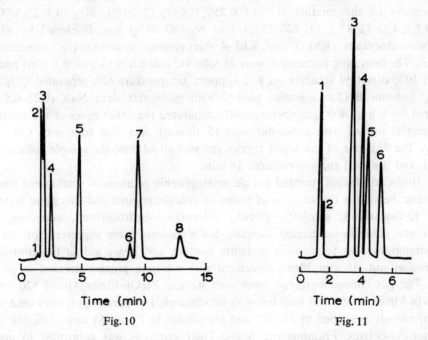

Fig. 10.   Fig. 11

Fig. 10. Fractovap GV (Carlo Erba), thermal conductivity detector, Synachrom 60–80 mesh, column 5 m × 2 mm stainless steel, temperature 26°, carrier gas He, 25 ml/min. 1, H_2; 2, $N_2 + O_2$; 3, CO; 4, CH_4; 5, CO_2; 6, C_2H_4; 7, C_2H_2; 8, C_2H_6.

Fig. 11. Synachrom 60–80 mesh; column 5 m × 2 mm stainless steel, temperature −78°, carrier gas He, 40 ml/min. 1, H_2; 2, Ne; 3, N_2; 4, O_2; 5, Ar; 6, CO.

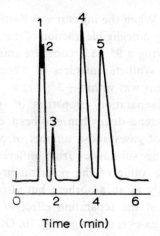

Fig. 12. Synachrom 60–80 mesh, column 5 m × 2 mm stainless steel, temperature 24°, carrier gas H_2, 27 ml/min. 1, N_2; 2, CO; 3, CH_4; 4, CO_2; 5, N_2O.

The chromatographic analysis of gases from pyrometallurgical processes in nonferrous metallurgy was reported by Dobrochiver et al.[6]. A gas chromatographic separation of the mixture H_2 (0.1–0.25%), CO_2 (2–20%), SO_2 (0.5–2%), CH_4 (3–4%), CO (2–3%), O_2 (20–25%), and N_2 (60–70%) was developed by which 2 chromatographs (KhL-3 and KhL-4 dual-column system) were connected in series. The operating parameters were as follows: column A (3.5 m × 6 mm) packed with 10% dimethyl sebacate on K-2 support, temperature 62°, separated CO_2 and SO_2; column B (2 m × 6 mm) packed with molecular sieve NaX (0.25–0.5 mm heated for 6 h at 400°), temperature 40°, separated the other gases of the mixture; column C packed with molecular sieve (5–10 mm), was used to absorb CO_2 and SO_2. The flow rate of the argon carrier gas was 60 ml/min, the sample volume was 1 ml, and the total analysis required 18 min.

Hollis and Hayes[7] utilized gas chromatographic columns of synthesized porous polymer beads for the analysis of water in volatile organic and inorganic systems, e.g. hydrocarbons, alcohols, glycols, chlorinated hydrocarbons, ammonia, and hydrogen chloride containing samples. Such columns were suggested for use for on-stream analysis to monitor moisture content and hence aid in the control of corrosion and other problems associated with water in process streams.

The gas chromatographic units used were a Perkin-Elmer Model 820 and a Perkin-Elmer Model 800 with hot-wire attachment. The porous polymers used were as previously described by Hollis[8] and are similar to Porapaks now available from Water Associates, Framingham, Mass. Their synthesis was controlled to give a rigid porous structure of relatively high surface areas, e.g. 100–700 m^2/g. Table 7 lists the retention times relative to water for air, methane, carbon dioxide, ethylene, acetylene, ethane, propylene, propane, and methyl chloride obtained on Porapak P, Q, R, S, and T columns at 32°C.

TABLE 6

KOVATS' RETENTION INDICES ON SYNACHROM

Column conditions: Synachrom 60–80 mesh; Fractovap GV (Carlo Erba); column 1.5 m × 5 mm, glass, temperature 150°; thermal conductivity detector, 150 mA, temperature 200°; carrier gas, 60 ml He/min

Compound	B.p. (°C)	Kovats' index
Water	100.0	250
Methyl alcohol	64.7	319
Acetaldehyde	20.2	365
Ethyl alcohol	78.4	412
Formic acid	100.6	444
Acetonitrile	81.8	446
Acrolein	52.5	459
Acetone	56.5	468
Isopropyl alcohol	82.5	472
Dichloromethane	40.7	480
Acrylonitrile	78.5	488
Ethyl ether	34.6	493
Methyl acetate	56.9	494
Allyl chloride	44.6	495
Propyl alcohol	96.6	501
Cyclopentane	97.8	510
Acetic acid	118.1	526
tert.-Butyl alcohol	82.9	531
Vinyl acetate	72.3	551
Methyl ethyl ketone	79.5	559
Trichloromethane	61.3	569
sec.-Butyl alcohol	99.5	570
Tetrahydrofuran	66.0	572
2-Methylpentane	60.3	580
Ethyl acetate	76.8	582
Isobutyl alcohol	108.0	586
3-Methylpentane	63.3	595
1,2-Dichloroethane	82.4	595
Methylcyclopentane	71.8	604
Butyl alcohol	117.5	609
1-Chlorobutane	77.8	612
Carbon tetrachloride	76.7	618
Benzene	80.1	622
Cyclohexane	80.7	625
Propionic acid	141.1	631
Ethylene glycol	197.3	641
Cyclohexene	83.0	643
Acrylic acid	141.1	645
Isopropyl acetate	88.9	655
Propyl acetate	101.6	682
Pyridine	115.2	699
Ethylcyclopentane	103.5	711
Toluene	110.6	721

TABLE 7

RETENTION TIMES RELATIVE TO WATER*

Compound	Columns				
	P	Q	R	S	T
Air	0.084	0.049	0.0073	0.0175	0.004
Methane	0.100	0.083	0.0126	0.0286	0.0064
Carbon dioxide	0.166	0.187	0.0335	0.0670	0.0327
Ethylene	0.262	0.388	0.0627	0.121	0.0354
Acetylene	0.299	0.388	0.0946	0.145	0.0982
Ethane	0.342	0.562	0.0862	0.175	0.0414
Propylene	1.02	2.75	0.375	0.783	0.247
Propane	1.62	3.58	0.454	0.873	0.215
Methyl chloride	1.77	3.23		1.03	0.445
Water	1.00	1.00	1.00	1.00	1.000

* Temperature approximately 32°C.

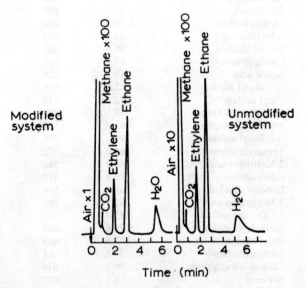

Fig. 13. Effect of instrument modification on water peak symmetry. Conditions: column, 6ft.× $\frac{3}{16}$ in. Porapak Q; temperature, −30°C; flow, 55 ml/min H_2.

Figure 13 illustrates the effect of instrument modification on water peak symmetry and Figs. 14 through 20 illustrate chromatograms of water in benzene, methyl chloride, hydrogen chloride, chlorine, ethylene oxide, ethylene glycol, and ammonia, respectively. Figure 21 shows the quantitative response of water when analyzed on a Porapak P column at 123°C.

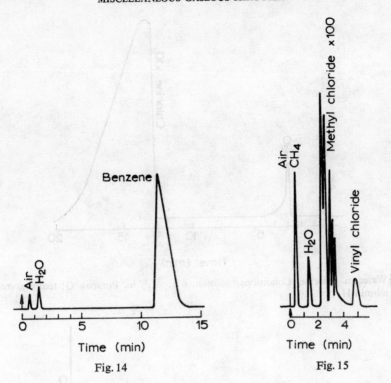

Fig. 14

Fig. 15

Fig. 14. Water in benzene. Conditions: column, 6 ft. $\times$ $\frac{3}{16}$ in. Porapak S; temperature, 165°C; flow, 80 ml/min He.

Fig. 15. Chromatogram of trace water in methyl chloride. Conditions: column, 6 ft. $\times$ $\frac{3}{16}$ in. Porapak Q; temperature, 98°C; flow, 55 ml/min H_2.

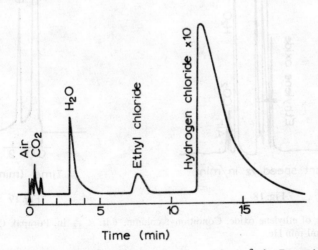

Fig. 16. Water in hydrogen chloride. Conditions: column, 6 ft. $\times$ $\frac{3}{16}$ in. Porapak R; temperature, 101°C; flow, 55 ml/min H_2.

References p. 414

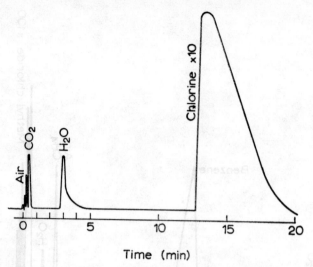

Fig. 17. Water in chlorine. Conditions: column, 6 ft. × $\frac{3}{16}$ in. Porapak Q; temperature, 101°C; flow, 50 ml/min H_2.

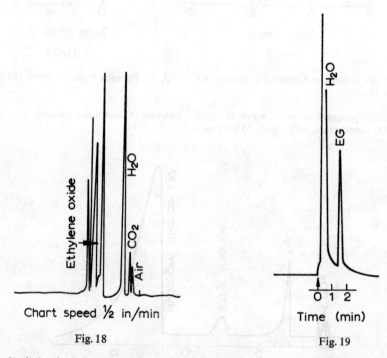

Fig. 18

Fig. 19

Fig. 18. Analysis of ethylene oxide. Conditions: column, 6 ft. × $\frac{3}{16}$ in. Porapak Q; temperature, 133°C; flow, 47 ml/min He.

Fig. 19. Trace quantity of ethylene glycol in water. Conditions: column, 3 ft. × $\frac{3}{16}$ in. Porapak Q; temperature, 200°C; flow, 90 ml/min H_2.

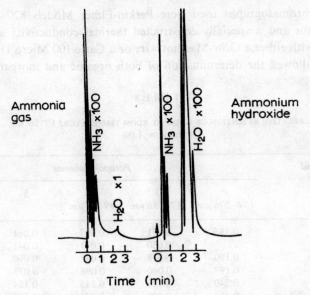

Fig. 20. Analysis of ammonia and water. Conditions: column, 2 ft. PE 1 on Porapak Q; temperature, 77°C; flow, 50 ml/min H_2.

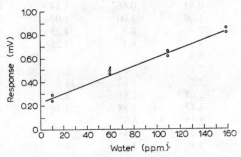

Fig. 21. Quantitative response of water. Conditions: column, 6 ft. $\times \frac{3}{16}$ in. Porapak P; temperature, 123°C; flow, 37 ml/min H_2.

Hollis and Hayes[9] reviewed the utility of microporous polymers for the GC separation of atmospheric gases, air pollution gases, reactive gases, and light petroleum gases. With a microporous polymer gel, the gas sample did not require drying before analysis to protect the column from deactivation by water as is required with solid adsorbents such as alumina, molecular sieves, and silica gel. Active gases such as the oxides of nitrogen, sulfur dioxide, and ammonia were resolved from one another as well as from the light hydrocarbon gases. The C_1–C_3 hydrocarbon gases were resolved from one another and from air, carbon dioxide, water, and carbonyl sulfide. The first separations by gas–gel chromatography were described by Hollis[8]. The synthesis of the polymers was designed to yield material in bead form with large surface areas ($\simeq$ 100–700 m^2/g) and strong physical structure.

The gas chromatographs used were Perkin-Elmer Models 820 and 800 with hot-wire detector and a specially constructed thermal conductivity gas chromatograph filtered with either a Gow-Mac hot wire or a Carle 100 Micro Detector. These modifications allowed the determination of both organic and inorganic gases. The

TABLE 8

RELATIVE RETENTION OF GASES AT ROOM TEMPERATURE ($\simeq 26°C$)
Ethane $= 1.00$

Compound	Porapak columns				
	P 104–206 μm	Q 117–180 μm	R 49–65 μm	S	T 132–166 μm
Air	0.185	0.071	0.087	0.061	0.071
Hydrogen		0.050	0.071	0.041	0.053
Carbon monoxide	0.190	0.078	0.095	0.070	0.075
Nitric oxide (NO)	0.197	0.080	0.098	0.073	0.086
Methane	0.240	0.128	0.145	0.124	0.127
Carbon dioxide	0.459	0.306	0.376	0.373	0.793
Nitrous oxide (N_2O)	0.578	0.433	0.466	0.467	
Ethylene	0.743	0.667	0.692	0.689	0.840
Acetylene	0.85	0.667	1.14	0.865	2.2
Ethane	1.00	1.00	1.00	1.00	1.00
Propylene	2.6	4.7	4.35	4.1	5.6
Propane	4.3	6.1	5.26	4.6	4.8
Propadiene	6.0				8.8
Methylacetylene	7.1				14.5
Water	2.5	1.67	11.6	5.2	22.0
Hydrogen sulfide	2.03	1.61	2.29	2.38	3.61
Carbonyl sulfide	3.47	3.38	3.46	3.75	4.65
Sulfur dioxide	8.55	6.92			
Hydrogen cyanide	6.27	5.44			
Methyl chloride	5.1	5.5		5.4	

relative retention of gas at room temperature ($\simeq 26°C$) obtained on Porapak columns P, Q, R, S, and T is shown in Table 8 and shows the separation of oxides of carbon and nitrogen. Carbon dioxide was well separated at room temperature on all of the gas–gel chromatographic 6 ft. columns tested. Table 9 lists the retention data obtained on 6 ft. columns of Porapak Q, R, S, T, and Q–T at 65°C. The C_1, C_2, and C_3 hydrocarbons can be separated on porous polymer gels at room temperature and at 65°C as shown in Table 9. Differences in the number and kinds of polar sites in the polymer grossly change the relative retention times of polar to non-polar molecules, such as the acetylenes, carbon dioxide, and water relative to the paraffins and inert gases. Table 10 shows the gas-retention data for polymers at 100°C on Porapak columns P, Q, R, S, and T. The chlorohydrocarbons gave excellent peak shape and were well reserved from porous polymer columns. The gaseous

TABLE 9

RETENTION DATA (min) AT 65°C

Column size, 6 ft. × $\frac{1}{10}$ in.; flow rate, 30 ml/ min; carrier gas, helium

Compound	Porapak columns				
	Q 117–180 μm	R 93–174 μm	S 124–134 μm	T 132–166 μm	Q and T* 121–148 μm
Air	0.48	0.55	0.50	0.44	0.56
Hydrogen	0.37	0.46	0.44	0.37	0.45
Nitric oxide	0.56	0.59	0.53	0.58	0.62
Methane	0.66	0.77	0.71	0.62	0.74
Carbon dioxide	1.15	1.37	1.21	1.50	1.75
Ethylene	1.91	2.15	1.82	1.60	2.03
Acetylene	1.91	2.75	2.05	3.73	3.45
Nitrous oxide	1.43	1.62	1.39		1.95
Ethane	2.58	2.75	2.34	1.77	2.45
Water	3.74	18.6	9.64	33.9	24.75
Hydrogen sulfide	3.62	4.75	4.38	5.10	5.14
Hydrogen cyanide	5.95	42.9	19.45	51.0	47.2
Carbonyl sulfide	6.0	6.76	6.14	6.0	6.9
Sulfur dioxide	7.87	26.9	41.56	33.6	29.4
Propylene	8.26	9.5	8.54	8.77	11.43
Propane	9.45	10.66	9.6	8.20	11.43
Propadiene	10.63	12.05	11.0	13.35	15.95
Methylacetylene	10.63	13.3	12.02	20.0	23.7
Methyl chloride	10.12	12.85	11.54	16.3	18.2
Vinyl chloride	18.5	24.86	22.7	30.7	36.5
Ethylene oxide	15.66	24.15	22.7	41.2	43.35
Ethyl chloride	31.6	53.1	48.5	76.6	86.0
Carbon disulfide	83.5			72	

* Column contained 33% Porapak Q and 67% Porapak T.

compounds of sulfur (H₂S, COS, and SO₂) were found to have excellent elution characteristics on Porapak Q as shown in Fig. 22. Fluorohydrocarbons were also well separated on Porapak Q and phosgene, sulfuryl chloride, and cyanogen chloride had reasonably good elution peaks from Porapak Q and Porapak S. Gas–gel chromatography has been shown to be particularly useful by Hollis and Hayes[9] for the analysis of toxic, "polar", and reactive gases such as hydrogen cyanide and the oxides of nitrogen and sulfur. Because of the rapid elution from porous polymer columns, trace amounts of these gases could be determined in hydrocarbons or other gas mixtures. Figure 23 is a chromatogram showing the elution of hydrogen cyanide from SO₂ and water at 70°C on a column of Porapak Q. The separation of acetylene, ethane, ethylene, and carbon dioxide on Porapak T at 35°C is shown in Fig. 24.

Obermiller and Charlier[10] described the gas chromatographic separation of nitrogen, oxygen, argon, carbon monoxide, carbon dioxide, hydrogen sulfide,

TABLE 10

GAS RETENTION DATA (min) FOR POLYMERS AT 100°C
Flow rate, 50–55 ml/min; carrier gas, hydrogen.

Compound	Porapak columns				
	P	Q	R	S	T
Air	0.31	0.28	0.27	0.41	0.40
Water	0.97	1.23	3.45	3.50	7.10
Methane	0.38	0.39	0.36	0.52	0.45
Ethylene	0.53	0.83	0.74	1.11	0.92
Ethane	0.53	1.02	0.90	1.31	1.00
Acetylene			0.90	1.17	1.40
Propylene			2.32	3.55	2.60
Propane	0.97	3.03	2.58	3.55	2.60
Butadiene	2.36	8.05	7.50	11.00	9.3
Methyl chloride	1.22	3.22	2.84	4.48	4.42
Vinyl chloride	1.61	5.51	4.75	7.39	7.07
Ethyl chloride	2.68	9.81	8.30	13.31	13.15
Ethylene oxide			4.32	7.87	9.00
Carbon dioxide			0.60	0.80	0.82

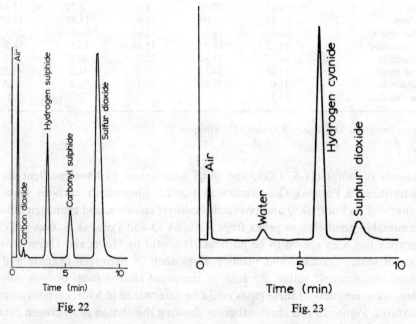

Fig. 22   Fig. 23

Fig. 22. Chromatogram showing separation of sulfur gases. Column 6 ft. × $\frac{1}{10}$ in. Porapak Q
(117–180 μm); temperature, 68°C; flow rate, 30 ml/min; carrier gas, helium.

Fig. 23. Chromatogram showing elution of hydrogen cyanide. Column, 6 ft. × $\frac{1}{10}$ in. Porapak Q;
temperature, 70°C; flow rate, 30 ml/min; carrier gas, helium.

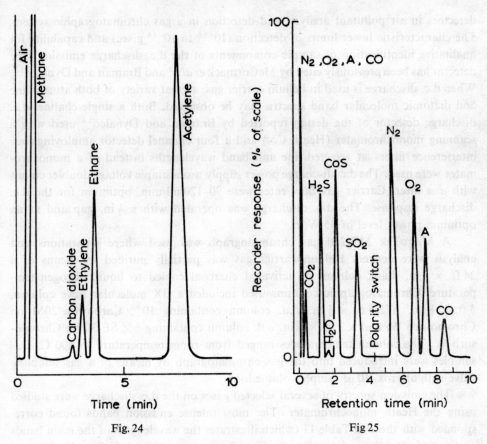

Fig. 24. Chromatogram showing separation of acetylene. Column, 6 ft × $\frac{3}{16}$ in. Porapak T (132 to 166 μm); temperature, 35°C; flow rate, 55 ml/min; carrier gas, hydrogen.

Fig. 25. Separation of N_2, O_2, A, CO, CO_2, H_2S, COS, and SO_2 on Porapak Q, 50–80 mesh. Column temperature "hot" +75°C, "cold" −65°C. Carrier gas, helium with 100 p.p.m. SO_2 at 15 ml/min; split flow 50–60 ml/min; sample size 700 μl.

carbonyl sulfide, and sulfur dioxide. Columns used were $\frac{1}{16}$ in. o.d. × 0.046 in. i.d. packed with 50–80 mesh Porapak Q. The "hot" column was 5 ft. long and the "cold" column 25 ft. long.

A splitter was used with the columns made from Swagelok $\frac{1}{8}$ in. union elbow drilled to accept a short piece of $\frac{1}{16}$ in. o.d. × 0.046 in. i.d. tubing so that the gas coming from the inlet was swept directly over it. The other end of the union elbow was fitted to a micrometer valve where the split ratio was adjusted. The short piece of $\frac{1}{16}$ in. o.d. tubing is connected directly to the "hot column". The carrier gas was helium with 100 p.p.m. SO_2 (J. T. Baker). Figure 25 illustrates the separation of N_2, O_2, A, CO, CO_2, H_2S, COS, and SO_2 on Porapak Q, 50–80 mesh.

Braman[11] investigated the use of direct current (d.c.) discharge emission type detectors in air pollutant analysis and detection in a gas chromatographic system. The characteristic lower limits of detection (10^{-10} to 10^{-16} g/sec) and capability for qualitative identification of sample components of the d.c. discharge emission type detector has been previously cited by McCormack et al.[12] and Braman and Dynalco[13]. When d.c. discharge is used in helium carrier gas, a great variety of both atomic line and diatomic molecular band spectra may be observed. Both a single channel d.c. discharge detector of the design reported by Braman and Dynalco[13] used with a scanning monochromater (Heath Co.) and a four channel detector employing four interference filters at selected line and band wavelengths instead of a monochromater were used. The d.c. discharge power supply was a simple voltage doubler circuit with a π filter. Carrier gas flow rates were 90–120 ml/min, optimum for the d.c. discharge response. The d.c. discharge was operated with a $\frac{1}{2}$ in. gap and at an optimum power level of 35 W/in.

A MicroTek MT-220 gas chromatograph was used where separations and analysis were desired. Helium carrier gas was partially purified by means of a 24 ft. $\times$ $\frac{3}{8}$ in. diam. column of activated charcoal cooled to liquid nitrogen temperature. Chromatographic columns used included a BX molecular sieve column, 3 ft. $\times$ $\frac{1}{8}$ in., a 2.5 ft. $\times$ $\frac{1}{4}$ in. o.d. column containing 10% Carbowax 20M on Chromosorb W, and a 3 ft. $\times$ $\frac{1}{8}$ in. o.d. column containing 5% SE-30 on Chromosorb W. The column temperatures ranged from room temperature to 100°C. Gas samples were introduced into the gas chromatograph by means of a gas sampling valve with approx. 10 μl sample volume loop.

The emission spectra of several selected gases on the d.c. discharge were studied using the Heath monochromater. The most intense emission bands found corresponded with those in Table 11 (which illustrates the wavelengths of the main bands or lines of interest in detection of air pollutant compounds). Sulfur dioxide and carbonyl sulfide produced S_2 and SO bands and S and O line spectra in a helium discharge. The S_2 and SO bands were extensive and somewhat diffuse. Nitrogen oxides produced O lines and N lines. Limits of detection for the nitrogen oxides and other permanent gases were determined from responses at C_2, CH, CN, H lines and N or O line wavelengths (Table 12). The notation "quenches" indicates that background radiation at the indicated wavelength was reduced by the sample component.

The limits of detection were in the 0.1–1 p.p.m. (vol.) range for the permanent gases, based upon the analysis of 1 ml samples of gas. For nitrogen oxides, this method is not as sensitive as the chemiluminescent technique of Fontion et al.[14] where limits of detection are approx. 4 p.p.b. Carbon-containing gases all gave good positive responses at C_2 and CH band wavelengths while the nitrogen oxides did not.

The d.c. discharge technique does have a lower limit of detection for sulfur dioxide than the method of West and Gaeke[15] which is suitable for analysis in the microgram range. The limits of detection indicated the d.c. discharge detector was

TABLE 11

SPECTAL EMISSION WAVELENGTHS OF AIR POLLUTANT COMPOUNDS

Compound	Wavelength (Å)	Band or line	Remarks
SO$_2$	3064.1	SO	Degraded to the red
	3164.8	SO	
	3271.0	SO	
	7771–7774	O	Oxygen triplet
	2769.4–6165.8	S$_2$	These bands merge with the SO bands to produce almost a continuum of background in discharges.
CO	5165.2	C$_2$	Most intense of C$_2$ series
	7771–7774	O	Oxygen triplet only if O$_2$ impurities are low
CO$_2$	5165.2	C$_2$	
	7771–7774	O	Oxygen triplet, stronger than for CO
COS	5165.2	C$_2$	
	3064.1	SO	Also other SO bands
	7771–7774	O	Less prominent
N$_2$O, NO, NO$_2$	3883.4	CN	
	4216.0	CN	Best, avoids interference from He line at 3888 Å
	6008.5	N	Other N lines are likely useful
	4358.3	N	
O$_2$	7771–7774	O	Good selectivity
H$_2$	6562	H	Good selectivity
Organic compounds in general	5165.2	C$_2$	
	4314.2	CH	
	3883.4	CN	
	6562	H	

TABLE 12

LIMITS OF DETECTION FOR SELECTED GASES BAND OR LINE SPECTRA OBSERVED

Gas	C$_2$	CH	CN	H	Others
N$_2$	1% of CN	10% of CN	1.4×10^{-10} l	3% of CN	N line
O$_2$	Quenches	Quenches	None	None	O line
H$_2$	None	None	Quenches	9.5×10^{-11} l	
NO	None	None	3.3×10^{-9} l	Quenches	O line N line
N$_2$O	None	20% of CN	1×10^{-9} l	Quenches	N line O line
NO$_2$	30% of CN	None	4×10^{-10} l	None	O line N line
CO	4.2×10^{-11} l	25% of C$_2$	Very weak	Very weak	O line
CO$_2$	2.9×10^{-10} l	2.9×10^{-10} l	Quenches	Quenches	O line
COS	7×10^{-10} l	7×10^{-10} l	Quenches	Weak	O line
SO$_2$	None	Very weak	Weak	None	O line
CF$_4$	30% of CH	7×10^{-10} l	Quenches	Very weak	

suitable for gas analysis of automobile exhaust. The four channels of the detector were found particularly useful for the confirmation of the identity of air peaks, hydrogen, carbon monoxide, hydrocarbons, and carbon dioxide.

REFERENCES

1 R. M. BETHEA AND M. C. MEADOR, *J. Chromatog. Sci.*, 7 (1969) 655.
2 N. BRENNER AND L. S. ETTRE, *Anal. Chem.*, 31 (1959) 1815.
3 D. R. DEANS, *et al.*, *Chromatographia*, 4 (1971) 279.
4 A. DRAVNIEKS, B. K. KROTOSZYNSKI, J. WHITFIELD, A. O'DONNELL AND T. BURGWALD, *Environ. Sci. Technol.*, 5 (1971) 1220.
5 O. DUFKA, J. MALINSKY, J. CHURACEK AND K. KOMAREK, *J. Chromatog.*, 51 (1970) 111.
6 I. G. DOBROCHIVER, S. P. ZHAROVA AND N. Y. ZUZYKINA, *Zavodsk. Lab.*, 35 (1969) 1053; *Chem. Abstr.*, 72 (1970) 38471d.
7 O. L. HOLLIS AND W. V. HAYES, *J. Gas Chromatog.*, 4 (1966) 235.
8 O. L. HOLLIS, *Anal. Chem.*, 38 (1966) 309.
9 O. L. HOLLIS AND W. V. HAYES, in A. B. LITTLEFIELD (Ed.), *Gas Chromatography, 1965*, Institute of Petroleum, London, 1967, p. 57.
10 E. L. OBERMILLER AND G. O. CHARLIER, *J. Chromatog. Sci.*, 7 (1969) 580.
11 R. S. BRAMAN, *Atmos. Environ.*, 5 (1971) 669.
12 A. J. MCCORMACK, S. C. TONG AND W. D. COOKE, *Anal. Chem.*, 37 (1965) 1470.
13 R. S. BRAMAN AND A. DYNALCO, *Anal. Chem.*, 40 (1968) 95.
14 A. FONTION, A. J. SABADELL AND R. J. RONCO, *Anal. Chem.*, 42 (1970) 575.
15 P. W. WEST AND G. C. GAEKE, *Anal. Chem.*, 28 (1956) 1816.

Chapter 19

POLYCYCLIC AROMATIC HYDROCARBONS

Polycyclic aromatic hydrocarbons (also referred to as polynuclear aromatic hydrocarbons) (PAH) are perhaps the most extensively studied components both in polluted air as well as in foodstuffs. This is a consequence of members of this chemical class having been shown to be carcinogenic to laboratory animals[1-3] as well as being suspected carcinogens for man[4]. The occurrence of polycyclic aromatic hydrocarbons in the environment is extensive and aspects of their occurrence, isolation and identification have been reviewed by Gunther and Buzzetti[5], Gunther et al.[6], Haenni[7], Schaad[8], Hoffman and Wynder[9], and Sawicki et al.[10-13].

Polycyclic hydrocarbons are found in air, soil and water, in many members of both animal and plant kingdoms, marine and non-marine sediments, as well as a spectrum of foods, including smoked foods, processed rice, roasted coffee, baked goods, barbecued meat, raw agricultural commodities such as oranges and products such as commercial waxes and paraffins, mineral oils, commercial solvents. The majority of polycyclic hydrocarbons in the environment is derived during incomplete combustion of organic matter at high temperatures and under pyrolytic con-

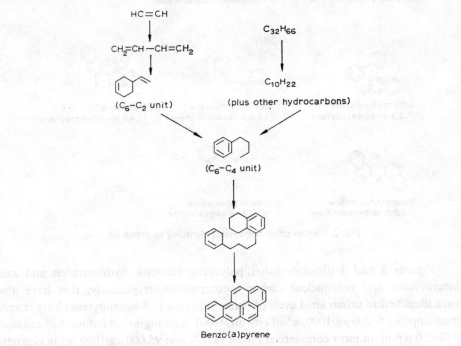

Fig. 1. Benzo[a]pyrene pyrosyntheses as suggested by Badger[14].

ditions involving two types of reaction, *viz.* pyrolysis and pyrosynthesis. For example, at temperatures above 400–500°C, organic components are partially cracked to smaller and relatively unstable molecules (radicals) which then recombine to form larger more stable polycyclic aromatic hydrocarbons. Figure 1 illustrates a scheme for benzo[*a*]pyrene pyrosynthesis as suggested by Badger *et al.*[14]. Besides benzo-[*a*]pyrene at least 11 carcinogenic polycyclics have been identified in urban air[15–19] (Fig. 2).

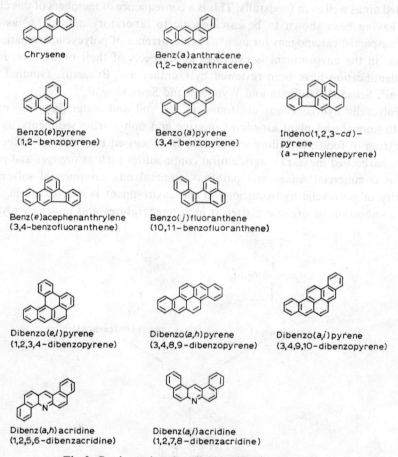

Chrysene

Benz(a)anthracene
(1,2-benzanthracene)

Benzo(e)pyrene
(1,2-benzopyrene)

Benzo(a)pyrene
(3,4-benzopyrene)

Indeno(1,2,3-*cd*)-
pyrene
(a-phenylenepyrene)

Benz(e)acephenanthrylene
(3,4-benzofluoranthene)

Benzo(*j*)fluoranthene
(10,11-benzofluoranthene)

Dibenzo(e,*l*)pyrene
(1,2,3,4-dibenzopyrene)

Dibenzo(a,h)pyrene
(3,4,8,9-dibenzopyrene)

Dibenzo(a,*i*)pyrene
(3,4,9,10-dibenzopyrene)

Dibenz(a,h)acridine
(1,2,5,6-dibenzacridine)

Dibenz(a,*i*)acridine
(1,2,7,8-dibenzacridine)

Fig. 2. Carcinogenic polycyclics identified in urban air.

Figures 3 and 4 illustrate other polycyclic aromatic hydrocarbons and aza-heterocyclics and polynuclear carbonyl compounds, respectively, that have also been identified in urban air. Levels of benzo[*a*]pyrene (3,4-benzopyrene) have ranged from approx. 6–200 μg/1000 m^3 of city air, 0.001–0.01 mg/m^3 in automobile exhaust, 0.01–2.0 p.p.m. in most commercial motor fuels[9], and 95,000 μg/1000 m^3 in cigarette smoke[20].

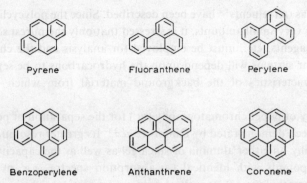

Pyrene   Fluoranthene   Perylene

Benzoperylene   Anthanthrene   Coronene

Fig. 3. Miscellaneous polycyclic aromatic (hydrocarbons in urban atmosphere).

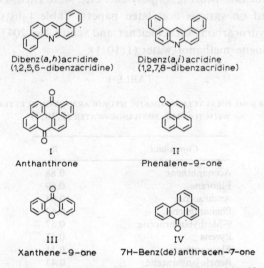

Dibenz(*a,h*)acridine   Dibenz(*a,i*)acridine
(1,2,5,6-dibenzacridine)   (1,2,7,8-dibenzacridine)

I   II
Anthanthrone   Phenalene-9-one

III   IV
Xanthene-9-one   7H-Benz(de)anthracen-7-one

Fig. 4. Aza-heterocyclics and polynuclear carbonyl compounds identified in urban air.

The analysis of polycyclic hydrocarbons is complicated because of both their extremely low concentrations as well as their admixture with large numbers and amounts of other classes of organic compounds. The methods used for the enrichment and isolation of polycyclic hydrocarbons from air and food sources include fractionation into neutral, basic, and acidic portions followed by chromatography of the neutral fraction on silica gel, alumina, or Florisil.

The enrichment is best achieved by distribution between organic solvents followed by column or thin-layer chromatography. It is generally agreed that unequivocal identification of the polycyclic hydrocarbons requires spectral characterization in addition to chromatographic R_F values and retention times. Polycyclic hydrocarbons are often characterized by their intense fluorescence. Direct ultraviolet[21,22], fluorimetric[23-26], and phosphorimetric[27] examinations of paper and thin-layer chromatograms as well as GLC eluents[12] have been described. Since the polycyclic hydrocarbons are isolated in very small amounts, it is stressed that only the purest solvents, adsorptive media, reagents, etc., must be employed for analysis and the choice of a particular system or systems will depend upon the hydrocarbons to be separated and the chemical characteristics of the background material from which they are to be isolated.

The utility of paper chromatography (PC) for the separation of polycyclic hydrocarbons has been demonstrated by Dubois *et al.*[28]. Its greater resolution than column chromatography on either alumina or silica gel as well as the capacity of PC to often separate compounds with identical UV absorption spectra, *e.g.* pyrene, 2-methylpyrene and 4-methylpyrene, as well as achieve separations which were impractical on a column, *e.g.* benzo[*a*]pyrene and benzo[*e*]pyrene, fluoranthene and benz[*a*]anthracene, benzo[*e*]pyrene and benzo[*g,h,i*]perylene, etc., were cited. Optimal separations have been achieved on various acetylated papers. Table 1 lists the R_F values of some polycyclic hydrocarbons on Schleicher and Schull No. 2043b acetylated paper developed with toluene–methanol–water (1:10:1).

TABLE 1

ABSOLUTE R_F VALUES OF SOME POLYCYCLIC AROMATIC HYDROCARBONS ON ACETYLATED PAPER DEVELOPED WITH TOLUENE–METHANOL–WATER (1:10:1)

Compound	R_F
Acenaphthene	0.88
Fluorene	0.69
Anthracene	0.64
Phenanthrene	0.62
9-Methylanthracene	0.57
Pyrene	0.52
Fluoranthene	0.51
Benz[*a*]anthracene	0.43
Benzo[*a*]pyrene	0.41
Perylene	0.28
Chrysene	0.22
Benzo[*g, h, i*]perylene	0.18
Dibenz[*a, h*]anthracene	0.16
Benzo[*e*]pyrene	0.10
Anthanthrene	0.07

Van Duuren[29] has demonstrated that R_F values for polycyclic hydrocarbons can be markedly different with different papers and these differences can be employed for sequential separations. Microgram quantities are often separated by paper chromatography and spot detection achieved by chromogenic reagents and ultraviolet and fluorescence spectrometry.

Useful combinations of techniques involving paper chromatography have been demonstrated by Hoffman and Wynder[30] which involved initial solvent partitioning of the polycyclic hydrocarbons followed by alumina column chromatography, then their resolution by acetylated paper chromatography. Table 2 lists partition coefficients of some polycyclic hydrocarbons.

TABLE 2

PARTITION COEFFICIENTS AT $20°$ OF SOME POLYCYCLIC HYDROCARBONS

Compound	$C_c/C_m{}^a$	$C_n/C_c{}^a$
Pyrene	150.0	4.40
Benzo[a]pyrene	14.9	1.53
Benzo[e]pyrene	14.5	1.68
Fluoranthene	140.0	1.80
Benzo[b]fluoranthene	14.7	1.94
Benzo[j]fluoranthene	14.8	1.75
Benzo[k]fluoranthene	14.9	1.76
Benz[a]anthracene	18.6	1.75
Dibenz[a, h]anthracene	14.9	1.80
Perylene	14.9	1.69
Benzo[g, h, i]perylene	14.6	1.77
Chrysene	14.6	1.65

^{a} C_c = concentration in cyclohexane, C_m = concentration in methyl alcohol:water (4:1), C_n = concentration in nitromethane. Spectroscopically determined in absence of other solutes.

Both single and two-dimensional thin-layer chromatography (because of the small amount of sample required and their speed of analysis) have been found to be particularly useful for the separation of complex mixtures into various classes or types of polycyclic compounds prior to their spectral characterization. Various absorbents including silica gel, alumina, cellulose, cellulose acetate, magnesium hydroxide, and Porapak Q with an equally large variety of developing solvents have been employed for the TLC analysis of environmental pollutant polycyclic hydrocarbons and related derivatives. Tables 3 and 4 list a variety of absorbents and developing solvents that have been employed for the single- and two-dimensional TLC, respectively, of polycyclic hydrocarbons and their various derivatives (*e.g.* aza, imino, carbonyl, and phenol).

Table 5 lists reagents and physical properties that have been employed for the characterization of the polycyclic hydrocarbons and derivatives after TLC.

The TLC separation and low-temperature luminescence measurement of polycyclic aromatic hydrocarbons was described by Hood and Winefordner[54]. The several distinct advantages in making fluorescence and phosphorescence measurements at low temperature have been cited by McGlynn *et al.*[55], Winefordner *et al.*[56], and Muel and Lacroix[57]. For example, the extremely sharp and numerous fluorescence bands which characterize the low-temperature fluorescence spectra of poly-

TABLE 3

TLC FOR ANALYSIS OF AIR POLLUTANTS[13]

Compounds	Layer material	Solvent	Ref.
Polynuclear aromatic hydrocarbons (arenes)	Alumina	Carbon tetrachloride	31
		Hexane	32, 33
		Pentane–ether (19:1, v/v)	32, 34
	Cellulose	Dimethylformamide–water (1:1, v/v)	34
	Cellulose[a]	Iso-octane[a]	32, 35
	Cellulose acetate	Ether–methanol–water (4:4:1, v/v)	33
		Toluene–ethanol–water (4:17:4, v/v)	32, 34
	Magnesium hydroxide	Benzene	36
	Porapak Q	Acetone	37
		Acetone–propanol (1:1, v/v)	37
		Ethanol	37
		Ethyl acetate	37
		n-Propanol	37
	Silica gel	Carbon tetrachloride	31, 38
		Cyclohexane–benzene (4:1, v/v)	32
		Hexane	31, 32
		Hexane–o-dichlorobenzene–pyridine (20:2:1, v/v)	32
		Heptane–benzene (4:1, v/v)	38
	Silica gel[b]	Carbon tetrachloride	38
		Heptane–benzene (4:1, v/v)	38
	Silica gel[c]	Heptane–benzene (4:1, v/v)	38
	Silica Gel GF$_{254}$	Petroleum ether, b.p. 50–70°	39
	Silica Gel GF$_{254}$[d]	Petroleum ether, b.p. 50–70°	39
Aza arenes	Alumina	Pentane–ether (19:1, v/v)	40
		Pentane–nitrobenzene (9:1, v/v)	41
		Pentane–2-nitropropane–triethylamine (9:1:0.05, v/v)	41
	Cellulose	Acetic acid–water (3:7, v/v)	40
		Dimethylformamide–water (35:65, v/v)	40
	Silica gel	Pentane–ether (7.3, v/v)	42
Imino arenes	Cellulose	Dimethylformamide–water (3:1, v/v)	43
		Dimethylformamide–water (35:65, v/v)	10
	Florisil	Pentane–ether (3:1, v/v)	43
Ring carbonyl arenes	Alumina	Pentane–ether (19:1, v/v)	44
		Toluene	44
	Cellulose	Dimethylformamide–water (35:65, v/v)	44
Polynuclear phenols	Polyamide	2-Butanone–2,4-pentanedione–ethanol–water (15:5:15:65, v/v)	45
			45
	Silica gel	Triethylamine	45
	Silica gel[e]	Toluene–ethyl formate–formic acid (5:4:1, v/v)	
Aromatic carbonyl compounds[f]	Silica gel	Benzene	46
		Benzene–methanol (19:1, v/v)	46
		Hexane–ethyl acetate (1:1, v/v)	46
		Methylene chloride	46

[a] Immobile phase is 20% dimethylformamide in ether.
[b] Impregnated with 2% 2,4,7-trinitrofluorenone in benzene and dried at 60°C for 20 min.
[c] Impregnated with 2% 1,3,5-trinitrobenzene in benzene and dried at 60°C for 20 min.
[d] Impregnated with a saturated solution of picric acid in benzene and dried.
[e] Saturated with 0.3 M sodium acetate.
[f] 4-Nitrophenylhydrazones prepared before separation.

TABLE 4

TWO-DIMENSIONAL CHROMATOGRAPHY OF POLYCYCLIC AROMATIC HYDROCARBONS AND THEIR DERIVATIVES[13]

Compounds	Layer material	Solvent	Ref.
Polycyclic hydrocarbons	Alumina–cellulose (2:1)	1. Pentane 2. Dimethylformamide–water (35:65, v/v)	12
	Alumina–cellulose acetate (2:1)	1. Pentane 2. Toluene–ethanol–water (4:17:4, v/v)	12, 47
Aza-polycyclic hydrocarbons	Alumina	1. Cyclohexane–ethyl acetate (19:1, v/v)	48
	Alumina–cellulose (2:1)	1. Pentane 2. Dimethylformamide–water (35:65, v/v)	12
		1. Cyclohexane–ethyl acetate (19:1, v/v) 2. Dimethylformamide–water (35:65, v/v)	12
	Silica gel–cellulose (2:1)	1. Pentane–ether (9:1, v/v) 2. Dimethylformamide–water (35:65, v/v)	42
Imino polycyclic hydrocarbons	Alumina	1. Cyclohexane–ethyl acetate (4:1, v/v)	12
		2. Pentane–toluene (1:1, v/v)	12
Ring-carbonyl polycyclic hydro-carbons	Alumina	1. Cyclohexane–ethyl acetate (4:1, v/v)	47
		2. Pentane–toluene (1:1, v/v)	12
		1. Cyclohexane–ethyl acetate (4:1, v/v)	47
		2. Pentane–toluene (1:1, v/v)	12
		1. Cyclohexane–ethyl acetate (19:1, v/v)	48
		2. Toluene	48
		1. Cyclohexane–ethyl acetate (1:1, v/v)	
		2. Toluene	49
	Alumina–cellulose acetate (2:1)	1. Pentane 2. Toluene–ethanol–water (4:17:4, v/v)	12
Polynuclear phenols	Silica Gel F_{254}	1. Toluene–methanol (1:1, v/v) 2. Cyclohexane–triethylamine (1:1, v/v)	45

nuclear aromatics enhance selectivity over room-temperature measurements and the increase in fluorescence intensity with decreasing temperature enhances sensitivity.

In the study of Hood and Winefordner[54] ethanol, which forms an ideal low-temperature rigid matrix[58], was used as the thin-layer extractant. A fluorescence spectrophometer (Fluorispec, Model SF-1, Baird-Atomic) with X–Y recorder was

TABLE 5

LOCATION AND CHARACTERIZATION OF SPOTS AFTER TLC[13]

Compound	Number of compounds	Reagent	Color		Detection limit (µg)	Ref.
			Visible	Fluorescence		
Polynuclear aromatic hydrocarbons (arenes)	20	None. Wet and dry layer		B, BG, Y, Pk		34
	17	None		B, V, YG, O	0.001–1	50
	23[a]	None. Low temperature fluorescence, phosphorescence			0.005–1	51
	30	Formaldehyde–sulfuric acid	Visible			31
	30	Tetracyanoethylene	Visible			31
	9	Piperonal	R, B, G		0.1–10	50
	3	7,7,8,8-Tetracyanoquinodimethane	OR-Bk		0.5–1	52
Aza arenes	17	None, and fluorescence with acid solution		YG-B	0.001–0.01	40
	22	None, and fluorescence with TFA fumes		B, Y, BG → G, O, B(TFA)		41
	11[a]	Nitrobenzene or 2-nitropropane		G, B, RO	<1	53
	26	None		G, P, B, O	<0.1	52
Imino arenes	3	7,7,8,8-Tetracyanoquinodimethane	V, P		0.8–3	42
	34	None		P, Y, B, G	0.2–1	50
	8	None		B		10
	27	None		B		52
	3	7,7,8,8-Tetracyanoquinodimethane	B, G, V		0.5	10, 50
	8	Acid, examination in uv	Y, R, G		0.2–0.4	10
	27	Tetraethylammonium hydroxide		B, G, Y		10
	27	3-Methylthioaniline, NO$_2$TFA or CS$_2$		Quenching		10
Aromatic carbonyl compounds	22	None		Y, G, B, O	0.0005–0.1	50
	22	TFA		Y, G, B, O, R	0.0005–1	50
	22	None		Y, G, B		50
Aromatic carbonyl p-nitrophenyl hydrazones	19	KOH	R, O, P	Y, B, G, R, O		50
Polynuclear phenols	45	None		B, G, O, P, Y		45
	45	Alkali		B, G, O, P, Y		45
	45	TFA		B, G, O, P, Y		45

[a] Fluorescence on wet plate, dry plate, and TFA-treated plate. TFA = trifluoroacetic acid.

Colors: B, blue; Bk, black; Br, brown; G, green; O, orange; P, purple; Ph, pink; R, red; Y, yellow.

TABLE 6

R_{FB}^a VALUES AND LOW-TEMPERATURE LUMINESCENCE CHARACTERISTICS OF POLYNUCLEAR AROMATIC HYDROCARBONS

Compound	R_{FB}^a	Fluorescence[b,d]			Phosphorescence[c,d]			
		λ_{ex}(nm)	λ_{em}(nm)	L. of D. (µg/ml)	λ_{ex}(nm)	λ_{em}(nm)	L. of D. (µg/ml)	τ (sec)
Pyrene	1.40	338	374	0.002	328	595	0.4	0.46
2,3-Benzfluorene	1.40	320	342	0.01	315	507	0.4	2.8
1,2-Benzanthracene	1.30	348	387	0.03	310	505	0.05	1.4
20-Methylcholanthrene	1.10	300	396	0.008				
Triphenylene	1.10	288	354	0.3	285	460	0.003	16.2
1,2-Benzpyrene	1.00	333	390	0.03	325	544	0.02	2.1
3,4-Benzpyrene	1.00	369	405	0.003	330	517	2.0	2.3
Perylene	0.90	438	470	0.002				
1,2,5,6-Dibenzanthracene	0.90	300	396	0.008	295	554	0.02	1.5
1,12-Benzperylene	0.80	386	421	0.005	295	556	0.09	0.67
1,2,3,4-Dibenzanthracene	0.77	290	377	0.007	287	565	0.09	0.90
1,2,3,4-Dibenzpyrene	0.73	331	504	0.07				
3,4,9,10-Dibenzpyrene	0.73	421	473	0.05	295	530	0.5	2.7
3,4,8,9-Dibenzpyrene	0.70	313	453	0.003				
1,2,4,5-Dibenzpyrene	0.67	305	397	0.007	379	609	0.6	0.67
Coronene	0.67	341	446	0.004	310	557	0.004	9.6

[a] R_F values relative to 3,4-benzpyrene.
[b] Spectral bandwidth 2 nm.
[c] Spectral bandwidth 8 nm.
[d] All limits of detection (L. of D.) were measured at excitation, λ_{ex}, and emission, λ_{em}, wavelengths given. The wavelengths are generally peak values uncorrected for instrumental response.

used for determining all fluorescence and phosphorescence spectra and for measuring limits of detection and analytical curves. Table 6 lists the R_F values and low-temperature luminescence characteristics of a number of polycyclic aromatic hydrocarbons.

The identification of polynuclear hydrocarbons in air pollutants was described by Matsushita et al.[59]. Particulates suspended in the air at Kawasaki (75 mg) were collected in an air sampler (Staplex Co.) mounted with a Gelman A glass filter. The filter was then inserted in a vacuum sublimation tube, heated at 300° under 0.01 torr, and polynuclear hydrocarbons condensed in a 2 × 200 mm side tube at room temperature, then eluted with 75 μl benzene to prepare a sample solution.

A 4 μl sample of this solution was first chromatographed on a 4 × 20 cm plate of Alumina G–Cellulose D-O (95:5) and developed with n-hexane–ether (19:1) for 35 min, 15 cm, and then second-dimensionally developed on a 26% acetylated cellulose plate, 16 × 20 cm, 300 μm thick, with methanol–ether–water (4:4:1) for 60 min, 10 cm. Seventy-six spots were revealed by UV light and were extracted from the plates with benzene and prepared for fluorescence measurements as n-hexane solutions. The purity and identification of the samples was established by fluorescence and excitation spectra obtained with a Hitachi fluorescence spectrometer MPF-2. From 76 spots, 28 were identified as 5,12-dihydrotetracene, 3-methylpyrene, 7H-benzo[c]fluorene, pyrene, fluoranthene, benzo(mono)fluoranthene, benz[a]anthracene, chrysene, 13H-naphtho[2,3-b]fluorene, benzo[e]pyrene, perylene, benzo[k2]fluoranthene, benzo[g,h,i]perylene, benz[c]acridine, 11H-naphtho-[2,1-a]fluorene, indeno-[1,2,3-Cd]pyrene, anthranthrene, coronene, picene, tetracene, benzo[b]chrysene, tribenzo[a,c,j]tetracene, tribenzo[a,e,i]pyrene, dibenzo[a,i]pyrene, dibenzo[a,h]pyrene, benzo[g]chrysene, 1,2,3,5,6,7-hexahydrotriangulene, naphtho[2,3-k]fluoranthene.

White and Howard[60] investigated the utility of cellulose and cellulose acetate plates for the separation of polycyclic aromatic hydrocarbons from vegetable oils. The TLC of 29 compounds was studied with the following systems: cellulose (immobile phase, dimethylformamide in ethyl ether; mobile phase, isooctane) and cellulose acetate with ethanol–toluene–water (17:4:4) as developer. The cellulose reverse phase system more effectively separated the compounds into groups according to their ring structure. The cellulose acetate multi-phase technique was superior in separating the individual 4-, 5-, and 6-ring compounds. For example, the pairs benzo[a]pyrene and benzo[e]pyrene and benz[a]anthracene and chrysene, difficult to separate on cellulose, were readily separated with the celluloid acetate system. When the two compounds were used in conjunction with one another, only 4 of the 29 compounds studied could not be adequately separated for subsequent quantitative analysis, viz. pyrene and fluoranthene and perylene and benzo[g,h,i]perylene. For the quantitative determination of the polycyclic aromatic hydrocarbons, it was estimated that a difference in R_F units of 0.03 was required for the reverse phase system and 0.05 units for the multi-phase system. Table 7 lists the R_F values of 29 polycyclic aromatic hydrocarbons determined using both systems.

TABLE 7

R_F VALUES[a] OF POLYCYCLIC AROMATIC HYDROCARBONS

Polycyclic hydrocarbons	Cellulose[b]		Acetylated cellulose[c]	
	Range	Ave.	Range	Ave.
7,12-Dimethylbenz[a]anthracene	0.71–0.77	0.74	0.49–0.51	0.50
4-Methylpyrene	0.64–0.72	0.70	0.67–0.72	0.70
5,6-Dimethylchrysene	0.68–0.71	0.69	0.41–0.43	0.42
3-Methylcholanthrene	0.67–0.71	0.69	0.53–0.61	0.57
Anthracene	0.63–0.67	0.65	0.53–0.60	0.57
12-Methylbenz[a]anthracene	0.62–0.64	0.63	0.34–0.35	0.34
7,8-Dimethylbenz[a]anthracene	0.62–0.64	0.63	0.42–0.42	0.42
Benzo[c]phenanthrene	0.59–0.63	0.62	0.52–0.56	0.54
1,2-Dihydrobenz[e]aceanthrylene	0.59–0.63	0.62	0.38–0.39	0.39
Cholanthrene	0.59–0.60	0.60	0.49–0.50	0.50
5-Methylchrysene	0.58–0.59	0.59	0.39–0.39	0.39
Pyrene	0.57–0.59	0.58	0.59–0.64	0.62
Fluoranthene	0.56–0.57	0.57	0.58–0.62	0.60
7-Methylbenz[a]anthracene	0.56–0.56	0.56	0.44–0.45	0.45
4-Methylbenzo[a]pyrene	0.50–0.52	0.51	0.37–0.37	0.37
Benz[a]anthracene	0.44–0.47	0.45	0.42–0.42	0.42
Chrysene	0.43–0.44	0.44	0.35–0.36	0.35
Triphenylene	0.43–0.44	0.44	0.48–0.50	0.49
Benzo[a]pyrene	0.39–0.41	0.40	0.23–0.23	0.23
Benzo[k]fluoranthene	0.39–0.41	0.40	0.39–0.41	0.40
Benzo[e]pyrene	0.37–0.40	0.39	0.55–0.59	0.56
Anthanthrene	0.35–0.37	0.36	0.32–0.33	0.32
Perylene	0.32–0.33	0.33	0.45–0.49	0.47
Benzo[g,h,i]perylene	0.32–0.33	0.33	0.50–0.52	0.51
Dibenz[a,h]anthracene	0.30–0.30	0.30	0.47–0.49	0.48
Dibenzo[a,i]pyrene	0.28–0.28	0.28	0.24–0.24	0.24
Dibenzo[a,i]phenanthrene	0.27–0.28	0.27	0.55–0.59	0.57
Dibenzo[a,e]pyrene	0.25–0.27	0.26	0.38–0.38	0.38
Dibenz[b,i]anthracene	0.18–0.19	0.19	0.51–0.56	0.54

[a] Average of five determinations.
[b] Solvent system: immobile phase, 20% DMF in ethyl ether; mobile phase, isooctane.
[c] Solvent system: ethanol–toluene–water, 17:4:4 (v/v/v).

The charge-transfer TLC of polycyclic hydrocarbons was described by Short and Young[61]. It was shown that the chromatographic behavior of these compounds was modified by the relative strength of charge-transfer interactions when an insoluble electron acceptor, such as pyromellitic dianhydride (30%) was dispersed in the solid phase (silica gel). Improved separations of mixture were obtained and brightly colored spots fluorescing in the visible region permitted ready identification of components. Table 8 lists the R_F values and ratios of 16 polycyclic hydrocarbons (with respect to anthracene) on untreated and treated charge-transfer plates.

Berg and Lam[62] described the separation of polycyclic hydrocarbons by TLC on alumina and silica gel plates impregnated with 2,4,7-trinitrofluorenone (TNF)

TABLE 8

R_F VALUES AND RATIOS RESPECT TO ANTHRACENE
Eluent cyclohexane, with plates prepared by Method (b).

Hydrocarbon	Treated plates		Untreated plates		Spot color	Ionization potential (eV)
	R_F value	R_F ratio	R_F value	R_F ratio		
Naphthalene	0.97	1.20±0.05	0.78	1.30±0.05	Yellow	8.1
Hexamethylbenzene	0.89	1.10	0.72	1.20	Yellow	7.9
Anthracene	0.81	1.00	0.6	1.0	Pink	7.4
Phenanthrene	0.81	1.00	0.54	0.90	Yellow	8.15
N-Phenylcarbazole	0.81	1.00	0.24	0.40	Brick red	
1,3,6,8-Tetraphenyl-pyrene	0.73	0.90	0.42	0.70	Green	
Benzo[a]fluorene	0.69	0.85	0.45	0.75	Brown	
Fluoranthene	0.57	0.70	0.54	0.90	Yellow-green	
Dibenzo[a, c]anthracene	0.53	0.65	0.24	0.40	Orange	7.52
Pyrene	0.53	0.65	0.6	1.0	Orange	7.5
Naphthacene	0.47	0.58	0.39	0.65	Olive green	7.0
Perylene	0.45	0.55	0.39	0.65	Olive green	7.2
Benzo[a]pyrene	0.32	0.40	0.36	0.60	Orange-brown	
Benzo[b]fluoranthene	0.32	0.40	0.39	0.65	Orange-brown	
Carbazole	0.16	0.20	0.06	0.10	Orange	
9,10-Dimethyl-anthracene	0	0	0.56	0.93	Purple	

Method (b) – Pyromellitic dianhydride (60 g) was mixed with 130 ml of dry carbon tetrachloride and milled for 3 h, when the viscosity of the suspension ceased to increase. Silica gel (120 g) was added, together with an additional 250 ml of carbon tetrachloride and the mixture milled again for 30 min. The resulting slurry was then used to prepare thin-layer chromatographic plates (12 × 2.5 × 0.4 cm) by a dipping method. Plates were dried for 20 min, spotted, dried for a further 10 min, eluted with cyclohexane, dried in darkness for 10–15 min, and heated with hot air to complete dryness. Some plates were sprayed with a saturated solution of pyromellitic dianhydride in acetone to produce extra color. The colors faded after about 5 days. The treated slurry was kept dry and away from light but lost its color-forming properties in about 5 days. Re-milling and treatment with acetic anhydride did not restore the activity of the slurry.

and caffeine, respectively. Mixtures of light petroleum with small amounts of polar solvents were used for development. Table 9 lists the fluorescent and colored spots of 21 hydrocarbons detected on alumina, silica, silica–caffeine, dimethylformamide–alumina, TNF–silica, and TNF–alumina plates.

TLC and spectral procedures for the analysis of aza heterocyclic hydrocarbons in complex mixtures from air pollution sources was described by Sawicki et al.[23]. The following separation procedures were added: column chromatography (alumina–pentane with increasing amounts of ether followed by increasing amounts of acetone; circular paper chromatography (formamide–water, 35:65), TLC (alumina with pentane–ether, 19:1), and TLC (cellulose with formamide–water, 35:65). The following characterization procedures were used: ultraviolet absorption spectrophotometry

TABLE 9

FLUORESCENCE AND VISIBLE SPOT COLORS FOR 21 POLYCYCLIC AROMATIC HYDROCARBONS ON TREATED
AND UNTREATED ALUMINA AND SILICA PLATES

	Hydrocarbon[a]	Fluorescent spots[b]				Coloured spots[b]	
		Al	Sil	Caf Sil	DMF Al	TNF Sil	TNF Al
1	Anthracene	V	V	V	V	V	R-Br
2	Pyrene	G	G	V	G	Br	Br
3	Chrysene[c]	B-V	V	V	V	Y-Br	Y-Br
4	3,4-Benzofluoranthene	B	B	B	B	V	Y-Br
5	3,4-Benzopyrene[d]	Y	Y	B (V)	Y	Y-Gr	Gr-G
6	Perylene[c,e]	B (Br)	B (Br)	B-G	B (Y)	Y	Y-G
7	1,12-Benzoperylene	Y	Y-G	V	Y-G	B-Gr	Br
8	Coronene[c]	V	V	V	V	Br	R-Br
9	Fluoranthene	B	lB	B	B		
10	1,2-Benzanthracene	lB	lB	V			
11	1,2-5,6-Dibenzanthracene[c]	B-G	B	V			
12	1,2-3,4-Dibenzanthracene	B	B	V			
13	1,2-3,4-Dibenzopyrene[d]	B	Y	Y			
14	1,2-4,5-Dibenzopyrene[c,d]	G	B	V			
15	3,4-9,10-Dibenzopyrene[c,d]	B	Y-Br	B-V			
16	Fluorene	—	—	—		Y	
17	Acenaphthylene	—	—	—		Y	
18	Phenanthrene	—	—	—		Y	
19	Triphenylene	—	—	—		Y	
20	1,2-Benzopyrene[c,d]	B	B	V		Br	
21	2,3-Benzanthracene	—	G	G		V	

[a] Hydrocarbon mixtures used were as follows: I = 1-3, II = 1-8; III = 9-15; IV = 16-21.
[b] Abbreviations: Al = alumina; Sil = silica gel; Caf = caffeine; DMF = dimethylformamide;
TNF = 2,4,7-trinitrofluorenone; B = blue; Br = brown; G = green; Gr = grey; l = light;
R = red; V = violet; Y = yellow. A dash indicates that the spot is non-fluorescent.
[c] A saturated solution (in benzene) was used for spotting the plates.
[d] Concerning the numbering of positions in pyrene see ref. 11.
[e] In most cases a perylene spot in UV light was bright blue with a distinct brown or yellow-brown
centre region.

of an eluent or an extract, quenchofluorometry, spectrophotofluorometry of an
extract or an eluent, spectrophotofluorometry on the thin-layer plate or on paper,
spectrophotophosphorimetry on paper, color tests, and observation of fluorescence
on the thin-layer plate or on paper before and after trifluoroacetic acid fuming. These
procedures above testify to the overall complexity of polycyclic hydrocarbon analysis
and the variety of chromatographic techniques that are required. Figure 5 illustrates
the overall separation and characterization scheme described above for the aza
heterocyclic hydrocarbons.

The TLC behavior of a variety of polycyclic aromatic amines and heterocyclic
imines was studied by Sawicki et al.[10]. Table 10 summarizes their R_F values and spot

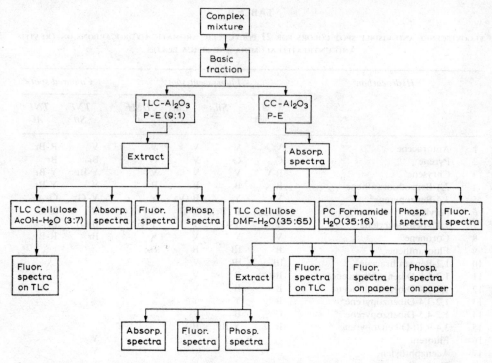

Fig. 5. Separation and characterization of aza-heterocyclic hydrocarbons. DMF=dimethylform-amide; P–E=pentane–ether.

colors obtained on cellulose powder developed with dimethylformamide–water (35:65) and detected with 5 systems.

Hancock et al.[63] recently studied the relationship of diesel locomotive exhaust to PAH concentrations on vegetation. By means of TLC and GLC, five polynuclear aromatic hydrocarbons were identified and quantified: anthracene, fluoranthene, pyrene, benz[a]anthracene, and benzo[a]pyrene. (3-Methylpyrene was not detected.) Individual PAH contents ranged from 5 to 110 µg/kg dry plant material and samples from the control area in most instances had higher concentrations than samples from the railroad right-of-way.

The leaf extracts were first chromatographed on Silica Gel 7 (250 µm) with hexane–ether (99.5:0.5) as the solvent and examined by short-wave ultraviolet. This separated the PAH from other compounds in the leaf extracts. PAH standards and sample spots having the same R_F values were scraped from the plates and extracted three times with 1 ml aliquots of hexane–benzene (1:1); the extracts were combined, evaporated to dryness at room temperature, and adjusted to 50 µl. A 10 µl aliquot was analyzed by GLC and the remaining 40 µl of the silica gel extract was chromatographed on cellulose plates using dimethylformamide–water (1:1). Table 11 lists the R_F values relative to benzo[a]pyrene (BrP) and fluorescent colors

TABLE 10

TLC[a] OF POLYNUCLEAR AROMATIC AMINES AND HETEROCYCLIC IMINES

Compound	R_F	Color[b]				
		D_4	D_5	D_6	D_7	D_9
1-Naphthylamine	0.81	B	YG	B	Q	Q
2-Naphthylamine	0.78	B	YG	B	Q	Q
1-Anthramine	0.61	lBG	dB	B	Q	Q
2-Anthramine	0.48	BG	dB	B	Q	Q
9-Anthramine	0.68	B	dB	lB	lB	Q
1-Aminofluorene	0.67	lB	BG	Q	lYV	lB
2-Aminofluorene	Streak	lB	BG	Q	lB	Q
3-Aminofluorene	Streak	lB	dY	Q	lB	Q
4-Aminofluorene	0.65	fV	BG	Q	Q	dRV
2-Aminodibenzofuran	0.66	fV	B	Q	Q	dV
2-Aminodibenzothiophene	0.38	fBV	P	Q	Q	Q
2-Aminodibenzoselenophene	Streak	fBV	RO	Q	Q	Q
3-Aminofluoranthene	0.43	G	R	G	Q	G
8-Aminofluoranthene	0.19	G	O	G	Q	G
6-Aminochrysene	0.35	B	YO	B	Q	BV
7-Aminobenz[a]anthracene	0.34	B	B	B	Q	Q
1-Aminopyrene	0.50	B	B	lB	Q	B
6-Aminobenzo[a]pyrene	0.17	G	RO	G	Q	Q
Carbazole	0.68	lB	B	Q	Q	Q
2-Methylcarbazole	0.46	B	B	Q	Q	Q
9-Methylcarbazole	0.39	B	B	Q	Q	Q
9-Phenylcarbazole	0.09	B	B	Q	Q	Q
9 H-Pyrido[2,3-]indole	0.69	B	B	Q	B	B
β-Carboline	0.54	B	B	Q	B	B
5,10-Dihydroindeno[1,2-b]indole	0.31	lB	B	Q	Q	Q
5-Methyl-5,10-dihydroindeno[1,2-b]indole	0.15	lB	B	Q	Q	Q
4 H-Benzo[d,e,f]carbazole	0.46	B	B	Q	Q	Q
11 H-Benzo[a]carbazole	0.34	B	B	Q	Q	Q
5 H-Benzo[b]carbazole	0.18	B	YG	lB	Q	Q
7 H-Benzo[c]carbazole	0.36	B	B	lB	Q	Q
13 H-Dibenzo[a,i]carbazole	0.10	B	B	Q	Q	Q
7 H-Dibenzo[b,g]carbazole	0.12	B	YG	lB	Q	Q
7 H-Dibenzo[c,g]carbazole	0.12	B	B	lB	Q	Q
13 H-Benzo[g]pyrido[2,3-a]carbazole	0.13	lB	G	Q	Q	Q
13 H-Benzo[g]pyrido[3,2-a]carbazole	0.26	B	G	Q	Q	Q
5 H-Naphtho[2,3-c]carbazole	0.11	B	Y	lB	Q	Q
7,8-Dimethyl-11 H-benzo[a]carbazole	0.16	lB	B	lB	Q	lB
1,2,4-Trimethyl-13 H-dibenzo[a,g]-carbazole	0	lB	B	lB	Q	lB
1,2,4-Trimethyl-13 H-dibenzo[a,i]-carbazole	0	lB	B	Q	Q	Q
2,5-Dimethyl-13 H-naphtho[2,1-a]carbazole	0	B	G	Q	Q	Q
11,12-Dimethyl-14 H-phenanthro[3,2-a]-carbazole	0	B	Y	lB	Q	Q
Phenothiazine	0.51	B	B	Q	Q	Q
N-Phenylphenothiazine	0.03	B	dB	Q	Q	Q
Acridone	0.75	B	B	lB	B	B
Phenanthridone	0.69	B	B	lB	lB	lB
Benzo[a]pyrene	0.07	B	B	Q	Q	Q
Fluoranthene	0.26	B	B	Q	lB	B

References pp. 445–447

TABLE 10. (Cont.)

TLC^a OF POLYNUCLEAR AROMATIC AMINES AND HETEROCYCLIC IMINES

Compound	R_F	Color				
		D_4	D_5	D_6	D_7	D_8
Perylene	0.07	B	B	Q	Q	B
Acridine	0.74	B	B	Q	B	fB
Benz[c]acridine	0.26	B	B	Q	BG	fB
Benzanthrone	0.41	B	B	Q	Y	B
1-Acetylpyrene		B	B	Q	Q	B
1,4-Dihydroxyanthraquinone		R	Q	Q	Y	R

^{a} TLC on cellulose powder developed with dimethylformamide–water (35:65).
^{b} Detection: D_1 = p-dimethylaminobenzaldehyde (5% reagent and 1% conc. hydrochloric acid in alcohol); D_2 = p-dimethylaminocinnamaldehyde (0.1% reagent and 1% conc. hydrochloric acid in alcohol); D_3 = 3-methyl-2-benzothiazolinone hydrazone (the plate was treated with 0.35% aq. reagent salt and then a few minutes later with 0.6% aq. ferric chloride solution); D_4 = UV light; D_5 = UV light (after treatment with one drop of 29% methanolic tetraethylammonium hydroxide); D_6 = UV light (after spraying with a 25% acetone solution of 3-methylthioaniline); D_7 = UV ligh t (after treatment with fumes of a nitrogen dioxide–trifluoroacetic acid (1:1) mixture); D_7 = the same as D_8, with a drop of CS_2 added. B = blue; d = dull; f = faint; G = green; l = light; O = orange; P = purple; Q = quenched; R = red; V = violet; Y = yellow.

TABLE 11

R_F VALUES RELATIVE TO BaP, AND FLUORESCENT COLORS OF THE PAH DEVELOPED BY THIN-LAYER CHROMATOGRAPHY USING BOTH SILICA GEL AND CELLULOSE

Compound	Silica gel hexane–ether (99.5:0.5)		Cellulose dimethylformamide–water (1:1)	
	R_F	Fluor, color dry	R_F	Fluor, color dry
Anthracene	1.31	Blue	2.28	Dark blue
Fluoranthene	1.21	Green	2.00	Green
Pyrene	1.34	Blue-green	1.90	Blue-green
Benz[a]anthracene	1.12	Light blue	1.36	Blue
Benzo[a]pyrene	1.00	Yellow-green	1.00	Yellow-green

of a number of polycyclic hydrocarbons obtained by TLC on silica gel and cellulose and Table 12 lists the GLC retention times of these hydrocarbons obtained on 5% SE-30 at 210°C.

The utility of GLC and TLC and subsequent characterization by fluorimetry of alkylated derivatives of the polycyclic aromatic hydrocarbons in urban atmospheres was elaborated by Sawicki et al.[12]. The R_F value and retention time, by themselves,

TABLE 12

GAS CHROMATOGRAPHIC RETENTION TIMES AND RELATIVE RETENTION VOLUMES OF PAH STANDARDS
IN LEAF WAXES OF LITTLE BLUESTEM AND POST OAK

Compound	5% Se 30 column at 210°C*	
	Retention time (min)	V_r
Anthracene	0.94	0.19
Fluoranthene	1.81	0.38
Pyrene	2.08	0.43
Benz[a]anthracene	4.76	1.00
Benzo[a]pyrene	12.12	2.54

* 5 ft × $\frac{1}{4}$ in. diameter glass column 60 ml/min N_2 gas.

are inadequate for characterization of the polycyclic compounds. GLC followed by fluorimetric examination of the eluent bands, and mixed-adsorbent two-dimensional TLC followed by direct fluorimetric examination was found to be exceedingly useful. A Perkin-Elmer Model 800 chromatograph was used with a 2 m × $\frac{1}{8}$ in. stainless steel column containing 10% Apiezon L on 60–80 mesh Chromosorb W (HMDS). The operating parameters were: column, injection port, and detector (FID) 240, 400 and 300°C, respectively, helium flow of 40 ml/min at the vent and 80 ml/min through the column, air flow and hydrogen flow at 400 and 20 ml/min, respectively. Spectrophotometric examination of column chromatographic effluents was done with a Cary Model II recording spectrophotometer, fluorimetric determinations with an Aminco-Bowman spectrophotofluorimeter with a solid state attachment and 1P21 phototube, and phosphorimetric determinations with an Aminco-Keirs spectrophotophosphorimeter with a 1P21 phototube.

Substantial amounts of alkylated polycyclic hydrocarbons were found to be present in the atmosphere, especially where a special form of industrial pollution was heavy (e.g. asphalt and petroleum industries).

The two-dimensional TLC separation and spectrophotofluorimetric identification and estimation of dibenzo[a,e]pyrene was reported by Bender[64]. The hexacyclic derivative was identified as being present in urban airborne particulate matter obtained from composites of samples from the city of Birmingham and from a combination of various other cities in Alabama and its concentration was estimated at 100 µg/g of benzene solubles. The two-dimensional TLC utilized aluminum oxide-cellulose acetate (2:1) plates developed first by cyclohexane or cyclohexane saturated with dimethylformamide, and in the 90° development with dimethylformamide-water (3:1) saturated with diethyl ether.

Majer et al.[65] described the utility of TLC and mass spectrometry for the rapid estimation of trace quantities of polycyclic aromatic hydrocarbons. Use of a microcentrifuge technique facilitated the transfer of eluted samples from the chromatogram to the mass spectrometer and enabled limits of detection in the picogram region to be obtained. In conjunction with such other techniques as paper and TLC, the integrated ion-current method offers a promising potential as an analytical technique. TLC employed cellulose powder (MN300, 250 μm) plates developed with dimethyl-formamide–water (70:30). The R_F values of the polycyclic compounds were fluoranthene 0.73, benzo[a]anthracene 0.68, benzo[a]pyrene 0.62, benzo[g,h,i]perylene 0.56, and coronene 0.45.

A GEC-AE1 MS9 mass spectrometer was equipped with a direct insert probe. The use of an integrated ion-current (IIC) method for the estimation of polycyclic compounds was facilitated by their characteristic properties in that they generally have low vapor pressures at room temperature and yet readily sublime without decomposition at temperatures between 200 and 500°C. In addition, fragmentation resulted in the molecular ion carrying a high percentage of the total ion current making its use ideal for analytical purposes.

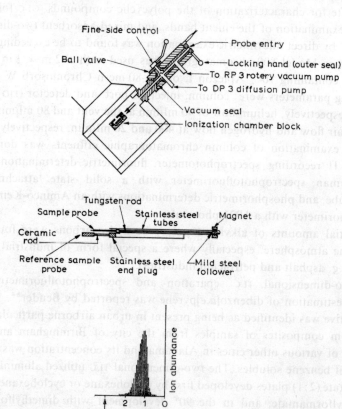

Fig. 6. MS 9 source and direct insertion probe. The probe used in this work was not fitted with a reference sample probe. Below, a typical integrated ion-current curve of a polycyclic hydrocarbon.

Figure 6 illustrates an MS9 source and direct insertion probe as well as a typical integrated ion-current curve of a polycyclic hydrocarbon. Table 13 lists the limits of detection of a number of polycyclic hydrocarbons by the integrated ion-current (IIC) method.

TABLE 13

LIMIT OF DETECTION OF POLYCYCLIC COMPOUNDS BY THE IIC METHOD

Polycyclic compound	m/e of molecular ion	Reference peak	Decade	Limit of detection (g)
Fluorene	166	164	1.012715	1×10^{-12}
Anthracene	178	169	1.053724	1×10^{-11}
Fluoranthene	202	181	1.116111	1×10^{-12}
Pyrene	202	181	1.116501	1×10^{-11}
Benzo[a]anthracene	228	219	1.041574	5×10^{-13}
Benzo[a]pyrene	252	231	1.091370	5×10^{-14}
Benzo[g,h,i]perylene	276	264	1.045721	1×10^{-14}
Coronene	300	264	1.136401	1×10^{-14}

The GC determination of polynuclear hydrocarbons in dust was described by Giberti et al.[66]. A Carlo Erba Fractovap C (Mode PAID1f) with FID and linear temperature programmer was used. The 35 m × 0.35 mm glass capillary column was coated with PE-30 and operated at 200° with nitrogen carrier at 0.25 atm and 1 ml/min flow. The injector temperature was 250° and samples were 0.5–2.0 μl of a solution containing 0.5–2 g of each component with a 1:100 splitter. The column used had approx. 10,000 theoretical plates as measured with n-hexacosane.

Dust samples (0.5–1.0 g) were originally extracted with 100 ml of cyclohexane in a Soxhlet, evaporated to 5 ml and extracted with 5 ml of methanol–water (4:1). This extract was further extracted 3 times with 5 ml portions of cyclohexane and the combined cyclohexane fractions were extracted 5 times with 5 ml portions of nitromethane. Both fractions were evaporated to dryness and made up to ca. 10 μl in ether. The cyclohexane fraction containing aliphatic hydrocarbons C_{18}–C_{30} were identified from retention volumes (V_r) by using a series of known alkanes. V_r (referred to chrysene = 1.00) of the nitromethane fraction for identified peaks varied from fluoranthene 0.31, pyrene 0.36, 1,2-benzofluorene 0.48, 1,2-benzanthracene 0.98, 2,3-benzofluoroanthene 2.63, perylene 3.21, 1,3,5-triphenylbenzene 4.15, 1,12-benzoperylene 5.92, and coronene 15.5. For quantitative determination, an internal standard of 1,3,5-triphenylbenzene with $V_r = 4.15$ was used and in synthetic mixtures the components could be determined to ±3%.

Eleven current techniques for the separation of benzo[a]pyrene from other hydrocarbons have been reviewed by Sawicki et al.[47]. Duncan[67] described an electron-capture GC procedure for the determination and separation of benzo[a]pyrene from benzo[e]pyrene, perylene and benzo[k]fluoranthene in 15 min. A Varian Aerograph Hy-Fi gas chromatograph was used with a 250 mCi tritium source. Benzo[a]pyrene was best separated from all other hydrocarbons by using a 10 ft. × $\frac{1}{8}$ in. stainless steel column packed with a mixture of 40% by volume 48–65 mesh sodium chloride and 60% 60–80 mesh Chromosorb G (A.W., DMCS treated) coated with 2% SE-30. The operating conditions were oven 215°C, injector 270°C, detector 190°C, nitrogen carrier gas at 200 ml/min, and the sample size was 20 μl. Further identification of benzo[a]pyrene and benzo[k]fluoranthene was made by ultraviolet fluorescence analysis after gas chromatographic separation. The response of electron-capture detection was linear over a range of 0–125 ng benzo[a]pyrene.

A method was described by Sawicki et al.[68] for studying atmospheric hydrocarbons quantitatively by GC. The air sample was introduced into a chromatographic column, 2.4 m × 1.60 mm i.d. containing activated Al_2O_3 coated with β,β'-oxydipropionitrile. The column was maintained at 0° with helium as the carrier gas. The detection limit of CH_4, C_2H_4, $CH_3CH=CH_2$, butane, and isobutane was 0.01 p.p.m. as detected by hydrogen flame ionization.

The utility of short capillary columns for the separation of polynuclear hydrocarbons, petroleum fractions and high-boiling waxes was demonstrated by Gouw et al.[69]. A 10 m × 0.010 in. capillary column coated with OV-101 was used to separate 1,2-benzopyrene from 3,4-benzopyrene, as well as separate the diastereomers of 1,3,5-triphenyldecane and resolve anthracene from phenanthrene. By temperature programming, it was possible to resolve C_4 from C_5 and yet obtain a linear relation between the retention times and the boiling points of the eluting n-hydrocarbons between C_{10} and C_{42} in a determination requiring less than 40 min.

The analysis of polycyclic aromatic hydrocarbons in tobacco smoke has been reviewed by Wynder and Hoffmann[70] and the chemical composition of tobacco and tobacco smoke by Stedman[71]. Davis[72] analyzed benzo[a]pyrene (BaP) in cigarette smoke by GC with an ionization detector utilizing a helium glow discharge as the electron source capable of operation at temperatures up to 400°C. As little as 1 ng of BaP could easily be measured with satisfactory precision. A Beckman GC-5 gas chromatograph was equipped with an electron-capture detector and a dual flame-ionization detector. The column was 9 ft. × $\frac{1}{4}$ in. o.d. stainless steel packed with 3% SE-30 on 60–80 mesh Chromosorb W. Typical operating conditions using the electron-capture detector were column temperature 245 or 250°C, detector temperature 275–280°, inlet temperature 350°.

Figure 7 depicts gas chromatograms of benzo[a]pyrene, benzo[e]pyrene and perylene showing response differences under identical GC conditions. A typical chromatogram of the benzo[a]pyrene fraction isolated from cigarette smoke is shown in Fig. 8. Benzo[a]pyrene values obtained by the GC method were found to be in good agreement with those obtained by the fluorimetric method of Davis et al.[73].

The detection and identification of polycyclic hydrocarbons in cigarette smoke,

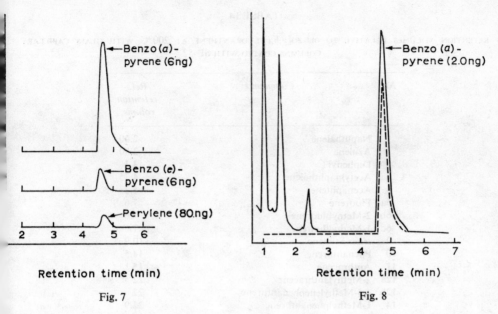

Fig. 7

Fig. 8

Retention time (min)

Retention time (min)

Fig. 7. Gas chromatograms of benzo[a]pyrene, benzo[e]pyrene, and perylene showing response differences under identical GC conditions.

Fig. 8. Gas chromatograms of benzo[a]pyrene fraction (solid line) isolated from cigarette smoke (fractionation method B) and pure benzo[a]pyrene, 1.46 ng (dotted line).

in airborne particles, and in petroleum waxes by GC with flame ionization or electron-capture detection has been reported[74-78].

The evaluation of polynuclear hydrocarbons in cigarette smoke by glass capillary columns with flame-ionization and electron-capture detectors was described by Carugno and Rossi[79]. A Fractovap Carlo Erba Model C was equipped with a linear tempera-ture programmer and a glass capillary column 65 m × 0.30 mm i.d. coated with SE-52. The chromatographic conditions used were: (a) isothermal: column temperature 200°C, injection temperature 320°C, carrier gas nitrogen at an inlet pressure of 0.67 atm and a flow rate of 1 ml/min, splitter-ratio 1:60, samples injected as solutions in tetrahydrofuran (size 1–2 μl); (b) linear-programmed temperature: 100–300°C, rate 1.8°C/min. All the other conditions were the same as those listed above. The relative retention volumes of 38 polynuclear hydrocarbons relative to benzo[g,h,i]-fluoranthene at 200°C are given in Table 14. The chromatogram obtained at this temperature is shown in Fig. 9. Thirty out of 38 components of the standard mixture have been identified under isothermal conditions. Retention indices on SE-52 at linear programmed temperature 100–300° are listed in Table 15. The retention indices were calculated as described by Vanden Dool and Kratz[78] from the retention tem-peratures and based on the n-paraffinic series from 10 to 36 carbon atoms. Figure 10 illustrates a chromatogram of a standard mixture of polynuclear hydrocarbons

TABLE 14

No.	Compound	Rel. retention volume
1	Naphthalene	2.00
2	Azulene	2.80
3	Diphenyl	3.3
4	Acetylnaphthalene	4.75
5	Acenaphtene	5.3
6a	Fluorene	7.6
6b	2-Methylfluorene	7.6
6c	1-Methylfluorene	7.6
9	trans-Stilbene	10.9
10	Phenanthrene	14.5
11	Anthracene	15.1
12a	2-Methylanthracene	22
12b	4,5-Methylenephenanthrene	22
14	1-Methylphenanthrene	22.9
15	9-Methylanthracene	26.3
16	Fluoranthene	36.3
17	Pyrene	43
18	1,2-Benzofluorene	58.4
19	2,3-Benzofluorene	61.5
19b	3,4-Benzofluorene	61.5
20	1-Methylpyrene	67.6
21	3-Methylpyrene	69.3
22a	Benzo[g,h,i]fluoranthene	100
22b	3,4-Benzophenanthrene	100
24	1,2-Benzanthracene	120
25a	Chrysene	124
25b	Triphenylene	124
26	Naphthacene	134
27	7-Methyl-1,2-Benzanthracene	209
28a	2,3-Benzofluoranthene	308
28b	3,4-Benzofluoranthene	308
29	11,12-Benzofluoranthene	313
30	7,12-Dimethyl-1,2-Benzanthracene	318
31	1,2-Benzopyrene	364
32	3,4-Benzopyrene	378
33	Perylene	404
34a	3-Methylcolanthrene	553
34b	20-Methylcolanthrene	553

obtained on a glass capillary column coated with SE-52 and linear programmed from 100 to 300°C at 1.8°C/min. The electron-capture detection was more sensitive and gave a better response than FID for compounds such as 1,2-benzopyrene and 3,4-benzopyrene.

TABLE 15

RETENTION INDICES ON SE-52 AT LINEAR PROGRAMMED TEMPERATURE $100°$ UP TO $300°$

No.	Compound	T_r	I_{pr}
1	Naphthalene	105	1150
2	Azulene	111.5	1292
3	Diphenyl	116	1373
4	Acetylnaphthalene	121	1437
5	Acenaphthene	123	1458
6	Fluorene	132.5	1555
7	2-Methylfluorene	145	1673
8	1-Methylfluorene	145.5	1677
9	trans-Stilbene	146.5	1686
10	Phenanthrene	152.5	1741
11	Anthracene	153.5	1750
12	2-Methylanthracene	166	1870
13	4,5-Methylenephenanthrene	166.5	1875
14	1-Methylphenanthrene	168	1890
15	9-Methylanthracene	171	1920
16	Fluoranthene	181	2020
17	Pyrene	186	2070
18	1,2-Benzofluorene	196.5	2179
19a	2,3-Benzofluorene	198	2195
19b	3,4-Benzofluorene	198	2195
20	1-Methylpyrene	200	2215
21	3-Methylpyrene	200.5	2220
22	Benzo[g,h,i]fluoranthene	211	2326
23	3,4-Benzophenanthrene	211.5	2332
24	1,2-Benzanthracene	217	2389
25a	Chrysene	217.5	2395
25b	Triphenylene	217.5	2395
26	Naphtacene	220	2425
27	7-Methyl-1,2-Benzanthracene	231.5	2575
28a	2,3-Benzofluoranthene	242.5	2700
28b	3,4-Benzofluoranthene	242.5	2700
29	11,12-Benzofluoranthene	243	2706
30	7,12-Dimethyl-1,2-benzanthracene	243.5	2713
31	1,2-Benzopyrene	247	2760
32	3,4-Benzopyrene	248	2773
33	Perylene	250	2800
34a	3-Methylcholanthrene	258	2906
34b	20-Methylcholanthrene	258	2906
35	1,2,7,8-Dibenzanthracene	270.5	3078
36a	1,2,3,4-Dibenzanthracene	273	3114
36b	1,2,5,6-Dibenzanthracene	273	3114
37	Benzotetraphene	274.5	3136
38a	Benzo[g,h,i]perylene	275.5	3150
38b	Picene	275.5	3150
39	Anthanthrene	278	3186
40	1,2,3,4-Dibenzopyrene	297	3477
41	Coronene	300.5	3544
42	1,2,4,5-Dibenzopyrene	301.5	3567
43	3,4,9,10-Dibenzopyrene	303	3600
44	3,4,8,9-Dibenzopyrene	303.5	3620

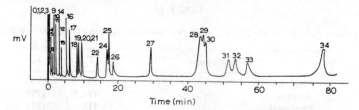

Fig. 9. Chromatogram of a standard mixture of hydrocarbons. Glass capillary column coated with SE 52, column temperature 200°C, flame-ionization detector. The numbers of the peaks correspond to the compounds reported in Table 14.

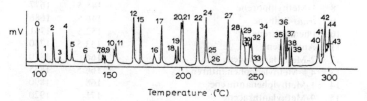

Fig. 10. Chromatogram of a standard mixture of polynuclear hydrocarbons. Glass capillary column coated with SE 52, linear programmed temperature 100–300°C, rate 1.8°C/min, electron-capture detector. The numbers of the leaks correspond to the compounds reported in table 15.

Brocco et al.[80] determined polynuclear hydrocarbons in atmospheric dust by a combination of TLC and GLC. The polyhydrocarbons were first separated by TLC on Silica Gel G plates with a cyclohexane–benzene (1:1.5) developing solvent, located by UV irradiation, extracted from the adsorbent with ether, and finally determined by gas chromatography using a Carlo Erba Fractovap Model C with a 30 m glass capillary column coated with silicone rubber SE-52, with 1,3,5-triphenylbenzene as the internal standard. The dust sample was applied to the thin-layer plates either directly or in a cyclohexane extract. Phenanthrene, fluoranthene, 3-methylpyrene, 3,4-benzofluoranthene, and 3,4-benzopyrene were determined in p.p.m. amounts.

Table 16 lists the R_F values of some typical polycyclic and paraffinic hydrocarbons on Silica Gel G and alumina plates developed with different solvents.

Vernon[81] studied the gas–solid chromatographic (GSC) behavior of aromatic hydrocarbons and polycyclic hydrocarbons on modified alumina stationary phases. A Pye series 104 gas chromatograph was equipped with FID and used with a 0.45 m × 0.32 mm i.d. copper tubing packed with 6.5 g of modified alumina (80–100 mesh). The stationary phase was 50 g of chromatography grade alumina, type 0, slurried with a solution of 0.5 g of NaOH and 5 g of NaCl in 50 ml of water; the slurry was dried at 400° for 24 h before column packing. Nitrogen carrier gas flow rate was maintained at 30 ml/min for alkyl benzenes and 60 ml/min for the polycyclic hydrocarbons.

The retention data and calculated chromatographic parameters are given in Fig. 11 and Table 17 for benzene and alkylbenzenes and in Fig. 12 and Table 18 for

TABLE 16

R_F VALUES OF SOME TYPICAL POLYCYCLIC AND PARAFFINIC HYDROCARBONS

Compounds	R_F values on silica gel				R_F values on Alumina G			
	Cyclo-hexane	Hexane	Benzene	Ether	Cyclo-hexane	Hexane	Benzene	Ether
Polycyclic								
Anthracene	0.31	0.25	0.74	0.95	0.55	0.42	0.95	1.00
3,4-Benzofluoranthene	0.17	0.12	0.64	0.95	0.30	0.078	0.71	1.00
1,2-Benzopyrene	0.20	0.13	0.75	0.95			0.89	1.00
3,4-Benzopyrene	0.20			0.95	0.38	0.078	0.86	1.00
Fluoranthene	0.23	0.19	0.70		0.46	0.34	0.85	1.00
3-Methylpyrene	0.26	0.20	0.66	0.93	0.49	0.27	0.93	1.00
Pyrene	0.27	0.18	0.67	0.95	0.51	0.30	0.91	1.00
Paraffins								
C_{18}	0.72	0.72	0.79	1.00	0.98	0.94	0.94	1.00
C_{20}	0.73	0.72	0.79	1.00	0.98	0.94	0.94	1.00
C_{22}	0.72	0.72	0.79	1.00				
C_{24}	0.72	0.72	0.79	1.00	0.86	0.94	1.00	1.00
C_{28}	0.72	0.72	0.79	1.00	0.98	0.94	0.94	1.00

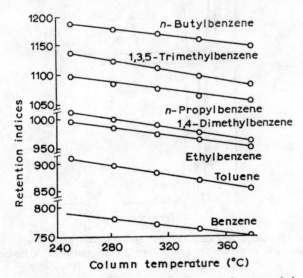

Fig. 11. Plot of retention index against column temperature for aromatic hydrocarbons.

polycyclic hydrocarbons. The chromatographic separation of high-boiling aromatic hydrocarbons was found to be quite feasible by GSC with the production of symmetrical peaks for hydrocarbons with boiling peaks as high as 350°.

Bhatia[82] described the gas chromatography of polycyclic aromatic mixtures

TABLE 17

GSC RETENTION INDICES FOR C_6–C_{10} AROMATIC HYDROCARBONS

Hydrocarbon	GSC retention index (I)			$\Delta I/°C$
	250°	300°	350°	
Benzene	790	774	759	−0.31
Toluene	910	890	869	−0.41
p-Xylene	1012	993	973	−0.39
1,3,5-Trimethylbenzene	1139	1117	1097	−0.42
Ethylbenzene	995	979	964	−0.31
n-Propylbenzene	1098	1082	1066	−0.32
n-Butylbenzene	1192	1176	1161	−0.31

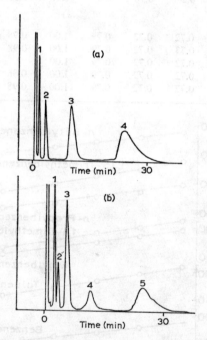

Fig. 12. Gas–solid chromatograms of polycyclic hydrocarbons. (a) 1, Naphthalene; 2, acenaphthalene; 3, phenanthrene; 4, pyrene. (b) 1, Biphenyl; 2, acenaphthylene; 3, fluorene; 4, anthracene; 5, fluoranthrene.

using columns packed with OV-7 (20% phenylmethyl silicone)-coated glass beads which were pre-treated with dimethyl chlorosilane. A Beckman Model GC-4 gas chromatograph equipped with a dual flame-ionization detector system and a linear temperature programmer was used with 20 ft. × $\frac{1}{8}$ in. columns packed with OV-7 coated glass beads (60–80 or 100–120 mesh) treated as above. The columns were

TABLE 18

GSC RETENTION INDICES FOR POLYCYCLIC HYDROCARBONS AT 400°

Hydrocarbon	B.p. (°C)	GSC retention index
Naphthalene	218	1342
Biphenyl	254	1507
Acenaphthene	278	1555
Acenaphthylene		1600
Fluorene	295	1732
Phenanthrene	340	1878
Anthracene	354	1935
Pyrene	260/60 mm Hg	2123
Fluoranthrene	250/60 mm Hg	2138

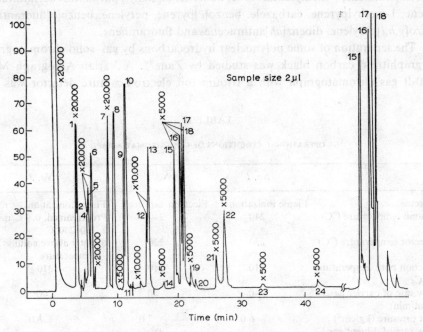

Fig. 13. Gas chromatograms of a synthetic mixture of polycyclic aromatic hydrocarbons and n-alkanes. Column, 60–80 mesh glass beads coated with OV-7; temperature program, 4 min at 170°C; 6°C/min rise to 260°C, at 260°C to end of analysis. (1) Phenanthrene, (2) anthracene, (3) carbazole, (4) 2-methylphenanthrene, (5) 3-methylphenanthrene, (6) 1-methylphenanthrene + 1-methylanthracene, (7) fluoranthene, (8) pyrene, (9) chrysofluorene, (10) 2,3-benzofluorene, (11) n-tetracosane, (12) 1,2-benzanthracene, (13) chrysene, (14) n-octacosane, (15) benzo[k]fluoranthene, (16) benzo[a]pyrene, (18) perylene, (19) 3-methylcholanthrene, (20) n-dotriacontane, (21) 1,2,5,6-dibenzanthracene, (22) 1,12-benzoperylene, (23) n-hexatriacontane, (24) dibenzo[a, i]pyrene. Portion of original chromatogram is shown on the right to indicate more clearly the resolution between peaks 15, 17, and 18.

References pp. 445–447

run for 4 min at 170°C, followed by a temperature rise of 6°C/min to 260°C and completed at 260°C until the end of the analysis. Peak areas were integrated by a Perkin-Elmer Model 154 printing integrator.

A chromatogram showing the resolution of benzo[k]fluoranthene, benzo[a]pyrene, benzo[e]pyrene, and perylene from each other as well as from other polycyclic aromatic hydrocarbons and n-alkanes in a 24-component synthetic mixture is shown in Fig. 13.

An electron-capture GLC procedure for the analysis of polycyclic hydrocarbons in cigarette smoke was described by Robb et al.[83] The particulate phase from cigarette smoke was initially fractionated over silica gel and the hexane and benzene–hexane (1:9) eluant fractions subsequently determined using a Wilkens Aerograph Hy-Fi 600-B gas chromatograph equipped with a 15 ft. column containing 15% SE-30 silicone on ABS Anakrom operated isothermally at 180, 200, 250, and 280°C with nitrogen as carrier gas. Benzo[a]pyrene as well as 15 polycyclic aromatic hydrocarbons were detected in ca. 0.01 μg amounts. The polycyclic hydrocarbons identified were naphthalene, 9-methylanthracene, fluorene, biphenyl, phenanthrene, anthracene, pyrene, 1-methylpyrene, carbazole, benzo[e]pyrene, perylene, benzo[b]fluoranthene, benzo[g,h,i]perylene, dibenz[ah]anthracene, and fluoranthene.

The separation of some polynuclear hydrocarbons by gas–solid chromatography on graphitized carbon black was studied by Zane[84]. A Varian Aerograph Model 1520-B gas chromatograph with a tritium foil electron-capture detector was used

TABLE 19

OPERATIONAL CONDITIONS OF GAS CHROMATOGRAPH

	No. 1	No. 2	No. 3
Detector	Flame ionization	Electron capture	Flame ionization
Column temperature (°C)	240	240	Programmed, 0.8°C/min, 190–250°C
Detector temperature (°C)	250	220[a]	Slightly above column temperature
Injection port temperature (°C)	310	310	310
Flow rate of carrier gas (ml/min)	40	40	40
Inlet pressure (kg/cm²)	6.0	7.0	6.0
Flow rate of diluent gas (ml/min)		480[b]	
D.c. polarizing voltage (V)		15	
Column packing	1% SE-30 silicone rubber on Diasolid H 60–80 mesh		
Column tubing	Stainless steel, 12 ft. in length and ⅛ in. o.d.		

[a] This is the maximum temperature allowable for tritium type detector.
[b] This flow rate produced better reproducibility of peak height and less tailing in this experiment, although the diluent flow should be 50–100 ml/min according to the instruction manual.

for studies ranging to 415°C. For higher oven temperatures, a Barber-Colman Model Selectra-System 5000 was used with FID. The operating conditions were injector temperature 320°, detector temperature 210° (for electron capture) and 340° (for flame ionization), nitrogen carrier gas with an inlet pressure of 5.7 atm and carrier gas flow rate of 60 ml/min at 30°. A 70 cm × 2.3 mm i.d. stainless steel column was packed with graphitized carbon black (100–120 mesh). A second column was packed with the same absorbent (80–100 mesh) and both columns were conditioned in the instrument for 2 h at 415° before use. The retention times found at 350° for the polycyclic aromatic hydrocarbons chromatographed on the 80 to 100 mesh graphitized carbon black column using a carrier gas flow of 70 ml/min were acenaphthene 1.8, acenaphthylene 1.9, fluorene 14.0, phenanthrene 15.0, 1-methylphenanthrene 19.2, anthracene 20.0, pyrene 80.0, and fluoranthene 80.0 min, respectively.

Arito et al.[85] described the GC determination of polycyclic hydrocarbons in particulate air pollutants. Vacuum sublimation was employed to extract organic materials from the particulates instead of employing Soxhlet extraction. The sublimate was dissolved in benzene and analyzed by FID or electron-capture GLC. A

TABLE 20

RETENTION TIMES OF POLYCYCLIC HYDROCARBONS UNDER ISOTHERMAL AND PROGRAMMED COLUMN CONDITIONS

Compound	Retention time (min)	
	Isothermal[a] 240°C	Programmed[b] 0.8°C/min 190–250°C
Fluorene	2.9	9.1
2-Methylfluorene	3.8	12.8
Anthracene	4.9	16.5
Fluoranthene	9.8	30.6
Pyrene	11.3	33.8
Benzo[b]fluorene	14.8	41.4
3-Methylpyrene	16.3	44.0
Benzo[mno]fluoranthene	22.1	52.5
Benzo[a]anthracene	25.5	57.0
Chrysene	25.9*	58.1*
7-Methylbenzo[a]anthracene	39.6	70.5
Benzo[b]fluoranthene	53.8	81.7
Benzo[e]pyrene	62.8	88.8
Benzo[a]pyrene	65.2	90.1*
Perylene	68.1	93.3

* The value was obtained by injecting a benzene solution of single component of the hydrocarbon.
[a] Gas chromatographic conditions were given in Nos. 1 and 2 of Table 19.
[b] Gas chromatographic conditions were given in No. 3 of Table 19.

TABLE 21

RESPONSES OF ELECTRON-CAPTURE DETECTOR AND FLAME-IONIZATION DETECTOR TO POLYNUCLEAR HYDROCARBONS

Compound	Detector				B/A
	Flame ionization (A)		Electron capture (B)		
	Attenuation	Response (cm/μg)	Attenuation	Response (cm/μg)	
Fluoranthene	50	12.0	100	15.8	1.32
Pyrene	50	11.8	100	7.6	0.64
3-Methylpyrene	20	18.4	100	11.4	0.62
Benzo[mno]fluoranthene	20	22.6	100	30.0	1.33
Benz[a]anthracene	20	11.7	100	6.8	0.58
Chrysene	20	8.2	100	0.6	0.07
7-Methylbenz[a]anthracene	10	14.5	100	7.9	0.54
Benzo[b]fluoranthene	10	11.2	100	10.2	0.91
Benzo[e]pyrene	10	16.3	100	3.8	0.23
Benzo[a]pyrene	10	6.0	100	7.2	1.20

* Each response was expressed as peak height in cm/μg of polynuclear hydrocarbon at the attenuation given above.
A and B were obtained under the gas chromatographic conditions given in Nos. 1 and 2 of Table 19 respectively.

Perkin-Elmer Model 800 gas chromatograph with the operating conditions described in Table 19 was used. Table 20 lists the retention times of 15 polycyclic hydrocarbons obtained during isothermal 240°C and programmed (0.8°C/min, 190–250°C) conditions. Table 21 lists the responses of electron-capture and flame-ionization detection to the polycyclic hydrocarbon. Benzo[e]pyrene and benzo[a]pyrene can be quantitatively evaluated using electron-capture detection.

Burchfield et al.[86] developed a gas-phase fluorescence detector for the analysis of mixtures of polycyclic hydrocarbons by gas chromatography. A Mikro-Tek Model MT-160 gas chromatograph equipped with a 13.5 mCi ^{63}Ni electron capture detector was connected in series to an Aminco-Bowman spectrophotofluorimeter. Glass columns (6 ft. × ¼ in. i.d.) were packed with 10% Dexsil 300 coated on 80–100 mesh Chromosorb W. Column temperatures were varied between 240 and 325°C using isothermal operation. The injection port and EC detector temperatures were maintained at 290 and 325°. Nitrogen was the carrier gas at 90 ml/min and the electrometer sensitivity was in the range of 3.2×10^{-9} to 8.0×10^{-10} AFS. The Aminco-Bowman spectrophotofluorimeter (SPF) was equipped with a 150 watt, 7.5 A, 17–23 V d.c. Hanoven 901 C1 Xeon lamp and 1P21 photomultiplier tube and used with a 3 mm i.d. × 5 mm o.d. × 20 mm quartz flow-through cell.

Of the liquid phases evaluated (OV-1, DC-200 and JXR (methyl silicone), and Dexsil 300 (a borane–silicone copolymer)), the latter was found to be the most stable (up to 450°C) for the separation of polycyclic hydrocarbons. Although it did separate fluorene from phenanthrene and anthracene, it could not be temperature programmed using the electron-capture detection and could not separate the pairs benzo[a]pyrene–perylene and chrysene–triphenylene. The excitation and emission spectra of the polycyclic hydrocarbons were very similar to those found in the liquid phase. However, the gas-phase measurements were more convenient to make and less susceptible to Raman and Rayleigh light scattering by the solvent than liquid phase measurements, although fluorescence intensity was lower. Measurement of fluorescence was more sensitive and specific than electron-capture detection and in addition permitted the analysis of mixtures of compounds which could not be separated on the present chromatographic columns.

Separation of polycyclic aromatic hydrocarbons from air samples on alumina (or silica gel columns) have all been described where the individual fractions were measured by ultraviolet absorption spectroscopy[87-92]. Popl et al.[93] separated a mixture of fluoranthene, triphenylene, 1,2-benzanthracene, chrysene, 3,4-benzo-pyrene, and coronene on a column packed with Al_2O_3 containing 2 wt.-% of water, by eluting with a gradient of cyclopentane–ether. By means of a flow-through cell ultraviolet spectrophotometer, the extractions of the eluted components were recorded at the wavelengths of 260, 275, and 296 nm at 30 sec intervals, thus considerably enhancing the sensitivity and selectivity of the determination. The column was 1 m × 4 mm i.d. containing alumina (Woelm, Neutral) (heated for 8 h at 400° and then deactivated by addition of 2% water), and used with a gradient elution pump (Dialagrad Model 190, Isco) and a Pye Unicam SP-800B uv spectrophotometer. The main advantages of the method are its speed with the entire analysis requiring only 2.5 h, as well as the sample size (0.1–0.5 mg) which is advantageous for determining polycyclic hydrocarbons in foodstuffs and in the atmosphere.

REFERENCES

1 H. L. Falk, P. Kotin and A. Miller, *Intern. J. Air Pollution*, 2 (1960) 273.
2 R. Schoental, in B. Clair (Ed.), *Polycyclic Hydrocarbons*, Vol. I, Academic Press, New York, 1964, p. 486.
3 H. Oettel, *Angew. Chem.*, 70 (1958) 532.
4 H. L. Falk, P. Kotin and E. Mehler, *Arch. Environ. Health*, 8 (1964) 721.
5 F. A. Gunther and F. Buzzetti, *Residue Rev.*, 9 (1964) 90.
6 F. A. Gunther, F. Buzzetti and W. E. Westlake, *Residue Rev.*, 17 (1967) 81.
7 E. O. Haenni, *Residue Rev.*, 24 (1968) 42.
8 R. E. Schaad, *Chromatog. Rev.*, 13 (1970) 61.
9 D. Hoffmann and E. L. Wynder, in A. C. Stern (Ed.), *Air Pollution*, Vol. II, Academic Press, New York, 2nd edn., 1968, p. 187.
10 E. Sawicki, H. Johnson and K. Kosinski, *Microchem. J.*, 10 (1966) 72.
11 E. Sawicki, *Natl. Cancer Inst. Monograph*, 9 (1962) 201.
12 E. Sawicki, T. W. Stanley, S. McPherson and M. Morgan, *Talanta*, 13 (1966) 619.
13 C. R. Sawicki and E. Sawicki, in E. Niederweser and G. Pataki (Eds.), *Progress in Thin-Layer Chromatography*, Vol. III, Ann Arbor Science, Ann Arbor, Mich., 1972, p. 233.

14 G. M. BADGER, J. K. DONNELLY AND T. M. SPOTSWOOD, *Australian J. Chem.*, 19 (1965) 1023.
15 E. SAWICKI, S. P. MCPHERSON, T. W. STANLEY, J. MEEKER AND W. C. ELBERT, *Intern. J. Air Water Pollution*, 9 (1965) 515.
16 J. M. COLUCCI AND C. R. BEGEMAN, *J. Air Pollution Control Assoc.*, 15 (1965) 113.
17 R. E. WALLER, B. T. COMMINS AND P. J. LAWTHER, *Brit. J. Ind. Med.*, 22 (1965) 128.
18 E. SAWICKI, T. R. HAUSER, W. C. ELBERT, F. T. FOX AND J. E. MEEKER, *Am. Ind. Hyg. Assoc. J.*, 23 (1962) 137.
19 E. SAWICKI, *Proc. Arch. Environ. Health*, 14 (1967) 46.
20 E. L. SYNDER AND D. HOFFMANN, *New Engl. J. Med.*, 262 (1960) 540.
21 G. E. MOORE, J. L. MONKMAN AND M. KATZ, *Natl. Cancer Inst. Monograph*, 9 (1962) 153.
22 E. SAWICKI, T. R. HAUSER AND T. W. STANLEY, *Intern. J. Air Pollution*, 2 (1960) 253.
23 E. SAWICKI, T. W. STANLEY AND W. C. ELBERT, *Occupational Health Rev.*, 16 (1964) 8.
24 E. SAWICKI, T. W. STANLEY AND W. C. ELBERT, *J. Chromatog.*, 18 (1965) 512.
25 J. L. MONKMAN AND T. J. PORRO, *Proc. Instr. Soc. Am.*, 6 (1960) D 41.
26 E. SAWICKI, W. ELBERT, T. W. STANLEY, T. R. HAUSER AND F. T. FOX, *Intern. J. Air Pollution*, 2 (1960) 273.
27 E. SAWICKI AND J. D. PFAFF, *Anal. Chim. Acta*, 32 (1965) 521.
28 L. A. DUBOIS, A. CORKERY AND J. L. MONKMAN, *Intern. J. Air Pollution*, 2 (1960) 236.
29 B. L. VAN DUUREN, *Natl. Cancer Inst. Monograph*, 9 (1962) 135.
30 D. HOFFMANN AND E. L. WYNDER, *Natl. Cancer Inst. Monograph*, 9 (1962) 91.
31 N. KUCHARCZYK, J. FOHL AND J. VYMETAL, *J. Chromatog.*, 11 (1963) 55.
32 H. MATSUSHITA AND Y. SUZUKI, *J. Chromatog.*, 29 (1964) 108.
33 G. M. BADGER, J. K. DONNELLY AND T. M. SPOTSWOOD, *J. Chromatog.*, 10 (1963) 397.
34 E. SAWICKI, T. W. STANLEY, W. C. ELBERT AND J. D. PFAFF, *Anal. Chem.*, 36 (1964) 497.
35 E. SAWICKI, T. W. STANLEY, J. D. PFAFF AND W. C. ELBERT, *Chemist-Analyst*, 53 (1964) 6.
36 L. K. KEEFER, *J. Chromatog.*, 31 (1967) 390.
37 J. JANAK AND U. KUBECOVA, *J. Chromatog.*, 33 (1968) 132.
38 R. G. HARVEY AND M. HALONEN, *J. Chromatog.*, 25 (1966) 294.
39 H. KESSLER AND E. MULLER, *J. Chromatog.*, 24 (1966) 469.
40 E. SAWICKI, T. W. STANLEY, J. D. PFAFF AND W. C. ELBERT, *Anal. Chim. Acta*, 31 (1964) 359.
41 E. SAWICKI, W. C. ELBERT AND T. W. STANLEY, *J. Chromatog.*, 17 (1965) 120.
42 C. R. ENGLE AND E. SAWICKI, *J. Chromatog.*, 31 (1967) 109.
43 D. F. BENDER, E. SAWICKI AND R. M. WILSON, JR., *Anal. Chem.*, 36 (1964) 1011.
44 E. SAWICKI, T. W. STANLEY, W. C. ELBERT AND M. MORGAN, *Talanta*, 12 (1965) 605.
45 E. SAWICKI, M. GUYER, R. SCHUMACHER, W. C. ELBERT AND C. R. ENGEL, *Mikrochim. Acta*, (1968) 1025.
46 E. D. BARBER AND E. SAWICKI, *Anal. Chem.*, 40 (1968) 984.
47 E. SAWICKI, T. W. STANLEY, W. C. ELBERT, J. MEEKER AND J. MCPHERSON, *Atmos. Environ.*, 1 (1967) 131.
48 E. SAWICKI AND H. JOHNSON, *J. Chromatog.*, 23 (1966) 142.
49 E. SAWICKI, T. W. STANLEY AND W. C. ELBERT, *Mikrochim. Acta*, (1965) 1110.
50 E. SAWICKI AND H. JOHNSON, *Mikrochim. Acta*, (1964) 435.
51 E. SAWICKI AND H. JOHNSON, *Mikrochim. J.*, 8 (1964) 85.
52 E. SAWICKI, C. R. ENGEL AND W. C. ELBERT, *Talanta*, 14 (1967) 1169.
53 E. SAWICKI, M. GUYER AND C. R. ENGEL, *J. Chromatog.*, 30 (1967) 522.
54 L. V. S. HOOD AND J. D. WINEFORDNER, *Anal. Chim. Acta*, 42 (1968) 199.
55 S. P. MCGLYNN, B. T. NEELY AND L. NEELY, *Anal. Chim. Acta*, 28 (1963) 472.
56 J. D. WINDFORDNER, W. J. MCCARTHY AND P. A. ST. JOHN, in D. GLICK (Ed.), *Methods of Biochemical Analysis*, Vol. 15, Interscience, New York, 1967.
57 B. MUEL AND G. LACROIX, *Bull. Soc. Chim. France*, (1960) 2139.
58 J. D. WINEFORDNER AND P. A. ST. JOHN, *Anal. Chem.*, 35 (1963) 2211.
59 H. MATSUSHITA, Y. ESUMI AND K. YAMADA, *Bunseki Kagaku*, 19 (1970) 951.
60 R. H. WHITE AND J. W. HOWARD, *J. Chromatog.*, 29 (1967) 108.
61 G. D. SHORT AND R. YOUNG, *Analyst*, 94 (1969) 259.
62 A. BERG AND J. LAM, *J. Chromatog.*, 16 (1964) 157.
63 J. L. HANCOCK, H. G. APPLEGATE AND J. D. DODD, *Atmos. Environ.*, 4 (1970) 363.

64 D. F. BENDER, *Environ. Sci. Technol.*, 2 (1968) 204.
65 J. R. MAJER, R. PERRY AND M. J. READE, *J. Chromatog.*, 48 (1970) 328.
66 A. GIBERTI, G. P. CARONTI AND V. CANTUTI, *J. Chromatog.*, 15 (1964) 141.
67 R. M. DUNCAN, *Am. Ind. Hyg. Assoc. J.*, 30 (1969) 624.
68 E. SAWICKI, R. C. COREY, A. E. DOOLEY, J. B. GISCLARD, J. L. MONKMAN, R. E. NELIGAN AND A. RIPPERTON, *Health Lab. Sci.*, 7 (1970) 23.
69 T. W. GOUW, F. M. WITTEMORE AND R. E. JENTOFT, *Anal. Chem.*, 42 (1970) 1394.
70 E. L. WYNDER AND D. HOFFMANN, *Tobacco and Tobacco Smoke Studies in Experimental Carcinogenesis*, Academic Press, New York, 1967.
71 R. L. STEDMAN, *Chem. Rev.*, 68 (1968) 153.
72 H. J. DAVIS, *Anal. Chem.*, 40 (1968) 1583.
73 H. J. DAVIS, L. A. LEE AND T. R. DAVIDSON, *Anal. Chem.*, 38 (1966) 1752.
74 A. LIBERTI, G. P. CARTONI AND V. CANTUTI, *J. Chromatog.*, 15 (1964) 141.
75 V. CANTUTI, G. P. CARTONI A. LIBERTI AND A. G. TORRI, *J. Chromatog.*, 17 (1965) 60.
76 J. A. WILMSHURST, *J. Chromatog.*, 17 (1965) 50.
77 W. LIJINSKY, I. I. DOMKY AND J. WARD, *J. Gas Chromatog.*, 3 (1965) 152.
78 M. VAN DEN DOOL AND P. D. KRATZ, *J. Chromatog.*, 11 (1963) 463.
79 N. CARUGNO AND S. ROSSI, *J. Gas Chromatog.*, 5 (1967) 103.
80 D. BROCCO, V. CANTUTI AND G. P. CARTONI, *J. Chromatog.*, 49 (1970) 66.
81 F. VERNON, *J. Chromatog.*, 60 (1971) 406.
82 K. BHATIA, *Anal. Chem.*, 43 (1971) 609.
83 E. W. ROBB, G. C. GUVERNATOR, M. D. EDMONDS AND A. BAVLEY, *Beitr. Tabakforsch.*, 3 (1965) 279.
84 A. ZANE, *J. Chromatog.*, 38 (1968) 130.
85 H. ARITO, R. SODA AND H. MATSUSHITA, *Ind. Health*, 5 (1967) 243.
86 H. P. BURCHFIELD, R. J. WHEELER AND J. B. BERNOS, *Anal. Chem.*, 43 (1971) 243.
87 G. J. CLEARY, *J. Chromatog.*, 20 (1965) 89.
88 G. GRIMMER AND A. HILDEBRANDT, *J. Chromatog.*, 20 (1965) 98.
89 G. S. SFORZOLINI, F. PASCASIO, E. MARCHESOTTI AND M. N. CHIUCCHIU, *Ann. Ist. Super. Sanita.*, 3 (1967) 45.
90 P. WEDGWOOD AND R. L. COOPER, *Analyst*, 79 (1954) 163.
91 R. L. COOPER, *Analyst*, 79 (1954) 573.
92 B. T. COMMINS, *Analyst*, 85 (1958) 386.
93 M. POPL, D. DOLANSKY AND J. MOSTECKY, *J. Chromatog.*, 59 (1971) 329.

Chapter 20

AROMATIC HYDROCARBONS

1. BENZENE

Benzene is obtained by a number of procedures generally involving its separation from petroleum sources. Petroleum-derived benzene is produced commercially by two types of process, *e.g.* reforming-separation and dealkylation. The former process generally employs hydroforming[1] and the more recently developed Udex solvent extraction process[2] for the production of benzene, while in the dealkylation processes, toluene or xylene is demethylated in the presence of hydrogen to yield benzene (and methane) and no solvent extraction of the product is necessary. Benzene is one of the principal compounds present in the light oil recovered from coal carbonization gases. Miscellaneous sources of benzene include the distillation of coal tar and from the manufacture of carburetted gas.

Benzene has been largely employed in the past for blending with gasoline. At the present time, chemical usage accounts for a large part of the benzene consumed in the U.S. with the following estimated percentage use in 1968: styrene (43), phenol (19), chemical intermediates for nylon manufacture (9), synthetic detergents (6), maleic anhydride (4), DDT (3), mono- and dichlorobenzene (2), aniline (2), benzene hexachloride (0.3), nitrobenzene (0.3), and miscellaneous usage including solvents (4.4). (The total U.S. production of benzene in 1971 was approx. 10.3 billion lb. from petroleum, coal, coke-oven, and tar distillers.)

Acute toxicity by benzene is due to its narcotic action and in many respects resembles that caused by other low molecular weight petroleum hydrocarbons[3]. Brief exposures to high concentrations of benzene in air (>100 p.p.m.) may cause chronic benzene intoxication, the symptoms, signs, and blood changes of which are similar to severe anemia and leukopenia[4]. Industrial benzene poisoning results almost exclusively from the inhalation of benzene in the atmosphere. The absorption and excretion of benzene has been described[5].

Benzene in air has been determined spectrophotometrically[6-10]. The GLC determination of the composition of crude benzene was described by Khazaimova *et al.*[11]. Crude benzene was separated into 5 fractions by distillation and the fractions analyzed on a 2.5 m × 4–6 mm column of 20% tricresyl phosphate on 1NZ-600 with a katharometer detector and helium as carrier gas.

Traces of solvent vapors present in air were determined by Omar[12] by first condensing the vapors on active carbon at room temperature, then desorbing the vapor adsorbates at elevated temperature, and finally determining their content by GC using thermal conductivity detection. The method was tested for air containing benzene, toluene, and *p*-xylene vapors ($\sim$0.3 mg/dm^3), which were adsorbed at 20°

449

in U-tubes (diam. 4 mm) filled with 0.5 g active carbon (fractions of particle diam. 0.3–0.4 up to 7.3–10.7 cm) at the gaseous mixture flow rate of 30 l/h and then desorbed at 250–300°C into a chromatographic column containing "Oktoil-S" on Chromosorb P with hydrogen as carrier gas at a flow rate of 60 ml/min. The amounts of the vapors measured chromatographically differed from those calculated as adsorbed on active carbon by 0.03 mg (0.12%).

The selectivity of GLC separation of benzene, toluene, ethylbenzene, and isomeric xylenes on 25 stationary phases (benzene and naphthalene derivatives) was studied by Gudkov and Teterina[13]. Best results were obtained at 50°C on a 16 m × 0.3 mm column containing α-naphthaldehyde with an evaporator temperature of 210–220°, nitrogen carrier gas at 5 ml/min and FID. Satisfactory results were also obtained at 38°C using a capillary column containing α-methylnaphthalene at 11.6 ml nitrogen/min.

An analytical procedure was described by Fett et al.[14] which determined benzene, toluene, ethylbenzene, and total aromatics in non-olefinic hydrocarbon solvents which boil between 100 and 450°F. The method was based on the addition of a known amount of internal standard (hexamethylbenzene), then segregation of the aromatic hydrocarbons in the sample by liquid chromatography[15] on silica followed by gas chromatographic analysis of this hydrocarbon fraction (plus solvent). The method was applicable to samples containing 0.1% or more of total aromatics with an estimated precision of ±3% relative and air accuracy of ±5% relative. The analysis cost per sample was 1.5–2.5 man hours, depending upon the number of samples determined simultaneously. The liquid chromatography separations were carried out in the glass–Teflon unit shown in Fig. 1. The manifold provided outlets for 8 separate samples. The GC analyses were performed on a Perkin-Elmer Model 900 with dual columns, linear temperature programming and differential flame-ionization detection. The columns used were 10 ft. × ⅛ in. o.d. packed with 10% Apiezon L on 60–80 mesh Chromosorb W (AW, DMCS). Integration was obtained with an

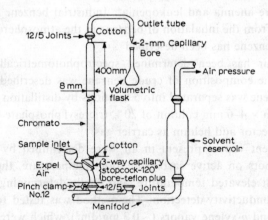

Fig. 1. Liquid chromatographic column and auxillary equipment.

Infotronics CRS-40 system. The GC analysis of the liquid chromatography fractions were performed using helium as carrier gas at 20 ml/min, the injection port and column to detector manifold at 300°C and the column held at 50° for 4 min (following injection of sample) then programmed at 5°/min to 220°C, holding this final temperature until the analysis was completed. A typical chromatogram of the determination of benzene, toluene, ethylbenzene, and total aromatics in hydrocarbon solvents is shown in Fig. 2.

The physical properties, selectivity, efficiency, and preparation of organic derivatives of clay materials (beidellite and montmoullonite type) as well as their possible applications were studied by Taramasso and Fuchs[16]. These clay minerals are phyllosilicates which have a lamellar structure built up from two planar layers of silica tetrahedrons and alumina octahedrons. One modified substance that has been employed in GC is Bentone-34 which is the dimethyldioctadecylammonium complex of montmoullonite. It has been employed as a stationary phase because of its high selectivity towards aromatic hydrocarbons and in particular towards *meta* and *para* isomers. Investigations by Taramasso and Fuchs[17] with clay minerals having cation exchange capabilities ranging from 24 to 150 mequiv./100 g of mineral have shown that "beidellite type" complexes were superior (for selectivity toward the aromatic hydrocarbons) to that of the "montmoullonite type" clay mineral complexes. Figure 3 illustrates the separation of aromatic hydrocarbons on a combined column of different clay mineral complexes. The interlamellar complexes are used as stationary phases deposited on the surface of a convenient support, 10 to 15 wt-% and can be used between the temperature limits of 50 and 150°. Above

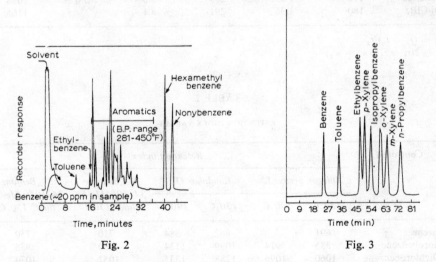

Fig. 2   Fig. 3

Fig. 2. Gas chromatogram of the separation of benzene, toluene, ethylbenzene, and total aromatics in hydrocarbon solvents.

Fig. 3. Separation of aromatic hydrocarbons on a combined column.

150°C some of the clay mineral complexes begin to decompose. Up to 150° all the complexes investigated could be used, even with the very sensitive ionization detectors. In order to separate mixtures of aromatic hydrocarbons which were very strongly retained by the complexes. the highest temperature possible is usually advisable in order to obtain good efficiency and reasonable retention times.

The chlorination of benzene yields a mixture which contains unreacted benzene, monochlorobenzene as well as the isomeric *para-*, *meta-*, and *ortho-*dichlorobenzenes. Karasek and Fong[18] described a 6 min isothermal GC method for the separation and quantitative analysis of these components. A Carle 8000 gas chromatograph was used with a microthermistor bead TCD. The column used was 10 ft. × 0.096 in. i.d.,

TABLE 1

RETENTION AND WEIGHT FACTOR DATA FOR CHLORINATED BENZENES

Chromatographed on a 10 ft. 3% Bentone-34–10% DC-200 silicone column at 150°C and 27 ml/min carrier flow

Compound	B.p. (°C)	Weight factors	Retention time (sec)	Capacity ratio[a] (k_1)	Resolution[b] (R)	Retention index
Air			32			
Benzene	80.1	1.00	58	0.8	6.2	710
Mono-ClBz	132	1.16	111	2.5	5.1	930
p-Di-ClBz	173.4	1.26	179	4.6	1.7	1052
m-Di-ClBz	172	1.33	208	5.5	3.6	1088
o-Di-ClBz	180	1.36	291	8.1		1165

[a] $k_1 = (t_1 - t_a)/t_a$
[b] $R = 2(t_j - t_1)/(w_j + w_1)$

TABLE 2

RETENTION INDEX VALUES

Compound	Retention index					
	Apiezon grease L^a		Emulphor O^a		3% Bentone-34–10% Me-Si-fluid[b] 150°C	5% Bentone-34–10% Me-Si-fluid[b] 150°C
	130°C	190°C	130°C	190°C		
Benzene	691		862	884	710	750
Chlorobenzene	885	914	1099	1134	930	988
p-Dichlorobenzene	1060	1096	1288	1335	1052	1074
m-Dichlorobenzene	1058	1097	1275	1320	1088	1148
o-Dichlorobenzene	1076	1117	1126	1375	1165	1257

[a] Reported by Wehrli and Kovats[19].
[b] Reported by Karasek and Fong[18].

stainless steel packed with a mixed phase of 3% Bentone-34 added to 10% DC-200 silicone oil on 80–100 mesh Chromosorb W. The column temperature was 150°C, with both the detector and inlet temperatures at 160°C, helium flow rate 27 ml/min detector output 1 mV, and sample size 0.5 µl. Table 1 lists the retention and weight factor data for chlorinated benzene obtained under the operating parameters above. Table 2 lists the retention index values compared with Wehrli and Kovats[19] indices. The indices in Table 2 show a much greater difference between the most difficultly separable pair, the *para* and *meta* isomers, for the Bentone-34 column than the others. Figure 4 illustrates a chromatogram of chlorinated benzene compounds obtained on a 10 ft. column of 5% Bentone-34–10% DC-200 silicone at 150°C and Table 3 lists the retention data for the chlorinated benzenes under these conditions. This column provided a higher capacity ratio and greater resolution between

TABLE 3

RETENTION DATA FOR CHLORINATED BENZENES

Chromatographed on a 10 ft. 5% Bentone-34–10% DC-200 silicone column at 150° and 38 ml/min carrier flow

Compound	Retention time (sec)	Capacity ratio (k_1)	Resolution (R)
Air	28		
Benzene	60	1.1	
Chlorobenzene	132	3.7	7.7
p-Dichlorobenzene	192	5.9	4.7
m-Dichlorobenzene	270	8.6	4.4
o-Dichlorobenzenes	450	15.1	5.2

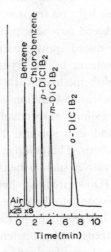

Fig. 4. Chromatogram of chlorinated benzene compounds. Flow 38 ml/min (He); temperature 180°C; column 10 ft. × ⅛ in.; 5% Bentone-34 with 10% DC-200 silicone.

the isomeric disubstituted compounds than the 3% Bentone-34 column indicated in Table 2.

Nabivach and Yatsenko[20] separated the contaminants (*e.g.* CS_2, thiophene, and hydrocarbons) in pure benzene (0.03–0.04 ml) using a 2-section column (two 2 m sections, 5 mm in. diam.) and thermal conductivity detection. The first section was filled with diatomite, INZ-600, previously heated to 700° and washed with HCl, impregnated with 20% polyethylene glycol adipate. The second section was filled with 20% triethylene glycol–*n*-butyrate on Chromosorb W (0.25–0.5 mm).

The course of the catalytic benzene oxidation to maleic anhydride was followed by pulsed gas chromatography[21] on a column of silicone oil MS200 supported on GC-22. The column temperature was 175° and air carrier gas was 40 ml/min. The retention times of CO_2, benzene, and maleic hydrazide were 1.30 min, 2.40 min, and 4.10 min, respectively.

The GLC analysis of bromo- and chlorobenzenes has been reported by Nakada *et al.*[22]. The columns used were (*a*) 4 mm × 2 m containing silicone SE-30, isothermal at 120 and 160°, temperature programming from 70 to 250°C at 5.8°/min, (*b*) 4 mm × 2 mm containing Bentone-24 and operated isothermally at 100 and 160°C and temperature programming from 70 to 200° at 5.8°/min. Compounds determined were C_2H_5X, 1,4-, 1,3-, 1,2-$C_6H_4X_2$, 1,3,5-, 1,2,4-, 1,2,3-$C_6H_3X_3$, 1,2,4,5-$C_6H_2X_4$, 1,2,3,5-, 1,2,3,4-$C_6H_2Br_4$, C_6HBr_5, C_6X_6 (X = Cl or Br).

2. TOLUENE

Toluene is principally produced from coal or from petroleum sources. From coal it is mainly a byproduct of metallurgical coke manufacture. In the coking operations gas, ammonia, light oil, and tar are obtained as byproducts, toluene being recovered largely from the light oil fraction, boiling up to 200°C. Production of toluene per ton of coke produced now is approx. 0.35–0.40 gal./ton and in 1968 coke-oven plants in the U.S. produced approximately 66 million gallons. Toluene is produced principally from petroleum by the hydroforming of selected petroleum naphthols which are rich in naphthenic hydrocarbons. The total U.S. production of toluene in 1966 from all sources was 584 million gallons (coke oven and petroleum sources 22.8 and 561.2 million gallons, respectively). Toluene has two direct uses, *i.e.* as a gasoline component and as a solvent. The amount used for solvents in 1966 was approx. 65 million gallons. The balance of toluene produced is used for the manufacture of various derivatives such as benzene, toluene diisocyanate (TDI), trinitrotoluene (TNT), phenol (Dow process), benzoic acid, benzyl chloride and toluene sulfonate. Toluene is second to benzene in the production of aromatic chemicals and in the entire field of organic chemical raw materials it is fourth in tonnage, following ethylene, benzene, and propylene in that order.

There are a number of derivatives of toluene accounting for approx. 5% of the total toluene consumed in chemical derivatives. These compounds including benzaldehyde, sulfonic acids, alkylated toluenes, and ring-chlorinated derivatives are used in such areas as dyes, medicinals, flavors, perfumes, sweetening agents, germicides, etc.

Toluene is a more powerful narcotic and more acutely toxic than benzene[23]. The acute[24] and chronic toxicity[25] of toluene in man as well as its absorption and excretion[24] has been reported.

Toluene in air has been determined largely by spectrophotometric techniques[6,8,9,26,27].

Gudkov and Teterina[28] studied the GC separation of benzene, toluene, ethylbenzene, and isomeric xylenes on 25 different stationary phases (benzene and naphthalene derivatives). Best results were obtained at 50°C on a 16 m × 0.3 mm capillary column containing α-naphthaldehyde with an evaporator temperature of 210–220°, nitrogen carrier at 5 ml/min and a flame-ionization detector.

The utility of high-resolution partition capillary columns was described by Liberti and Zoccolillo[29]. By roughening the glass surface either with a layer of carbon or with polymerized material (polytrifluorochloroethylene or polybutadiene), a stationary phase of any polarity could be successfully coated. The capillary after being coated with polymeric material according to Grob[30] was then treated with a solution of Carbowax squalene or trimer acid to yield columns of enhanced efficiency. The high resolution of these columns permitted the separation of polar isotopic molecules such as dimethyl sulfoxide from deuterium dimethyl sulfoxide using 28 m Carbowax 1450 glass column at 70°C with nitrogen carrier gas at an inlet pressure of 0.5 kg/cm² and a flow rate of 1.0 ml/min.

Toluene and deuterotoluenes were separated on a 200 m squalene capillary column at 16°C with inlet pressure of nitrogen at 1.05 kg/cm² and a flow rate of 0.57 ml/min.

3. STYRENE

Styrene (phenylethylene, vinylbenzene) is manufactured commercially by the dehydrogenation of ethylbenzene although a number of other methods have been used or considered for commercial production. These include sidechain chlorination of ethylbenzene followed by dehydrochlorination, pyrolysis of petroleum and recovery from various petroleum processes, oxidative conversion of ethylbenzene to α-phenylethanol via acetophenone, and subsequent dehydration of the alcohol. Ethylbenzene itself is principally prepared by the liquid-phase alkylation (of benzene and ethylene) using aluminum chloride as catalyst with ethyl chloride or hydrogen chloride as promotor.

Styrene is extensively used for the manufacture of plastics, including polystyrene, rubber-modified impact polystyrene, acrylonitrile–butadiene–styrene terpolymer (ABS), styrene–acrylonitrile copolymer (SAN) and for the production of styrene–butadiene synthetic rubber.

References pp. 468–469

The production of styrene monomer was 4.4 billion lb. in 1971. Its utility is based both on its ease of handling (b.p. 145°C) and the activation of the vinyl group by the benzene ring which makes styrene easy to polymerize and co-polymerize under a variety of conditions. The major impurities in styrene are of an oxidative origin and include aldehydes (benzaldehyde and formaldehyde) and peroxides.

Overexposure to the vapors of styrene may cause central nervous system depression (anesthesia). The threshold-limit value for styrene vapor in air is 100 p.p.m. Styrene is detoxified in the body by oxidative cleavage of the double bond.

Styrene plastics rank third in volume in the plastics field behind polyethylene and polyvinyl chloride. Styrene homopolymer accounts for approximately one-quarter of the monomer production and is widely used in packaging, toys, appliances, housewares, etc. Expanded forms of polystyrene are used as foam plastics in construction and refrigeration. Chemically modified copolymers of styrene–divinyl-benzene are the basis of many ion-exchange resins.

The acute and chronic toxicity of styrene in experimental animals[31] as well as its absorption and excretion has been described[31,32]. The absorption of styrene is primarily through the respiratory tract in both animals[31] and man[32]. Styrene is metabolized to benzoic acid, conjugated with glycine and excreted as hippuric acid. Styrene vapors occur not only at industrial sites where styrene and polystyrene are produced but can also be found in the form of depolymerization products in the thermal processing of polystyrene.

Styrene in air has been determined by ultraviolet[33,34] and infrared spectroscopy[33,35], spectrophotometrically[6,34], and by the nitration method[33].

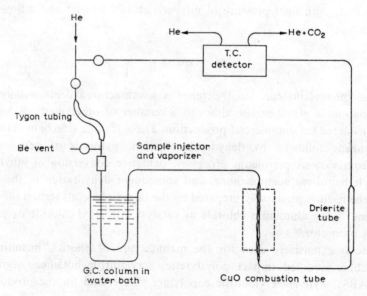

Fig. 5. Schematic diagram of apparatus.

Nelsen *et al.*[36] determined styrene in aqueous emulsions and styrene–butadiene latexes (20–25% wt.-% solid) by gas chromatography. The apparatus (Fig. 5) consisted of a sample injector and vaporizer, a short separating GC column, the oxidative–dehydration train, and a thermal conductivity detector. The detector was the hot-wire type (Gow-Mac, Model 9285) operated at 50°C with a bridge current of 200 mA. The separating column, 12 in. × $\frac{5}{16}$ in. was packed with C-22 insulating brick (30–60 mesh) containing 20% silicone oil SF-96 (General Electric). The oxidizer consisted[37] of 8 in. × $\frac{1}{4}$ in. stainless steel tubing containing 14–18 mesh copper oxide maintained at 675°C.

The sample injector and vaporizer are shown in Fig. 6. The disposable glass capillary sampling tube was approx. 65 mm × 1 mm i.d. and constricted at the mid-portion to half its original diameter. The vaporizer section was packed with glass wool and heated electrically. The GC column was operated at 75°C, the vaporizer at 140°C, and the helium flow was 60 ± 0.5 ml. Essentially complete elimination of

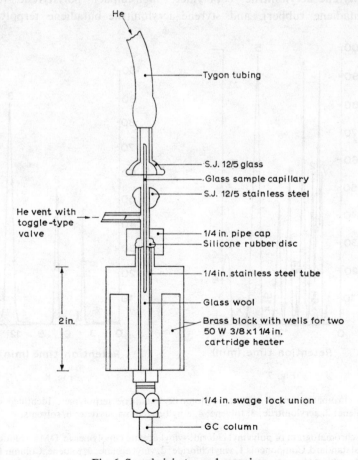

Fig. 6. Sample injector and vaporizer.

free styrene from the polymer was achieved in the vaporizer and the minimum
detectable amount of styrene was of the order of a few hundredths of 1 %.

A gas chromatographic method was developed by Shapras and Claver[38] for
the direct determination of residual monomers, non-polymerizable volatiles, and
volatile additives in styrene-based polymers. Solutions or dispersions of polymers
were injected in dimethylformamide using toluene as an internal standard. A Wilkens Model 600 Hy-Fi gas chromatograph equipped with a straight bore injection
block and hydrogen flame detector was used. Two columns in series were used,
3 ft. $\times$ $\frac{1}{8}$ in. o.d. containing 20% Tween 81 followed by a 10 ft. $\times$ $\frac{1}{8}$ in. o.d. column
of 10% Resoflex 446 on Chromosorb W, 30–60 mesh. The columns were held at
120° and the injection block at 210°C. Nitrogen was the carrier gas at 30 ml/min and
a hydrogen flow rate of 23 ml/min was used for the detector which was operated at an
impedence of $10^9\,\Omega$, an output sensitivity of $1\times$ and attenuated between $1\times$ and
$128\times$ depending on monomer concentration. The polymers analyzed were polystyrene, styrene–acrylonitrile copolymer, high-impact polystyrene (containing
styrene–butadiene rubber), and styrene–acrylonitrile–butadiene terpolymer. The

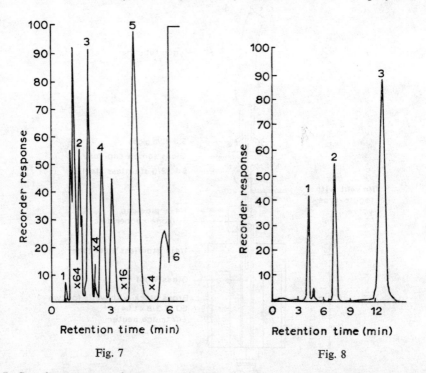

Fig. 7   Fig. 8

Fig. 7. Gas chromatogram of styrene–acrylonitrile–butadiene terpolymer. Identified components
are 1, butadiene; 2, acrylonitrile; 3, toluene; 4, ethylbenzene; 5, styrene; 6, solvents.

Fig. 8. Gas chromatogram of polyvinyl chloride–vinyl acetate copolymer in DMF solution. Toluene
as an internal standard. Components 1, vinyl chloride; 2, vinyl acetate; 3, toluene. Column 10ft $\times$ $\frac{1}{4}$ in.
o.d., 10% Resoflex 446; carrier gas N_2; flow rate 25 ml/min; detector, H_2 flame; sample size 5 μl.

relative retention (toluene = 1) of the components were butadiene 0.40, acrylo-nitrile 0.76, toluene 1.00, ethylbenzene 1.40, styrene 2.25, and dimethylformamide 3.75.

Figure 7 shows a chromatogram of a styrene–acrylonitrile–butadiene terpolymer. Less than 10 p.p.m. of monomer could be detected and the resolution of minor components could be improved by using temperature programming. The GLC method was applicable to other than styrene-based polymers. Figure 8 shows a chromato-gram of volatiles in a sample of polyvinyl chloride–vinyl acetate copolymer.

Since the development of polystyrene food cartons, much interest has been engendered concerning the possibility of residual styrene monomer in food. Two gas chromatographic methods were described by Finley and White[39] to determine if styrene monomer was present in milk. One technique had a sensitivity of 5 p.p.m. and the other a sensitivity of 0.01 p.p.m. following inoculation of concentration styrene monomer ranging from 100 p.p.m. to 0.1 p.p.m. and 10 p.p.m. to 0.01 p.p.m, re-spectively, in milk. Standard samples of styrene monomer were obtained from polysty-rene by distillation at 145°C and purified by preparatory gas chromatography on an Aerograph Autoprep instrument equipped with a 20 ft. × $\frac{3}{8}$ in. diam. stainless steel column packed with 20% Carbowax 20M on Chromosorb W maintained at 150°C with a helium flow of 100 ml/min. The material collected was analyzed on an F & M Model 609 analytical gas chromatograph equipped with a Model 400 oven and a flame-ionization detector, and a 6 ft. glass column packed with 20% Carbowax 20M plus 2.5% KOH on 60–80 mesh Gas Chrom R. The oven temperature was 100°C with a helium flow of 60 ml/min.

The extraction procedure of choice for removing styrene from milk involved initial centrifugation at 19,500 rev./min for 40 min. For the first 30 min the temperature was held at 30° and for the last 10 min it was held at 3°C to allow for solidification of the fat phase. The fat layer was removed and immersed in 0.1 N alcoholic KOH and saponified for 7 min, cooled, and extracted with carbon tetra-chloride. The CCl₄ fraction was analyzed by GC on an F & M Model 609 chromato-graph with a hydrogen flame detector. The column of choice was 10% Apiezon L on 80–100 mesh firebrick at 100°C with helium carrier gas at 60 ml/min. Linear detection was obtained down to 0.1 p.p.m. of styrene monomer. Milk which had been stored in polystyrene cartons for up to 8 days was analyzed by this technique and no styrene monomer was detected.

The determination of styrene monomer in polyester resins by GLC was reported by Esposito and Swann[40]. Pentane was used for the solvent separation of the resin prior to the analysis on a Triton X-305 column with methylisobutyl ketone as the internal standard. An F & M Model 500 Linear Programmed Temperature gas chromatograph was used with a Brown Electronic recorder (Minneapolis Honey-well) and a disc integrator. The column was 6 ft. × $\frac{1}{4}$ in. copper tubing packed with 20% Triton X-305 on acid-washed Chromosorb W. The operating conditions were column temperature 85°C, injection port temperature 300°C, detector cell temperature

300°C, detector cell current 160 mA, and helium flow at exit 85 ml/min. Figure 9 shows a chromatogram obtained from a typical polyester resin containing styrene. The time required was 15 min and the complete procedure including extraction was approx. 30 min.

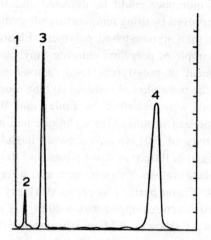

Fig. 9. Analysis of polyester resin containing styrene monomer. 1, Pentane; 2, methylethyl ketone; 3, methylisobutyl ketone; 4, styrene.

Jones and Reynolds[41,42] employed pyrolysis gas chromatography for the analysis of copolymers of styrene–methylmethacrylate and styrene–ethylacrylate in the presence of azodiisobutyronitrile and a stabilizer in the ratio of 68:32 styrene:MMA and 99:1 styrene:EA. Lehmann and Brauer[43] analyzed the pyrolyzates of polystyrene and polymethylmethacrylate. The optimum temperature for maximum yield of styrene was 730°C and no higher than 450°C for MMA.

Pyrolysis gas chromatography was used by Spagnolo[44] to detect and estimate small amounts (0.25–4%) of polystyrene in polymethylmethacrylate. The sample was pyrolyzed in a helium stream and the volatile products, essentially methylmethacrylate and styrene, chromatographed on a diisodecyl phthalate column. The polystyrene content was calculated from the peak area ratio of styrene to methylmethacrylate, compared to a calibration curve obtained from synthetic mixtures of the two polymers. A Perkin-Elmer Model 800 gas chromatograph with Hot Wire Detector Accessory was used. Pyrolysis was accomplished using a Perkin-Elmer pyrolyzer with quartz tube and porcelain sample boats. The chromatographic column was a 6 ft. × $\frac{1}{4}$ in. aluminum tube containing 5% diisodecyl phthalate on Teflon 6. The operating conditions were pyrolysis temperature 600°C, column temperature 125°C, injector temperature 300°C, hot wire oven temperature 200°C, filament current +210 mA, attenuation 128× for methylmethacrylate peak and 2× for styrene peak, helium flow rate 40 ml/min and approximate sample size was 5 mg.

The retention times for the methylmethacrylate monomer and styrene were 2.2 and 6.0 min, respectively.

Raley and Kaufman[45] described the gas chromatography of chlorostyrenes, ethylbenzene, and xylenes. The best separation of the xylenes was obtained with a Bentone-34–dinonyl phthalate column while the six isomeric chlorostyrenes were best resolved on a 4P3E-modified column. A Podbelniak detector Type 9981 thermal conductivity detector with Gow-Mac W filament was used with a Sargent Model S-72180-35 1 mV recorder. The columns were 12 ft. × $\frac{3}{16}$ in. o.d. stainless steel tubing packed with the following: 10 p.p.h. Bentone-34 (FGM), 5% Bentone-34 and 5% diisodecyl phthalate (DDP), 10 p.p.h. Bentone-34 and 10 p.p.h. tricresyl phthalate (TCP), 10 p.p.h. Bentone-34 and 10 p.p.h. dinonyl phthalate (DNP). The columns of Bentone-34 and Bentone-34/4P3E used 70–80 mesh Anakrom AS support and the last three were on 60–100 mesh Chromosorb W. The operating conditions for both the chlorostyrenes and alkylbenzenes were injection port and detector temperatures, 170 and 250°C, respectively, and helium flow 50 ml/min. The column temperatures for chlorostyrenes and alkylbenzenes were 150 and 100°C,

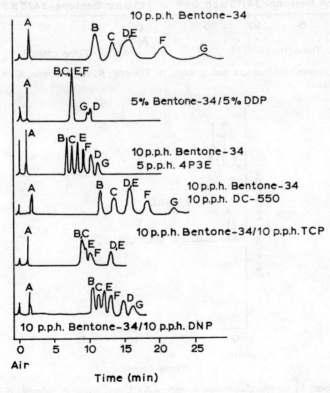

Fig. 10. Separation of chlorostyrenes. A, Toluene; B, *p*-Cl styrene; C, *m*-Cl styrene; D, *trans-β*-Cl styrene; E, *o*-Cl styrene; F, *α*-Cl styrene; G, *cis-β*-Cl styrene.

respectively, and the sample size in both cases was 0.3 µl. Figure 10 shows the separations of chlorostyrenes on the various Bentone-34 modified columns. With all columns, peaks B, C, E, F, and G eluted in the same order and only the relative portion of peak D (*trans-β*-chlorostyrene) was shifted. Figure 11 illustrates the reso-

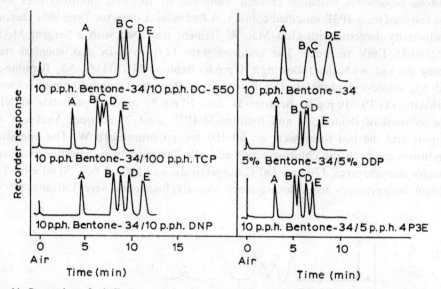

Fig. 11. Separation of ethylbenzene and xylenes. A, Toluene; B, ethylbenzene; C, *p*-xylene; D, *m*-xylene; E, *o*-xylene.

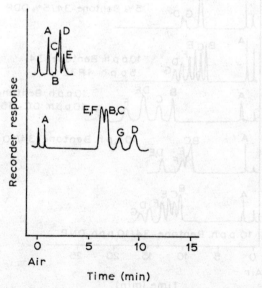

Fig. 12. Separations by octadecylamine on Anakrom As. Upper trace: A, toluene; B, ethylbenzene; C, *p*-xylene; D, *m*-xylene; E, *o*-xylene. Lower trace: A, toluene; B, *p*-Cl styrene; D, *trans-β*-Cl styrene; E, *o*-Cl styrene; F, *α*-Cl styrene; G, *cis-β*-Cl styrene.

lution of ethylbenzene–xylenes on the same six columns. In every case the various components eluted in the same order. A column packing of Bentone-34 and 4P3E offered significant advantages in the resolution of the chlorostyrenes and, to a lesser extent, of ethylbenzene–xylene mixtures. It was suggested that the selectivity of Bentone-34 was possibly due to its stable, highly adsorptive, oleophilic surface coating, the solubility parameter of which may be rather precisely modified by the addition of stationary phases of varying degrees of polarity. The major advantage of Bentone-34 over a support coated with an oleophilic stationary phase, *e.g.* silica coated with octadecylamine, was considered to be the degree of attachment, *e.g.* ionic *vs.* coordinate bonding, preventing loss of the coating by volatilization. A 12 ft. × $\frac{3}{16}$ in. column containing 70–80 mesh Anakrom AS coated with 3.6 p.p.h. octadecylamine (equal in % C to 10 p.p.h. Bentone-34) at 100°C separated alkylbenzenes in about the same fashion as 5% Bentone-34–5% DDP on Anakrom AS and somewhat better than 10 p.p.h. Bentone-34 on Anakrom AS as shown in Fig. 12. The gas rate was 25 ml/min and the order of elution was the same.

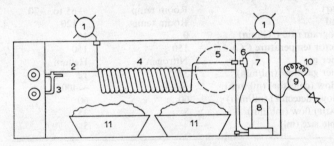

Fig. 13. Gas chromatograph. 1, Gas sample valves; 2, column support; 3, sample splitter; 4, capillary column; 5, oven fan; 6, "Swagelok" "Tee" (for ADF); 7, insulated detector oven; 8, flame-ionization detector; 9, six-port valve assembly (see Fig. 14); 10, packed column; 11, trays (aluminum) of dry ice.

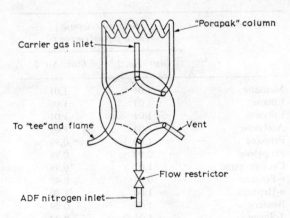

Fig. 14. Six-port valve assembly for C_1–C_{12} determination. ═, C_1–C_2 determination position; - - - -, backflush position.

Papa *et al.*[46] presented a GC method for the determination of C_1–C_{12} hydro-carbons in automotive exhaust. The minimum detectable concentration of each hydrocarbon was 1 p.p.b. The method has detected about 200 individual peaks in chromatograms of automotive exhaust which represents over 200 paraffinic, ole-finic, and aromatic hydrocarbons. The total analysis time was 25–30 min. A Perkin-Elmer Model gas chromatograph was used with several modifications (Fig. 13)

TABLE 4

CHROMATOGRAPHIC CONDITIONS

	C_1–C_2 Detn.	C_3–C_{12} Detn.
Column substrate	Porapak T and Q	DC-200
Column length (ft.)	3.5	150
Column diameter (in.)	⅛ o.d.	0.02 i.d.
Column material	Stainless steel	Stainless steel
Column temperature (°C)		
Start	Room temp.	−65 to −70
End	Room temp.	+120
Program rate (°C/min)	0	10
Detector temperature (°C)	150	150
Carrier gas	Nitrogen	Helium
Carrier gas flow (ml/min)	35	12
Air flow (detector) (ml/min)	∼400	∼400
H_2 flow (detector) (ml/min)	60	60
N_2 (ADF) flow (ml/min)	5	35
Sample size (ml)		5

TABLE 5

RELATIVE RESPONSES FOR GASEOUS DILUTED HYDROCARBONS

Hydrocarbons	Relative response (area p.p.m. carbon)	
	Instr. No. 1	Instr. No. 2
Methane	1.01	1.01
Ethane	1.01	1.00
Ethylene	1.01	1.01
Acetylene	1.04	1.03
Propane	1.00	0.99
Propylene	0.99	0.98
Cyclopropane	1.01	0.98
n-Pentane	0.97	0.98
n-Heptane	1.00	1.00
Benzene	1.00	0.99
Toluene	1.01	0.98
Ethylbenzene	0.97	0.97

TABLE 6

RELATIVE RESPONSES FOR LIQUID HYDROCARBONS

Hydrocarbons	Relative response (area/mmole carbon)	
	Instr. No. 1	Instr. No. 2
n-Pentane	0.96	0.97
n-Heptane	1.00	1.00
Isooctane	1.02	1.00
2-Methyl-1-pentene	0.96	0.97
Octene-1	1.03	0.97
Decene-1	1.00	1.05
Toluene	1.00	0.96
Ethylbenzene	1.00	0.98
o-Xylene	1.00	0.99
Benzene	0.97	1.00

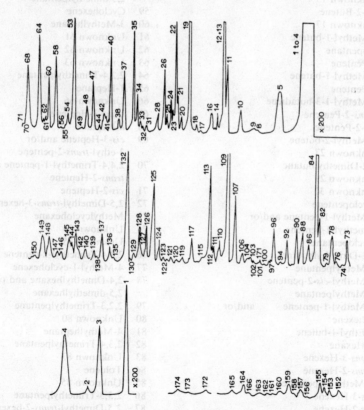

Fig. 15. Gas chromatogram of typical exhaust sample.

TABLE 7

IDENTIFICATION OF HYDROCARBON PEAKS

Peak number	Hydrocarbons	Peak number	Hydrocarbons
1	Methane	46	3-Methyl-*trans*-2-pentene and/or 3-methyl-*cis*-2-pentene
2	Ethane	47	Methylcyclopentane
3	Ethylene	48	2,4-Dimethylpentane
4	Acetylene	49	2,2,3-Trimethylbutane
5	Propylene	50	3,4-Dimethyl-1-pentene
6	Propane	51	4,4-Dimethyl-*cis*-2-pentene
7	Cyclopropane	52	3,3-Dimethylpentane
8	Propadiene	53	Benzene
9	Methylacetylene	54	Cyclohexane
10	Isobutane	55	3-Ethyl-1-pentene
11	Isobutylene and/or 1-butane	56	5-Methyl-1-hexene
12	1,3-Butadiene	57	4-Methyl-1-hexene
13	*n*-Butane	58	2-Methylhexane and/or 2,3-dimethylpentane
14	*trans*-2-Butene	59	Cyclohexene
15	Unknown 15	60	3-Methylhexane
16	*cis*-2-Butene	61	Unknown 61
17	Unknown 17	62	Unknown 62
18	3-Methyl-1-butene	63	Unknown 63
19	Isopentane	64	2,2,4-Trimethylpentane
20	1-Pentene	65	1-Heptane
21	2-Methyl-1-butene	66	Unknown 66
22	*n*-Pentene	67	*trans*-3-Heptene
23	2-Methyl-1-3-butadiene	68	*n*-Heptane
24	*trans*-2-Pentene	69	*cis*-3-Heptene and/or 3-ethyl-*trans*-2-pentene
25	*cis*-2-Pentene	70	2,4,4-Trimethyl-1-pentene and/or *trans*-2-Heptene
26	2-Methyl-2-butene	71	*cis*-2-Heptene
27	Unknown 27	72	2,5-Dimethyl-*trans*-3-hexene
28	2,2-Dimethylbutane	73	Methylcyclohexane
29	Unknown 29	74	Unknown 74
30	Unknown 30	75	Unknown 75
31	Cyclopentene	76	2,4,4-Trimethyl-2-pentene
32	4-Methyl-1-pentene and/or 3-methyl-1-pentene	77	4-Methyl-1-cyclohexene
33	Cyclopentane	78	2,4-Dimethylhexane and/or 2,5-dimethylhexane
34	2,3-Dimethylbutane	79	2,2,3-Trimethylpentane
35	2-Methylpentane	80	Unknown 80
36	4-Methyl-*cis*-2-pentene	81	4-Methylheptane
37	3-Methylpentane	82	2,3,4-Trimethylpentane
38	2-Methyl-1-pentene and/or 1-hexene	83	Unknown 83
39	2-Ethyl-1-butene	84	Toluene
40	*n*-Hexane	85	Unknown 85
41	*trans*-3-Hexene	86	2,3,3-Trimethylpentane
42	*trans*-2-Hexene	87	2,5-Dimethyl-*trans*-2-hexene
43	2-Methyl-2-pentene		
44	*cis*-3-Hexene		
45	*cis*-2-Hexene		

TABLE 7 (cont.)

Peak number	Hydrocarbons	Peak number	Hydrocarbons
88	2-Methyl-3-ethylpentane and/or 2,3-dimethylhexane	121	Unknown 121
		122	Unknown 122
89	Unknown 89	123	Unknown 123
90	3,4-Dimethylhexane and/or 3-Methylheptane	124	n-Propylbenzene
		125	1-Methyl-4-ethylbenzene and/or 1-methyl-3-ethylbenzene
91	Unknown 91		
92	2,2,5-Trimethylhexane	126	1,3,5-Trimethylbenzene
93	1-Octene	127	Unknown 127
94	trans-1,2-Dimethylcyclohexane	128	1-Methyl-2-ethylbenzene
95	Unknown 95	129	Unknown 129
96	n-Octane	130	Unknown 130
97	trans-2-Octene	131	tert.-Butylbenzene
98	Unknown 98	132	1,2,4-Trimethylbenzene
99	Dimethylheptane	133	Unknown 133
100	cis-2-Octene	134	Isobutylbenzene
101	cis-1,2-Dimethylcyclohexane	135	Unknown 135
102	Unknown 102	136	Unknown 136
103	Ethylcyclohexane	137	sec.-Butylbenzene
104	Unknown 104	138	1-Methyl-3-isopropylbenzene
105	Unknown 105	139	n-Decane
106	Unknown 106	140	1,2,3-Trimethylbenzene
107	Ethylbenzene	141	1-Methyl-4-isopropylbenzene
108	Unknown 108	142	1,3-Diethylbenzene
109	m + p-Xylene	143	Unknown 143
110	Unknown 110	144	n-Butylbenzene and/or 1-methyl-4-n-propylbenzene
111	Unknown 111		
112	Unknown 112	145	1,3-Dimethyl-5-ethylbenzene and/or 1,2-diethylbenzene
113	o-Xylene		
114	2-Methyloctane	146	1-Methyl-2-n-propylbenzene
115	Unknown 115	147–153	Unknown
116	Unknown 116	154	Durene
117	n-Nonane	155–172	Unknown
118	Unknown 118	173	1-Dodecene
119	Isopropylbenzene	174–200	Unknown
120	Unknown 120		

including two gas sampling valves mounted on the top of the column oven, a gas sample valve piped with a six-port flow switching valve and the Porapak column (Fig. 14), an auxiliary detector flow (ADF) of nitrogen added at the column outlet by passing the gas through an appropriate restrictor (packed column, valve, etc.) to the six-port valve (Fig. 14) and then into the detector oven. The instrument pyrometer was replaced with a −60 to 250°C (Assembly Products, Model No. 429-8396). A 3.5 ft. × $\frac{1}{8}$ in. o.d. column packed with 1:1 mixture of Porapak Q and Porapak T (80–100 mesh) (Waters Associates) was used for C_1–C_2 determinations. A 150 ft. × 0.02 in. i.d. capillary column coated with Dow Corning DC-200 silicone oil was used for the C_3–C_{12} determinations. The gas chromatographic conditions for the determination of the C_1–C_{12} hydrocarbons are listed in Table 4. The capillary column oven was cooled with powdered dry ice contained in two aluminum trays (8 × 1.5 × 1.25 in.) placed inside the column oven (Fig. 13). Tables 5 and 6 list the relative response for gaseous and liquid hydrocarbons, respectively. Figure 15 illustrates a gas chromatogram of a typical exhaust sample the peaks of which are tabulated in Table 7.

REFERENCES

1 J. J. SWIFT, S. R. STILES, E. W. HOWARD AND M. TARNPOLL, Petrol. Refiner, 32 (1953) 105.
2 R. L. TOLLETTI, Calif. Oil World, 40 (1953) 18.
3 F. FLURY, Arch. Exptl. Pathol. Pharmakol., 138 (1928) 65.
4 L. GREENBERG, M. R. MAYERS, L. GOLDWATER AND A. R. SMITH, J. Ind. Hyg. Toxicol., 21 (1939) 395.
5 H. H. SCHRENK, W. P. YANT, S. J. PEARCE, F. A. PATTY AND R. R. SAYERS, J. Ind. Hyg. Toxicol., 23 (1941) 20.
6 H. H. SCHRENK, S. J. PEARCE AND W. P. YANT, U.S. Bur. Mines Rept. No. 3287, 1935.
7 B. H. DOLIN, Ind. Eng. Chem., Anal., Ed., 15 (1943) 242.
8 P. A. MOFFETT, T. F. DOHERTY AND J. L. MONKMAN, Ind. Hyg. Assoc. Quart., 17 (1956) 186.
9 N. W. HANSON, D. A. REILLY AND H. E. STAGG (Eds.), The Determination of Toxic Substances in Air, Heffer, Cambridge, 1965, p. 51.
10 M. B. JACOBS, The Analytical Chemistry of Industrial Poisons, Hazards and Solvents, Interscience, New York, 1944, p. 400.
11 A. M. KHAZAIMOVA, M. F. KOVALENKO AND E. A. ZHERDEVA, Koks. Khim., 6 (1970) 35; Chem. Abstr., 73 (1970) 94394p.
12 A. K. OMAR, Pr. Cent. Inst. Ochr. Pr., 20 (1970) 23; Chem. Abstr., 73 (1970) 28539.
13 S. F. GUDKOV AND N. A. TETERINA, Tr. Vses. Nauchn. Issled. Inst. Prirodn. Gazov, 40 (1969) 133; Chem. Abstr., 73 (1970) 457.
14 E. R. FETT, D. J. CHRISTOFFERSEN AND L. R. SNYDER, J. Gas Chromatog., 6 (1968) 572.
15 L. R. SNUDER, Anal. Chem., 33 (1961) 1535.
16 M. TARAMASSO AND P. FUCHS, J. Chromatog., 49 (1970) 70.
17 M. TARAMASSO AND P. FUCHS, Chromatographia, (1969) 239.
18 F. KARASEK AND I. FONG, J. Chromatog. Sci., 9 (1971) 497.
19 A. WEHRLI AND E. KOVATS, Helv. Chim. Acta, 42 (1959) 2709.
20 V. M. NABIVACH AND S. S. YATSENKO, Khim. Teknol., 12 (1969) 132.
21 W. CELLER, R. POTOCKI AND J. F. BERAK, Przemysl Chem., 49 (1970) 230; Chem. Abstr., 73 (1970) 2851h.
22 M. NAKADA, S. FUKUSHI, H. TOMITA AND Y. MAIHIKO, Kogyo Kagaku Zasshi, 73 (1970) 929.
23 H. F. SMYTH AND H. F. SMYTH, JR., J. Ind. Hyg., 10 (1928) 261.
24 W. F. VON OETTINGEN, P. A. NEAL AND D. D. DONAHUE, J. Am. Med. Assoc., 118 (1942) 579.
25 L. GREENBURG, M. R. MAYERS, H. HERMANN AND S. MOSKOWITZ, J. Am. Med. Assoc., 118 (1942) 573.
26 W. P. YANT, S. J. PEARCE AND H. H. SCHRENK, U.S. Bur. Mines Rept. Invest. No. 3323, 1936.
27 T. DAMBRAUSKAS AND W. A. COOK, Am. Ind. Hyg. Assoc. J., 24 (1963) 568.
28 S. F. GUDKOV AND N. A. TETERINA, Tr. Vses. Nauchn. Issled. Inst. Prirodn. Gazov, 40/48 (1969) 133; Chem. Abstr., 73 (1970) 1266926.
29 A. LIBERTI AND L. ZOCCOLILLO, J. Chromatog., 49 (1970) 18.
30 K. GROB, Helv. Chim. Acta, 51 (1968) 119.
31 H. C. SPENSER, D. D. IRISH, E. M. ADAMS AND V. K. ROE, J. Ind. Hyg. Toxicol., 24 (1942) 295.
32 C. P. CARPENTER, C. B. SHAEFFER, C. S. WEST AND H. F. SMYTH, JR., J. Ind. Hyg. Toxicol., 26 (1944) 69.
33 V. K. ROWE, P. J. ATCHISON, E. N. LUCE AND E. M. ADAMS, J. Ind. Hyg. Toxicol., 25 (1943) 348.

34 H. Ley and H. Dirking, *Chem. Ber.*, 67 (1934) 1331.
35 N. W. Hanson, D. A. Reilly and H. E. Stagg (Eds.), *The Determination of Toxic Substances in Air*, Heffer, Cambridge, 1965, p. 181.
36 F. M. Nelsen, F. T. Eggertsen and J. J. Holts, *Anal. Chem.*, 33 (1961) 1150.
37 F. T. Eggertsen, S. Groennings and J. J. Holtz, *Anal. Chem.*, 32 (1960) 904.
38 P. Shapras and G. C. Claver, *Anal. Chem.*, 36 (1964) 2282.
39 J. W. Finley and J. C. White, *Bull. Environ. Contam. Toxicol.*, 2 (1967) 41.
40 G. C. Esposito and M. H. Swann, *J. Gas Chromatog.*, 3 (1965) 192.
41 C. E. R. Jones and G. E. J. Reynolds, *J. Gas Chromatog.*, 5 (1967) 25.
42 C. E. R. Jones and G. E. J. Reynolds, *J. Soc. Chem. Ind.*, 25 (1967) 260.
43 F. A. Lehmann and G. M. Brauer, *Anal. Chem.*, 33 (1961) 673.
44 F. Spagnolo, *J. Gas Chromatog.*, 6 (1968) 609.
45 C. F. Raley and J. W. Kaufman, *Anal. Chem.*, 40 (1968) 1371.
46 L. J. Papa, D. L. Dinsel and W. C. Harris, *J. Gas Chromatog.*, 6 (1968) 270.

Chapter 21

CHLORINATED ALIPHATIC HYDROCARBONS

1. CHLOROFORM

The principal manufacturing route to chloroform is now methane chlorination, although some chloroform is also produced by various processes by limited reduction of carbon tetrachloride. Until comparatively recently, chloroform was made almost entirely from acetone or ethanol by reaction with chlorine and alkali. Chloroform can also be produced by decomposition of pentachloroethane with aluminum chloride. Technical grade chloroform generally contains one or more stabilizers which vary according to the requirements of specifications. Common stabilizers that have been used include industrial methylated spirit (0.2%) and absolute alcohol (0.6–1%), thymol, tert.-butylphenol, or n-octylphenol (0.0005–0.01%). A representative grade chloroform generally contains not more than 300 p.p.m. water, 200 p.p.m. methylene chloride, 550 p.p.m. bromochloroethane, 1500 p.p.m. carbon tetrachloride, and 2 p.p.m. acid (as HCl). The production of chloroform in the U.S. in 1962 was 98,209 lb. One of the most important uses of chloroform is as a starting material in the production of chlorodifluoromethane, $CHClF_2$, an important refrigerant and coolant in air conditioning. Chlorodifluoromethane is also an intermediate in the production of polytetrafluoroethylene (PTFE) and of bromochlorofluoromethane, $CBrClF_2$, which is used as a vaporizing fire-fighting agent against fires involving flammable liquids. Chloroform is also used as a mildew preventative for tobacco seedlings, for soil fumigation, and in mixtures with methylene chlorine as a heat-transfer medium. Chloroform is widely used as an extractant of nicotine from aqueous solutions as well as an extractant for penicillin, essentials oils, and alkaloids. The anesthetic grade of chloroform is still employed for that purpose as well as in the manufacture of analgesics, anthelmintics, expectorants, liniments, and carminatives.

Chloroform slowly decomposes on prolonged exposure to sunlight in the presence or absence of air, and in the dark when air is present. The products of oxidation include phosgene, hydrogen chloride, chlorine, carbon dioxide, and water·

The powerful anesthetic action of chloroform is associated with cumulative toxic effects, primarily on the liver. Inhalation of concentrations of 0.03–0.04% chloroform can prove fatal in 30 min or less. A maximum safe concentration is considered to be 0.001%. The most pronounced effect from acute exposure of chloroform is depression of the central nervous system. Aspects of the acute[1] and chronic[2] effects of chloroform in man as well as its absorption, excretion, and metabolism have been described[3].

Chlorinated hydrocarbons are most frequently determined colorimetrically by

471

the Fujiwara reaction in which a red color is produced when the hot solution is treated with sodium hydroxide in the presence of pyridine[4-8].

Gas chromatographic determination of the chlorinated hydrocarbons is readily accomplished because of the differences in their boiling points. Flame ionization is generally not employed because of its lack of sensitivity to chlorinated hydrocarbons. Although electron-capture detection is highly sensitive to this group of compounds, the effect varies considerably according to the individual substance, hence complicating quantitative analysis.

Mixtures of chloroform, carbon tetrachloride and trichloroethylene have been separated from an air sample (10–20 ml) without enrichment by introducing the sample directly onto a 1600 cm × 6 mm U-tube containing 28% paraffin oil on Sterchamol (0.3 mm)[9]. The column temperature was 77°C, helium was the carrier gas at 90 ml/min, and a thermal conductivity cell was used for detection. The retention times were chloroform 6, carbon tetrachloride 10, and trichloroethylene 13 min, respectively.

Gutsche and Hermann[10] described the utility of a combination of gas chromatography and flame photometry (in forensic toxicological analysis) for the detection of chlorine with detection limit of 0.003 μg and precision of Cl peaks of ±2.2%. (The chlorine compounds analyzed were carbon tetrachloride, dichloromethane, dichloroethylene, and chloroform in admixture.) A Perkin-Elmer Model $F_{20}H$ gas chromatograph was used with a thermal conductivity detector and a column containing 15% polyethylene glycol 1500 on Celite 5454 at 50°C with argon as carrier gas at 16 ml/min. The filter flame photometer was specific for chlorine at 359.9 nm.

Byproducts in the manufacture of chloroform (e.g. C_2H_5Cl, C_2H_5OH, CH_3Cl, CH_2Cl_2, Cl_2CHCH_3, CCl_4, and $ClCH_2CH_2Cl$) were separated on a column of tricresyl phosphate coated on Celite. The column temperature was 68° and nitrogen was the carrier gas[11].

The isolation and gas chromatographic analysis of volatile substances (e.g. chloroform, dichloromethane, carbon tetrachloride, 1,1-dichloroethane, 1,2-dichloroethane, and trichloroethylene) from serum, brain, and liver was described by Berlet[12]. A cold spot condenser was devised to isolate small quantities of volatile material from aqueous solutions and biological material for subsequent analysis by gas chromatography. Recovery from biological material ranged from 70 to 100% and concentrations as low as 0.001% were detectable in 0.1–0.2 ml specimens. The use of this method was demonstrated by measuring 1,1-dichloroethane in blood and homogenates of brain and liver of rats treated with a single cutaneous dose of a topical analgesic containing 85% of 1,2-dichloroethane as active ingredient. A Perkin-Elmer Model F6/F7 gas chromatograph with flame-ionization detector was used with two 2 m × 4 mm columns containing 4% polyethylene glycol 1500 K on 80–100 mesh Chromosorb G and 15% polypropylene glycol on 60–100 mesh Celite 545, respectively. The columns were operated at 50–70°C with helium as carrier gas at 45 ml/min.

2. CARBON TETRACHLORIDE

Carbon tetrachloride is manufactured principally by the chlorination of hydrocarbons, particularly methane, although carbon disulfide chlorination is still used to some extent. Chlorination of aliphatic or aromatic hydrocarbons at pyrolytic or near pyrolytic temperatures (chlorinolysis) generally results in the production of some byproducts such as chloromethanes and higher chlorination derivatives. When methane is used in the Hüls process. the principal co-product is perchloroethylene. When ethylene is substituted for methane in this process, perchloroethylene is the main product and carbon tetrachloride one of a number of coproducts including hexachlorobutadiene, hexachloroethane, and hexachlorobenzene. A number of products has been used as effective stabilizers for carbon tetrachloride and they include 0.34–1 % diphenylamine, ethyl acetate, resins and amines, diethylcyanamide, and hexamethylene tetramine.

The production of carbon tetrachloride in the U.S. in 1971 was 1009 million lb. The principal uses of carbon tetrachloride are in fluorinated refrigerant and propellant manufacture (80 % of annual output), fire extinguishers, and formerly in degreasing and dry cleaning trades. Other uses of carbon tetrachloride include solvent for the reactants in organic chlorination processes such as in the manufacture of chlorinated rubber, the recovery of tin from tin-plating waste, grain fumigant alone or in admixture with 2–25 % of ethylene bromide or dichloride or with 20 % carbon disulfide, in insecticidal formulations with DDT and benzene hexachloride, in the bleaching of flour with or without benzoyl peroxide, in the separation of coal from shale, in the synthesis of nylon-7 monomer (7-aminoheptanoic acid), and as a machining lubricant. Carbon tetrachloride is used pharmaceutically in the treatment of hookworm and liver fluke in cattle and sheep and hookworm in man.

Carbon tetrachloride's narcotic and toxic properties have long been recognized. Inhalation of high concentrations of CCl_4 vapor can produce unconsciousness in 1–2 min and continued exposure can produce respiratory paralysis and even death. Repeated exposure to concentrations of about 100 p.p.m. can produce chronic toxic effects. The administration of carbon tetrachloride can lead to liver necrosis[13], accumulation of neutral lipids[14], and decreased activity of microsomal enzymes which catalyze the oxidation and reduction of drugs[15-18]. Aspects of acute[19-21] and chronic[19,22-24] toxicity of carbon tetrachloride in animals as well as its absorption, excretion, and metabolism have been described[24-26].

The utility of gas chromatography in the determination of carbon tetrachloride in a case of accidental poisoning was reported by Fischl and Labi[27]. Separation of CCl_4 was obtained on an F& M Model 500 gas chromatograph equipped with a thermal conductivity detector and a column of 16 ft. × $\frac{1}{4}$ in. copper tubing packed with 25 % diethyleneglycol adipate on 30–60 mesh Chromosorb W. The column temperature was 90°C, helium carrier flow 40 ml/min, and detector bridge current 150 mA. A 1 ml sample of freshly drawn serum was extracted with an equal volume of diethyl ether, centrifuged, and the solvent phase dried over sodium sulfate. Aliquots of the ether extract in amounts of 10–30 μl were analyzed. Under the above conditions the retention times of the ether and carbon tetrachloride were 2 and 4 min, respectively. The minimal detectable amount of CCl_4 was 500 p.p.m.

Martur and Kozlova[28] described the GLC of CCl_4 using a 600 cm × 0.5 cm column containing silicone rubber on firebrick at 200°C with helium as carrier gas at 33 ml/min. Impurities emerged from the column in the following order: CH_3Cl, C_2H_5Cl, CH_2Cl_2, $CHCl_3$, $CHCl_2CH_2Cl$, C_2Cl_4, CCl_3CH_2Cl, $CHCl_2CHCl_2$, C_2HCl_5, and C_2Cl_6.

The separation and determination of a number of organochlorine compounds in sewage by GLC was described by Mikhailyuk and Murzakaev[29]. CCl_4, C_2HCl_3, C_2Cl_4, C_2HCl_5, C_2Cl_6, and $(CCl{=}CCl_2)_2$ were separated at 130° on a column filled with a support coated with 5% tritolyl phosphate. Nitrogen was the carrier gas at 64.9 l/min at 0.75 atm pressure on leaving the column. The sample was originally collected by extraction of 1 liter of waste water with 2 × 10 ml samples of pentadecane followed by injection of 10 μl of the extract into the chromatograph and detection by flame ionization.

3. METHYLENE CHLORIDE

Methylene chloride is produced by chlorination of either methane or methyl chloride or by the reduction of chloroform or carbon tetrachloride. Other chloromethanes are coproducts of methylene chloride, their relative proportions depending on the process operating conditions.

The production of methylene chloride in the U.S. in 1962 was 143,782 lb. Commercial grades of methylene chloride may contain added stabilizers such as 0.0001–1.0% of phenolic compounds (*e.g.* phenol, hydroquinone, *p*-cresol, resorcinol, thymol, 1-naphthol), small quantities of amines, or a mixture of 1,4-dioxane and nitromethane. Methylene chloride is principally used in solvent applications, *e.g.* as a paint stripper, in removal of lacquer from metal surfaces, extraction of grease from raw wood, extraction of naturally occurring heat-sensitive substances such as edible fats, cocoa, butter, and beer flavoring in hops; as a solvent for cellulose triacetate in rayon yarn manufacture, in the production of PVC fiber. Methylene chloride is also used as a propellant in aerosol mixtures, a refrigerant in low-pressure refrigerators and air-conditioning installations, and as a low-temperature heat-transfer medium.

Methylene chloride vapor does not form explosive mixtures with air at ordinary temperatures but forms toxic products such as phosgene when exposed to hot surfaces or an open flame.

Methylene chloride is the least toxic of the chloromethanes, *e.g.* about half as toxic as chloroform[30]. It is used to some extent as a local anesthetic in Europe.

Divencenzo et al.[31] reported the gas chromatographic analysis of methylene chloride in breath, blood, and urine. Breath samples from human subjects exposed to CH_2Cl_2 were collected in Saran bags, while blood and urine were placed directly into sealed flasks. A structural modification of the gas sample valve enabled a heated flask to be inserted into the gas sampling system resulting in the direct measurement of the head-space vapors. The analyses were performed on a Varian 2100 gas chromatograph equipped with a Carle gas sample valve. To permit the sampling of head space vapors in Erlenmeyer flasks, one gas loop was removed from the gas valve and replaced by two $\frac{1}{16} \times 21$ in. lengths of metal tubing. A 16-gauge stainless steel needle was soldered to each free end of the tubing and both needle points were inserted simultaneously through the serum cap of the sample flask. The valve was then switched to sweep the carrier gas through the flask and transfer the contents of the head space volume to the chromatographic column. The gas chromatographic conditions for the analysis of the chamber air, breath, blood, and urine are listed in Table 1. The methods which were described for the measurement of methylene chloride are rapid and simple and should be generally applicable to several volatile organic compounds.

TABLE 1

GAS CHROMATOGRAPHIC CONDITIONS FOR THE ANALYSIS OF METHYLENE CHLORIDE IN BREATH, BLOOD, AND URINE

Detector	Flame ionization
Mode	Single column U-shaped, all glass
Liquid phase	20% Carbowax 20M on Gas Chrom Q, 80–100 mesh
Column size	6 ft. × mm i.d.
Carrier gas	Nitrogen
Column conditioning	200°C for 24 h with N_2 flow rate of 5 ml/min
Carrier gas flow rates (ml/min)	
(a) For chamber air and breath analysis	70
(b) For blood and urine analysis	90
H_2 flow rate (ml/min)	70
Air flow rate (ml/min)	435
Detector temperature (°C)	230
Injection port temperature (°C)	90
Column temperature	25
Methylene chloride retention time (min)	4.5

4. TRICHLOROETHYLENE AND TETRACHLOROETHYLENE

Both trichloro- and tetrachloroethylene are important industrial solvents used extensively for dry cleaning as well as for fumigation. Trichloroethylene is used extensively in cold degreasing applications and as a solvent for agricultural chemicals, hair sprays, and in dry cleaning. Tetrachloroethylene is increasingly replacing trichloroethylene in dry cleaning because of its lower vapor pressure.

Both compounds, however, are hepatoxic and aspects of the effects of human exposure have been reported[32-37].

Trichloroethylene has been determined by spectrophotometric[38,39] and infrared spectroscopy[40,41]. Chlorinated aliphatic hydrocarbons can also be determined (following the formation of chloride ion) by nephelometry, titration[9,42], and colorimetry[43]. Trichloroethanol and trichloroacetic acid (the major metabolites of trichloroethylene and tetrachloroethylene) have been determined in urine by colorimetry[44-46].

Trichloroacetic acid has been identified by paper[47] as well as by gas chromatography[48].

Laham[49], using a combination of gas and paper chromatographic techniques, identified di- and trichloroacetic acids and the corresponding alcohols in the urine of rats dosed orally with trichloroethylene (1 g/kg).

Daniel[50] studied the oral metabolism of uniformly labeled ^{36}Cl-trichloroethylene and tetrachloroethylene in the rat. The partition of radioactivity in urine, feces, and expired air was measured and the nature and amount of individual metabolites determined. The specific activities of metabolic trichloroacetic acid and trichloroethanol were shown to be the same as that of the administered trichloroethylene hence suggesting an intramolecular rearrangement of chloride and no exchange of chloride with the body chloride pool. About 15% of the dose of trichloroethylene was excreted in the urine while tetrachloride underwent little metabolism and only 2% of the radioactivity was excreted in the urine. Both trichloroethylene and tetrachloroethylene were largely excreted through the lungs although there was an appreciable difference between the rates of excretion of the two compounds.

The purification of ^{36}Cl-labeled trichloroethylene and tetrachloroethylene was achieved by gas chromatography at 50° on a 2 m column of Celite (60–80 mesh) containing 25% silicone grease (Dow-Corning) as the liquid phase. Analysis of expired air from dosed animals was achieved by first drawing the air through two absorbers containing anhydrous calcium chloride and soda lime, respectively, then through two traps each containing $\frac{3}{16}$ in. Lessing-rings kept at the temperature of liquid air. The chlorinated metabolites were retained in the traps under these conditions, subsequently recovered by vacuum distillation, and the distillate examined by gas chromatography at 50° on a 1 m column of Celite (60–80 mesh) containing 28.4% Reoplex 400 as liquid phase and a katharometer as detector.

A reaction mechanism suggested by Daniel to operate *in vivo* and which would account for the known metabolites of trichloroethylene would be

The formation[51] of an hypothetical oxide was originally proposed by Powell[51] and was suggested by Daniel[50] to be probably due to a non-specific enzyme system.

(Trichloroethylene oxide is also believed to be formed when trichloroethylene is oxygenated in the presence of actinic radiation.) Rearrangement of the oxide would yield trichloroacetaldehyde, the formation of which in men exposed to trichloroethylene vapor has been quantitatively determined[52]. The oxidation of chloral to trichloroacetic acid can be carried out by an enzyme present in the liver of a variety of experimental animals[53] while reduction of chloral to trichloroethanol would involve alcohol dehydrogenase.

The metabolism of tetrachloroethylene was suggested by Daniel[50] possibly to involve the following series of reactions.

$$\underset{Cl}{\overset{Cl}{>}}C=C\underset{Cl}{\overset{Cl}{<}} \longrightarrow \underset{Cl}{\overset{Cl}{>}}C\overset{O}{-}C\underset{Cl}{\overset{Cl}{<}} \longrightarrow Cl_3C \cdot COCl \longrightarrow Cl_3C\,COOH \; + \; Cl^-$$

The acid chloride would be rapidly hydrolyzed to trichloroacetic acid and neither trichloroethanol or oxalic acid would be formed. The excretion of organic chlorine compounds in the urine of persons exposed to vapors of trichloroethylene and tetrachloroethylene was studied by Ogata et al.[54].

The urinary levels of trichloroethanol, trichloroacetic acid, and total trichloro compounds were almost proportional to the environmental concentration of trichloroethylene. The total excretion of organic chloride following tetrachloroethylene exposure was only 2.8% of the tetrachloroethylene retained as against 75% of the trichloroethylene retained. This indicated that tetrachloroethylene was mostly eliminated through expiration rather than in urine and that its metabolites are retained for a longer period in the body.

The excretion kinetics of urinary metabolites in a patient addicted to trichloroethylene were elaborated by Ikeda et al.[55]. The analyses of urine revealed the presence of up to 160 μg/ml of trichloro compounds (mostly trichloroacetic acid) which gradually disappeared in 3 weeks as the psychotic symptoms cleared up. The excretion half-lives of trichloroethylene metabolites for the initial rapid phase (succeeding slow phase in parentheses) were 5.8 (49.7) h for trichloroethanol, 22.5 (72.6) h for trichloroacetic acid, and 7.5 (72.6) h for total trichloro compounds.

The analysis of air and breath for chlorinated hydrocarbons by infrared spectroscopic and gas chromatographic techniques was reported by Boettner and Dallos[56]. Three gas chromatographs were used: (1) Research Specialties Model 600 gas chromatograph with an argon ionization detector, (2) Wilkens Hy-Fi Model 600 chromatograph with hydrogen flame and electron-capture detectors, and (3) Beckman Model GC2A chromatograph with thermal conductivity detectors. Table 2 lists the types of columns and detectors used for the separation of chlorinated aliphatic solvents. The limits of detection for the compounds with the various detectors are tabulated in Table 3. Table 4 lists the limits of detection for chlorinated aliphatic compounds using infrared long-path gas cell techniques. Table 5 depicts a comparison of the sensitivities of infrared and gas chromatographic techniques for various chlorinated solvents. (The IR and GC samples were 10 μl and 10 ml, respectively.)

Although GC techniques are considerably more sensitive than infrared absorption on an absolute basis, the ability of the IR method to handle a much larger sample makes it comparable on a weight per unit volume basis. In accessing the various detectors it was found that the thermal conductivity detector had poor sensitivity but a large dynamic range and its sensitivity was rather independent of molecular weight. The sensitivity of GC using argon or flame-ionization detectors decreased slightly with the higher molecular weight chlorinated aliphatic compounds. The electron-capture detector was by far the most sensitive detector for compounds having three or more chlorine atoms, but it had a limited dynamic range and in addition it was easily "poisoned" either by an excess of compound being analyzed or by other compounds in the sample.

Beroadschi and Stefan[57] described the GC of a trichloroethylene–chloroacetic acid mixture. Chloroacetic acid, obtained by hydrolysis of trichloroethylene at 180 to 185°C, was determined on a 1.8 m × 0.1 cm² cross-section column containing 25% transformer oil and 0.5% glycerol on Celite, with hydrogen carrier gas at 0.5 kg/cm² at 82°. The detector was a thermal conductivity cell and chloroacetic acid and tri-

TABLE 2

COLUMN COATINGS AND SUPPORTS FOR THE SEPARATION OF CHLORINATED ALIPHATIC COMPOUNDS
USING GAS CHROMATOGRAPHS WITH VARIOUS DETECTORS

Compound	Detector			
	Thermal conductivity	Argon ionization	Flame ionization	Electron capture
Methyl chloride	C	A	A	A
Dichloromethane	C	A	C	B
Chloroform	C	A	A	B
Carbon tetrachloride	A	A	A	B
Ethyl chloride	A	A	A	A
1,2-Dichloromethane	A	A	A	B
1,1,1-Trichloroethane	C	A	A	B
1,1,2-Trichloroethane	C	A	C	B
1,1,2,2-Tetrachloroethane	C	A	C	E
1,2-Dichloropropane	C	A	C	C
1,2,3-Trichloropropane	C	A	C	E
Chloroethylene	A	A	A	A
cis-1,2-Dichloroethylene	A	A	C	B
trans-1,2-Dichloroethylene	A	A	C	B
Trichloroethylene	C	A	A	B
Tetrachloroethylene	D	A	C	B

A: 20% Carbowax 600 on C-22 Firebrick.
B: 5% Silicone 550 and 5% Ucon (water-insoluble) on Chromosorb P.
C: 10% Silicone 550 on C-22 Firebrick.
D: 20% Carbowax 20M on C-22 Firebrick.
E: 10% Silicone SE 30 on Chromosorb W.

TABLE 3

LIMITS OF DETECTION FOR CHLORINATED ALIPHATIC COMPOUNDS USING GAS CHROMATOGRAPHS
WITH VARIOUS DETECTORS

Compound	Detector			
	Thermal conductivity $\times 10^{-1}$ (µg)	Argon ionization $\times 10^{-2}$ (µg)	Flame ionization $\times 10^{-3}$ (µg)	Electron capture $\times 10^{-3}$ (µg)
Methyl chloride	1.2	2.0	3.0	8.5
Dichloromethane	4.2	5.0	1.3	8.6
Chloroform	6.0	4.3	20.0	0.08
Carbon tetrachloride	4.8	5.0	20.0	0.002
Ethyl chloride	1.4	6.0	1.6	11.0
1,2-Dichloromethane	3.4	4.1	13.0	13.0
1,1,1-Trichloroethane	2.6	5.2	6.0	0.03
1,1,2-Trichloroethane	2.8	4.0	8.6	0.07
1,1,2,2-Tetrachloroethane	5.0	8.0	16.0	0.008
1,2-Dichloropropane	0.9	5.5	8.8	23.0
1,2,3-Trichloropropane	2.8	3.8	4.0	0.07
Chloroethylene	0.2	1.9	2.2	2.3
cis-1,2-Dichloroethylene	4.0	6.5	2.6	13.0
trans-1,2-Dichloroethylene	2.2	3.5	2.5	8.4
Trichloroethylene	2.5	10.0	8.5	0.02
Tetrachloroethylene	3.2	5.3	21.0	0.003

TABLE 4

LIMITS OF DETECTION FOR CHLORINATED ALIPHATIC COMPOUNDS USING INFRARED LONG-
PATH GAS CELL TECHNIQUES

Compound	Analytical wavelength (µm)	Scale Expansion (µg/l)	
		1×	5×
Methyl chloride	13.45	2.4	0.7
Dichloromethane	13.35	0.9	0.5
Chloroform	8.2	1.5	0.4
Carbon tetrachloride	12.55	0.15	0.03
Ethyl chloride	10.15	11.0	5.0
1,2-Dichloroethane	8.1	2.0	0.9
1,1,1-Trichloroethane	9.2	1.2	0.4
1,1,2-Trichloroethane	13.5	1.4	0.6
1,1,2,2-Tetrachloroethane	12.4	1.2	0.5
1,2-Dichloropropane	13.47	3.1	1.8
1,2,3-Trichloropropane	13.75	3.1	0.7
Chloroethylene	10.63	15.0	3.1
cis-1,2-Dichloroethylene	11.82	0.4	0.2
trans-1,2-Dichloroethylene	12.25	0.6	0.4
Trichloroethylene	11.82	1.2	0.5
Tetrachloroethylene	11.0	0.5	0.2

TABLE 5

COMPARISON OF THE SENSITIVITIES OF INFRARED AND GAS CHROMATOGRAPHIC TECHNIQUES

Compound	Limit of detection ($\mu g/l$)				
	Infrared absorption	Thermal conductivity	Argon ionization	Flame ionization	Electron capture
Methyl chloride	4.2	12	0.2	0.3	0.9
Dichloromethane	3.0	42	0.5	0.1	0.9
Chloroform	2.4	60	0.4	2.0	0.008
Carbon tetrachloride	0.2	48	0.5	2.0	0.0002
Ethyl chloride	30.0	14	0.6	0.2	1.1
1,2-Dichloromethane	5.4	34	0.4	1.3	1.3
1,1,1-Trichloroethane	2.4	0.9	0.5	0.6	0.003
1,1,2-Trichloroethane	3.6	1.1	0.4	0.9	0.007
1,1,2,2-Tetrachloroethane	3.0	50	0.8	1.6	0.0008
1,2-Dichloropropane	11.0	9.0	0.6	0.9	2.3
1,2,3-Trichloropropane	4.2	28	0.4	0.4	0.007
Chloroethylene	19.0	2.0	0.2	0.2	0.2
cis-1,2-Dichloroethylene	1.2	40	0.7	0.3	0.3
trans-1,2-Dichloroethylene	2.4	22	0.4	0.3	0.8
Trichloroethylene	3.0	25	1.0	0.9	0.002
Tetrachloroethylene	1.2	32	0.5	2.1	0.0003

chloroethylene yielded 2 sharp and distinct peaks which appeared at 107.5 and 178 theoretical plates, respectively.

Mixtures of chlorine derivatives of methane, ethane, and ethylene were analyzed by GLC[58] using a 4.90 m × 4 mm copper tube containing 15% Apiezon M on Chromosorb R (80–100 mesh); the carrier gas was helium at 66.7 ml/min and 2.5 atm inlet pressure. The column temperature was 150°. A 6 μl sample of a mixture of 15 chlorinated hydrocarbons showed the retention time as a constant function of the boiling point. Commercial C_2Cl_4 contained very little C_2HCl_3 and no other impurities, but after it had been used and distilled several times, it contained more C_2HCl_3 and other impurities.

Stewart et al.[59] described the detection of halogenated hydrocarbons in the expired air of human subjects by electron-capture gas chromatography. A series of timed human exposures to known concentrations of trichloroethylene, 1,1-trichloroethane, and tetrachloroethylene vapor in a room was initially monitored by (a) infrared using a 10 m path-length gas cell, (b) a modified Davis halide meter[60], and (c) Volhard titration[61].

Samples of air expired were collected in 6 l volume "Saran" bags at appropriate intervals in the post-exposure period and analyzed by GLC using an Aerograph Hy-Fi Model A600B. Gas-tight syringes were used to transfer the breath aliquot from the Saran bags and introduce the sample into the instrument. A stainless steel 6 ft. × ⅛ in. o.d. column was packed with Apiezon L or Carbowax 20M on Chromosorb R. The limits of detection for trichloroethylene in air ranged from 2 to 10 p.p.b. and for 1,1,1-trichloroethane or tetrachloroethylene in air from 1 to 5 p.p.b. Trichloroethylene was detectable in breath for 330 h following exposure and 1,1,1-trichloroethane and tetrachloroethylene was detectable for 800 h following exposure.

A method for the GLC determination of methylene chloride, ethylene dichloride, and trichloroethylene in spice oleoresins has been reported[62]. The method employed a microcoulometric gas chromatograph equipped with a Porapak Q column and is based on the volatility of the solvent residues, their relative retention times on a poly-aromatic bead chromatographic column and their detectability by a halide-specific microcoulometric method. A Dohrmann Model C-200 gas chromatograph was used with a microcoulometric detector and a 6 ft. × 6 mm o.d. aluminum column packed with 150–200 mesh Porapak Q. (The column was conditioned 1 h at 230°, passing nitrogen through at 20 ml/min.) The operating conditions were damping range 200 Ω, removable inlet temperature 140°, column temperature 160°, furnace temperature ca. 825°, and nitrogen flow 90 ml/min. The chlorinated solvents eluted as follows: methylene chloride ca. 7 min, ethylene dichloride ca. 17 min, and trichloroethylene ca. 24 min. Relative retention times were ethanol 0.52, methylene chloride 1, ethylene chloride 2.58, and trichloroethylene 3.52.

A cryogenic trapping technique for introducing pure trace samples into a mass spectrometer or gas chromatograph with specific reference to the determination of trichloroethylene and benzene in air was described by Snyder[63]. The trapping system

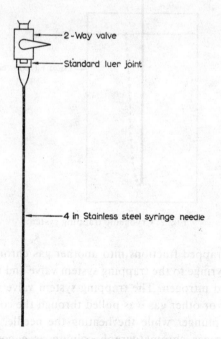

Fig. 1. Trapping system with Teflon seated two-way valve.

consisted of a standard 4 in. 22 gauge stainless steel syringe needle equipped with a Teflon seated two-way valve (Fig. 1). Fractions were collected at the effluent of a post column stream splitter (1:1 ratio) attached to an F & M Model 810 gas chromatograph equipped with FID. A Bendix Model 12 time-of-flight mass spectrometer was used for analyzing the fractions. The post column strean splitter was equipped with a standard three-way syringe valve which is used to direct the flow of the stream splitter, the trapping syringe needle trap or to the atmosphere, depending on whether a particular sample constituent is to be trapped. The cryogenic trapping system as shown in Fig. 2 was attached to the 3-way valve at the effluent end of the stream splitter. Three inches of the trapping needle was then immersed in liquid nitrogen (Fig. 2).

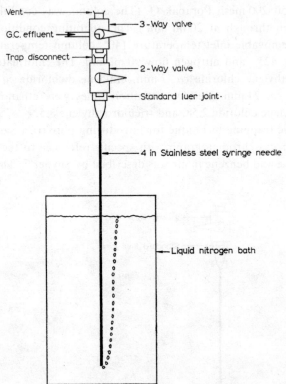

Fig. 2. Cryogenic trapping system.

     Re-injection of trapped fractions into another gas chromatograph required the attachment of a gas syringe to the trapping system valve and removal of the trapping needle from the liquid nitrogen. The trapping system valve was then opened and a desired volume of air or other gas was pulled through the collection needle by withdrawing the syringe plunger while the heating the needle. The fraction was then injected into another gas chromatograph column as a conventional gas sample.

An example of this procedure is shown in Fig. 3 which represents the chromatograms from the gas chromatograph injections of 3 ml of 100 p.p.m. trichloroethylene in air. Figure 4 shows the chromatograms of 3 ml of 100 p.p.m. of benzene in air blend.

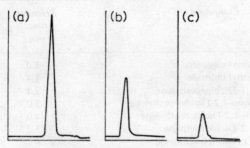

Fig. 3. 3 ml of 100 p.p.m. trichloroethylene in air blend. 16 ft. Carbowax 20 M; $\frac{3}{16}$ in. o.d., 150°C; 30 ml/min; range 10; attenuation 32. (a) Trichloroethylene injection with stream splitter closed; (b) trichloroethylene injection with fraction being collected; (c) reinjection of B with splitter closed.

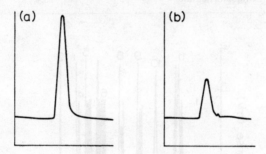

Fig. 4. 3 ml of 100 p.p.m. benzene in air blend. 16 ft. Carbowax 20 M, $\frac{3}{16}$ in. od..; 100°C; 30 ml/min; range 10; attenuation 64. (a) Benzene injection with stream splitter closed; (b) benzene injection with fraction being collected.

A mixture of eleven chlorinated derivatives of C_1 and C_2 hydrocarbons in 1,2-dichloroethane was resolved and analyzed by programmed-temperature gas chromatography by Hinshaw[64]. A Beckman Thermotrac temperature programmer was connected to a Gow-Mac thermal conductivity cell (TR-11B) heated to 150°C by heating tapes. The cell was operated at 230°C and a filament current of 150 mA. A Minneapolis Honeywell 0–1 mV strip chart recorder was used in conjunction with a Ridgefield Attenumatic integrator (Esterline Angus Instrument, Indianapolis). Dual 6 ft. × $\frac{1}{4}$ in. copper columns were packed with 30–60 mesh Firebrick coated with 15–18% SE-30. A helium flow rate of 90 ml/min was used at a column inlet pressure of 30 p.s.i.g. A thermal conductivity detector was used at 230°C.

All compounds present in the catalytic oxychlorination of ethylene to 1,2-dichloroethane reaction were identified from retention data on silicone rubber and diisodecyl phthalate columns and mass spectrometric identification of fractions

TABLE 6

RETENTION DATA AND SENSITIVITY FACTORS FOR COMPOUNDS PRESENT
IN REACTION MIXTURES

Peak no.*	Compound	Retention time (min)	Sensitivity factor
1	Vinyl chloride	1.0	
2	Ethyl chloride	1.4	0.751
3	1,1-Dichloroethylene	2.4	
4	trans-1,2-Dichloroethylene	3.0	1.007
5	cis-1,2-Dichloroethylene	4.0	1.008
6	1,2-Dichloroethane	5.2	1.000
7	Carbon tetrachloride	6.2	1.114
8	Trichloroethylene	7.3	1.149
9	1,1,2-Trichloroethane	8.9	1.179
10	Perchloroethylene	10.3	1.312
11	1,1,2,2-Tetrachloroethane	12.5	1.345
12	Pentachloroethane	15.4	1.498

*See Fig. 5.

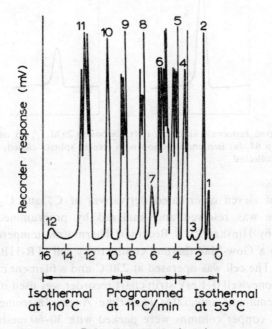

Fig. 5. Gas chromatogram of chlorinated hydrocarbons. 1, Vinyl chloride; 2, ethyl chloride; 3, 1,1-dichloroethylene; 4, trans-1,2-dichloroethylene; 5, cis-1,2-dichloroethylene; 6, 1,2-dichloroethane; 7, carbon tetrachloride; 8, trichloroethylene; 9, 1,1,2-trichloroethane; 10, perchloroethylene; 11, 1,1,2,2-tetrachloroethane; 12, pentachloroethane.

TABLE 7

PARAMETERS FOR DETERMINATION OF SEVERAL HALOGENATED AND AROMATIC HYDROCARBONS[a]

Compound	Chromatographic settings				Attenuation	Minimum air sample (l)	Efficiency[b] (%)	Retention time (min)	p.p.m. Conversion factor
	Oven temp. (°C)	Detector temp. (°C)	Injection block temp. (°C)	Helium flow rate (ml/min)					
Benzene	50	50	100	175	×2	10	89	5.8	313
Carbon tetrachloride	50	50	100	175	×1	10	95	4.1	158.8
Ethylene dichloride	75	75	150	156	×1	4	96	6.3	247
Methyl chloroform	50	50	100	175	×2	4	100	4.2	183.3
Methylene chloride	50	50	100	175	×2	4	97	4.8	288
Perchloroethylene	75	75	150	156	×1	4	95	4.6	147.4
Toluene	75	75	150	156	×2	4	93	5.7	266
Trichlorotrifluoroethane	50	50	100	175	×4 and ×16	2	92	0.8	130.5
Trichloroethylene	75	75	150	156	×1	4	95	4.2	186
Xylene	100	100	200	140	×1	4	95	5.4	230.7

[a] In all cases the column inlet pressure is 24 p.s.i. and the detector set at 8 V.
[b] Efficiencies are based on 0.5–2 TLV concentrations and sampling rates of 1–2 l/min.

collected from the gas chromatograph. Four microliter liquid samples were injected at a starting temperature of 53°C. After 4 min, the column temperature was programmed at about 11°C/min for 5 min. The temperature was then held constant at 110°C for the remainder of the analysis. A typical chromatogram obtained as described is shown in Fig. 5. The identified compounds (*e.g.* vinyl chloride, ethyl chloride, 1,1-dichloroethylene, *trans*-1,2-dichloroethylene, *cis*-1,2-dichloroethylene, 1,2-dichloroethane, carbon tetrachloride, trichloroethylene, 1,1,2-trichloromethane, perchloroethylene, 1,1,2,2-tetrachloroethane, and pentachloroethane) and their retention times on the SE-30 column are shown in Table 6. Sensitivity factors were obtained by dividing the actual weight percent of each component by the relative area percent. Factors relative to that for 1,2-dichloroethane were calculated by dividing these values by the factor for 1,2-dichloroethane. These relative sensitivity factors, which correct for any differences in thermal conductivity of the components, are also shown in Table 6.

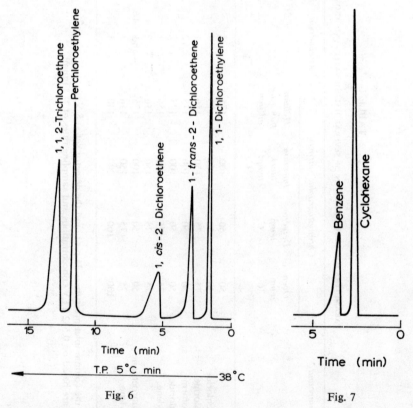

Fig. 6

Fig. 7

Fig. 6. Separation of chlorinated hydrocarbons on 80–100 mesh unsilanized ceramic nodules coated with 3.8 w/w Apiezon L. Volume injected 2 λ.

Fig. 7. Separation of cyclohexane–benzene on 80–100 mesh unsilanized ceramic nodules coated with 3.8% w/w Apiezon L, carried out isothermally at 60 °C. Volume injected 1 λ.

Reid and Halpin[65] described the GLC of 7 halogenated and 2 aromatic hydrocarbons. Table 7 lists the operating parameters and retention times for the determination of carbon tetrachloride, ethylene dichloride, methyl chloroform, methylene
chloride, perchloroethylene, trichloroethylene, trichlorotrifluoroethane, benzene,
toluene, and xylene.

Castellucci and Eisaman[66] compared cellular nodules (low density ceramic
material) with smooth glass beads and diatomaceous earth for the solid phase in
GLC of chlorinated aliphatics, chlorinated olefins, normal hydrocarbons, etc. The

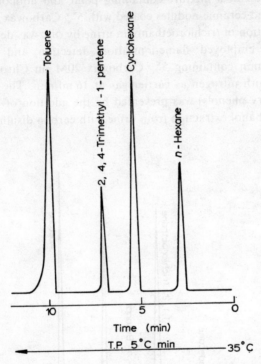

Fig. 8. Separation of hydrocarbons on 80–100 mesh unsilanized ceramic nodules coated with 3.8 %w/w
Apiezon L. Volume injected 1 λ.

points of comparison were separating power, pressure drop, flow rates, theoretical
plates, and the general structure of chromatographic curves. The nodules were made
(Pittsburgh Corning Corp.) primarily from a high alumina soda lime silica glass and
have a highly spherical geometry. Gas chromatographic data were obtained with a
Micro-Tek Model 2500 instrument equipped with a thermal conductivity detector
and using columns of 1219 mm × 5.2 mm i.d. stainless steel under isothermal or
linear-programmed conditions. All column packings were prepared by adding 0.6 g
of substrate dissolved in a suitable solvent to 34 ml of the support. Figure 6 shows
the programmed-temperature separation of chlorinated hydrocarbons on 80 to

100 mesh unsilanized ceramic nodules coated with 3.8% Apiezon L. The separation of cyclohexane–benzene on the same support at 60°C is shown in Fig. 7 which illustrates the programmed-temperature separation of a mixture of polar and non-polar compounds on 60–80 mesh silanized ceramic nodules coated with 5% Carbo-wax 20M–TPA and Fig. 8 shows the separation of hydrocarbons on 80–100 mesh unsilanized ceramic nodules coated with 3.8% Apiezon L obtained with temperature programming. It was shown in these studies that cellular ceramic nodules silanized and unsilanized performed well as a solid support in gas chromatography. Figure 9 shows the separation of a mixture containing polar and nonpolar compounds on 60–80 mesh silanized ceramic nodules coated with 5% Carbowax 20M–TPA.

The determination of trichloroethanol in urine by GLC was described by Sedivek and Flek[67] who employed flame-ionization detection and a stainless steel (2 m × 2 mm) column containing 5% Carbowax 20M on Chromosorb W (60 to 80 mesh) at 130° with nitrogen as carrier gas at 16 ml/min. The main interference (co-extracted urinary phenols) was prevented by the addition of paraformaldehyde prior to trichloroethanol extraction from urine with carbon disulfide.

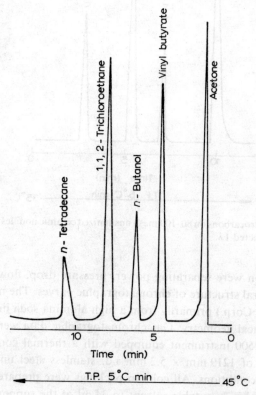

Fig. 9. Separation of a mixture containing polar and nonpolar compounds on 60–80 mesh silanized ceramic nodules coated with 5% w/w Carbowax 20M-TPA. Volume injected 2 λ.

REFERENCES

1 K. B. LEHMAN AND F. FLURY, *Toxicology and Hygiene of Industrial Solvent*, Williams and Wilkins, Baltimore, 1943.
2 P. J. R. CHALLEN, D. E. HICKISH AND J. BEDFORD, *Brit. J. Ind. Med.*, 15 (1958) 243.
3 W. F. VON OETTINGEN, C. C. POWELL, N. E. SHARPLESS, W. C. ALFORD AND J. PECORA, *Natl. Inst. Health Bull.*, 191 (1949).
4 J. H. BRUMBAUGH AND D. E. STALLARD, *J. Agr. Food Chem.*, 6 (1958) 465.
5 G. W. ROGERS AND K. K. KAY, *J. Ind. Hyg. Toxicol.*, 29 (1947) 8229.
6 A. O. GETTLER AND H. BLUME, *Arch. Pathol.*, 17 (1931) 555.
7 G. A. LUGG, *Anal. Chem.*, 38 (1966) 1532.
8 G. A. HUNGOLD AND B. SCHUHLEIN, *Z. Anal. Chem.*, 179 (1961) 81.
9 W. LEITHE, *The Analysis of Air Pollutants*, Ann Arbor Science, Ann Arbor, Mich., 1971, p. 235.
10 B. GUTSCHE AND R. HERMANN, *Z. Anal. Chem.*, 245 (1969) 274.
11 Z. S. SMOLYAN, N. D. DEMINA AND M. I. VLASOVA, *Gaz Kromatogr. Akad. Nauk, SSSR, Tr. Vtoroi Vses. Konf. Moscow, 1962*, 1964, p. 276; *Chem. Abstr.*, 62 (1965) 4588.
12 H. H. BERLET, *Z. Anal. Chem.*, 243 (1968) 335.
13 C. GALLAGNER, *Australian J. Exptl. Med. Sci.*, 40 (1962) 241.
14 T. F. SLATER, *Nature*, 209 (1966) 36.
15 H. LAL, S. K. PURI AND G. C. FULLER, *Toxicol. Appl. Pharmacol.*, 16 (1970) 35.
16 J. V. DINGELL AND M. HEIMBERG, *Biochem. Pharmacol.*, 17 (1968) 1269.
17 D. NEUBERT AND O. MAIBAUER, *Arch. Exptl. Pathol. Pharmacol.*, 235 (1959) 291.
18 R. KATO, E. CHIESARA AND P. VASARELLY, *Biochem. Pharmacol.*, 11 (1962) 211.
19 E. M. ADAMS, H. C. SPENCER, V. K. ROWE, D. D. MCCOLLISTER AND D. D. IRISH, *Arch. Ind. Hyg. Occupational Med.*, 6 (1952) 50.
20 K. B. LEHMANN, *Arch. Hyg.*, 74 (1911) 1.
21 B. D. NINMAN AND A. BERNSTEIN, *Arch. Environ. Health*, 16 (1968) 777.
22 W. F. VON OETTINGEN, *U.S. Publ. Health Serv. Publ. No. 414*, 1955.
23 D. D. MCCOLLISTER, W. H. BEAMER, G. J. ATCHISON AND H. C. SPENCER, *J. Pharmacol. Exptl. Therap.*, 102 (1951) 1112.
24 D. D. MCCOLLISTER, R. C. HOLLINGSWORTH, F. OYEN AND V. K. ROWE, *Arch. Ind. Health*, 13 (156) 1.
25 A. B. ESCHENBRENNER AND E. MILLER, *Gew. Natl. Res. Inst.*, 6 (1946) 325.
26 T. H. BRODY, D. N. CALVERT AND A. F. SCHNEIDER, *J. Pharmacol. Exptl. Therap.*, 131 (1961) 341.
27 J. FISCHL AND M. LABI, *Israel J. Med. Sci.*, 2 (1966) 84.
28 V. G. MARTUR AND V. S. KOZLOVA, *Metody Anal. Khim. Reaktivov Prep.*, 15 (1968) 31; *Chem. Abstr.*, 69 (1968) 92770B.
29 Y. I. MIKHAILYUK AND F. G. MURZAKAEV, *Gigiena i Sanit.*, 35 (1970) 73; *Chem. Abstr.*, 73 (1970) 123345f.
30 L. T. FAIRHALL, *Industrial Toxicology*, Williams and Wilkins, Baltimore, 1957.
31 G. D. DIVINCENZO, F. Y. YANO AND B. D. ASTILL, *Am. Ind. Hyg. Assoc. J.*, 32 (1971) 387.
32 Am Conf. Govt, Ind. Hygienists, *Arch. Environ. Health*, 1 (1960) 140.
33 R. D. STEWART, H. H. GAY, D. S. ERLEY, C. L. HAKE AND A. W. SCHAFFER, *Arch. Environ. Health*, 2 (1961) 516.
34 R. D. STEWART, H. H. GAY, D. S. ERLEY, C. L. HAKE AND A. W. SCHAFFER, *Am. Ind. Hyg. Assoc. J.*, 22 (1961) 252.
35 B. SOUCEK AND D. VLACHOVA, *Brit. J. Ind. Med.*, 17 (1960) 60.
36 S. TANAKA AND M. IKEDA, *Brit. J. Ind. Med.*, 5 (1968) 214.
37 R. FRANT AND J. WESTENDORP, *Ind. Hyg. Occupational Med.*, 1 (1946) 308.
38 V. T. STACK, JR., D. E. FOREST AND J. K. WAHL, *Am. Ind. Hyg. Assoc. J.*, 22 (1961) 184.
39 N. W. HANSON, D. A. REILLY AND H. E. STAGG (Eds.), *Determination of Toxic Substances in Air*, Heffer, Cambridge, 1965, p. 101.
40 R. D. STEWART, H. H. GAY, D. S. ERLEY, C. L. HAKE AND J. E. PETERSON, *J. Am. Ind. Hyg. Assoc.*, 23 (1962) 167.

41 S. BINASCHI, G. GAZZANIGA, S. RIZZO AND M. RIVA, *Boll. Soc. Ital. Biol. Sper.*, 45 (1968) 94.
42 F. A. PATTY, H. H. SCHRENK AND W. P. YANT, *Ind. Eng. Chem. Anal. Ed.*, 4 (1949) 259.
43 M. B. JACOBS, *The Analytical Chemistry of Industrial Poisons, Hazards and Solvents*, Interscience, New York, 2nd edn., 1949, p. 562.
44 M. OGATTA, Y. TAKATSUKA AND K. TOMOKUNI, *Brit. J. Ind. Med.*, 27 (1970) 378.
45 T. A. SETO AND M. O. SCHULTZE, *Anal. Chem. (Washington)*, 28 (1956) 1625.
46 G. GAZZANIGA. S. BINASCHI AND A. SPORTELLI, *Boll. Soc. Ital. Biol. Sper.*, 45 (1968) 97.
47 M. OGATA, K. SUGIYAMA AND Y. KURODA, *Okayama Igakkai Zasshi*, 74 (1962) 247.
48 M. OGATA AND K. TOMOKUNI, *Abst. 16th Congr. Occupational Health*, 1969, p. 98.
49 S. LAHAM, *Proc. Can. Physiol. Soc.*, 36 (1957) 72.
50 J. W. DANIEL, *Biochem. Pharmacol.*, 12 (1963) 795.
51 J. F. POWELL, *Brit. J. Ind. Med.*, 2 (1945) 142.
52 G. SCANSETTI, G. F. RUBINO AND G. TROMPEO, *Med. Lavoro*, 50 (1959) 743.
53 J. R. COOPER AND P. J. FRIEDMAN, *Biochem. Pharmacol.*, 1 (1958) 76.
54 M. OGATA, Y. TAKATSUKA AND K. TOMOKUNI, *Brit. J. Ind. Med.*, 28 (1971) 386.
55 M. IKEDA, H. OHTSUKI, H. KAWAI AND M. KUNIYOSHI, *Brit. J. Ind. Med.*, 28 (1971) 203.
56 E. A. BOETTNER AND F. C. DALLOS, *J. Am. Ind. Hyg. Assoc.*, 26 (1965) 289.
57 D. BERSADSCHI AND V. STEFAN, *Rev. Chim (Bucharest)*, 17 (1966) 309.
58 J. KURZ AND M. SCHUIERER, *Faerber-Ztg.*, 17 (1964) 328; *Chem. Abstr.*, 64 (1966) 11835.
59 R. D. STEWART, J. D. SWANK, C. B. ROBERTS AND H. C. DODD, *Nature*, 198 (1963) 696.
60 A. W. SCHAFFER AND H. R. HOYLE, *Am. Ind. Hyg. Assoc. J.*, 22 (1961) 93.
61 J. E. PETERSON, H. R. HOYLE AND E. J. SCHNEIDER, *Am. Ind. Hyg. Assoc. Quart.*, 17 (1956) 429.
62 L. A. ROBERTS, *J. Assoc. Offic. Anal. Chemists*, 51 (1968) 825.
63 R. E. SNYDER, *J. Chromatog. Sci.*, 9 (1971) 638.
64 L. D. HINSHAW, *J. Gas Chromatog.*, 4 (1966) 300.
65 F. H. REID AND W. R. HALPIN, *Am. Ind. Hyg. Assoc. J.*, 29 (1968) 391.
66 N. T. CASTELLUCCI AND P. R. EISAMAN, *J. Gas Chromatog.*, 6 (1968) 599.
67 V. SEDIVEK AND J. FLEK, *Collection Czech. Chem. Commun.*, 34 (1969) 1533.

Chapter 22

PHENOLS

Phenol has been derived from a number of sources including "natural" phenol obtained from coal tar. All syntheses of phenol utilize benzene as the raw material except for one method starting from toluene. The cumene process accounts for approx. 48.5% of U.S. production and basically involves the cleavage of cumene hydroperoxide into phenol and acetone, *viz.*

$$C_6H_5C(CH_3)_2COOH \rightarrow C_6H_5OH + (CH_3)_2CO$$

Cumene hydroperoxide is formed by first alkylating benzene with propylene, then oxidation with air. The only process for the manufacture of phenol by vapor-phase reactions is the Raschig–Hooker process. The reactions involved are the oxychlorination of benzene followed by hydrolysis of the chlorobenzene. This process accounts for approximately one-sixth of the U.S. capacity. Other processes include the sulfonation process (still in use where the location of the plant is in close proximity to paper manufacturing to provide a ready market for the sodium sulfite byproduct), hydrolysis of monochlorobenzene with aqueous sodium hydroxide, and toluene oxidation *via* benzoic acid. The total U.S. production of phenol production in 1966 was approx. 1.33 billion lb. of which 1.2 billion lb. were synthetic (*e.g.* cumene 685, Raschig–Hooker 240, sulfonation 220, Dow chlorination 220, and toluene oxidation 36 million lb., respectively). Estimates of worldwide production capacity indicate that the U.S. produces approximately half the world total. (Practically all synthetic phenol is used and sold at a purity in excess of 98%.) Commercial grades of phenol as a rule contain 90–92% or 80–82% of phenol with the remainder consisting of cresols and water ("liquified phenol").

The most important reaction of phenol is its condensation with formaldehyde in the manufacture of resins, accounting for the consumption of approx. 60% in the U.S. phenol usage. Other important uses of phenol include the formation of azo dyes, synthesis of caprolactam leading to nylon-6, preparation of a variety of products prepared *via* oxidation of phenol (*e.g.* benzenediols, benzenetriols, diphenyls), preparation of phenolic aldehydes, ethers, alkyl phenols, chlorophenols, cresols which are of importance as pharmaceutical, pesticidal, and organic intermediates. Phenol traces are contained in automobile exhaust and in tobacco smoke.

Human exposure in industry has been generally limited to accidental contact of phenol with the skin or to inhalation of phenol vapors. Prolonged oral or subcutaneous administration of phenol can cause damage to the lungs, liver, kidneys, heart, and G.I. tract[1,2]. The toxicity of phenol in laboratory animals[3-5] and its chronic effects in man[6] as well as its absorption, excretion, and metabolism[3,7,8] have been described.

Phenol in air has been largely determined by colorimetry as an azo dye[9-12], with 4-amino antipyrene[13,14], with 2,6-dibromoquinone-4-chlorimide[15], and with diphenylpicrylhydrazyl[16].

The GC determination of phenol and cresols in waste gases was studied by Naito et al.[17]. Stack gases containing phenol and cresols were passed into 0.1 N NaOH solution, the solution acidified with 5 N H_2SO_4 and extracted with ether. The ether layer was analyzed on a 2.1 m column packed with 30% dioctyl sebacate on Chromosorb WAW, and a hydrogen flame-ionization detector. The column temperature was 152°C, the carrier gas nitrogen, and isopropyl benzene was used as an internal standard for quantitative analysis. Phenol and cresols in waste gases were removed quantitatively when their concentrations in air were 50–1000 p.p.m. by washing with NaOH solution in a column packed with Rashig rings.

Spears[18] described the quantitative determination of phenol in cigarette smoke. Phenol was determined by a method involving solvent partition, steam distillation, and gas chromatography. The purity of the isolated phenol and cresols was demonstrated by spectrophotometry and the accuracy of the method was established by isotope dilution technique. The gas chromatography was carried out on a Perkin-Elmer Model 154-0 Vapor Fractometer equipped with a 1 mV recorder and approx. 8000 Ω thermistor. The detector voltage was 8 V and peak areas were determined by a disc integrator.

TABLE 1

PHENOLIC COMPOUNDS OF CIGARETTE SMOKE

Component	Retention time (min)	Major ultraviolet absorption maxima[a] (mμ)
2-Methoxyphenol	9.5	281.5, 275.5
2,6-Dimethylphenol	13.5	277.5, 272
2,4,6-Trimethylphenol	18.2	284, 279, 277
2-Methylphenol	21.2	278.5, 272
Phenol	23.8	278, 271.5, 265
2,4-Dimethylphenol	27.2	285, 279
2,5-Dimethylphenol	27.2	281, 275.5
4-Methylphenol	30.0	286, 279.5, 276.5, 273
3-Methylphenol	31.2	280, 273
3-Ethylphenol	42.0	279, 272
4-Ethylphenol	42.0	285, 278.5, 275, 272
4-Methoxyphenol	90	301, 294, 290
3-Methoxyphenol	108	280.5, 273.5
Internal standard (2-hydroxyacetophenone)	7.8	

[a] Solvent for all ultraviolet spectra was cyclohexane.

TABLE 2

BOILING POINTS[1], HEAT OF VAPORIZATION AT BOILING POINT[2], COLUMN[a] CAPACITY RATIO AND HETP
FOR THE METHYL AND POLYMETHYL PHENOLS

Peak no. (see figs.)		Abbreviation	T_B (°C)	ΔH_B (cal/g)	k'^a	H^a (cm)
1	Phenol	P	182	116	10.7	0.16
2	2,6-Dimethylphenol	2,6-DMP	210	92	12.0	
3	2-Methylphenol	2-MP	191	99	14	0.12
4	4-Methylphenol	4-MP	202	105	18	0.11
5	3-Methylphenol	3-MP	202	105	19.5	
6	2,4,6-Trimethylphenol	2,4,6-TMP	222		22.5	0.09
7	2,3,6-Trimethylphenol	2,3,6-TMP	226		24	
8	2,4-Dimethylphenol	2,4-DMP	210	91	25	
9	2,5-Dimethylphenol	2,5-DMP	210	91	26.5	0.10
10	2,3-Dimethylphenol	2,3-DMP	218	93	32	0.11
11	3,5-Dimethylphenol	3,5-DMP	219	99	37	0.09
12	3,4-Dimethylphenol	3,4-DMP	225	101	42	0.11
13	2,4,5-Trimethylphenol	2,4,5-TMP	232		52	
14	2,3,5-Trimethylphenol	2,3,5-TMP	233		55	
15	2,3,5,6-Tetramethylphenol	2,3,5,6-TeMP	247		56	
16	2,3,4,6-Tetramethylphenol	2,3,4,6-TeMP			58	
17	2,3,4-Trimethylphenol	2,3,4-TMP	236		63	
18	3,4,5-Trimethylphenol	3,4,5-TMP	248		83	
19	2,3,4,5-Tetramethylphenol	2,3,4,5-TeMP	266		134	
20	Pentamethylphenol	PMP	267		151	

[a] Phase: 1 part tri-2,4-xylenyl phosphate (TXP), 2 parts di-(3,3,5-trimethylcyclohexyl)-o-phthalate (TMCP). Temperature 115°C, outlet carrier gas velocity 56 cm/sec hydrogen, inlet pressure 2.5 kg/cm².

The gas chromatographic column was a 6 ft. × ¼ in. stainless steel column containing 8% UCON oil 50HB 2000 on 42–60 mesh firebrick containing 25% nylon by weight. The column was conditioned at 168°C for 24 h, operated at 168°C and helium flow of 78 ml/min with a sample which did not contain more than 0.5 mg of phenol. (The same column was in use 17 months and allowed approx. 200 chromatograms with little or no loss in resolution of the individual phenols.) Table 1 lists the phenolic compounds of cigarette smoke and their retention times on major ultraviolet absorption maxima.

Packed capillary columns, using silanized Chromosorb P coated with a mixture of di-(3,3,5-trimethyl cyclohexyl)-o-phthalate and tri-2,4-xylenylphosphate allowed the rapid analysis of the mixtures of phenol, methyl phenols, and polymethyl phenols resulting from the pyrolysis of phenol–formal polycondensates[19]. Although the practical upper temperature limit was 115°C, the high permeability and efficiency of the packed capillary columns allowed a complete analysis of 10 compounds (phenol, the cresols, and xylenols) in less than 15 min, and the analysis of 19 of the 20 compounds (phenol and all the methyl and polymethyl phenols) in 55 min.

The experiments were carried out with a gas chromatograph of conventional design constructed in the author's laboratory. It was fitted with a flame-ionization detector and an injection port with a splitting system. The injection port and the splitting system were heated to about 300°C to insure a sufficiently fast vaporization of the mixture. Pyrolysis was carried out with a platinum coil and heating of the injection port and splitting system at 250°C was then convenient because the phenols were produced in the vapor phase.

Table 2 lists the boiling points, heats of vaporization, column capacity ratio, and HETP for the methyl and polymethyl phenols studied. Table 3 lists the retention times relative to phenol of the methyl and dimethyl phenols using columns of tri-2,4-xylenyl phosphate (TXP) and di-(3,3,5-trimethylcyclohexyl)-o-phthalate (TMCP).

Figures 1–4 show the analysis of a mixture of phenol, the methyl and dimethyl phenols and 2,4,6-trimethyl phenol on the three columns used. The column made with the mixture of the two phases gave the best results. Only the resolution of 2,4- and 2,5-dimethyl phenol was less than 1 (ca. 0.80) and all other compounds could be quantitatively determined with good accuracy in less than 15 min. Figure 4 shows that on this column a mixture of all the polymethyl phenols could be resolved in less than 55 min with the exception of 2,3,5-trimethyl phenol and 2,3,5,6-tetra-methyl phenol.

Mono- and dihydric phenolic materials of natural and man-made origin constitute an important class of water pollutants. The gas chromatography of phenolics by aqueous injection combined with FID has been studied by Baker[20-25]. The advantages of direct aqueous injection GLC are that it eliminates pretreatment or extraction

TABLE 3

RETENTION TIMES OF METHYL AND DIMETHYLPHENOLS RELATIVE TO PHENOL

Compound	Relative retention time				
	TXP (115°C)	TMCP (115°C)	Calculated on the mixed phase[a]	Observed (115°C)	
				Column 1	Column 2
P	1.0	1.0	1.0	1.0	1.0
2,6-DMP	0.87	1.27	1.13	1.08	1.11
2-MP	1.17	1.27	1.25	1.25	1.24
4-MP	1.53	1.64	1.60	1.59	1.56
3-MP	1.63	1.74	1.70	1.70	1.64
2,4-DMP	1.87	2.09	2.02	2.01	2.13
2,5-DMP	1.90	2.24	2.12	2.08	2.22
2,3-DMP	2.39	2.70	2.58	2.54	2.61
3,5-DMP	2,66	3.06	2.92	2.89	3.03
3,4-DMP	3.04	3.42	3.28	3.23	3.47

[a] 1 part TXP–2 parts TMCP.

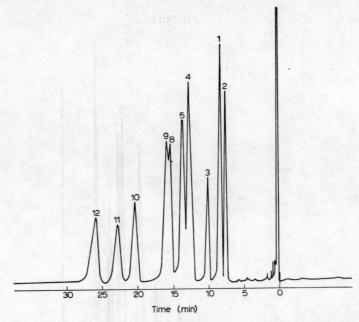

Fig. 1. Analysis of a mixture of phenol, methyl-, and dimethylphenols on a packed column, 5 m long, ilanized Chromosorb P 80–100 μm, tri-2,4-xylenyl phosphate, 115°C, carrier gas hydrogen, outlet elocity 120 cm/sec. Numbers refer to peaks named in Table 2.

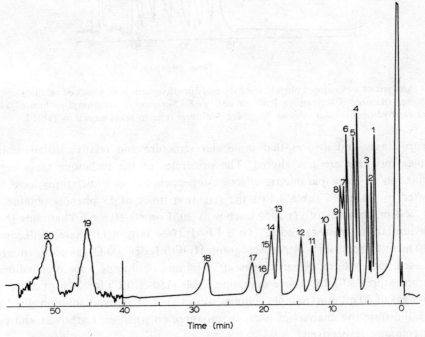

Fig. 2. Analysis of a mixture of phenol, methyl-, and dimethylphenols on a packed capillary column, 5 m long, silanized Chromosorb P 80–100 μm, mixed liquid phase, 115°C, carrier gas hydrogen, outlet velocity 65 cm/sec. Numbers refer to peaks named in Table 2.

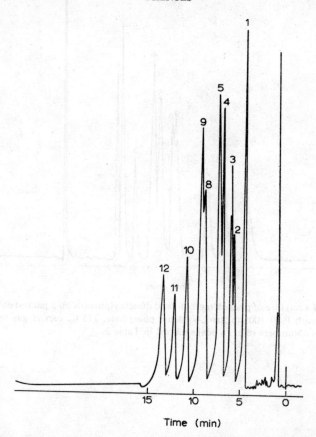

Fig. 3. Analysis of a mixture of phenol, methyl-, and dimethylphenols on a packed capillary column, 5 m long, silanized Chromosorb P 80–100 μm, di(3,3,5-trimethyl cyclohexyl)phthalate, 140°C, carrier gas hydrogen, outlet velocity 30 cm/sec. Numbers refer to reaks named in Table 2.

in many cases and assures that molecular structure and relative distribution of complex mixtures are not altered. The principles of the technique including an explanation of some parametric effects using phenolics as illustrations have been reported by Baker[22]. Table 4 lists the retention times of 17 phenols obtained on 10 ft. × ⅛ in. columns of (1) 20% Carbowax 20M on 60–80 mesh Diatomite (hexamethyldisilazone treated) and (2) 10% FFAP (free fatty acid phase, Wilkens) on 60–80 mesh Chromosorb T. An Aerograph Hy-Fi Model 600 D gas chromatograph was used with helium as carrier gas at 20 ml/min. Columns were preconditioned at temperatures 30–50°C above operating levels (180–210°C) for 24 h or more with carrier gas at 20 ml/min. Sample volumes were 1–3 μl of aqueous solution. Figures 5 and 6 illustrate the analysis of phenolic mixtures obtained on Carbowax and polyester columns, respectively.

Baker and Malo[26] recently examined various column substrate and support combinations and GLC operating parameters to determine the most suitable procedure for phenolic analyses by direct aqueous injection. The chromatographs used were a Varian Aerograph single-column Hy-Fi Model 600D equipped with a Model 328 isothermal temperature controller and a Varian Aerograph dual-column, linear temperature-programmed Model 204-1B each equipped with flame-ionization detectors. All columns were packed in 0.125 in. diam. stainless steel and preconditioned at 30–50°C in a Varian Model 550 chromatographic oven. Table 5 lists the retention and calibration values for phenol and 14 monohydric phenolic compounds using 10 substrates.

The actual elution interval of phenol corresponding to the data of Table 5 is shown in Table 6. The polyethylene glycols, Carbowaxes 1540, 4000 and 20M and their chemical modifications were determined to be among the best of the available substrates for aqueous phenolic analysis. FFAP is a reaction product between Carbowax 20M and 2-nitroterephthalic acid. Substrate loadings of 5–10% FFAP provided reasonably good separation efficiency and long column life. The combination of FFAP supported on Chromosorb T was found to be ideally suited to aqueous phenolic analyses at temperatures up to 200°C and provided maximum

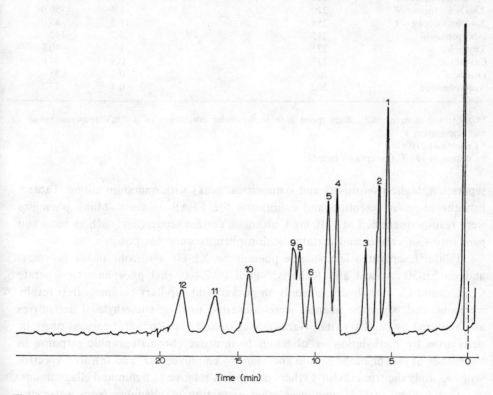

Fig. 4. Analysis of a mixture of phenol and all polymethylphenols. Same column as for Fig. 2 115°C, carrier gas velocity 62 cm/sec. Numbers refer to peaks named in Table 2.

TABLE 4

PHENOLIC RETENTION

Phenol	B.p. (°C)	Stainless steel 10 ft. × ⅛ in. Carbowax–Diatomite		Stainless steel 10 ft. × ⅛ in. Polyester–polyethylene	
		Relative retention	Calibration[a,c]	Relative retention	Calibration[a,c]
Phenol	182	1.0	45.2	1.0	28.8
o-Cresol	192	1.0	51.0	1.0	27.0
m-Cresol	203	1.3	51.9	1.3	30.5
p-Cresol	202	1.3	51.9	1.3	31.3
o-Chlorophenol	176	0.8	86.7	0.6	44.8
m-Chlorophenol	214	3.6	106.0	3.6	45.4[b]
p-Chlorophenol	217	3.6	125.0	3.6	46.5[c]
2,3-Dichlorophenol		1.8	76.5	1.9	46.8
2,4-Dichlorophenol	210	1.8	117.0	1.9	55.3
2,5-Dichlorophenol	210	1.8	133.0	1.9	50.4
2,6-Dichlorophenol	220	1.6	71.5	1.5	50.0
3,4-Dichlorophenol	254			11.5	43.0[b]
o-Nitrophenol	215			0.7	140.
Thymol	233			1.8	30.9
Carvacrol	237			0.9	94.0
Guaiacol	205			0.7	55.6
Acetophenone	202			0.4	27.2

[a] Calibration is in ng/in.^{2}, chart speed is 90 in./h, under conditions of a 1 mV response range 1 and attenuation 1.
[b] Column at 210°C.
[c] Column at 187°C (exceptions noted).

separation, highest sensitivity and symmetrical peaks with minimum tailing. Table 7 lists the phenolic retention and calibration for FFAP columns. Many phenolics were readily quantified at 1–10 mg/l, although certain separations such as *meta* and *para* forms of cresols and certain dichlorophenols were not possible.

Pillion[27] separated less volatile phenols on XE-60, a silicone nitrile polymer, and on SE-30. Adlard and Roberts[28] used tri-2,4-xylenyl phosphate to separate C_6, C_7, and C_8 monohydric phenols on packed and capillary columns. Best results were obtained when the phenols were converted to their trimethylsilyl derivatives and analyzed by capillary GLC. Narasimhachari and von Rudloff[29] formed phenolic derivatives by methylation or silylation to improve chromatographic response in their study of polyphenols. Nelson and Smith[30] employed GLC and infrared spectroscopy to study the trimethylsilyl ethers of phenols relative to lignin and ellagitannins.

Goren–Strul et al.[31] employed other extraction of phenolics from water and subsequent GLC analysis on the tri-2,4-xylenyl phosphate column.

TABLE 5

PHENOLIC RETENTION AND CALIBRATION VALUES

Compound	B.p. (°C)	15% STAP (70–80 W) r[a] / c	10% STAP (70–80 W) r / c	10% Carbowax 1540 (60–80 HMDS W) r / c	20% Carbowax 4000 TPA (60–80 HMDS W) r / c	20% Carbowax 20M TPA (60–80 A/W HMDS-W) r / c	20% Carbowax 20M TPA (60–80 W) r / c	20% Carbowax 20M (60–80 A/W W) r / c	15% UCON 50LB550X (60–80 HMDS W) r / c	15% UCON 50LB550X (60–80 DMCS-W) r / c
Length (ft.) of 0.125 in. diam. stainless steel column		10	5	10	10	10	10	10	5	5
Aerograph Model No.		600D/328	204-1B	600D/328	600D/328	600D/328	600D/328	600D/328	600D/328	204-1B
Recommended max. temp. (°C)		225	255	200	200	250	250	250	200	200
Phenol	182	1.0 / 43	1.0 / 44	1.0 / 49	1.0 / 32	1.0 / 45	1.0 / 29	1.0 / 48	1.0 / 45	1.0 / 45
o-Cresol	192	1.0 / 39	1.0 / 38	1.0 / 45	1.0 / 31	1.0 / 51	1.0 / 27	1.0 / 42	1.0 / 39	1.0
m-Cresol	203	1.3 / 43	1.3 / 38	1.3 / 49	1.3 / 37	1.3 / 52	1.3 / 28	1.3 / 45	1.3 / 45	1.3
p-Cresol	202			1.3 / 54	1.4 / 34	1.3 / 51	1.3 / 28	1.3 / 41	1.3 / 46	1.3
o-Chlorophenol	176	0.7 / 67	0.7	0.6 / 67	0.6 / 46	0.8 / 87	0.7 / 39	0.6 / 54	0.6 / 54	0.6
m-Chlorophenol	214	3.9 / 96	3.0	b		3.6 / 106	4.1 / 48	3.8 / 88		
p-Chlorophenol	217	3.9 / 88	3.9	b		3.6 / 125	4.1 / 50	3.8 / 84		
2,3-Dichlorophenol		1.8 / 85	1.9	2.1[c] / 138		1.8 / 77	1.8 / 53	1.8 / 76		
2,4-Dichlorophenol	210	1.8 / 89	1.8	2.0[c]		1.8 / 117	1.8 / 77	1.8 / 95	1.9	1.9
2,5-Dichlorophenol	210			2.1[c]		1.8 / 133	1.8 / 68		1.9	1.9
2,6-Dichlorophenol	220	1.5 / 60	1.5	1.5[c]		1.6 / 72	1.5 / 52	1.5 / 83	1.5	1.5
3,4-Dichlorophenol	254			b						
o-Nitrophenol	215	b	b	0.5 / 148						
Thymol	233	b	b	1.9[c]			1.8 / 25			
Guaiacol	205	0.7 / 56	0.6 / 56	0.6 / 92	0.7 / 49	0.7 / 44	0.7 / 44	0.7 / 66	0.8 / 69	0.8

[a] r = relative retention, c = calibration, ng/in.² at 90 in./chart speed, 1 mV response with range 1, attenuation 1.
b No peak.
c Smear estimated.

References pp. 516–517

Crouse *et al.*[32] employed two columns, a polar, polyethylene glycol succinate (DEGS) and a nonpolar Apiezon N column to determine various phenols contained in steam-distilled cigarette smoke volatiles.

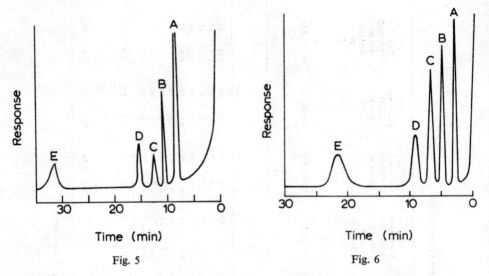

Fig. 5   Fig. 6

Fig. 5. Phenolic analyses. Analyses were made by aerograph with 10 in. × ⅛ in. stainless steel column coated with 20% carbowax–terephthalic acid on 60–80 mesh diatomite, HMDS treated. Column temperature was 210°C and injection temperature 250°C. Hydrogen and helium flow rates were each 20 ml/min at electrometer range 1 and attenuation 1, with chart speed at 12 in./h, 1 mV full scale response, and 1 μl sample of approximately 100 mg/l solutions of each phenolic. A, *o*-Chlorophenol; B, phenol; C, *m*-cresol; D, 2,4-dichlorophenol; E, *p*-chlorophenol.

Fig. 6. Chromatographic analysis of phenolics in aqueous solution. Analyses were made by aerograph with 10 in. × ⅛ in. stainless steel column coated with 10% polyester on 60–80 mesh fluorocarbon resin medium. Column temperature was 188°C and injection temperature 250°C; flow rate for nitrogen was 60 ml/min, for hydrogen 30 ml/min, at electrometer range 1 and attenuation 1, with chart speed at 12 in./h, 1 mV full scale response, and 1 μl sample of approximately 100 mg/l solutions of each phenolic. A, *o*-Chlorophenol; B, phenol; C, *m*-cresol; D, 2,4-dichlorophenol; E, *p*-chlorophenol.

Nonaka[33] has studied the utility of steam as the carrier gas for the gas–solid chromatography of organic compounds. This technique, called Steam–Solid Chromatograph, ssc, has permitted the analysis of samples which are liquid and solid at room temperature and also samples of extremely dilute aqueous solutions[34–37]. The distinctive feature of ssc is that it permits direct analysis of aqueous solutions of organic substances without any pretreatment such as extraction, concentration or derivatization since the chromatographic system is filled with water vapor before injecting the aqueous samples. Steam carrier gas can be used in combination with ordinary adsorbents as a stationary phase and detection of the components is possible even for extremely dilute aqueous samples with a hydrogen flame-ionization detector which functions in the steam carrier gas as well as in the ordinary carrier gas.

An Ohkura 2000 gas chromatograph was modified by the addition of a steam generator and a cupric oxide column as shown schematically in Fig. 7. The steam generator was a stainless steel boiler having dimensions of 45 mm i.d. and 90 mm in height (120 ml in capacity). This boiler was capable of being adjusted to various steam temperatures within a range from 120 to 140°C and of keeping the temperature at a constant level with an accuracy of better than ±0.1°C. (The pressure of the steam was also maintained at 2 and 3.5 atm.) Before being introduced into the analytical column, the steam passed through a heated 50 cm × 5 mm i.d. column packed with CuO pellets in order to remove organic impurities. The CuO column was kept at 500–600°C during the operation. The analytical columns generally used were 1.5–4.5 m long × 1.8–4.0 mm i.d. borosilicate glass tubes for samples other than amines (which utilized aluminum tubes). The glass tubes were washed with mineral acids and the aluminum tubes were etched by a 5% aqueous solution of potassium hydroxide.

TABLE 6

REPRESENTATIVE PHENOL ELUTION INTERVALS

Column[a]	Temp. (°C)	Carrier flow (ml/min)	Time (min)
10 ft. 15% STAP on 70–80 W	190	20	10
5 ft. 10% STAP on 70–80 T	190	20	6.5
5 ft. 10% STAP on 60–80 W	190	20	6.5
10 ft. 10% Carbowax 1540 on 60–80 W-HMDS	170	40	18.5
10 ft. 20% Carbowax 4000-TPA on 60–80 W-HMDS	170	40	12.5
10 ft. 20% Carbowax 20M-TPA on A/W 60–80 W-HMDS	210	20	8.0
10 ft. 20% Carbowax 20M-TPA on 60–80 W	210	20	10.9
10 ft. 20% Carbowax 20M on A/W 60–80 W	210	20	18.6
5 ft. 15% UCON 50LB550X on 60–80 W-HMDS or DMCS	170	20	7.2
10 ft. 10% FFAP on 60–80 T	170	60	7.5
5 ft. 10% FFAP on 60–70 T	170	30	7.5
5 ft. 5% FFAP on A/W 70–80 W-DMCS	150	40	8.0
5 ft. 10% FFAP on A/W 60–80 W-DMCS	170	20	6.2
5 ft. 2% FFAP on A/W 60–80 W	170	20	3.0
5 ft. 5% FFAP on A/W 70–80 W	150	40	8.0
5 ft. 10% FFAP on A/W 60–80 W	170	20	8.5
5 ft. 10% FFAP on 60–80 W	190	20	4.8
10 ft. 10% FFAP on 60–80 W	170	20	18.5
5 ft. 5% FFAP on JM porous polymer	190	20	12

[a] All columns 0.125 in. diam. stainless steel with Chromosorb supports.

TABLE 7

PHENOLIC RETENTION AND CALIBRATION FOR FFAP COLUMNS

Compound	B.p. (°C)	10% 60-80 T		10% 60-80 T		10% 60-80 T		2% 60-80 A/W W		5% 70-80 A/W W		10% 60-80 A/W W-DMCS		5% 70-80 A/W W-DMCS		10% 60-80 A/W W-DMCS	
		10		5		10		5		5		5		5		5	
		600D/328		204-1B		600D/328		600D/328		600D/328		600D/328		600D/328		600D/328	
		200		200		275		275		275		275		275		275	
		r	c	r	c	r	c	r	c	r	c	r	c	r	c	r	c
Phenol	182	1.0	29	1.0	18	1.0	42	1.0	65	1.0	46	1.0	33	1.0	47	1.0	37
o-Cresol	192	1.0	27	1.0	18	1.0	43	1.0	60	1.0	53	1.0	32	1.0	57	1.0	32
m-Cresol	203	1.3	30	1.4	19	1.3	40	1.3	75	1.3	53	1.3	36	1.4	46	1.4	37
p-Cresol	202	1.3	31	1.4	19	1.3	40	1.3	75	1.3	53	1.3	36	1.4	49	1.4	37
o-Chlorophenol		0.6	45	0.6	23	0.6	62	0.6	104			0.6	65			0.6	54
2,3-Dichlorophenol	210	1.9	47	2.0	39	1.9	79	2.0	140			1.8	65			1.9	67
2,4-Dichlorophenol	220	1.9	55	2.0	38	1.9	98	2.0	137			1.8	68			1.9	76
2,6-Dichlorophenol		1.5	50	1.5	38	1.5	78	1.4	130			1.4	69			1.5	74
Guaiacol	205	0.7	56	0.7	26	0.7	55	0.7	95			0.6	44			0.7	45

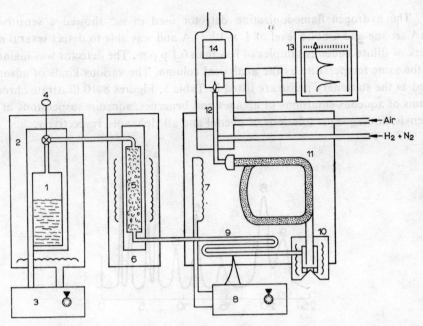

Fig. 7. Schematic diagram of ssc. (1) Steam boiler, (2) electric furnace for the boiler, (3) thermoregulator, (4) needle valve, (5) CuO column, (6) furnace for the CuO column, (7) regulated air bath, (8) thermoregulator for the air bath, (9) stream buffer for carrier steam, (10) injection port, (11) analytical column, (12) hydrogen flame-ionization detector, (13) d.c. amplifier and chromatogram recorder, (14) dew point hygrometer for measuring the steam flow rate.

TABLE 8

ADSORBENTS USED IN THE PRESENT STUDY OF SSC

Samples	Adsorbents
Low-boiling hydrocarbons	Activated alumina[a]
High-boiling hydrocarbons	Diatomaceous firebrick[b,c]
Monohydric alcohols	Diatomaceous firebrick[b,c] modified by HF
Polyhydric alcohols	Diatomaceous firebrick[b] modified by KF or KHF$_2$
Aldehydes, ketones, ethers, and esters	Diatomaceous firebrick[b,c] modified by HF
Carboxylic acids and phenols	Diatomaceous firebrick[b,c] baked after addition of H$_3$PO$_4$
Amines	Sintered magnesia[d]

[a] Gas chromatographic use (specific surface area, 400 m^2/g; Gasukuro Kôgyo Co., Tokyo).
[b] For example Isolite LBK-28 (SiO$_2$ 50%, Al$_2$O$_3$ 45%, CaO 0.7%, bulk density 0.80, Isolite Kôgyo Co., Osaka).
[c] Chromosorb P AW (Johns-Manville Products Corp., New York) can be used in place of firebrick-Isolite.
[d] M-1 (electrically sintered; MgO 96%, bulk density 2.4, Nippon Kagaku Kôgyo Co., Osaka).

The hydrogen flame-ionization detector used in ssc showed a sensitivity of $7 \mu A$ sec/mg and a noise level of 1×10^{-13} A and was able to detect several microliters of dilute aqueous samples of less than 0.1 p.p.m. The detector was maintained at the same temperature as the analytical column. The various kinds of adsorbents used as the stationary phase are listed in Table 8. Figures 8–10 illustrate chromatograms of aqueous emulsions of alkanes and benzenes, aqueous suspensions of polyphenyls, and aqueous solutions of phenol and alkylphenols, respectively.

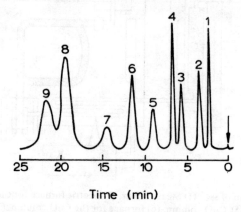

Time (min)

Fig. 8. Chromatogram of an aqueous emulsion of alkanes and benzene derivatives. Column, aluminum tube of 3 m × 2 mm, activated alumina, 40–60 mesh; column temperature, 125°C; carrier gas, steam 12 ml/min; detector, hydrogen flame ionization. 1, n-Pentane; 2, n-hexane; 3, n-heptan; 4, benzene; 5, n-octane; 6, toluene; 7, n-nonane; 8, p- and m-xylene; 9, o-xylene.

Lower alkylamines, pyridine, aniline, and toluidine were also examined in the form of 0.1–0.01% aqueous solutions. The aromatic amines were separated on a magnesia column at 101°C while the alkylamines showed insufficient separation and fairly marked tailing.

The microdetermination of phenol by mono- and polychloroacetylation followed by microcoulometric gas chromatography was studied by Chin et al.[38]. A Dohrmann Model 200 MCGC equipped with a Vycor tube in the injection port was used. The fractionating column was a 4 ft. × ¼ in. o.d. aluminum tube packed with 8.69% ethylene glycol adipate on 70–80 mesh Anakrom ABS. The titrating cell was Model T-300-S. The inlet, column and furnace temperatures were 200, 130, and 800°C, respectively. Dry nitrogen was the carrier gas at the rate of 11 on the MicroTek flow meter and 1.5 on the sweep flow meter steel ball. The oxygen rate was 30 on the oxygen flow meter steel ball and all the analyses were conducted at a setting of 30 Ω. Acetylations were made with mono-, di-, and trichloroacetyl chlorides and the base solution used was 0.05 N NaOH. Because of the high chlorine content of trichloroacetyl chloride, the sensitivity was much greater than when mono- and dichloroacetyl chlorides were used (Fig. 11). The rate of mono- and polychloroacetylation by monochloroacetyl chloride was the fastest and by trichloroacetyl chloride the slowest.

Cohen *et al.*[39] studied the trace determination of phenols as their 2,4-dinitro-phenyl ethers by gas chromatography. The ethers of some 40 phenols were prepared by the reaction of 1-fluoro-2,4-dinitrobenzene with the requisite phenol according to the procedure of Reinheimer *et al.*[40]. Table 9 shows the retention time data and sensitivity to electron-capture detection of the 2,4-dinitrophenyl ethers on a silicone GE XE-60 column. Linear calibration curves were obtained over the range 1–10 ng. The elution sequence of 2,4-dinitrophenyl ethers was influenced by the nature of the phenolic portion of the molecule and under the conditions specified spanned a sufficiently wide time range to be of value in phenol characterization. The *ortho*,

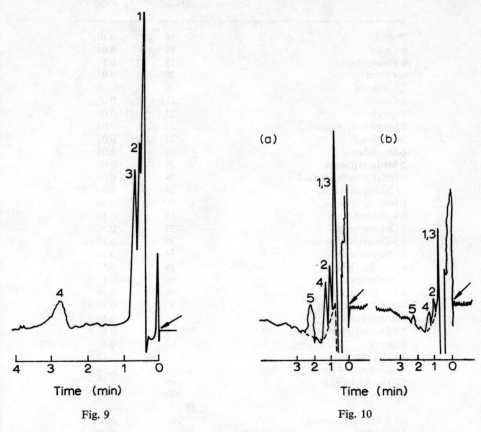

Fig. 9

Fig. 10

Fig. 9. Chromatogram of a 0.01% aqueous suspension of polyphenyls. Column, aluminum tube of 2 m × 4 mm, chromosorb P AW, 30–60 mesh, column temperature, 260°C; carrier gas steam, 45 ml/min; detector, hydrogen flame ionization. 1, *o*-Terphenyl; 2, *p*-terphenyl; 3, triphenylene; 4, *p*-quarterphenyl.

Fig. 10. Chromatograms of 1.0 p.p.m. (a) and 0.3 p.p.m. (b) aqueous solution of alkylphenols. Column, glass tube of 3 m × 3.7 mm, chromosorb P AW baked after addition of 3% H_3PO_4, 30–60 mesh; column temperature, 144°C; carrier gas, steam 46 ml/min; detector, hydrogen flame ionization; sample size, 5 μl. 1, Phenol; 2, *o*-cresol; 3,2,4-dimethylphenol; 4, *p*-ethylphenol; 5, *p*-tert-butylphenol; dotted curve, water only.

TABLE 9

RETENTION TIME DATA AND ELECTRON-CAPTURE SENSITIVITY OF 2,4-DINITROPHENYL
DERIVATIVES OF PHENOLS

Conditions: 1% GE XE-60 and 0.1% Epikote 1001 on Chromosorb G, acid-washed, dimethyl-chlorosilane coated, 60–80 mesh. Temperature: 215°. Column material: glass 140 cm in length

Parent phenol	Retention time relative to 1-naphthol derivative	Sensitivity[a] to electron-capture detection $(g \times 10^{-9})$
Phenol	16	0.10
o-Cresol	17	0.05
4-Fluorophenol	19	0.10
m-Cresol	20	0.05
2,5-Xylenol	20	0.05
p-Cresol	21	0.05
Thymol	23	0.05
2-Isopropoxyphenol	23	0.05
3,5-Xylenol	23	0.05
4-Ethylphenol	28	0.05
2-Methoxyphenol	29	0.05
3-Ethyl-5-methylphenol	29	0.05
2-Chlorophenol	29	0.05
4-Isopropylphenol	31	0.10
3,4-Xylenol	31	0.10
3-Chlorophenol	33	0.10
3,5-Di-tert.-butylphenol	36	0.10
4-Chlorophenol	38	0.10
4-tert.-Butylphenol	38	0.10
4-Allylphenol	38	0.1
4-sec.-Butylphenol	40	0.10
4-tert.-Pentylphenol	50	0.10
4-Bromophenol	56	0.10
2,4-Dichlorophenol	58	0.10
Eugenol	62	0.10
4-Iodophenol	92	0.10
3,5-Dimethyl-4-methylthiophenol	94	0.10
[a]ftert.-Octylphenol	95	0.20
4-Naphthol	100 (8.2 min)	0.20
2,4,5-Trichlorophenol	102	0.20
2-Naphthol	141	0.20
4-Cyclohexylphenol	153	0.30
2-Nitrophenol	168	0.50
3-Nitrophenol	178	0.50
4-Nitrophenol	238	0.50
4-Benzylphenol	285	0.40

[a] Sensitivity expressed as the weight of derivative producing a peak with height equivalent to 10% full-scale deflection at an amplification producing a noise level of 5% f.s.d.

meta, para sequence of elution (the "ortho effect" normally associated with the GC of free isomeric phenols) was also displayed by their 2,4-dinitrophenyl ethers although the separation of the isomeric ethers was poorer than the corresponding phenols. The electron-capturing properties of the derivatives were associated with the presence of the aromatic nitrogens and were suggested to compare favorably with phenol derivatives prepared with chloracetic anhydride[41] or α-bromo-2,3,4,5,6-pentafluoro-toluene[42].

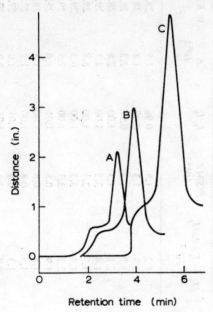

Fig. 11. Chromatograms of microcoulometric gas chromatography (MCGC) of the mono- and poly-chlorophenyl acetates. A, $C_6H_5OCOCH_2Cl$; B, $C_6H_5OCOCHCl_2$; C, $C_6H_5OCOCCl_3$.

Pettitt *et al.*[43] studied the temperature dependency on the sensitivity of electron-absorbing derivatives of amines, alcohols, phenols, and anilines. The mode of the capture process, *i.e.* dissociated or non-dissociative, was determined by plotting ln $KT^{3/2}$ *vs.* $1/T$ where K is the capture coefficient. It was shown that electron-absorbing derivatives and operating conditions could be preselected to yield optimum resolution and sensitivity. The anhydrides chosen for derivative formation were the chloroacetyl, chlorodifluoracetyl, pentafluoropropionyl, and heptafluorobutyryl because the resultant derivative could be made in macroquantities and chromatographed under mild conditions.

The electron-capture detector was of the design of Simmonds *et al.*[44] and incorporated a 15 mCi [63]Ni source plated on gold foil (New England Nuclear Corp.). The detector was mounted in tandem with a Model 5D-6 Statham precision oven. Ionization currents were measured with a Cary Model 31 vibrating reed electrometer (Applied Physics Corp.). The signal from the electrometer was monitored with a

TABLE 10

PROPERTIES OF PHENOL DERIVATIVES

Derivative	Molarity ($\times 10^{-5}$)	Detector temperature (°C)	$\int (I_b - I_e)\, dx/I_e$ ($\times 10^{-2}$)	F/SMZ ($\times 10^{10}$)	I_o/I_b	K ($\times 10^8$)	$\ln KT^{3/2}$	Capture mechanism
Pentafluoropropionyl	1.0	350	1.2	1.3	0.96	1.5	28.47	Non-dissociative
	1.0	300	1.4	1.3	0.97	1.8	28.56	
	1.0	250	1.7	1.3	0.97	2.2	28.58	
	1.0	200	2.1	1.3	1.00	2.8	28.66	
Heptafluorobutyryl	0.5	350	1.4	3.3	1.00	4.7	29.60	Non-dissociative
	0.5	300	1.7	3.3	1.00	5.8	29.67	
	0.5	250	1.9	3.3	1.00	6.3	29.61	
	0.5	200	2.2	3.3	1.00	7.5	29.63	
Chloroacetyl	5.0	350	1.9	0.8	1.01	1.6	28.48	Dissociative
	5.0	300	1.6	0.8	1.00	1.3	28.19	
	5.0	250	1.5	0.8	1.00	1.2	27.97	
	5.0	200	1.3	0.8	1.01	1.0	27.66	
Chlorodifluoroacetyl	1.0	350	0.9	3.3	1.08	3.2	29.20	Dissociative
	1.0	300	0.83	3.3	1.08	3.0	29.00	
	1.0	250	0.82	3.3	1.10	3.0	28.86	
	1.0	200	0.11	3.3	1.09	3.9	28.99	

TABLE 11

PROPERTIES OF ANILINE DERIVATIVES

Derivative	Molarity ($\times 10^{-5}$)	Detector temperature (°C)	$\int (I_b - I_e)\,dx/I_e$ ($\times 10^{-2}$)	F/SMZ ($\times 10^{10}$)	I_o/I_b	K ($\times 10^{8}$)	$\ln KT^{3/2}$	Capture mechanism
Pentafluoropropionyl	1.0	350	6.0	1.1	1.20	8.2	30.14	Non-dissociative
	1.0	300	7.1	1.1	1.20	9.6	30.17	
	1.0	250	8.3	1.1	1.23	12	30.22	
	1.0	200	9.5	1.1	1.24	13	30.21	
Heptafluorobutyryl	1.0	350	7.0	1.1	1.20	9.5	30.28	Non-dissociative
	1.0	300	7.7	1.1	1.23	10	30.25	
	1.0	250	8.9	1.1	1.23	12	30.29	
	1.0	200	10	1.1	1.24	14	30.28	
Chloroacetyl	10	350	4.5	0.15	1.07	0.73	27.72	Dissociative
	10	300	4.5	0.15	1.03	0.71	27.58	
	10	250	4.4	0.15	1.04	0.71	27.43	
	10	200	4.3	0.15	1.03	0.68	27.23	
Chlorodifluoroacetyl	1.0	350	9.4	3.3	1.01	3.2	29.19	Non-dissociative
	1.0	300	1.3	3.3	1.01	4.5	29.42	
	1.0	250	1.9	3.3	1.04	6.7	29.67	
	1.0	200	2.5	3.3	1.02	8.5	29.76	

Leeds and Northrup Model G potentiometer recorder. A Datapulse Model 102 pulse generator provided the polarizing potential for the EC detector. The settings used were amplitude 30 V, width 1.0 μsec, and pulse interval 100 μsec. The carrier gas was a mixture of 10% methane and 90% argon which was passed through a 5X molecular sieve trap prior to entering the chromatographic column. Three columns were used: (A) 100 ft. × 0.03 in. i.d. stainless steel capillary coated with 10% STAP, (B) 200 ft. × 0.02 in. i.d. stainless steel capillary coated with 10% Apiezon L, and (C) 6 ft. × 2.5 mm i.d. glass coiled column packed with 10% DC-200 on 100 to 120 mesh Gas Chrom G.

All compounds were initially examined by FID to establish purities and relative retention times. Before each analysis the electron-capture detector was allowed to purge overnight at 350°C. The temperature dependency of the compounds was observed over the range 350–200°C, decreasing at 50° intervals. The majority of the derivatives examined capture dissociatively, thus the maximum response occurred at 350°C. Most of the phenol and aniline derivatives were acceptions, only the chlorodifluoroacetate of phenol and the chloroacetyls of both aniline and phenol captured dissociatively. All derivatives of phenol and aniline, as expected, had correspondingly greater responses than their saturated analogs due to the enhanced effect of conjugation with the aromatic ring. The properties of phenol and aniline derivatives are shown in Tables 10 and 11, respectively.

The temperature dependency of the capture coefficient, K, for these compounds was determined by plotting $\ln KT^{3/2}$ vs. $1/T$. This results in a linear function and the plot will have a positive slope if the electron attachment is non-dissociative; a negative slope is dissociative. The capture coefficient is related to the detector response by

$$S(I_b - I_e)/I_e\, dx = KSMZ/F$$

as derived by Wentworth et al.[45,46] where I_b = standing current, I_e = decreased current from the introduction of a capturing species into the detector, S = chart speed in in./min, M = molar concentration of injected solution, Z = sample size, and F = gas flow in l/min. To correct for column bleed, a correction factor of I_0/I_b, where I_0 = standing current when pure carrier gas is passed through the cell, was utilized to adjust K.

The microdetermination of pentafluorobenzyl derivatives of phenols and mercaptans by electron-capture gas chromatography was described by Kawahara[42]. A Perkin-Elmer Model 811 with tritium foil electron-capture detector of parallel plate design was used. An 8 ft. × ¼ in. aluminum column containing 50–50% mixture of 5% F.S. 1265 and 3% DC-200 on Chromosorb P (60–80 mesh) was employed at 200°C, with nitrogen flow at 40 ml/min. The sensitivity was set at an attenuation of ×50 and ×100 while the power supply was operated on 10 V d.c.

Phenols and mercaptans were converted to their respective derivatives by reaction with α-bromo-2,3,4,5,6-pentafluorotoluene (Aldrich) by the procedure of

TABLE 12

PROPERTIES OF ETHERS AND THIOETHERS

Pentafluorobenzyl aryl (or alkylthio) ether	M.p. (°C)	R.I. (n_D^{240})	Yield (%)	Retention time (min)	Relative retention (to phenyl ether)
tert.-Butylthio		1.4744	99	1.61	0.67
n-Butylthio		1.4676	98	1.85	0.77
Phenyl	75–76		96	2.40	1.00
2-Tolyl	78–79		88	2.72	1.13
4-Tolyl	74–75		85	2.91	1.21
2,4-Xylyl	50–52		94	3.27	1.36
2-Chlorophenyl	91–93		99	3.74	1.56
4-Chlorophenyl	66–67		94	4.11	1.72
2,4-Dichlorophenyl	73–74		86	5.91	2.46
n-Decylthio		1.4662	86	7.36	3.06
n-Undecylthio		1.4665	100	9.68	4.03
Naphthyl-2	132–133		84	10.46	4.35
n-Dodecylthio		1.4668	96	12.91	5.37
2-Nitro, 4-chlorophenyl	99–99.5		92	16.26	6.76

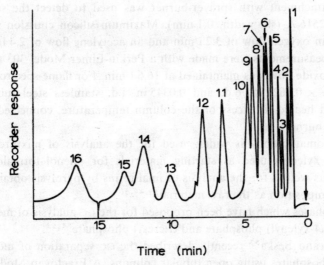

Fig. 12. Separation of various phenols and mercaptans as their pentafluorobenzyl ethers and thioethers, respectively. Equal quantities by weight were use for the 3rd through 11th and 16th compounds. The 13th through 15th contain twice the quantity while the 12th contains three times the quantity. 1, Solvent; 2, α-bromo-2,3,4,5,6-pentafluorotoluene; 3, 2-methyl-2-propanethiol; 4, butanethiol-1; 5, phenol; 6, o-cresol; 7, p-cresol; 8, 2,4-xylenol; 9, o-chlorophenol; 10, p-chlorophenol; 11, 2,4-dichlorophenol; 12, n-decanethiol-1; 13, n-undecanethiol-1; 14, β-naphthol; 15, n-dodecanethiol-1; 16, 4-chloro-2-nitrophenol.

Claisen et al.[47]. Table 12 lists the melting points, refractive indices and retention times of these ether and thioether derivatives. Figure 12 shows the separation of 14 pentafluorobenzyl ethers and thioethers of phenols and mercaptans, respectively, under conditions as described above. These derivatives provided the advantages of excellent stability in aqueous solution and excellent response to electron-capture detection and specificity in the presence of impurities and hence were suggested particularly for the GC analysis of phenols and mercaptans from water or sewage samples.

A silica-specific detector based on interfacing gas chromatograph and a flame emission or atomic absorption spectrometer was described by Morrow et al.[48]. Either the emission of the silicon line at 2516 Å from an oxygen–acetylene flame or the absorption of this line in a nitrous oxide–acetylene flame could be used. Mixtures of silated alcohols and phenols were identified in the presence of the non-silylated species. The silicone flame detector possesses another advantage in that it permits the estimation of the number of silicone atoms per molecule of silylated compound through a comparison of peak areas obtained by flame emission or atomic absorption and thermal conductivity detection.

A Beckman Model GC-2A gas chromatograph was equipped with a 6 ft. × $\frac{1}{4}$ in. o.d. stainless steel column packed with 20 % SE-30 on 30–60 mesh Chromosorb W. The thermal conductivity cells were operated at 200 mA. A Beckman Model DU spectrometer equipped with a 1P28 multiplier phototube (operated at 67 V/dynode) and flame attachment with sprayer-burner was used to detect the silicon atomic emission at 2516 Å (slit width, 0.1 mm). Maximum silicon emission intensity was achieved at an oxygen flow of 3.2 l/min and an acetylene flow of 2.4 l/min. Atomic absorption measurements were made with a Perkin-Elmer Model 303 spectrometer. The nitrous oxide flow was maintained at 16.6 l/min. For flame-spectrometric detection a 14 in. × 0.0625 in. o.d. and 0.0345 in. i.d. stainless steel tube, thermally insulated and heated in excess of the column temperature, connected the column outlet to the burner.

Gas chromatography is widely used for the analysis of mixtures of phenol, cresols, and xylenols used as starting materials for phenol–formaldehyde poly-condensates as well as for the analysis of final resins by pyrolysis obtaining phenols of related composition as the raw material[19,49,50].

Liquid phases which have been proposed for the GC analysis of methyl phenols include tri-(2,4-xylenyl) phosphate and tricresyl phosphate[51,52].

Hrivnak and Beska[53] recently described the GC separation of methyl phenols on tricresylphosphates using open tubular columns. A Fractovap Model G1 (Carlo Erba) gas chromatograph was equipped with FID and used with 20 m × 0.01 in. i.d. stainless steel open tubular columns. Nitrogen was the carrier gas. Table 13 lists the net retention times (relative to 3-methylphenol) of phenol, cresols, and xylenols at 120° using columns of tri-ortho- (TOCP), tri-meta- (TMCP), and tri-para-cresylphosphate (TPCP) and shows that the column with TPCP afforded the best separation for closely related compounds. A chromatogram of a mixture of phenol, cresols, and xylenols using TPCP as the liquid phase is illustrated in Fig. 13. The high efficiency of the open tubular column and the good selectivity of the liquid phase used allowed a complete separation of the 10 compounds in less than 10 min. This column had a good thermal stability up to 135°, and the undesirable effect of the metallic wall of the column usually resulting in tailing was eliminated by addition of phosphoric acid to the liquid phase[54].

Aly[55] described a TLC method for the separation and identification of phenols in waters as their antipyryl or *p*-nitrophenylazo dyes. The antipyryl dyes were prepared by the condensation of phenols with 4-amino-antipyrine in the presence of

TABLE 13

THE NET RETENTION RELATIVE TIMES[a] OF PHENOL, CRESOLS, AND XYLENOLS AT 120°

No.	Compound	TOCP	TMCP	TPCP	TPP
1	2,6-Dimethylphenol	0.57	0.54	0.55	0.56
2	Phenol	0.59	0.60	0.58	0.61
3	2-Methylphenol	0.71	0.71	0.69	0.69
4	4-Methylphenol	0.93	0.93	0.92	0.94
5	3-Methylphenol	1.00	1.00	1.00	1.00
6	2,4-Dimethylphenol	1.11	1.10	1.10	1.07
7	2,5-Dimethylphenol	1.16	1.15	1.16	1.11
8	2,3-Dimethylphenol	1.46	1.53	1.45	1.41
9	3,5-Dimethylphenol	1.68	1.65	1.71	1.65
10	3,4-Dimethylphenol	1.92	1.92	1.94	1.92

[a] Relative to 3-methylphenol.

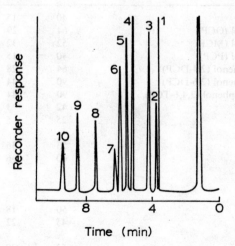

Fig. 13. Gas chromatogram of phenol and methylphenols on TPCP liquid phase at 130°. Numbers refer to peaks named in Table 13.

alkaline oxidizing agent. The extracted dyes were separated on neutral silica gel plates. The azo dyes were prepared by coupling of phenols and p-nitrobenzene diazonium fluoborate and chromatographed on basic silica gel plates. A mixture of phenol, o-, m-, p-chlorophenol, 2,4-, and 2,6-dichlorophenol was separated in a single run by this method and the procedure was successfully applied to the separation and identification of chlorination products of phenols in waters.

Table 14 lists the $R_F \times 100$ values for 4-amino-antipyrine derivatives of some phenols on neutral silica gel plates developed with 3 solvent systems: (1) methylene chloride–acetyl acetone (70:30), (2) chloroform–ethyl acetate (80:20), and (3) ethyl acetate–hexane–acetic acid (70:20:10). Table 15 lists the R_F values of p-nitrophenyl-azo dyes formed from different phenols on basic silica gel plates and developed with chloroform–acetone (9:1), and Table 16 illustrates the TLC identification of chlorination products of phenol, 2-chloro, and 4-chlorophenols as antipyryl, and p-nitrophenyl-azo dyes.

In waste water, the phenol content is usually determined by spectrophotometry after coupling with diazotized p-nitroaniline. Since every phenol coupling product has a different extraction coefficient, it is generally the practice to separate the phenols

TABLE 14

$R_F \times 100$ VALUES FOR 4-AMINOANTIPYRINE DERIVATIVES OF SOME PHENOLS
ON NEUTRAL SILICA GEL PLATES
Quantities applied: 1–2 µg of phenol as derivative

Compound	R_F ($\times$ 100) in solvent system		
	1	2	3
Phenol	40	15	35
2-Chlorophenol (OCP)	64	29	45
3-Chlorophenol (MCP)	52	32	52
4-Chlorophenol (PCP)	40	15	35
2,4-Dichlorophenol (2,4-DCP)	64	28	45
2,6-Dichlorophenol (2,6-DCP)	90	44	53
2,4,6-Trichlorophenol (2,4,6-TCP)	90	44	53
2-Nitrophenol	42	3	49
3-Nitrophenol	25	5	40
4-Nitrophenol			
1-Naphthol	53	36	56
2-Naphthol	60	66	80
Resorcinol			
Catechol			
Hydroquinone			
2-Cresol	50	18	43
3-Cresol	45	22	45
4-Cresol			
2,5-Xylenol	53	30	54
2,6-Xylenol	47	21	44

and determine their quantitites separately. Heier[56] separated a number of phenols by TLC on a slurry of 2.8g Silica Gel G in 5.6 ml of a solution of 10 g $Na_2CO_3 \cdot 10\ H_2O$ in 90 ml water. The developing solvent was toluene–ethanol–acetone (60:6:0.5) and freshly diazotized p-nitroaniline was the detecting reagent. The phenols were identified by their color on the plate and the R_p value [$R_p(X) = R_F(X)/R_F$ (phenol) where X is the compound to be determined]. The extractions of the spots were measured with an ERI-10 extraction recording apparatus. The sensitivity was 0.04–0.3 μg/spot and the precision was ±18%, depending on the phenol and its amount.

Thielmann[57] separated a mixture of phenol, cresols, dimethylphenols, and naphthols as their Fast Blue Salt BB addition compounds by TLC on Silica Gel G–K_2CO_3 (1:2) plates. Thielman[57] also reported the TLC separation of phenol, isomeric cresols, dimethylols, and naphthols on Silica Gel G after reaction with

TABLE 15

$R_F \times 100$ VALUES OF p-NITROPHENYLAZO DYES OF SOME PHENOLS ON BASIC SILICA GEL PLATES

Compound	R_F ($\times\ 100)^a$	Color of azo dye	
		Before exposure to ammonia	After exposure to ammonia
Phenol	45	Orange	Reddish-orange
2-Chlorophenol (OCP)	21	Orange	Reddish-orange
3-Chlorophenol (MCP)	30	Orange	Reddish-orange
4-Chlorophenol (PCP)	82	Yellow	Purple
2,4-Dichlorophenol (2,4-DCP)	79	Pink	Pink
2,6-Dichlorophenol (2,6-DCP)	5	Orange	Red
2,4,6-Trichlorophenol (2,4,6-TCP)			
2-Nitrophenol	10	Orange	Orange
3-Nitrophenol	20	Orange	Orange
4-Nitrophenol	45	Orange	Orange
1-Naphthol	86	Pink	Blue
	55	Yellow	Blue
2-Naphthol	83	Orange	Reddish-orange
Catechol	34	Dark green	Dark green
	4	Pink	Pink
Resorcinol	85	Yellow	Yellow
	8	Orange	Orange
Hydroquinone	30	Dark green	Dark green
2-Cresol	54^b	Yellow	Mauve
3-Cresol	43^b	Yellow	Magenta
4-Cresol	94^b	Yellow	Purple
2,5-Xylenol	62^b	Yellow	Lilac
2,6-Xylenol	74^b	Yellow	Pale lilac

^{a} Solvent system, chloroform–acetone (9:1). Quantities applied, 1–2 μg of the phenol as derivative.
^{b} After Crump using Silica Gel G (Merck) treated with 0.5 N sodium hydroxide.

References pp.516–517

1-phenyl-2,3-dimethyl-4-amino-5-pyrazolone and after formation of azo dyes. The developing solvents were benzene, butyl acetate and butyl acetate–benzene (1:9). Azo compounds were prepared by coupling the phenolic compounds with anthraquinone-1-diazonium chloride (using stabilized form of Fast Red Salt AL).

TABLE 16

THIN LAYER CHROMATOGRAPHIC IDENTIFICATION OF CHLORINATION PRODUCTS OF PHENOL, 2-CHLOROPHENOL, AND 4-CHLOROPHENOL AS ANTIPYRYL AND p-NITROPHENYLAZO DYES

Components	Chlorination products			
	p-Nitrophenylazo dyes		Antipyrine dyes	
	No. of spots	Components detected	No. of spots	Components detected
Phenol	5	Phenol, OCP, PCP[a], 2,4-DCP 2,6-DCP	3	Phenol and PCP, OCP and 2,4-DCP 2,6-DCP and 2,4,6-TCP
2-Chlorophenol	3	OCP[a], 2,4-DCP, 2,6-DCP	2	OCP and 2,4-DCP, 2,6-DCP and 2,4,6-TCP
4-Chlorophenol	2	PCP, 2,4-DCP	3	PCP, 2,4-DCP 2,4,6-TCP

[a] Present in traces.

REFERENCES

1   W. B. DEICHMANN AND P. OESPER, *Ind. Med.*, 9 (1940) 296.
2   M. BIEBL, *Z. Ges. Exptl. Med.*, 93 (1934) 515.
3   W. B. DEICHMAN AND J. WITHERUP, *J. Pharmacol. Exptl. Therap.*, 80 (1944) 233.
4   M. I. SMITH, E. ELVOVE AND W. H. FRAZIER, *U.S. Public Health Rept.*, 45 (1930) 2509.
5   K. TOLLENS, *Arch. Exptl. Pathol. Pharmakol.*, 52 (1905) 220.
6   H. ZANGGER, F. FLURY AND H. ZANGGER, *Lehrbuch der Toxikologie*, Springer, Berlin, 1928.
7   W. B. DEICHMANN, *Federation Proc.*, 2 (1943) 35.
8   W. B. DEICHMANN, *Arch. Biochem.*, 3 (1944) 345.
9   E. LAHMANN, *Staub*, 26 (1966) 530.
10 H. BUCHWALD, *Ann. Occupational Hyg.*, 9 (1966) 7.
11 J. CHALUPA AND J. DUORAKOVA, *Sb. Vysoke Skoly Chem. Technol. Praze*, 5 (1962) 267.
12 M. M. BRAVERMAN, S. HOCHHEISER AND M. B. JACOBS, *Ind. Hyg. J.*, 18 (1957) 132.
13 R. G. SMITH, J. D. MCEWEN AND R. E. BARROW, *Ind. Hyg. J.*, 20 (1959) 142.
14 D. C. ABBOTT, *Proc. Soc. Water. Treat. Exam.*, 13 (1964) 153.
15 H. D. GIBBS, *J. Biol. Chem.*, 72 (1949) 649.
16 G. J. PAPARIELLO AND M. A. JANISH, *Anal. Chem.*, 38 (1966) 211.
17 S. NAITO, M. KANEKO, S. SETSUDA, J. MATSUZAKI, S. FUKUL AND S. KANNO, *Eisei Kagaku*, 16 (1970) 41.

18   A. W. SPEARS, *Anal. Chem.*, 35 (1963) 320.
19   L. LANDAULT AND G. GUIOCHON, *Anal. Chem.*, 39 (1967) 713.
20   R. A. BAKER, *Intern. J. Air Water Pollution*, 10 (1966) 591.
21   R. A. BAKER, *J. Am. Water Works Assoc.*, 55 (1963) 913.
22   R. A. BAKER, *J. Am. Water Works Assoc.*, 58 (1966) 751.
23   R. A. BAKER, *Water Res.*, 1 (1967) 61.
24   R. A. BAKER, *Water Res.*, 1 (1967) 97.
25   R. A. BAKER, *Water Res. Standards*, 2 (1962) 983.
26   R. A. BAKER AND B. A. MALO, *Environ. Sci. Technol.*, 1 (1967) 997.
27   E. PILLION, *J. Gas Chromatog.*, 3 (1965) 238.
28   E. R. ADLARD AND G. W. ROBERTS, *J. Inst. Petrol.*, 51 (1965) 376.
29   N. NARASIMHACHARI AND E. VON RUDLOFF, *Can. J. Chem.*, 40 (1962) 1123.
30   P. F. NELSON AND J. G. SMITH, *Tappi*, 49 (1966) 215.
31   S. GOREN-STRUL, H. F. W. KLEIJN AND A. E. MOSTAERT, *Anal. Chim. Acta*, 34 (1966) 322.
32   R. H. CROUSE, J. W. GERNER AND H. J. O'NEILL, *J. Gas Chromatog.*, 1 (1963) 18.
33   A. NONAKA, *Anal. Chem.*, 44 (1972) 271.
34   A. NONAKA, *Bunseki Kagaku*, 16 (1967) 260.
35   A. NONAKA, *Bunseki Kagaku*, 16 (1967) 1169.
36   A. NONAKA, *Bunseki Kagaku*, 17 (1968) 91.
37   A. NONAKA, *Bunseki Kagaku*, 17 (1968) 944.
38   W. T. CHIN, T. E. CULLEN AND R. P. STANOVICK, *J. Gas Chromatog.*, 6 (1968) 248.
39   I. C. COHEN, J. NORCUP, J. H. A. RUZICKA AND B. B. WHEALS, *J. Chromatog.*, 44 (1969) 251.
40   J. D. REINHEIMER, J. P. DOUGLAS, H. LEISTER AND M. B. VOELKEL, *J. Org. Chem.*, 22 (1957) 1743.
41   R. J. ARGAUR, *Anal. Chem.*, 40 (1968) 122.
42   F. K. KAWAHARA, *Anal. Chem.*, 40 (1968) 1009.
43   B. C. PETTITT, P. G. SIMMONDS AND P. ZLATKIS, *J. Chromatog. Sci.*, 7 (1969) 645.
44   P. G. SIMMONDS, D. C. FENNIMORE, B. C. PETTITT, J. E. LOVELOCK AND A. ZLATKIS, *Anal. Chem.*, 39 (1967) 1428.
45   W. E. WENTWORTH AND R. S. BECKER, *J. Am. Chem. Soc.*, 84 (1962) 4263.
46   W. E. WENTWORTH, E. CHEN AND J. E. LOVELOCK, *J. Phys. Chem.*, 70 (1966) 445.
47   L. CLAISEN, O. EISLEB AND F. KRAMERS, *Ann. Chem.*, 418 (1919) 69.
48   R. W. MORROW, J. A. DEAN, W. D. SHULTS AND M. R. GUERIN, *J. Chromatog. Sci.*, 7 (1969) 572.
49   S. T. PRESTON, *A Guide to the Analysis of Phenols by Gas Chromatography*, Polyscience, Evanston, Ill., 1966.
50   J. ZULAICA AND G. GUIOCHON, *J. Polymer Sci.*, 4 (1966) 567.
51   V. T. BROOKS, *Chem. Ind. (London)*, 1959, 1317.
52   A. R. PATERSON, in H. J. NOEBELS, R. F. WALL AND N. BRENNER (Eds.), *Gas Chromatography, Instr. Soc. Am. Symp., June 1959*, Academic Press, 1961, p. 323.
53   J. HRIVNAK AND E. BESKA, *J. Chromatog.*, 54 (1971) 277.
54   J. HRIVNAK, *J. Chromatog. Sci.*, 8 (1970) 602.
55   O. M. ALY, *Water Res.*, 2 (1968) 587.
56   H. HEIER, *Fortschr. Wasserchem. Ihrer Grenzgeb.*, 12 (1970) 20; *Chem. Abstr.*, 73 (1970) 59094u.
57   H. THIELMANN, *Pharmazie*, 25 (1970) 365.

Chapter 23

ANILINE AND DERIVATIVES

Aniline is produced commercially by the reduction of nitrobenzene, either in a continuous, vapor-phase process, or by a batch solution method. It may also be prepared by the amination of chlorobenzene. U.S. production of aniline in 1961 was 122,702,000 lb. The two most important uses of aniline are in the manufacture of dyes and rubber chemicals. The rubber industry consumes approximately two-thirds of the aniline production in the manufacture of derivatives such as vulcanization accelerators and antioxidants. Minor amounts of aniline are used in the textile, paper, and metallurgical industries, in the preparation of surfactants, and photographic chemicals. In the pharmaceutical industry aniline is important in the manufacture of sulfur drugs and in the animal feed industry in the production of arsanilic acid.

The toxicity of aniline in man is characterized by its ability to form methemoglobin[1]. Aniline has been analyzed by colorimetry[2-5], as well as by a number of

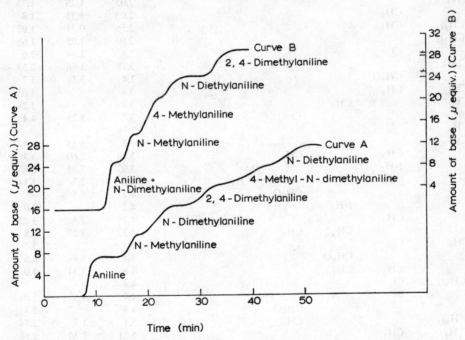

Fig. 1. Separation of aromatic amines at 137°C. A, Column length 4 ft.; stationary phase liquid paraffin; nitrogen pressure 47.2 cm Hg; nitrogen flow rate 75 ml/min. Recording of zones by automatic titration; B, on column capable of showing hydrogen bonding with primary and secondary amines only; column length 4 ft.; stationary phase Lubrol MO; nitrogen pressure 75 cm Hg; nitrogen flow rate 120 ml/min.

References p. 528

TABLE 1

RETENTION VOLUMES OF AROMATIC AMINES RELATIVE TO ANILINE IN THREE TYPES OF COLUMN AT 137°C

N atom	Substituents in position					Column stationary phase		
	2	3	4	5	6	Paraffin wax	Lubrol MO	Benzyl-diphenyl
	F					0.72		0.63
	F		F			0.74	0.75	
	F			F		0.79	1.08	
			F			0.96	1.28	1.1
		F					0.99	1.52
		F	CH$_3$			1.88	2.28	
			CH$_3$			1.95	1.63	1.75
	CH$_3$		F			2.0	2.28	
		CH$_3$				2.0	1.56	1.86
CH$_3$						2.05	1.28	1.73
	CH$_3$					2.05	1.53	1.8
(CH$_3$)$_2$		CH$_3$				2.56	0.71	1.07
(CH$_3$)$_2$						2.60	1.05	1.68
	Cl					2.8	2.34	2.6
C$_2$H$_5$						3.18	1.66	2.33
		CH$_3$O				3.4	2.7	
CH$_3$	CH$_3$					3.4		3.05
CH$_3$		CH$_3$				3.4		3.28
			Cl			3.7	5.25	4.4
CH$_3$						3.76		
C$_2$H$_5$		Cl				3.8	5.35	4.4
	CH$_3$		CH$_3$			3.84	2.60	3.36
	NH$_2$					3.92	7.95	6.6
	CH$_3$			CH$_3$		4.0	2.7	3.14
			CH$_3$O			4.1	5.5	
			NH$_2$			4.2		10.0
	CH$_3$				CH$_3$	4.25	2.5	3.22
	CH$_3$	CH$_3$				4.27	3.27	3.8
CH$_3$	Cl					4.3		3.7
		CH$_3$O	CH$_3$			4.4	5.8	6.6
		CH$_3$	CH$_3$			4.6	3.34	4.0
(CH$_3$)$_2$	Cl					4.8		
	Br					4.9	4.5	5.0
			CH$_3$O			4.9		5.7
(CH$_3$)$_2$				CH$_3$		5.1	1.7	2.74
C$_2$H$_5$	CH$_3$					5.21	2.34	3.66
(CH$_3$)$_2$		CH$_3$				5.4	1.8	2.9
(C$_2$H$_5$)$_2$				CH$_3$		5.45	1.9	3.0
n-C$_3$H$_7$						5.82	2.7	4.0
C$_2$H$_5$		CH$_3$				5.9	3.02	4.4

TABLE 1 (cont.)

N atom	Substituents in position					Column stationary phase		
	2	3	4	5	6	Paraffin wax	Lubrol MO	Benzyldi-phenyl
C_2H_5			CH_3			6.0	2.9	4.4
			NH_2			6.2		12.4
			C_2H_5O			6.2		8.8
			Br			6.4	10.0	8.8
		Br				6.4	10.0	9.0
CH_3			Cl			7.35		6.9
		Br	CH_3			8.8		
		NO_2				9.8		
	Cl	Cl				10.0		
	Cl		Cl			10.0		
	I					10.1		
$(C_2H_5)_2$			CH_3			10.5	3.2	5.0
$(C_2H_5)_2$		CH_3				10.8	3.1	5.1
n-C_4H_9						11.0	4.7	7.0
			I			13.0		23.2
		I				13.1	22.4	23.2
	CH_3O			CH_3O		14.0		
		Cl	Cl			14.6		
	CH_3O			NH_2		16.4		
iso-C_5H_{11}						16.8	6.4	
$(CH_3)_2$			Br			17.3	6.8	13.3

methods *via* diazotization with nitrous acid and coupling with a suitable phenol; the azo dye formed is then photometrically determined[6].

The separation and microestimation of volatile aromatic amines by GLC was initially described by James and Martin[7] and James[8]. The columns used consisted of 4 ft. × 3–4 mm i.d. glass tubing coated with stationary phases such as paraffin wax (m.p. 49°C), liquid paraffin, Lubrol MO (ICI), a polyethylene oxide condensate, and benzyldiphenyl. Detection was achieved with a gas-density meter[9]. Figure 1, curve A shows a typical isothermal separation of aniline, N-methylaniline, N-dimethylaniline, 2,4-dimethylaniline, 4-methyl-N-dimethylaniline and N-diethylaniline using a 4 ft. column at 137°C with liquid paraffin or paraffin wax as the stationary phase. Each step denotes a separate substance and the horizontal line between steps denotes a period in which no titrable material emerged from the columns. Figure 1, curve B shows a separation obtained at 137°C on a 4 ft. column containing Lubrol MO (a polyether stationary phase) which allowed hydrogen bonding by primary and secondary amines, but not by tertiary amines. Table 1 lists the retention volumes of a number of aromatic amines relative to aniline in three types of columns at 137°C (paraffin wax, Lubrol MO, and benzyldiphenyl).

The determination of aniline and toluidine isomers in aqueous and non-aqueous media by gas chromatography of their formanilide and formyl toluidine

derivatives was reported by Umeh[10]. A Perkin-Elmer Model F11 gas chromatograph equipped with dual flame-ionization detector was used with 3 columns of
sodium dodecyl benzenesulfonate (PBS) coated on Chromosorb G. The instrument
conditions and the columns employed for the various separations and determinations
are listed in Table 2. Figure 2 shows a chromatogram of the separation of formyl
derivatives of aniline, N-methylaniline, N-methyl-o-toluidine and the isomeric
toluidines on a 12 ft. 2.5% DBS–Chromosorb G column. Figure 3 illustrates the
separation of (a) the acetyl derivatives of the isomeric toluidines and (b) acetyl
derivatives of N-methyl aniline, N-methyl-p-toluidine and the isomeric toluidines
on 12 ft. and 6 ft. columns of 2.5% DBS columns, respectively. The formyl deri-

TABLE 2

INSTRUMENTAL CONDITIONS FOR THE ANALYTICAL SEPARATIONS

Chromatography of	Instrumental conditions
Formanilide (or any formyl derivative of any of the toluidines) for formic acid determination using methyl stearate as internal standard	Column A: oven temperature 200°, nitrogen inlet pressure 10 lb./in.^{2}
Formyl derivatives of aniline, o-, m, and p-toluidine, N-methyl aniline, and N-methyl p-toluidine using methyl stearate as internal standard	Column B: oven temperature 210°, nitrogen inlet pressure 15 lb./in.^{2}
Acetyl derivatives of o-, m-, and p-toluidine, N-methyl aniline, and N-methyl p-toluidine	Column B: oven temperature 200°, nitrogen inlet pressure 12 lb./in.^{2}

Columns: A, 6 ft. × 3 mm i.d. glass column packed with 2.5% (w/w) sodium dodecylbenzene-
sulphonate (DBS) on NAW 60–80 mesh Chromosorb G; B, the 12 ft. × 3 mm i.d. version of
column A; C, 6 ft. × 3 mm i.d. glass column packed with 2.5% (w/w) DBS on NAW 60–80 mesh
1% (w/w) base-loaded Chromosorb G.
The oxygen and hydrogen inlet pressures and the injection port temperature were the same for all
the separations, viz. 26 and 18 lb./in.^{2} and 250°, respectively.

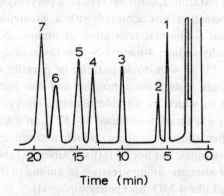

Fig. 2. Separation of formyl derivatives of (1) N-methyl aniline, (2) N-methyl o-toluidine, (4) o-
toluidine, (5) aniline, (6) m-toluidine, (7) p-toluidine, on a 12 ft. 2.5% DBS on Chromosorb G column
using methyl stearate (3) as internal standard.

vatives of the amines were quantitatively separated on Column B (12 ft.) but not on Column C (6 ft. of DBS). The acetyl derivatives, however, were separated on both columns B and C, although *m*- and *p*-toluidines were not quantitatively separated on Column C [Figs. 3(a) and (b)].

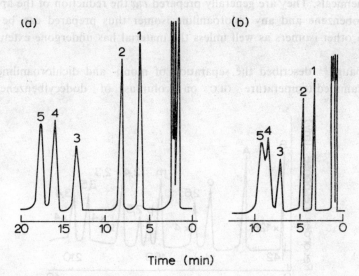

Fig. 3. (a) Separation of acetyl derivatives of *o*-, *m*-, and *p*-toluidines, N-methyl aniline and N-methyl *p*-toluidine on a 12 ft. 2.5% DBS on Chromosorb G column. 1, N-methyl aniline; 2, N-methyl *p*-toluidine; 3, *o*-toluidine; 4, *m*-toluidine; 5, *p*-toluidine. (b) Separation of acetyl derivatives of (1) N-methyl aniline, (2) N-methyl *p*-toluidine, (3) *o*-toluidine, (4) *m*-toluidine, (5) *p*-toluidine on a 6 ft. 2.5% DBS on Chromosorb G column coated 1% with potassium hydroxide.

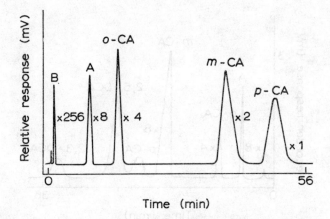

Fig. 4. Chromatogram showing separation of monochloroaniline isomers. Column: 2 m of Siponate DS 10 at 15 wt.% on acid-washed white Chromosorb precoated with 2% NaOH. Carrier, helium at 60 ml/min; temperature, 180°C; plates, 990; relative volatility α(p/m) = 1.28; resolution, 2.2.

The separation and determination of aniline and toluidine and other related amines *via* the GC of their trifluoroacetyl derivatives on a stationary phase mixture of 9.5% Apiezon L and 3.5% Carbowax 20M coated on 80–100 mesh Aeropak 30 was described by Dove[11].

Chloroanilines are important intermediates in the production of a variety of organic chemicals. They are generally prepared *via* the reduction of the appropriate chloronitrobenzene and any chloroaniline isomer thus prepared can be expected to contain other isomers as well unless the material has undergone extensive purification.

Bombaugh[12] described the separation of mono- and dichloroaniline isomers by programmed-temperature GLC on columns of dodecylbenzene sodium

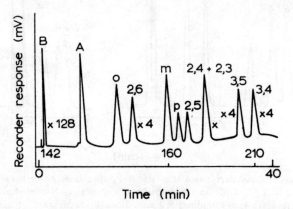

Fig. 5. Separation of mono- and dichloroaniline mixture using programmed-temperature operation. Column: 1.3 m Siponate DS 10 at 15 wt.% plus 0.7 m of silicone oil 550 at 20 wt.% on acid-washed white chromosorb precoated with 2% NaOH. Sample size, 10 μl; temperature, programmed from 142° to 210°C at 4°C/min.

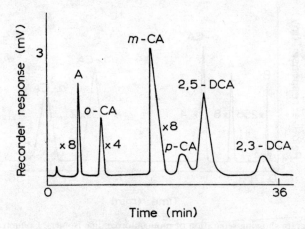

Fig. 6. Isothermal chromatograms of typical *m*-chloroaniline synthesis product. Column, same as Fig. 5; temperature, 145°C; carrier, helium at 58 ml/min; sample size, 5 μl.

TABLE 3

RELATIVE RETENTION DATA BY SEVERAL SUBSTRATES

Chloroaniline isomers	B.p. (°C)	PEG 20M	Silicone	Tide	Siponate	SipSo
o-	208	1.94	2.41	2.19	1.79	1.97
2,6-		2.21	4.23	3.52	1.93	2.32
m-	229	4.28	3.84	4.96	4.20	4.34
p-	231	4.18	3.62	5.78	6.10	5.66
2,5-	245	6.16	7.04	8.22	6.39	6.20
2,4-	251	5.89	6.82	10.07	9.94	8.37
2,3-	252	7.12	8.48	10.83	3.73	8.83
3,5-	260	13.85	11.61	18.74	24.85	16.74
3,4-	272	16.49	13.97	24.88	28.15	22.20
Aniline = 1						

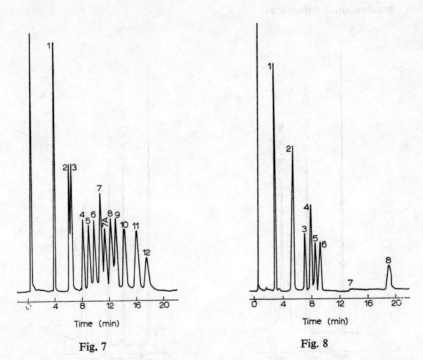

Fig. 7   Fig. 8

Fig. 7. Alkylanilines. 1, N,N-dimethylaniline; 2, aniline; 3, N-methylaniline; 4, o-methylaniline; 5, m-methylaniline; 6, p-methylaniline; 7, 2,6-dimethylaniline; 7a, 2,4-dimethylaniline; 8, o-ethylaniline; 10, p-ethylaniline; 11, 2,3-dimethylaniline. Column A, 120°C, isothermal operation, amplification $10^2 \times 16$.

Fig. 8. Chloroanilines. 1, o-Chloroaniline; 2, m- and p-chloroanilines; 3, 2,4,6-trichloroaniline; 4, 2,4-dichloroaniline; 5, 2,5-dichloroaniline; 6, 2,3-dichloroaniline; 7, 2,4,5-trichloroaniline; 8, 3,4-dichloroaniline. Column A, 180°C, isothermal operation, amplification $10^2 \times 16$.

sulfonate and silicone oil. A Consolidated Electrodynamics Corp. Mode 201A and Aerograph A-100-C gas chromatograph were used to obtain comparative thermal response data. Columns were prepared from $\frac{1}{4}$ in. copper tubing. The analytical column used to resolve 8 of the 9 isomers was a 2 m column with 1.3 m containing a packing of dodecylbenzene sodium sulfonate on 15% white (Siponate DS-10, American Alcolac Corp., Baltimore) Chromosorb, followed by 0.7 m containing silicone oil 440 on 20% white Chromosorb. (The white Chromosorb was acid-washed, water-washed, then loaded to 2% with NaOH prior to loading with the stationary phase.)

The separation of monochloroaniline isomers on 2 m of Siponate DS-10 on Chromosorb at 180°C is shown in Fig. 4. Figure 5 shows a temperature-programmed separation of mono- and dichloroaniline on a 1.3 m column of Siponate DS-10 plus 0.7 m of silicone oil 550. Figure 6 illustrates an isothermal separation at 145°C of a typical *m*-chloroaniline synthesis product on a 1.3 m column of Siponate plus 0.7 m of silicone oil 550. Table 3 lists relative retention data of 9 chloroaniline derivatives obtained on 5 substrates.

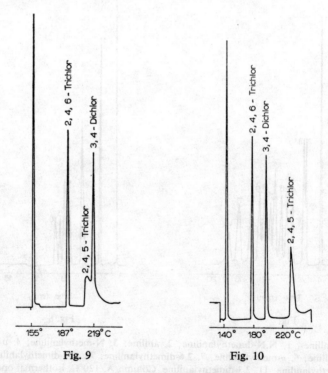

Fig. 9

Fig. 10

Fig. 9. Chloroanilines. Column B, temperature, 155°C at the start and programmed at a rate of 8°C/min. Amplification 10 × 128.

Fig. 10. Chloroanilines. Column C, Temperature, 140 °C at the start and programmed at a rate of 10 °C/min. Amplification 10 × 128.

The GC of alkylated and chlorinated anilines has been described by Henkel[13]. An F & M Model 810 gas chromatograph was used with FID and three columns: A, 2.9 m × ¼ in. o.d., 4.5 m i.d. copper tubing packed with 1.5% Ucon 50-HB2000 on 70–80 mesh Chromosorb G preheated for 15 h at 230°C under a helium flow of 100 ml/min; B, 1.70 m × ¼ in. o.d., 4.5 mm i.d. copper tubing packed with 1.25% Ucon 50-HB2000 on 70–80 mesh Chromosorb G, preheated for 15 h at 230°C under a helium flow rate of 110 ml/min; C, length and diameter as column B but packed with 1% Versamid 900 on 70–80 mesh Chromosorb G, preheated for 15 h at 280°C under a helium flow rate of 110 ml/min. The operating conditions were carrier gas helium at an inlet pressure of 40 p.s.i.g. for column A, flow rate of 100 ml/min for Column B and 110 ml/min for column C, hydrogen flow at 16 p.s.i.g. 50 ml/min, air flow at 15 p.s.i.g. 220 ml/min, detector and injection block temperatures at 310 and 270°C, respectively, and sample size 5.0 µg.

Figure 7 illustrates a chromatogram of the isothermal separation of 11 alkylanilines on column A at 120°C. Figure 8 shows the isothermal chromatographic separation of 8 chloroanilines on column A.

Figures 9 and 10 show chromatograms of the temperature-programmed separation of chloroanilines on columns B and C, respectively. Table 4 lists the relative retention times for the alkyl anilines on column A at 110 and 120°C, as well as the relative retention times for the chloroanilines on column A at 180°C.

TABLE 4

RELATIVE RETENTION TIMES OF ALKYLANILINES OBTAINED AT 110° AND 120°C AND RELATIVE RETENTION TIMES OF CHLOROANILINES AT 180°C
Retention times obtained on column A

Alkylaniline	Relative retention time		Chloroaniline	Relative retention time
	110°C	120°C		180°C
Aniline	1.00	1.00	2-Chloro-	1.00
2-Methyl-	1.38	1.34	3-Chloro-	1.90
3-Methyl-	1.55	1.47	4-Chloro-	1.90
4-Methyl-	1.71	1.61	2,3-Dichloro-	3.29
2-Ethyl-	2.17	2.01	2,4-Dichloro-	2.81
4-Ethyl-	2.58	2.39	2,5-Dichloro-	3.00
2,3-Dimethyl-	2.90	2.61	3,4-Dichloro-	6.86
2,4-Dimethyl-	2.02	1.86	2,4,6-Trichloro-	2.50
2,5-Dimethyl-	2.32	2.12		
2,6-Dimethyl-	1.85	1.76		
3,4-Dimethyl-	3.22	2.87		
N-Methyl-	1.05	1.05		
N,N-Dimethyl-	0.58	0.64		

REFERENCES

1 E. E. EVANS, R. C. SHARSTA AND A. L. LYNCH, *Arch. Ind. Health*, 18 (1958) 452.
2 W. A. RIEHL AND K. F. HAGER, *Anal. Chem.*, 27 (1955) 1768.
3 E. A. RIEHL AND K. F. HAGER, *Anal. Chem.*, 27 (1955) 1768.
4 N. W. HANSON, D. A. REILLY AND H. E. STAGG (Eds.), *The Determination of Toxic Substances in Air*, Heffer, Cambridge, 1965, p. 63.
5 J. T. STEWART, T. D. SHAW AND A. B. RAY, *Anal. Chem.*, 41 (1969) 360.
6 J. L. CLIPSON AND L. C. THOMAS, *Analyst*, 88 (1963) 1971.
7 A. T. JAMES AND A. J. P. MARTIN, *Brit. Med. Bull.*, 10 (1954) 170.
8 A. T. JAMES, *Anal. Chem.*, 28 (1956) 1564.
9 A. J. P. MARTIN AND A. T. JAMES, *Biochem. J.*, 63 (1956) 138.
10 E. O. UMEH, *J. Chromatog.*, 51 (1970) 139.
11 R. A. DOVE, *Anal. Chem.*, 39 (1967) 1188.
12 K. J. BOMBAUGH, *Anal. Chem.*, 37 (1965) 73.
13 H. G. HENKEL, *J. Gas Chromatog.*, (1965) 320.

Chapter 24

POLYCHLORINATED BIPHENYLS

Despite the fact that polychlorinated biphenyls (PCB's) have been available commercially for over 40 years, it is only within the last 5 years that they have been recognized to be of environmental concern. PCB's are manufactured in the United States, Great Britain, France, Germany, USSR, Japan, Spain, Italy, and Czechoslovakia and are marketed under a number of commercial trade names, *e.g.* Aroclor, Clophen, and Phenoclor. The series of Aroclors (Monsanto) are marketed under various numbers and consist of mixtures of chlorinated biphenyls and terphenyls. The first two digits represent the molecular type: 12 = chlorinated biphenyls, 25 and 44 = blends of chlorinated biphenyls and chlorinated terphenyls (75% biphenyl and 60% biphenyl, respectively), 54 = chlorinated terphenyls. The last two digits give the weight percent of chlorine, *e.g.* Aroclor 1242 is a chlorinated biphenyl containing 42% chlorine.

The viscosity of the PCB's increases in direct proportion to the chlorine content from very fluid liquids to viscous products and solids. In the commercial process for PCB manufacture, biphenyls are chlorinated with anhydrous chlorine with either iron filings or ferric chloride as the catalyst, the byproduct is hydrogen chloride and the product is a mixture of several PCB's. In the process of replacing hydrogen atoms with those of chlorine, a large number of substitution combinations arise, *viz.*

For example, three monochlorobiphenyl isomers are possible, 12 dichlorobiphenyl isomers, 21 trichlorobiphenyl isomers, and so on. Theoretically, 210 compounds can be prepared by this substitution process (a typical PCB example would be 2,4,6,2'4'-pentachlorobiphenyl). Mass spectroscopic studies[1] of Aroclor 1260 revealed the presence of 11 isomers; five containing six chlorine atoms, five containing seven chlorine atoms, and one containing eight, while Aroclor 1254 was found to contain 18 compounds, *e.g.* one containing three chlorine atoms, four containing four chlorines, four containing five chlorines, five containing six chlorines, and four containing seven chlorines[2].

Monsanto[3], the sole U.S. producer, has recently revealed figures for the production and sales for the Aroclors from 1960 through estimated 1971. Approximately 353,000 short tons of PCB's was produced during this decade. Closed system electrical uses ranged from 11,000 to 20,000 tons/year during the period. The report showed that Monsanto voluntarily reduced its non-electrical sales of PCB from the high of 16,000 tons in 1970 to 4400 tons in 1971. Both production and domestic sales of PCB's roughly doubled in the decade. The cumulative sales in North America between 1930 and 1970 is suggested to be on the order of 500,00 tons, assuming a constant rate of growth of domestic sales from 1930. Although corresponding data on production of PCB's outside the U.S. are not available, current estimates suggest that the total U.S. production represents approx. one-half of the total world production[4]. Aroclor 1242 and grades with lower percentages of chlorine have composed one-half or more of the total production between 1963 and 1970.

The chemical properties that make PCB's desirable industrial chemicals are their excellent thermal stability, their strong resistance to both acidic and basic hydroxides and action of corrosive chemicals, and their *general* inertness. They are insoluble in water but possess a low finite vapor pressure. Their boiling points range from 278° for Aroclor 1221 to 415°C for Aroclor 1268. All are stable to prolonged heating at 150°C and the lower Aroclors can be distilled at atmospheric pressure without appreciable decomposition.

The largest categories of use of PCB's have been in capacitors and transformers (as dielectrics) and in certain "plasticizer" applications including carbonless duplicating paper. The major uses for PCB's prior to 1970, in order of importance in terms of volume of material used were capacitors, plasticizer applications, transformer fluids, hydraulic fluids and lubricants, and heat transfer fluids. Miscellaneous uses include machine tool cutting oils, high vacuum oils, specialized lubricants and gasket sealers, formulations in epoxy paints, resins and chlorinated rubber, printing inks, waxes, formulations in epoxy paints, resins and chlorinated rubber, synthetic adhesives, textile dyes, protective coatings for wood, metal and concretes, as sealers in water-proofing compounds and putty. PCB's have also been incorporated into pesticide formulations, especially with such insecticides as DDVP, Lindane[5,6], Chlordane, Aldrin, Dieldrin and Toxaphene, to suppresss their vaporization and hence extend their "kill-life" and have also been shown to increase the insecticidal properties of DDT[7].

The first identification of polychlorinated biphenyls in regard to ecology was by Jensen[8] who identified PCB's in the bodies of 200 pike taken from different parts of Sweden, in other fish, and in eagle feathers collected in 1944. PCB's along with DDE [1,1-dichloro-2,2-bis(p-chlorophenyl)ethane] are now reported to be the most abundant of the chlorinated aromatic pollutants in the global ecosystem[9]. Extracts of sea eagles, pike, and salomon[9] as well as in British[10] and Canadian[11-13] wildlife contained PCB's, and in the former instance it was found that in birds' livers and eggs the PCB residues were greater than the organochlorine pesticide residues. PCB's have also been found in fish, mussels, and birds from the River Rhine and the Netherlands coastal areas[1] and in marine animals and wildlife in Sweden, England, and the U.S.A.[1,9,12,14].

Polychlorinated biphenyls have also been found in human adipose tissue[15], samples of human milk[16], and in foods (margarine, vegetable oils, and particularly fish)[17].

Essentially the same type of residue pattern is becoming apparent for the polychlorinated biphenyls that has been found for the persistent organochlorine insecticides. The PCB's are extremely stable, chemically fat soluble and hence persistent in the environment. The four possible pathways by which PCB's could be dispersed in the environment include (a) because of their numerous manufacturing applications and general inertness, PCB's could be flushed as wastes into rivers, lakes, etc., to pollute fish and other wildlife, (b) via industrial smoke, in exhaust from aircraft engines, incineration and combustion of PCB-containing products (incineration at 2000°F or above for 2 sec will destroy PCB's, but poorly operated incinerators or open burning may result in PCB's being released to the atmosphere unchanged), and leaching of plasticizer from plastic objects in waste disposal areas, (c) via insecticidal formulations to enhance kill ratios, and (d) via direct contamination of feeds and foodstuffs through leakage of heat transfer fluids and recycled paper (e.g. PCB's in carbonless paper being used as chipboard or liners in containers for breakfast foods, cereals, etc.[14] or migration from surface coatings (paints, etc.) into food and feeds.

Information relative to the biological decomposition of the PCB's is scant and it was suggested[18] that they are possibly more stable than DDT and its metabolites since the PCB's lack the ethane component between the aromatic rings which is the site of action of most of the transformations of DDT. This factor coupled with their physical and chemical characteristics for persistence indicates that these materials are capable of biological magnification up the food chain. (Although the concentrations of the PCB's are generally in the parts-per-billion range in the environment, their high lipid solubility results in their accumulation in fatty tissues of lower animals and marine life.) Because of the very low aqueous solubility of the PCB's, when they are discharged into a river or lake, they will accumulate on the sediment in relatively nigh concentration and redissolve very slowly.

It has been estimated by Nisbet and Sarofin[19] that the total loss of PCB's into the U.S. environment over the last 40 years would approach 30,000 tons to the atmosphere, 60,000 tons to water, and 300,000 tons to dumps. Of this total, remaining residues might be 20,000 tons from the air (which would be distributed on land or water), 30,000 tons in water, and perhaps 250,000 tons in dumps.

Compared with the chlorinated pesticides, definitive aspects of acute, subacute and chronic toxicity in man still remain rather poorly known. Chloracnegen effects have been reported as early as 1936, following industrial exposure to the PCB's[20-24]. Occupational chloracne, however, has not been a problem with recent usage of the PCB's. Approximately 10 cases of fatal intoxication involving persons who handled or were exposed to chlorinated biphenyls or naphthalenes in their occupations have been described[25-27]. In all cases histological examination revealed fatty liver degeneration, necrosis, and cirrhosis. It is important to note that chlorinated naphthalene as well as chlorinated dibenzofurans have been recently identified in two commercial biphenyls samples (Phenoclor DP6 and Clophen A60) by Vos et al.[28,29].

Human intoxication with the heat exchanger Kanachlor 400, a Japanese manufactured PCB with 48% chlorine and containing as its main components 2,4,3',4'-,
2,5,3',4'-, 2,3,4,4'-, and 3,4,3',4'-tetrachlorobiphenyl, and 2,3,4,3',4'-pentachloro-
biphenyl, was reported[30] in Western Japan in 1968 and has been referred to as the
"Yusho" disease.

More than 900 people were eventually affected following the consumption of
contaminated rice oil containing levels estimated to range from 2000 to 3000 p.p.m.
PCB's (average of about 2500 p.p.m.). These levels were derived from the known
organic chlorine content of the rice oil related to the known organic chlorine content
of Kanachlor 400. Exposure levels to the oil were calculated to approximate
15,000 mg/day. The minimum dose necessary to produce positive clinical effects
was estimated to be 0.5 g. The clinical aspects associated with "Yusho" included
chloracne, blindness, systemic gastrointestinal symptoms with jaundice, edema, and
abdominal pain. Chloracne is very persistent with some patients showing evidence
of it after three years.

Newborn infants born from poisoned mothers had skin discoloration due to
the presence of PBC *via* placental passage. (The dark skin discoloration regressed
after a period of 2–5 months.) Gingival hyperplasia with pigmentation was seen in
several cases. Decreased birth weights were also noted but no evidence could be
obtained in regard to the possible retardation in physical and mental activities of
the babies.

The skin of stillborn infants showed hyperkeratosis and atrophy of the epidermis
and cystic dilation of the hair follicles. Residues of PCB have been found in fetal
tissue[31,32].

Examination of autopsy tissues of two Yusho fatalities revealed the presence
of chlorobiphenyls in all of the examined organs, especially mesenterial fatty tissues,
skin, and bone marrow[33]. Kojima *et al.*[34,35] found PBC's with longer retention
times (probably pentachloro- and higher chlorinated biphenyls) in autopsy tissues and
it was assumed that their presence might have been responsible for the observed
long duration of the intoxication symptoms.

Hypopericardium and abdominal edema have been observed in chicken[36,37]
and Japanese quail after PCB ingestion. Toxic and teratogenic effects of PCB's
in chick embryo have also been reported by McLaughlin *et al.*[38].

PCB's as inducers of hepatic enzymes (*e.g.* increasing the rate of circulating
estradiol in the liver) may be responsible for aberrations in calcium metabolism in
certain species of birds[9] and are generally considered to be more of a potent threat
than DDT to our declining bird populations, especially for predatory birds that
accumulate fairly high levels of PCB's. Anderson *et al.*[12] have suggested evidence
that PCB's affected eggshell thickness, although to a lesser extent than DDE.

A number of the more salient biological and toxicological aspects of the PCB's
are summarized in Table 1. It is possible to distinguish two actions of the PCB's
on mammals, *e.g.* liver damage and skin lesions. Liver damage has been predomi-

TABLE 1

SOME BIOLOGICAL AND TOXICOLOGICAL EFFECTS IN THE PCB'S

(1) Acute oral LD_{50} in mammals varies from approximately 2–10 mg/kg. (Apparent increase in mammalian toxicity with decrease in chlorine content).

(2) Generally, enzyme induction increases with increase in chlorination of PCB's.

(3) Induction of hepatic hydroxylating microsomal enzymes and increased estrogenic activity in the rat.

(4) Enlargement of the liver and vacuolar or fatty degeneration of liver cells in rats, guinea pigs, and monkeys.

(5) Production of hydropericardial edema in chickens and Japanese quail.

(6) Teratogenic effect in chick embryo.

(7) Adverse reproductive effects in rats at levels of *ca.* 100 p.p.m. in diet.

(8) Possible adverse reproductive effects in mink.

(9) Possible implications in aberrations in calcium metabolism and reproduction in ring doves.

(10) Effects on hatchability in chickens, Japanese quail.

(11) Skin, liver, and kidney lesions in rabbits following dermal exposure.

(12) Possible immunosuppressive effects in rabbits.

(13) Chemical porphyrogenic effects in many species.

(14) Chloracnegenic and hepatotoxic effects in man.

(15) Hyperglyceridemic effects in man.

(16) Human miscarriages, stillbirths and transplacental transmission in abnormal pigmentation from "rice-oil disease" ("Yusho").

(17) PCB residues in human adipose tissue, serum, and milk.

(18) Hepatoxic, chloracnegenic, and porphyrogenic effects of chlorinated dibenzofuran contaminants in several species.

(19) Chloracnegenic effects of chlorinated napthalene contaminants in man.

nantly manifest in mice, rats, guinea pigs, rabbits, and monkeys in feeding and to a lesser extent in inhalation studies. Liver damage, as well as transplacental transmission as shown in abnormal pigmentation and miscarriages, have also been observed in the "Yusho" poisonings in Japan. Skin lesions have been observed in rabbits in dermal toxicity studies. In addition, chloracnegenic and porphyrogenic as well as edema effects in many species have been caused by commercial PCB preparations. It is believed[28,29,39,40] that the liver damage and skin lesions are caused primarily by chlorinated dibenzofuran contaminants and to a minor extent by PCB itself. These contaminants are also responsible for the edema formation observed (in fowl) while the PCB's are suggested to be the causative agents for the hepatic porphyria.

The human[41] and non-mammalian[42] effects of the PCB's have been reviewed.

Because of their similarity in structures and chemical properties, PCB's, if present in a sample, are carried through the usual pesticide extraction and screening procedures and are frequently mistaken for DDT and analogs in monitoring tests, *viz.*

Reynolds[13,43] described an activated Florisil column technique which separated Heptachlor, Aldrin, DDE and PCB with the first elution (60 ml n-hexane) from Lindane; and Heptachlor epoxide, DDD and DDT with the second elution (40 ml 50% ethyl ether in hexane).

Koeman et al.[1] eluted a number of apolar compounds DDE and PCB on activated Florisil columns using n-hexane, then Dieldrin and Endrin with 10% diethyl ether in hexane.

The separation of PCB's from DDT and its analogs and other common pesticides by column chromatography on silicic acid–Celite was reported by Armour and Burke[44]. Aldrin and PCB were eluted with the first fraction (250 ml petroleum ether) and Lindane, Heptachlor, Heptachlor epoxide, Dieldrin, Endrin, p,p'-DDE, o,p'-DDT, p,p'-DDT, and p,p'-DDD with the second fraction (200 ml acetonitrile–hexane–methylene chloride (1:19:80)). Determination of PCB's and pesticides could also be made on separate column eluates without cross-interference with recoveries of Aroclors 1254 and 1260 and of several chlorinated pesticides ranging from 76 to 100% and 80 to 100%, respectively. The effect of water content of silicic acid on elution and separation of Aroclor 1260 and DDT analogs is shown in Figure 1.

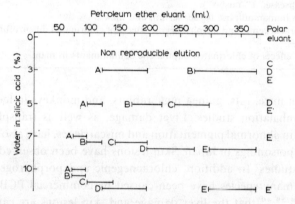

Fig. 1 Effect of silicic acid water content on elution and separation of Aroclor 1260 and DDT and analogs. 25 ml fractions of petroleum ether eluate analyzed by GLC. Respective compounds found present in volumes crossed by horizontal line. Various polar eluants used to elute pesticides C, D and E. A, Aroclor 1260; B, p,p'-DDE; C, o,p'-DDT; D, p,p'-DTE, E, p,p'-TDE.

The maximum margin of separation of PCB from pesticides with good reproducibility of elution pattern was obtained with silicic acid containing 3% water. A more polar eluant was required for the complete elution of p,p'-DDT and p,p'-TDE, when the water content was 5% less.

A semi-quantitative determination of PCB's in tissue samples by TLC was described by Mulhern et al.[45]. Cleanup by hexane–acetonitrile partitioning and Florisil column chromatography was performed on samples before oxidative treatment[46] to convert DDE to DCBP (4,4-dichlorobenzophenone), then the PCB components were determined semiquantitatively (with a lower limit of sensitivity of 0.2 μg) by TLC with no prior separation from chlorinated pesticides required. TLC separation was accomplished on silver nitrate-incorporated Aluminum Oxide G plates developed with 5% benzene in hexane. The spots were detected using a spray consisting of 5 ml of water and 10 ml of 2-phenoxy-ethanol diluted to 200 ml with acetone and containing 1–5 drops of 30% hydrogen peroxide. followed by exposure to germicidal UV light. The R_F values of Aroclor mixtures chromatographed by this TLC system are listed in Table 2. The lower limit of detection of PCB by this procedure is 0.2 μg.

TABLE 2

R_F VALUES FOR AROCLOR MIXTURES AND PESTICIDES ON AgNO$_3$-INCORPORATED ALUMINUM OXIDE G PLATES DEVELOPED WITH 5% BENZENE IN HEXANE

Compound	R_F value
1242	0.91
1248	0.93
1254	0.93
1260	0.93
1262	0.94
DDE	0.93
DDT	0.88
DDD	0.74
Heptachlor epoxide	0.58
Dieldrin	0.48
DCBP	0.30

The separation and identification of DDT analogs in the presence of poly-chlorinated biphenyls by two-dimensional TLC was described by Fehringer and Westfall[47]. Separation resulted from the different migration patterns and migration distances relative to p,p'-DDE that mixtures of PCB's exhibit under two sets of conditions. Two-dimensional TLC was employed using MN-Kieselgel G-HR as the absorbent with n-heptane alone and n-heptane–acetone (98:2) as developing solvents and silver nitrate incorporated into the absorbent layer as the chromogenic agent. First-dimensional development was with n-heptane and subsequent 90° development was performed with n-heptane–acetone (98:2).

The synthesis and TLC of 23 chlorobiphenyls has been described by Hutzinger et al.[48]. All chlorobiphenyls prepared were homogeneous on silica gel thin-layer chromatograms developed in heptane and gave one peak when chromatographed on a 4% SE-30 column. Table 3 lists the name, structural formula, starting material sources, method of purification, melting points, and R_F values of the individual chlorobiphenyls.

The discovery of PCB's in the environment was not accomplished until 1966 for three basic reasons: (1) they were not deliberately distributed about the ecosystem, (2) their presence was not immediately evident due to their relatively low acute

TABLE 3

PHYSICAL AND TLC PROPERTIES OF CHLOROBIPHENYLS

Compound		Preparation	Purification[a]	Physical data	
Name	Formula	Starting material (sources[c,b])		M.p. (°C)[e]	R_F (× 100)[f]
Monochlorobiphenyls					
2-		2-Aminobiphenyl (I)	TLC, MeOH aq.	33–34	35
3-		3-Chloroaniline (I)	Dist., MeOH aq.	16–17	49
4-			MeOH	77–78	43
Dichlorobiphenyls					
2,2'-		2-Chloroiodobenzene (V)	TLC, MeOH	59–60	38
3,3'-		3,3'-Dinitrobiphenyl (II) → 3,3'-Diaminobiphenyl	TLC, EtOH aq.	26–27	45
4,4'-			MeOH	148–149	41

Substitution	Starting material / preparation	Method	M.p. (°C)	Yield (%)
2,4-	2,4-Dichloroaniline (I)	TLC, MeOH aq.	24–25	50
3,4-	3,4-Dichloroaniline (III)	TLC, MeOH	45–46	42
2,6-	2,6-Dichloroaniline (I)	Col., MeOH aq.	35–36	48
Trichlorobiphenyls 2,4,6-	2,4,6-Trichloroaniline (I)	Col., TLC, EtOH aq.	62–63 / 62.5	51
2,4,4'-	4,4'-Dichloro-2-nitrobiphenyl (18) → 2-Amino-4,4'-dichlorobiphenyl	MeOH aq.	57–58	53
	2,2'-Dichlorobenzidine (20) or 2,4-Dichloroiodobenzene (20)	Dist., col., EtOH aq.	41	58
Tetrachlorobiphenyls 2,4,2',4'-				
3,4,3',4'-	3,4,3',4'-Tetraaminobiphenyl (VI) or 3,3'-Dichlorobenzidine (VIII)	TLC, EtOH aq.	173	44

| Compound[a] | | Preparation | Purification[d] | Physical data | |
Name	Formula	Starting material (sources[c,b])		M.p. (°C)[e]	R_F (×100)[b]
3,5,3',5'-		3,5,3',5'-Tetrachlorobenzidine (20)	TLC, MeOH, hexane	164	65
2,6,2',6'-		2,6-Dichloroaniline (I) → 2,6-Dichloroiodobenzene	Col., EtOH, aq.	198	40
2,3,4,5-		2,3,4,5-Tetrachloronitrobenzene (IV) → 2,3,4,5-Tetrachloroaniline	Col., MeOH	91	48
2,3,5,6-		2,3,5,6-Tetrachloronitrobenzene (I) → 2,3,5,6-Tetrachloroaniline	Col., hexane	79	50
Pentachlorobiphenyl 2,3,4,5,6-		Pentachloronitrobenzene (I) → Pentachloroaniline	Col., MeOH	123	51

Compound	Structure	Prepared from (source)	Method[d]	m.p. (°C)	Ref.
Hexachlorobiphenyls 2,4,6,2',4',6'-		2,6,2',6'-Tetrachlorobenzidine (20) or 2,4,6-Trichloroiodobenzene (21)	Col., MeOH/EtOH	112–113	69
3,4,5,3',4',5'-		3,3',5,5'-Tetrachlorobenzidine (20)	TLC, EtOH aq.	201–202	54
Octachlorobiphenyls 2,3,4,6,2',3',4',6'-		3,3'-Diamino-2,4,6,2',4',6'-tetrachlorobiphenyl (25)	TLC,EtOH	132	70
2,3,5,6,2',3',5',6'-		2,3,5,6-Tetrachloronitrobenzene (I) → 2,3,5,6-Tetrachloroaniline → 2,3,5,6-Tetrachloro-1-iodobenzene	TLC, EtOH aq.	160–161	68
Decachlorobiphenyl		Aroclor 1268 (VII)	Benzene	305–306	76

[a] The following chlorobiphenyls are commercially available (source[b]): 2-chlorobiphenyl (IV, V, VIII), 3-chlorobiphenyl (IV, V, VIII), 4-chlorobiphenyl (I, IV, V, VIII), 2,2'-dichlorobiphenyl (IV, VIII), and 4,4'-dichlorobiphenyl (I, IV, VIII).

[b] Code for commercial suppliers: I, Aldrich Chem. Co.; II, Sapon Lab.; III, Eastman; IV, Chemical Procurement Labs: V, K & K Lab.; VI, Burdick & Jackson Lab.; VII, Monsanto Chem. Co.; VIII, Pfaltz & Bauer Chemicals.

[c] Prepared by method given in reference or commercially available.

[d] TLC = preparative thin-layer chromatography (hexane); col. = silica column (hexane); MeOH = methanol; EtOH = ethanol; dist. = distillation at 0.5 mm.

[e] New compound, when no literature m.p. given.

[f] In heptane on commercially prepared thin-layer plates (Merck Silica Gel F-274).

toxicities, and (3) the difficulty of analytical detection and separation. The poly-chlorinated biphenyls were first detected as interfering peaks in the GLC analysis of environmental samples being analyzed for chlorinated pesticide residues. This difficulty is shown in Fig. 2 which illustrates gas chromatograms of a standard pesticide mixture, Aroclor 1254, and a mixture of both Aroclor 54 and chlorinated pesticides.

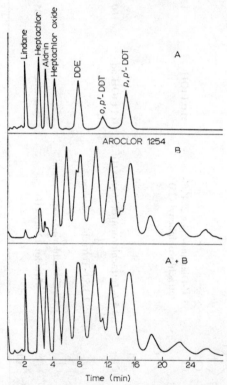

Fig. 2. Gas chromatogram A is of a standard pesticide mixture, B is typical of Aroclor 1254, and A + B is the chromatogram of a 50:50 mixture of both. The concentration of Aroclor 1254 is approximately 10 times those of the pesticides in A. Gas chromatograph (Packard 7620) conditions; gas flow 80 ml/min nitrogen; temperatures: oven, 210°C, inlet 230°C, detector 218°C; [63]Ni electron-capture detector; 5 ft. × ⅛ in. glass column packed with 8% SE-30 on Chromosorb Q 80–90 mesh.

Reynolds[13,43] reviewed the problem of pesticide residue analysis in the presence of PCB's. The reported GLC patterns indicating PCB interferences have shown marked resemblances[1,10,43,49] with Aroclors 1254 and 1260. This type of interference with pesticide residue analysis is shown with a standard mixture of organochlorine pesticides and a commercial PCB mixture (Aroclor 1254) in the chromatogram of Fig. 3. The results confirm that the presence of PCB's in the sample extracts will cause interference with pesticide residue analysis under these or similar operating conditions.

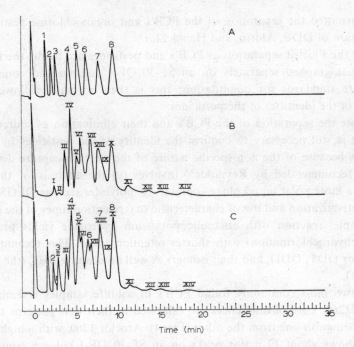

Fig. 3. PCB interference with organochlorine pesticide residue analysis on 4% SE-30/6% QF-1 on 60–80 mesh Chromosorb W, 6 ft. × ⅛ in. borosilicate column. Chromatogram A, standard mixture of organochlorine pesticides; 1, 0.08 ng Lindane; 2, 0.10 ng Heptachlor; 3, 0.10 ng Aldrin; 4, 0.14 ng Heptachlor epoxide; 5, 0.20 ng DDE; 6, 0.20 ng Dieldrin; 7, 0.30 ng DDD, 8, 0.50 ng DDT. Chromatogram B: 5 ng of Aroclor 1254 (the 14 major peaks are numbered I to XIV). Chromatogram C: combination of above organochlorine standard pesticide mixture and Aroclor 1254. Injector 250°C, column 200°C, detector base 250°, N₂ at 20–30 ml/min.

The GLC was carried out using a Varian Model 1200 gas chromatograph fitted with a tritium electron-capture detector and a spiral glass column (6 ft. × ⅛ in. o.d.) packed with 6% QF-1 and 4% SE-30 on Chromosorb W (AW). (The number of theoretical plates for DDT = 2227.) The operating conditions were column 190°, injector 245°, detector base 240°C, nitrogen flow rate 40 ml/min, Varian Aerograph Model 20 recorder 1 mV fullscale deflection, and chart speed ⅔ in./min.

It is of note that the peaks of the commonly found pesticides all have a corresponding PCB peak that would interfere if present in the same extract. The problem of the interferences of PCB's with the analysis of chlorinated pesticides has been largely overcome by prior column chromatography[1,13] or *via* an initial extraction with hexane as in regular pesticide analysis, partitioning with acetonitrile to remove fats, passage of cleaned-up extract through both Florisil and silicic acid columns, then finally analyzed by GLC with preferably a chloride-specific detector. The pesticides are retained on the silicic column for later elution and identification.

The use of column (30 cm × 2.5 cm, o.d.) packed with 40 ml (*ca.* 19 g or 10 cm in height) Florisil (60–80 mesh, stored at 130°C until used) and elution with 20 ml hexane permitted the separation of the PCB's and organochlorine pesticides (with the exception of DDE, Aldrin, and Heptachlor[13].

After the Florisil separation of PCB's and pesticides are made, the two eluates are chromatographed separately on an SE-30–QF-1 column and compared with appropriate standards for quantitation; this is then normally followed by confirmation of the identities of the pesticides.

Despite the separation of the PCB's and their elimination as sources of interference, it is still necessary to confirm the identity of the pesticides by additional techniques because of the non-specific nature of the electron-capture detector. The sequence recommended by Reynolds[13] involves (a) the analysis of the pesticide eluate on a more polar liquid phase column, e.g. polyester such as DEGS or DEGA and (b) derivatization and use of characteristic GLC retention times of the derivatives. For example, reaction with ethanolic potassium hydroxide yields products (via mainly dehydrochlorination) with shorter retention times. This technique is most effective for DDT, DDD, and their isomers as well as α- and γ-BHC (the β isomer is unaffected).

The two most commonly found PCB's in wildlife samples resemble Aroclor 1254 and 1260. Certain differences are exhibited in their GLC patterns that can be used to distinguish one from the other (Fig. 4). Aroclor 1260, with a higher chlorine content, shows about 17 major peaks on an SE-30–QF-1 column compared with 11 with Aroclor 1254. Another major difference is the peak height ratio of peaks 10 and 13, e.g. this ratio is approx. 9.3 and 1.0 for Aroclors 1254 and 1260, respectively.

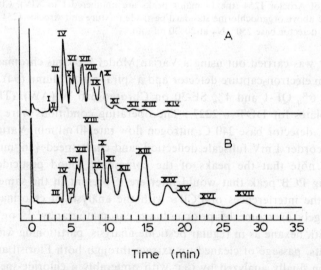

Fig. 4. Comparison of Aroclors 1254 and 1260. Chromatogram A: 5 ng of Aroclor 1254; the 14 major peaks are numbered I to XIV as in chromatogram B of Fig. 3. Chromatogram B: 5 ng of Aroclor 1260; the 17 major peaks are numbered I to XVII (the early peaks correspond to those in Aroclor 1254). GLC parameters as in Fig. 3.

Figure 5 illustrates GLC chromatograms of a number of Aroclor mixtures determined on SE-30–QF-1 columns. The more highly chlorinated biphenyls (Aroclors 1254 and 1260) are easily detected while the compounds of the lower chlorinated mixtures, *e.g.* Aroclors 1221, 1232, and 1242, and the higher molecular weight mixtures (Aroclor 5460), are less responsive with the usual operating parameters and are thus more likely to go undetected.

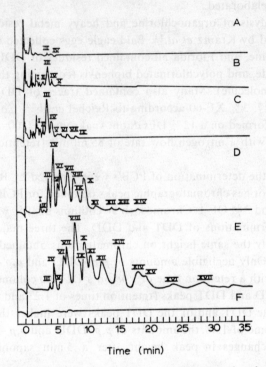

Time (min)

Fig. 5. GLC profiles of the more popular Aroclor mixtures under normal analytical conditions. Note: all peak numbers correspond to those of Aroclor 1254. Chromatograms: A, 5 ng Aroclor 1221; B, 5 ng. Aroclor 1232; C, 5 ng Aroclor 1242; D, 5 ng Aroclor 1254; E, 5 ng Aroclor 1260; F, 5 ng Aroclor 5460. GLC parameters as in Fig. 3.

The estimation of PCB's following GLC analysis in the past has been generally achieved *via* the following techniques: (*1*) the PCB estimation[1] based on the peak of Phenoclor DP6 having r_x (relative retention time with Dieldrin = 1) equal to 1.45, (*2*) the PCB's are estimated on the assumption that they have similar electron-capture responses to *p,p'*-DDE and a factor is applied to fit the assumed 54% chlorine content of the PCB's[50], (*3*) estimates of the PCB's are reported as the *sum* of all the PCB components[14] and (*4*) estimation of PCB's based on an average of two peaks[1].

A GLC study[44] of Aroclors 1221, 1232, 1254, 1260, 1262, 4465, 5442, and 5460 on columns of 10% DC-200 and 15% QF-1/10% DC-200 with electron-capture detection showed the Aroclors to be multicomponent mixtures with retention times throughout and beyond the retention time range of common chlorinated pesticides. With a detector sensitivity producing 1/9 fullscale recorder response to 1 ng Heptachlor epoxide, significant responses were obtained with about 10–20 ng of the various Aroclors.

In the above study of Armour and Burke[44] the efficacy of silicic acid columns for the separation of Aroclors and chlorinated pesticides added to trout and salmon extracts was also elaborated.

The GLC analysis of organochlorine and heavy metal residues in bald eagle eggs was described by Krantz et al.[51]. Bald eagle eggs collected in 1968 from nests in Wisconsin, Maine, and Florida all contained residues of DDE, DDD, Dieldrin, Heptachlor epoxide, and polychlorinated biphenyls (containing three to eight atoms of chlorine per molecule). Many also contained traces of DDT. Fractions were analyzed on OV-17, 3% XE-60 according to Reichel et al.[52], Confirmatory residue analyses were performed on a 12% DEGS on Anakrom SO (100–110 mesh), column temperature 190° with a nitrogen flow rate of 85 ml/min (retention time of Dieldrin was 9.5 min).

Methods for the determination of PCB's were surveyed by Risebrough et al.[50]. Several of the major gas chromatographic peaks obtained for PCB on 10% DC-200–Chromosorb W and 3% QF-1–Chromosorb W columns at 195° were found to interfere with the determinations of DDT and DDD. The three major peaks generally have approximately the same height on chromatograms obtained with an electron-capture detector. Only negligible amounts of p,p'-DDD and p,p'-DDT are present, if the PCB peak with a retention time of 1.48 on a DC-200 column is approximately as high as the DDD and DDT peaks (retention times of 1.27 and 1.68, respectively). However, when the DDT and/or the DDT peaks were higher than the PCB peak with a retention time of 1.48, the amounts of p,p'-DDD and p,p'-DDT could be estimated from the changes in peak height after a 5 min saponification with 5% KOH in ethanol.

Gas chromatographic analytical conditions for the PCB's have usually been those employed for chlorinated pesticides including such liquid phases as SE-30 (or OV-1, OV-101)[15,48,53–55], DC-200–QF-1[56], or XE-60[57]. Apiezon L has been reported to give greater resolution of the many components in the Arochlor series of commercial PCB mixtures than a large number of other liquid phases tested[58]. Even with Apiezon L, however, long capillary or SCOT columns are required for resolution of the large number of similar isomers occurring in the Arochlors[58].

Quantitative analysis for PCB's is complicated by the number of peaks appearing on chromatograms and by the variation in detector response to each component. This variation may be 3000-fold using an electron-capture detector[55]. Although a correlation between total electron-capture peak area and amount of a particular mixture of PCB's can be made[55], any variation in the relative composition of the mixture will cause an unpredictable change in the total peak area response. This is particularly troublesome in analyzing PCB's extracted from biological materials, since the PCB mixtures obtained often do not exactly match any of the commercial PCB mixtures.

Stalling[59] has recommended programmed-temperature gas chromatography on SCOT columns using electrolytic conductivity or microcoulometric detectors for reliable quantitative analysis of PCB mixtures. In conjunction with mass spectrometry to both confirm the identification of a given peak as being due to a PCB, and determine the number of chlorine atoms per molecule (necessary for quantitation by microcoulometry), this recommended procedure should indeed cope adequately with the major problems in PCB analysis where large numbers of PCB isomers are involved.

Increasing interest in environmental aspects of PCB's and development of specific methods for the synthesis of many individual isomers[48,60] may be expected to result in an increasing amount of research involving less complex mixtures of PCB's than those discussed above. Preparative fractionation of a small number of very similar isomers will become increasingly necessary to those involved in synthetic work.

Albro and Fishbein[61] recently described the quantitative and qualitative analysis of polychlorinated biphenyls by GLC and flame ionization. Retention indices on six liquid phases were presented for the mono-, di-, and trichlorobiphenyls and the additivity of $\frac{1}{2}$ (R.I) value in predicting retention indices was confirmed for these compounds. A method was given for predicting the relative molar response of the hydrogen flame-ionization detector to PCB's for which no pure standard is available. Solute–solvent interaction between these "non-polar" compounds and a variety of liquid phases was also elaborated.

Varian Aerograph Model 2100 and Hewlett-Packard Model 5750 gas chromatographs, both equipped with hydrogen flame-ionization detectors and 1 mV recorders, were used for these studies. Injection port and detector temperatures were 260 and 300°, respectively, hydrogen and air flows were 30 and 320 ml/min, respectively, and samples were injected in less than 5 μl of n-heptane or methylene chloride.

The columns used in this study were 0.318 cm o.d. stainless steel and the column packings were: (1) OV-101, 10% on 80–100 mesh High Performance Chromosorb W, column length 3 m, (2) Versilube F-50, 5% on 80–100 mesh High Performance Chromosorb W, column length 2.75 m, (3) Apiezon L, 6% on 100–120 mesh acid-washed Chromosorb W, column length 2 m, (4) OV-17, 10% on 80–100 mesh Supelcoport, column length 3 m, (5) OV-225, 10% on 80–100 mesh Supelcoport, column length 2.5 m, (6) cyclohexanedimethanol succinate (CHDMS), 3% on 100–120 mesh Gas Chrom Q, column length 4 m. Some samples were run on two additional columns, a 1.5 m × 0.2 cm (i.d.) column packed with 5% "purified" Apiezon L on 100 to 120 mesh Gas Chrom Q. For the latter, Apiezon L (1 g) was passed through a column of Florisil (50 g) in n-heptane. 0.75 g of colorless grease was eluted with 300 ml of heptane, after which 0.23 g of orange wax was eluted with 250 ml of 5% methyl acetate in heptane. The white grease showed only aliphatic hydrocarbon absorption in the infrared and had the same thermal stability as a GLC liquid phase as Apiezon L.

The source of PCB contamination in the marine environment (Glasgow and Clyde areas) was studied by Holden[62]. Although the extraction of organochlorine compounds from sewage solids using a mixture of n-hexane and isopropanol is believed to be less efficient than extraction from water, it was found by Holden[62] that many sludge samples gave gas–liquid chromatograms similar to those of PCB mixtures. The GLC column used was packed with 10% DC-200 silicone on Chromosorb W. PCB compounds were initially separated from all major organochlorine pesticide residues, except p,p'-DDE on silica columns[63].

The peak pattern for sludge A closely resembled a PCB formulation containing about 50 wt.-% of chlorine (Aroclor 1254) but the pattern for sludge B suggested Aroclor 1260. On the basis of wet weight of crude sludge, the estimated concentrations of PCB·s using Aroclor 1254 as reference in 15 samples were in the range <0.1–14 p.p.m. In terms of dry matter, the values ranged from 1 to 185 p.p.m. It was estimated that an average concentration of PCB of the order of 1 p.p.m. would be equivalent to a discharge in the Clyde estuary of the order of 1 ton of PCB's/year. PCB compounds found in wildlife, using GLC at about 200°C, have been shown to contain mostly six to seven chlorine atoms[1], although compounds with more chlorine atoms may not be sufficiently volatile to be detectable at this temperature.

Veith and Lee[64] studied the PCB sources in the Milwaukee River drainage basin. The GLC analyses of water from the river indicated that isomers of chlorinated biphenyl similar to those used in industry were present. Analyses of municipal sewage treatment plant effluents, industrial discharges and the Milwaukee River water near combined sewer outfalls presented evidence that PCB's were discharged to natural waters through municipal and industrial wastes. PCB concentrations at the μg/l level suggested that PCB's in large ecosystems such as Lake Michigan have resulted in part through water transport from metropolitan areas. The cleanup of extracts for GC analysis was conducted with liquid chromatography on Florisil as described by Reynolds[43] and Hughes et al.[65].

The analyses of water extracts were conducted on an Aerograph 1745-20 gas chromatograph equipped with concentric tube electron-capture detectors (^{3}H, 250 mCi) and a 50:50 effluent splitter for simultaneous analysis with electron-capture and flame-ionization detectors. The GLC columns consisted of 2.0m × 1.8 mm glass coils which were packed with either 3% OV-101, OV-101/XE-60 (3:3 percent) or OV-101–QF-1 (3:4.5 percent) coated on Gas Chrom Q (720–140 mesh). The carrier gas, purified nitrogen, was maintained at 21 ml/min and the injector, column, and detector temperatures were 250, 180, and 220°, respectively. The PCB's present in the Milwaukee River were seldom comparable to a single Aroclor mixture and the data were reported as mixtures of Arochlor 1260 and Aroclor 1242. The relative variations of the composition of the PCB's in the river suggested that the lesser chlorinated isomers may be removed through preferential vaporization or co-distillation arising from their greater volatility than the heavier isomers and/or through the more rapid degradation of the lesser chlorinated isomers. The chemical stability of the PCB isomers toward nucleophilic and electrophilic substitution has been found by Veith[66] to increase with increasing chemical content. Veith and Lee[64] suggested that the above studies indicated that similar trends exist toward microbial degradation whereby the lesser-chlorinated isomers of Aroclors 1232 and 1242 are selectively removed from natural waters.

Ahling and Jensen[67] described a new method based on an application of the common reversed liquid–liquid partition method for the extraction of PCB's and chlorinated pesticides from water. In this method the water was passed through a filter (3 g) containing a mixture of n-undecane and Carbowax 4000 monostearate on Chromosorb W and the absorbed pesticides were eluted with petroleum ether (10 ml). When detected by means of electron-capture GC, the sensitivity was 10 ng/m^3 of Lindane with a sample size of 200 l. The recovery of added pesticides was 50 to 100% (DDT, 80%) and of PCB (93–100%).

The filter column was 30 cm × 1 cm i.d. with a glass-filter disk G1 (100 to 120 μm). The gas chromatograph was a Varian Aerograph 204 with EC detector and a Speedomax G-1 mV recorder. The column was borosilicate glass 160 cm × 0.20 cm containing either 4% methyl silicone oil (SF-96) or 8% QF-1 on 80–100 mesh HMDS-treated Chromosorb W. Nitrogen carrier gas was purified with a 6 in. molecular sieve and was used at a flow rate of 30 ml/min. The detector and injector temperatures were 205 and 220°C, respectively, and the column temperature was chosen to give a retention time of 20 min (about 190°C).

Figure 6(a) and (b) shows gas chromatogram extracts from distilled water and natural water samples determined on SF-96 columns. These peaks are responsible for the recoveries over 100% when a very small amount (180 p.p.b. or lower) of standard was added to distilled water. Figure 7 illustrates a gas chromatogram of extract from sewage sludge obtained from an SF-96 column.

Keil et al.[68] studied the effects of uptake of Aroclor 1242 on growth, nucleic acids, and chlorophyll of a marine diatom Cylindrotheca closterium. Following exposure of the cells with 0.1 and 0.01 p.p.m. levels of the PCB, the PCB was extracted from a dried lyophilized pellet with acetone and analyzed by GLC using two column systems: (a) a 6 ft. × ¼ in. o.d. glass column packed with 4% SE-30–2% QF-1 on Chromoport XXX, and (b) a 6 ft. × ¼ in. o.d. glass column packed with 1.5% OV-17-1–1.95% QF-1 on Chromoport XXX. The operating parameters were inlet temperature 235°C, column temperature 200°C, detector temperature 350°C, and carrier gas flow 60 ml/min. Identification of Aroclor 1242 was accomplished by measuring relative retention times of the five major peaks along with relative peak heights. Quantification was accomplished by comparison of the area under the curve of the five late eluters with the standard Aroclor. Verification of the peak identities by TLC positively identified Aroclor 1242 as the major PCB components of the sample. Some PCB materials not common to the known Aroclor 1242 mixture were

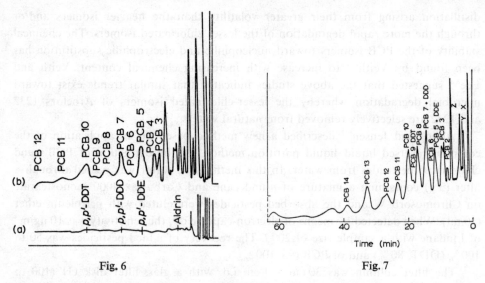

Fig. 6   Fig. 7

Fig. 6. (a) Gas Chromatogram from SF-96 column of extract from distilled water. (b) Gas chromatogram from SF-96 column of extract from natural water.

Fig. 7. Gas chromatogram from SF-96 column of extract from sewage sludge.

isolated. These materials, in all cases early GLC eluters, were believed to be possible metabolic products of Aroclor 1242.

Cylindrotheca closterium were found to absorb and concentrate Aroclor 1242 900–1000 times above the level in sea water. At 0.1 p.p.m. concentration in sea water PCB's inhibited growth and diminished levels of RNA and chlorophyll of this marine diatom.

The occurrence of Aroclor 1254 in the water, sediment, and biota of Escambia Bay, Florida, was described by Duke et al.[69]. Tissue of fish, crabs, oysters, and shrimp were extracted for 4 h with petroleum ether in a Soxhlet apparatus following initial mixing with anhydrous sodium sulfate. Extracts were concentrated, partitioned with acetonitrile, the acetonitrile then evaporated just to dryness, the residue transferred to a Florisil column[70] with petroleum ether and the Aroclor 1254 eluted from the column with 6% ethyl ether in petroleum ether.

Sediment samples were dried at room temperature and extracted for 4 h with 10% acetone in petroleum ether in a Soxhlet apparatus. The extracts were evaporated to dryness and the residues eluted from a Florisil column as described above. The extracts of all substrates were identified and measured by electron-capture gas chromatography using three columns of different polarity, e.g. DC-200, QF-1 and mixed DC-200 and QF-1 to confirm identification. In a few samples, TLC was employed for additional confirmation and to assess the amount of DDT present. Interference from DDT was negligible due to the relatively high residues of Aroclor 1254 found in the samples. Quantitation of this multiple peak compound was approximated by averaging the peak height of five major peaks indicating recovery rates above 80%.

Additional acute studies indicated that juvenile shrimp were the most sensitive to Aroclor 1254 and were killed when exposed to 5.0 p.p.b. in flowing sea water. The Aroclor content in water from Escambia Bay even near the mouth of the river contained less than 1 p.p.b. and shrimp collected from the bay contained a maximum of 2.5 p.p.m. The above study illustrated the need for both conducting continued surveillance of estuaries in order to preserve these nursery grounds for valuable fishery resources as well as conducting long-term tests on the effect of sublethal concentrations of PCB's on estuarine organisms in sensitive stages of their life history.

Westöö and Noren[71] reported the determinations by GLC and TLC of organochlorine pesticides in animal foods in the presence of polychlorinated biphenyls. The extraction of organochlorine pesticides and PCB's from animal foods, routine cleanup of the extract as well as main GLC and TLC procedures were performed according to Noren and Westöö[72].

Both Varian Aerograph Hy-Fi Models 600 and III-1200 gas chromatographs were used equipped with electron-capture detectors. Column I consisted of a 5 ft. × $\frac{1}{8}$ in. Pyrex glass packed with 5% DC-11 on Chromosorb W 60–80 mesh, nitrogen flow rate 75 ml/min, temperatures of column, injector, and detector 195, 210, and 195°, respectively. Column II was a 5 ft. × $\frac{1}{8}$ in. Pyrex glass packed with a mixture of equal parts of 5% DC-11 on Chromosorb W 60–80 mesh and 15% QF-1 on Chromosorb W 60–80 mesh, nitrogen carrier gas flow rate 30 ml/min. Column, injector, and detector temperatures were 171, 200, and 200°C, respectively.

For TLC[72] Aluminum Oxide G plates were developed with n-hexane–anhydrous ethyl ether (40:0.8). Detection was accomplished using a silver nitrate spray consisting of 0.10 g silver nitrate in 1 ml of water and 10 ml of 2-phenoxyethanol diluted to 200 ml with acetone and treated with 1 drop of 30% hydrogen peroxide[73].

Grant et al.[74] described the oral metabolism of Aroclor 1254 in male normal and carbon tetrachloride-treated rats. Hexane extracts of tissues were analyzed by GLC–electron capture analysis on a Varian Aerograph Model 600D gas chromatograph fitted with a coiled 4 ft. × $\frac{1}{4}$ in. o.d. glass column containing 4% SE-30 and 6% QF-1 on Chromosorb W (80–100 mesh). The nitrogen flow rate was 120 ml/min with column and injection temperatures of 193 and 225°C.

Residues were found in all tissues analyzed (blood, heart, kidney, brain, liver, and fat) with the greatest concentration in the fat. The GLC–EC pattern of the residues was different from the standard mixture administered, indicating that all components were not metabolized at the same rate. Higher residues were found in the carbon tetrachloride-treated rats. Aroclor 1254 residues in the brain, spleen, blood, testes, heart, kidney, liver, and fat were reduced by 90, 84, 80, 79, 78, 76, 64, and 33%, respectively, in 20 days. Aroclor 1254 significantly increased the size of the liver and also the percent lipid in the liver as well as to potentiate the toxicity of carbon tetrachloride in the rat.

Figure 8 shows the chromatograms from 24 ng of Aroclor 1254 and from the residue found in the liver of a rat. It was shown that the components of the Aroclor 1254 mixture with the shorter retention times, peaks 1, 2, and 3 and presumably with the lowest chlorine contents[2], were metabolized to a greater degree than those with the longer retention times. This observation agrees with that reported in studies with Phenoclor DP6 fed to Japanese quail[1].

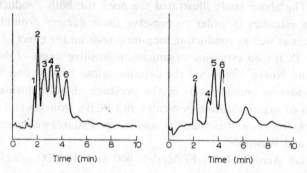

Fig. 8. GLC–EC tracings from 24 ng of Aroclor 1254 (left) and from the residue found in the liver of a group 2 rat.

A comparative toxicologic study with PCB's in chickens with special reference to porphyrin, edema formation, liver necrosis, and tissue residues was described by Vos and Koeman[29]. Three 60% chlorinated PCB's were used in a comparative 60-day oral toxicity test (400 p.p.m.) in chickens, e.g. Phenoclor DP6 (I), Clophen A60 (II), and Aroclor 1260 (III). Using GLC analysis of liver and brain of dead chicks gave PCB levels that varied from 120 and 2900 p.p.m. Tissues were extracted with petroleum ether in a Soxhlet extraction apparatus after drying with anhydrous sodium sulfate. Cleanup of the samples was performed by liquid–liquid partition with dimethylformamide and column chromatography using activated Florisil. The PCB preparations in the final extract were measured by GLC using a Varian Aerograph Model 204-1B with electron-capture detection and a Pyrex glass column (5 ft. × $\frac{1}{8}$ in.) filled with 10% DC-200 on Gas Chrom Q (80–100 mesh) operated at 200°C with nitrogen as carrier gas at a flow of 50 ml/min. For quantitation one peak, considered to be representative, was selected (the peak with $r_x = 1.45$ relative to Dieldrin). The approximate amount of total PCB residue was calculated from the height of this peak with compound I (Phenoclor DP6) as the standard.

In a number of tissue extracts the amounts of PCB were measured also by a Dohrmann C 250A microcoulometric detection system equipped with a T-300-S-halogen cell coupled to a Microtek MT-220 gas chromatograph and employing a 6 ft. × $\frac{1}{4}$ in. Pyrex column filled[75] with a mixed-bed column packing QF-1 plus DC-200. With this system 0.6 μg of Phenoclor DP6 gave about half-scale response on a 1 mV recorder (microcoulometer range 200 Ω). By this method, the content of chlorine in the extract is determined by addition of the chlorine contents of the different PCB peaks. Since the average chlorine contents of the compounds used were 60%, the total residue could be calculated.

Kojima et al.[31] described the utility of GLC for the analysis of chlorobiphenyls from the skin of a stillborn infant who was born from a patient suffering from chloro-biphenyl poisoning and also showed that chlorobiphenyls passed through the placenta barrier and accumulated in the fetus. Kaneclore 400 is a mixture of di-, tri-, tetra-, and pentachlorobiphenyls (chlorine content approx. 48%). Approximately 2300 p.p.m was concluded to be the concentration of Kaneclore 400 in the patient—rice oil. Histologically, many melamine pigments were demonstrated in the darkish skin of the stillborn infant and this pigmentation could possibly be caused by chloro-biphenyls since some components of Kaneclore 400 were detected in the darkish skin of the infant by GLC. Parallel experiments with mice confirmed the fact that chlorobiphenyls passed through the placenta barrier and accumulated in the fetus. GLC analyses were carried out using a Shimadzu GC-1C instrument equipped with an electron-capture detector of pulse type. The glass column (4 × 262.5 mm) was packed with 1.5% SE-30 on 60–80 mesh Chromosorb W. The operating temperature for the column and detector was 200°C; the carrier gas was nitrogen with a flow rate of 65.5 ml/min. Figures 9 and 10 illustrate the method of extraction and cleanup of chlorobiphenyls from the rice oil and biological materials, respectively.

The biological interaction between a series of polychlorinated biphenyl com-pounds and Dieldrin and DDT was studied by Lichtenstein et al.[7]. Many PCB's were found toxic to *Drosophila melanogaster* Meigen and houseflies *Musca domestica* L, but to a lesser extent than Dieldrin or DDT. Their toxicity increased with a *decrease* in their chlorine content and moreover, sublethal dosages of several of the PCB plasticizers increased the toxicity of Dieldrin and DDT. Since the PCB's are in the environment, their potential effects on biological systems, especially in combination with other synthetic chemicals, should thus be considered. The chromatographic behavior of 11 PCB's was examined by both TLC and GLC. The latter utilized two systems. One instrument was a Packard gas chromatograph, Model 7834, equipped with a 150 mCi tritium affinity ionization detector operated at 50 V and a 1.83 m × 4.0 mm i.d. glass column packed with 5% SE-30 on 60–80 mesh Chromosorb W AW-DMCS (conditioned for 7 days at 250°C before use) with a column pressure of 25 p.s.i. of nitrogen (flow rate of 125 ml/min). The injection, oven, and detector cell temperatures were 240, 190, and 215°C, respectively.

The second instrument was a Jarrell-Ash gas chromatograph Model 28-700 equipped with a 100 mCi tritium electron affinity ionization detector and operated at 20 V. A 1.22 m glass column (4.0 mm i.d.) contained a 1:1 mixture of 5% QF-1 and 5% DC-200 coated on Anakrom AS (80–90 mesh) and conditioned for 7 days at 250°C before use (column pressure of 16 p.s.i., flow rate of nitrogen was 100 ml/min). The injection port, oven, and detector temperatures were 250, 190, and 210°C, respec-tively. For TLC, Aluminum Oxide G plates were developed with *n*-heptane, sprayed with the Mitchell reagent[75] and exposed to UV light for 30 min.

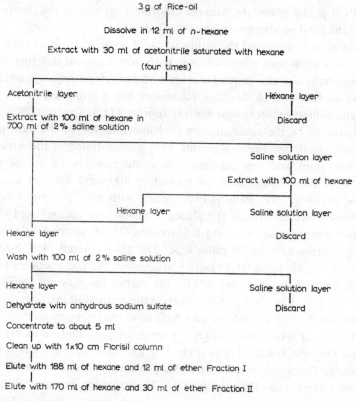

Fig. 9. Method of extraction and cleanup of chlorobiphenyls from rice oil.

The input of PCB's into California coastal waters from urban sewage outfalls has been described by Schmidt et al.[76]. Aliquots of concentrated extracts were analyzed by GLC using a Microtek 220 gas chromatograph equipped with a ^{63}Ni electron-capture detector and a 3% QF-1 column on Chromosorb (80–100 mesh, acid-washed and DMCS treated). The column, injection port, and detector temperatures were 190, 230, and 250°, respectively. Nitrogen was used as a carrier and purge gas; flow through the column was 95 ml/min and flow through the purge was 45 ml/min. PCB was identified by the presence of profile on chromatograms identical to those of one of the commercial PCB preparations (Aroclor 1254). Other chromatograms were identical except for minor peaks to chromatograms of Aroclor 1242, mixtures of Aroclor 1242 and 1254 or of Aroclor 1260. PCB was quantified by direct comparison with standard preparations of these commercial mixtures. The profile of PCB peaks was unchanged by either saponification or passage through a sulfuric acid –Celite column[77] used for additional cleanup of extracts.

It is noted that daily outputs of PCB in the order of two kilograms are equivalent to a ton a year and because of the low solubility of these compounds in water and their high affinity for lipids, a large fraction of the PCB can be expected to

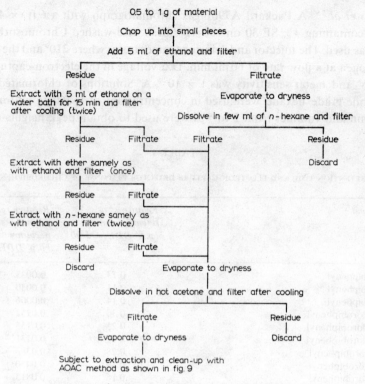

0.5 to 1 g of material
|
Chop up into small pieces
|
Add 5 ml of ethanol and filter

Residue — Filtrate

Residue:
Extract with 5 ml of ethanol on a water bath for 15 min and filter after cooling (twice)

Filtrate:
Evaporate to dryness
|
Dissolve in few ml of *n*-hexane and filter

Residue — Filtrate

Filtrate — Residue

Residue:
Extract with ether samely as with ethanol and filter (once)

Residue: Discard

Residue — Filtrate

Residue:
Extract with *n*-hexane samely as with ethanol and filter (twice)

Residue — Filtrate

Residue: Discard

Evaporate to dryness
|
Dissolve in hot acetone and filter after cooling

Filtrate — Residue

Filtrate:
Evaporate to dryness
|
Subject to extraction and clean-up with AOAC method as shown in fig. 9

Residue: Discard

Fig. 10. Extraction method from biological materials.

pass into marine food chains. The highest output was recorded from Los Angeles County, one of the most industrialized areas of the state of California.

Zitko[78] described the solubilization of PCB in water by non-ionic surfactants as well as the determination of concentration of the solubilized PCB by UV spectrophotometry and fluorescence and gas–liquid chromatography and the toxicity of PCB to Atlantic salmon (*Salmo salar* parr.).

Gas chromatography was carried out on a Varian 600D instrument equipped with a 250 mCi tritium electron-capture detector, using a 5 ft. × $\frac{1}{8}$ in. glass column containing 4% SE-30 on Chromosorb W (100–120 mesh) operated at 180°C. Ultraviolet and fluorescence spectra were recorded on a Beckman DK-2A and on a Perkin-Elmer MPF-2A instrument, respectively. Retention times and electron-capture responses have been reported for mono- and dichlorobiphenyls[79,80]. Emery and Gasser[81] reported the GLC separation on polyphenyl thioethers of mono- and some di- and trichlorobiphenyls. As a general rule, chlorine substitution in position 2 shortens and vicinal disubstitution lengthens the retention time. The reported relative electron-capture detector responses are 1, 0.2, 2.3, 7.6, 5.8, and 3.5 for 4-, 3-, 2-mono-, 4,4'-, 3,3'-, and 2,2'-dichlorobiphenyl, respectively[80]. The retention times and electron-capture detector responses of a number of PCB's has been reported by Zitko et al.[55]. A Packard A7901 gas chromatograph with a 6 ft. × 4 mm glass column containing 4 % SE-30 on 100–120 mesh acid-washed Chromosorb operated 200°C was used. The injector and detector temperatures where 210° and the carrier gas was nitrogen at a flow rate of 60 ml/min. D.c. voltage in the electron-capture detector was 95 V and meter sensitivity was 1 × 10^{-8} A. Solutions of chlorinated biphenyls in pesticide-grade hexane were used in concentrations of 1–260 µg/ml for injections on column and volumes from 2 to 5 µl were used to obtain peak heights of 30–70 %

TABLE 4

RETENTION TIME AND ELECTRON-CAPTURE DETECTOR RESPONSE OF CHLOROBIPHENYLS

Compound	Relative retention time (p,p'-DDE = 1.00)	Relative response per ng + standard deviation (p,p'-DDE = 1.00)
4-Chlorobiphenyl	0.17	0.0033 ± 0.00005
2-Chlorobiphenyl	0.11	0.0030 ± 0.00004
3-Chlorobiphenyl	0.14	0.0006 ± 0.00001
4,4'-Dichlorobiphenyl	0.30	0.0152 ± 0.0002
3,3'-Dichlorobiphenyl	0.25	0.0155 ± 0.0001
2,2'-Dichlorobiphenyl	0.15	0.0131 ± 0.0001
3,4-Dichlorobiphenyl	0.28	0.0388 ± 0.0002
2,4-Dichlorobiphenyl	0.19	0.0450 ± 0.0004
2,6-Dichlorobiphenyl	0.16	0.0815 ± 0.0049
2,4,4'-Trichlorobiphenyl	0.39	0.298 ± 0.010
2,4,6-Trichlorobiphenyl	0.24	0.276 ± 0.006
2,2',4,4'-Tetrachlorobiphenyl	0.51	0.206 ± 0.003
3,3',4,4'-Tetrachlorobiphenyl	0.96	0.770 ± 0.027
2,2',6,6'-Tetrachlorobiphenyl	0.33	0.0403 ± 0.0016
3,3',5,5'-Tetrachlorobiphenyl	0.70	0.625 ± 0.045
2,3,4,5-Tetrachlorobiphenyl	0.69	0.715 ± 0.010
2,3,5,6-Tetrachlorobiphenyl	0.51	0.505 ± 0.010
2,3,4,5,6-Pentachlorobiphenyl	1.00	1.30 ± 0.017
2,2',4,4',6,6'-Hexachlorobiphenyl	0.78	0.545 ± 0.029
3,3',4,4',5,5'-Hexachlorobiphenyl	2.98	1.15 ± 0.064
2,2',3,3',4,4',6,6'-Octachlorobiphenyl	2.62	1.58 ± 0.020
2,2',3,3',5,5',6,6'-Octachlorobiphenyl	2.45	1.53 ± 0.076
Decachlorobiphenyl	8.20	1.61 ± 0.086
Aroclor 1254		0.910 ± 0.008
Aroclor 1260		1.35 ± 0.010

of the fullscale pen deflection. Relative retention times and electron-capture detector responses of chlorinated biphenyls and Aroclor 1254 and 1260 are shown in Table 4. Chlorine substitution in positions 2 and 6 shortens and vicinal substitution lengthens the retention time also in tri- and tetrachlorobiphenyls. In some instances these effects overshadow the effect of the number of chlorines in the molecule which increases the retention time. The detector response increases strongly with increasing number of chlorine atoms in the molecule (*e.g.* the response of decachlorobiphenyl is 500 times stronger than that of 4-chlorobiphenyl). PCB's found in wildlife are generally of the Aroclor 1254 or 1260 type[44,50,71]. In the study of Zitko *et al.*[55], these Aroclors gave 14 and 16 peaks with relative retention times from 0.48 to 3.28 and from 0.72 to 6.80, respectively. The presented data indicates that the electron-capture response of all chlorinated biphenyls with 4–9 atoms of chlorine/molecule is not likely to exceed the value of 1.6 relative to p,p'-DDE under the described conditions.

The application of carbon-skeleton chromatography to the qualitative differentiation between PCB's and DDT was demonstrated by Asai *et al.*[82]. PCB's and biphenyl yield identical carbon-skeleton chromatograms that are strikingly different from that of DDT. Comparisons of the relative retention times of products yielded by the PCB's with known compounds, plus partial identification by their UV spectra suggested that the products formed at 300°C catalyst temperature were cyclohexylbenzene and biphenyl, and at 260°C were cyclohexylbenzene and a small amount of bicyclohexyl. The apparatus consisted of a NIL–Beroza carbon-skeleton determination (National Instrument Laboratories, Inc., Rockville Md.) attached to the injection port of an Aerograph Model A-600-B gas chromatograph equipped with a flame-ionization detector. For the electron-capture gas chromatograms, the same gas chromatograph was equipped with a tritium detector. A sodium chloride–neutral palladium catalyst (1 wt.-% as the metal on DMCS-treated Gas Chrom Q, 80 to 100 mesh)[83] was used to change the carbon-skeleton determinator. The temperature of the catalyst bed was maintained at either 260 or 300°C. The carbon-skeleton chromatograms were obtained using a 3 ft. × ⅛ in. stainless steel column packed with 5% DC-200 on DMCS-treated Gas Chrom Q, 80–100 mesh. The column temperature was held at 105°C and the hydrogen flow rate was 20 ml/min. The carbon-skeleton chromatograms of Aroclor 1260, biphenyl, bicyclohexyl-, and cyclohexylbenzene were obtained on a 3 ft. × ⅛ in. stainless steel column packed with 80–100 mesh Carbowax 400 on Porasil S. The column temperature was 178°C and the hydrogen flow rate was 20 ml/min. For the electron-capture determinations, 2 ft. × ⅛ in. stainless steel column packed with 10% DC-200 on DMCS-treated Gas Chrom Q 80–100 mesh was used with a column temperature of 180°C and nitrogen flow rate of 85 ml/min. Figure 11 illustrates carbon-skeleton chromatograms of biphenyl and Aroclor 1260 at two catalyst temperatures, *viz.* 260 and 300°C. The two peaks (cyclobenzene and biphenyl) observed at the higher catalyst temperature was due to incomplete hydrogenation of the biphenyl (hydrogenation is favored at lower temperatures)[84].

Gas chromatograms for Aroclor 1260 and a mixture of p,p'-TDE and p,p'-DDT obtained with the electron-capture detector are shown in Fig. 12 and illustrate how the PCB's can interfere with the analysis of the DDT pesticide group. Carbon-skeleton chromatograms obtained at 260 and 300° for the samples, at much higher concentrations, are depicted in Fig. 13 and suggest that this technique can be an

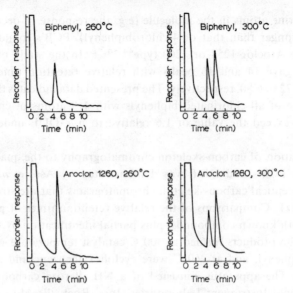

Fig. 11. Carbon-skeleton chromatograms of biphenyl and Aroclor 1260 at two catalyst temperatures.

extremely useful means for ascertaining the presence or absence of PCB's in samples which from gas chromatographic responses show the presence of the DDT pesticide group.

The qualitative identification of PCB's in metabolism studies and some surveillance situations by carbon-skeleton chromatography should require samples of a microgram or less. The response of this technique for a specific Aroclor depends upon its chlorine content; the greater the chlorine content, the larger will be the amount required for detection. For 1 µg of Aroclor 1260 with the electrometer attenuator at 8× a response of 13 % of fullscale deflection was obtained. For 1 µg quantities of Aroclors with less chlorine contents than Aroclor 1260, responses should increase as the chloride content decreases.

The gas chromatographic–mass spectrometric behavior of PCB's in human adipose tissue was described by Biros et al.[15]. The gas chromatographic column was a stainless steel capillary, 100 ft. × 0.020 in. i.d. coated with OV-silicone oil. Programmed-temperature analyses were made both for the six Aroclor 1200 series standards and the tissue extracts. Figures 14 and 15 illustrate total ion current monitor chromatograms obtained for Aroclors 1254 and 1260 and detail the programming conditions used. The molecular separator and gas inlet temperatures were maintained at 210°C and 215°C, respectively. The mass spectra were recorded at 80 eV electron energy with 2300 V accelerating voltage and the filament emission current was 100 µA. Helium carrier gas flow rate was 4 ml/min and the injector temperature was 175°C. The mass spectra were scanned magnetically over the range m/e 5 to m/e 500 in 6 sec. Figures 16 and 17 illustrate the total ion current monitor chromato-

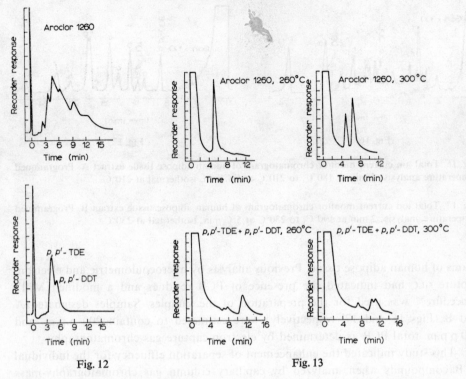

Fig. 12

Fig. 13

Fig. 12. Gas chromatograms of Aroclor 1260 and a mixture of *p,p'*-TDE and *p,p'*-DDT using the electron-capture detector.

Fig. 13. Carbon-skeleton chromatograms of two catalyst temperatures of Aroclor 1260 and a mixture of *p,p'*-TDE and *p,p'*-DDT.

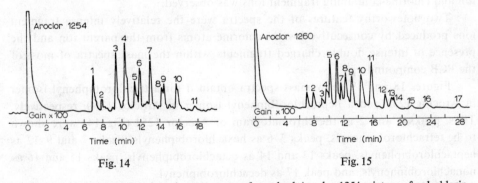

Fig. 14

Fig. 15

Fig. 14. Total ion current monitor chromatogram of standard Aroclor 1254 mixture of polychlorinated biphenyls. Programmed-temperature analysis: 2 min at 185°C, to 210°C at 5°C/min, isothermal at 210°C. See text for remaining instrumental parameters and partial peak identification.

Fig. 15. Total ion current monitor chromatogram of standard Aroclor 1260 mixture of polychlorinated biphenyls. Programmed-temperature analysis: 2 min at 200°C, to 230°C at 5°C/min, isothermal at 230°C. See text for remaining instrument parameters and partial peak identification.

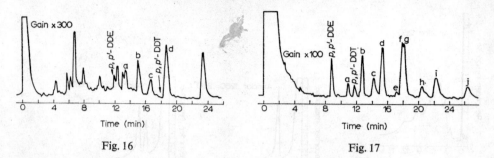

Fig. 16   Fig. 17

Fig. 16. Total ion current monitor chromatogram of human adipose tissue extract A. Programmed temperature analysis: 5 min at 180°C, to 210°C at 5°C/min, isothermal at 210°C.

Fig. 17. Total ion current monitor chromatogram of human adipose tissue extract B. Programmed temperature analysis: 2 min at 190°C, to 230°C at 5°C/min, isothermal at 230°C.

grams of human adipose tissue. Previous analysis by microcoulometric and electron-capture GLC had indicated the presence of PCB residues and a modified Mill's procedure[85] was used for the preparation of the samples. Samples designated A and B (Figs. 16 and 17, respectively) were estimated to contain 200 p.p.m. and 600 p.p.m. total PCB as determined by electron-capture gas chromatography.

This study indicated the enhancement of separation effluency for the individual PCB compounds when analyzed by capillary column gas chromatography-mass spectrometry with no evidence of thermal degradation of any of the PCB compounds. All components of the Aroclors gave molecular ion groups of high intensity as would be expected from highly chlorinated biphenyl structures. The characteristic isotopic distribution pattern[86] corresponding to the number of chlorine atoms in the parent ion and chlorine-containing fragment ions was observed.

Two noteworthy features of the spectra were the relatively intense fragment ions produced by consecutive loss of chlorine atoms from the parent ion and the presence of intense doubly charged fragments within the mass spectra of most of the PCB compounds.

Figures 18 and 19 show mass spectra obtain d for a trichlorobiphenyl isomer in Aroclor 1232 and a heptachlorobiphenyl isomer in Aroclor 1260, respectively. Thus, peaks 1 and 2 in the chromatogram of Aroclor 1254 (Fig. 14) were shown to be tetrachlorobiphenyls, peaks 3–6 as hexachlorobiphenyls, peaks 7 and 9–12 as heptachlorobiphenyls, peaks 13 and 14 as octachlorobiphenyls, peaks 15 and 16 as nonachlorobiphenyls, and peak 17 as decachlorobiphenyl.

The mass spectra of the components of adipose tissue of sample extract A (Fig. 16) revealed that peaks a, b, c, and d were polychlorinated biphenyls identical with those obtained respectively for peaks 6, 7, 9, and 10 of Aroclor 1254 (Fig. 14).

The mass spectrum of peak d (an isomer of hexachlorobiphenyl, m.w. 358) is shown in Fig. 20.

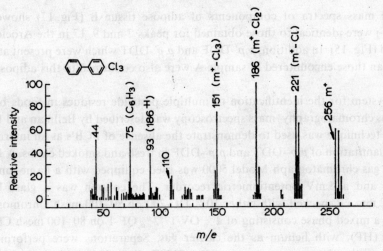

Fig. 18. Mass spectrum of trichlorobiphenyl contained in standard Aroclor 1232.

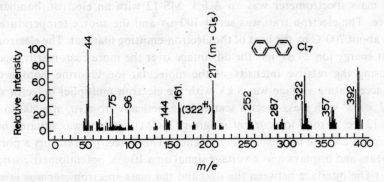

Fig. 19. Mass spectrum of a heptachlorobiphenyl found in standard Aroclor 1260.

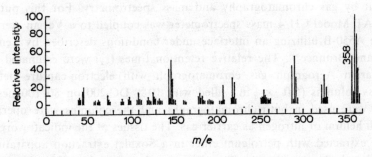

Fig. 20. Mass spectrum of a hexachlorobiphenyl contained in human adipose tissue sample A (peak d, Fig. 16).

The mass spectra of components of adipose tissue B (Fig. 17) showed that peaks a–j were identical to those obtained for peaks 7 and 9–17 in the Aroclor 1260 standard (Fig. 15). In addition, p,p'-DDE and p,p'-DDT which were present at higher levels than those encountered in sample A were also confirmed in this adipose tissue sample.

A system for the identification of multiple pesticide residues in foods by combined gas chromatography–mass spectroscopy was described by Bellman and Barry[87] and this technique was used to demonstrate the absence of PCB's as an intermediate in the quantitation of p,p'-DDT and p,p'-DDE in fresh and smoked chubs. A Barber-Colman gas chromatograph Model 5000 was used equipped with a flame-ionization detector and a 5 mV potentiometric recorder. The column was a glass U-tube, 6 ft. × 2 mm i.d. packed with either 10% DC-200 on 80–100 mesh Chromosorb W (HP) or a mixed phase consisting of 2% OV-17/2% QF-1 on 80–100 mesh Chromosorb W (HP); with helium as the carrier gas. Separations were performed isothermally with column temperatures of 180–215°C, depending on the pesticides being separated with the GLC column effluent split between the mass spectrometer and the flame-ionization detector.

The mass spectrometer was an A.E.I. MS-12 with an electron bombardment ion source. The electron trap was set at 100 μA and the source temperature maintained at about 200°C by the heat of the electron-emitting filament. The electron bombardment energy ion 24 eV has the advantage over the more customarily used 70 eV of increasing the relative intensity of the molecular ion chlorine isotope cluster. The ion-accelerating voltage was 8 kV with the electron multiplier detector set at a gain of 7×10^4 with mass spectra scanned magnetically from ca. m/e 500 to m/e 12 in about 12 sec. A total ion-current monitor (TICM) detector was situated between the ion source and the mass analyzing magnet. (This detector intercepts a portion of the ion beam and displays it as a voltage signal on a 10 mV potentiometric strip chart recorder.) The interface between the GLC and the mass spectrometer was essentially as reported by Markey[88].

The occurrence and distribution of PCB's in the River Rhine and the coastal areas in the Netherlands has been reported by Koeman et al.[1]. The identification was carried out by gas chromatography and mass spectrometry. For this purpose a Varian MAT Model CH 4 mass spectrometer was coupled to a Varian Aerograph Oven type A550-B utilizing an interface under conditions described by Tennoever deBrauw and Brunnee[89]. The relative retention times (r_x) were estimated with a 204-1B Varian Aerograph gas chromatograph with electron-capture detection. Pyrex glass columns (5 ft. × $\frac{1}{8}$ in.) filled with 10% DC-200 on 80–100 mesh Gas Chrom Q were used in both chromatographs and the columns were operated at 200°C with helium or nitrogen as carrier gas. The tissues of the indicator organisms used were extracted with petroleum ether in a Soxhlet extraction apparatus after drying with anhydrous sodium sulfate. The samples were cleaned up by dimethylformamide partition[90] and column chromatography over activated Florisil. The

apolar compounds including the PCB's were eluted with hexane and then a 10%
diethyl ether in hexane solution was used to elute Dieldrin and Aldrin. Tables 5
and 6 list the mass numbers and numbers of chlorine atoms per molecule of peaks
with relative retention times (relative to Dieldrin = 1) longer than 0.20 and shorter
than 1.00 and of peaks longer than 1.00, respectively. Mass spectra were taken
from all peaks present in the total ion current chromatogram obtained with the
double ion source of the CH4 mass spectrometer. In Fig. 21(A) and (B) gas chromatograms are shown on an eider extract and a technical PCB mixture (Phenoclor DP6)
illustrating a strong resemblance between the electron-capture chromatograms and
the ion-current chromatograms. Both from the mass numbers and the numbers of
chlorine atoms/molecule, it was concluded that most of the compounds present in
the extracts could be identified as PCB's (with a strong probability of Phenoclor DP6,
Clophen A60, and Aroclor 1260 being present). Other chlorinated compounds
found were hexachlorobenzene (HCB), p,p'-DDE, and Telodrin.

The lower chlorinated PCB's were found more frequently in the tissues of the
roach than in the seabirds. This suggests the possibility that the lower chlorinated
PCB's are less persistent, and this was confirmed by an experiment with Japanese
quail which were fed a diet containing 2000 p.p.m. of Phenoclor DP6. (Chromato-

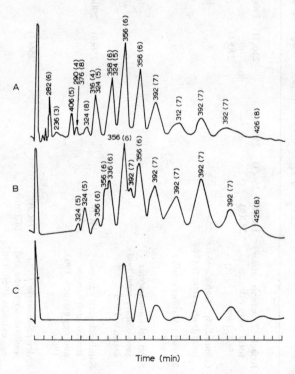

Time (min)

Fig. 21. Gas chromatograms of (a) an eider extract, (b) a technical PCP mixture (Phenochlor DP 6)
and (c) a quail liver and brain extract.

References pp. 574–577

TABLE 5

MASS NUMBERS AND NUMBERS OF CHLORINE ATOMS/MOLECULE (IN PARENTHESES) OF PEAKS WITH RELATIVE RETENTION TIMES (RELATIVE TO DIELDRIN = 1) LONGER THAN 0.20 AND SHORTER THAN 1.00

Sampling area	Type of sample	r_x												
		0.25	0.32	0.33	0.34	0.40	0.48	0.57	0.61	0.65	0.71	0.72	0.83	0.86
1	Roach (*Leuciscus rutilus*) total body extracts 1967, 1968	HCB 282(6)	256(3)	256(3)		256(3) + 290(4)	290(4)	290(4)	290(4)	290(4)	290(4)		324(5)	342(5)
1	Groundling (*Gobio gobio*) total body extracts 1968	+	+			+	+					+		
2	Mussel (*Mytilus edulis*) 1968	+	+		+	+								
3	Sand eel (*Ammodytes lanceolatus*) total body extracts 1966	+		+		+	+							
3	Sandwich tern (*Sterna sandvicensis*) tissue and egg extracts 1965, 1966	HCB 282(6)				256(3)	290(4)	290(4)	290(4) + 376(8)	Telodrin 406(8)			342(5)	
3	Elder (*Somateria mollissima*) tissue and egg extracts 1966, 1967	HCB 282(6)				256(3)			290(4) + 376(8)	Telodrin 406(8)			324(5)	
	Phenochlor DP6											324(5)	324(5)	
	Arochlor 1260											324(5)	324(5)	
	Clophen A60											324(5)	324(5)	

With the exception of HCB, Telodrin, and compound 376(8) all compounds identified are conformable to PCBs. +, At the given retention time a peak is present in the chromatogram. A 10% DC-200 on Gaschrom Q (80–100 mesh) column was used at 200°C, carrier gas N_2 (±50 ml/min). Mass number calculations are based on Cl = 35. HCB = Hexachlorobenzene; Telodrin = 1,3,4,5,6,7,8,8-octachloro-1,3,3a,4,7,7a-hexahydro-4,7-methano-isobenzofuran.

TABLE 6

MASS NUMBERS AND NUMBERS OF CHLORINE ATOMS/MOLECULE (IN PARENTHESES) OF PEAKS WITH RELATIVE RETENTION TIMES (RELATIVE TO DIELDRIN =1) OF PEAKS LONGER THAN 1.00

Sampling area	Type of sample	r_x 1.01	1.17	1.23	1.45	1.57	1.69	1.96	2.27	2.67	3.20	3.58
1	Roach (Leuciscus rutilus) total body extracts 1967, 1968	DDE 316(4) + 324(5)	+	+	358(6)	392(7)	358(6) 392(7)	392(7)	392(7)	392(7)	392(7)	426(8)
1	Groundling (Gobio gobio) total body extracts 1968	+		+							+	
2	Mussel (Mytilus edulis) 1968	+	+				+					
3	Sand eel (Ammodytes lanceolatus) 1966	+	+	+	+		+	+		+	+	+
3	Sandwich tern (Sterna sandvicensis) tissue and egg extracts 1965, 1966	DDE 316(4) + 324(5)		358(6) 324(5)	358(6)	392(7)	358(6)	392(7)	392(7)	392(7)	392(7)	426(8)
3	Eider (Somateria mollissima) tissue and egg extracts 1966, 1967	DDE 316(4) + 324(5)		358(6) + 324(5)	358(6)	392(7)	358(6)	392(7)	392(7)	392(7)	392(7)	426(8)
	Phenoclor DP6	358(6)		358(6)	358(6)	392(7)	358(6)	392(7)	392(7)	392(7)	392(7)	426(8)
	Arochlor 1260	358(6)		358(6)	358(6)	392(7)	358(6)	392(7)	392(7)	392(7)	392(7)	426(8)
	Clophen A60	358(6)		358(6)	358(6)	392(7)	358(6)	392(7)	392(7)	392(7)	392(7)	426(8)

With the exception of DDE, all compounds identified are conformable to PCBs. +, At the given retention time a peak is present in the chromatogram. See notes under Table 5.

gram in Fig. 21(B)). A chromatogram representing the residue in liver and brain of these birds is shown in Fig. 21(C) and shows that a number of the compounds present in the original mixture were metabolized, particularly the lower chlorinated ones. Residues in the brains and livers of the quail were measured on a semiquantitative scale by peak height comparison of peak $r_x = 1.45$ using the Phenoclor DP6 mixture as a standard. The residue levels calculated in this manner were about 20 times higher in the quail tissues than in the livers and brains of the eiders. The quail developed hydropericardia with a dose of 2000 p.p.m. Flick et al.[91], also, have reported hydropericardia in White Leghorn cockerels dosed with 400 p.p.m. in their diet.

Bagley et al.[2] described the identification of polychlorinated biphenyls in two bald eagles by combined GLC–mass spectrometry. The separation and identification of the unknown components were obtained with an LKB Model 9000 GLC-MS equipped with stainless steel molecule separator system of Ryhage[92]. The 9 ft. × 0.25 in. spiral glass column was packed with 1% SE-30 on 100–120 mesh Gas Chrom Q. The operating temperatures (°C) were flask heater 220, GLC oven 180, separator 240, and ion source 270. The carrier gas was helium at a rate of 35 ml/min, the ionization potential was 70 eV, trap current 60 μA, accelerating voltage 3.5 kV and the scan time was 12 sec from m/e 2–400. Silica gel TLC as described by Mulhern[93] was used to separate interfering organochlorine pesticides. The major peaks as shown in the total ion current chromatogram for Aroclor 1254 (Fig. 22(a)) are numerically labeled in order of their retention time. The peaks in total ion current

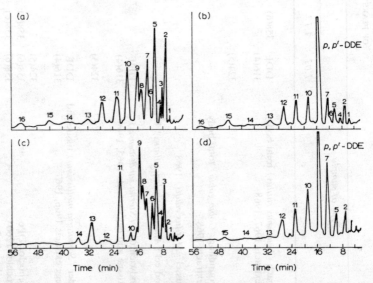

Fig. 22. Total ion current chromatogram of (a) standard PCB mixture (Arochlor 1254, 1.5 μg); (b) eagle carcass 68-056, TLC zone IV; (c) eagle carcass 68-056, TLC zone III; and (d) eagle carcass 68-050, TLC zone IV. GLC conditions: 9 ft. × 0.25 in. 1% SE-30 column, oven temperature 180°C, and helium flow 35 ml/min.

chromatograms for eagle samples (Fig. 22(d)) are labeled as they relate to the Aroclor standard and show that peaks 1–16 have approximately the same retention time as identically number peaks in Aroclor 1254.

The identification of PCB's and DDT in mixtures by mass spectrometry was reported by Hutzinger et al.[94]. Figure 23 illustrates a 70 eV mass spectra of a 1:1 mixture of Aroclor 1242 and DDT at different sample temperatures. The peaks arising from the individual major components of the Aroclor mixture can easily be distinguished from major peaks found in the most common chlorinated insecticides. A useful spectrum of Aroclor 1242 could be obtained at low temperatures by direct introduction with the aid of a temperature-controlled probe[95], and fractional sublimation from many chlorinated pesticides directly in the ion source seems possible (see Fig. 23).

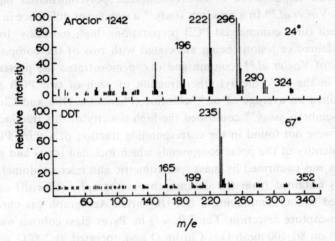

Fig. 23. 70 eV mass spectra (C.E.C. 21-110 B instrument) of a 1:1 mixture of Aroclor 1242 and DDT at different sample temperatures. Isotope clusters at m/e 222, 256, 290, and 324 result from individual components of Aroclor; the group of peaks at m/e 186 are mainly fragment ions from this substance.

The GLC–MS characterization of PCB's and ^{36}Cl-labeling of Aroclors 1248 and 1254 was recently described by Stalling and Huckins[96].

The isomer composition of Aroclor 1200 series was characterized by GLC–MS using temperature programming and SE-30 support coated open-tubular (SCOT) capillary columns (50 ft. × 0.020 in.). Neutron irradiation of the commercial material was used to prepare the ^{36}Cl-labeled material and purification of this material was accomplished by silicic acid column chromatography. Yields of the purified product were between 63 and 99 % with no detectable alteration of the isomer composition; 10 % of the ^{36}Cl produced was associated with the unchanged PCB isomers and the remaining radioactivity was contained in polychlorinated terphenyls. The terphenyls were produced by irradiation polymerization and were easily separated from the PCB components. Mass spectrometry utilizing either temperature-programmed GLC or direct probe sample introduction was used to characterize the irradiation products.

The analysis of PCB's by GC–MS was described by Murata et al.[97]. PCB's (PCB-300, 400, 500, and 600) were separated on a 3 m × 3 mm glass column packed with 1% OV-17 on Chromosorb W (60–80 mesh) at 170°C. The chromatograph was connected to a mass spectrometer mounted with a Ryhage separator. The operating parameters were separator temperature 200°C, ionization chamber temperature 230°C, accelerator voltage 3.5 kV, ionization potential 70 eV, trap current 60 μA. PCB's 300 to 600 yielded 15, 15, 16, and 18 components, respectively, with the main components being $C_{12}H_7Cl_3$, $C_{12}H_6Cl_4$, $C_{12}H_5Cl_5$, $C_{12}H_4Cl_6$, and $C_{12}H_3Cl_7$.

The identification and toxicological evaluation of chlorinated dibenzofuran and chlorinated naphthalene in two commercial polychlorinated biphenyls was described by Vos et al.[28] In a previous study[29] a significant difference in toxicity was found between three commercial PCB preparations, high mortality, liver necrosis, and chick edema-like lesions being associated with two of the compounds tested. In the study of Vos et al.[28], column and GLC demonstrated the presence of polar compounds in the 25% diethyl ether fraction of each of two PCB preparations which contained an average of 60% chlorine (Phenoclor DP6 and Clophen A60), and a chick-embryo assay[98] confirmed the high toxicity of this fraction. The polar compounds were not found in the corresponding fraction of a third PCB (Aroclor 1260). The identity of the polar compounds which included tetra- and pentachlorodibenzofuran was confirmed by mass spectrometric and microcoulometric analyses.

Aliquots obtained from the column chromatography (Florisil) of the PCB's were analyzed by GLC employing a 204-1B Varian Aerograph gas chromatograph with electron-capture detection. The 5 ft. × $\frac{1}{8}$ in. Pyrex glass column was filled with 10% DC-200 on 80–100 mesh Gas Chrom Q and operated at 200°C with nitrogen as carrier gas at 50 ml/min. Mass spectrometry was used to identify compounds present in the polar fraction (25% diethyl ether–hexane) from the column fractionation of PCB's. For this purpose a CH4 mass spectrometer (Varian MAT) was coupled to a Varian Aerograph oven, type A550B, with a 5 ft. × $\frac{1}{8}$ in. Pyrex glass column filled with 10% DC-200 on 80–100 mesh Gas Chrom Q. The design of the interface between the above gas chromatograph and the mass spectrometer has been described by Tennoever deBrauw and Brunnee[89].

High-resolution mass spectra were taken on a double-focus spectrometer type SM (Varian MAT). Exact mass measurements were made by the peak matching method and measurements were also made from spectra which were recorded on photographic plates processed by Varian MAT on a Leitz comparator SAM 1.

Quantitative GLC information was obtained via microcoulometric analysis of the third polar Florisil column fractions using a Dohrmann microcoulometer coupled to a Microtek MT220 gas chromatograph with a 6 ft. × $\frac{1}{4}$ in. glass column containing the same 10% DC-200 columns packing described above. The column effluents were oxidized in a "Dohrmann" combustion furnace and the HCl produced by chlorinated compounds was detected and measured quantitatively in a T-300-S titration cell.

The microcoulometric gas chromatogram of the third fraction (25% diethyl ether–hexane eluates) obtained with a 10 mm Florisil column is shown in Fig. 24 and reveals the presence of polar compounds in Phenoclor and Clophen. (Data for peaks 1–5 are given in Table 7.)

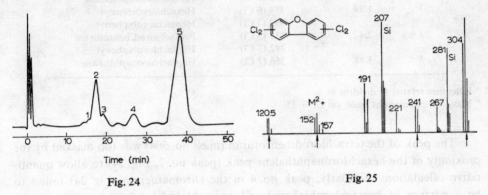

Fig. 24   Fig. 25

Fig. 24. Gas chromatogram of the third fraction of Clophen A60 (showing pattern identical to that for the third fraction of Phenoclor DP6). Data for peaks 1–5 are given in Table 5.

Fig. 25. Mass spectrum of tetrachlorodibenzofuran (mass no. 304) obtained from the third fractions of Clophen A60 and Phenoclor DP6 (peak 1 in Fig. 24).

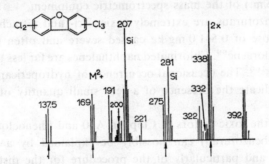

Fig. 26. Mass spectrum of pentachlorodibenzofuran (mass no. 338) obtained from the third fractions of Clophen A60 and Phenoclor DP6 (peak 4 in Fig. 24). The spectrum of heptochlorobiphenyl (mol. wt. 392) is also present.

Figure 25 shows the mass spectrum of tetrachlorodibenzofuran (mass no. 304) obtained from the third fraction of Clophen A60 and Phenoclor DP6 (peak 1 in Fig. 24). Figure 26 depicts a mass spectrum of pentachlorodibenzofuran (mass no. 338) obtained from the third fractions of Clophen A60 and Phenoclor DP6 (peak 4 in Fig. 24). The spectrum of heptachlorobiphenyl (m.w. 392) is also present.

TABLE 7

RETENTIONS AND COLLECTED MASS SPECTROMETRIC DATA OF PEAKS SHOWN IN FIG. 24

Peak no.	Relative retention[a]	Mass no.[b] and no. of chlorine atoms/mol	Identity of compounds
1	1.40	304 (4 Cl)	Tetrachlorodibenzofuran
2	1.58	332 (6 Cl)	Hexachloronaphthalene
3	1.74	358 (6 Cl)	Hexachlorobiphenyl
		392 (7 Cl)	Heptachlorobiphenyl
4	2.42	338 (5 Cl)	Pentachlorodibenzofuran
		392 (7 Cl)	Heptachlorobiphenyl
5	3.48	366 (7 Cl)	Heptachloronaphthalene

[a] Retention related to dieldrin $= 1$.
[b] Mass no. calculations base on Cl $= 35$.

The peak of the tetrachlorodibenzofuran (mass no. 304) was too masked by the proximity of the hexachloronaphthalene peak (peak no. 2, Fig. 24) to allow quantitative calculations. Similarly, peak no. 4 in the chromatogram (Fig. 24) found to be a mixture of heptachlorobiphenyl and pentachlorodibenzofuran could not be measured directly. However, by assuming that the whole peak was produced by the dibenzofuran compound alone, a certain maximum level could be indicated. This level was found to be 5 p.p.m. of pentachlorodibenzofuran in Clophen and 20 p.p.m. in Phenoclor. The minimum level could be indicated with the estimated lower limit of detection (1 p.p.m.) of the mass spectrometric equipment.

The halodibenzofurans are extremely toxic. Tri- and tetrachlorodibenzofuran in a single oral dose of 0.5–1.0 mg/kg caused severe and often lethal necrosis in rabbits[91,99] and chloracne[99]. Chlorinated naphthalenes are far less toxic then chlorinated dibenzofurans[100]. The occasional occurrence of hydropericardia in chicks fed Aroclor[91] may indicate the presence of a very small quantity of a toxic factor in this preparation.

The origin of the toxic factors in Clophen A60 and Phenoclor DP6 (e.g. tetra- and pentachlorodibenzofuran) can possibly be explained by a consideration of their manufacture and particularly of the procedure for the distillation of crude PCB in which sodium hydroxide can be used[101].

For example, PCB can react with sodium hydroxide at elevated temperatures to yield phenolic compounds which may be saponified to polychlorohydroxy biphenyls by sodium hydroxide in a polyhydric alcohol medium[102]. Dibenzofuran derivatives could then be formed via a further loss of hydrochloric acid.

Williams and Blanchfield[103] described the TLC separation of two chlorodibenzo-p-dioxins from some polychlorinated biphenyls and organochlorine pesticides. After isolation by TLC, the chlorodibenzo-p-dioxins were analyzed by electron-capture gas chromatography. A Varian Aerograph Model 2100 gas chromatograph equipped with tritium source electron-capture detector (250 mCi) was used with a 6 ft. × ¼ in. o.d. glass column packed with 3% silicon gum XE-60 on 80–100 mesh Gas Chrom Q. The operating conditions were detector and injector temperatures 200°C, column oven temperature 200 or 175°C, and nitrogen carrier gas at 60 ml/min. TLC was accomplished using Aluminum Oxide DFO-plates developed with solvent A tetra-chloroethylene–acetone (1:4) and solvent B tetrachloroethylene–methanol–water (5:45:1). In solvent A, all of the chlorodibenzo-p-dioxins moved at or near the solvent front. In solvent B, the tetra- and octachlorodibenzo-p-dioxins did not move, whereas the di- and trichloro-p-dioxins (as well as Aroclors 1221, 1254, 1260, and p,p'-DDT, p,p'-DDD, p,p'-DDE, and Dieldrin) all moved at or near the solvent front. A clear separation was thus obtained between these compounds and the tetra- and octachlorodibenzo-p-dioxins. Since these latter two compounds remained at the point reached by solvent A, they were easily located and after elution from the aluminum oxide, were analyzed simultaneously in a single GLC analysis. When mixtures of the tetrachlorodibenzo-p-dioxin and Aroclor 1254 were analyzed, a clear separation was obtained and the GLC chromatogram of the tetrachlorodibenzo-p-dioxin eluate showed no peaks from the Aroclor 1254. Table 8 shows the GLC retention times (relative to Aldrin) of some chlorodibenzo-p-dioxins, PCB's (Cl, Cl_2, Cl_4, Cl_6, Cl_8, and Cl_{10}) and chlorinated pesticides at 175 and 200°C.

The separation of three chlorodibenzo-p-dioxins from some polychlorinated biphenyls (Aroclors 1254, 1260, and 1262) by chromatography on an aluminum oxide column was described by Porter and Burke[104]. PCB's were eluted first with 1% methylene chloride–hexane and chlorodioxins were then eluted with 20% methylene chloride–hexane. Recoveries of both PCB's and chlorodioxins approximated 100% (as analyzed by electron-capture GLC). A 6 ft. × 4 mm i.d. glass column packed with 3% OV-101 on 80–100 mesh Chromosorb W (HP) was used with nitrogen carrier flow of 120 ml/min. The column and detector temperatures were 200°C and the injection port 225°C. The concentric design electron-capture detector was operated at d.c. voltage to produce half-scale recorder deflection for 0.4 ng Heptachlor epoxide when full-scale deflection was 1×10^{-9} A. The retention times of the chlorodioxins relative to Aldrin (approx. 1.2 min) were 2,3-dichloro- 0.93, 2,3,7-trichloro- 1.67, and 2,3,7,8-tetrachloro- 3.0.

Sissons and Welti[58] characterized the major constituent polychlorinated biphenyls in Aroclor 1254 by high-resolution nuclear magnetic resonance and mass spectrometry following separation by liquid–solid and gas–liquid chromatography. The Kovats retention indices of these compounds together with those of 40 synthesized PCB's have been used to predict the structures of the remaining Aroclor 1254 constituents, together with all those of Aroclor 1242 and 1260. Analyses were performed on a Pye Model 74 instrument fitted with a [63]Ni electron-capture detector and a 7 ft. × ¼ in. Pyrex column of 2% Apiezon L + 0.02% Epikote resin on 80–100 mesh Varoport 30. Argon was the carrier gas at an inlet pressure of 25 p.s.i. and a flow of 80 ml/min. The column and injector were maintained at 230°C and the

TABLE 8

GLC RETENTION TIMES (RELATIVE TO ALDRIN (R_A)) OF SOME CHLORODIBENZO-p-DIOXINS, PCB, AND CHLORINATED PESTICIDES AT 2 COLUMN TEMPERATURES

Compound	R_A (175°C)	R_A (200°C)
Chlorodibenzo-p-dioxins		
2,7-Di-	1.12	
2,3,7-Tri-	2.45	
1,2,3,4-Tetra-	4.7	3.86
2,3,7,8-Tetra-	4.96	4.06
Hexa-		11.3
		12.7
		16.3
Hepta-		23.2
		26.8
Octa-		48.0
Polychlorobiphenyls		
Cl_1	0.03	
Cl_2	0.75	
Cl_4	1.14	1.06
Cl_6	1.47	1.28
Cl_8	6.25	4.72
Cl_{10}	21.5	14.6
Chlorinated pesticides		
p,p'-DDE	2.99	2.45
p,p'-DDT	7.24	4.8
p,p'-DDD	7.5	4.85
Dieldrin	3.48	2.85
Aldrin	1.00[a]	1.00[a]

[a] The absolute retention times for Aldrin were 0.96 and 0.24 min on the 175 and 200°C columns, respectively.

detector at 325°C. For combined GLC–MS analysis, Aroclor mixtures were separated using a Perkin-Elmer 50 ft. × 0.02 in. i.d. support coated open tubular (SCOT) Apiezon L column. The column effluent was split between the FID and an AEI MS12 mass spectrometer.

Stalling et al.[105] described a gel permeation chromatographic (PGC) technique for the cleanup of pesticide and PCB residues in fish extracts. A chromatographic system using cyclohexane and Bio Beads S-X2 (Bio-Rad Labs) gave the most satisfactory separation of lipids from DDT, DDD, DDE, Methoxychlor, Lindane, Dieldrin, Endrin, Malathion, Parathion with recoveries of the pesticides greater than 95%. Pesticide concentrations were equivalent to 0.01–1.0 p.p.m. based on the original fish tissue. Less than 0.5% of the lipids originally present in the fish extract was recovered in the pesticide-containing eluate. Polychlorinated biphenyls (as ^{36}Cl-labeled Aroclors 1248 and 1254) and phthalate esters were also recovered quantitatively by the same gel permeation technique. For example, the recoveries of ^{36}Cl-PBC and di-2-ethylhexyl phthalate were 96 and 98%, respectively.

The components of the chromatographic apparatus (Fig. 27) were connected with 0.125 in. o.d., 0.063 in. i.d. polytetrafluoroethene tubing (Polymer Corp., Polypenco Div., Reading, Pa.) using Cheminert fittings (Chromatronix, Inc., Berkeley, Cal.). A solvent pulse damper was constructed with a 24 in. capped tee line after the solvent pump to minimize pressure surges. The solvent pump was a Model 15, 501-11 chemical feed pump (Precision Control Products Corp., Waltham, Mass.), the column (Series 3500, 2.5 × 60 cm) was fitted with an adjustable plunger (Glenco Scientific, Houston, Tex.), and the sample introduction valve was a Model SV-8031 (Chromatronix).

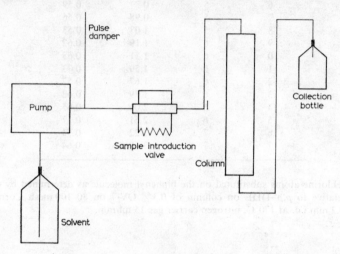

Fig. 27. Manual gel permeation apparatus for lipid–pesticide cleanup.

GLC pesticide and PCB analyses were made with a Model 804 Packard gas chromatograph equipped with a ^{63}Ni electron-capture detector. A coiled glass column (1.8 m × 2 mm i.d.) was packed with 0.3% OV-7 coated on 80–100 mesh Corning-110 glass beads and maintained at 170°C. Nitrogen was used as carrier gas at a flow rate of 15 ml/min.

An important advantage of gel permeation chromatography in pesticide cleanup is that it did not significantly alter the isomer ratio in multicomponent pesticides. In the liquid–liquid partition step prior to adsorption chromatography, the various isomers in multicomponent pesticide residues and industrial contaminants, *e.g.* toxaphene, chlordane, and PCB, as well as DDT, DDD, and DDE isomers, are not partitioned in an identical manner. Hence the ratios of components in the partitioned sample must or should be compared with a partitioned standard and thus another source of variability is introduced into the analysis. This is an important consideration in the cleanup of samples containing PCB residues since the partition coefficients for the components of Arochlor 1254 varied from 0.25 to 0.73 in acetonitrile–hexane (Table 9).

TABLE 9

PARTITION VALUES FOR AROCLOR 1254 CONSTITUENTS IN ACETONITRILE–HEXANE

Peak no.	Number Cl[a]	Retention value[b]	p-Value
1	4	0.27	0.39
2	4	0.37	0.47
3	4	0.45	0.25
4	5	0.50	0.55
5	5	0.63	0.60
6	5	0.77	0.59
7	5	0.98	0.56
8	5	1.07	0.53
9	6	1.19	0.62
10	6	1.31	0.63
11	6	1.59	0.73
12	6	1.67	0.57
13	6	1.79	0.62
14	6	2.04	0.65
15	6	2.61	0.57
16	7	3.22	0.53
17	7	3.68	0.64

[a] Number of chlorine atoms substituted on the biphenyl molecule as determined by GLC-MS.
[b] Retention relative to p,p'-DDE on column of 0.3% OV-7 on 80–100 mesh Corning-110 glass beads 1.8×0.2 mm i.d. at 170°C, nitrogen carrier gas 15 ml/min.

It was found that in general, after gel permeation fractionation, samples required little or no additional cleanup. However, additional cleanup with Florisil was suggested for samples in which whole body pesticide residues are less than 0.1 µg/g. Many fish samples contain PCB, and the separation of PCB from the pesticide is necessary. In this case, the cyclohexane effluent from the GPC is concentrated to an appropriate volume and an aliquot is applied to a silicic acid–Celite 545 column[44,96]. The PCB and pesticide fractions are eluted from the silicic acid–Celite column and each fraction is analyzed by GLC or GLC–mass spectroscopy[96,105].

Burne et al.[106] described a multifunctional gradient elution device for use in high-speed liquid chromatography (LC) which could produce a variety of shapes and time durations without hardware changes. Liquids from two reservoirs were mixed automatically by two electronically controlled proportioning valves. The device also provided a means of rapidly changing solvent concentrations during optimization of separations for subsequent sample analysis under constant composition conditions.

A DuPont Model 820 High Speed LC was used equipped with both a UV photometer and a differential refractometer. Figure 28 illustrates a schematic diagram of a multifunctional gradient device.

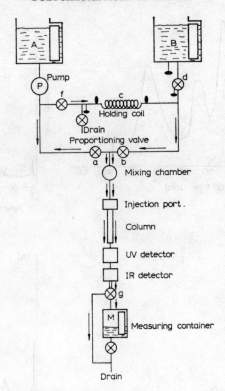

Fig. 28. Schematic diagram of a multifunctional gradient device.

Eight exponential gradients were provided, four had fast rates of change at the beginning of the gradient [eqn. (1)] with continuously decreasing rate and four had slow starting rates with continuously increasing rate [eqn. (2)].

$$C = K(1 - (1 - t)^n)$$ (1)

$$C = Kt^n$$ (2)

where C = fraction of B in A; t = reduced time = time/total time for complete gradient; $n = 2, 3, 4$, and 5, and K = constant. With all the gradients, the total time, flow rate and initial and final concentration could be varied. Separations of a mixture of 60% chlorinated biphenyls chromatographed under both constant composition and gradient conditions are shown in Fig. 29. A comparison of two other gradients for the separation of this same mixture are shown in Fig. 30.

The reproducibility of retention times in gradient elution chromatography was found dependent on variables such as initial solvent concentration gradient rate, final solvent, concentration, gradient profile reproducibility, column stability and column inlet pressure.

References pp. 574–577

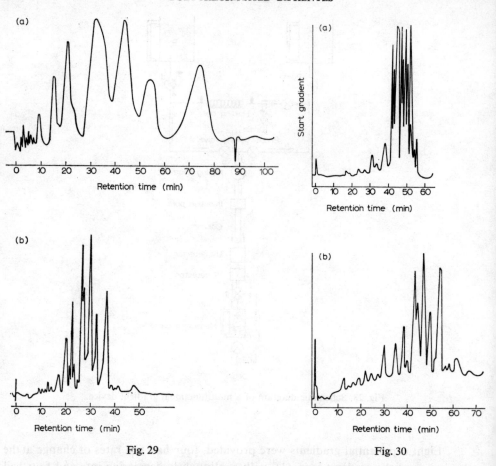

Fig. 29   Fig. 30

Fig. 29. (a) Separation of a mixture of 60% chlorinated biphenyls constant composition. Column, ODS "Permaphase"; mobile phase, 60% methanol/40% water; column temperature, 50°C; flow rate, 2 ml/min; column pressure, 120 p.s.i.g. (b) Chlorinated biphenyls separated with an exponential gradient ($n = 3$ in eqn. (2)). Gradient from 20% methanol/25% water at 3% min.

Fig. 30. (a) Separation of chlorinated biphenyls. Operating conditions as for Fig. 29 except for the linear gradient. (b) Operating conditions as for Fig. 29 except for the exponential gradient ($n = 2$ in eqn. (2)).

REFERENCES

1   J. H. KOEMAN, M. C. TENNOEVER DEBRAUW AND R. H. DE VOS, *Nature*, 221 (1969) 1126.
2   G. E. BAGLEY, W. L. REICHEL AND E. CRUMARTIE, *J. Assoc. Offic. Anal. Chemists*, 53 (1970) 251.
3   *Monsanto News Release*, Washington, D. C., Nov. 30, 1971.
4   *Federal Govt. Task Force Rept., PCB's and the Environment*, National Technical Information Service, Springfield, Va. 1972.
5   I. HORNSTEIN AND W. N. SULLIVAN, *J. Econ. Entomol.*, 46 (1953) 937.

6 C. H. TSAO, W. N. SULLIVAN AND I. HORNSTEIN, *J. Econ. Entomol.*, 46 (1953) 882.
7 E. P. LICHTENSTEIN, K. R. SCHULZ, T. W. FUHREMANN AND T. T. LIANG, *J. Econ. Entomol.*, 62 (1969) 761.
8 S. JENSEN, *New Scientist*, 32 (1966) 612.
9 R. W. RISEBROUGH, P. REICHE, D. B. PEAKALL, S. G. HERMAN AND M. N. KIRVEN, *Nature*, 220 (1968) 1098.
10 D. C. HOLMES, J. H. SIMMONS AND J. O'G. TATTON, *Nature*, 216 (1967) 227.
11 A. V. HOLDEN AND K. MARSDEN, *Nature*, 216 (1967) 1274.
12 D. W. ANDERSON, J. J. HICKEY, R. W. RISEBROUGH, D. F. HUGHES AND R. E. CHRISTENSEN, *Can. Field. Nutr.*, 83 (1969) 89.
13 L. M. REYNOLDS, *Residue Rev.*, 34 (1971) 27.
14 S. JENSEN, A. G. JOHNELS, M. OLSSON AND G. OTTERLIND, *Nature*, 224 (1969) 247.
15 F. J. BIROS, A. C. WALKER AND A. MEDBURY, *Bull. Environ. Contam. Toxicol.*, 5 (1970) 317.
16 R. RISEBROUGH AND V. BRODINE, *Environment*, 12 (1970) 16.
17 G. WESTÖÖ, K. NOREN AND M. ANDERSON, *Var Foeda*, 2–3 (1970) 10.
18 D. B. PEAKALL AND J. L. LINCER, *Bioscience*, 20 (1970) 958.
19 I. C. T. NISBET AND A. F. SAROFIN, Paper presented at *PCB Meeting, National Institute of Environmental Health Sciences, Quail Roost, N.C., Dec. 16–18, 1971.*
20 L. SCHWARTZ, *Am. J. Public Health*, 26 (1936) 58.
21 L. SCHWARTZ AND F. A. BARLOW, *U.S. Public Health Rept.*, 57 (1942) 1742.
22 L. SCHWARTZ AND S. M. PECK, *N.Y. State Med. J.*, 43 (1943) 1711.
23 J. W. JONES AND H. S. ALDEN, *Dermatol. Syphilol.*, 33 (1936) 1022.
24 J. K. MEIGS, J. J. ALBOM AND B. L. KARTIN, *J. Am. Med. Assoc.*, 154 (1954) 1417.
25 F. B. FLINN AND J. E. JARVIL, *Proc. Soc. Exptl. Biol. Med.*, 35 (1936) 118.
26 L. M. GREENBURG, M. R. MAYERS AND A. R. SMITH, *J. Ind. Hyg. Toxicol.*, 21 (1939) 29.
27 C. K. DRINKER, W. F. WARREN AND G. A. BENNETT, *J. Ind. Hyg. Toxicol.*, 19 (1937) 283.
28 J. G. VOS, J. H. KOEMAN, H. L. VANDER MAAS, M. C. TENNOEVER DEBRAUW AND H. DE VOS, *Food Cosmet. Toxicol.*, 8 (1970) 625.
29 J. G. VOS AND J. H. KOEMAN, *Toxicol. Appl. Pharmacol.*, 17 (1970) 656.
30 S. JAEKI, A. TSUTSUI, K. OGURI, H. YOSHIMURA AND M. HAMANA, *Fukuoka Acta Med.*, 62 (1971) 20; *Chem. Abstr.*, 74 (1971) 146294.
31 T. KOJIMA, H. FUKUMOTO AND J. MAKISUMI, *Japan. J. Legal Med.*, 23 (1969) 415.
32 K. INAGAMI, T. KOGA AND Y. TOMITA, *Shokuhin Eiseigaku Zasshi*, 10 (1969) 312; *Chem. Abstr.*, 72 (1970) 120110.
33 M. KIKUCHI, Y. MIKAGI, M. HASHIMOTO AND T. KOJIMA, *Fukuoka Acta Med.*, 62 (1971) 89.
34 T. KOJIMA, *Fukuoka Acta Med.*, 62 (1971) .
35 T. KOJIMA AND H. FUKUMOTO, *Nippon Hoigaku Zasshi*, 24 (1970) 314.
36 E. L. MCCUNE, J. E. SAVAGE AND B. L. O'DELL, *Poultry Sci.*, 41 (1962) 295.
37 D. F. FLICK, C. D. DOUGLASS AND L. GALLO, *Poultry Sci.*, 42 (1963) 855.
38 J. MCLAUGHLIN, J. P. MARLIAC, M. J. VERRETT, M. K. MUTCHLER AND G. G. FITZHUGH, *Toxicol. Appl. Pharmacol.*, 5 (1963) 760.
39 J. G. VOS, Paper presented at *PCB Meeting, National Institute of Environmental Health Sciences, Quail Roost, N.C., Dec. 16–18, 1971.*
40 J. G. VOS AND R. B. BEEMS, *Toxicol. Appl. Pharmacol.*, 19 (1971) 617.
41 L. FISHBEIN, in *PCB's and the Environment*, Natl. Tech. Inform. Serv., Springfield, Va., 1972.
42 L. STICKEL, in *PCB's and the Environment*, Natl. Tech. Inform. Serv., Springfield, Va, 1972.
43 L. M. REYNOLDS, *Bull. Environ. Contam. Toxicol.*, 4 (1969) 128.
44 J. A. ARMOUR AND J. A. BURKE, *J. Assoc. Offic. Anal. Chemists*, 53 (1970) 761.
45 B. M. MULHERN, E. CROMARTIE, W. L. REICHEL AND A. A. BELISLE, *J. Assoc. Offic. Anal. Chemists*, 54 (1971) 548.
46 H. L. HALLER, *J. Am. Chem. Soc.*, 67 (1945) 1591.
47 N. V. FEHRINGER AND J. E. WESTFALL, *J. Chromatog.*, 57 (1971) 397.
48 O. HUTZINGER, S. SAFE AND V. ZITKO, *Bull. Environ. Contam. Toxicol.*, 6 (1971) 209.
49 S. JENSEN AND G. WIDMARK, *Uni Residues in the Environment*, Rept. of OECD Pesticide Conf., Stockholm, 1967.

50 R. W. RISEBROUGH, P. REICHE AND H. S. OLCOTT, *Bull. Environ. Contam. Toxicol.*, 4 (1969) 192.
51 W. C. KRANTZ, B. M. MULHERN, G. E. BAGLEY, A. SPRANT, F. J. LIGAS AND W. B. ROBERTSON, JR., *Pesticides Monitoring J.*, 4 (1970) 136.
52 W. L. REICHEL, T. G. LAMONT AND E. CROMARTIE, *Bull. Environ. Contam. Toxicol.*, 4 (1969) 24.
53 M. L. PORTER AND J. A. BURKE, *J. Assoc. Offic. Anal. Chemists*, 54 (1971) 1426.
54 A. C. TAS AND R. H. DE VOS, *Environ. Sci Technol.*, 5 (1971) 1216.
55 V. ZITKO, O. HUTZINGER AND S. SAFE, *Bull. Environ. Contam. Toxicol.*, 6 (1971) 160.
56 J. W. ROTE AND P. G. MURPHY, *Bull. Environ. Contam. Toxicol.*, 6 (1971) 160.
57 D. T. WILLIAMS AND B. J. BLANCHFIELD, *J. Assoc. Offic. Anal. Chemists*, 54 (1971) 1429.
58 D. SISSONS AND D. WELTI, *J. Chromatog.*, 60 (1971) 15.
59 D. L. STALLING, in A. S. TAHORI (Ed.), *Workshop on Pesticide Residue Analysis, 2nd International Congress of Pesticide Chemistry, Feb. 22–26, 1971 Tel Aviv, Israel*, Gordon and Breach, London, 1972, p. 413.
60 B. MELVAS, *PCB Conf., Natl. Swedish Environ. Protection Board, Sept. 29, 1970, Stockholm.*
61 P. W. ALBRO AND L. FISHBEIN, *J. Chromatog.*, 69 (1972) 273.
62 A. V. HOLDEN, *Nature*, 228 (1970) 1220.
63 A. V. HOLDEN AND K. MARSDEN, *J. Chromatog.*, 44 (1969) 481.
64 G. D. VEITH AND G. G. LEE, *Water Res.*, 5 (1971) 1107.
65 R. A. HUGHES, G. D. VEITH AND G. F. LEE, *Water Res.*, 4 (1970) 547.
66 G. D. VEITH, *Ph. D. Thesis*, Univ. of Wisconsin, Madison, Wisc., 1970.
67 B. AHLING AND S. JENSEN, *Anal. Chem.*, 42 (1970) 1483.
68 J. E. KEIL, L. E. PRIESTER AND S. H. SANDIFER, *Bull. Environ. Contam. Toxicol.*, 6 (1971) 156.
69 T. W. DUKE, J. I. LOWE AND A. J. WILSON, JR., *Bull. Environ. Contam. Toxicol.*, 5 (1970) 171.
70 P. A. MILLS, J. F. ONLEY AND R. A. GAITHER, *J. Assoc. Offic. Agr. Chemists*, 46 (1963) 182.
71 G. WESTÖÖ AND K. NOREN, *Acta Chem. Scand.*, 24 (1970) 1639.
72 K. NOREN AND G. WESTÖÖ, *Acta Chem. Scand.*, 22 (1968) 2289.
73 M. F. KOVACS, *J. Assoc. Offic. Agr. Chemists*, 46 (1963) 884.
74 D. L. GRANT, W. E. J. PHILLIPS AND D. C. VILLENEUVE, *Bull. Environ. Contam. Toxicol.*, 6 (1971) 102.
75 L. C. MITCHELL, *J. Assoc. Offic. Agr. Chemists*, 40 (1957) 999.
76 T. T. SCHMIDT, R. W. RISEBROUGH AND F. GRESS, *Bull. Environ. Contam. Toxicol.*, 6 (1971) 235.
77 R. I. STANLEY AND H. T. LEFAVOURE, *J. Assoc. Offic. Anal. Chemists*, 48 (1965) 666.
78 V. ZITKO, *Bull. Environ. Contam. Toxicol.*, 5 (1970) 279.
79 H. WEINGARTEN, W. D. ROSS, J. M. SCHLABER AND G. G. WHEELER, *Anal. Chim. Acta*, 26 (1962) 391.
80 N. L. GREGORY, *J. Chem. Soc. B*, (1968) 295.
81 E. M. EMERY AND G. M. GASSER, *U.S. Pat. 3,520,108*, July 14, 1970.
82 R. I. ASAI, F. A. GUNTHER, W. E. WESTLAKE AND Y. IWATA, *J. Agr. Food Chem.*, 19 (1971) 396.
83 R. I. ASAI, F. A. GUNTHER AND W. E. WESTLAKE, *Residue Rev.*, 19 (1967) 57.
84 M. BEROZA AND R. SARMIENTO, *Anal. Chem.*, 36 (1964) 1744.
85 P. MILLS, *J. Assoc. Offic. Anal. Chemists*, 42 (1959) 734.
86 J. H. BEYNON, *Mass Spectrometry and Its Applications to Organic Chemistry*, Elsevier, New York, 1960, p. 298.
87 S. W. BELLMAN AND T. L. BARRY, *J. Assoc. Offic. Anal. Chemists*, 54 (1971) 499.
88 S. P. MARKEY, *Anal. Chem.*, 42 (1970) 306.
89 M. C. TENNOEVER DEBRAUW AND C. BRUNNEE, *Anal. Chem.*, 229 (1967) 321.
90 M. J. DE FAUBERT MAUDER, H. EGAN, E. W. GODLY, E. W. HAMMOND, J. ROBURN AND J. THOMSON, *Analyst*, 89 (1964) 168.
91 D. F. FLICK, R. G. O'DELL AND V. A. CHILDS, *Poultry Sci.*, 44 (1965) 1460.
92 R. RYHAGE, *Anal. Chem.* 36 (1964) 759.
93 B. M. MULHER, *J. Chromatog.*, 34 (1968) 556.
94 O. HUTZINGER, W. D. JAMIESON AND V. ZITKO, *Nature*, 226 (1970) 664.

95 W. D. JAMIESON AND F. G. MASON, *Rev. Sci. Instr.*, in press.
96 D. L. STALLING AND J. N. HUCKINS, *J. Assoc. Offic. Anal. Chemists*, 54 (1971) 801.
97 T. MURATA, S. TAKAHASHI AND R. TACHIKAWA, *Shimadzu Hyoron*, 28 (1971) 93; *Chem. Abstr.*, 76 (1972) 30414z.
98 M. VERRETT, J. P. MARLIAC AND J. MCLAUGHLIN, *J. Assoc. Offic. Anal. Chemists*, 47 (1964) 1003.
99 H. BAUER, K. H. SCHULZ AND Y. SPIEGELBERG, *Arch. Gewerbepath. Gewerbehyg.*, 18 (1961) 538.
100 H. T. HOFMANN, *Arch. Exptl. Pathol. Pharmacol.*, 232 (1958) 228.
101 J. W. J. FAY AND J. M. RICHARDS, *Office Tech. Serv. P. B. Rept. 75859*, Bios Final Rept., 1947, p. 893.
102 Progil SA, *Brit. Pat. 779,221*, July 17, 1957.
103 D. T. WILLIAMS AND B. J. BLANCHFIELD, *J. Assoc. Offic. Anal. Chemists*, 54 (1971) 1429.
104 M. L. PORTER AND J. A. BURKE, *J. Assoc. Offic. Anal. Chemists*, 54 (1971) 1426.
105 D. L. STALLING, R. C. TINDLE AND J. L. JOHNSON, *J. Assoc. Offic. Anal. Chemists*, 55 (1972) 32.
106 S. H. BURNE, J. A. SCHMIT AND P. E. JOHNSON, *J. Chromatog. Sci.*, 9 (1971) 592.

Chapter 25

PHTHALATE ESTERS

The aromatic dicarboxylic esters such as the phthalic acid esters are among the most important industrial chemicals employed. Although they have enjoyed extensive utility for three decades, primarily as plasticizers for a variety of films and plastics, it has only been within the last several years that their migration from plastics into human tissue as well as their increasing occurrence in the ecology have been reported. Hence, there is increasing concern regarding the consequences in man of chronic ingestion, absorption and/or inhalation of low levels of a variety of phthalate esters.

Phthalates are made industrially by esterification of the requisite alcohol with phthalic anhydride in the presence of a catalyst such as sulfuric acid or *p*-toluene-sulfonic acid or non-catalytically at high temperature. Phthalic anhydride itself is made by oxidizing naphthalene in air using vanadium pentoxide as a catalyst and is purified by distillation. The phthalate esters are, in most cases, liquids with very high boiling points and very low vapor pressures. (The low vapor pressure is important in contributing to their general stability in plastics.)

In terms of quantity, use applications and concomitant environmental considerations, the area of plasticizers is by far the most important aspect of the phthalate esters. Plasticizer production of phthalate esters in the U.S. has grown from approx. 300 million lb. in 1960 to approx. 1.30 billion lb. in 1970. In 1971 di(2-ethylhexyl)-phthalate (DOP, DEHP) production as a general purpose plasticizer for poly(vinyl chloride) (PVC) was 350 million lb. or a fourth of the total plasticizer production. The closely related phthalates, diisooctyl phthalate (DIOP) and diisodecyl phthalate (DIDP) accounted for another fourth of the market, *e.g.* 85 and 123 million lb., respectively. Other phthalates added another 296 million lb. or approx. 20%. These include 59 million lb. of N-octyl-N-decyl, 23 million lb. of dibutyl, and 21 million lb. of diethyl phthalate. It is of interest to note that since PVC began to be commercialized in the early 1930's, by 1966 over 2 billion lb. had been produced with forecast of 6 billion by the mid 1970's.

Phthalate ester plasticizers are extensively used in PVC for food packaging, refrigerator gasketing, luggage, handbags, coated cloth, vinyl floor coverings (tile and sheet goods), waterproof boots and shoes, electrical insulation, industrial hose, and tank liners. Besides PVC, the phthalates are used to plasticize polyvinyl acetate, polyvinylidene chloride, polystyrene, ethyl cellulose, cellulose nitrate, acetate, and acetate butyrate, chlorinated rubber, high styrene–butadiene protein compounds, shellac, acrylic-type resins, polyamides, polyesters, epoxy, phenolic, alkyds, poly-urethan, nitrile and neoprene rubber, and chloroethylene resins.

Other phthalate esters of importance are the octyldecyl esters where low-temperature properties and low volatility are important. The lower members of the series, particularly the methyl, ethyl, and butyl derivatives, find large-volume application as plasticizers for polar polymers such as polyvinyl acetate and the cellulosics. The methyl and ethyl esters also serve as insect repellants[1] when mixed with indalone and a variety of alkyl phthalates have been used extensively as stationary phases in gas chromatography. Other suggested areas of utility have included the use of bis(methoxyethyl)phthalate mixed with polypropylene–poly(methylacrylate) for cigarette filters[2] and in aerosol pesticide formulations.

An increasing number of reports have been cited concerning the appearance of phthalate esters in the environment (including the food chain as well as in human tissue). The widespread occurrence of the phthalates in aquatic ecosystems was recently announced by the U.S. Bureau of Sport, Fish, and Wildlife[3]. The compounds identified in various species of fish were di-*n*-butyl- and di-2-ethylhexyl phthalate in concentrations ranging from 0.2 to 3.2 p.p.m. These two plasticizers are quite stable, are stored in fish tissue and are suggested to be probably concentrated in food chains. Phthalate esters have been found in water, sediments, and in fish and other aquatic organisms in industrial and heavily populated areas in the U.S.[4,5]. The presence of dioctyl phthalate[6] and di-ethylhexyl phthalate[7] in various crude oils and petroleum has been reported. (The detection of phthalic anhydride in crude oils from North Africa has also been cited[8].)

Phthalic acid as well as its short-chain alkyl esters have been found in lipid extracts of plant materials, microorganisms, and in tobacco smoke[9]. Diheptyl phthalate has been produced by *Alternaria kikachiana* Tanaka fungus which produces a black spot on pears[9]. Phthalic acid can be synthesized in rat liver and some bacteria can make diesters of phthalic acid[10].

It has been increasingly apparent that the phthalate esters which are lipid-soluble substances can migrate from plastic packaging into foods, particularly into fatty foods[11]. Feofanov[12] has reported up to 150 mg of dibutyl phthalate/kg in cheese with 15% fat content. Phthalate esters have also been identified in fat used in deep-fryers[13] and in milk[14].

There has been evidence reported of the presence of phthalate esters in animal tissues such as beef pineal gland[15] and heart[16]. Perhaps the most dramatic recent finding has been that of Jaeger and Rubin[17] who reported that di-2-ethylhexyl phthalate (DEHP) and butylglycolbutyl phthalate (BGBP), often used in the PVC used in bags to store human blood and also in tubing through which blood is passed in heart–lung machines and kidney machines, are leached out of the plastic and into the blood. It was found that at the end of the maximum storage period of 21 days, blood stored in PVC bags contained 5–7 mg of DEHP/100 ml of blood (50–70 p.p.m.). It was calculated by the authors[18] that under these conditions an average size man requiring a total of 14 pints (not uncommon in the treatment of severe hemorrhage) could receive intravenously as much as 350 mg of plasticizer.

Neerguard[19] reported the leaching of diethyl phthalate from PVC tubing which leached 10–20 mg of plasticizer into every liter of dialysis fluid. However, despite this evidence, no cause and effect relationship between patients' hepatitis symptoms and plasticizer was shown. It has also been reported in some preliminary work at the National Institutes of Health in the U.S. that phthalate esters may be present in concentrations of 20 p.p.m. and higher in blood taken from human patients[20].

Physicians in the U.S. are reporting an increasing incidence of shock lung, a condition characterized by an impeded circulation of blood in the lungs. Although the cause of shock lung (a sometimes fatal disorder) is not definitely known, one theory suggests that it is related to the presence in the blood of microemboli consisting of aggregates of blood platelets. Rubin[21] has suggested that the mounting incidence of shock lung may be caused by the increasing medical use of plasticized PVC.

Nazir et al.[16] have recently reported the specific localization of DEHP in bovine, dog, rabbit, and rat heart mitochondria. It was noted that the mitochondrial compound differed from commercial DEHP in showing some evidence of optical activity and being crystallizable from hexane at low temperatures. It is not yet clear whether (and to what extent) the phthalate ester occurs naturally in heart mitochondria or is ingested by the animals as an environmental pollutant[10]. The specific localization of DEHP in heart muscle mitochondria was considered to be significant in that this substance may possibly influence the bioenergetics of the myocardial cell.

The acute and chronic toxicities of the phthalate esters in various species of laboratory animals has been described[22-25]. (The acute and LD_{50} in rats is about 30–34 g/kg.) The pathology produced by the phthalate esters is usually non-specific unless the alcohol portion released on hydrolysis in vivo has activity[26]. The teratogenicity of phthalate esters in chick embryo[27,28] and rat[29] have been reported. Guess et al.[28,30-32] have stressed the "subtle toxicities" of plasticizers (including phthalate esters) and stabilizers used in the manufacture of polyvinyl plastics. Butyloctyl phthalate (as well as dibutylethyl sebacate and dioctyl adipate) stimulated the growth of human amnion (Wish strain) cells and the K-B (Eagle-human cancer line) tissue culture. The observed increased growth, especially with the human cancer line, might be considered as a potential subtle change that could be undesirable.

Jaeger and Rubin[17] described the extraction, metabolism and accumulation by some biological systems of plasticizers from plastic devices. One such plasticizer, butylglycolylbutyl phthalate (BGBP) was metabolized by the isolated perfused rat liver glycolyl phthalate (GP) (I) while another plasticizer (di-2-ethylhexyl) phthalate

COOH
C–O–CH$_2$–COOH
‖
O
Glycolyl phthalate

References pp. 605–607

(DEHP) was found to be accumulated in the liver unchanged. DEHP, in addition, was identified in samples of human tissues taken from patients who had received transfusions of blood stored in plastics bags.

In Fig. 1 are shown the elution patterns from the chromatographic fractionation of acid-soluble extracts of perfusion plasma which had either perfused a liver, or had circulated in the apparatus in the absence of a liver. The first peak, I, was common to both experimental conditions and was identified as uric acid (based on its UV spectrum and elution pattern). Peak, II, a compound formed only when a liver was present, has not yet been identified. Peak III which also appeared only when a functioning rat liver was present was subsequently identified as glycolyl phthalate (based on comparison of its infrared spectrum of the methyl ester with an analogous synthetic glycolyl phthalate derivative.)

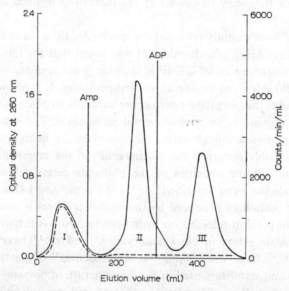

Fig. 1. Circulation of perfusion liquid, —— in the presence and - - - - in the absence of a rat liver. An amount of [^{14}C]adenosine diphosphate and [^{14}C]adenosine monophosphate was added to a neutralized extract of plasma following perfusion to act as a marker during further chromatographic fractionation. The total extract was applied to a 0.7 × 10 cm column of Dowex-1 (formate form) anion exchange resin. Elution of the column was with a non-linear gradient of ammonium formate (0–2 N, pH 5.5) and the absorbance at 260 nm was monitored continuously in a Gilford spectrophotometer. Portions of each fraction were counted in a Packard Tricarb liquid scintillation counter. In order to simplify this figure, only the peak radioactive fractions are displayed.

It was also found that DEHP, unlike BGBP, was not de-esterified by the isolated rat liver, but rather was accumulated by that organ, primarily in unmetabolized form. Jaeger and Rubin[33] also reported that although simple saline solutions were unable to extract DEHP from PVC tubing, even after 6 h of recirculation in the tubing, a 4% bovine serum albumin (BSA) solution was found to extract approx. 40% of the amount of plasticizer extractable by whole blood.

Fractionation of the blood indicated that the plasticizer was located almost entirely in the plasma fraction and was specifically associated with the lipoprotein fraction of plasma. Preliminary experiments with tissue obtained from two patients who had received blood transfusions (from DEHP plasticized blood bags) indicated that spleen, liver, lung, and abdominal fat all contained significant quantities of DEHP ranging from 0.025 mg/g (dry wt.) in spleen to 0.270 mg/g (dry wt.) in abdominal fat[19].

The contamination of blood stored in plastic packs was also studied by Marcel and Noel[34] who identified the presence of dihexyl phthalate in lipid extracts of plasma from the transfusion packs. The dihexyl phthalate content of the lipid extracts (obtained *via* extraction of plasma with chloroform–methanol mixtures according to Folch *et al*.[35] were enriched by chromatography on thick layers of Silica Gel H[36] and final purification was achieved by TLC on the same medium and developed with hexane–diethyl ether–acetic acid (90:10:1). The concentrations of dihexyl phthalate (mg/100 ml of plasma) were 0, 4, 7, 11.5, and 11.5 for storage periods in plastic packs of 0, 4, 8, 15, and 21 days, respectively. The identity of the plasma extracted phthalate ester was confirmed by infrared spectrometry and GLC as dihexyl phthalate.

The isolation, identification and specific localization of di-2-ethylhexyl phthalate in bovine heart muscle mitochondria was described by Nazir *et al*.[16]. Lipids were separated by a modified silicic acid procedure[37].

The composition of methylated fatty acids was determined by gas–liquid chromatography using argon beta-ionization detectors containing 20 mCi of ^{90}Sr sources (Barber-Colman Model 10 and Electronic Instruments for Research Model AU-8). Siliconized, U-shaped glass columns, 8 ft. × 4 mm i.d. and 6 ft. × 5 mm i.d. were used containing Gas-Chrom P, 80–100 mesh, previously acid-washed and deactivated with 2% (v/v) dimethyl dichlorosilane[38] and coated with either diethyleneglycol succinate (200 ml of a 8% solution in acetone/15 g of support) or Apiezon L (AE1, Manchester) (200 ml of a 4% solution in chloroform/15 g of support). The diethylene glycol succinate and Apiezon L columns were operated at temperatures of 180 and 200° and outlet flow rate of 75 and 300 ml/min, respectively with a detector polarizing voltage of 1000 V. Di-2-ethylhexyl phthalate was quantitatively determined following its initial isolation from mitochondrial fractions on a modified silicic acid column[39]. (The fraction eluted with 4% ether in hexane was evaporated to a small volume under partial vacuum and aliquots of 5–10 μl were analyzed by GLC on a two-component SE-52/XE-60 column as described by Nair *et al*.[39] at a temperature of 200° and outlet flow rate of 250 ml/min.) Peak areas obtained by triangulation were compared with those obtained from authentic DEHP.

The retention values characteristic of the isolated compounds from heart muscle mitochondrial triglyceride relative to methyl octadecanote and compared with DEHP are shown in Table 1. (The retention values were obtained by GLC on diethyleneglycol succinate, Apiezon L and SE-52/XE-60 columns.) The isolated compound was subsequently isolated in pure form by preparative GLC on diethyleneglycol succinate (10 ft. × 8 mm column operated at 200° and an outlet flow rate of 750 ml/min). Microhydrogenation[40] and microozonization[41] of the di-2-ethylhexyl phthalate isolated by preparative GLC was carried out and showed it to be unaltered indicating the absence of aliphatic double bonds, thus indicating that the isolated compound could not be an unsaturated fatty acid.

TABLE 1

RETENTION TIME DATA ON THE UNIDENTIFIED COMPONENT FROM BOVINE HEART MUSCLE MITOCHONDRIAL TRIGLYCERIDE RELATIVE TO METHYL OCTADECANOATE[a]

Stationary phases	Unknown compound	DEHP[b]
Diethylene glycol succinate[c]	13.6	13.5
Apiezon L[d]	4.4	4.4
SE52-XE60 (2:1, v/v)[e]	4.6	4.6

[a] Methyl octadecanoate = 1.00.
[b] DEHP = di-2-ethylhexylphthalate.
[c] Column temperature 180°, outlet flow rate 75 ml/min.
[d] Column temperature 200°, outlet flow rate 300 ml/min.
[e] Column temperature 200°, outlet flow rate 250 ml/min.

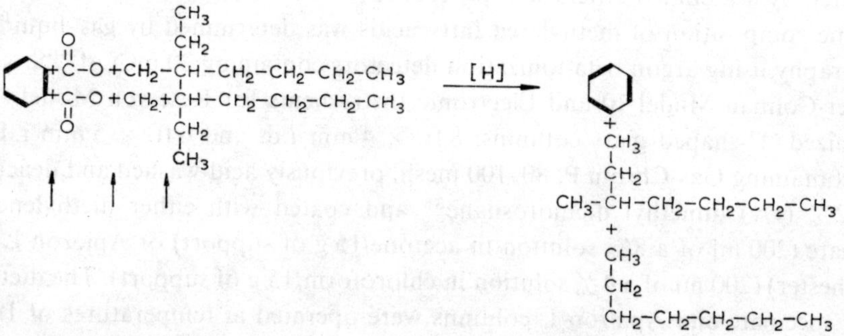

Fig. 2. Diagramatic representation of structural subunits derived from di(2-ethylhexyl)phthalate upon carbon skeleton chromatography.

Figure 2 shows a diagramatic representation of structural subunits derived from di-2-ethylhexyl phthalate upon carbon skeleton chromatography. In this technique[40] aliphatic multiple bonds are saturated and functional groups containing oxygen are stripped from compounds giving as products the parent hydrocarbon and/or the next lower homolog which are identified by their retention times. The unknown compound subjected to carbon skeleton chromatography gave peaks with retention times corresponding to n-heptane (11.9 min), a branched-chain hydrocarbon (3-methyl heptane which is 2-ethyl hexane, 20.0 min), and a weak peak for benzene (9.5 min). This breakdown pattern (Fig. 2) and relative proportions in which each of the hydrocarbons were obtained suggested a disubstituted benzene

TABLE 2

RETENTION TIME DATA FOR SEVERAL CLOSELY RELATED ALCOHOLS RELATIVE TO ISOAMYL ALCOHOL
ON THREE DIFFERENT STATIONARY PHASES

Stationary phase	Relative retention time[a]							
	1-Butanol	Isobutyl alcohol	Isoamyl alcohol	4-Heptanol	2-Ethyl-1-hexanol	1-Octanol	2-Octanol	Unknown
Diethylene glycol succinate	0.73	0.51	1.00[b]	1.58	4.77	7.15	3.42	4.77
Apiezon L			1.00[c]	5.49	14.82	20.75	12.60	14.82
SE52-XE60			1.00[d]	3.19	8.72	12.45	6.82	8.72

[a] Isoamyl alcohol = 1.00.
[b] Isoamyl alcohol = 4.3 min.
[c] Isoamyl alcohol = 1.7 min.
[d] Isoamyl alcohol = 2.0 min. Experimental conditions are described in the text. Unknown sample
represents the alcohol moiety of the compound isolated from bovine heart muscle mitochondria.

TABLE 3

RELATIVE RETENTION TIME DATA OF o-, m-, AND p-DIMETHYL ESTERS OF BENZENE DICARBOXYLIC ACIDS
ON THREE DIFFERENT STATIONARY PHASES

Stationary phase	Relative retention time[a]			
	Ortho	Meta	Para	Unknown
Diethylene glycol succinate	1.21	1.10	1.00[b]	1.21
Apiezon L	0.63	1.00	1.00[c]	0.63
SE52-XE60	0.92	1.07	1.00[d]	0.92

[a] Dimethyl terephthalate = 1.00.
[b] Dimethyl terephthalate = 11.7 min.
[c] Dimethyl terephthalate = 3.8 min.
[d] Dimethyl terephthalate = 2.5 min.
Unknown sample represents the methylated acid moiety of the compound isolated from bovine
heart muscle mitochondria.

ring with a 2-ethylhexyl side chain. (Identical results were obtained when authentic
DEHP was subjected to carbon-skeleton chromatography.) Following drastic
alkaline hydrolysis, the alcohol portion upon gas chromatography in diethyleneglycol
succinate, Apiezon L and SE-52/XE-60 columns revealed the presence of a single
component having retention times identical with those of authentic 2-ethyl-1-hexanol
(Table 2). When the acidic portion was methylated and subjected to gas chromatography, only one component having the same retention characteristics as those of

dimethyl phthalate (*e.g. ortho*) was detected (Table 3). The separation of several diesters of phthalic acid was achieved on a 6 ft. × 6 mm o.d. stainless steel column containing 5% SE-30 on Chromosorb W at a column temperature of 225° with a flow rate of 30 ml/min of nitrogen (Fig. 3).

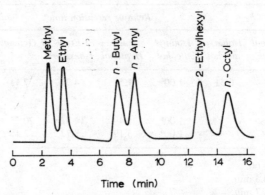

Time (min)

Fig. 3 Gas-liquid chromatographic separation of diesters of phthalic acid on a 5% SE30 column.

The mass spectrum of the isolated compound was identical with that of an authentic sample of DEHP when determined on a Consolidated Electrodynamics Corp. Model 21-110B mass spectrometer. The molecular formula established by peak matching the molecular ion $C_{24}H_{38}O_4^+$ was found to be 390.2765 compared with a theoretical value of 390.2770. The m/e values of fragments with highest relative abundance corresponded to rearrangement ions of phthalic anhydride (149; M-241), the diacid fragment (167; M-233) (15), and the one derived by the loss of 2-ethyl-hexyl group from the parent molecular ion (279; M-111).

Wandel and Tengler[42] described the GLC determination of bis(2-ethylhexyl)-phthalate and tributyl-o-acetyl citrate in foods in contact with plastic film. Dry or non-fatty foods such as rice, meal, herbs, instant coffee or cocoa, were analyzed by a direct method. For example, a 50 g sample of food was mixed sand and extracted in a Soxhlet extraction apparatus with methylene chloride. The methylene chloride extract was concentrated and the residue dissolved in methanol and analyzed on a 50 cm × 3 mm column containing either 5% SE-52 on Celite 545 or 10% Resoflex LAC-2-R-446 on Chromosorb at 230°C and a helium flow of 42 ml/min using dioctyl adipate as internal standard and flame ionization for detection. Tributyl-o-acetyl citrate was determined on the same column at 200° with dibutyl phthalate as internal standard. The foods examined by the above method had 4–16 p.p.m. of plasticizer or 0.06–0.29 mg/dm² of packaging surfaces. The relative error was about 10%. As little as 0.1 mg of bis(2-ethylhexyl)phthalate or tributyl-0-acetyl citrate (Citroflexa 4) in the 50 g sample (corresponding to 2 p.p.m.) could be detected.

Samples of cheese or sausage were analyzed by an indirect method. The evapor-ated residues from the methylene chloride extraction were shaken 15 times with 10 ml portions of methanol, the combined methanol extracts were concentrated to

3–5 ml and quantitatively saponified with *p*-toluenesulfonic acid for 3 h at 130°C. The saponified mixture was analyzed for alcohols using a flame-ionization detector and a 2 m × 4.65 mm i.d. copper column packed with 15% polyethylene glycol P1500 on Celite 545 or 10% Resoflex LAC-2-R-446 on Chromosorb. 2-Ethylhexanol from bis-(2-ethylhexyl)phthalate was determined at a column temperature of 100°C and a helium flow of 92 ml/min using heptanol as an internal standard. Butanol from tributyl-O-acetyl citrate was determined at a column temperature of 90° and a helium flow of 45 ml/min using propanol as an internal standard.

A method for the isolation and detection of DEPH from milk lipids was described by Cerbulis and Ard[43]. Milk was dialyzed and evaporated to dryness and the residue extracted with petroleum ether. The petroleum ether extract of dried milk was then chromatographed on an alumina column as described by Hanahan[44] and the fraction containing free fatty acids and DEHP then separated by TLC on Silica Gel G. DEHP was distinguished from other phthalates by TLC and infrared spectroscopy. The solvent systems used were (*1*) for TLC (*a*) petroleum ether (b.p. 30–60°C)–diethyl ether–acetic acid (90:10:1), (*b*) chloroform–methanol–water (65:25:4), and (*c*) benzene and (*2*) for alumina column (activated, chromatographic grade, Matheson, Coleman and Bell) (*a*) chloroform–methanol (1:1), (*b*) and (*c*) ethanol–chloroform–water (5:2:1) and (5:2:2), respectively.

TABLE 4

CHROMATOGRAPHY OF MILK LIPIDS ON AN ALUMINA COLUMN

Solvent	Volume of solvent (ml)	Yield (g)	Lipid components
Chloroform	2000	53.0	Glycerides, cholesterol, pigments
Chloroform–methanol (1:1)	500	Traces	Lecithin, sphingomyelin
Ethanol–chloroform–water (5:2:1)	800	Traces	Inositide (?)
Ethanol–chloroform–water (5:2:2)	900	0.8	Free fatty acids and DOP
Methanol	900	No residue	

Table 4 shows the results of chromatography of milk lipids on an alumina column. The residue of the ethanol–chloroform–water (5:2:2) fraction containing free fatty acids and DEHP was chromatographed on Silica Gel G. The migration of bis(2-ethylhexyl)phthalate and Mesamoll (ethane sulfonate of phenol and cresol) from nitrite rubber into milk was studied by gas chromatography[45]. Potassium oxalate was added to the milk to prevent coagulation, after which the plasticizers were extracted with a methanol–ether mixture. The extract was then concentrated and the products then saponified to form 2-ethylhexanol and phenol and cresol (from Mesamoll). The saponified mixture was steam distilled and the 2-ethyl hexanol, phenol, and cresol in the distillate were extracted with ether and quantitatively determined by GLC using a P-2000 column and adding triethylphenol as an internal standard. The relative error for the range 2–40 p.p.m. was 20–30%.

The TLC analysis of plasticizers has been described by Peereboom[46,47]. The plasticizer or the residue obtained by extraction of a food-packing material was dissolved in 5% diethyl ether, concentrated and applied on Silica Gel G plates containing 0.005% of a water-soluble fluorescence indicator Ultraphor (Badische Anilin- & Soda-Fabrik). On irradiation with UV light (365 nm) all plasticizers became visible as fluorescent spots or in the case of phthalates as dark spots. The following solvent mixtures were preferably used: isooctane + 10% ethyl acetate, benzene + 5% ethyl acetate, and dibutyl ether + 20% hexane.

TABLE 5

RELATIVE R_A VALUES OF SOME PLASTICIZERS WITH MICRO-ADSORPTION CHROMATOGRAPHY IN THREE SOLVENT MIXTURES

Identification with 0.005% Ultraphor in UV light

R_A value = R_F value with reference to dibutyl sebacate

Plasticizer	R_A value in the solvent mixtures		
	Isooctane + 10% ethyl acetate	Benzene + 5% ethyl acetate	Dibutyl ether + 20% hexane
1. Triacetin	0.18	0.34	0.17
2. Ethylphthalyl ethyl glycolate	0.22	0.66	0.30
3. Acetyltriethyl citrate	0.26	0.51	0.29
4. Triphenyl phosphate	0.33	0.80	0.50
5. Tricresyl phosphate	0.42	0.86	0.69
6. Butylphthalyl butyl glycolate	0.43	0.90	0.65
7. 2-Ethylhexyl-diphenyl phosphate	0.46	0.77	0.58
8. Diethyl phthalate	0.51	0.79	0.60
9. Acetyltributyl citrate	0.53	0.85	0.70
10. Di-n-butyl phthalate	0.74	1.03	0.84
11. Diisobutyl adipate	0.83	0.86	0.85
12. Dibutyl sebacate	≡1.00	≡1.00	≡1.00
13. Dinonyl phthalate	1.01	1.18	1.14
14. Di-2-ethylhexyl phthalate	1.14	1.16	1.15
15. Butyl stearate	1.61	1.23	1.28
16. Paraflex G 62 (epoxidized natural glycerides)	0.09	0.06	0.06
	0.15	0.13	0.14
	0.25	0.20	0.28
	0.33	0.26	0.42
	0.60	0.40	
		0.47	
		0.65	
		0.72	
		1.04	

TABLE 6

COLOR REACTIONS OF SOME PLASTICIZERS (500 µg) ON CHROMATOPLATES

Plasticizer	Color reaction[a]								
	1	2	3	4	5	6	7	8	9
1. Triacetin	Yellow	Purple	Yellow-green		Blue-black		Purple		
2. Ethylphthalyl ethyl glycolate		Yellow	Purple						Yellow
3. Acetyltriethyl citrate	Orange-yellow			+	Blue	Blue	±Yellow	±Pink	Orange-purple
4. Triphenyl phosphate					Red	Blue	Yellow		Orange-purple
5. Tricresyl phosphate	Red				Red	Blue	Yellow		
6. Butylphthalyl butyl glycolate	Yellow	Yellow	Purple						
7. 2-Ethylhexyl diphenyl phosphate	Blue	Purple	Green		Blue	Blue	Yellow	Purple-brown	Orange-purple
8. Diethyl phthalate	Yellow	Yellow		+	Blue		Purple-red		
9. Acetyltributyl citrate	±Yellow	—	Blue				±	±Pink	
10. Di-n-butyl phthalate	±Yellow		±Yellow		Green-blue		±Yellow	±Pink	
11. Diisobutyl adipate	±Yellow		±Blue				±Yellow	±	
12. Dibutyl sebacate	Blue	±Orange	Brown-red		Blue			Purple-brown	
13. Dinonyl phthalate	Blue	±Orange					±Yellow	Purple-brown	
14. Di-2-ethylhexyl phthalate	Blue	Green			Black		Yellow	Brown-red	
15. Butyl stearate	Blue		Green				Yellow	Brown-red	
16. Paraflex G 62, $R_F \sim 0.00$	Blue	±Brown	Brown	+	Black		Yellow	Brown	

[a] Color reactions in which only a faint color is obtained are indicated as ±.

The R_F values of the plasticizers were calculated with reference to dibutyl sebacate (Table 5). Only the groups of plasticizers 5–6 (tricresyl phosphate and butylphthalyl butyl glycolate), 5–7 (tricresyl phosphate, butylphthalyl butyl glycolate, and 2-ethylhexyl diphenyl phosphate), and 7–9 (2-ethylhexyl diphenyl phosphate, diethyl phthalate, and acetyl tributyl citrate) were not sufficiently separated in any of the systems and had to be identified by means of the specific reagents mentioned in Table 5. Table 6 contains a survey of some more-or-less specific color reactions. The procedure of these spot test color reactions was as follows.

(1) Spray with a solution of phosphomolybdic acid (10 % in ethanol). Heat for about 20 min at 100°.

(2) Spray with a solution of resorcinol (20 % in ethanol, add some $ZnCl_2$). Heat for 10 min at 150°, spray with 4 N H_2SO_4, heat for approx. 20 min at 120°, spray with 40 % KOH.

(3) Spray with a solution of thymol (20 % in ethanol), heat for 10 min at 90°, spray with 4 N H_2SO_4, heat for approx. 10–15 min at 120°.

(4) Spray with 1 N alcoholic KOH solution, heat for 15 min at 60°. Only the citrates become visible as yellow spots. Spray with a 50 % solution of urea. The citrates are visible in uv light (365 μm) as strong fluorescent spots.

(5) Spray with a solution of vanillin (20 % in ethanol), heat for 10 min at 80°, spray with 4 N H_2SO_4, heat for approx. 30 min at 110°.

(6) Spray with a solution of 2,6-dichloroquinone-chloroimide (2 % in ethanol), spray after 2–3 h with 2 % borax.

(7) Spray with an alkaline $KMnO_4$ solution (a mixture 1:1 of 1 % $KMnO_4$ and 2 % Na_2CO_3). The colors shown after 1–2 h are noted.

(8) Spray with a mixture of acetic anhydride–50 % H_2SO_4 (1:2), heat for approx. 10 min at 80°.

(9) Spray with a 0.5 N ethanolic solution of KOH; heat for 15 min at 60°. Spray with a diazonium reagent (prepared from 0.8 g p-nitroaniline in 250 ml water and 20 ml 25 % HCl; a 5 % $NaNO_2$ solution is added till the reagent mixture is entirely colorless.

The TLC of dimethyl- and dibutyl phthalates and tricresyl phosphate on Aluminum Oxide D (VEB) was described by Pietsch and Mayer[48]. Benzene–chloroform (50:2.5) was used for development (10 cm, *ca.* 20 min) with detection of phthalate esters accomplished with a spray reagent consisting of 20 % resorcinol, 40 % $ZnCl_2$, and 3 % sulfuric acid. Heating the plates at 150°C revealed the phthalate esters as yellow spots. Detection of tricresyl phosphate was achieved *via* sequential use of (a) 15 % aq. sodium hydroxide, (b) 0.1 g p-nitroaniline in 5 ml of 25 % hydrochloric acid diluted to 100 ml, and (c) 6 % aq. sodium nitrite. The plates were first sprayed with solution (a) and kept in an oven at 110° for 30 min; 10 ml of solution (b) and 1 ml of solution (c) were mixed and sprayed on the cooled plate revealing tricresyl phosphate as a violet spot. The R_F values of dimethyl- and dibutyl phthalates were 0.21 and 0.33 and for tricresyl, 0.23. Any antioxidants added to the plastic material as stabilizers appeared as blue or brown spots when sprayed with phosphomolybdic acid in ethanol.

Braun[49,50] separated a variety of plasticizers on Silica Gel G with methylene dichloride, after they had been extracted from the plastic material with benzene or ether (provided the polymer itself was not soluble). Detection was accomplished using an antimony pentachloride reagent ($SbCl_5$ + $CHCl_3$ or CCl_4 (1:4), freshly prepared before use), which yielded brown spots with most of the plasticizers after the plate was heated to 210°. Phthalate esters were also detected using a resorcinol–zinc chloride–sulfuric acid reagent[46]. Table 7 lists the R_F values of 26 common plasticizers on Silica Gel G developed with methylene chloride.

TABLE 7

R_F VALUES OF IMPORTANT PLASTICIZERS ON SILICA GEL G
Solvent: methylene dichloride

Compound	R_F	Compound	R_F
Dimethyl phthalate	51	Di-(2-ethylhexyl) adipate	44
Dibutyl phthalate	69	Dinonyl adipate	44
Dihexyl phthalate	80	Adipic acid polyester	2
Dioctyl phthalate	86	Dibutyl sebacate	41
Di-(2-ethylhexyl)phthalate	85	Dioctyl sebacate	61
Didecyl phthalate	85	Di-(2-ethylhexyl) sebacate	61
Diisodecyl phthalate	84	Sebacic acid polyester	2
Trioctyl phosphate	23	Triethyl citrate	12
Diphenyloctyl phosphate	42	Tributyl citrate	14
Triphenyl phosphate	47	Triethylacetyl citrate	15
Diphenylcresyl phosphate	51	Tributylacetyl citrate	24
Tricresyl phosphate	53	Tri(2-ethylhexyl) acetyl citrate	46
Dioctyl adipate	42	Glycerol triacetate	13

For gas chromatography, columns of 10% Resoplex 446 on silica gel have been generally successful[51]. In isothermal operations, a column temperature of 190° was used for the separation of dimethyl-, diethyl-, and dibutyl phthalates and 230° for the higher boiling phthalates except diisodecyl phthalate. For a positive identification, 10–100 mg of plasticizer was sealed in a small tube with approx. 1 ml of dry methanol containing 3 wt.-% of p-toluenesulfonic acid and heated for 3 h at 130°. The alcohol was identified by GLC using a column containing Reoplex 400 on 80–100 mesh silica gel with temperature-programmed from 80 to 160° at 125°C/min, and a carrier flow rate of 42 ml/min.

Both TLC and GLC methods were examined by Diemair and Pfeilsticker[52] for the rapid isolation and concentration of monomeric ester plasticizers. The extraction with nitromethane as solvent and TLC on Silica Gel G with nitromethane at 50° proved successful for the separation and analysis of plasticizers from food products. The plasticizers studied were dioctyl-, dibutyl glycol-, didecyl-, and dinonyl phthalates, tricresyl phosphate, tributyl citrate, acetyl-tri(2-ethylhexyl) citrate, and dioctyl adipate.

GLC was carried out with a Perkin-Elmer Fraktometer F6 chromatograph equipped with a flame-ionization detector. A 30 cm × 6.35 mm U-column was packed with 3% Silopren-R (Bayer) on Chromosorb R, 50–60 mesh. The injection port and detector temperatures were 350 and 400°C and the column was run under temperature programming from an initial temperature of 160° at 10°C/min. Helium was the carrier gas at 120 ml/min.

A TLC and GLC technique for the analysis of plasticizers in poly(vinylidene chloride) mixed polymer films was described by Groebel[53]. TLC on silica gel using methylene chloride for development separated tributyl acetyl citrate (R_F 0.32), dibutyl sebacate (0.38), and bis(2-ethylhexyl) phthalate (0.58). Use of a column containing 20% poly(diethylene glycol succinate) permitted the GLC separation of dibutyl sebacate, tributyl acetylcitrate and bis(2-ethylhexyl) phthalate (3.80, 7.14, and 11.25 min, respectively).

The identification and determination of seven plasticizers in nitrocellulose, vinyl, and acrylic type lacquers by programmed-temperature gas chromatography was reported by Esposito[54]. The analysis was conducted on lacquer samples after treatment to remove the resin. The equipment used was an F & M Model 500 gas chromatograph equipped with a Brown Electronic recorder (Minneapolis-Honeywell) and a disc integrator (Disc Instruments). The column consisted of a 6 ft. × ¼ in. copper tubing packed with 20 wt.-% of silicone grease on acid- and alkali-washed Chromosorb W. The detector cell temperature was set at 300°C, the detector cell current at 160 mA, injection port temperature at 330°C, and helium flow at the exit was 120 ml/min. The chromatographic column was heated at 210° then programmed at a heating rate of 4°C/min to 290°C. The plasticizers were identified by calculating the retention times relative to butyl sebacate (Table 8).

TABLE 8

RELATIVE RETENTION TIME DATA FOR PLASTICIZERS

Plasticizer	Relative retention time (dibutyl sebacate = 1)
Dimethyl phthalate	0.16
Diethyl phthalate	0.26
Dibutyl phthalate	0.67
Butylbenzyl phthalate	1.32
Di-(2-ethylhexyl) phthalate (DOP)	1.65
Tricresyl phosphate	1.78–1.97[a]
Dioctyl sebacate	2.20

[a] Several peaks produced in this range.

GLC analysis of plasticizer esters has been accomplished using a Perkin-Elmer Model 116E chromatograph with a thermistor detector[55]. Two stationary phases were used, 0.5% silicone SE-30 and 0.5% neopentylglycol polysebacate. Analysis at 200° was possible for liquids with boiling points <400°C. Retention indices were reported for phthalic and adipic esters of normal and branched chain alcohols on 2 columns of different polarity.

The GLC determination of plasticizers such as dioctyl phthalate or dioctyl adipate and related compounds was effected on gas chromatographic columns at 200–240°C using SE-30 or poly(neopentyl glycol adipate) on glass beads[56].

A special technique of mild pyrolysis of the sample in the injection port permitted analysis of the plasticizers without a preliminary extraction step. The chromatograph used was a Perkin-Elmer Model 116R with a thermistor katharometer detector. Two columns were used, both copper tubing, 4 mm i.d., 6 mm o.d. One 2m column was packed with 0.5% silicone gum SE-30 on glass beads, previously etched for a few minutes by concentrated hydrofluoric acid and washed. The sieved beads were 125–160 μm. The other column (3 m long) was packed with 0.5% poly(neopentyl glycol adipate) on the etched glass beads. Hydrogen was used as carrier gas with a flow rate of about 60 ml/min. The main part of the electrical circuit for the pyrolysis unit is a platinum coil welded to two copper wires very similar to those used in the pyrolysis of polymers[57]. The device is fitted to the chromatograph in place of the conventional silicone rubber cap. The platinum wire is 35 mm × 0.4 mm in diameter and the coil is 2 mm o.d. (resistance 0.02 Ω). For about the same quantity of solute, peaks were narrower and higher when the liquid was injected than when the plasticizer was obtained on pyrolyzing the polymer. The pyrolysis peaks tail somewhat, while with liquid injection they were Gaussian. PVC plasticized with diesters may be identified only by the comparison of their retention data on the two columns. Table 9 shows the retention indices of some plasticizers. In the first row the values were computed according to the earlier described technique of Zulaica and Guio-

TABLE 9

COMPUTED AND MEASURED RETENTION INDICES

Compounds	I_R			$I_R(pyr.) - I_R(inj.)$
	Computed	Injection	Pyrolysis	
Dimethyl phthalate	1457	1450	1469	19
Diethyl phthalate	1587	1583	1594	11
Di-n-butyl phthalate	1926	1932	1920	−16
Bis (2-ethylhexyl) phthalate	2482	2470	2496	26
Tri-n-butyl phosphate	1663	1648	1633	−15
Di-n-butyl sebacate	2169	2150	2156	−6
Bis (2-ethylhexyl) adipate	2412	2391	2400	9

References pp. 605–607

chon[58]. In the second row they were measured from pure compounds injected as liquids and in the third row they were measured from the pyrolyses of plastic products containing known plasticizers at different concentrations. The difference between the last two values are less than $\pm 1\%$ of the indices.

Dialkyl phthalates at concentrations of 1–10 p.p.m. in aqueous solutions have been separated and determined by GLC after extraction into hexane[59]. The instrument used was an Aerograph 1520 equipped with a 1:1 stream splitter. The two equal effluent gas streams were monitored by flame-ionization and electron-capture detectors to demonstrate the difference in response to both sample and solvent. The column (5 ft. $\times \frac{1}{8}$ in. o.d.) comprised 1% Carbowax 20M on DMCS Chromosorb G, 80–100 mesh, coated with 1% polyvinyl pyrrolidone. The column was conditioned overnight at 230°C in a stream of nitrogen, followed by 10 days at 190°C. Nitrogen used as carrier gas at a flow rate of 20 ml/min was dried by passing through molecular sieve 5A.

Figure 4 illustrates a chromatogram (at a column temperature of 190°C) of a 4 µl mixture of diethyl-, dibutyl-, di-(2-ethylhexyl)-, and di-(3,5,5'-trimethylhexyl)-phthalates, each at a concentration of 1 p.p.m.

Figure 5 illustrates an example of the type of chromatogram obtained from an actual sample by using only the electron-capture detector and shows that di-(3,4,5'-trimethylhexyl)- and di- (2-ethylhexyl) phthalates are easily separated. A better separation of the lower boiling dialkyl phthalates could be affected by GLC at a lower column temperature than 190°C.

The GLC of a number of commercial plasticizers was elaborated by Courtier[60]. Commercial samples of n-octyl, 2-ethylhexanyl, and isooctyl phthalates showed the presence of several C_8 alcohols produced in the oxo process. The chromatogram of the phthalates made from C_7, C_8, and C_9 alcohols from the oxo process showed 20 peaks. Chromatograms of ethylhexyl and octyl azelates showed the presence of acids from adipic to undecylenic acid.

A studied was described by Courtier[61] of the control of purity of bis(2-ethyl-hexyl) phthalate, the separation and identification of light impurities as well as the analysis of plasticizers of higher molecular weight, $e.g.$ trimellitic esters. A high-temperature apparatus was used and a CuO furnace operated at 700° was included which converted the eluted compounds to CO_2 and H_2O. CO_2 was detected by a thermistor catharometer and the water was absorbed in $CaCl_2$. The columns used were 2 m $\times$ 4 mm i.d. copper tubing containing silicone grease or SE-30 on Chromosorb W (60–80 mesh) or Chromosorb G. Isooctyl phthalates from 3 different sources were resolved and the impurities detected. Impurities found in bis(2-ethylhexyl)-phthalate were air, ethylhexene, ethylhexanol, phthalic anhydride, ethylhexyl ether, and ethylhexyl benzoate. Analogous results were obtained for trimellitic esters.

The determination of dimethyl-, diethyl-, and dibutyl phthalates in small arms double-base propellants has been reported by Norwitz and Abatoff[62]. The phthalates were extracted from the propellant with methylene chloride and an aliquot passed

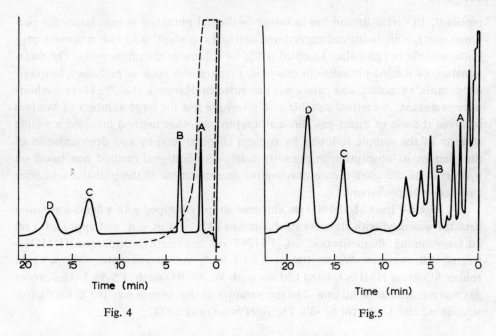

Fig. 4   Fig.5

Fig. 4. Chromatogram of 4 µl of a mixture of A,diethyl phthalate, B, dibutyl phthalate, C, di-(2-ethylhexyl) phthalate, and D di-(3,5,5'-trimethylhexyl) phthalate, each at a concentration of 1 p.p.m., showing relative response to electron-capture, ——, and flame-ionization, - - - -, detectors. Column, 1 % Carbowax 20 M on DMCS Chromosorb G, 80–100 mesh, coated with 1 % polyvinyl-pyrrolidone (5 ft. × ⅛ in. o.d.); temperature, 190°C; flow rate, 20 ml/min; instrument, Aerograph 1520; sensitivity of electron-capture and flame-ionization detectors, 1 × 8 and 0.1 × 16, respectively.

Fig. 5. Typical example of the type of chromatogram obtained after extraction of a polyvinyl chloride-type plastic. A, diethyl phthalate; B, dibutyl phthalate; C, di(2-ethylhexyl) phthalate; D, di(3,5,5'-trimethylhexyl) phthalate.

through a silicone rubber SE-30 column at 200°C for dimethyl and diethyl phthalates and 230°C for dibutyl phthalate. Triacetin was used as the internal standard in the determination of dimethyl and diethyl phthalates while dimethyl sebacate was used as the internal standard for the determination of dibutyl phthalate.

A single column gas chromatograph was used equipped with a thermal conductivity cell (Aerograph Model A90) and a 6 ft. × ¼ in. o.d. column containing 20 % SE-30 on acid-washed Chromosorb W (60–80 mesh). The injection port and detector temperatures were both 275°C and the carrier gas was helium at 35 ml/min. The retention times obtained at a column temperature of 210° for the analyzed phthalates were dimethyl- 2.75, diethyl- 4.0, and dibutyl- 12.0 min, respectively. (The internal standard triacetin was 1.75 min.)

Diethyl phthalate is a commonly employed denaturant. It is widely used since it seldom interferes with the intended use of the preparation and may be a preferred component because of its utility as a perfume fixative, a plasticizer or an insect repellant. In Great Britain the inclusion of diethyl phthalate is mandatory for per-fumes made with industrial methylated spirits (α-grade)[63] and the minimum pro-portion of diethyl phthalate accepted is 1 % by volume of the preparation. The deter-mination of diethyl phthalate in ethanolic preparations such as perfumes, lacquers, deodorants, varnishes, and paints was described by Hancock et al.[64]. Three methods were evaluated. A method suitable as a screening test for large numbers of samples involved the use of direct gas chromatography. Another method involved a simple cleanup of the sample followed by column chromatography and determination of the ester by its absorption in the ultraviolet. An additional method was based on hydrolysis of the ester thence gravimetric determination of the phthalic acid after conversion to phthalanil.

A Perkin-Elmer Model 800 gas chromatograph equipped with a flame-ionization detector was used with stainless steel columns (9 ft. × $\frac{1}{8}$ in. o.d. and approx. 0.1 in. i.d.) containing fluorosilicone oil, FS1265 on hexamethyl disilazone (HMDS)-treated Chromosorb W, 80–100 mesh, 1.5–98.5, or alternatively, silicone gum rubber SE-30 on HMDS-treated Chromosorb W, 80–100 mesh, 1.5–98.5. The carrier gas was nitrogen at 30 ml/min. The temperature of the column was 165°C for fluoro-silicone oil and 150°C for SE-30. The inlet block was 250°C.

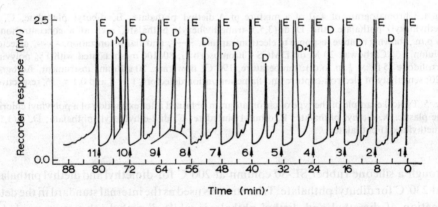

Fig. 6. Successive chromatograms obtained with a Perkin-Elmer model 800 gas chromatograph of samples. Column, FS 1265: Chromosorb W (hexamethyldisilazane) 1$\frac{1}{2}$: 98$\frac{1}{2}$; nitrogen flow; 30 ml/min; temperature, 165°C. 1,2,3, Perfumes; 4, after-shave lotion containing isopropyl myristate; 5, 6, hair lacquer; 7, deodorant; 8, nail varnish remover; 9, cologne; 10, surgical spirit containing methyl salicylate; 11, cologne with diethyl phthalate absent. E = ethanol; D = diethyl phthalate; I = isopropyl myristate; M = methyl salicylate.

Figure 6 illustrates successive chromatograms of a variety of ethanol preparations including perfumes, lacquers, deodorants, and varnishes. Relative retention times of dimethyl-, diethyl-, dibutyl-, and diethylhexyl phthalates, isopropyl myristate, and methyl salicylate on Fluorosilicone oil and SE-30 are listed in Table 10.

Column chromatography utilized 100–200 mesh alumina, Brockman 1–2 with a mixture of 60 % v/v of light petroleum and 40 % v/v of diethyl ether used for elution.

TABLE 10

RELATIVE RETENTION TIMES

Substance	FS1265	Relative retention time, SE30
Dimethyl phthalate	0.65	0.59
Diethyl phthalate	1.00	1.00
Dibutyl phthalate	3.20	5.70
Diethylhexyl phthalate	38.2	
Isopropyl myristate	0.98	3.0[a]
Methyl salicylate	0.02	

[a] Commercial samples of isopropyl myristate often contain an impurity with a relative retention time on SE30 of 1.15.

TABLE 11

COMPARISON OF RESULTS FROM GAS CHROMATOGRAPHIC, GRAVIMETRIC, AND TITRATION METHODS

Type of preparation	Diethyl phthalate by			Results
	Gas chromatography (%)	Gravimetric method (%)	Titration method (%)	
Perfume spray	0.7	0.71	0.9	
Perfume spray	1.0	1.10	1.0	
Perfume	1.0	1.00	2.4	
("Honey and Flowers")	⎰ nil	nil		Top layer
Hair oil	⎱ 0.8	1.00		Lower layer
Pre-electric shave lotion	1.3	1.39	1.4	Contains isopropyl myristate
Eau de cologne	0.9	0.80	1.2	
Hair dressing	0.95	0.90	1.2	
Aerosol perfume	1.2	1.2	1.9	
Perfume	0.6	0.74	1.1	
Hair lacquer	0.7		2.0	Contains diethyl and dibutyl phthalates
Hair tonic	nil	nil	nil	Contains castor oil
After-shave	0.8	0.84		Contains isopropyl myristate
Hair tonic	0.85	0.80		Contains glycerin
Hair lacquer	1.7	1.9		Contains poly(vinyl pyrrolidone)
Eau de cologne	0.90	0.86		Contains benzyl acetate, 2-phenyl-ethanol, methyl naphthyl ketone
Bay rum	0.75	0.86		
Eau de cologne	0.25	0.38		
Hair dressing	0.70	0.66		Contains xylol, salicylic acid, resorcinol, hexachlorophene
Lavender water	0.80	0.78		

References pp. 605–607

The eluate fraction containing the diethyl phthalate was evaporated to dryness, taken up in ethanol, and measured at 227 nm.

Table 11 compares the results from gas chromatographic, gravimetric and titration methods of the determination of diethyl phthalate in a variety of ethanolic preparations.

Gas chromatographic analyses of the phthalate esters have generally employed isothermal conditions[65-67] with only limited use of temperature programming for a few of the plasticizers. One of the major handicaps of isothermal column operations has been the necessity of employing either two different columns or two different temperatures for the identification of both the original plasticizer and the products obtained from it by hydrolysis and esterification.

Krishen[68] recently described the programmed gas chromatography for identification of a variety of ester plasticizers used in PVC and other polymers. The gas chromatographic unit was a Hewlett-Packard 5750B dual flame-ionization chromatograph equipped with a Moseley–Hewlett–Packard 7127A 1 mV recorder. Two stainless steel columns each 6 ft. × ⅛ in. o.d. packed with 10% UCW-98 on 60–80 mesh Diatoport S were employed in the dual operation mode. The initial column oven temperature was 100°C and after 4 min of isothermal operation, the temperature was programmed at a rate of 8°C/min to a maximum of 330°C, and the final temperature held constant for 8 min. The injection block and the detector were maintained at 270°C. The helium, hydrogen, and air pressures were 60 (Flowrator 0.8), 14, and 30 p.s.i.g., respectively.

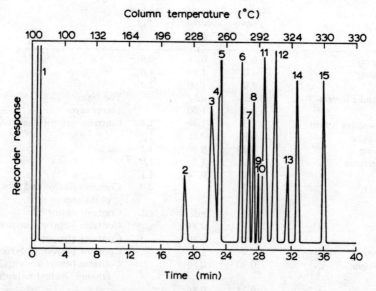

Fig. 7. Gas chromatography of ester plasticizers. 1, Tetrahydrofuran; 2, triethyl citrate; 3, methylphthalylethyl glycolate; 4, ethylphthalylethyl glycolate; 5, dibutyl phthalate; 6, dibutyl sebacate; 7, acetyltributyl citrate; 8, butylphthalylbutyl glycolate; 9, butylbenzyl phthalate; 10, trioctyl phosphate; 11, di(2-ethylhexyl) adipate; 12, di(2-ethylhexyl) phthalate; 13, di(2-ethylhexyl) azelate; 14, di(2-ethylhexyl) sebacate; 15, di-n-decyl phthalate.

Samples of the plasticizers were dissolved in tetrahydrofuran and then injected into the gas chromatogram. When the plasticizers were present in PVC, the polymer sample was first dissolved in tetrahydrofuran, insoluble components allowed to settle out, and a sample of the clear solution chromatographed. A 1 % solution of the polymer suitable for this purpose, when the plasticizer content of the polymer was between 10 and 40%.

Figure 7 shows a gas chromatogram of a mixture of ester plasticizers. Most of the commonly used materials were easily separated by programmed GC and could be identified by their retention times. The retentions calculated relative to di(2-ethyl-hexyl)phthalate (DOP) are shown in Table 12.

TABLE 12

RELATIVE RETENTION DATA FOR PLASTICIZERS

Plasticizer	Retention relative to DOP (100.00)
Phthalates	
Dibutyl	76.9
Butyl benzyl	92.8
Di-(2-ethylhexyl)	100.0
Di-*n*-decyl	120.0
Adipates	
Di-(2-ethylhexyl)	95.0
Azelates	
D-(2-ethylhexyl)	105.4
Citrates	
Triethyl	62.5
Acetyltributyl	89.1
Sebacates	
Dibutyl	85.9
Di-(2-ethylhexyl)	108.7
Glycolates	
Methylphthalylethyl	73.6
Ethylphthalylethyl	76.4
Butylphthalylbutyl	91.3
Phosphates	
Trioctyl	92.8

Although the relative retentions of the plasticizers are helpful for the identi-fication of plasticizers, the complexity of mixtures normally encountered generally necessitates hydrolysis and esterification to obtain information about the components of plasticizers. Simple hydrolysis (using lithium metal in methanol and refluxing the

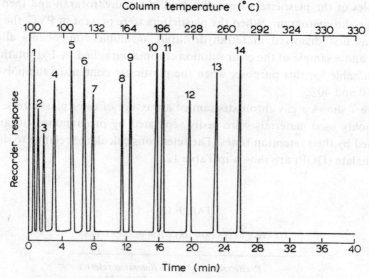

Fig. 8. Gas chromatography of alcohols and methyl esters of acids. 1, Methanol; 2, butanol; 3, pentanol; 4, hexanol; 5, heptanol; 6, 2-ethylhexanol; 7, octanol; 8, dimethyl adipate; 9, decanol; 10, dimethyl *o*-phthalate; 11, dodecanol; 12, tetradecanol; 13, methyl palmitate; 14, methyl stearate.

mixture for 2 h) followed by esterification to the respective methyl esters, thence gas chromatography permitted the facile identification of these products.

Figure 8 depicts a gas chromatogram of alcohols and methyl esters of acids of the commonly used plasticizers. The identification of unknown alcohols and methyl esters of acids facilitated by their relative retentions are given in Table 13.

TABLE 13

RELATIVE RETENTION DATA FOR HYDROLYSIS PRODUCTS

Component	Retention relative to DOP (100.00)
Methanol	0.034
Butanol	1.82
Pentanol	3.98
Hexanol	8.04
Heptanol	15.5
2-Ethylhexanol	21.0
Octanol	24.5
Dimethyl adipate	37.0
Decanol	40.0
Dimethyl-*o*-phthalate	51.1
Dodecanol	53.0
Tetradecanol	64.3
Methyl palmitate	76.1
Methyl stearate	84.8

Phthalic anhydride is manufactured by catalytic oxidation of naphthalene and o-xylene, which process allows the possible formation of impurities such as naphthalene, maleic, and citraconic anhydrides, benzoic and o-toluic acids, phthalide as well as unreacted o-xylene. The identification and quantitation of impurities in phthalic anhydride is of importance for quality control because of stringent specifications for phthalic anhydride plasticizer and resin use.

The quantitative analysis of the impurities usually found in phthalic anhydride was achieved by Cucarella and Crespo[69] using a silicone SE-30 and Carbowax 20M column and a hydrogen flame detector. The phthalic anhydride was esterified with methanol and chromatograms were obtained using phenoxy acetic acid as an internal standard. An F & M Model 700 gas chromatograph was used equipped with an F & M Model 240 temperature programmer and a Minneapolis-Honeywell 1 mV recorder. The column employed was a stainless steel tubing of 240 cm length × 6.3 mm diameter, packed with 28% SE-30 and 2% Carbowax 20M on 60–80 mesh acid-washed Chromosorb W. The column was conditioned at 220° with nitrogen flowing until a stable baseline was obtained. The operating conditions were detector and injection port temperatures 250°C and column temperature 175°C isothermal or programmed as indicated. The flow rates were nitrogen 25 ml/min, hydrogen 30 ml/min, and air 300 ml/min.

The programmed-temperature procedure for the separation of o-xylene, maleic acid dimethyl ester, citraconic acid dimethyl ester, benzoic acid methyl ester, o-toluic acid methyl ester, naphthalene, phenoxyacetic acid methyl ester, phthalide, and phthalic acid methyl ester was 13 min isothermal at 175°C, then increasing 2°/min until 200°C, and later isothermal. Attenuation was necessary, usually × 100 or × 200. Table 14 lists the retention times, temperatures, and the area correction factors for each of the above substances.

TABLE 14

RELATIVE RETENTION DATA

Substance	Retention temperature (°C)	Retention time (min)	Relative retention time[a]	F
o-Xylene	175	5.5	0.23	0.48
Maleic acid, dimethyl ester	175	8.5	0.35	1.65
Citraconic acid, dimethyl ester	175	10.5	0.44	0.95
Benzoic acid, methyl ester	175	12	0.50	0.60
o-Toluic acid, methyl ester	181	16	0.67	0.54
Naphthalene	186	18.5	0.77	0.42
Phenoxyacetic acid, methyl ester	197	24	1.00	1.00
Phthalide	200	33	1.38	0.88
Phthalic acid, dimethyl ester	200	38	1.58	

[a] Relative to phenoxyacetic acid, methyl ester.

The qualitative and quantitative determination of the various related substances commonly found in phthalic anhydride (PAA) (obtained from the naphthalene conversion process) is of importance because of the many use applications of the phthalate esters. For example, plasticizers for PVC medical application (*e.g.* blood bags, heart and kidney machine tubing, etc.) as well as for food wraps must contain a minimum of impurities.

Trachman and Zucker[70] described the quantitative determination of maleic anhydride, benzoic acid, naphthalene, and 1,4-naphthoquinone in phthalic anhydride by GLC. Separation of all the related components was achieved by isothermal analysis using a Silicone SF-96 column and a hydrogen flame detector. An F & M Model 1609 gas chromatograph was employed with a Brown Electronic recorder (Minneapolis-Honeywell) and a Disc chart integrator (Disc Instruments). The operating conditions were injection port temperature 250°C, column temperature 220°, detector cell temperature 200°C, helium flow rate 60 ml/min (30 p.s.i. at inlet). The column (6 ft. × $\frac{1}{4}$ in. stainless steel) was packed with 30% Silicone SF-96 on 60–80 mesh acid-washed Chromosorb W. Retention data for phtahlic anhydride and possible impurities is shown in Table 15.

TABLE 15

RETENTION DATA FOR PHTHALIC ANHYDRIDE AND POSSIBLE IMPURITIES

Sample	Retention time (min measured from injection)	Relative retention[a]
Acetone (if used)	1.1	0.24
Maleic anhydride	2.4	0.52
o-Dichlorobenzene	4.6	1.00
Benzoic acid	6.2	1.35
Naphthalene	8.4	1.83
Phthalic anhydride	11.4	2.48
1,4-Naphthoquinone	16.3	3.54

Relative to solvent peak *o*-dichlorobenzene = 1.00.

A considerable amount of phthalic anhydride is also prepared *via* an *o*-xylene conversion process. The column and operating parameters used above were also applicable for the analysis of phthalic anhydride and expected impurities from the *o*-xylene process (*e.g.* phthalate, *o*-xylene, maleic anhydride, benzoic acid).

The use of vapor-phase chromatography in the production analysis of phthalic anhydride was described by Moreno and Mendoza[71]. Acidic impurities were esterified with diazomethane and analyzed on a column of 30% SE-30 on Chromosorb W (60–80 mesh) at a programmed temperature from 80 to 200°C at 12°/min. Adequate separations and quantitative estimation of impurities such as naphthalene, benzoic acid, maleic and phthalic acids, and 1,4-naphthoquinone were obtained in 8 min.

Recent advances have suggested that the speed and efficiency of liquid chromato-graphy (LC) are rapidly approaching that of gas chromatography. A number of high-efficiency LC supports have recently been introduced, including Zipax (DuPont's CSP support)[72,73], Corasil I and II and Durapak (Water Associates)[74]. With the exception of Durapak, these materials, in the micron particle range, consist of particles with a solid core and thin porous coating. This unique combination permits very high coefficients of mass transfer. Durapak consists of conventional liquid phases such as β,β'-oxydipropionitrite, chemically bonded to a rigid porous bead. The Zipax material has been reported to have a relatively inert surface[72].

Majors[75] has recently described the high-speed liquid chromatography of a number of antioxidants and plasticizers using solid core supports. Figure 9 shows a schematic of the high-pressure liquid chromatographic system employed. The unit

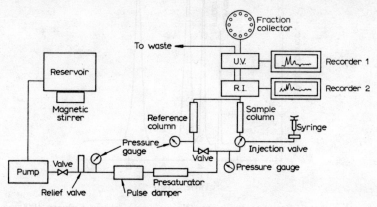

Fig. 9. Schematic diagram of liquid chromatographic system.

employed a Whitey Micro-Regulating High Pressure feed pump (Whitey Research Tool Co., Oakland, Calif.) equipped with an 11 mm plunger, capable of a maximum flow rate of 7.3 l/h and output pressure of 5000 p.s.i. For most experiments up to 2500 p.s.i. output pressure, an ALC-100 LC pulse damper (Waters Associates) was used. To insure saturation of the carrier liquid when doing liquid–liquid chromato-graphy, a pre-saturator column was placed before the chromatographic column. It consisted of an 8 in. × 3/8 in. o.d. stainless steel tube packed with 62–100 μm Porasil A (Waters Associates) loaded with 5 wt.-% of the same liquid phase used in the chromatographic column. The detectors employed in the system were (1) a Refracto-Monitor Model 1103 (Laboratory Data Control (LDC), Danbury, Conn.), with a cell volume of 3 μl and interchangeable prisms to cover solvent refractive indices from 1.31 to 1.55 and (2) UV Monitor Ultraviolet Absorbance (LDC) with an 8 μl cell volume and minimum absorbance range of 0.02 o.d. units. The detector output was monitored by a 10 mV recorder. The columns employed were 2.1 mm i.d. × 0.125 in. o.d. precision-bore stainless steel.

Figures 10 and 11 show the separation of didecyl-, dibenzyl-, and decylbenzyl phthalates using Zipax and Corasil supports, respectively. These plasticizers cannot be directly determined in less than 8 min on Zipax with HETP values less than a millimeter as summarized in Table 16. At the same flow rate, the separation times on Corasil were considerably longer and the plate heights greater although not excessive.

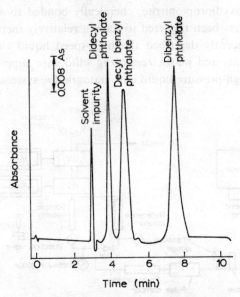

Fig. 10. Separation of phthalate plasticizers using Zipax support column. Column, 1000 mm × 2.1 mm i.d.; packing, 0.5% β,β′-oxydipropionitrile on 20–37 μm Zipax support; carrier, isooctane; flow rate, 0.50 ml/min; sample, 10.6 μl of a mixture of 0.40 μl/ml each of didecyl phthalate and decylbenzyl phthalate and 0.35 mg/ml of dibenzyl phthalate in heptane.

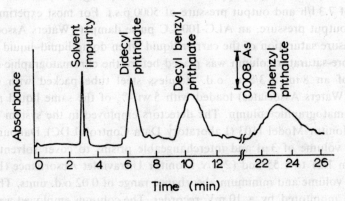

Fig. 11. Separation of phthalate plasticizers using Corasil I support. Column, packing, carrier and flow rate as in Fig. 10.

TABLE 16

EVALUATION OF SOLID CORE SUPPORTS FOR PHTHALATE PLASTICIZERS

	Zipax		Corasil I		Deactivated Corasil	
	V_R (ml)	HETP (mm)	V_R (ml)	HETP (mm)	V_R (ml)	HETP (mm)
Didecyl phthalate	2.0	0.38	3.1	1.9	3.1	1.8
Decylbenzyl phthalate	2.4	0.93	5.1	3.3	4.7	2.7
Dibenzyl phthalate	3.8	0.51	11.6	1.9	8.4	2.5

The analysis of phthalate esters has also been carried out by polarographic[76], fluorescence[77], gravimetric[64], and spectrophotometric (of phthalic acid at 284 nm)[78] and mass spectroscopic techniques[79].

Because of the apparent facility of the phthalate esters to be leached from a variety of plastic devices (e.g. tubing, gloves, vials, caps, bottles, etc.), it is well to stress the stringent precautions required both for its analyses as well as its biological elaboration. A further precaution is raised in regard to the purity of reagent solvents. For example, Asakawa and Genjida[80] isolated and identified dibutyl- and di-2-ethylhexyl phthalates from benzene and chloroform and diisobutyl- and isobutyl phthalates from hexane and petroleum ether. In addition, they extracted dimethyl-, diethyl-, and dibutyl phthalates from polyethylene buckets and dibutyl phthalate from plastic tape at the neck of chloroform bottles.

REFERENCES

1 *Chemical Economics Handbook*, Standford Res. Inst., Monlo Park, Calif., 1966.
2 G. D. TOUEY AND R. C. MUMPOWER, *Brit. Pat. 989,181*, April 14, 1965; *Chem. Abstr.*, 63 (1965) 1964d.
3 F. L. MAYER, JR., D. L. STALLING AND J. L. JOHNSON, Paper presented at *Midwest Regional American Chemical Society Meeting, St. Louis, Mo., Oct. 28–29, 1971.*
4 Anon, *Environ. Health Newsletter*, Dec. 15 (1971).
5 Anon, *Chem. Eng. News*, Nov. 1 (1968) 8.
6 H. F. PHILLIPS AND I. A. BREGER, *Geochim. Cosmochim. Acta*, 25 (1958) 51.
7 I. A. BREGER, *J. Am. Oil Chemists Soc.*, 43 (1966) 197.
8 I. A. BREGER, *Chem. Eng. News*, Jan. 17 (1972).
9 N. SUGIYAMA, C. KASHIMA, M. YANAMOTO, T. SUGAYA AND R. MOHRI, *Bull. Chem. Soc. Japan*, 39 (1966) 1573.
10 P. P. MAIR, *Chem. Eng. News.* Nov. 8 (1971).
11 M. WANDLE AND H. TENLOER, *Deut. Lebensm. Rundschau*, 59 (1963) 326.
12 V. D. FEOFANOV, N. F. TOLIKINA, O. N. BELYAJSKAYA AND U. E. GUL, *Izv. Vysshikh Uchebn. Zavedenii Pishchevaya Tekhnol.*, (1971) 156.
13 E. G. PERKINS, *J. Am. Oil Chemists Soc.*, 44 (1967) 197.
14 P. W. PARODI AND R. J. DUNSTAN, *Australian J. Dairy Technol.*, 23 (1968) 20.
15 R. S. TABORSKY, *J. Agr. Food Chem.*, 15 (1967) 1073.

16 D. J. NAZIR, A. P. ALCAREZ, B. A. BIERL, M. BEROZA AND P. P. NAIR, *Biochemistry*, 10 (1971) 4228.
17 R. J. JAEGER AND R. J. RUBIN, *Science*, 170 (1970) 460.
18 R. J. JAEGER AND R. J. RUBIN, *Lancet*, ii (1970) 151.
19 J. NEERGUARD, *Scand. J. Urol. Nephrol.* 5 (1971) 141.
20 Anon, *Chem. Eng. News*, 49 (1971) 14.
21 W. RUBIN, *Chem. Eng. News*, 49 (1971) 12.
22 C. B. SHAFFER, C. P. CARPENTER AND H. J. SMYTH, JR., *J. Ind. Hyg. Toxicol.*, 27 (1945) 130.
23 H. C. HODGE, *Proc. Soc. Exptl. Biol. Med.*, 53 (1943) 20.
24 C. P. CARPENTER, C. S. WEIL AND H. F. SMYTH, JR., *Arch. Ind. Hyg. Occupational Med.*, 8 (1953) 219.
25 D. K. HARRIS, *Brit. J. Ind. Med.*, 10 (1953) 255.
26 D. W. FASSETT, in F. A. PATTY (Ed.), *Industrial Hygiene and Toxicology*, Interscience, New York, 2nd edn., 1967, p. 1771.
27 R. K. BOWER, S. HABERMAN AND P. D. MINTON, *J. Pharmacol. Exptl. Therap.*, 171 (1970) 314.
28 W. L. GUESS, S. HABERMAN AND P. D. MINTON, *J. Pharmacol. Exptl. Therap.*, 171 (1970) 314.
29 A. R. SINGH, W. H. LAWRENCE AND J. AUTIAN, *J. Pharm. Sci.*, 61 (1972) 51.
30 D. CALLEY, J. AUTIAN AND W. C. GUESS, *J. Pharm. Sci.*, 55 (1966) 15B.
31 W. L. GUESS AND S. HABERMAN, *J. Biomed. Mater Res.*, 2 (1968) 313.
32 J. NEMATOLLAHI AND W. L. GUESS, *J. Pharm. Sci.*, 56 (1967) 1446.
33 J. J. JAEGER AND R. RUBIN, *Federation Proc.*, 29 (1970) 411.
34 Y. L. MARCEL AND S. P. NOEL, *Lancet*, i (1970) 35.
35 J. FOLCH, M. LEES AND G. H. SLOANE-STANLEY, *J. Biol. Chem.*, 226 (1957) 497.
36 H. H. O. SCHMID, L. L. JONES AND H. K. MANGOLD, *J. Lipid Res.*, 8 (1967) 692.
37 L. W. WHEELDON, Z. SCHUMERT AND D. A. TURNER, *J. Lipid Res.*, 6 (1965) 481.
38 P. P. NAIR AND D. A. TURNER, *J. Am. Oil. Chemists. Soc.*, 40 (1963) 352.
39 P. P. NAIR, I. SARLOS AND J. MACHIZ, *Arch. Biochem. Biophys.*, 114 (1966) 488.
40 M. BEROZA AND R. SARMIENTO, *Anal. Chem.*, 38 (1966) 1040.
41 M. BEROZA AND B. A. BIERL, *Anal. Chem.*, 38 (1966) 1976.
42 M. WANDLE AND H. TENGLER, *Deut. Lebensm. Rundschau*, 60 (1964) 335.
43 J. CERBULIS AND J. S. ARD, *J. Assoc. Offic. Anal. Chemists*, 50 (1967) 646.
44 D. J. HANAHAN, *Lipid Chemistry*, Wiley, New York, 1960, p. 115.
45 A. REICHLE AND H. TENGLER, *Deut. Lebensm. Rundschau*, 64 (1968) 142.
46 J. W. C. PEEREBOOM, *J. Chromatog.*, 4 (1960) 323.
47 J. W. C. PEEREBOOM, *J. Chromatog.*, 11 (1963) 55.
48 H. P. PIETSCH AND R. MAYER, *Nahrung*, 9 (1965) 151.
49 D. BRAUN, *Chimia (Aarau)*, 19 (1965) 77.
50 D. BRAUN, *Kunststoffe*, 52 (1962) 2.
51 M. WANDEL AND H. TENGLER, *Kunststoffe*, 55 (1965) 655.
52 W. DIEMAIR AND K. PFEILSTICKER, JR., *Z. Anal. Chem.*, 212 (1965) 53.
53 W. F. GROEBEL, *Z. Lebensm- Untersuch.-Forsch.*, 137 (1968) 7.
54 G. G. ESPOSITO, *Anal. Chem.*, 35 (1963) 1439.
55 J. ZUCAICA AND G. GUIOCHON, *Compt. Rend.*, 255 (1962) 254.
56 J. ZULAICA AND G. GUIOCHON, *Anal. Chem.*, 35 (1963) 1724.
57 J. STRASSBURGER, G. M. BRAUER, M. TRYON AND A. F. FORZJATI, *Anal. Chem.*, 32 (1960) 454.
58 F. J. ZULAICA AND G. GUIOCHON, *Bull. Soc. Chim. France*, (1963) 1242.
59 W. BUNTING AND E. A. WALKER, *Analyst*, 92 (1967) 575.
60 J. C. COURTIER, *Plastiques Mod. Elastomeres (Paris)*, 17 (1965) 132; *Chem. Abstr.*, 64 (1966) 5254.
61 J. C. COURTIER, *Chromatog. Methods Immed. Separ. Prod. Mtg. Athens (1965)*, (publ. 1966), Vol. 1, p. 185.
62 G. NORWITZ AND J. B. ABATOFF, *J. Chromatog. Sci.*, 9 (1971) 682.
63 *The Methylated Spirits Regulations, Statutory Instruments No. 2230*, H.M.S.O., London, 1952.
64 W. HANCOCK, B. A. ROSE AND D. D. SINGER, *Analyst*, 91 (1966) 499.
65 W. FISCHER AND G. LEUKROTH, *Plastverarbeiter*, 20 (1969) 107.
66 H. HAASE, *Kautschuk Gummi*, 20 (1967) 501.

67 F. I. H. TUNSTALL, *Anal. Chem.*, 42 (1970) 542.
68 A. KRISHEN, *Anal. Chem.*, 43 (1971) 1130.
69 M. C. M. CUCARELLA AND F. CRESPO, *J. Gas Chromatog.*, 6 (1968) 39.
70 H. TRACHMAN AND F. ZUCKER, *Anal. Chem.*, 36 (1964) 266.
71 A. M. MORENO AND V. MENDOZA, *Bol. Inst. Quim. Univ. Natl. Auton. Mex. Bull.*, 17 (1965) 67; *Chem. Abstr.*, 63 (1965) 17958f.
72 J. J. KIRKLAND, *Anal. Chem.*, 41 (1969) 7.
73 J. J. KIRKLAND, *J. Chromatog. Sci.*, 7 (1969) 7.
74 I. HALASZ AND P. WALKING, *J. Chromatog. Sci.*, 7 (1969) 129.
75 R. E. MAJORS, *J. Chromatog. Sci.*, 8 (1970) 338.
76 A. F. FORZIATI, in *Analytical Chemistry of Polymers*, Part II, Interscience, New York, 1962.
77 G. M. BRAUER, in *Analytical Chemistry of Polymers*, Part II, Interscience, New York, 1962.
78 G. WILDBRETT, K. F. EVERS AND F. KIERMEFER, *Z. Lebensm. Untersuch.-Forsch.*, 137 (1968) 356.
79 J. C. TOU, *Anal. Chem.*, 42 (1970) 1381.
80 Y. ASAKAWA AND F. GENJIDA, *J. Sci. Hiroshima Univ.*, Serv. A-2, 34 (1970) 103.

Chapter 26

NITRILOTRIACETIC ACID

Increasing concern for the effects of phosphorus in the eutrophication of water bodies has resulted in a search for a substitute for the sodium tripolyphosphate used in laundry detergents. (About 2 billion lb. of phosphate builders are consumed per year.) Trisodium nitrilotriacetate (nitrilotriacetic acid, NTA), I, has until recently

$$NaOOCCH_2 -N \begin{cases} CH_2COONa \\ CH_2COONa \end{cases} \quad I$$

been considered the most promising substitute in both liquid and granular detergent formulations. NTA is produced *via* the interaction of ammonia, formaldehyde, and hydrogen cyanide.

$$NH_3 + 3\ CH_2O + 3\ HCN \xrightarrow{H_2SO_4} N(CH_2CN)_3 + 3\ H_2O$$

$$N(CH_2CN)_3 + NaOH + 3\ H_2O \rightarrow N(CH_2COONa)_3 + 3\ NH_3$$

NTA is an effective chelating agent but does not by itself possess all the necessary properties for a detergent builder and hence is used in combination with the polyphosphates. Some test-marketed formulations in the U.S. contained as much as 10% NTA. However, animal studies[1] have shown that NTA may enhance the teratogenic effect of such heavy metals as cadmium and mercury. As a result the U.S. has suspended the use of NTA in detergents pending a conclusive determination of its safety. Diverse toxicity studies[2] are continuing on NTA (as well as other chelates) in a spectrum of environments with the strong prospect that NTA might still be used to some degree in detergents. An indication of the possible extent of use could be gleaned from the estimated worldwide NTA capacity of 750 million lb.[3] (before the above U.S. ban).

A recent sub-acute and acute oral toxicity evaluation by Nixon[4] has indicated that the sodium and calcium salts of NTA have a low order of toxicity to several species (rats, monkeys, and dogs). However, extremely high concentrations of Na_3NTA (7500 p.p.m. or greater) incorporated into the diet of rats for a period of 90 days produced kidney lesions of varying severity. The oral LD_{50} of Na_3NTA (20–50% solution) was 1.1–1.68 g/kg in rats; about 750 mg/kg (as a 50% solution) in monkeys and more than 5 g/kg (as an 80% solution) in dogs. The metabolism of NTA was determined in rabbit, monkey, and dog by Michael and Wakim[5] and it was shown that NTA did not have any demonstrable effect on bone development following its oral administration. (NTA was absorbed readily by the dog, relatively poorly by the rabbit, and poorly by the monkey.)

Reproduction and teratology studies of NTA in rats and rabbits have been most recently described by Nolen et al.[6]. NTA fed to two generations of rats. either continuously or from days 6 to 15 of pregnancy (throughout the period of organogenesis) at dietary levels of 0.1 or 0.5% produced no abnormalities in reproduction or in the fetuses. Doses of 2.5–250 mg/kg/day given by intubation to rabbits during organogenesis for one pregnancy also had no detectable effects on reproduction or the fetuses. Pollard[7] has reported on the physical characteristics, chemical properties, and some aspects of the biochemical and toxicological properties of NTA. Chromosome breakage has been demonstrated at high dose levels of NTA with *in vitro* cytogenic systems[8] and mutagenic effects have been demonstrated in *Drosophila melanogaster* with high concentrations of NTA[9]. However, in preliminary studies, NTA has been reported to be non-mutagenic in *E. coli* K-12[10] and Ophiostoma[11].

Aspects of the biodegradability of NTA have been reviewed by Thom[12], Thompson and Duthie[13], and Shumate et al.[14].

Experiments have been reported, mostly in relation to the activated sludge process, which show that NTA is biodegradable (to natural end-products, CO_2, H_2O, and inorganic nitrogen) at summer temperatures after adequate acclimatization periods (often of several weeks). The degree of removal also appears to decrease at low temperatures as well as with increasing concentrations of NTA.

As a mechanism of NTA biodegradation, Thompson and Duthie[13] suggested an initial cleavage and immediate metabolism of two acetate groups followed by metabolism of the glycine fragment. The rate and extent of NTA degradation was said to be comparable with those of glucose and citric acid.

A number of unresolved problems relating to the degradability of NTA resolve about the degradability of NTA under anerobic conditions, *e.g.* approx. 3.0% of sewage in rural areas are disposed by septic tanks. Under these conditions, high concentrations of non-degraded NTA may be anticipated in wells or receiving waters. The degradability of NTA chelates has not been completely elaborated. Thus, while the ferric chelate is degradable, there is no apparent available data concerning the degradability of a spectrum of other chelates that could be produced. It is possible that degradation of chelates such as that of lead may result in accumulation of high levels of heavy metals in receiving waters.

NTA in water has been analyzed by a variety of techniques including zinc–zircon procedure[13], polarographic[15-20], and colorimetric[21-23] as well as by an automated procedure involving the analysis of a bismuth–NTA complex at pH 2 using twin-cell oscillographic d.c. polarography[18].

A gas chromatographic method of determination of nitrilotriacetic acid (NTA) in lake water at the microgram level was described by Chau and Fox[24]. The NTA was first concentrated on a Dowex 1 (50–100 mesh) anion exchange resin, eluted with formic acid and converted to a propyl ester (by treatment with *n*-propanol saturated with hydrogen chloride). The method was found to be sensitive, specific, and free from the interferences of metals and common fatty and amino acids. The sensitivity of the method was in the order of 0.01 μg (injected).

A MicroTek GC Model 220 equipped with dual columns and hydrogen FID was used. The columns were 6 ft. × ¼ in. stainless steel packed with 3% OV-1 on Chromosorb WHP (80–100 mesh). The chromatographic parameters were nitrogen carrier gas at 65 ml/min, hydrogen flow 60 ml/min, air flow 566 ml/min, temperature program 180–225°C at 3°C/min, detector temperature 250°C, sensitivity 4×10^{-11} to 6.4×10^{-10}. A Disc Chart Integrator Model 229 (Disc Instruments) was used to measure the peak areas. A Nuclear-Chicago Liquid Scintillation System Unilux II was used for radiometric assay of the ^{14}C-NTA.

Figure 1 is a chromatogram showing relative retention of propyl esters of NTA, heptadecanoic acid (internal standard) and several fatty acids.

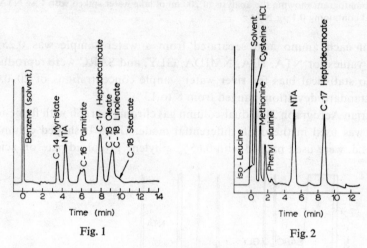

Fig. 1   Fig. 2

Fig. 1. Typical chromatogram showing relative retention of propyl esters of NTA, heptadecanoic acid, and several fatty acids.

Fig. 2. Typical chromatogram showing relative retention of propyl esters of NTA, heptadecanoic acid, and several amino acids.

Figure 2 is a chromatogram showing relative retention of propyl esters of NTA, heptadecanoic acid, and several amino acids. Figure 3 illustrates a chromatogram of the analysis of 200 ml of lake water spiked with 1 μg NTA. (The injection volume was 5 μl containing 0.1 μg NTA.)

Warren and Malec[25] developed a quantitative GC method for the determination of NTA and related aminopolycarboxylic acids in samples of river water, well water, primary sewage effluent, and secondary sewage efflient. NTA, N-methyl imino-diacetic acid (NMIDA), glycine (GLY), sarcosine (SARC), N-oxalyl iminodiacetic acid (N-oxalyl-IDA) were converted to n-butyl- and N-trifluoroacetyl-n-butyl esters and analyzed on ethylene glycol adipate columns. For quantitative analysis, the

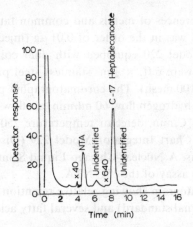

Fig. 3. Chromatogram showing the analysis of 200 ml of lake water spiked with 1 μg NTA. Injection volume, 5 μl containing 0.1 μg NTA.

amount of each amino acid separated from a water sample was 0.25–1000 μg. Response values for NTA, IDA, NMIDA, GLY, and SARC were reproducible and showed no statistical bias for river water sample concentrations of 20–0.025 mg/l. Relative standard deviations ranged from 8 to 13%.

A Varian Aerograph 2100 dual-column gas chromatograph with flame-ionization detectors was used in the dual, differential mode. Glass U-shaped columns, 1.9 m × 2 mm i.d. were used packed with 0.65% ethylene glycol adipate on acid-washed

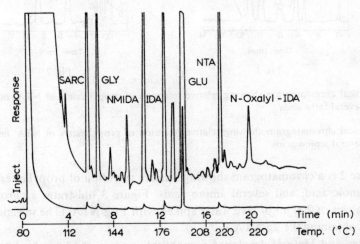

Fig. 4. Chromatogram of n-butyl esters of N-oxalyl-IDA, NTA, and NMIDA and N-TFA n-butyl esters of SARC, GLY, IDA, and GLU. Sample, N-oxalyl-IDA, NTA, NMIDA, IDA, GLY, and SARC each at the 10 μg level in 10 ml of Meramec River water; GLU, 1 μmole (147.1 μg), added as internal standard; derivatized amino acids in 300 μl of Freon 113, 1 μl of sample injected. Column, 1.9 m × 2 mm i.d. glass U column packed with 0.6 % w/w stabilized grade EGA on 80–100 mesh. Chromosorb W: initial temperature, 80°C programmed at 8°/min to 220°C; attenuation, upper pen at 1×10^{-11} A/mV, lower pen at 50×10^{-11} A/mV.

Chromosorb W, 80–100 mesh. Mass spectra were obtained with a GC–MS system composed of a Varian Aerograph 1700 gas chromatograph interfaced with a Varian CH 7 mass spectrometer. For a typical GC scan, the following conditions were used: injection size 1–3 μl, sensitivity 1×10^{-11} A/mV, column oven temperature 80 to 220° at 8°/min, injection block temperature 230°, detector block temperature 250°C, nitrogen carrier gas 60 p.s.i., 20 ml/min, H_2 25 p.s.i., 35 ml/min, air 10 p.s.i., 400 ml/min. Figure 4 illustrates a temperature-programmed chromatogram of n-butyl esters of N-oxalyl-IDA, NTA and NMIDA, and N-TFA on butyl esters of SARC, GLY, IDA, and GLU.

The determination of NTA by high-speed ion exchange chromatography has been reported by Longbottom[26]. The method was applied to the analysis of sewage samples and solutions of detergent formulations with a sensitivity of 1.0 mg/l. A DuPont 820 Liquid Chromatograph was operated at room temperature and a pressure of 1000 p.s.i. to maintain a flow rate of 0.5 ml/min. The flow was maintained with a UV photometric detector that measured the absorbance of unfiltered light emitted at 254 nm by a low-pressure mercury lamp. The separating column was 1 m long by $\frac{1}{4}$ in. o.d. stainless steel packed with Sax (DuPont), a strong anion exchange resin coated onto Zipax[27]. The mobile phase was 0.02 M $Na_2B_4O_7 \cdot 1OH_2O$ (pH 9.0). The response of the system was linear through the range of 0.1–1.5 μg. Figure 5 illustrates a chromatogram of 1.0 μg each of the monoamine chelating agents diethanol glycine (DEG), ethanol diglycine (EDG), iminodiacetic acid (IDA), and NTA. Figure 6 illustrates a chromatogram of 100 μl of a filtered sewage sample spiked with 3.0 mg/l NTA. Other UV absorbing materials in the sample eluted under the conditions of the test, but none inteferred with the NTA peak.

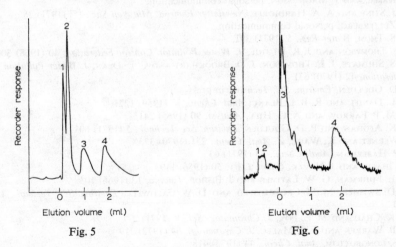

Fig. 5

Fig. 6

Fig. 5. Chromatogram of monoamine chelating agents (1 μg of each). 1, DEG; 2, EDG; 3, IDA; 4, NTA.

Fig. 6. Chromatogram of 100 μl of a filtered sewage sample dosed with 3.0 mg/l NTA. 1, Flow stopped for injection; 2, flow restored; 3, unidentified material; 4, NTA.

A number of solutions of metallic salts (*e.g.* $BaCl_2Sr(NO_3)_2$, $Mg(NO_3)_2$, $CaCl_2$, $MnCl_2$, $ZnCl_2$, $HgCl_2$, and $Fe(NO_3)_2$ and NTA were tested and had no effect as to the formation of their chelates on the NTA chromatogram while $Pb(NO_3)_2$, $CuSO_4$, and $FeSO_4$ were found to have low-response chelates with ferric ions when available and the method successfully reseparates NTA^{3-} from Fe^{3+}; all metal interferences were overcome by the addition of a ferric salt in excess of the NTA to samples before injection. After the addition of the iron, all the metal–NTA mixtures gave 100% of the theoretical response.

NTA has also been separated and identified in urine by paper chromatography following the administration of the ^{14}C-labeled chelate[5]. The solvent systems used were isopropanol–2 N hydrochloric acid (1:1) (R_F NTA, 0.82) and phenol–water (4:1) with 1 g KCN and 2 ml of 88% ammonium hydroxide (R_F NTA, 0.25).

REFERENCES

1 W. D. RUCKELSHAUS AND J. L. STEINFELD, *Joint Statement on NTA on Behalf of Environmental Protection Agency Press Release*, Washington, D.C., Dec. 18, 1970.
2 C. M. TARZWELL, *Quarterly Report of Research, Jan.–March, 1971*, National Marine Water Quality Laboratory, West Kingston, R. I.
3 Anon, *Chem. Eng. News*, Jan. 4 (1971) 15.
4 G. A. NIXON, *Toxicol. Appl. Pharmacol.*, 18 (1971) 398.
5 W. R. MICHAEL AND J. M. WAKIM, *Toxicol. Appl. Pharmacol.*, 18 (1971) 407.
6 G. A. NOLEN, L. W. KLUSMAN, D. L. BACK AND E. V. BUEHLER, *Food Cosmet. Toxicol.*, 9 (1971) 509.
7 R. R. POLLARD, *Soap Chem. Specialities*, 42 (1966) 58; 130.
8 B. KIHLMAN, personal communication.
9 C. RAMEL AND J. MAGNUSSON, personal communication.
10 G. J. STINE AND A. A. HARDIGREE, *Newsletter Environ. Mutagen Soc.*, [5] (1971) 38.
11 G. ZETTERBERG, personal communication.
12 N. S. THOM, *Water Res.*, 5 (1971) 391.
13 J. E. THOMPSON AND J. R. DUTHIE, *J. Water Pollution Control Federation*, 40 (1968) 306.
14 K. S. SHUMATE, J. E. THOMPSON, J. D. BROOKHART AND C. L. DEAN, *J. Water Pollution Control Federation*, 42 (1970) 631.
15 P. D. GOULDEN, *Environ. Sci. Technol.*, in press.
16 R. L. DANIEL AND R. B. LEBLANC, *Anal. Chem.*, 31 (1959) 1221.
17 R. M. P. FARROW AND A. G. HILL, *Analyst*, 90 (1965) 241.
18 B. K. AFGHAN AND P. D. GOULDEN, *Environ. Sci. Technol.*, 5 (1971) 601.
19 J. WERNET AND K. WAHL, *Z. Anal. Chem.*, 251 (1970) 373.
20 J. P. HABERMAN, *Anal. Chem.*, 43 (1971) 63.
21 J. CIHALIK AND J. NOVAK, *Chem. Listy*, 50 (1956) 1193.
22 D. L. FUHRMAN, G. W. LATIMER AND J. BISHOP, *Talanta*, 13 (1966) 103.
23 R. D. SWISHER, M. M. CRUTCHFIELD AND D. W. CALDWELL, *Environ. Sci. Technol.*, 1 (1967) 820.
24 Y. K. CHAU AND M. E. FOX, *J. Chromatog. Sci.*, 9 (1971) 271.
25 C. B. WARREN AND E. J. MALEC, *J. Chromatog.*, 64 (1972) 219.
26 J. E. LONGBOTTOM, *Anal. Chem.*, 44 (1972) 418.
27 J. K. KIRKLAND, *J. Chromatog. Sci.*, 8 (1970) 72.

Chapter 27

MISCELLANEOUS INDUSTRIAL CHEMICALS

1. DIMETHYL SULFOXIDE

Dimethyl sulfoxide (DMSO) is manufactured from dimethyl sulfide (DMS) which is obtained either by processing spent liquors from the kraft pulping process or by the reaction of methanol or dimethyl ether with hydrogen sulfide. The synthesis from methanol is accomplished by a vapor-phase reaction over a catalyst at temperatures above 300°C. In the preparation of DMSO *via* the oxidation of DMS, nitrogen dioxide or oxygen containing minor amounts of nitrogen dioxide are used. DMSO, because it is the most polar of the common aprotic solvents, is widely used as a solvent for displacement, base-catalyzed, and electrolytic reactions. It is also extensively employed as a polymerization and spinning solvent (particularly for polymerization of acrylonitrile) and is a particularly good solvent for polysaccharides and other polymers having active hydrogens or polar groups. DMSO has also been employed on a limited scale as a pesticide solvent for fungicides, insecticides, and herbicides[1,2] and as a carrier for iron chelates in citrus production[3-5].

DMSO is a relatively stable solvent of low toxicity, *e.g.* LD_{50}, for single-dose oral administration to rats is about 40,000 mg/kg. However, DMSO has the ability to penetrate the skin and thus represents a hazard in that it may also carry with it certain chemicals with which it is combined.

DMSO has been used to facilitate the percutaneous absorption of drugs[6-9], as a medication for the treatment of pain associated with musculoskeletal disorders[10,11], as a cryoprotective agent[12,13], and as a radioprotective compound[14-17].

DMSO has been shown to reversibly inhibit amino acid transport[18] and DNA[19] and RNA[20] synthesis in tumor cells[21], and also to inhibit thyroidal transport and organification in hamsters[22,23], to increase fetal resorption in rats[24], and to increase the mutagenic effect of ethylmethanesulfonate in *Arabidopsis thaliana*[25].

Aspects of toxicity of DMSO have been described for many species including mice[26,27], rats[26,27], dogs[27], and primates[28]. The oral or dermal administration of DMSO to several species of laboratory animals has been reported to cause ocular abnormalities characterized by a change in lenticular relucency and progressive lenticular myopia. These changes have been observed in dogs[29-31], rabbits[32,33], and pigs[32]. The metabolism of DMSO has been studied in man[34,35] and rat[34,36] and its mechanism of action elaborated[37].

Wing *et al.*[38] studied the adsorption, excretion and biotransformation of DMSO 80% gel (Demasorb) in man and in miniature pigs following topical administration. Both man and miniature pig transformed DMSO to dimethyl sulfone ($DMSO_2$) and dimethyl sulfide (DMS). DMSO and $DMSO_2$ were excreted in the expired air. In man, the relative amounts of DMSO and $DMSO_2$ in the plasma were similar to those found in the urine. The biological half-life of $DMSO_2$ in both the plasma and urine of man was 2.5–3 days. Urinary excretion of DMSO plus $DMSO_2$ ranged from 9 to 35% of the dose in both man and miniature pigs while only 1.6% of the dose was present in the feces of miniature pigs. Whereas $DMSO_2$ was the main excretory product in the urine of man, DMSO was the major component in the urine of miniature pigs. Concentrations of DMSO and $DMSO_2$ and DMS were determined by GLC by a modification of the procedure of Wallace and Mahon[39]. A Victoreen Model 400 gas chromatograph was equipped with two columns (6 ft. × $\frac{1}{8}$ in. diam. stainless steel tubing) packed with 20% Carbowax 20M on Chromosorb W. For the measurement of DMSO and $DMSO_2$, the system was operated isothermally at an oven temperature of 170°C, with the injector and the flame-ionization detectors at 210 and 220°C, respectively. DMS was measured isothermally at an oven temperature of 40°C with the injector and flame-ionization detector temperatures at 180 and 210°C, respectively. DMSO, $DMSO_2$, and DMS had retention times of 2, 6, and 0.5 min, respectively. Each urine and plasma sample was diluted with 4 volumes of methanol and, after the precipitated protein and other insoluble substances had been removed by centrifugation, 2 μl of the supernatant fluid from each sample was analyzed by GLC. Similarly, a volume of the methanol solution (2–4 μl) used to trap DMS from the expired air was analyzed.

Paulin et al.[40] described the GLC of DMSO in plasma and cerebrospinal fluid. An F & M Model 700 dual column, dual flame-ionization detector chromatograph was used equipped with a 1.0 mV range Honeywell recorder. The columns used were 4 ft. × $\frac{1}{4}$ in. o.d. copper tubing packed with 25% Carbowax 20M on 80–100 mesh Chromosorb P. Samples of 5 μl of plasma or cerebrospinal fluid were injected directly onto the column. The chromatographic parameters were injection port temperature 225°, column temperature 160°C, detector temperature 205°C, carrier helium flow rate 60 ml/min, hydrogen flow rate 60 ml/min, and air flow rate 500 ml/min. The column was heated at 220°C for approx. 24 h prior to the analysis of any samples. Subsequent to the conditioning of the column, 5 μl aliquots of standard solution of DMSO in distilled water were used to obtain a calibration curve. The lower limit of sensitivity of the overall procedure for the direct analysis of DMSO in dog cerebrospinal fluid and plasma was found to be 22 μg/ml with linearity of the curve within the range of 22–110 μg/ml. DMSO was eluted in 10 min while dimethyl sulfide chromatographed under the above conditions was eluted with the solvent front.

The GLC quantitative determination of DMSO in butanol was described by Stuart[41]. A Perkin-Elmer Model 800 gas chromatograph with dual 10 ft. × $\frac{1}{8}$ in. o.d. copper columns containing 20:80 ratio of Carbowax 20M and SE-30 on equal parts of Chromosorb P (80–100 mesh) (MHDS treated) and Teflon 6 (40–60 mesh) was used. The operating conditions were column temperature 100°C followed by programming at 5°C/min to 150°C, injection temperature 200°C, detector temperature 150°C, carrier gas helium at 40 ml/min, flame-ionization detector with hydrogen at 20 ml/min, air at 250 ml/min, attenuation ×1000, and sample size 0.5 μl. Restrictor tubes with one-to-one splitting ratio were installed in the instrument so that only part of the column flow entered the detector. A Bristol 0–2 mV recorder was used at a chart speed of 0.5 in./min. Although the work reported was done using temperature programming, the samples could be analyzed under isothermal conditions. The method was considered adequate for the analysis of DMSO to within ±2% in various solvent mixtures.

Variali[42] described the quantitative analysis of toluene–DMSO and benzene–DMSO mixtures by GLC. Two columns were employed: (A) 2 m × 4 mm i.d. packed with 20% 1:4 Carbowax 20M-SE-30 on 1:1 silanized Chromosorb P–Chromosorb T (Teflon 6) and (B) 5 m × 3 mm i.d. packed with 1:1 100–120 mesh glass beads–Chromosorb T (30–60 mesh), the beads coated with 0.2% of 2:3 Carbowax 20M-SE-30 and the Chromosorb T with 10% of the same mixture.

The chromatographic analysis of DMSO in non-aqueous solutions has been reported[41–44]. The quantitative determination of DMSO in water by GLC was reported by Seager and Stone[45]. A Perkin-Elmer Model 810 gas chromatograph was used equipped with both dual hydrogen flame-ionization and dual hot-wire detectors. A stream splitter at the end of the column diverted 30% of the column flow through the flame detector and the remainder through the hot-wire detector. The column was 2 m × ⅛ in. o.d. stainless steel tubing packed with 80–100 mesh Porapak T (Waters Assoc.). The carrier gas was helium at a flow rate of 50 ml/min measured at the exhaust port of the hot-wire detector. The operating temperatures for all measurements were 190°C for the column and 200°C for the injection port and both detectors. A column preconditioned at 250°C for 1 h before use was found to give retention times of 10 sec for water and 170 sec for DMSO using the operating conditions given above. Figure 1 illustrates the separation of 2.17% DMSO obtained using the column and conditions described above.

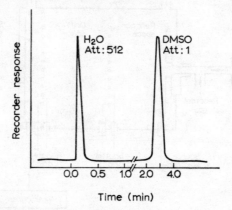

Fig. 1. Separation of DMSO from water on Porapak T. Column temperature, 190°C; injection port and detector temperatures, 200°C; carrier gas, 50 ml/min helium.

References pp. 629–630

Cates and Meloan[44] separated a series of alkyl and aryl sulfoxides by GLC using Carbowax 1500 or 20M on Gas Chrom Z, a silanized diatomaceous solid support. An F & M Model 609 flame-ionization gas chromatograph was used with 3 ft. × $\frac{1}{4}$ in. o.d. aluminum columns containing 20% Carbowax 1500 or 20M on Gas Chrom Z and helium as carrier gas at 15 p.s.i.g. inlet pressure. The sulfoxides were separated isothermally at column temperatures of 200 and 250°C.

Cohen and Karasek[46] recently described the operating aspects and utility of plasma chromatography (PC). The plasma chromatograph is a positive and negative ion–molecule reactor with an ion drift spectrometer which operates with an atmospheric pressure gas and can be employed with a gas chromatograph for detection and peak characterization. The plasma chromatograph was also suggested to serve the GC as an ion interface to the mass spectrometer in which principally molecular ions are formed. In both the PC and PC–MS many organic molecule types were said to be observed at ultra-trace concentration levels down to 10^{-12} mole fraction.

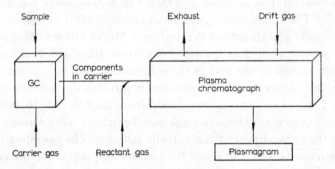

Fig. 2. Functional block diagram of gas chromatograph with plasma chromatograph detector.

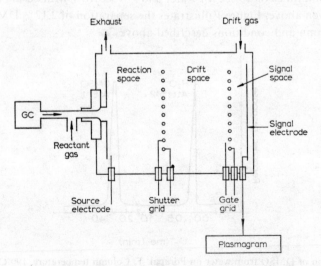

Fig. 3. Schematic diagram of plasma chromatograph cell.

Dimethyl sulfoxide was used as an example to describe the positive and negative PC plasmagram signals as well as the characteristic PC–MS mass spectrogram.

Figures 2 and 3 illustrate the functional block diagram of gas chromatograph with plasma chromatograph detector and a schematic diagram of a plasma chromatograph cell.

Williams et al.[47] analyzed dimethyl sulfone in urine by GLC. A Research Specialties Series 600 instrument was used with a flame-ionization detector and a 6 ft. stainless steel column containing 30% butanediol succinate on 45–60 mesh Chromosorb W. The column and injection block temperatures were 115 and 130°C, and the carrier gas was nitrogen at a flow rate of 60 ml/min.

Jucker et al.[48] isolated and identified DMSO (by paper chromatography) in urine of rats following intraperitoneal administration of ^{35}S-DMSO. The developing solvent was n-butanol–acetic acid–water (65:25:10) and the R_F of DMSO was 0.59.

Malinin et al.[49] described the distribution of ^{14}C-labeled DMSO in tissues of guinea pigs. The uptake of ^{14}C-DMSO serum fractions (following i.p. administration) was determined by gel filtration of 10ml serum samples at 4°C for approx. 18 h with a 36 × 2.5 cm column of Sephadex G-200 in 0.15 M NaCl. The absorbance of 5 ml eluted fractions was determined spectrophotometrically at 280 nm. (No DMSO was bound to serum proteins.) Autoradiographic evidence indicating an overwhelming tendency of DMSO solvated in body fluids to be confined to the interstitial spaces.

2. KETENE

Ketene is a very reactive gaseous basic starting material manufactured primarily by pyrolysis of acetic acid at 700–800°C. For practical conversions, any acidic catalyst is generally used, e.g. ca. 0.3% triethyl phosphate. It is also formed by pyrolysis of virtually any compound containing an acetyl group, e.g. acetone. Ketene is used in many important industrial processes as an acetylating agent for hydroxy and amino compounds. The major use of ketene is in the manufacture of acetic anhydride which in turn is used primarily for production of cellulose acetate. Another important application of ketene is in the preparation of diketene which is an important intermediate for the preparation of dihydroacetic acid and acetoacetic esters. Ketene also has utility in the acetylation of viscose rayon fiber and in textile finishing, as a rodenticide, as an additive for non-corrosive hydrocarbon fuels as well as serving as a photochemical source (as well as methyl ketene) for the generation of methylene and ethylidene radicals.

Ketene is a highly toxic gas with symptoms resulting from inhalation of excessive amounts resembling those caused by phosgene. Amounts as low as 1 p.p.m. produce definite toxic symptoms in laboratory animals[50]. The toxicity of ketene has been studied by Cameron and Neuberger[51], Wooster et al.[52], and Treon et al.[50] in mice, rats, cats, and guinea pigs. Ketene has been reported to be mutagenic in Drosophila melanogaster[53] but not in Neurospora[54].

Ketene reacts with hydroxy and amine groups to form acetyl derivatives which has formed the basis of its analysis *via* gravimetric[55], titrimetric[51], and photometric techniques[56,57]. Ketene has also been determined by ultraviolet spectroscopy[58].

Non-quantitative gas chromatographic analyses of ketene have been reported by Young[59] and Dehara[60] with a dioctyl phthalate-coated column used in the latter case. A method for the quantitative analysis of ketene and other low-boiling organic and inorganic impurities by gas chromatography using Porapak-R has been developed by Kikuchi *et al.*[61]. The method gave reproducible results in the analysis of diketene pyrolyzate and of ketene from acetic acid. A relationship between the pyrolysis conditions and the composition of the pyrolyzate was established from this data as well as suggesting the reaction mechanisms of pyrolysis.

A Perkin-Elmer Model 154 was modified with a tungsten filament detector of a Shimadzu GC-2A gas chromatograph and used with a dual column system as a Hitachi QPD-33 recorder. Figures 4–6 illustrate the gas flow diagram of the apparatus,

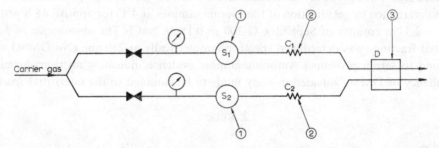

Fig. 4. The gas flow diagram of the apparatus. S_1, S_2, sampling valves; C_1, C_2, columns; D, detector.

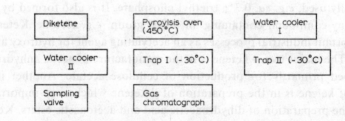

Fig. 5. Gas flow diagram.

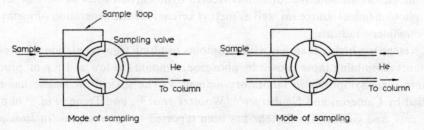

Fig. 6. Gas sampling equipment.

the flow diagram from ketene generator to gas sampling valve (Perkin-Elmer), and
details of the gas sampling equipment, respectively. Table 1 lists the operating
parameters for the dual column system. The first column (C_1) (Porapak R) separated
the hydrogen–air–carbon monoxide mixture, methane, ethane, ketene, allene, and

TABLE 1

OPERATING PARAMETERS FOR DUAL COLUMN ANALYSIS OF KETENE AND OTHER GASES

	I	II
Column	C_1: "Porapak" R. 1.5 m	C_1: DOP 20 wt.-% on Microsorb W, 2.0 m
	C_2: Silica gel 2.0 m	C_2: Silica gel 2.0 m
Column temperature	C_1: 95°C	C_1: 48°C
	C_2: 30°C	C_2: room temperature
Carrier gas	He, C_1: 60 ml/min	He, C_1: 100 ml/min
	C_2: 100 ml/min	C_2: 100 ml/min
Detector current	190 mA	
Chart speed	Start 5 min. 40 mm/min	
	5 min. 10 mm/min	

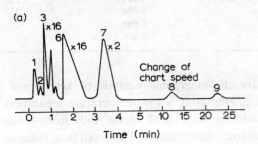

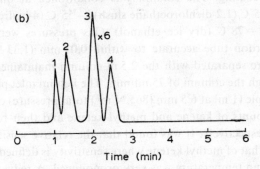

Fig. 7. Gas chromatographic separations of ketene and other gases obtained on (a) Poropak R and
(b) DOP columns. (a) 1, Carbon monoxide and air; 2, methane; 3, carbon dioxide; 4, ethylene;
5, ethane; 6, ketene; 7, allene; 8, 9, unidentified. (b) 1, Air, carbon monoxide, methane, ethane,
ethylene; 2, carbon dioxide; 3, ketene; 4, allene.

two unidentified peaks. The hydrogen, air, and carbon monoxide were analyzed on the silica gel (C) system. (Normally, however, hydrogen and air were negligible and it was not necessary to use the silica gel column.) Figures 7(a) and (b) illustrate chromatograms of gaseous mixtures obtained on Porapak R and dioctyl phthalate (DOP) columns, respectively. Relative sensitivities of each component against air were obtained and are listed in Table 2. In comparing the utility of Porapak R and DOP columns [Figs. 7(a) and (b)] it was found that ketene could be separated on a Porapak column without inhibiting the separation of inorganic gases and lower hydrocarbons. The DOP column separated only carbon dioxide, ketene, and allene.

TABLE 2

RELATIVE SENSITIVITIES

Component	C_1 ("Porapak" R column)	C_2 (Silica gel column)
Air (standard)	1.000	1.000
Carbon monoxide	0.988	0.988
Methane	0.830	0.832
Carbon dioxide	1.211	1.228
Ethylene	1.204	1.220
Ketene	1.068	
Allene	1.476	

The quantitative chromographic analysis of ketene and methyl ketene was accomplished by Laufer[62] using a low-temperature separation of the components on a Haloport F column. An F & M Model 1609 flame-ionization gas chromatograph was used without modification. The analytical column was 2.55 m × ¼ in. polyethylene tubing packed with 30–60 mesh Haloport F (F & M Scientific) without any liquid-phase coating. The column was conditioned at one of the following temperatures: 0–35°C (1,2-dichloroethane slush), −55°C (4-hydroxy-4-methyl-2-pentanone slush), or −78°C (dry ice–ethanol). Gas pressures were measured with a quartz spiral bourdon tube accurate to within 0.01 mm (1.33 N/m²). Ketene and methyl ketene were separated with the 2.5 m column maintained at −35°C with a helium flow through the column of 75 ml/min. The helium inlet pressure was 50 p.s.i. The injection sample (1 ml at 6.5 mm (865 N/m²) total pressure) consisted of approximately equal amounts of ketene and methyl ketene and their retention times were 0.9 and 1.8 min, respectively. It was found that the relative sensitivity of ketene was 25% greater than that of methyl ketene where sensitivity is defined as peak area/mole. The effect of column temperature was very pronounced. A reduction in temperature resulted in longer retention times, e.g. at −78°C ketene was eluted after 2 min and methyl ketene was retained on the column. Loss of methyl ketene within the chromatographic column was considered to be due to rapid polymer formation.

Jenkins[63] and Blake and Hole[64] have reported rapid dimerization of methyl ketene in the gas phase. Jenkins[63] observed a higher rate of dimerization at 0°C than at 20°C which is in agreement with a negative ΔH of dimerization.

3. DIISOCYANATES AND ISOCYANATES

Phosgenation (the reaction of amines with phosgene) is used almost exclusively for the manufacture of isocyanate. The problems are multiplied in the manufacture of a diisocyanate where the simple byproducts may be intermolecular, *e.g.* a mixed carbamyl chloride/amine hydrochloride, and the urea byproduct may be polymeric.

The overall reaction sequence for the manufacture of 80/20 toluene diisocyanate (TDI) can be illustrated as follows.

The estimated production of TDI in the U.S. in 1965 was 260 million lb. TDI is used primarily for production of urethan foams and pre-polymer coatings based on its reaction with a hydroxyl-terminated polyester and water. The purity of commercial TDI (consisting of $80 \pm 2\%$ of 2,4 isomer and $20 \pm 1\%$ of 2,6 isomer) is approx. 95.5%.

In the process where TDI is used on a very large scale in the manufacture of flexible foams, the highly exothermic reaction is controlled by the addition of appropriate catalysts (*e.g.* blowing agents such as azodicarbonamide). As a result of the exotherm, significant quantities of TDI vapor may appear in the atmosphere and in concentrations substantially above the present threshold limit value of 0.02 p.p.m.[65]. Prolonged exposure of workers to low air concentrations of TDI (0.1 p.p.m.) has been reported to produce a variety of acute and chronic respiratory effects[66-71].

Crude MDI (methylene diphenylisocyanate) used in rigid urethane foam formulations is prepared by treating the condensation product of aniline and formaldehyde with phosgene. Any uncondensed aniline initially present will be converted to phenylisocyanate, which because of its volatility is more toxic than the diisocyanates.

The hazards involved in the decomposition of polyurethanes have been suggested by Paisley[72]. For example, the decomposition of polyurethanes used in wire insulation occurs at 220–275°C producing isocyanates and N-oxides. (The tempera-

ture of a soldering iron in normal soldering operations is approx. 300°C.) The possibility of a serious incipient hazard in combating fires involving buildings and refriger-
ated compartments where large quantities of polyurethane foams are used was also
raised by Paisley[72].

The identification and analysis of urethane foams by pyrolysis–gas chromatography was reported by Takeuchi *et al.*[73]. Pyrolysis was performed at a series of
temperatures between 650 and 1000°C in a quartz tube attached to the inlet port of the
separation column. The identification of the pyrolysis product was also carried out
by infrared spectroscopy and mass spectrometry. The determination of isocyanates
utilized as raw material was carried out using the pyrograms obtained at 850°C.

A Yanagimoto GP-1000 Pyrolyzer was used as illustrated in Fig. 8. The tube of
the pyrolyzer was made of quartz (10 mm i.d. × 136 mm). The pyrolysis heater,
B, could be heated up to 1200°C, and the precut heater, C, could be regulated in-

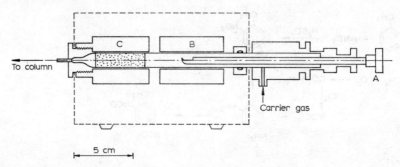

Fig. 8. Pyrolysis apparatus. A, Sample charging rod; B, pyrolysis heater; C, precut heater.

TABLE 3

FORMULATION FOR FLEXIBLE FOAMS

| | Parts by weight | | | | |
	Foam 1	Foam 2	Foam 3	Foam 4	Foam 5
Isocyanates					
PAPI[a]	50	35	25	15	0
TDI[b]	0	9.9	16.5	23.1	33.6
Polypropylene glycol-based					
polyether	95	95	95	95	95
Distilled water	0.4	0.4	0.4	0.4	0.4
Crosslinking agents					
(triethanolamine)	10	10	10	10	10
Additives					
(stabilizer, catalysts, etc.)	2	2	2	2	2

[a] PAPI = Polyaryl polyisocyanate.
[b] TDI = Toluene diisocyanate.

dependently up to 400°C. A Yanagimoto Gas Chromatograph Model GCG-500F with double flame-ionization detectors and a 2 m × 3 mm i.d. column containing 10% Carbowax 6000 coated on Diasolid L (80–100 mesh) was used. Temperature programming was carried out from 50 to 210°C at 12°C/min, carrier gas nitrogen at 0.7 kg/cm² inlet pressure (20 ml/min at 50°C), hydrogen 25 ml/min, air 0.7 l/min, sample size 0.3–0.5 mg, and pyrolysis temperature range 650–1000°C. For infrared spectroscopy, pyrolysis products were trapped by a U-shaped capillary collector containing a small amount of CCl_4 (10 µl) which was attached to the outlet of the pyrolyzer.

Table 3 lists the formulations for the flexible foams analyzed. As shown in this table, the synthesized foams have similar compositions except for the isocyanates used. The isocyanates, TDI and PAPI, have the following structures.

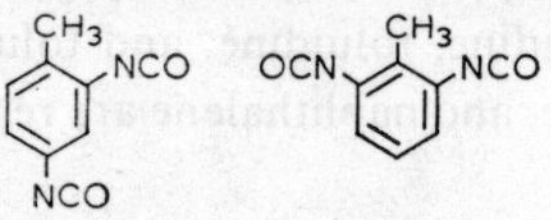

TABLE 4

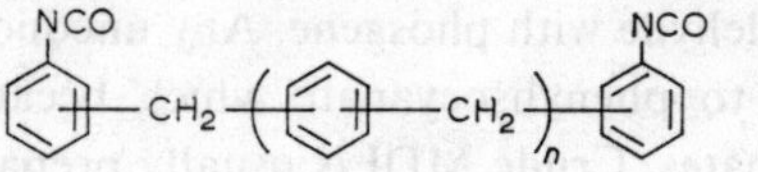

MASSES OF THE PYROLYSIS PRODUCTS AT 850°C

Peak no.	Identified peaks	Chemical formula	Calculated mass[a]	Determined mass
1	Acetonitrile	C_2NH_3	41.02655	41.032
2	Acrylonitrile	C_3NH_3	53.02655	53.029
3	Pyrrole	C_4NH_5	67.04220	67.046
4	Benzene	C_6H_6	78.04695	78.042
5	Pyridine	C_5NH_5	79.04220	79.046
6	Toluene	C_7H_8	92.06260	92.062
7	Aniline	C_6NH_7	93.05785	93.067
8	Benzonitrile	C_7NH_5	103.04220	103.040
9	Styrene	C_8H_8	104.06260	104.025
10	Ethylbenzene	C_8H_{10}	106.07825	106.075
11	m-Toluidine	C_8NH_9	107.07350	107.054
12	p-Toluidine	C_8NH_9	107.07350	107.054
13	Phenylisocyanate	C_8NOH_5	119.03711	119.080
14	o-, m-, p-Toluylenediamine[b]	$C_7N_2H_{10}$	122.08440	122.101
15	Naphthalene	$C_{10}H_8$	128.06260	128.061

[a] Calculated as $^{12}C = 12.00000$.
[b] Identified on Apiezon Grease-L column.

On decomposition of the respective foams, the TDI segment will yield toluylene-diamine and diaminobenzene while PAPI yields toluidine, aniline and phenyl-isocyanate. Table 4 lists the peak number and masses of the pyrolysis products at 850°C. Figures 9 and 10 illustrate typical pyrograms at 850 and 1000°C, respectively. The structure of the synthesizer urethane foam is divided into the following bondings: polyether, urethane, urea, biuret, and allophanate. The polyether segment may be decomposed at a relatively low temperature because of its linear structure. The latter four bondings are concerned in the three-dimensional structure of the foams and hence their their thermal decomposition may be slower than the linear portion. As shown in Figs. 9 and 10, fairly different pyrograms are obtained when urethane foams are decomposed at elevated temperatures. The main pyrolysis products at 850°C are ring compounds such as benzene, benzonitrile, aniline, phenyl-isocyanate, alkyl benzene derivatives, naphthalene, and pyrrole. Smaller quantities of acrylonitrile and acetonitrile as well as traces of pyridine, toluidine, and toluylene-diamine were also identified. Nitriles, pyrrole, pyridine, and naphthalene are regarded as second-order thermal reactions.

Crude MDI used in rigid urethane foam formulations is prepared by treating the condensation product of aniline and formaldehyde with phosgene. Any uncondensed aniline initially present will be converted to phenylisocyanate which because of its volatility is more toxic than the diisocyanates. Crude MDI is usually prepared in chlorinated aromatic solvent and traces of these may also give a slight odor to the product.

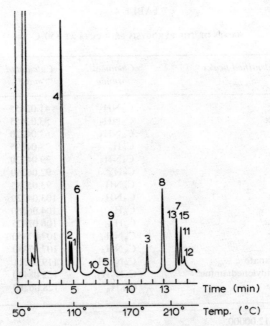

Fig. 9. Pyrogram of foam 3 at 850°C. The peak numbers correspond to those given in Table 4.

The gas chromatographic determination of phenylisocyanate and monochlorobenzene in crude methylenediphenyl isocyanate (MDI) covering the range from 0.05 to 0.5% was described by Hanneman and Robinson[74].

An F & M Model 810 gas chromatograph was equipped with FID and used with a 10 ft. × ¼ in. stainless steel column packed with 2.5% SE-30 on 70–80 mesh Chromosorb G, acid-washed and dimethyldichlorosilane-treated. The operating con-

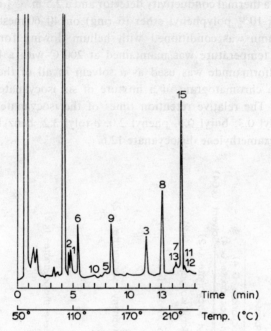

Fig. 10. Pyrogram of foam 3 at 1000°C. The peak numbers correspond to those given in Table 4.

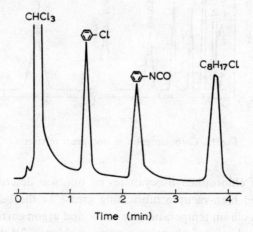

Fig. 11. Typical chromatogram for phenylisocyanate and monoclorobenzene analysis: 0.144% phenylisocyanate in crude MDI.

ditions were column temperature 135°C, detector temperature 300°C, injector temperature 270°C, carrier gas helium at 60 ml/min, attenuation 8×10^2.

Figure 11 illustrates a chromatogram of the analysis for phenyl isocyanate and monochlorobenzene (0.144% phenylisocyanate in crude MDI).

Ruth[75] described the GC analysis of various isocyanates in the presence of reactive organic halides. An F & M Model 720 dual-column gas chromatograph was equipped with a thermal conductivity detector and a 2.5 m × ¼ in. o.d. aluminum column containing 10% polyphenyl ether (6 ring) on 40–60 mesh Chromosorb T (Teflon). The column was conditioned with helium flowing for 16 h at 200°C. The column oven temperature was maintained at 200°C with a bridge current at 150 mA. Dimethylformamide was used as a solvent in all of the determinations. Figure 12 shows a chromatogram of a mixture of six isocyanates and one diisocyanate at 200°C. The relative retention times of the isocyanates (with reference to DMF) were allyl 0.3, butyl 0.5, phenyl 2.0, o-tolyl 3.2, benzyl 4.9, p-methoxyphenyl 7.9, and hexamethylene diisocyanate 12.6.

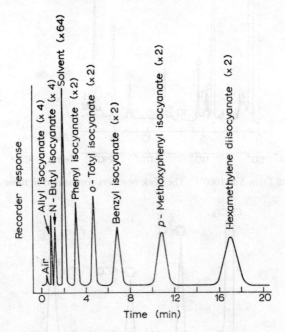

Fig. 12. Chromatogram of isocyanate mixture.

The analysis of chlorophenylisocyanates by GLC was described by Strepikheev et al.[76] who utilized high-vacuum lubricating grease as the best stationary phase on Celite 545. The column temperature was 125° and argon carrier gas was 550 mm Hg. The retention volumes of phenylisocyanate, p-chloro-, 2,4-dichloro-, and 2,4,6-trichlorophenylisocyanates were 0.5, 1.0, 2.25, and 4.3, respectively.

REFERENCES

1 G. A. Bean, *Plant Disease Rept.*, 49 (1965) 810.
2 H. L. Keil, *Agr. Chem.*, 20 (1965) 23.
3 I. Stewart and C. D. Leonard, *Soil Sci.*, 84 (1957) 89.
4 C. D. Leonard, *Ann. N.Y. Acad. Sci.*, 141 (1967) 148.
5 B. C. Smale, *Sulfur Inst. J.*, August (1969).
6 A. M. Kligman, *J. Am. Med. Assoc.*, 193 (1965) 796.
7 R. B. Stoughton and W. Fritsch, *Arch. Dermatol.*, 90 (1964) 512.
8 M. B. Sulzberger, A. Cortese, Jr., L. Fishman, H. S. Wiley and P. S. Peyakovich, *Ann. N.Y. Acad. Sci.*, 141 (1967) 437.
9 S. W. Jacobs, M. Bischel and R. J. Herschler, *Current Therap. Res.*, 6 (1964) 134.
10 E. E. Rosenbaum, R. J. Herschler and S. W. Jacob, *J. Am. Med. Assoc.*, 192 (1965) 309.
11 E. E. Rosenbaum and S. W. Jacob, *Northwest Med.*, 63 (1964) 167.
12 J. E. Lovelock and M. W. H. Bishop, *Nature*, 183 (1959) 1394.
13 M. J. Ashwood-Smith, *Ann. N.Y. Acad. Biol. Sci.*, 141 (1967) 41.
14 M. J. Ashwood-Smith, *Intern. J. Radiation Biol.*, 3 (1961) 41.
15 M. J. Ashwood-Smith, *Ann. N.Y. Acad. Sci.*, 141 (1967) 45.
16 S. E. Kim and W. S. Moos, *Health Phys.*, 13 (1967) 601.
17 W. S. Moos and S. E. Kim, *Experientia*, 22 (1966) 814.
18 R. F. Hagemann and T. C. Evans, *Nature*, 218 (1968) 583.
19 R. F. Hagemann and T. C. Evans, *Proc. Soc. Exptl. Biol. Med.*, 128 (1968) 648.
20 R. F. Hagemann, *Experientia*, 25 (1969) 1298.
21 R. F. Hagemann and T. C. Evans, *Proc. Soc. Exptl. Biol. Med.*, 128 (1968) 1008.
22 V. H. Ferm, *J. Embryol. Exptl. Morphol.*, 16 (1966) 49.
23 V. H. Ferm, *Lancet*, i (1966) 208.
24 M. B. Juma and R. E. Staples, *Proc. Soc. Exptl. Biol. Med.*, 125 (1967) 567.
25 C. R. Bhatia, *Mutation Res.*, 4 (1967) 375.
26 E. R. Smith, Z. Hadidian and M. M. Mason, *J. Clin. Pharmacol.*, 8 (1968) 315.
27 J. E. Willson, D. E. Brown and E. K. Timmens, *Toxicol. Appl. Pharmacol.*, 7 (1965) 104.
28 E. E. Vogin, S. Carson, G. Cannon, C. R. Linegar and L. F. Rubin, *Toxicol. Appl. Pharmacol.*, 16 (1970) 606.
29 L. F. Rubin and P. A. Mattis, *Science*, 153 (1966) 83.
30 K. R. Kleberger, *Ann. N.Y. Acad. Sci.*, 141 (1967) 381.
31 E. R. Smith, Z. Hadidian and M. M. Mason, *Ann. N.Y. Acad. Sci.*, 141 (1967) 96.
32 L. F. Rubin and K. C. Barnett, *Ann. N.Y. Acad. Sci.*, 141 (1967) 333.
33 R. Esila and T. Tenhunen, *Acta Ophthalmol.*, 14 (1967) 530.
34 E. Gerhards and G. Gibian, *Ann. N.Y. Acad. Sci.*, 141 (1967) 65.
35 K. H. Kolb, G. Jaenicki, M. Kramer and P. E. Schulze, *Ann. N.Y. Acad. Sci.*, 141 (1967) 85.
36 C. W. Denki, R. M. Goodman, R. Miller and T. Donovan, *Ann. N.Y. Acad. Sci.*, 141 (1967) 77.
37 T. J. Franz and J. T. Van Bruggen, *Ann. N.Y. Acad. Sci.*, 141 (1967) 73.
38 K. K. Wing, G. M. Wang, J. Dreyfuss and E. C. Schreiber, *J. Invest. Dermatol.*, 56 (1971) 44.
39 T. J. Wallace and M. J. Mahon, *Nature*, 201 (1964) 817.
40 H. J. Paulin, J. B. Murphy and R. E. Larson, *Anal. Chem.*, 38 (1966) 651.
41 R. A. Stuart, *J. Gas Chromatog.*, 4 (1966) 388.
42 G. Variali, *Rass. Chim.*, 22 (1970) 47.
43 T. Wallace and J. J. Mahon, *J. Am. Chem. Soc.*, 86 (1964) 4099.
44 V. E. Cates and C. E. Meloan, *Anal. Chem.*, 35 (1963) 650.
45 S. L. Seager and C. C. Stone, *J. Chromatog. Sci.*, 9 (1971) 438.
46 M. J. Cohen and F. W. Karasek, *J. Chromatog. Sci.*, 8 (1970) 330.
47 K. I. H. William, S. H. Burstein and D. S. Layne, *Arch. Biochem. Biophys.*, 113 (1966) 251.
48 H. B. Jucker, P. M. Ahmad and E. A. Miller, *Federation Proc.*, 24 (1965) 546.
49 G. I. Malinin, D. J. Fontana and D. C. Braungart, *Cryobiology*, 5 (1969) 328.

50 J. F. Treon, H. E. Sigmon, K. V. Kitzmiller, F. F. Heyroth, F. F. Younker and J. Cholak, *J. Ind. Hyg.*, 31 (1949) 209.
51 G. R. Cameron and A. Neuberger, *J. Pathol. Bacteriol.*, 45 (1937) 653.
52 H. A. Wooster, C. C. Lushbaugh and C. E. Redemann, *J. Ind. Hyg. Toxicol.*, 29 (1947) 56.
53 I. A. Rapoport, *Dokl. Akad. Nauk. SSSR.*, 58 (1947) 119.
54 K. A. Jensen, I. Kirk, G. K. Olmark and M. Westergaard, *Coldspring Harbor Symp. Quant. Biol.*, 16 (1951) 245.
55 A. M. Potts, F. P. Simon and R. W. Gerard, *Arch. Biochem.*, 24 (1949) 329.
56 W. M. Diggle and J. S. Gage, *Analyst*, 78 (1953) 473.
57 R. M. Mehdenhall, *Am. Ind. Hyg. Assoc. J.*, 21 (1960) 201.
58 A. R. Harrison and J. S. Lake, *J. Phys. Chem.*, 63 (1959) 1489.
59 J. R. Young, *J. Chem. Soc.*, (1958) 2909.
60 M. Dehara, *J. Soc. Org. Syn. Chem. Japan*, 20 (1962) 730.
61 Y. Kikuchi, T. Kikkawa and R. Kato, *J. Gas Chromatog.*, 5 (1967) 261.
62 A. H. Laufer, *J. Chromatog. Sci.*, 8 (1970) 677.
63 A. D. Jenkins, *J. Chem. Soc.*, (1952) 2563.
64 P. G. Blake and K. J. Hole, *J. Phys. Chem.*, 70 (1966) 1464.
65 H. G. Parkes, *Proc. Roy. Soc. Med.*, 63 (1970) 10.
66 H. Eckins, G. W. McCarl, H. G. Brughsh and J. P. Fahy, *J. Am. Ind. Hyg. Assoc.*, 23 (1962) 265.
67 A. Munn, *Trans. Assoc. Ind. Med. Off.*, 9 (1960) 134.
68 G. M. Hama, *Arch. Ind. Health*, 16 (1957) 232.
69 A. Munn, *Ann. Occupational Hyg.*, 8 (1965) 163.
70 K. S. Williamson, *Trans. Assoc. Ind. Med. Off.*, 15 (1965) 29.
71 J. M. Peteri, *Proc. Roy. Soc. Med.*, 63 (1970) 14.
72 D. P. G. Paisley, *Brit. J. Ind. Med.*, 26 (1969) 79.
73 T. Takeuchi, S. Tsuge and T. Okumoto, *J. Gas Chromatog.*, 6 (1968) 542.
74 W. W. Hanneman and C. C. Robinson, *J. Gas Chromatog.*, 6 (1968) 256.
75 G. W. Ruth, *J. Gas Chromatog.*, 6 (1968) 1513.
76 Y. A. Strepikheev, R. A. Semenova and A. N. Ushakov, *Zh. Analit. Khim.*, 20 (1965) 757.

SUBJECT INDEX